AF352872

Computational Approaches: Drug Discovery and Design in Medicinal Chemistry and Bioinformatics

Computational Approaches: Drug Discovery and Design in Medicinal Chemistry and Bioinformatics

Editors

Marco Tutone
Anna Maria Almerico

MDPI • Basel • Beijing • Wuhan • Barcelona • Belgrade • Manchester • Tokyo • Cluj • Tianjin

Editors

Marco Tutone
Dipartimento di Scienze e
Tecnologie Biologiche Chimiche
e Farmaceutiche
University of Palermo
Palermo
Italy

Anna Maria Almerico
Dipartimento di Scienze e
Tecnologie Biologiche Chimiche
e Farmaceutiche
University of Palermo
Palermo
Italy

Editorial Office
MDPI
St. Alban-Anlage 66
4052 Basel, Switzerland

This is a reprint of articles from the Special Issue published online in the open access journal *Molecules* (ISSN 1420-3049) (available at: www.mdpi.com/journal/molecules/special_issues/comput_appr_drug_dis_des_med_chem_bio).

For citation purposes, cite each article independently as indicated on the article page online and as indicated below:

LastName, A.A.; LastName, B.B.; LastName, C.C. Article Title. *Journal Name* **Year**, *Volume Number*, Page Range.

ISBN 978-3-0365-2779-6 (Hbk)
ISBN 978-3-0365-2778-9 (PDF)

Contents

MDPI
St. Alban-Anlage 66
4052 Basel
Switzerland
Tel. +41 61 683 77 34
Fax +41 61 302 89 18
www.mdpi.com

Molecules Editorial Office
E-mail: molecules@mdpi.com
www.mdpi.com/journal/molecules

About the Editors

Marco Tutone

Marco Tutone graduated from the University of Palermo, Italy, in Medicinal Chemistry and Pharmaceutical Technology with summa cum laude in 2001. He received a Ph.D. in Pharmaceutical Sciences from the University of Palermo, Italy, in 2006. From 2006 to 2008, he was a Post-doc research fellow funded by MIUR (Italian Ministry of University and Research) at the University of Palermo. In 2008, he was an Assistant Professor at the University of Palermo. He was a lecturer of Advanced Methodologies in Medicinal Chemistry, and currently, he is a lecturer of Advanced Medicinal Chemistry and Drug Design. In October 2019, he became an Associate Professor of Medicinal Chemistry at the University of Palermo. Marco Tutone is the owner of one international patent: US20210002238A1 *"Oxadiazole derivatives for the treatment of genetic diseases due to nonsense mutations"*. The research activity of Marco Tutone was oriented to studies and applications of molecular modeling in the field of Medicinal Chemistry (Molecular Dynamics, QSAR and 3D-QSAR, Pharmacophore Modeling, Molecular Docking, HTVS of compounds with potential antiviral, antitumor, anti-protozoarian activity). In recent years, his activity has been also oriented to the study of substances of natural origins, such as the betalains. He was involved in the design, synthesis, and biological evaluation of compounds with activity at the CNS level, in particular, dopaminergic modulators. Since 2014, he has been involved in the study of translational readthrough-inducing drugs (TRIDs) within the FFC projects.

Anna Maria Almerico

Anna Maria Almerico graduated in chemistry with full marks and honors in 1977 at the University of Palermo. During her academic career, she has been a Visiting Professor at the School of Chemical Sciences, University of East Anglia (UK, 1983–1984), at the College of Pharmacy, Ohio State University (USA, 1986–1987) and at Masaryk University, Brno (Czech Republic, 1993). Currently, she is a Full Professor of Medicinal Chemistry in the Faculty of Pharmacy, University of Palermo. Her scientific interests are mainly devoted to the design and synthesis of new nitrogen heterocycles of biological interest. In particular, her most recent studies are devoted to the preparation and evaluation of the biological properties of compounds related to well known anticancer/antiviral/antiparasitic drugs.

Preface to "Computational Approaches: Drug Discovery and Design in Medicinal Chemistry and Bioinformatics"

To date, computational approaches have been recognized as a key component in drug design and discovery workflows. Developed to help researchers save time and reduce costs, several computational tools have been developed and implemented in the last twenty years. At present, they are routinely used to identify a therapeutic target, understand ligand–protein and protein–protein interactions, and identify orthosteric and allosteric binding sites, but their primary use remains the identification of hits through ligand-based and structure-based virtual screening and the optimization of lead compounds, followed by the estimation of the binding free energy. The repurposing of an old drug for the treatment of new diseases, helped by in silico tools, has also gained a prominent role in virtual screening campaigns.

Moreover, the availability and the decreasing cost of hardware and software, together with the development of several web servers on which to upload and download computational data, have contributed to the success of computer-assisted drug design. These improved, accurate, and reliable methods should help to add new and more potent molecules to the group of approved drugs. Nevertheless, the ease of access of computational tools in drug design (software, databases, libraries, and web servers) should not encourage users with little or almost no knowledge of the underlying physical basis of the methods used, who could compromise the interpretation of the results. The role of the computational (medicinal) chemist should be recognized and included in all research groups. These considerations led us to promote a volume collecting original contributions regarding all aspects of the computational approaches, such as docking, induced-fit docking, molecular dynamics simulations, free energy calculations, and reverse modeling. We also include ligand-based approaches, such as molecular similarity fingerprints, shape methods, pharmacophore modeling, and QSAR. Drug design and the development process strive to predict the metabolic fate of a drug candidate to establish a relationship between the pharmacodynamics and pharmacokinetics and highlight the potential toxicity of the drug candidate. Even though the use of computational approaches is often combined, we tried to identify which of these play a central role in each manuscript.

In this Special Issue, the use of molecular dynamics simulations, both unbiased and biased, cover a major part of the contributions. The non-covalent inhibition of the immunoproteasome was investigated in-depth through MD binding and binding pose metadynamics. MD simulations provided insight into the structural features of hTSPO (Translocator Protein) and the previously unknown interplay between PK11195, a molecule routinely used in positron emission tomography (PET) for the imaging of neuroinflammatory sites, and cholesterol. The interaction of certain endogen substrates, drug substrates, and inhibitors with wild-type MRP4 (WT-MRP4) and its variants, G187W and Y556C, were studied to determine differences in the intermolecular interactions and affinity related to SNPs using several approaches, but particularly all-atom, coarse-grained, and umbrella sampling molecular dynamics simulations (AA-MDS and CG-MDS, respectively). Natural sodium–glucose co-transporter 2 (SGLT2) inhibitors were selected to explore their potential against an emerging uropathogenic bacterial therapeutic target, i.e., FimH, which plays a critical role in the colonization of uropathogenic bacteria on the urinary tract surface, and molecular dynamics

simulations were carried out to study the potential interactions. Doxorubicin encapsulation in carbon nanotubes with haeckelite or Stone–Wales defects as drug carriers were investigated using a molecular dynamics approach. The combined use of different approaches has been reported in a series of papers associated with the virtual screening of libraries. Almeelebia and co. screened 224,205 natural compounds from the ZINC database against the catalytic site of the Mtb proteasome. Pharmacophore-based virtual screening and molecular docking were carried out to identify potential Src inhibitors starting from a total of 891 molecules. Finally, MD simulations identified two molecules as potential lead compounds against Src kinase. An in silico study identified a methotrexate analog as a potential inhibitor of drug-resistant human dihydrofolate reductase for cancer therapeutics. A structure-based method for high-throughput virtual screening aimed to identify new dual-target hit molecules for acetylcholinesterase, and the 7 nicotinic acetylcholine receptor was reported and confirmed in vitro. A new complementary computational analysis called "dock binning" evaluates antibody–antigen docking models to identify why and where they might compete in terms of possible binding sites on the antigen. Interesting drug repurposing strategies have been reported. Hudson and Samudrala presented a computational analysis of a novel drug opportunities (CANDO) platform for shotgun multitarget repurposing. It implements several pipelines for the large-scale modeling and simulation of interactions between comprehensive libraries of drugs/compounds and protein structures. Qi and co. data-mined the crowd extracted expression of differential signatures (CREEDS) database to evaluate the similarities between gene expression signature (GES) profiles from drugs and their indicated diseases for GES-guided drug-repositioning approaches. In late 2019, the SARS-CoV-2 pandemic focused the attention of many researchers on not only vaccines but also new antiviral drugs. These reasons boosted the use of computational approaches to explore large libraries of natural compounds, already approved drugs, and in-house and commercial compounds. In this Special Issue, Baig and co. studied the efficacy of the Mpro inhibitor PF-00835231 against Mpro and its reported mutants in clinical trials. Several in silico approaches were used to investigate and compare the efficacy of PF-00835231 and five drugs previously documented to inhibit Mpro. Li and co. computationally investigated the MPD3 phytochemical database along with the pool of reported natural antiviral compounds to be used against SARS-CoV-2. Pedretti and co., exploiting the availability of resolved structures, designed a structure-based computational approach. The innovative idea of their study was to exploit known inhibitors of SARS-CoV 3CL-Pro as a training set to perform and validate multiple virtual screening campaigns. In the context of antiviral drugs, Regad and co. investigated the emergence of HIV-2 resistance. They proposed a structural analysis of 31 drug-resistant mutants of HIV-2 protease (PR2), an important target against HIV-2 infection. A wide series of contributions regarding the use of QSAR, machine learning, and deep learning has reported interesting outcomes. A multiple-molecule drug design based on systems biology approaches and a deep neural network to mitigate human skin aging was developed by Yeh and co. With the proposed systems medicine design procedure, they not only shed light on the skin-aging molecular progression mechanisms, but they also suggested two multiple-molecule drugs to mitigate human skin aging. The construction of quantitative structure–activity relationship (QSAR) models was used to predict the biological activities of the compounds obtained with virtual screening and identify new selective chemical entities for the COX-2 enzyme. The three-dimensional QSAR model, employing a common-features pharmacophore as an alignment rule, was built on 20 highly active/selective HDAC1 inhibitors. The predictive power of the 3D QSAR model represents a useful filtering tool for screening large chemical databases, finding novel derivatives with improved HDAC1 inhibitory

activity. Different machine learning (ML) and deep learning (DL) algorithms using various integer and binary-type fingerprints were evaluated to develop quantitative structure–activity relationship (QSAR) models, which are important for hERG potassium channel blocker prediction.

Throughout this book, all the recent aspects of the computational approaches applied to several research fields are reported. We express our deep gratitude to all the contributors to this Special Issue for their commitment, hard work, and outstanding papers. We also thank all the reviewers involved in the manuscript revisions for their unpaid contributions to improve any aspects of the submitted works. Last but not least, we deeply thank Mrs. Jessie Zhang for her assistance during the period in which we served as guest editors.

Marco Tutone, Anna Maria Almerico
Editors

Article

Immunoproteasome and Non-Covalent Inhibition: Exploration by Advanced Molecular Dynamics and Docking Methods

Giulia Culletta [1,2] , Maria Zappalà [2] , Roberta Ettari [2], Anna Maria Almerico [1] and Marco Tutone [1,*]

[1] Dipartimento di Scienze e Tecnologie Biologiche Chimiche e Farmaceutiche (STEBICEF), Università degli Studi di Palermo, Via Archirafi 32, 90123 Palermo, Italy; giulia.culletta@unime.it (G.C.); annamaria.almerico@unipa.it (A.M.A.)

[2] Dipartimento di Scienze Chimiche, Biologiche, Farmaceutiche ed Ambientali, Università di Messina, Viale Annunziata, 98168 Messina, Italy; maria.zappala@unime.it (M.Z.); roberta.ettari@unime.it (R.E.)

* Correspondence: marco.tutone@unipa.it

Abstract: The selective inhibition of immunoproteasome is a valuable strategy to treat autoimmune, inflammatory diseases, and hematologic malignancies. Recently, a new series of amide derivatives as non-covalent inhibitors of the β1i subunit with K_i values in the low/submicromolar ranges have been identified. Here, we investigated the binding mechanism of the most potent and selective inhibitor, N-benzyl-2-(2-oxopyridin-1(2H)-yl)propanamide (**1**), to elucidate the steps from the ligand entrance into the binding pocket to the ligand-induced conformational changes. We carried out a total of 400 ns of MD-binding analyses, followed by 200 ns of plain MD. The trajectories clustering allowed identifying three representative poses evidencing new key interactions with Phe31 and Lys33 together in a flipped orientation of a representative pose. Further, Binding Pose MetaDynamics (BPMD) studies were performed to evaluate the binding stability, comparing **1** with four other inhibitors of the β1i subunit: N-benzyl-2-(2-oxopyridin-1(2H)-yl)acetamide (**2**), N-cyclohexyl-3-(2-oxopyridin-1(2H)-yl)propenamide (**3**), N-butyl-3-(2-oxopyridin-1(2H)-yl)propanamide (**4**), and (S)-2-(2-oxopyridin-1(2H)-yl)-N,4-diphenylbutanamide (**5**). The obtained results in terms of free binding energy were consistent with the experimental values of inhibition, confirming **1** as a lead compound of this series. The adopted methods provided a full dynamic description of the binding events, and the information obtained could be exploited for the rational design of new and more active inhibitors.

Keywords: immunoproteasome; non-covalent inhibitors; molecular dynamics; MD binding; metadynamics; induced-fit docking

check for **updates**

Citation: Culletta, G.; Zappalà, M.; Ettari, R.; Almerico, A.M.; Tutone, M. Immunoproteasome and Non-Covalent Inhibition: Exploration by Advanced Molecular Dynamics and Docking Methods. *Molecules* **2021**, *26*, 4046. https://doi.org/10.3390/molecules26134046

Academic Editor: Brullo Chiara

Received: 20 May 2021
Accepted: 29 June 2021
Published: 2 July 2021

Publisher's Note: MDPI stays neutral with regard to jurisdictional claims in published maps and institutional affiliations.

1. Introduction

Protein turnover is essential for cellular function and homeostasis; in eukaryotic cells, the ubiquitin-proteasome system (UPS) is the central non-lysosomal pathway devoted to protein degradation. Whereas the lysosomal pathway mainly degrades membrane proteins or extracellular proteins imported into the cell by endocytosis, UPS, present both in the cytoplasm and nucleus, controls the degradation of damaged, incorrectly synthesized, or no longer useful intracellular proteins. Proteins are firstly tagged with several ubiquitin units; then, the polyubiquitinated proteins are rapidly hydrolyzed to small peptides by the proteasome, whereas ubiquitin is released and recycled [1]. The 26S constitutive proteasome consists of a barrel-shaped 20S catalytic core and two 19S regulatory caps. The catalytic core is constituted of four packed rings, each composed of seven different subunits, the two outer α, and the two inner β, respectively. The proteolytic activities reside in the β1c, β2c, and β5c subunits that are responsible for caspase-like (C-L), trypsin-like (T-L), and chymotrypsin-like (ChT-L) activities, respectively.

Immunoproteasome is a specialized form of proteasome present in the vertebrates, constitutively expressed in lymphocytes and monocytes and induced by cytokines, such as IFN-α and TNF-α, in many other cell types. In immunoproteasomes, the constitutive catalytic subunits (β1c, β2c, and β5c) are replaced by the corresponding immunosubunits: β1i, β2i, and β5i. While β2i and β5i maintain the same type of activities as the β2c and β5c subunits, β1i mainly performs a ChT-L activity, thus cleaving peptides after hydrophobic amino acids [2]. High levels of immunoproteasomes have been found in a wide number of inflammatory diseases, such as Crohn's disease or inflammatory bowel disease, and autoimmune diseases like rheumatoid arthritis or systemic lupus erythematosus [3]. Furthermore, immunoproteasomes are overexpressed in hematologic malignancies, including multiple myeloma or acute myeloid leukemia [4]. Therefore, the discovery of selective immunoproteasome inhibitors is pivotal to bring new chances for the treatment of the above-mentioned diseases. Exhaustive reports on selective covalent and non-covalent immunoproteasome inhibitors have been recently published [5,6]. The main class of covalent immunoproteasome inhibitors is that of peptide derivatives bearing an electrophile warhead able to interact with the nucleophilic hydroxyl group of catalytic Thr1. Just to give some examples, ONX-0914, a tripeptide α',β'-epoxyketone, was the first β5i-selective inhibitor identified; another α',β'-epoxyketone, UK-101, and the peptidyl aldehyde IPSI-001 showed a selective activity on the β1i subunit [5,6]. However, the covalent irreversible inhibition of a human enzyme is not always desirable in medicinal chemistry, as it can be responsible for potential toxicity due to off-target binding. Another drawback is that a single mutation in the catalytic amino acid (i.e., Thr1) could cause a loss of activity and acquired resistance. [7]. Non-covalent inhibition is therefore strongly desirable, because it is free of these disadvantages. Lacking a reactive warhead, non-covalent inhibitors may have an improved selectivity and less reactivity and instability and, therefore, may not exhibit the side effects that occur in covalent inhibitor therapies (e.g., liver toxicity and idiosyncratic adverse reactions) [8,9]. Furthermore, the enzyme–inhibitor complexes have reduced lifetimes, and this promotes an extensive tissue distribution of the drug [10]. To date, few non-covalent immunoproteasome inhibitors show selectivity towards the β1i and/or β5i subunits. One of them is Argyrin B, a natural cyclic peptide that is a reversible, noncompetitive inhibitor of β5i and β1i [8]. Other compounds are N,C-capped dipeptides, such as PKS2279 and PKS2252, in which the insertion of a β-amino acid markedly reduces the inhibitory potency against constitutive proteasomes, yet maintain potent inhibitory activity against immunoproteasomes [11]. Recently, some of us identified a panel of selective non-covalent inhibitors of the β1i and/or β5i subunits, characterized by a 2(1H)-pyridone scaffold linked to an amide function [12]. *N*-Benzyl-2-(2-oxopyridin-1(2H)-yl)propanamide (**1**) proved to be the most potent and selective inhibitor, with a Ki = 21 nM against the β1i subunit. Four other compounds of this series, *N*-benzyl-2-(2-oxopyridin-1(2H)-yl)acetamide (**2**), *N*-cyclohexyl-3-(2-oxopyridin-1(2H)-yl)propanamide (**3**), *N*-butyl-3-(2-oxopyridin-1(2H)-yl)propanamide (**4**), and (*S*)-2-(2-oxopyridin-1(2H)-yl)-*N*,4-diphenylbutanamide (**5**), showed remarkable inhibitory activity towards the β1i subunit (Figure 1). Derivatives **3–5** were also active against the β5i subunit, whereas none of the compounds **1–5** proved to affect the constitutive catalytic subunits.

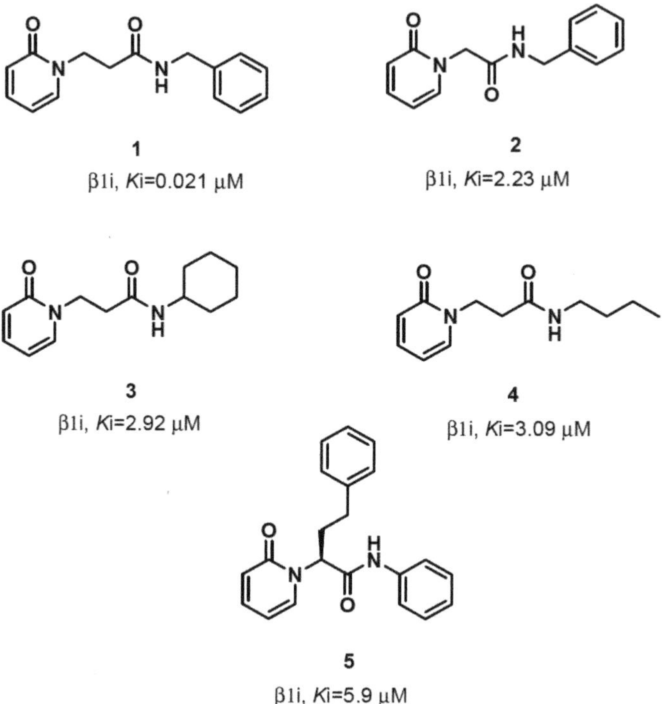

1

βli, *Ki*=0.021 μM

2

βli, *Ki*=2.23 μM

3

βli, *Ki*=2.92 μM

4

βli, *Ki*=3.09 μM

5

βli, *Ki*=5.9 μM

Figure 1. Structures and K_i values of the selective β1i inhibitors **1–5**.

The available experimental structures of immunoproteasomes provided the basis for several computational investigations. In the recent past, most of these studies made use of molecular docking methods. In particular, the binding mode of the non-covalent amide derivatives **1** and **2** was investigated at this level [12], while, to the best of our knowledge, the most accurate computational investigations were performed just on the β1i subunit (Figure 2A,B) and the peptide α',β'-epoxyketone UK101 (Figure 2C) using molecular dynamics (MD) simulations. The observed selectivity of UK101 for the β1i subunit is rationalized by the requirement for both a linear hydrocarbon chain at the N-terminus and a bulky group at the C-terminus of the inhibitor [13]. In recent years, the constant update of hardware capabilities allowed the development of enhanced MD methods able to provide a full dynamical description of the target–ligand-binding events [14]. These methods are usually employed given that the sampling issue is fundamental to describing these slow processes while docking methods continue to be pivotal to screening large libraries, also assisted by MD [15,16].

In this manuscript, we investigated the binding mechanism of the previously identified most active non-covalent amide **1** in the β1i subunit. For this purpose, we employed advanced molecular dynamics methods, such as MD binding (MDB) [17] and Binding Pose MetaDynamics (BPMD) [18]. In particular, we used the MDB tools implemented in the BiKi suite [19] to analyze the binding mechanism and gain insights into the ligand entrance pathway. Then, plain MD was performed to study the stability and conformational space into the immunoproteasome–ligand complex, thus allowing elucidation of the compound dynamic behaviors within the binding cavity. Successively, a cluster analysis provided representative poses that were used to evaluate the binding stability using the BPMD protocol. To have a comparative point of view, we also carried out induced-fit docking (IFD) and BPMD studies of the other four compounds (**2–5**) that showed high inhibitory

activities towards the β1i subunit. The results obtained could provide further information to develop the most selective and active immunoproteasome inhibitors.

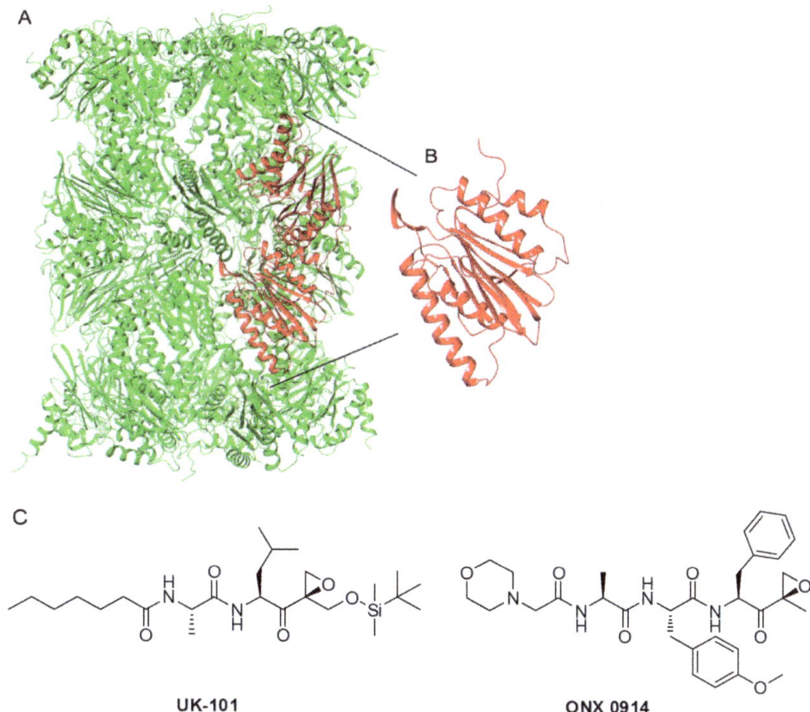

Figure 2. (**A**) 3D structure of the immunoproteasome, the two β1i subunits in red, (**B**) 3D structure of the β1i subunit, and (**C**) the structure of inhibitors UK-101 and ONX-0914.

2. Results

2.1. MD-Binding (MDB) Analysis

We began the study using the crystal structure of the murine immunoproteasome in complex with the inhibitor ONX-0914 (Figure 2C) bound to the β5i subunit (PDB ID: 3UNF) [20]. Murine and human immunoproteasomes share a sequence identity of more than 90%, and the few nonidentical residues were external to the active sites. In the literature, a crystal structure of human immunoproteasome was recently released in complex with a boronic acid derivative [21], but the docking of compound 1 was previously studied on the β1i subunit structure derived from the PDB ID:3UNF that do not bind any ligand. For these reasons, we used it as a starting point to carry out our simulations. To gain insights into the ligand-binding mode, we employed the MDB method to predict the path of ligand entrance into the cavity. This method has the advantage of describing complex events without incurring prohibitive time and computational costs. It is based on an adaptive, electrostatics-inspired bias and a campaign of trivially parallel short MD simulations to identify a near-native binding pose from the sampled configurations. At a reasonable computational cost, this method also provides accurate predictions when the starting protein conformation is different from that of the crystal complex, a known hurdle for traditional molecular docking [22]. The advanced proposed MDB protocol would require the identification of the binding pocket with NanoShaper software [23], which can identify the atoms facing the lumen pocket entrance in the protein target. According to NanoShaper software, the attractive protein residues selected were Thr1, Val20, Ser21, Phe31, Lys33, Leu45, Ser46, Gly47, Ser48, Ala49, Ala52, Ser129, and Ser168 (Figure 3).

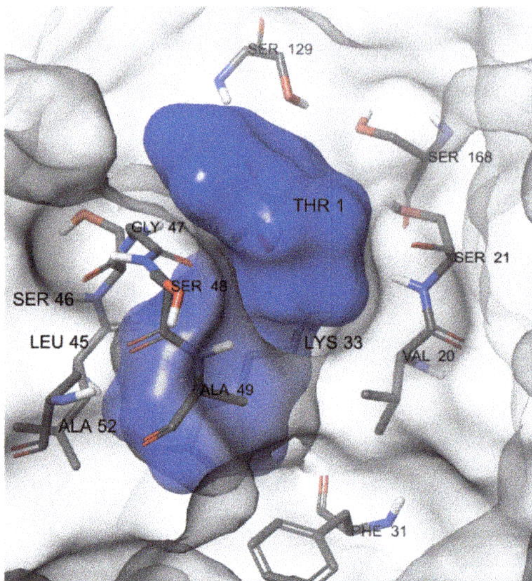

Figure 3. Identification of the binding cavity of the β1i subunit (solid blue) by NanoShaper software with the residues involved in the binding pocket.

Compound **1** is positioned with a random orientation at a predetermined distance, measured in terms of the thickness of the solvation shell around the ligand. From the set, we started 20 replicas of 20 ns for each entrance starting from the apo structure, thus collecting a total of 400 ns of MDB simulations.

The analysis of the results showed that the simulations overcome the energetic barrier in an average time of 2 ns, reaching the binding site. The unavailability of crystallographic structures for non-covalent ligand 1 did not allow the comparison of the conformations, employing the RMSD of the bound ligand. For these reasons, the RMSD of the protein backbone was used as a reference for any uncommon behaviors. All replicas showed a protein backbone RMSD average <2 Å, decreasing when the ligand arrived at the binding site (Supplementary Materials, Figures S1–S7). In most replicas, the ligand entered into the active site in the following 8 ns of the simulations, and in the last 10 ns, its refinement at the binding site was registered (Figure 4).

After the first 20 ns, the electrostatic bias was removed, and the sampling time was increased starting from the final frames of each MDB replica to enhance the sampling conformational changes and interactions of the ligand inside the binding site. For each replica, 10 ns more of the simulation was carried out, collecting a total of 600 ns of MD simulations. The plain MD simulations performed after the bias switch-off provided the local refinement of the binding mode. Once the binding simulation campaign was completed, the replicas ending without the ligand into the binding site were pruned out, and the remaining replicas were analyzed. The major part of the simulations showed a high stability, with the ligand strictly bound to the binding pocket, and in a few simulations, the ligand rapidly drifted away. Twelve replicas maintained a high stability at the binding site, as shown by averaged value of RMSD 1.5 Å of the complex (Supplementary Materials, Figures S8–S10). Then, the 12 replica trajectories were clustered. Each trajectory was recorded in 1000 frames, and these frames were clustered considering the RMSD of the binding site backbone (12,000 frames total). Each replica returned three representative clusters for a total of 36 MD representative poses. These last ones were further clustered and took into consideration the conformations of the ligand into the binding site and the most occurred interactions. In the end, it was possible to identify three final representative poses (pose 1, pose 2, and pose 3).

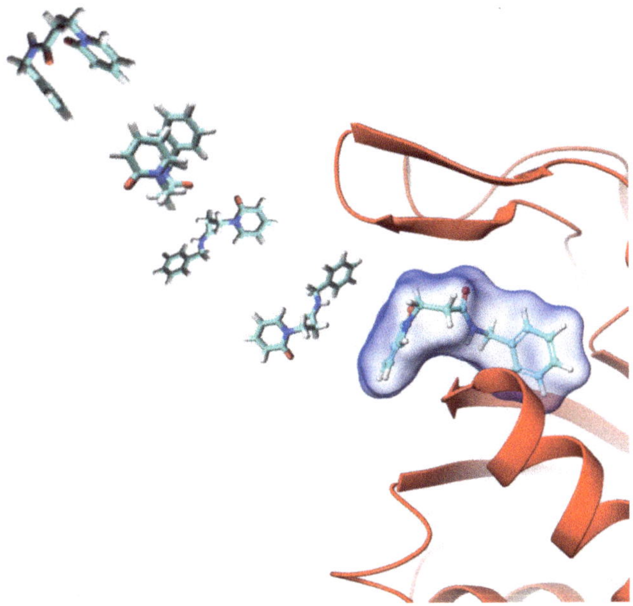

Figure 4. Time sequence of compound **1** approaching the active site of the β1i subunit studied by MD binding. The figure is representative of the 20 replicas.

The poses obtained from the clustering analysis were characterized by the important features observed during the simulations. In particular, in pose 1, two H-bonds were formed between the oxygen of the amide group and Ser21 and between the hydrogen of the amide group and Gly47 (Supplementary Information, Figure S11A). The binding of the ligand was strengthened by several van der Waals contacts between the benzyl group and the residues Val20, Phe31, Lys33, Gly47, Ala49, and Ala52. Val20, Ser21, Ser46, Gly47, Ala49, and Ala52 interacted with the linker between the two rings.

The identified pose 2 showed a series of noteworthy interactions that have not been previously identified. The benzyl group of pose 2 interacted by a pi-stacking interaction with Phe31. This pose was stabilized by several van der Waals contacts. The 2-pyridone moiety showed a series of contacts different from pose 1 (Lys33, Leu45, Ser46, Gly47, Ser48, Ala49, Ala52, Ser129, and Ser168). Ser21, Phe31, Ser46, and Gly47 interacted with the ethylene linker. It is worthy to note the absence of H-bonds in this pose (Supplementary Information, Figure S11B).

Pose 3 was characterized by the same H-bonds network observed in pose 1, with Ser21 and Gly47 residues. The 2-pyridone moiety formed one cation-pi-stacking interaction with the epsilon amino group of Lys33 (Supplementary Information, Figure S12A). As observed for the other poses, van der Waals contacts strengthened the ligand binding in pose 3. The benzyl group interacted with Ser21, Ala22, Leu45, and Ser46. The 2-pyridone moiety showed contacts with several residues: Phe31, Lys33, Gly47, Ser48, Ala49, and andAla52. The linker showed interactions with Thr1, Ser46, Gly47, and Ser168.

The major differences observed for these poses concerned the orientations of pose 1 and pose 3 related to their interactions with the residues of the binding site. In particular, besides the same H-bonds, a flipped orientation of the 2-pyridone and the benzyl moieties was observed. This evidence could reveal that the entrance mode of the ligand occurred in different ways without affecting the binding capability during the MD runs. The folded conformation assumed by the ligand in pose 2 seemed to represent an intermediate conformation. Concerning previous studies [12], two pi-stacking interactions and van der

Waals contacts between the rings and the residues Thr1, Val20, Phe31, Lys33, Leu45, Ser46, and Ala52 were identified (Figure 5).

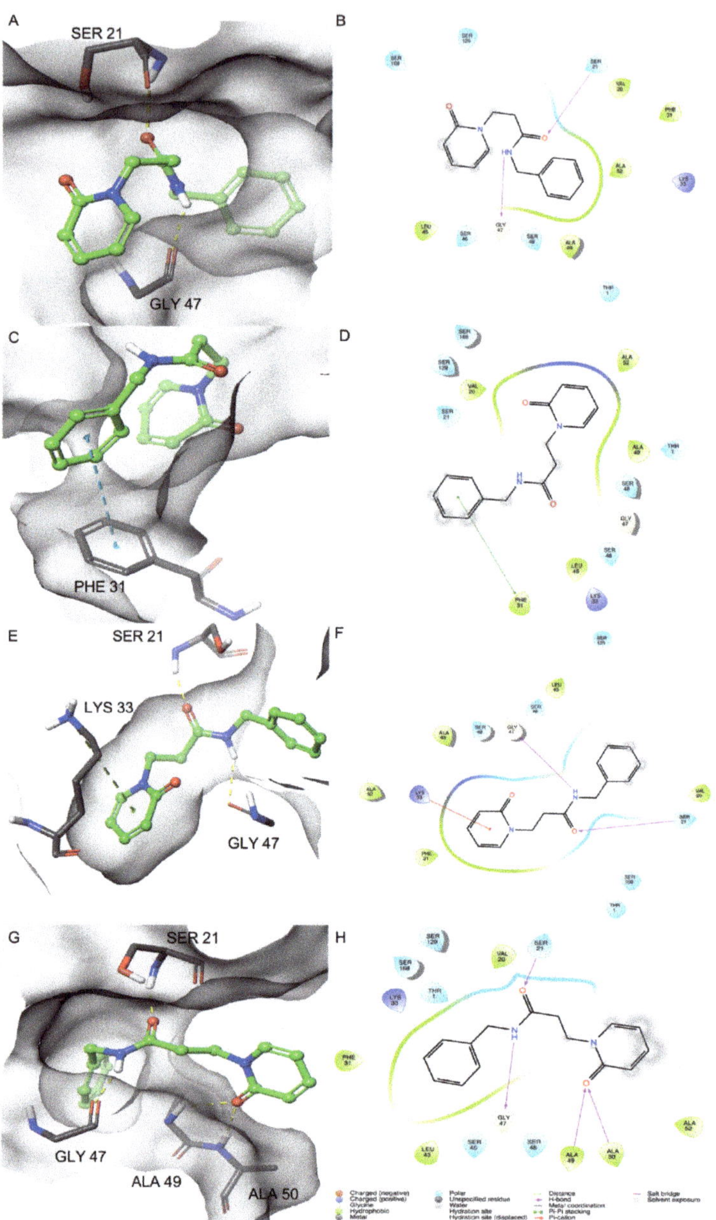

Figure 5. 3D and 2D-binding modes of compound 1: pose 1 (**A**,**B**), pose 2 (**C**,**D**), and pose 3 (**E**,**F**) after the MDB simulations and after IFD (**G**,**H**) into the β1i active site of murine immunoproteasome (PDB ID: 3UNF). In the 3D figures, the H-bonds are represented in yellow dashes, the cation-pi-stacking interactions in green dashes, and the pi-pi stacking in blue dashes.

2.2. Induced Fit Docking (IFD)

To add more information concerning the previous docking studies and compare the results obtained by the MDB, we used the more accurate protocol induced-fit docking (IFD) [24]. IFD predicts the ligand-binding modes and concomitant structural changes in the receptor. It confers flexibility to the protein side chains, allowing the ligand to adjust and optimize the binding interactions within the active site. IFD was carried out for compound 1 and for the other four compounds **2–5** that showed encouraging inhibitory activity on β1i. In the previous studies, classical docking was performed for compound **2**, while computational studies have not been performed yet for compounds **3–5**.

The best IFD pose of **1** reported the same two interactions observed after the MD simulation: the H-bonds between the residues Gly47 and Ser21 of the protein and NH amide and the carbonyl group of the molecule (Figure 5G,H). Besides, other hydrogen bonds were found. In particular, the carbonyl of 2-pyridone moiety formed an H-bond network with Ala49 and Ala50. Concerning the other MDB poses, this was a peculiar difference of the IFD pose that was not observed in the docking study. These residues, together with Ser48, stabilized the ligand binding by van der Waals contacts, such as observed in the previous docking study. Other van der Waals contacts were formed between the benzyl group and Val20, Phe31, Lys33, Leu45, Gly47, and Ala52. Finally, the ethylene linker between the rings interacted with Thr1, Val20, Ser21, Gly47, and Ala49. It is interesting to note that the pi-stacking interactions observed in pose 2 between the benzyl group and Phe31 and the cation-pi-stacking interaction in pose 3 between the 2-pyridone and Lys33 were not evidenced in the IFD pose but only as van der Waals contacts (Figure 5).

The other four analogs of amide 1 were characterized by structural variations at the *N*-substituent and the methylene/ethylene linker between the 2-pyridone scaffold and the amide function. Compound 2 showed a methylene linker between the 2-pyridone scaffold and the amide function, and the experimental activity was recorded with a *Ki* value of 2.23 μM on the β1i subunit. The best IFD pose for 2 showed three H-bonds: Ser21 with the carbonyl of amide and Gly47 with the NH amide and the carbonyl of 2-pyridone. The benzyl moiety of the molecule formed a cation-pi-stacking interaction with Lys33, as also evidenced for 1 in pose 3 (Figure 6A,B).

The cyclohexyl derivative **3** (*Ki* = 2.92 μM) formed four H-bonds. The residue Thr1 made two H-bonds with the carbonyl of amide and the carbonyl of 2-pyridone. Gly47 formed two H-bonds with NH amide and carbonyl of 2-pyridone (Figure 6C,D). The interactions of the best IFD pose of *n*-butyl derivative **4** (*Ki* = 3.09 μM) were characterized by two H-bonds between the carbonyl and NH of the amide of the molecule with Ser21 and Gly47, respectively. The 2-pyridone moiety formed pi-pi stacking with the Phe31 (Figure 6E,F). The last compound, (*S*)-2-(2-oxopyridin-1(2H)-yl)-*N*,4-diphenylbutanamide (**5**) (*Ki* = 5.9 μM), showed two H-bonds, one between Ser21 and carbonyl of amide and the other between Ala49 and carbonyl of 2-pyridone (Figure 6G,H). Additionally, for these molecules, the recurrent interactions were between the residues Ser21, Gly47, and the amide group, but it underlined the pi-stacking interactions with Phe31 and Lys33, which could constitute clear evidence of the key role of these residues in the inhibition pattern.

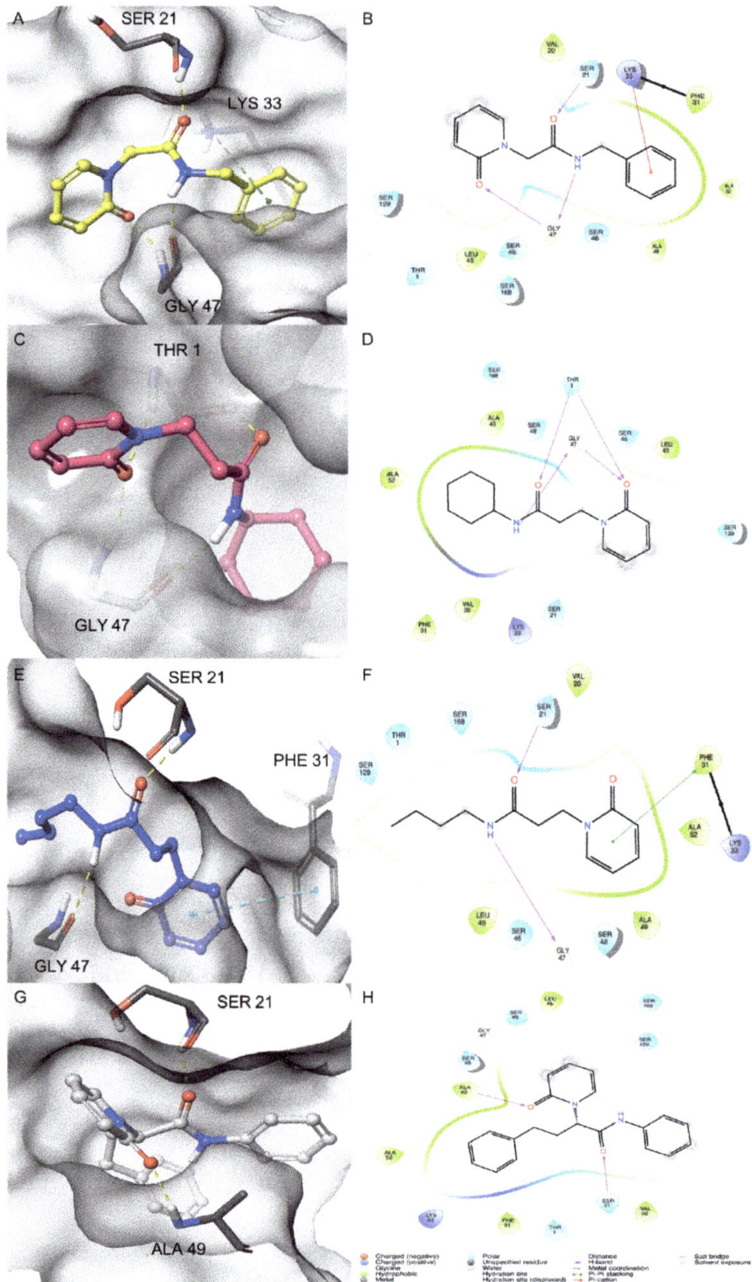

Figure 6. 3D and 2D-binding modes of compound **2** (**A,B**), compound **3** (**C,D**), compound **4** (**E,F**), and compound **5** (**G,H**) into the β1i active site of murine immunoproteasome (PDB ID: 3UNF) after the IFD study. In the 3D figures, the H-bonds are represented in yellow dashes, the cation-pi-stacking interactions in green dashes, and the pi-pi stacking in blue dashes.

2.3. Binding Pose MetaDynamics Analysis

Binding Pose MetaDynamics (BPMD) is an automated, enhanced sampling, metadynamics-based protocol, in which the ligand is forced to move around its binding pose. The possible higher mobility of the ligand under a biasing potential is a mark of the binding mode instability. This method showed the ability to reliably discriminate between the ligand-binding pose retrieved by MDB and a plausible alternative generated with IFD studies [22].

We decided to use BPMD to evaluate the affinity of the three representative poses obtained from MDB and the pose of IFD into the binding site for compounds **1** and **2–5**. The results were defined in terms of the pose stability based on the PoseScore, which is the RMSD of the ligand related to the starting coordinates of the heavy atoms of the ligand. Moreover, to evaluate the results, another metric is used, such as the PersScore, which is a clue of the H-bond formed between the ligand and the target during the simulations. The linear combination of these two scores provides a third score, the CompScore, which is calculated with Equation (1):

$$\text{CompScore} = \text{PoseScore} - 5 \times \text{PersScore} \tag{1}$$

Lower values of the CompScore indicate more stable complexes.

During the metadynamics simulations, pose 1 reached a steady PoseScore = 1.397, considered stable, while the PersScore showed that the hydrogen bonds identified at the start of the metadynamics run were kept for 60% of the averaged time (Figure 7A). In particular, the H-bond between the NH amide group of the ligand and Gly47 was kept for 88% of the simulation time, while the H-bond between the carbonyl of the ligand and Ser21 for 36% (Figure 7B). The CompScore value was -1.694. Due to the absence of H-bonds recorded, pose 2 with recorded pi stacking and van der Waals interactions showed the same value for the PoseScore and CompScore, 3.129 (Figure 7C), while, for pose 3, the scores were PoseScore = 3.349, PersScore = 0.223, and CompScore = 2.235, respectively (Figure 7E). As for pose 1, pose 3 kept the H-bond between NH amide and Gly47 as 26% and 18% between carbonyl and Ser21 (Figure 7F).

The PoseScore for the pose of amide 1 obtained by the IFD was 4.576, and the PersScore showed that the hydrogen bonds identified at the start of the MetaDynamics run were kept for 13% of the averaged time. The value of the CompScore was 3.917 (Figure 7G). It is interesting to point out that, of the four H-bonds detected by IFD, the two interactions between the amide group and Ser21 and Gly47 were maintained—in particular, the interaction between NH amide and Gly47 for 43% and 9% between carbonyl and Ser21 (Figure 7H).

The RMSD values and the percentage of the H-bonds retrieved from BPMD studies for the amide 1 in the three MDB poses and in the IFD pose showed that pose 1 could be considered more stable. Pose 1, pose 3, and the IFD pose adopted the same plain conformation and H-bonds between Ser21, Gly47, and the amide group. The differences were in the additional interactions between Ala49, Ala50, and the carbonyl of 2-pyridone, which led to a rotation of 2-pyridone, causing the ring to be specular in the IFD pose and showed a high value of RMSD (4.02 Å).

The BPMD analysis was also carried out for compounds **2–5** to evaluate their binding stability with respect to the most active compound of the series, **1**. The results of the BPMD calculations are reported in Figure 8. As can be evidenced from the plots, all showed PoseScore values higher than the averaged PoseScore for **1**. The hydrogen bonds identified at the start of the MetaDynamics run were maintained for 10-30% of the averaged time (Figure 8B,D,F,H) The CompScore values for compounds **2–5** were 4.750, 4.276, 5.979, and 1.728, respectively. Moreover, MM-GBSA-binding free energy calculations for all the complexes were performed. The plot of the calculated ΔG binding vs. the Ki values is reported in Figure 9, and it shows an R^2 = 0.8071 (compound **1** ΔG = -52.912 Kcal/mol, compound **2** ΔG = -41.684 Kcal/mol, compound **3** ΔG = -41.355 Kcal/mol, compound **4** ΔG = -36.701 Kcal/mol, and compound **5** ΔG = -35.340 Kcal/mol).

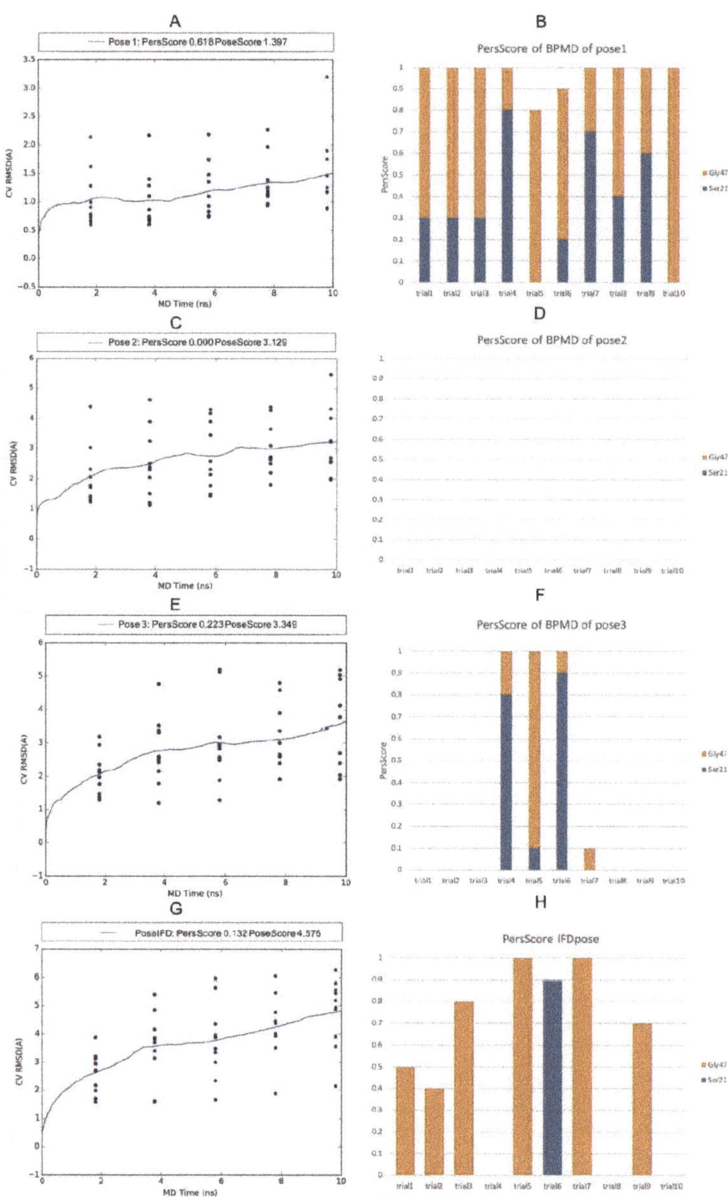

Figure 7. Plots of the RMSD estimate averaged over all 10 trials vs. the simulation time for the BindingPoseMetaDynamics runs: pose 1 (**A**), pose 2 (**C**), pose 3 (**E**), and IFD pose (**G**). Persistence Score Plots: pose 1 (**B**) pose 2 (**D**), pose 3 (**F**), and IFD pose (**H**). The blue dots represent the values of the CV RMSD at different times (2 ns, 4 ns, 6 ns, 8 ns, and 10 ns) for each simulation trial. The blue lines represent the mean CV RMSD values along the 10×10 ns of the simulation trials. The orange and blue bars represent the fraction of H-bonds maintained during the simulation for each trial.

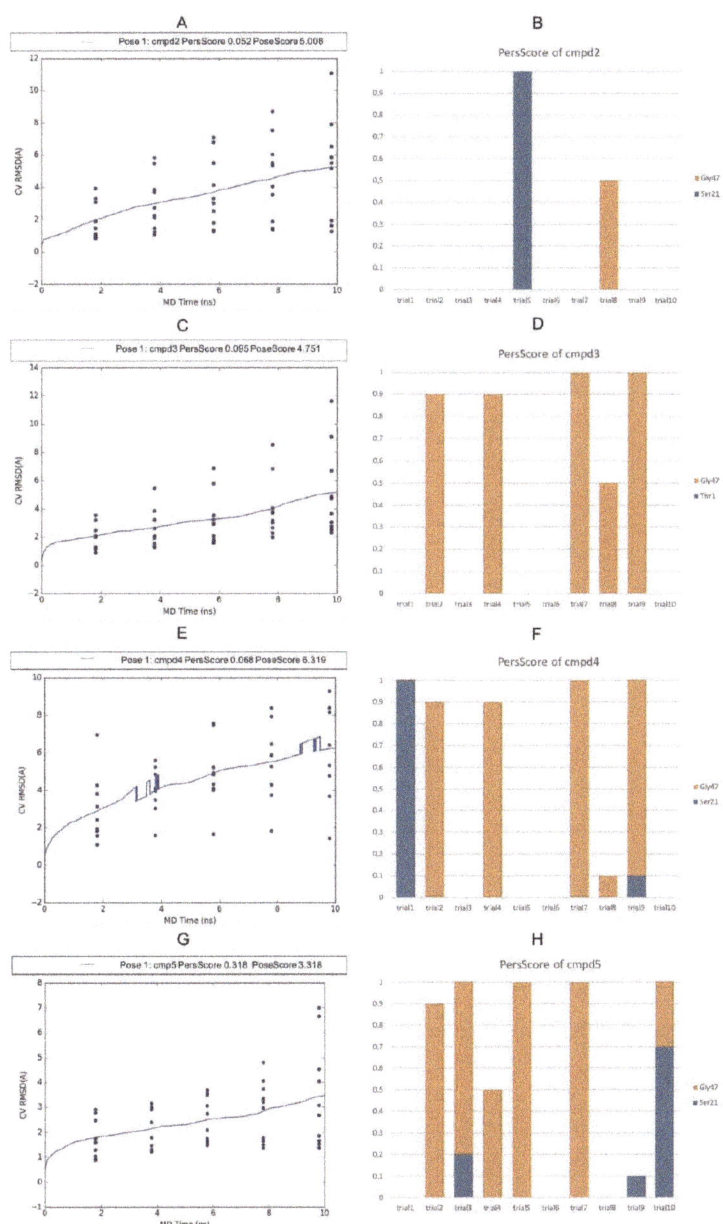

Figure 8. Plots of the RMSD estimate averaged over all 10 trials vs. the simulation time for the BindingPoseMetaDynamics runs: compound **2** (**A**), compound **3** (**C**), compound **4** (**E**), and compound **5** (**G**). Persistence Score plots of compounds **2** (**B**), compound **3** (**D**), compound **4** (**F**), and compound **5** (**H**). The blue dots represent the values of the CV RMSD at different times (2 ns, 4 ns, 6 ns, 8 ns, and 10 ns) for each simulation trial. The blue lines represent the mean CV RMSD values along the 10 × 10 ns of the simulation trials. The orange and blue bars represent the fraction of H-bonds maintained during the simulation for each trial.

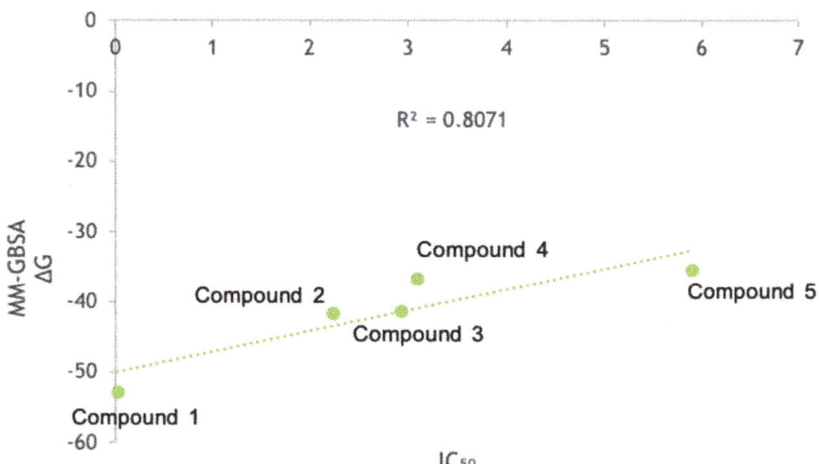

Figure 9. The plot of the MM-GBSA ΔG binding vs. the K_i values of compounds **1–5**. The binding free energy is expressed in Kcal/mol, and the IC_{50} is expressed in µM.

3. Discussion

The inhibition of the human immunoproteasome is a hot topic of recent years in medicinal chemistry due to its involvement in a wide range of diseases. Promising immunoproteasome inhibitors, both covalent and non-covalent, have been recently identified. In covalent inhibitors, the presence of a reactive warhead may cause significant off-target activities against other proteins, which may result in side effects (e.g., liver toxicity and idiosyncratic adverse reactions) and reduced selectivity over time [8,9]. For these reasons, the attention was focused on non-covalent immunoproteasome inhibitors. In this context, a series of amide derivative β1i subunit inhibitors with K_i values in the low micromolar or submicromolar range have been recently identified [12]. The use of computational approaches could characterize the binding process of these inhibitors—in particular, the use of advanced molecular dynamics approaches able to explore the dynamic features of the protein/ligand complex could overcome the limitations of semiflexible molecular docking methods in which the protein target is treated as a rigid body. Several advanced methods have been proposed in the last years for computing association and dissociation mechanisms, and all of them were shown to be promising in the interpretation of such mechanisms. With the aim to gain more insights into non-covalent inhibitors of the immunoproteasome, we decided to exploit these enhanced sampling methods.

Here, we investigated the dynamic binding mechanism of compound **1**, the most active of a series of non-covalent amide derivatives. With the aim of collecting mechanistic insight on the binding process, we performed the MDB protocol implemented in BiKi software to simulate the events that elapsed among the ligand unbound and the ligand entrance in the binding pocket. Successively, plain MD simulations were performed to extend the sampling of the bound states. The clustering of the survived complexes trajectories allowed identifying three representative poses (pose 1, pose 2, and pose 3) observed during the simulation. The most important interactions for the inhibition pattern were, in pose 1, two H-bonds between the amide group and Ser21 and Gly47 and, in pose 2, the benzyl group interacting by pi-pi stacking with Phe31. The residues Ser21 and Gly47 of pose 3 formed H-bonds with carbonyl and NH amide, and at the same time, the 2-pyridone moiety made a cation-pi-stacking interaction with the epsilon amino group of Lys33. Moreover, pose 1 showed a different orientation of the 2-pyridone moiety with respect to the docking and IFD studies. The 2-pyridone moiety was stabilized in the binding pocket by van der Waals contacts, as observed in MDB, while, in docking and IFD studies, it was stabilized by

H-bonds with Ala49 and Ala50 beyond van der Waals contacts. In pose 3, the peculiarity is represented by the 2-pyridone moiety interacting with the pi-stacking interaction with Lys33, which determined a flipped orientation with respect to pose 1. Finally, in pose 2, a different folded conformation with respect to pose 1 and pose 3 was observed, with a new pi-stacking interaction between the benzyl group and Phe31. The flipped orientations obtained for pose 1 and pose 3 suggested a different entrance mode of the ligand into the active site. Through the BPMD studies, it was possible to observe that both poses were stable, but, according to the RMSD value, the conformation of pose 1 showed major stability compared to pose 3 in the active site. Beyond the van der Waals contacts observed, the conformation of pose 1 could be strengthened by the pi-stacking interactions shown in pose 2 and pose 3 to improve the potency and selectivity of the β1i subunit. To pursue the matter, we also carried out IFD calculations for the other four amide derivatives **2–5** that showed an appreciable inhibitory activity on β1i. These studies revealed that Phe31 and Lys33 residues could play a key role in the inhibition pattern, in addition to the already known Ser21 and Gly47 ones, showing the importance not only of the hydrogen bonds but also of the pi-stacking interactions for the stabilization of the binding of the inhibitors. Moreover, the BPMD analysis confirmed the higher binding stability of inhibitor 1 with respect to the inhibitors 2–5, as evidenced by in vitro tests. Compound **1** showed the best CompScore (-1.694) with respect to the other compounds. The consistency of the computational analysis with the experimental data was further confirmed by the MM-GBSA-binding free energy calculations. These outputs were plotted against the experimental *Ki* values, and the $R^2 = 0.8071$ confirmed compound **1** as the best derivatives of this series.

4. Materials and Methods

4.1. System and Ligand Preparation

For the purposes of this study, we selected the catalytic subunit β1i (LMP2 and PSMB9) extracted from the murine i20S in complex with the inhibitor ONX-0914 bound to the β5i subunit (PDB ID: 3UNF) [20]. Both 20S subunits, murine and human, share a sequence identity of more than 90%, and the few nonidentical residues are external to the active sites. As reported in the literature, in the case of covalent cocrystallized inhibitors [25,26], the reactive residue at the catalytic site was rebuilt after removing the covalent inhibitor by breaking the covalent bond and filling in the open valence. In this case, the involved residue was Thr1. The protein was prepared with the Protein Preparation Wizard [27] included in the Maestro suite (Maestro, Schrödinger, LLC, 2021, New York, NY, USA): adding bond orders and hydrogen atoms to the crystal structure using the OPLS2005 force field. Next, Prime [28] was used to fix missing residues or atoms in the protein and to remove cocrystallized water molecules. The protonation states at pH 7.2 ± 0.2 of the protein and the ligand were evaluated using Epik 3.1 [29]. The hydrogen bonds were optimized through the reorientation of hydroxyl bonds, thiol groups, and amide groups. In the end, the systems were minimized with the value of convergence of the RMSD of 0.3 Å. The ligands were drawn using Marvin Sketch 19.25 [30]. Amide 1 was parameterized using the BiKi suite [19] at the AM1-BCC [22] level of theory. Partial charges were derived using the RESP method [23] in Antechamber [24]. Compounds **2–5** were prepared using Schrödinger LigPrep v. 2021-1 (LigPrep, Schrödinger, LLC, 2021, New York, NY, USA). The force field adopted was OPLS2005, and Epik was selected as the ionization tool at pH 7.0 ± 2.0. Tautomers generation was flagged, and the maximum number of conformers generated was set at 32.

4.2. MD-Binding Simulations

The MD-binding method [17] within the BiKi suite [19] (BiKi Technologies s.r.l., Genova, Italy) exploits an additive external force that is summed as the regular potential energy of the system to enhance the probability of observing the binding event. The bias is represented by external electrostatic-like forces acting between a subset of the residues of the binding site and the ligand. The intensity of the bias is controlled by the adaptivity

rules and gradually switches off as the ligand moves torward the subset of residues; after the conjectured passing of the transition state has occurred, it slowly recovers the behavior of classical unbiased MD [31].

The protocol for MD-binding consists of crucial steps: characterization of the binding pocket using NanoShaper [23] (BiKi Technologies s.r.l., Genova, Italy). NanoShaper calculations provide a characterization of the binding pocket, which identifies the atoms facing the pocket entrance in the protein structure. This information was then used by BiKi software for the initial ligand positioning outside the binding cavity. Subsequently, an additive external force was made to enhance the sampling of the binding event. Once the ligand was positioned through the "Residue Placement" tool in BiKi, the system was solvated in an orthorhombic box using the TIP3P water model [32]. A suitable number of counterions were added to neutralize the overall system. Then, the whole system underwent energy minimization by using the Amber99SB-ildn force field [33]. According to the standard protocol [17], four different consecutive equilibration steps were performed: (1) 100 ps in the NVT ensemble at 100 K with both the protein backbone and ligand restrained (1000 kJ/mol nm^2), (2) 100 ps in the NVT ensemble at 200 K with both the protein backbone and the ligand restrained, (3) 100 ps in the NVT ensemble at 300 K with the free protein and the ligand restrained, and (4) 1000 ps in the NPT ensemble at 300 K with the free protein and the ligand restrained. Electrostatic interactions were treated with the cutoff method for short-range interactions and with the particle mesh Ewald method for long-range interactions (rlist = 1.1 nm, cutoff distance = 1.1 nm, vdW distance = 1.1 nm, and PME order = 4). The constant temperature conditions were provided using the velocity rescale thermostat [34], which is a modification of Berendsen's coupling algorithm [35]. The coordinate output from the last simulation was then used as the input to produce the biased molecular dynamics. Finally, 20 replica production runs, 20-ns-long in the NVT ensemble at 300 K, were performed for each complex using C = 0.1 (the fraction of the felt force, here 10%), a smoothing window size of 1000 samples, and a maximal K(t) of 0.001 (maximal steering constant).

4.3. Plain MD Simulations

The plain MD simulations were carried out using Desmond 6.5 [36] using the OPLS2005 force field [37] (Desmond Molecular Dynamics System, D. E. Shaw Research, New York, NY, USA). The complexes were solvated in orthorhombic boxes using the TIP3P water model. Ions were added to neutralize the charges. The systems were minimized and equilibrated at a temperature of 303.15 K and a pressure of 1.013 bar. The system was simulated as an NPT ensemble; a Nose–Hover thermostat and Martyna–Tobia–Klein barostat were used. The integration time step was chosen to be 2 fs. To keep the hydrogen–heavy atom bonds rigid, the SHAKE algorithm was used. A 9 Å cutoff radius was set for the short-range Coulomb interactions, and smooth particle mesh Ewald was used for the long-range interactions. For each replica, we carried out 10-ns MD simulations for a total of 200 ns, with 1.2-ps detection ranges for energy and 4.8 ps for the trajectory frames. The stability of the systems was evaluated using the root mean square deviation (RMSD) of the aligned protein and ligand coordinate set calculated against the initial frame. Visualization and analysis of the MD trajectories were performed using the Desmond simulation analysis tools in Maestro. The trajectories frames were clustered according to the hierarchical cluster linkage method. The 1000 frames recorded in each simulation were clustered considering the binding site conformations into 10 clusters based on the atomic RMSDs.

4.4. Binding Pose MetaDynamics (BPMD)

Binding pose MetaDynamics (BPMD) is an automated, enhanced sampling, metadynamics-based protocol in which the ligand is forced to move around its binding pose. This method showed the ability to reliably discriminate between the correct ligand binding pose and plausible alternatives generated with docking or plain MD studies [18].

According to the protocol, 10 independent metadynamics simulations of 10 ns were performed using as a collective variable (CV) the measure of the root mean square deviation (RMSD) of the ligand heavy atoms with respect to their starting positions. The alignment before the RMSD calculations was done by selecting protein residues within 3 Å of the ligand. The Cαs of these binding site residues were then aligned to those of the first frame of the metadynamics trajectory before calculating the heavy atom RMSD to the ligand conformation in the first frame. The hill height and width were set to 0.05 kcal/ mol (about 1/10 of the characteristic thermal energy of the system, kBT) and 0.02 Å, respectively. Before the actual metadynamics run, the system was solvated in a box of SPC water molecules [38], followed by several minimizations and restrained MD steps that allow the system to slowly reach the desired temperature of 300 K, as well as releasing any bad contacts and/or strain from the initial starting structure. The final snapshot of the short unbiased MD simulation of 0.5 ns was then used as the reference for the following metadynamics production phase. After the simulation, the stability of the ligand during the course was represented by three scores: PoseScore, PersistenceScore (PersScore), and CompositeScore (CompScore). The PoseScore is indicative of the average RMSD from the starting pose. A steep increase of this value is a symptom that the ligand is not in a well-defined energy minimum and, probably, it might not have been accurately modeled. PersScore is a measure of the hydrogen bond persistence calculated in the last 2 ns of the simulation that have the same number of hydrogen bonds as the input structure, averaged over all the 10 repeated simulations. It covers a range between 0 and 1, where 0 indicates that either the starting ligand pose did not have any interactions with the target or that the interactions were lost during the simulations, while 1 indicates that the interactions between the staring ligand pose and the last 2 ns of the simulations were retained. CompositeScore is the linear combination of the PoseScore and PersScore; lower values equate to more stable complexes. Each complex, previously obtained, was run, Country) on a single node with a 1 GPU card NVIDIA GeForce RTX2070.

4.5. Induced-Fit Docking

The induced-fit protocol (IFD)—developed by Schrödinger [24]—is a method for modeling the conformational changes induced by ligand binding. This protocol models induced-fit docking of one or more ligands using the following steps, as also reported in references [39–42]. The protocol starts with an initial docking of each ligand using a softened potential (van der Waals radii scaling). Then, a side-chain prediction within a given distance of any ligand pose (5 Å) is performed. Subsequently, a minimization of the same set of residues and the ligand for each protein/ligand complex pose is performed. After this stage, any receptor structure in each pose reflects an induced fit to the ligand structure and conformation. Finally, the ligand is rigorously docked, using Glide XP (Glide, Schrödinger, LLC, 2021, New York, NY, USA), into the induced-fit receptor structure. IFD was performed using a standard protocol, and the OPLS2005 force field was chosen. The receptor box was centered on the active site of β1i, according to the NanoShaper calculations. During the initial docking procedure, the van der Waals scaling factor was set at 0.5 for both the receptor and ligand. The Prime refinement step was set on the side chains of residues within 5 Å of the ligand. For each ligand docked, a maximum of 20 poses was retained to then be redocked in XP mode.

4.6. MM-GBSA-Binding Free Energy Calculations

Prime/MM-GBSA was used for the estimation of ΔG binding. The MM-GBSA approach employs molecular mechanics, the generalized Born model, and the solvent accessibility method to elicit free energies from structural information, circumventing the computational complexity of free energy simulations, wherein the net free energy is treated as a sum of a comprehensive set of individual energy components, each with a physical basis [25]. In our study, the VSGB solvation model was chosen using the OPLS2005 force field with a minimized sampling method [28].

5. Conclusions

In this study, we investigated the mechanism of non-covalent inhibition of the potent and selective immunoproteasome inhibitor 1. For this purpose, we employed advanced molecular dynamics methods such as MD binding (MDB) and Binding Pose MetaDynamics (BPMD) and advanced docking methods such as induced-fit docking (IFD). MD binding allowed analyzing the binding mechanisms and gained insights into the ligand entrance pathway. Then, plain MD was performed to study the stability and conformational space in the immunoproteasome–ligand complex, thus allowing elucidation of the compound dynamic behavior within the binding cavity. These results were compared with the IFD poses of the other four inhibitors, revealing new key residues in the binding pattern, and confirmed the different binding stability of **1** with respect to the other compounds, **2–5**. The collected information and outcome could be useful in ameliorating the activity of compound 1 and providing a dynamical point of view for the definition of the pharmacophoric features that could be exploited through dynamic pharmacophore modeling approaches, such as the Common Hits Approach (CHA) [43] or MYSHAPE [44], for the scaffold-hopping of new non-covalent inhibitors through a virtual screening campaign.

Supplementary Materials: The following supplementary materials are available online, Figures S1–S12. Figure S1: MD-binding: Ligand RMSD calculated from the centroid of binding pocket and protein backbone RMSD (20 ns) for Replica 1 (**A,B**) and Replica 2 (**C,D**). Figure S2. MD-binding: Ligand RMSD calculated from the centroid of binding pocket and protein backbone RMSD (20 ns) for Replica 3 (**A,B**), Replica 4 (**C,D**), Replica 5 (**E,F**). Figure S3. MD-binding: Ligand RMSD calculated from the centroid of binding pocket and protein backbone RMSD (20 ns) for Replica 6 (**A,B**), Replica 7 (**C,D**), Replica 8 (**E,F**). Figure S4. MD-binding: Ligand RMSD calculated from the centroid of binding pocket and protein backbone RMSD (20 ns) for Replica 9 (**A,B**), Replica 10 (**C,D**), Replica 11 (**E,F**). Figure S5. MD-binding: Ligand RMSD calculated from the centroid of binding pocket and protein backbone RMSD (20 ns) for Replica 12 (**A,B**), Replica 13 (**C,D**), Replica 14 (**E,F**). Figure S6. MD-binding: Ligand RMSD calculated from the centroid of binding pocket and protein backbone RMSD (20 ns) for Replica 15 (**A,B**), Replica 16 (**C,D**), Replica 17 (**E,F**). Figure S7. MD-binding: Ligand RMSD calculated from the centroid of binding pocket and protein backbone RMSD (20 ns) for Replica 18 (**A,B**), Replica 19 (**C,D**), Replica 20 (**E,F**). Figure S8. Ligand and protein RMSD during MD-plain(10 ns). Replica 1 (**A**); Replica 2 (**B**); Replica 8 (**C**); Replica10 (**D**). Figure S9. Ligand and protein RMSD during MD-plain(10 ns). Replica 11 (**A**); Replica 12 (**B**); Replica 13 (**C**); Replica 14 (**D**). Figure S10. Ligand and protein RMSD during MD-plain(10 ns). Replica 15 (**A**); Replica 16 (**B**); Replica 17 (**C**); Replica 20 (**D**). Figure S11. Ligand Interaction Diagram of pose1 (**A**) and pose2 (**B**). Purple arrows show H-bond interactions and green line Pi-Pi stacking. Figure S12. Ligand Interaction Diagram of pose3 (**A**) and IFD pose (**B**). Purple arrows show H-bond interactions and red line Pi-cation.

Author Contributions: Conceptualization, G.C. and M.T.; methodology, G.C.; formal analysis, G.C.; investigation and data curation G.C. and M.T.; writing draft preparation, M.T.; writing—review and editing, G.C., M.T., M.Z., R.E. and A.M.A. All authors have read and agreed to the published version of the manuscript.

Funding: This research received no external funding.

Institutional Review Board Statement: Not applicable.

Informed Consent Statement: Not applicable.

Data Availability Statement: Not applicable.

Conflicts of Interest: The authors declare no conflict of interest.

Sample Availability: Samples of the compounds are not available from the authors.

References

1. DeMartino, G.N.; Gillette, T.G. Proteasomes: Machines for all Reasons. *Cell* **2007**, *129*, 659–662. [CrossRef]
2. Ferrington, D.A.; Gregerson, D.S. Immunoproteasome: Structure, Function, and Antigen Presentation. *Prog. Mol. Biol. Transl. Sci.* **2012**, *109*, 75–112. [CrossRef]
3. Ettari, R.; Previti, S.; Bitto, A.; Grasso, S.; Zappalà, M. Immunoproteasome-Selective Inhibitors: A Promising Strategy to Treat Hematologic Malignancies, Autoimmune and Inflammatory Diseases. *Curr. Med. Chem.* **2016**, *23*, 1217–1238. [CrossRef] [PubMed]
4. Kuhn, D.J.; Orlowski, R.Z. The Immunoproteasome as a Target in Hematologic Malignancies. *Semin. Hematol.* **2012**, *49*, 258–262. [CrossRef] [PubMed]
5. Ettari, R.; Zappalà, M.; Grasso, S.; Musolino, C.; Innao, V.; Allegra, A. Immunoproteasome-Selective and Non-Selective Inhibitors: A Promising Approach for the Treatment of Multiple Myeloma. *Pharmacol. Ther.* **2018**, *182*, 176–192. [CrossRef]
6. Xi, J.; Zhuang, R.; Kong, L.; He, R.; Zhu, H.; Zhang, J. Immunoproteasome-Selective Inhibitors: An Overview of Recent Developments as Potential Drugs for Hematologic Malignancies and Autoimmune Diseases. *Eur. J. Med. Chem.* **2019**, *182*, 111646. [CrossRef] [PubMed]
7. Ghosh, A.K.; Samanta, I.; Mondal, A.; Liu, W.R. Covalent Inhibition in Drug Discovery. *ChemMedChem* **2019**, *14*, 889–906. [CrossRef] [PubMed]
8. Allardyce, D.J.; Bell, C.M.; Loizidou, E.Z. Argyrin B, a Non-Competitive Inhibitor of the Human Immunoproteasome Exhibiting Preference for β1i. *Chem. Biol. Drug Des.* **2019**, *94*, 1556–1567. [CrossRef]
9. Zhan, W.; Singh, P.K.; Ban, Y.; Qing, X.; Ah Kioon, M.D.; Fan, H.; Zhao, Q.; Wang, R.; Sukenick, G.; Salmon, J.; et al. Structure–Activity Relationships of Noncovalent Immunoproteasome β5i-Selective Dipeptides. *J. Med. Chem.* **2020**, *63*, 13103–13123. [CrossRef]
10. Villoutreix, B.O.; Miteva, M.A.; Khatib, A.-M.; Cheng, Y.; Marechal, X.; Reboud-Ravaux, M.; Vidal, J. Blockade of the malignant phenotype by β-subunit selective noncovalent inhibition of immuno- and constitutive proteasomes. *Oncotarget* **2017**, *8*, 10437–10449. [CrossRef]
11. Singh, P.K.; Fan, H.; Jiang, X.; Shi, L.; Nathan, C.F.; Lin, G. Immunoproteasome β5i-Selective Dipeptidomimetic Inhibitors. *Chem. Med. Chem.* **2016**, *11*, 2127–2131. [CrossRef] [PubMed]
12. Ettari, R.; Cerchia, C.; Maiorana, S.; Guccione, M.; Novellino, E.; Bitto, A.; Grasso, S.; Lavecchia, A.; Zappalà, M. Development of Novel Amides as Noncovalent Inhibitors of Immunoproteasomes. *Chem. Med. Chem.* **2019**, *14*, 842–852. [CrossRef] [PubMed]
13. Lei, B.; Abdul Hameed, M.D.; Hamza, A.; Wehenkel, M.; Muzyka, J.L.; Yao, X.J.; Kim, K.B.; Zhan, C.G. Molecular Basis of the Selectivity of the Immunoproteasome Catalytic Subunit LMP2-Specific Inhibitor Revealed by Molecular Modeling and Dynamics Simulations. *J. Phys. Chem. B* **2010**, *114*, 12333–12339. [CrossRef]
14. De Vivo, M.; Masetti, M.; Bottegoni, G.; Cavalli, A. Role of Molecular Dynamics and Related Methods in Drug Discovery. *J. Med. Chem.* **2016**, *59*, 4035–4406. [CrossRef] [PubMed]
15. Culletta, G.; Almerico, A.M.; Tutone, M. Comparing molecular dynamics-derived pharmacophore models with docking: A study on CDK-2 inhibitors. *Chem. Data Coll.* **2020**, *28*, 100485. [CrossRef]
16. Tutone, M.; Culletta, G.; Livecchi, L.; Almerico, A.M. A Definitive Pharmacophore Modelling Study on CDK2 ATP Pocket Binders: Tracing the Path of New Virtual High-Throughput Screenings. *Curr. Drug Discov. Technol.* **2020**, *17*, 740–747. [CrossRef]
17. Spitaleri, A.; Decherchi, S.; Cavalli, A.; Rocchia, W. Fast Dynamic Docking Guided by Adaptive Electrostatic Bias: The MD-Binding Approach. *J. Chem. Theory Comput.* **2018**, *14*, 1727–1736. [CrossRef]
18. Fusani, L.; Palmer, D.S.; Somers, D.O.; Wall, I.D. Exploring Ligand Stability in Protein Crystal Structures Using Binding Pose Metadynamics. *J. Chem. Inf. Model.* **2020**, *60*, 1528–1539. [CrossRef]
19. Decherchi, S.; Bottegoni, G.; Spitaleri, A.; Rocchia, W.; Cavalli, A. BiKi Life Sciences: A New Suite for Molecular Dynamics and Related Methods in Drug Discovery. *J. Chem. Inf. Model.* **2018**, *58*, 219–224. [CrossRef]
20. Huber, E.M.; Basler, M.; Schwab, R.; Heinemeyer, W.; Kirk, C.J.; Groettrup, M.; Groll, M. Immuno- and constitutive proteasome crystal structures reveal differences in substrate and inhibitor specificity. *Cell* **2012**, *148*, 727–738. [CrossRef]
21. Ladi, E.; Everett, C.; Stivala, C.E.; Daniels, B.E.; Durk, M.R.; Harris, S.F.; Huestis, M.P.; Purkey, H.E.; Staben, S.T.; Augustin, M.; et al. Design and Evaluation of Highly Selective Human Immunoproteasome Inhibitors Reveal a Compensatory Process that Preserves Immune Cell Viability. *J. Med. Chem.* **2019**, *62*, 7032–7041. [CrossRef]
22. Lexa, K.W.; Carlson, H.A. Protein flexibility in docking and surface mapping. *Q. Rev. Biophys.* **2012**, *45*, 301–343. [CrossRef] [PubMed]
23. Decherchi, S.; Rocchia, W. A general and Robust Ray-Casting-Based Algorithm for Triangulating Surfaces at the Nanoscale. *PLoS ONE* **2013**, *8*, e59744. [CrossRef] [PubMed]
24. Sherman, W.; Day, T.; Jacobson, M.P.; Friesner, R.A.; Farid, R. Novel Procedure for Modeling Ligand/Receptor Induced Fit Effects. *J. Med. Chem.* **2006**, *49*, 534–553. [CrossRef] [PubMed]
25. Culletta, G.; Gulotta, M.R.; Perricone, U.; Zappalà, M.; Almerico, A.M.; Tutone, M. Exploring the SARS-CoV-2 Proteome in the Search of Potential Inhibitors via Structure-Based Pharmacophore Modeling/Docking Approach. *Computation* **2020**, *8*, 77. [CrossRef]

26. Selvaraj, C.; Panwar, U.; Dinesh, D.C.; Boura, E.; Singh, P.; Dubey, V.K.; Singh, S.K. Microsecond MD Simulation and Multiple-Conformation Virtual Screening to Identify Potential Anti-COVID-19 Inhibitors Against SARS-CoV-2 Main Protease. *Front. Chem.* **2021**, *8*, 595273. [CrossRef] [PubMed]

27. Madhavi Sastry, G.; Adzhigirey, M.; Day, T.; Annabhimoju, R.; Sherman, W. Protein and Ligand Preparation: Parameters, Protocols, and Influence on Virtual Screening Enrichments. *J. Comput.-Aided Mol. Des.* **2013**, *27*, 221–234. [CrossRef] [PubMed]

28. *Prime*; Schrödinger, LLC: New York, NY, USA, 2021.

29. *Epik*; Schrödinger, LLC: New York, NY, USA, 2021.

30. Marvin Sketch 19.25, ChemAxon. Available online: https://www.chemaxon.com (accessed on 2 July 2021).

31. Motta, S.; Callea, L.; Giani Tagliabue, S.; Bonati, L. Exploring the PXR Ligand-Binding Mechanism with Advanced Molecular Dynamics Methods. *Sci. Rep.* **2018**, *8*, 16207. [CrossRef]

32. Jorgensen, W.L.; Chandrasekhar, J.; Madura, J.D.; Impey, R.W.; Klein, M.L. Comparison of Simple Potential Functions for Simulating Liquid Water. *J. Chem. Phys.* **1983**, *79*, 926–935. [CrossRef]

33. Lindorff-Larsen, J.K.; Piana, S.; Palmo, K.; Maragakis, P.; Klepeis, J.L.; Dror, R.O.; Shaw, D.E. Improved Side-Chain Torsion Potentials for the Amber ff99SB Protein Force Field. *Proteins Struct. Funct. Genet.* **2010**, *78*, 1950–1958. [CrossRef] [PubMed]

34. Bussi, G.; Donadio, D.; Parrinello, M. Canonical Sampling through Velocity Rescaling. *J. Chem. Phys.* **2007**, *126*, 014101. [CrossRef]

35. Berendsen, H.J.C.; Postma, J.P.M.; van Gunsteren, W.F.; DiNola, A.; Haak, J.R. Molecular Dynamics with Coupling to an External Bath. *J. Chem. Phys.* **1984**, *81*, 3684–3690. [CrossRef]

36. Bowers, K.J.; Chow, E.; Xu, H.; Dror, R.O.; Eastwood, M.P.; Gregersen, B.A.; Klepeis, J.L.; Kolossvary, I.; Moraes, M.A.; Sacerdoti, F.D.; et al. Scalable Algorithms for Molecular Dynamics Simulations on Commodity Clusters. In *Proceedings of the 2006 ACM/IEEE SC|06 Conference (SC'06)*; Association for Computing Machinery: Tampa, FL, USA, 2006; ISBN 978-0-7695-2700-0.

37. Robertson, M.J.; Tirado-Rives, J.; Jorgensen, W.L. Improved Peptide and Protein Torsional Energetics with the OPLS-AA Force Field. *J. Chem. Theory Comput.* **2015**, *11*, 3499–3509. [CrossRef]

38. Dyer, K.M.; Perkyns, J.S.; Stell, G.; Pettitt, B.M. Site-renormalised Molecular Fluid Theory: On the Utility of a Two-site Model of Water. *Mol. Phys.* **2009**, *107*, 423–431. [CrossRef]

39. Almerico, A.M.; Tutone, M.; Pantano, L.; Lauria, A. Molecular Dynamics Studies on Mdm2 Complexes: An Analysis of the Inhibitor Influence. *Biochem. Biophys. Res. Commun.* **2012**, *424*, 341–347. [CrossRef] [PubMed]

40. Tutone, M.; Pibiri, I.; Lentini, L.; Pace, A.; Almerico, A.M. Deciphering the Nonsense Readthrough Mechanism of Action of Ataluren: An in Silico Compared Study. *ACS Med. Chem. Lett.* **2019**, *10*, 522–527. [CrossRef] [PubMed]

41. Pibiri, I.; Lentini, L.; Melfi, R.; Tutone, M.; Baldassano, S.; Ricco Galluzzo, P.; Di Leonardo, A.; Pace, A. Rescuing the CFTR protein function: Introducing 1,3,4-oxadiazoles as translational readthrough inducing drugs. *Eur. J. Med. Chem.* **2018**, *159*, 126–142. [CrossRef] [PubMed]

42. Almerico, A.M.; Tutone, M.; Lauria, A. Docking and multivariate methods to explore HIV-1 drug-resistance: A comparative analysis. *J. Comput.-Aided Mol. Des.* **2008**, *22*, 287–297. [CrossRef]

43. Wieder, M.; Garon, A.; Perricone, U.; Boresch, S.; Seidel, T.; Almerico, A.M.; Langer, T. Common Hits Approach: Combining Pharmacophore Modeling and Molecular Dynamics Simulations. *J. Chem. Inf. Model.* **2017**, *57*, 365–385. [CrossRef]

44. Perricone, U.; Wieder, M.; Seidel, T.; Langer, T.; Padova, A.; Almerico, A.M.; Tutone, M. A Molecular Dynamics-Shared Pharmacophore Approach to Boost Early-Enrichment Virtual Screening: A Case Study on Peroxisome Proliferator-Activated Receptor α. *Chem. Med. Chem.* **2017**, *12*, 1399–1407. [CrossRef] [PubMed]

Article

The Interplay of Cholesterol and Ligand Binding in *h*TSPO from Classical Molecular Dynamics Simulations

Hien T. T. Lai [1], Alejandro Giorgetti [2,3,4], Giulia Rossetti [2,3,5,6], Toan T. Nguyen [1,*], Paolo Carloni [2,3,7,8] and Agata Kranjc [2,3,9,10,*]

1   VNU Key Laboratory for Multiscale Simulation of Complex Systems, VNU University of Science, Vietnam National University, Hanoi 11416, Vietnam; laithithuhien_t60@hus.edu.vn

2   Institute of Neuroscience and Medicine (INM-9), Forschungszentrum Jülich, D-52425 Jülich, Germany; alejandro.giorgetti@univr.it (A.G.); g.rossetti@fz-juelich.de (G.R.); p.carloni@fz-juelich.de (P.C.)

3   Institute for Advanced Simulation (IAS-5), Forschungszentrum Jülich, D-52425 Jülich, Germany

4   Department of Biotechnology, University of Verona, Strada Le Grazie 15, 37134 Verona, Italy

5   Jülich Supercomputing Center (JSC), Forschungszentrum Jülich, 52428 Jülich, Germany

6   University Hospital Aachen, RWTH Aachen University, 52078 Aachen, Germany

7   Department of Physics, RWTH Aachen University, 52078 Aachen, Germany

8   JARA-BRAIN Institute "Molecular Neuroscience and Neuroimaging" INM-11, Forschungszentrum Jülich, 52428 Jülich, Germany

9   Laboratoire de Biochimie Théorique, UPR 9080 CNRS, Université de Paris, 13 rue Pierre et Marie Curie, F-75005 Paris, France

10   Institut de Biologie Physico-Chimique-Fondation Edmond de Rotschild, PSL Research University, 75005 Paris, France

*   Correspondence: toannt@hus.edu.vn (T.T.N.); a.kranjc.pietrucci@fz-juelich.de (A.K.)

Citation: Lai, H.T.T.; Giorgetti, A.; Rossetti, G.; Nguyen, T.T.; Carloni, P.; Kranjc, A. The Interplay of Cholesterol and Ligand Binding in *h*TSPO from Classical Molecular Dynamics Simulations. *Molecules* **2021**, *26*, 1250. https://doi.org/10.3390/molecules26051250

Academic Editor: Marco Tutone and Anna Maria Almerico

Received: 11 December 2020
Accepted: 3 February 2021
Published: 26 February 2021

Publisher's Note: MDPI stays neutral with regard to jurisdictional claims in published maps and institutional affiliations.

Abstract: The translocator protein (TSPO) is a 18kDa transmembrane protein, ubiquitously present in human mitochondria. It is overexpressed in tumor cells and at the sites of neuroinflammation, thus representing an important biomarker, as well as a promising drug target. In mammalian TSPO, there are cholesterol–binding motifs, as well as a binding cavity able to accommodate different chemical compounds. Given the lack of structural information for the human protein, we built a model of human (*h*) TSPO in the apo state and in complex with PK11195, a molecule routinely used in positron emission tomography (PET) for imaging of neuroinflammatory sites. To better understand the interactions of PK11195 and cholesterol with this pharmacologically relevant protein, we ran molecular dynamics simulations of the apo and holo proteins embedded in a model membrane. We found that: (i) PK11195 stabilizes *h*TSPO structural fold; (ii) PK11195 might enter in the binding site through transmembrane helices I and II of *h*TSPO; (iii) PK11195 reduces the frequency of cholesterol binding to the lower, N–terminal part of *h*TSPO in the inner membrane leaflet, while this impact is less pronounced for the upper, C–terminal part in the outer membrane leaflet, where the ligand binding site is located; (iv) very interestingly, cholesterol most frequently binds *simultaneously* to the so-called CRAC and CARC regions in TM V in the free form (residues L150–X–Y152–X(3)–R156 and R135–X(2)–Y138–X(2)–L141, respectively). However, when the protein is in complex with PK11195, cholesterol binds equally frequently to the CRAC–resembling motif that we observed in TM I (residues L17–X(2)–F20–X(3)–R24) and to CRAC in TM V. We expect that the CRAC–like motif in TM I will be of interest in future experimental investigations. Thus, our MD simulations provide insight into the structural features of *h*TSPO and the previously unknown interplay between PK11195 and cholesterol interactions with this pharmacologically relevant protein.

Keywords: *h*TSPO; PK11195; cholesterol; homology modeling; molecular dynamics (MD) simulation

1. Introduction

The translocator protein (TSPO) is a transmembrane protein (18kDa), evolutionary conserved and expressed in different organisms, from bacteria to humans [1]. Its biological

functions are conserved throughout the phylogenetic spectrum, like tetrapyrrole biosynthesis and/or sterol metabolism [1,2]. Indeed, the bacterial TSPO homology in *Rhodobacter sphaeroides* can be functionally replaced by rat TSPO [3], despite that these proteins share only about 30% sequence identity. The human protein (*h*TSPO) is expressed in all tissues and located in the outer mitochondrial membrane [4,5]. Its highest expression levels are found in steroid-synthesizing cells of endocrine organs indicating that it may play an important role in steroid synthesis from cholesterol [6]. Mammalian TSPO binds cholesterol with high affinity by the cholesterol recognition/interaction amino acid consensus (CRAC) motif (residues 150–156) [7,8]. This motif is preceded by a short three amino acids sequence L144–A145–F146 (LAF) that is highly conserved in mammalian TSPO. In the experiment with the *Rs*TSPO mutant, where the three amino acids (A136–T137–A138) preceding the CRAC region were replaced by mammalian LAF sequence, it was shown that the LAF motif greatly increased the binding affinity for cholesterol with respect to the original bacterial sequence [9]. An additional binding prediction sequence for cholesterol was found in TSPO—the inverse version of CRAC, the CARC motif (residues 135–141) [10]. While in the case of the nicotine acetylcholine receptor, functional studies clearly show that a substitution of a specific amino acid in CARC slows the kinetics of cholesterol binding, in the case of TSPO, it is not yet known whether CARC binds cholesterol as well [11]. In addition, TSPO has been proposed to play an important role in other cellular processes like porphyrin transport [12,13], mitochondrial respiration [4,14], and immunomodulation [15].

TSPO expression is highly upregulated in cancer and at the sites of neuroinflammation processes in cerebral ischemia, Alzheimer's, Parkinson's, and Huntington's diseases, and multiple sclerosis (reviewed in [16]). In addition, a human single nucleotide polymorphism of TSPO (A147T) is associated with different psychiatric disorders, like bipolar disorder, anxiety, and panic attacks [17–19], along with cancer. Thus, TSPO is an interesting target for the development of diagnostic and therapeutic ligands [16,20]. TSPO is overexpressed in the outer mitochondrial membrane of activated microglia [21–23] and reactive astrocytes [24]. Chronic activation of microglia leads to the release of neurotrophic and proinflammatory factors that are neurotoxic and cause neuronal damage and neurodegeneration [25–27]. The microglial activation is imaged in human brain in vivo by positron emission tomography (PET) of TSPO radiolabeled ligands. PK11195 (1-(2-chlorophenyl)-*N*-methyl-*N*-(1-methylpropyl)-3-isoquinolinecarboxamide) is one of the most commonly used, high affinity TSPO ligands for studying the diagnostics and treatment of brain inflammation and of other inflammatory diseases [28].

Second- (PRB28, PBR06, DAA1106) and third- (ER176 and GE-180) generation ligands have been developed (reviewed in [29,30]). PRB28, PBR06, and DAA1106 have higher binding affinity for *h*TSPO than PK11195. They provide a better signal-to-typical positron emission tomography (PET) noise ratio. ER176 and GE-180 are being developed to overcome differences in binding affinities for the WT or the A147T mutant *h*TSPO. This mutant emerges in 30% of Caucasians, 25% of Africans, 4% of Japanese, and 2% of Han Chinese according to the Hapmap database (http://hapmap.ncbi.nlm.nih.gov (accessed on 25 January 2020)). It binds PRB28 [31], PBR06 [32], and FEPPA [33]. This consequently leads to the lower PET signal intensity and can provide misleading results (absence of neuroinflammation) for the carrier of the A147T mutation.

TSPO folds into a bundle of five transmembrane (TM) helices and a short extramembrane helix placed in the cytoplasmic loop between helices TM I and TM II [34–36]. Bacterial and mouse TSPO can exist as monomers, dimers, and other oligomers as shown by experiments [34,35,37,38]. Different computational molecular modeling studies were carried out to further shed light on the dimerization of mouse and bacterial TSPO and on how ligands (small chemical compounds, porphyrin, cholesterol) interact with dimers or influence their stability [39–42]. In contrast, the oligomerization state of *h*TSPO has not been established. An in vitro study showed that it can adopt dimeric or trimeric forms under the inflammation conditions reproduced by high concentrations of reactive oxygen species (ROSs) [43]. Different experimental TSPO structures have been solved to date, notably the

NMR structure of mouse TSPO (*Mo*TSPO, PDB ID: 2MGY [36]) and the X-ray structures of *Rhodobacter sphaeroides* (*Rs*TSPO, PDB ID: 4UC1 [35]) and *Bacillus cereus* (*Bc*TSPO, PDB ID: 4RYI [34]) TSPOs.

The experimental structure of *h*TSPO has not yet been solved. Here, we built a monomeric structural model of *h*TSPO, alone and in complex with PK11195 by homology modeling and docking, and we ran molecular dynamics (MD) simulations of the protein in a lipid environment. We analyzed in detail the interactions of PK11195 and cholesterol with *h*TSPO and how PK11195 alters cholesterol interactions with this protein. Our structural model and results can be a valuable source for future studies of *h*TSPO and its interactions with cholesterol and/or other pharmacological ligands.

2. Results and Discussion

2.1. The hTSPO Structural Model

The sequence of the *h*TSPO was aligned with those of *Rs*TSPO, *Bc*TSPO, and *Mo*TSPO (the sequence identity is of 29%, 24%, and 81%, respectively; Figure 1). We decided to use *Rs*TSPO (PDB code: 4UC1 [35]) as a template in structural modeling of the human translocator protein [44,45].

Indeed, the latter is currently the best template choice in comparative modeling of mammalian TSPOs [40,41] for the following reasons: (i) the structural folds of the TSPOs on passing from the bacterial to the mammalian proteins are conserved [34–36]; (ii) the NMR *Mo*TSPO structure is affected by the ionic detergents used for the purification during the measurements [36]; as a result, the positions of highly conserved amino acids and of the transmembrane helices are altered [40]; (iii) the *Rs*TSPO crystal structure was resolved at a relatively good resolution (1.8 Å).

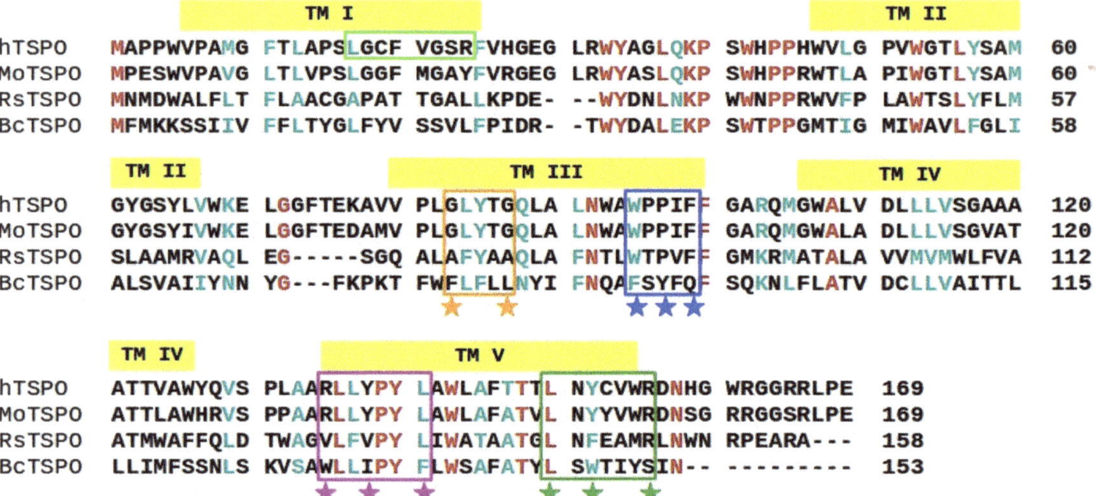

Figure 1. Multiple sequence alignment of TSPO from different organisms: human TSPO (*h*TSPO), *Mus musculus* (*Mo*TSPO) [36], *Rhodobacter sphaeroides* (*Rs*TSPO) [35], and *Bacillus cereus* (*Bc*TSPO) [34]. For the last three proteins, experimental structural information is available. Semi-conserved positions with more than 50% consensus according to ClustalO [46] are highlighted in cyan, while highly conserved positions with more than 90% consensus are shown in red. The oligomerization motif G83XXXG87 is indicated by the orange stars and rectangle. The W95XPXF99 motif is depicted with the blue stars and rectangle. The cholesterol-binding motif CRAC and its "mirror code" CARC are marked by dark green and violet rectangles and stars, respectively. The CRAC-like motif in TM I is highlighted by the light green rectangle.

We modeled the *h*TSPO based on the canonical *h*TSPO sequence reported in the UniProtKB database, Entry P30536 [47,48] (https://www.uniprot.org/uniprot/P30536 (accessed on 1 February 2019)). The model contains five transmembrane helices (TMs) arranged in the clockwise order TM I–TM II–TM V–TM IV–TM III. The four loops are located either in the cytoplasm, LP I and LP III, or in the cytosol, LP II and LP IV, respectively (Figure 2). LP I is composed of residues V26–H46 (Figure 1), and it was earlier suggested to play a role as the gate of the *h*TSPO ligand binding pocket [49]. Almost half of the residues composing LP I are fully conserved between *h*TSPO and *Rs*TSPO (Figure 1), which has a short α-helix in the middle of its loop. LP I in the *Mo*TSPO structure is shorter (10 residues) than in the bacterial TSPO structures (19 residues), and it lacks this α-helix.

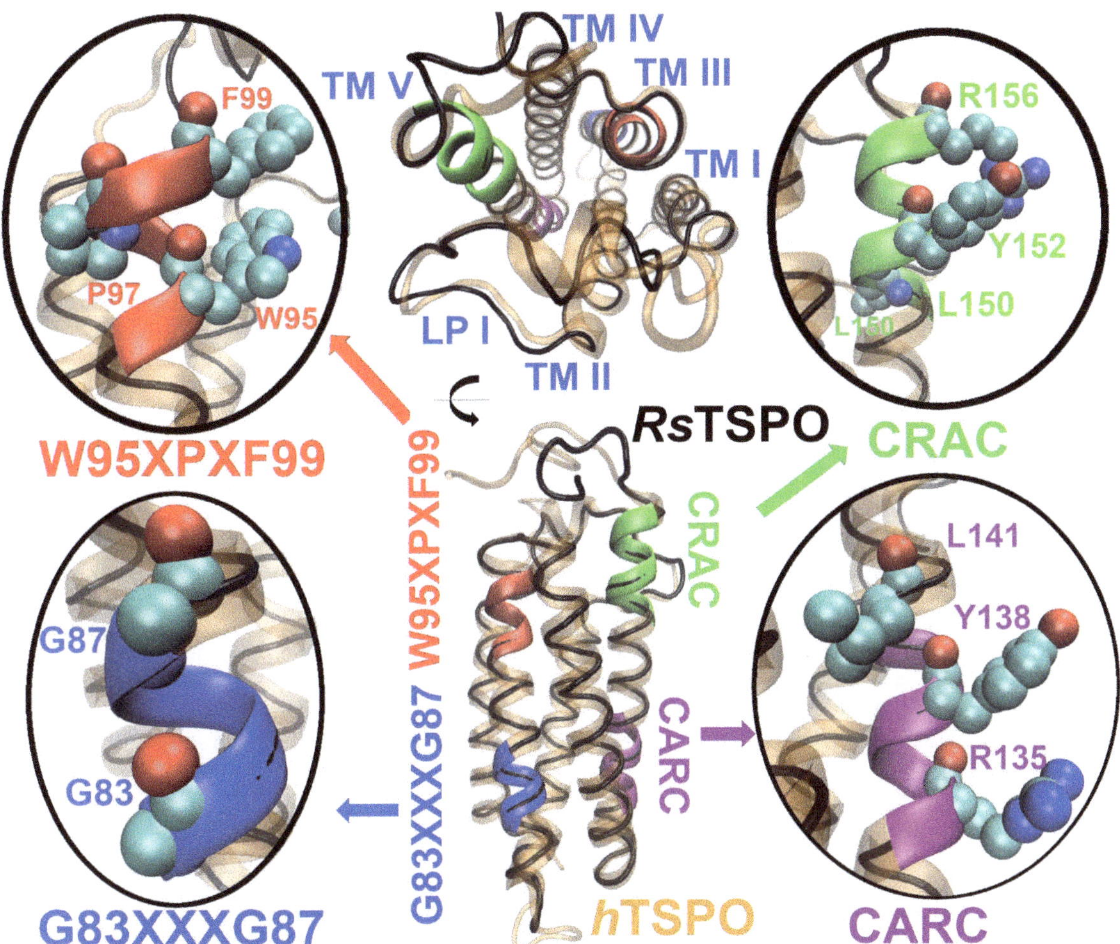

Figure 2. Top and side view of the structural model of the *h*TSPO monomer (orange cartoon representation) aligned with the *Rs*TSPO template (black tube representation). Four important functional regions are highlighted: cholesterol recognition/interaction amino acid consensus region (CRAC, in green color) and reverse region of the CRAC (CARC, in purple color), both involved in cholesterol binding, the G83XXXG87 motif (blue color) relevant for monomer–monomer interactions, and the W95XPXF99 motif (shown in red color) important for ligands binding. The figure was prepared with the VMD program [50].

Different structurally and/or functionally important motifs are conserved among *Rs*TSPO and *h*TSPO:

(i) An important feature of the mammalian TSPO is its capability of binding cholesterol molecules [7]. Two motifs are involved in cholesterol binding: the cholesterol recognition/interaction amino acid consensus motif (CRAC) represented as L150–X–Y152–X(3)–R156 (Figure 1, dark green rectangle) [7] and the reverse region of CRAC (named CARC) described as R135–X(2)–Y138–X(2)–L141 (Figure 1, purple rectangle) [7,10]. One helical turn before CRAC, there is a short sequence L144–A145–F146 that enhances the binding affinity of mammalian TSPO for cholesterol [9]. Even though cholesterol is absent in bacterial membranes, the CRAC motif is conserved in *Rs*TSPO (Figure 1, dark green rectangle). This is most probably in order to accommodate hopanoids, which have a structure and function similar to that of cholesterol in higher organisms [51].

In order to interact with cholesterol, side chains of the key CRAC, CARC, and LAF residues need to face the membrane. In our model, CRAC Y152 and R156 side chains are membrane exposed, while L150 is oriented towards the ligand binding cavity in the center of the TSPO (Figure 2). Different mutagenesis studies showed the importance of Y152 and of R156 for cholesterol binding. If one of these two residues is mutated to serine or leucine, respectively, the binding of cholesterol is abolished [7,8]. CARC, in contrast to CRAC, is not conserved in *Rs*TSPO (Figure 1, purple box). However, in our *h*TSPO model, all key residues from this motif, R135, Y138, and L141, are membrane exposed and can interact with cholesterol (Figure 2). Among the LAF residues, L144 is facing the interior of the protein in our model, while A145 and F146 are membrane exposed and available for cholesterol binding.

(ii) The G83XXXG87 motif from *h*TSPO (Figure 1, orange rectangle) coincides with the A75XXXA79 motif in *Rs*TSPO. These motifs represent widespread helix–helix interactions across different membrane proteins [52–56]. Both motifs are located in the third TM domain (TM III) of the respective proteins and represent a binding interface for TSPO monomer–monomer interactions. The G83 and G87 residues are exposed to the membrane in our *h*TSPO model and can interact with the second monomer (Figure 2).

(iii) The W95XPXF99 motif is fully conserved among human, mouse, and *Rs*TSPO, while the *Bc*TSPO sequence differs significantly in this region (Figure 1, blue rectangle). This motif is conserved also in other prokaryotic and eukaryotic TSPO sequences [40,57], and it was suggested to play a role in oligomerization processes [49], as well as in ligand binding [34,58]. In the *Mo*TSPO experimental structure, W95 points into the binding cavity and F99 is oriented toward the membrane, whereas in the *Rs*TSPO structure, both residues point into the binding pocket. In the *Bc*TSPO–PK11195 complex, residues F90 and Q94, which correspond to W95 and F99 in mammalian TSPO, respectively, interact with PK11195. This is also the case for our *h*TSPO model (Figure 2).

2.2. PK11195 Interactions with the hTSPO Model

We docked PK11195 to our *h*TSPO model. Two 3D structures of the TSPO in complex with PK11195 exist, the NMR structure of *Mo*TSPO [36] and the X-ray structure of *Bc*TSPO [34]. The ligand binding cavity has the same location in both proteins, but PK11195 adopts different binding poses. In our studies, we used the *h*TSPO–PK11195 complex where the ligand binding pose was similar to the one observed in the *Bc*TSPO, since currently, there are no experimental structures available for the *Rs*TSPO–PK11195 complex, which is a template of our model. Other docking poses differed from the one used as the starting configuration for the MD simulations (Figure S1), showing a small degree of convergence. Indeed, these poses are already cluster representatives, since AutodockVina [59], used here, does the clustering automatically.

However, many residues crucial for PK11195 binding in the *Bc*TSPO crystal structure are fully conserved in *Rs*TSPO, like Y32(31), P42(41), W51(50), N87(84), W138(135), A142(139), and L145(142) (numbering is for *Bc*TSPO and in parentheses for *Rs*TSPO) (Figure 1). Furthermore, the main structural differences between *Bc* and the *Rs*TSPOs appear at the monomer–monomer interface in the dimer structure and not in the ligand binding pocket [49].

The binding cavity in our model is lined with residues belonging to four TM domains and to LP I: G18, C19, V21, G22, F25 (in TM I), Y34, H43 (in LP I), H46, L49, G50, W53 (in TM II), N92, W95, P96, F99, F100 (in TM III), W143, T147, L150 (in TM V). The residues L49, W53, W95, W143, A147, and L150 are involved in the binding of PK11195 in both the *Mo* and *Bc*TSPOs experimental structures [34,36]. Beside these residues, others bind PK11195 in *Mo* and *Bc*TSPO, but they are specific to each structure. One third of the binding residues in our model are fully conserved among mammalian [60] and other prokaryotic and eukaryotic species [57], like Y34, W53, N92, W143, L150, and A/T147 (Figure 1, for more complete alignments see [57,60]). It was shown that either threonine or alanine at position 147 has no impact on PK11195 binding to the TSPO since it binds to both polymorphs with the same binding affinity, in contrast to other radioligands, which bind with significantly smaller binding affinity to the protein with threonine [31,60,61].

Next, we ran a 1 µs long MD simulation and analyzed the interactions between our *h*TSPO model and the PK11195 ligand. Additional MD simulation replicas of 450 ns were later run for holo and apo *h*TSPOs (two for each protein), and the results are reported in the Supplementary Materials. During the MD simulation, at around 450 ns, we observed quite sudden movement of the PK11195 chlorophenyl-*N*-isoquinoline part towards different binding pose, while the alkyl part of PK11195 remained at its initial position. The rings horizontally slide in a way that the chlorophenyl ring, which initially faces TM I, is placed between TM I and TM II, closer to the latter helix (Figure 3a, center, and Figure S2). This movement indicates that the initial binding pose may not be optimal and/or that the TSPO binding pocket possesses certain plasticity allowing for different binding poses of the ligand. Interestingly, the pose of PK11195 after 450 ns was similar to one docking position (Figure S1, pose 5; the RMSD between the two poses is 1.8 Å). We compared the binding pose that PK11195 adopts during the first 450 ns (Figure 3b) with the one it takes up after the movement (Figure 3c). We observed that residues binding constantly PK11195 during the whole length of the MD run are: G22, F25 (in TM I), Y34, H43 (in LP I), L49, W53 (in TM II), W95, P96 (in TM III), and T147, L150 (in TM V). Residues Y34, W53, W95, A/T147, and L150 are well conserved among TSPOs from different species, while G22, F25, and L49 are semi-conserved [57]; this indicates their importance for the structure and/or function of the protein.

The F25 side chain is initially facing towards the membrane, later it moves inside the binding pocket, establishing the π-stacking and hydrophobic interactions with F100, Y34, and PK11195. In the *Mo*TSPO (PDB code: 2MGY [36]) and *Bc*TSPO (PDB code: 4RYI [34]) structures, this residue points out of the binding site in the same orientation as it does in our model at the beginning of the MD simulation.

In our model, Y34 forms hydrophobic interactions with PK11195 throughout the MD simulation. This residue is fully conserved among TSPOs from different species, from mammalians to bacteria [57]. However, in the crystal structure of *Bc*TSPO–PK11195, this residue binds PK11195, while in the *Mo*TSPO–PK11195 complex, it faces the cytosol. Despite this ambiguity in the TSPO–PK11195 experimental structures, our result is in very good agreement with mutational studies showing that mutations Y34F, Y34F/F100A, and Y34F/F99A cause a large decrease in the binding affinity for PK11195 with respect to the WT TSPO [58]. These results indicate that the aromatic phenyl rings are crucial at this place for PK11195 binding.

A stable hydrogen bond (H-bond) is formed between the W53 indole amino group and the carbonyl-oxygen atom of PK11195. The H-bond between W53 and PK11195 was observed also in the *Bc*TSPO–PK11195 crystal structure [34], but not in the *Mo*TSPO–PK11195 NMR structure, which lacks any H-bond interaction [36]. Another H-bond observed in the *Bc*TSPO–PK11195 structure was formed between W143 and the PK11195 ligand [34]. In our model, this H-bond is formed occasionally during the first 450 ns; after this time, W143 constantly interacts with PK11195 through VdW interactions.

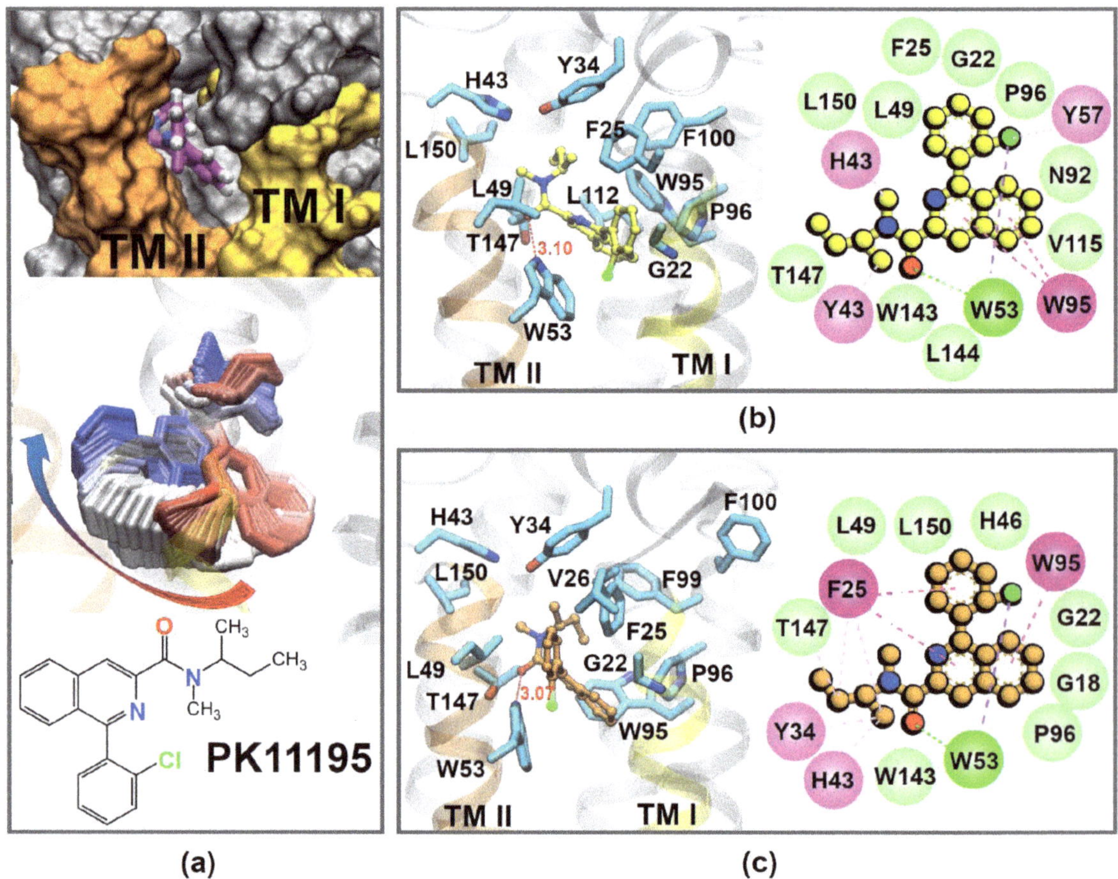

Figure 3. PK11195 binding interactions with the *h*TSPO model. (**a**) Top: Open access to the binding pocket between TM I and TM II of the *h*TSPO structural model that persists throughout the MD simulation. We do not observe other openings, nor the change in the LP I conformation. Center: PK11195 moves in the binding site of the *h*TSPO model from its initial state observed in the first 450 ns (red color) to the new pose (blue color, 450–1000 ns). The image of every hundredth frame is shown smoothed with a five frame window. Bottom: Chemical formula of PK11195. (**b**,**c**) 3D and 2D representations of the PK11195 binding pocket during the first 450 ns (top) and after the ligand movement, from 450 ns till the end of the MD run (bottom). 3D plots show PK11195 (yellow and orange balls and sticksrepresentation for 0–450 ns and for 450–1000 ns, respectively) and the residues binding it for more than 90% of the simulation time; the backbone and hydrogen atoms were omitted for clarity reasons. F100 was kept in (**c**), despite that it does not bind PK11195 anymore, to show the change in its side chain conformation. The most constant interactions, formed for more than 75% of the simulation time between PK11195 and the *h*TSPO model, are shown in the 2D plots obtained by the Discovery tool [62]. Legend: green circles—hydrogen bonds, light green circles—VdW interactions, light pink circles—π-alkyl, and dark pink circles—π-π interactions.

PK11195 is additionally bound through VdW, hydrophobic, or stacking interactions by H43, L49, P96, W95, T147, and L150. Some of these residues interact with PK11195 also in the *Bc*TSPO structure (F90, S91, A142, and L145, respectively) and in the *Mo*TSPO structure.

In addition to all the above described residues, F100 and L112 steadily bind PK11195 during the first 450 ns. At this time, the ligand moves to a new binding pose, and these two interactions are lost; however, the interactions with V26 and F99 are established (Figure 3b,c). Interestingly, the F100 side chain flips out of the binding pocket at around 720 ns (Figure 3c). This residue is oriented towards the binding site in *Bc*TSPO, as it is in our

model at the beginning of the MD simulations, while in *Mo*TSPO, it is facing the membrane like in our model at the end of the MD run. According to our results, we suggest that F100 spontaneously changes its position from inward to outward of the binding pocket and that the experimental structures captured it in one of these two different conformations.

The spontaneous change that we observe in the orientation of the F100 side chain may indicate that it is involved in placing the ligand inside the binding pocket, but not crucial for its binding, and that this role is left to F99 and Y34, as suggested by [58]. Deeper studies will of course need to be done to explore in detail the exact role of these three residues.

Different hypothesis were made in the literature about the possible binding pathways for PK11195: one suggested that the ligand enters the binding pocket from the cytosol and that LP I plays a role of the gate [49,63]; the second one proposes binding between the crevices in TM helices [34,63,64].

In the present work, we observed the opening between TM I and TM II of our *h*TSPO–PK11195 model, which during the whole simulation time gave direct access to the binding pocket through the membrane (Figure 3a). Furthermore, PK11195 adopts a position where its Cl-phenyl ring is placed in this crevice, enveloped by C19, G22, F25, and V26 from TM I and by H46, L49, and W53 from TM II. In contrast, in the apo *h*TSPO model, the TM I and TM II helices maintain the closed position, without any openings, during the full length of the MD simulation (Figure S3).

Furthermore, we noted that LP I is stabilized through a patch of interactions: cation-π interactions formed between K39 (LP I) and Y34 (LP I) and the stacking interactions between Y34, F25 (TM I), F99, F100 (TM III), and PK11195. These interactions prevent—within the time scale of our simulation—LP I from moving in a way to open the access to the binding site from the top of the protein and therefore from the cytosol. This result is consistent with the previous study [64]. However, we cannot exclude that on a longer time scale, LP I is able to perform larger movements, as was proposed by [63].

2.3. PK11195 Stabilizes hTSPO Structural Fold

The apo and holo *h*TSPO structural models were embedded into the membrane and evaluated by a 1 μs long MD simulation. We evaluated and compared the structural stability of the apo and holo *h*TSPOs by calculating the root mean squared deviations (RMSD) of backbone atoms, the root mean squared fluctuations (RMSF) of Cα atoms (Figure 4), and the helices' flexibility (Figure 5).

The RMSD of apo *h*TSPO fluctuates more than that of the holo protein (Figure 4a); however, both systems reach a plateau at around 400 ns. The RMSD fluctuations in the apo protein (from 400 ns on) are principally due to the bending of the TM I helix and due to the flexibility of the loops LP I, II, and III. In contrast, the RMSD of the *h*TSPO–PK11195 complex is lower than for the apo protein and becomes steady from around 400 ns on, indicating that PK11195 stabilizes the *h*TSPO structural model. The higher structural stability of the holo protein can be observed as well from the principal component analysis (PCA) (Figure 6). These results are in line with experimental data for *Mo*TSPO, where the interactions with its cognate ligand PK11195 stabilize its structural fold [36,65]. Similar observations were obtained for *Rs*TSPO, which also showed an important flexibility, especially around the ligand binding site [35,66]. It was shown that the quality of *Rs*TSPO crystals was significantly improved by adding cholesterol and PK11195 to the crystallization medium [35], suggesting that PK11195 can have a positive impact also on the *Rs*TSPO and not only on *Mo*TSPO structural stability.

The RMSD values of backbone atoms for each transmembrane helix (Figure 4c,d) clearly show that PK11195 increases the stability of TM I and TM II, while TM IV and TM V are stable regardless of the absence/presence of the ligand. In both—apo and holo—proteins TM I is the most flexible helix and TM IV the most stable one.

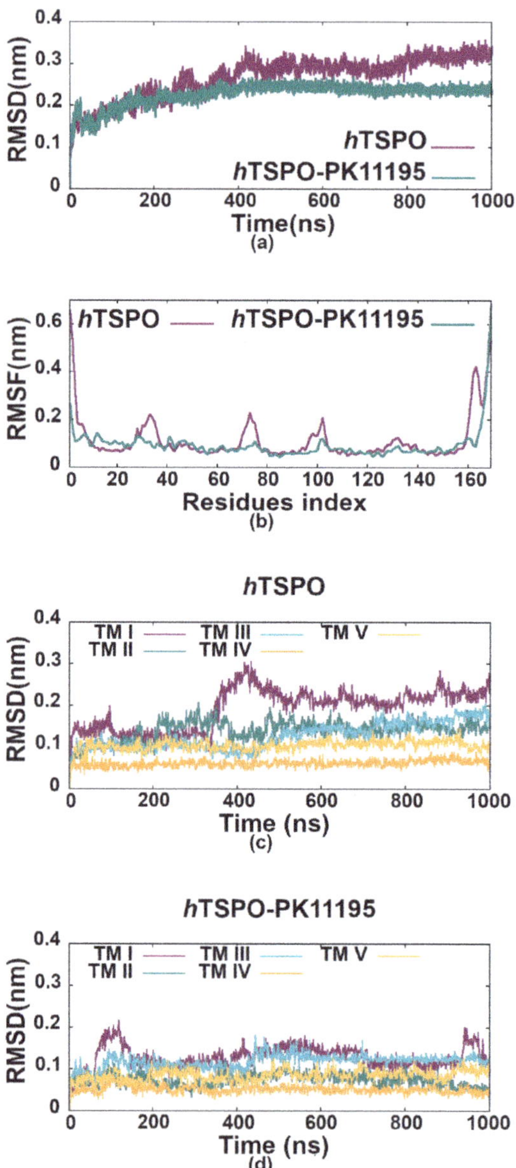

Figure 4. (**a**) Evolution of the root mean squared deviation (RMSD) values of the apo (violet graph) and holo (green graph) *h*TSPOs during the 1 μs long MD simulation. RMSD values were calculated for the backbone atoms of residues W5 to N158, excluding the N– and C–termini and H atoms. (**b**) Root mean squared fluctuations (RMSF) of the Cα atoms in the apo (violet graph) and holo (green graph) *h*TSPOs. RMSF values were calculated for equilibrated proteins (in the MD simulation range of 400 ns–1 μs). (**c,d**) RMSD values for each of the five transmembrane helices (TM I–TM V) in the apo (*h*TSPO) and holo (*h*TSPO–PK11195) proteins, respectively.

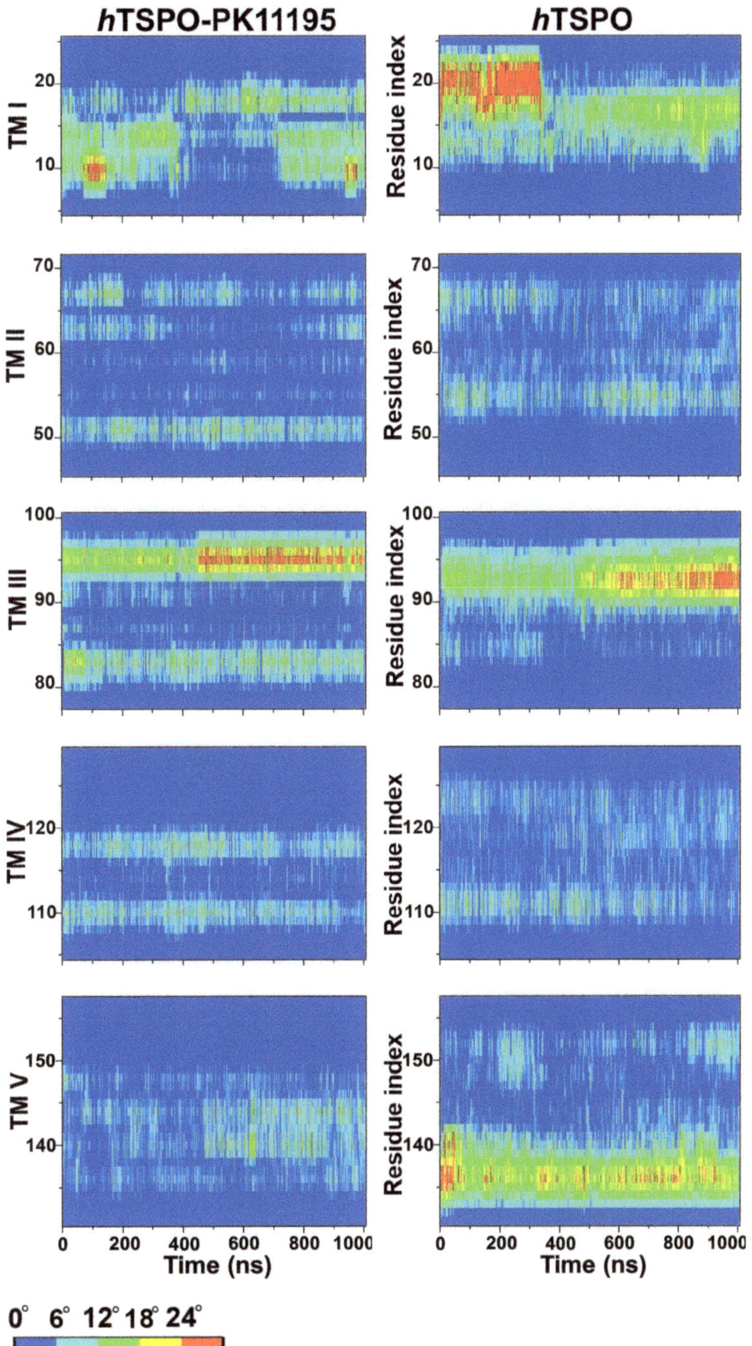

Figure 5. Analysis of the flexibility of each TM domain (TM I–TM V) in *h*TSPO–PK11195 and *h*TSPO structural models by means of Bendix [67]. y-axis: residue index number corresponding to the residues composing individual TM domain; x-axis: simulation time. The color scale indicates changes in helix angle/bending during the MD simulations, from blue: <6° to red: >24°.

*h*TSPO

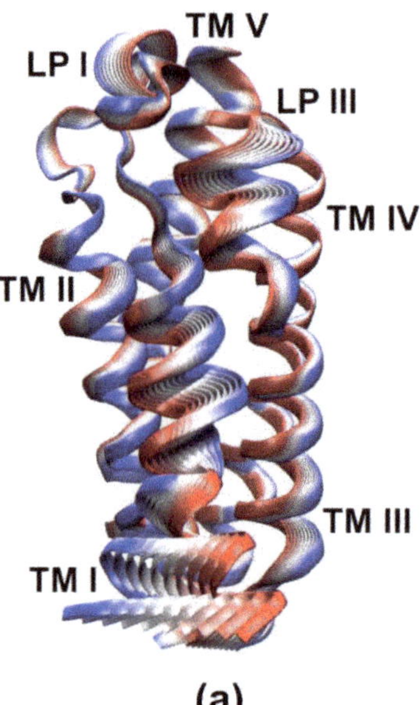

(a)

*h*TSPO-PK11195

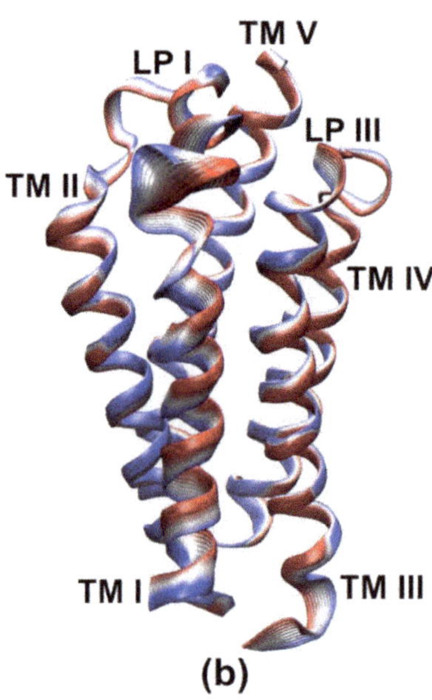

(b)

Figure 6. The principal component analysis of the (**a**) *h*TSPO and (**b**) *h*TSPO–PK11195 models showing the flexible parts of the protein. The image of every hundredth frame is shown, spanning from the beginning (red color) to the end (blue color) of the MD simulation.

The RMSD for TM I in the apo system has two plateau levels indicating a conformational change. Indeed, there is a kink in the α-helix due to the P15 residue (Figure 5, TM I in *h*TSPO). Helix kinks are a common feature of long α-helices, which are frequent in transmembrane proteins, and proline residues are strongly associated with the helix being kinked [68,69]. TM I becomes straight at around 350 ns. The alteration between the kinked and straight form of TM I is the reason for the RMSD change and also for the higher RMSD values with respect to the other helices in our model (Figure 4c). TM III (P96–P97) and TM V (P139) are also slightly kinked. However, TM V in the holo protein is very stable most probably owing it to the presence of PK11195, though its binding site is distant from P139.

The root mean squared fluctuations (RMSF) of Cα atoms were calculated in the range of 400 ns to 1000 ns, when both systems reach equilibration. The RMSF for both systems are very similar, but one can note important peaks at residues A35, E70, and F100 in the apo model (Figure 4b). These regions correspond to the first three loops fluctuating more in the apo than in the holo protein.

In the *h*TSPO–PK11195 complex, LP I with the small α-helix (residues G28 to G36) in the middle of it is stable during the whole MD simulation time. A crucial role in its high stability is played by a patch of interactions (described in details in the previous section) that hinders the free movement of LP I and contributes to the α-helix retaining its conformation. In contrast, LP I in the apo model varies in length (between F25–P45 and S23–P45), and the small α-helix is rarely formed. Due to its random coil structure and the absence of PK11195, LP I is more flexible than in the holo protein. LP II in the holo model is stable during the MD simulation, while in the apo protein, one helix turn at the C–terminus of TM II unfolds (data not shown), extending the length of LP II (W68–A78), which consequently fluctuates more than in the holo protein. Here, again, PK11195 seems to play an important role in the stability of TM II (Figure 4c,d).

LP III in the holo protein consists of residues G102–L109, and despite its length, it is more stable than in the apo protein where it is composed of residues F99–N104 (Figure 6). We observed that in both proteins, the parallel cation-π interactions are formed between R103 (LP III) and W33 (LP I α-helix) for more than 75% of the simulation time. Hydrophobic interaction between W33 and F100 (TM III) further stabilize the previous interaction in the holo protein (existent for 90% of the simulation time), but much less in the apo protein (existent for 40% of the simulation time). In addition, in the holo protein, F100 binds with Y34 (LP I) for 730 ns and for about 450 ns also with PK11195. This cascade stabilizes LP I, LP III, and the whole upper, C–terminal part of the holo *h*TSPO, namely the part in the outer membrane leaflet, where the ligand binding site is present. In the apo protein, this same cascade is not stabilized by PK11195, and indeed, LP III fluctuates more (Figure 6).

Finally, LP IV and LP V remain stable without changes in both models during all MD simulations, in line with our results that TM IV and TM V are the most stable helices regardless of the ligand's presence, and their termini do not unfold, as seen for some other helices described above (Figure 4c,d).

Taken together, the PK11195 ligand appears to reduce the fluctuations of the loops and to stabilize the overall structural fold of the TSPO protein.

2.4. Cholesterol Interactions with hTSPO

Our analyses show that the apo *h*TSPO model binds 1.5 times more cholesterol molecules than the holo one. This result is in good agreement with other studies postulating that PK11195 reduces the cholesterol binding to the TSPO [7].

We analyzed the average simulation time during which cholesterol interacts with each of the five TM helices (Table 1). In the holo protein, cholesterol interacts most often with TM I (47% of the simulation time) and TM V (48% of the simulation time), while in the apo protein, it interacts for 100% of the time with TM V and much less with other helices. Among other helices, TM II stands out, binding cholesterol for about 50% of the simulation time.

Table 1. The percentage (%) of the simulation time during which the individual transmembrane helices (TM I–TM V) bind the cholesterol molecule(s). The total simulation time is 1 µs.

	TM I	TM II	TM III	TM IV	TM V
Apo system	32	51	23	27	100
Holo system	47	26	19	15	48

Next we calculated the average number of cholesterol molecules bound to a single TM domain per frame (i.e., per ns). The holo TM I binds on average slightly more cholesterol than TM I in the apo protein, while for TM II, the result is inverted (Figure 7a). TM III and TM IV bind on average the same amount of cholesterol per frame, but holo TM V binds less than one cholesterol per frame, while apo TM V binds on average 1.5 cholesterol molecules per frame during the full length of the simulation time (Figure 7a). Indeed, the high affinity cholesterol binding motifs, CRAC and CARC, are located in TM V (Figures 1 and 2).

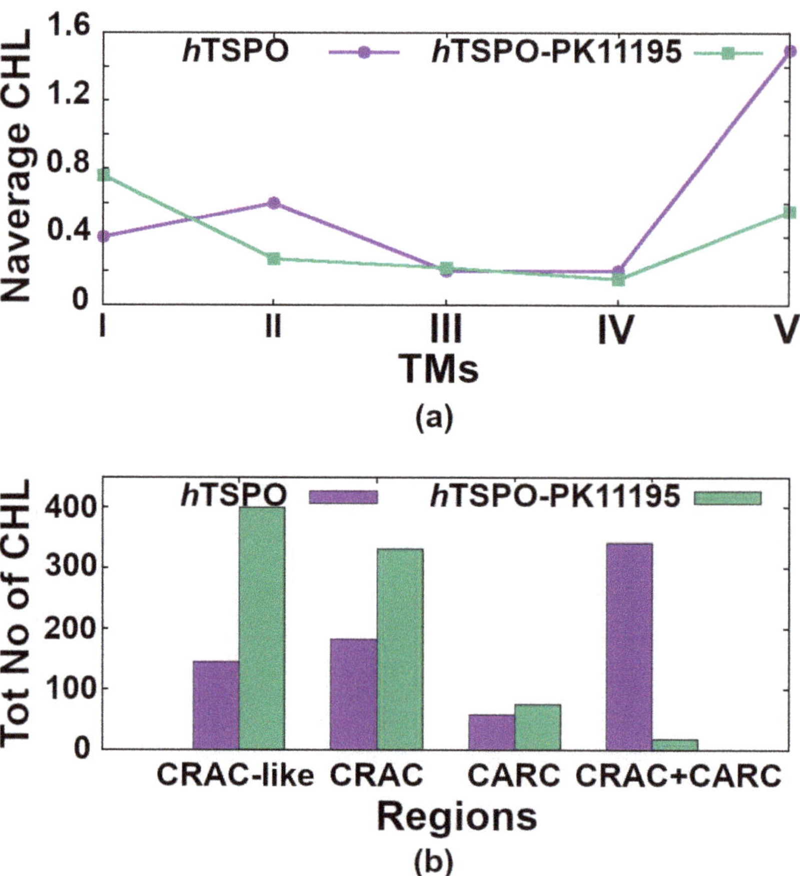

Figure 7. (**a**) The average number of cholesterol molecules (Naverage CHL) binding to the individual helix (TM I–TM V) in the apo (violet line) and holo (green line) *h*TSPOs at each frame of the 1 µs MD trajectory. (**b**) The total number of all cholesterol molecules (Tot No of CHL) binding either to the CRAC-like motif in TM I or to the CRAC and/or CARC in TM V during our 1 µs long MD simulation of apo *h*TSPO (violet) and holo *h*TSPO (green).

We analyzed why TM I of the holo protein binds cholesterol more frequently than other helices and why it binds more cholesterol than TM I in the apo protein (Figure 7b). We found that TM I has a CRAC–resembling motif, namely L17–X(2)–F20–X(3)–R24 (Figure 1), that attracts cholesterol. This motif is located in the upper, C–terminal part of the TSPO, in the outer membrane leaflet, next to the PK11195 binding site. We suggest that cholesterol binds more often to this motif in the holo than in the apo protein due to the higher stability of the TM I helix in the former protein (Figure 4c,d). The higher stability of holo TM I (i.e., it is less kinked than apo TM I (Figure 5, TM I) allows for the more optimal orientation of cholesterol binding residues L17, F20, and R24 and of cholesterol molecules with respect to the apo protein.

For TM II, the frequency of cholesterol binding is inverted: it interacts more often with the apo than with the holo protein. Cholesterol binds to both proteins only in the first half of the simulation time. Two cholesterols interact with apo TM II, one in the upper, C–terminal part (binding to residues P45–W47–V48–P51–V52) and one in the lower, N–terminal part of the protein (binding to residues T55-A59–Y62–L66). Cholesterol interacts almost twice more frequently with the lower part of the helix that is in the inner membrane leaflet than with the upper part of the helix that is in the outer membrane leaflet. In the holo protein instead, cholesterol binds exclusively to the upper part of TM II. Since apo TM II binds more cholesterol than holo, it is clear that cholesterol has higher binding affinity for the lower part of this helix (i.e., it rather binds to the N–terminal part of the *h*TSPO that is in the inner membrane leaflet than to its C–terminal part). We suggest that PK11195 reduces the frequency of cholesterol binding to the lower, N–terminal part of the *h*TSPO and in this particular case to TM II.

TM III and TM IV bind cholesterol equally frequently in both systems.

Finally, we determined the frequency of cholesterol binding to TM V, precisely to the CRAC (L150–X–Y152–X(3)–R156) and CARC (R135–X(2)–Y138–X(2)–L141) motifs. We defined how many cholesterol molecules bind individually to CRAC or CARC, as well as the frequency of cholesterol binding to both regions simultaneously (Figure 7b). For individual binding, we counted cases where only one motif at a time is occupied by cholesterol. For simultaneous binding, we counted only the cases when both motifs are occupied at the same time and with two different cholesterol molecules. Cases where one cholesterol molecule is bridging the two regions were excluded. Our results show that in the holo protein, cholesterol binds most often to CRAC and much less often to the CARC motif. Interestingly, the presence of PK11195 almost abolishes the simultaneous binding of cholesterol to both motifs.

In the apo protein, cholesterol binds more often to CRAC than to CARC. The number of cholesterol binding to these motifs is lower than in the holo protein, owing to the fact that cholesterol in the apo protein preferentially binds to both motifs at the same time.

Taken together, the apo protein binds more cholesterol molecules than the holo protein. In the apo protein indeed, cholesterol binds with about 50% frequency to TM II and with 100% frequency to the CRAC and CARC regions in TM V. In TM II, cholesterol binds more readily to the lower part of the helix, so to the N–terminal part of *h*TSPO present in the inner membrane leaflet. Very interestingly, in the apo protein, cholesterol binds most frequently to CRAC and CARC simultaneously (Figure 8, right panel), while in the holo protein, simultaneous binding to these motifs is almost abolished. Indeed, cholesterol in the holo protein binds mostly to the newly described motif in TM I and to the CRAC motif in TM V (Figure 8, left panel). Both motifs are present in the upper, C–terminal part of the *h*TSPO, next to the PK11195 binding site and in the outer membrane leaflet. In our study, the binding of cholesterol to the lower, N–terminal part of the holo *h*TSPO model is rarely observed.

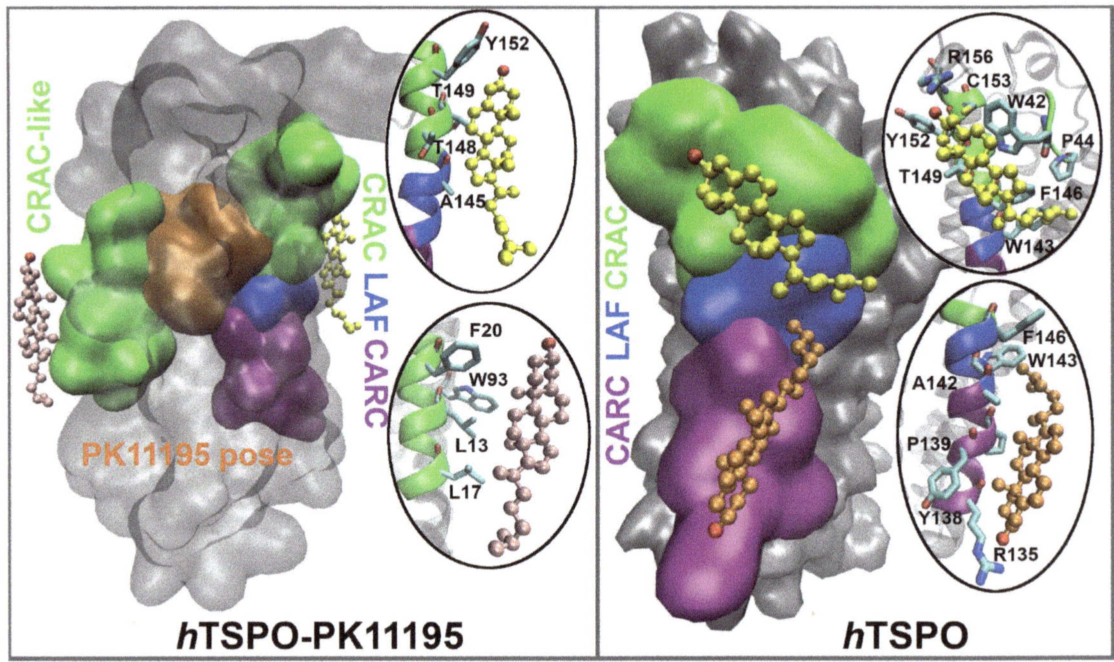

Figure 8. Cholesterol molecules bind most frequently to CRAC and CRAC–like regions (green surface representation) that are in the vicinity of the PK11195 binding site (orange surface representation) in the *h*TSPO–PK11195 system and to CRAC, LAF(blue surface representation), and CARC (purple surface representation) motifs in the *h*TSPO system. Corresponding residues from each region that interact with cholesterol (color coded, respectively) are represented within ellipses.

3. Materials and Methods

3.1. Building the 3D Structural Model of hTSPO

We used the sequence reported in the UniProtKB [47,48] database with the ID: P30536. It contains 169 amino acids, spanning from M1 to E169. This sequence was aligned with those of: *Mus musculus* TSPO (*Mo*TSPO; UniProt ID: P50637), *Rhodobacter sphaeroides* TSPO (*Rs*TSPO; UniProt ID: Q9RFC8), and *Bacillus cereus* TSPO (*Bc*TSPO; UniProt ID: Q81BL7), using the Multalin [70] and ClustalO [46] web-servers. 3D structural models of the *h*TSPO were built based on the *Rs*TSPO template (PDB ID: 4UC1 [35]). Twenty models of *h*TSPO were generated using the MODELLER program, Version 9.19 [71]. All models were analyzed according to the Discrete Optimized Protein Energy (DOPE) score using the built-in script of the MODELLER package [71,72]. In addition, the local structural quality of the *h*TSPO models in the biological membrane were examined using the QMEANBranescoring function [73] from the Swiss-model server [74–78]. All models were visually inspected and compared to the template and to the available mutagenesis data. The model corresponding best to the available experimental data, having the lowest DOPE score according to the MODELLER program [71,72] and the appropriate local structural quality as defined by the QMEANBrane tool [73], was chosen for docking and molecular dynamics (MD) simulation [79,80] studies. We checked that the orientation and tilt angles of the helices—once the model is inserted in a membrane—were appropriate. These parameters were computed by the Positioning the Proteins in Membranes (PPM) server [81] for the model and the *Rs*TSPO template and compared between them (Table 2).

Table 2. Comparison of the hydrophobic thickness and protein tilt angles for the *Rs*TSPO template and the *h*TSPO structural model. All values were obtained from the Positioning the Proteins in Membranes (PPM) server [81].

Model/Template	Hydrophobic Thickness (Å)	ΔG Transfer (kcal/mol)	Tilt Angle (o)
*h*TSPO	30.4 ± 4.1	−23.9	10.0 ± 1.0
*Rs*TSPO	30.2 ± 1.6	−40.8	7.0 ± 3.0

3.2. Docking of the PK11195 Ligand

We docked the PK11195 ligand (*N*-[(2R)-butan-2-yl]-1-(2-chlorophenyl)-*N*-methylisoquinoline-3-carboxamide) to the *h*TSPO structural model. The initial 3D structure of PK11195 was obtained from the PubChem database (https://pubchem.ncbi.nlm.nih.gov/compound/1345 (accessed on 1 February 2019)). Molecular docking was performed using the UCSF Chimera program [82] and AutoDock Vina package [59]. Protein and ligand input files were prepared by AutoDockTools. The ligand had fully flexible torsion of freedom, while the receptor side chains were kept rigid. Non-polar hydrogen atoms of the protein and the ligand were merged. The center of the grid was placed at X = −14.451 Å, Y = 25.618 Å, and Z = 25.297 Å. The grid dimensions were 76 × 72 × 66 Å, and the spacing between the grid points was set to 0.375 Å. The exhaustiveness parameter of the global search was set to 8 (default). Ten ligand binding modes were generated in search for a ligand pose with the lowest binding affinity.

We selected the model where the ligand had the lowest, i.e., the most negative docking binding affinity score, and it interacted with the two conserved residues W53 and W95 [57] shown to be important for binding [34].

3.3. Molecular Dynamics Simulations

The apo (*h*TSPO) and holo (*h*TSPO–PK11195 complex) models were then inserted into the lipid bilayer composed of phosphatidylcholine (POPC)—phosphatidylethanolamine (POPE)—cholesterol (CHL) with the ratio of 3:3:1 for POPC:POPE:CHL, respectively. The choice of the membrane composition was made based on the experimental studies of the mitochondrial membrane and the protein–lipid monolayers [83,84]. The membrane thickness was 3.04 nm and was built by the Mem-Builder web-server [85,86]. The *h*TSPO models were placed and properly oriented at the center of the membrane box by the Lambada and InflateGRO2 tools [87] (the tilt angle of all TM helices is 10°). Principally, these values correspond to those of the *Rs*TSPO template (Table 2).

The apo and holo models of *h*TSPO inserted in the POPC–POPE–CHL membrane were solvated with 12,758 water molecules enclosed in a solvation box with dimensions of 10.5 nm × 10.5 nm × 11.0 nm. 161 sodium (Na^+) and 166 chloride (Cl^-) ions were added to neutralize the system net charge and to reproduce the physiologic electronic strength of 0.15M. The MD simulations were run using the GROMACS 2018.6 package [88,89] and applying the SLIPIDSforce field [90] for the membrane, the AMBER99SB-ILDNforce field [91] for the *h*TSPO model and ions, and the TIP3P [92] force field for water. The force field parameters of PK11195 were prepared using the General Amber force field (GAFF) [93,94], introducing the RESPatomic charges and electrostatic potential (ESP) as calculated based on the B3LYP/6-31G* basis set using the Gaussian09 package [95]. The topology file of the PK11195 ligand was converted to GROMACS format using the ACPYPE tool [96]. The geometry of the *h*TSPO models was optimized by steepest descent minimization performed for 50,000 steps with a maximum force constant value of 1000 kJ/mol/nm. After the geometrical optimization, the systems underwent NPTequilibration for 10 ns with a time step of 2 fs. The systems were maintained at the reference pressure of 1 bar by coupling to the Parrinello–Rahman barostat [97,98] with uniform scaling of x-y box vectors and independent scaling for the z-axis (i.e., perpendicular to the membrane). The systems were coupled to the Nose–Hoover thermostat [99–101] to maintain the temperature at 310 K. A 1.2 nm

cut-off was set for the short-range non-bonded interaction. The LINCSalgorithm [102] was chosen to constrain all bonds involving hydrogen atoms. The holo and apo structural models of *h*TSPO were then simulated for 1 µs to detect the stable structure of *h*TSPO in the mitochondrial membrane. MD simulation parameters were the same as in the NPT equilibration run; only the thermostat was changed to the V-rescale thermostat [101].

3.4. Analysis

The root mean squared deviation (RMSD) for the entire *h*TSPO model and for the individual helices was calculated for the backbone atoms omitting the hydrogen atoms. The RMSD for the entire protein was calculated for the sequence from W5 to N158, excluding the N– and C–termini and the hydrogen atoms.

The root mean squared fluctuation (RMSF) was calculated for the Cα atoms of each *h*TSPO residue. They were calculated for equilibrated proteins, that is in the MD simulation range 400 ns–1 µs. For these calculations, the *g_rms* and *g_rmsf* tools from the GROMACS package were used [88,89].

Helices flexibility analyses was done by the Bendix plugin [67] in the VMD program [50]. To define the color code, we saved the values of the angle changes along each helix during the MD simulation time. The average values ranged from 0° to 24°; therefore, we divided the color code into 5 parts with the corresponding 6 angle values' extents. The highest change in the helical angle was observed for the TM I helix in the apo *h*TSPO model, i.e., 51°. The color code is described as blue: <6°, cyan: 6–12°, green: 12–18°, yellow: 18–24°, and red: >24°.

The principal component analysis (PCA) was performed on the apo and holo *h*TSPO including residues W5 to N158. The C– and N–termini, as well as hydrogen atoms were ignored. We used the *g_covar* and *g_anaeig* tools in the GROMACS package [88,89].

The analysis of PK11195 interactions with the *h*TSPO model (Figure 3) was carried out by home-made TCLand AWKscripts. To define a binding site of the ligand, we searched for all residues that were within 4.5 Å of any PK11195 heavy atom. We defined residues that bind PK11195 ligand for more than 90% of the simulation time as constant or principal binders. All of them form hydrophobic or stacking interactions with the ligand. For W53 that is H-bonding the carbonyl oxygen of PK11195, we calculated the frequency of H-bond formation using the distance criteria of 3.5 Å. The residues binding the PK11195 ligand for more than 90% of the simulation time are shown in the VMD representation in Figure 3. Residues that interact with the PK11195 for at least 75% were determined as frequent binders. All residues binding PK11195 for at least 75% are shown in the 2D plots in Figure 3. The subfigures in Figure 3 were made using the Visual Molecular Dynamics (VMD) [50] and Discovery Studio Visualizer [62] programs.

The cholesterol analysis was done by in-house written TCL and AWK scripts. The cholesterol was counted as bound to the *h*TSPO model, to the defined TM helix, or to different motifs (CRAC-like in TM I, CRAC/CARC in TM V) if it was found within 5 Å from any residue belonging to the analyzed region, respectively. Only contacts between heavy atoms were taken into account.

The total number of cholesterols bound to the apo and holo *h*TSPO systems throughout the 1 µs were counted and expressed as the ratio of the cholesterol molecules interacting with each system.

The data in Table 1 were obtained by counting all frames where the cholesterol interacts with the transmembrane helix in question. The percentage of the simulation time was calculated as the number of frames divided by the total number of frames (1000).

To obtain the average number of cholesterol molecules that bind the individual helix at each frame (as reported in Figure 7a), we counted the total number of cholesterol molecules interacting with the individual TM helix, and we divided this number by the total number of frames (1000).

For Figure 7b, we just counted the total number of cholesterol molecules that bind only to the CRAC–like motif in TM I, only to CRAC/CARC in TM V, or to both motifs (CRAC+CARC in TM V) at the same time.

4. Conclusions

The interplay between PK11195 and cholesterol interactions with *h*TSPO were studied by MD simulations of a homology model of the protein based on *Rs*TSPO. The ligand increases the stability of the protein in terms of RMSD, PCA, and Bendix analyses. During the MD simulation, PK11195 slides to a new position, in which its CL-phenyl ring initially facing TM I is placed between TM I and TM II, closer to the latter helix (Figures 3 and S2).The two helices detach from one other, while they stay close to each other in the apo protein. The ligand forms mostly hydrophobic and stacking interactions with the protein. Its carbonyl oxygen forms an H-bond with the W53 side chain. Two and three cholesterol molecules per ns bind, on average, to the holo and apo *h*TSPO, respectively. Hence, the presence of the ligand reduces the frequency of cholesterol binding to the protein. In the apo protein, the cholesterol molecules bind most of the time simultaneously to two well-known cholesterol binding motifs, CRAC and CARC in TM V. In the holo protein instead, cholesterol interacts with the CRAC–like motif in TM I and with the CRAC motif in TM V. Cholesterol binds much more rarely to the lower, N–terminal part of the holo *h*TSPO, that is in the inner membrane leaflet. Thus, PK11195 reduces cholesterol binding to this latter region, but it favors cholesterol interactions with the upper, C–terminal part of the protein in the outer membrane leaflet. Further studies are required to understand more in detail why this is the case.

Supplementary Materials: The following data are available online: Table S1: Experimentally defined dissociation constants (K_d) of the PK11195 molecule for different TSPOs. Figure S1: 3D and 2D representations of the PK11195 binding poses obtained by docking. Figure S2: MD snapshots showing the PK11195 ligand in its binding pocket at 0 ns, 250 ns, 500 ns, 750 ns and 1000 ns. Figure S3: Apo hTSPO does not exhibit an open access channel to the ligand binding pocket (located between the TM I and TM II helices in the holo protein). Figure S4: The number of cholesterol molecules binding to each helix of the apo and holo *h*TSPO proteins, averaged over our three MD simulations. Figure S5: The number of cholesterol molecules binding to apo and holo *h*TSPO averaged over our MD trajectories. Figure S6: Ligands motion in the binding site in our three MD simulations. Figure S7: Open access channels to the ligand binding pocket in our three MD trajectories.

Author Contributions: Conceptualization, T.T.N., G.R., A.G., and P.C.; methodology, T.T.N.; investigation, H.T.T.L, analyses, H.T.T.L. and A.K.; writing, original draft preparation, H.T.T.L. and A.K.; writing, review and editing, all authors; supervision, T.T.N., P.C., G.R., A.G., and A.K.; project administration, T.T.N.; funding acquisition, T.T.N. All authors read and agreed to the published version of the manuscript.

Funding: This research was funded by the World Bank and the Ministry of Science and Technology of Vietnam joint project "Fostering Innovation through Research, Science and Technology", Subproject Number 13/FIRST/1.a/VNU1, and the financial support of the Vietnam National University, Hanoi, for the Key Laboratory of Multiscale Simulations of Complex System, Annual Grant Number TXTCN.20.03.

Data Availability Statement: MD trajectories and protein models are available upon request to the corresponding authors.

Conflicts of Interest: The authors declare no conflict of interest.

Abbreviations

The following abbreviations are used in this manuscript:

TSPO	translocator protein
TM	transmembrane helix
LP	loop
PK11195	1-(2-chlorophenyl)-*N*-methyl-*N*-(1-methylpropyl)-3-isoquinolinecarboxamide molecule
*h*TSPO	the human TSPO
*h*TSPO-PK11195	the *h*TSPO-PK11195 complex
POPC	phosphatidylcholine lipid
POPE	phosphatidylethanolamine lipid
CHL	cholesterol
RMSD	root mean squared deviation
RMSF	root mean squared fluctuation
VMD	Visual Molecular Dynamics
MD	molecular dynamics
CRAC	cholesterol recognition/interaction amino acid consensus motif (L150–X–Y152–X(3)–R156)
CARC	reverse region of CRAC (R135–X(2)–Y138–X(2)–L141)

References

1. Fan, J.; Lindemann, P.; GJ Feuilloley, M.; Papadopoulos, V. Structural and functional evolution of the translocator protein (18 kDa). *Curr. Mol. Med.* **2012**, *12*, 369–386. [PubMed]
2. Bonsack, F.; Sukumari-Ramesh, S. TSPO: An evolutionarily conserved protein with elusive functions. *Int. J. Mol. Sci.* **2018**, *19*, 1694. [CrossRef]
3. Yeliseev, A.A.; Krueger, K.E.; Kaplan, S. A mammalian mitochondrial drug receptor functions as a bacterial "oxygen" sensor. *Proc. Natl. Acad. Sci. USA* **1997**, *94*, 5101–5106. [CrossRef] [PubMed]
4. Anholt, R.; Pedersen, P.L.; De Souza, E.; Snyder, S.H. The peripheral-type benzodiazepine receptor. Localization to the mitochondrial outer membrane. *J. Biol. Chem.* **1986**, *261*, 576–583. [CrossRef]
5. Yasin, N.; Veenman, L.; Singh, S.; Azrad, M.; Bode, J.; Vainshtein, A.; Caballero, B.; Marek, I.; Gavish, M. Classical and novel TSPO ligands for the mitochondrial TSPO can modulate nuclear gene expression: Implications for mitochondrial retrograde signaling. *Int. J. Mol. Sci.* **2017**, *18*, 786. [CrossRef]
6. Papadopoulos, V.; Baraldi, M.; Guilarte, T.R.; Knudsen, T.B.; Lacapère, J.J.; Lindemann, P.; Norenberg, M.D.; Nutt, D.; Weizman, A.; Zhang, M.R.; et al. Translocator protein (18 kDa): New nomenclature for the peripheral-type benzodiazepine receptor based on its structure and molecular function. *Trends Pharmacol. Sci.* **2006**, *27*, 402–409. [CrossRef] [PubMed]
7. Li, H.; Papadopoulos, V. Peripheral-type benzodiazepine receptor function in cholesterol transport. Identification of a putative cholesterol recognition/interaction amino acid sequence and consensus pattern. *Endocrinology* **1998**, *139*, 4991–4997. [CrossRef]
8. Li, H.; Yao, Z.x.; Degenhardt, B.; Teper, G.; Papadopoulos, V. Cholesterol binding at the cholesterol recognition/interaction amino acid consensus (CRAC) of the peripheral-type benzodiazepine receptor and inhibition of steroidogenesis by an HIV TAT-CRAC peptide. *Proc. Natl. Acad. Sci. USA* **2001**, *98*, 1267–1272. [CrossRef] [PubMed]
9. Li, F.; Liu, J.; Valls, L.; Hiser, C.; Ferguson-Miller, S. Identification of a key cholesterol binding enhancement motif in translocator protein 18 kDa. *Biochemistry* **2015**, *54*, 1441–1443. [CrossRef]
10. Fantini, J.; Di Scala, C.; Evans, L.S.; Williamson, P.T.; Barrantes, F.J. A mirror code for protein-cholesterol interactions in the two leaflets of biological membranes. *Sci. Rep.* **2016**, *6*, 21907. [CrossRef]
11. Fantini, J.; Di Scala, C.; Baier, C.J.; Barrantes, F.J. Molecular mechanisms of protein-cholesterol interactions in plasma membranes: Functional distinction between topological (tilted) and consensus (CARC/CRAC) domains. *Chem. Phys. Lipids* **2016**, *199*, 52–60. [CrossRef] [PubMed]
12. Wendler, G.; Lindemann, P.; Lacapère, J.J.; Papadopoulos, V. Protoporphyrin IX binding and transport by recombinant mouse PBR. *Biochem. Biophys. Res. Commun.* **2003**, *311*, 847–852. [CrossRef] [PubMed]
13. Verma, A.; Nye, J.S.; Snyder, S.H. Porphyrins are endogenous ligands for the mitochondrial (peripheral-type) benzodiazepine receptor. *Proc. Natl. Acad. Sci. USA* **1987**, *84*, 2256–2260. [CrossRef]
14. Hirsch, J.D.; Beyer, C.F.; Malkowitz, L.; Beer, B.; Blume, A.J. Mitochondrial benzodiazepine receptors mediate inhibition of mitochondrial respiratory control. *Mol. Pharmacol.* **1989**, *35*, 157–163.
15. Da Pozzo, E.; Tremolanti, C.; Costa, B.; Giacomelli, C.; Milenkovic, V.M.; Bader, S.; Wetzel, C.H.; Rupprecht, R.; Taliani, S.; Da Settimo, F.; et al. Microglial Pro-Inflammatory and Anti-Inflammatory Phenotypes Are Modulated by Translocator Protein Activation. *Int. J. Mol. Sci.* **2019**, *20*, 4467. [CrossRef]

16. Rupprecht, R.; Papadopoulos, V.; Rammes, G.; Baghai, T.C.; Fan, J.; Akula, N.; Groyer, G.; Adams, D.; Schumacher, M. Translocator protein (18 kDa)(TSPO) as a therapeutic target for neurological and psychiatric disorders. *Nat. Rev. Drug Discov.* **2010**, *9*, 971. [CrossRef]

17. Colasanti, A.; Owen, D.R.; Grozeva, D.; Rabiner, E.A.; Matthews, P.M.; Craddock, N.; Young, A.H. Bipolar Disorder is associated with the rs6971 polymorphism in the gene encoding 18 kDa Translocator Protein (TSPO). *Psychoneuroendocrinology* **2013**, *38*, 2826–2829. [CrossRef]

18. Costa, B.; Pini, S.; Martini, C.; Abelli, M.; Gabelloni, P.; Landi, S.; Muti, M.; Gesi, C.; Lari, L.; Cardini, A.; et al. Ala147Thr substitution in translocator protein is associated with adult separation anxiety in patients with depression. *Psychiatr. Genet.* **2009**, *19*, 110–111. [CrossRef]

19. Nakamura, K.; Yamada, K.; Iwayama, Y.; Toyota, T.; Furukawa, A.; Takimoto, T.; Terayama, H.; Iwahashi, K.; Takei, N.; Minabe, Y.; et al. Evidence that variation in the peripheral benzodiazepine receptor (PBR) gene influences susceptibility to panic disorder. *Am. J. Med Genet. Part Neuropsychiatr. Genet.* **2006**, *141*, 222–226. [CrossRef]

20. Dimitrova-Shumkovska, J.; Krstanoski, L.; Veenman, L. Diagnostic and Therapeutic Potential of TSPO Studies Regarding Neurodegenerative Diseases, Psychiatric Disorders, Alcohol Use Disorders, Traumatic Brain Injury, and Stroke: An Update. *Cells* **2020**, *9*, 870. [CrossRef]

21. Venneti, S.; Lopresti, B.J.; Wiley, C.A. The peripheral benzodiazepine receptor (translocator protein 18 kDa) in microglia: From pathology to imaging. *Prog. Neurobiol.* **2006**, *80*, 308–322. [CrossRef]

22. Kreutzberg, G.W. Microglia: A sensor for pathological events in the CNS. *Trends Neurosci.* **1996**, *19*, 312–318. [CrossRef]

23. Batchelor, P.E.; Liberatore, G.T.; Wong, J.Y.; Porritt, M.J.; Frerichs, F.; Donnan, G.A.; Howells, D.W. Activated macrophages and microglia induce dopaminergic sprouting in the injured striatum and express brain-derived neurotrophic factor and glial cell line-derived neurotrophic factor. *J. Neurosci.* **1999**, *19*, 1708–1716. [CrossRef]

24. Lavisse, S.; Guillermier, M.; Hérard, A.S.; Petit, F.; Delahaye, M.; Van Camp, N.; Haim, L.B.; Lebon, V.; Remy, P.; Dollé, F.; et al. Reactive astrocytes overexpress TSPO and are detected by TSPO positron emission tomography imaging. *J. Neurosci.* **2012**, *32*, 10809–10818. [CrossRef]

25. Kettenmann, H.; Kirchhoff, F.; Verkhratsky, A. Microglia: New roles for the synaptic stripper. *Neuron* **2013**, *77*, 10–18. [CrossRef]

26. Nakajima, K.; Tohyama, Y.; Maeda, S.; Kohsaka, S.; Kurihara, T. Neuronal regulation by which microglia enhance the production of neurotrophic factors for GABAergic, catecholaminergic, and cholinergic neurons. *Neurochem. Int.* **2007**, *50*, 807–820. [CrossRef]

27. Hanisch, U.K.; Kettenmann, H. Microglia: Active sensor and versatile effector cells in the normal and pathologic brain. *Nat. Neurosci.* **2007**, *10*, 1387–1394. [CrossRef]

28. Myers, R.; Manjil, L.G.; Cullen, B.M.; Price, G.W.; Frackowiak, R.S.; Cremer, J.E. Macrophage and astrocyte populations in relation to [3H] PK 11195 binding in rat cerebral cortex following a local ischaemic lesion. *J. Cereb. Blood Flow Metab.* **1991**, *11*, 314–322. [CrossRef]

29. Werry, E.L.; Bright, F.M.; Piguet, O.; Ittner, L.M.; Halliday, G.M.; Hodges, J.R.; Kiernan, M.C.; Loy, C.T.; Kril, J.J.; Kassiou, M. Recent developments in TSPO PET imaging as a biomarker of neuroinflammation in neurodegenerative disorders. *Int. J. Mol. Sci.* **2019**, *20*, 3161. [CrossRef]

30. Chauveau, F.; Boutin, H.; Van Camp, N.; Dollé, F.; Tavitian, B. Nuclear imaging of neuroinflammation: A comprehensive review of [11 C] PK11195 challengers. *Eur. J. Nucl. Med. Mol. Imaging* **2008**, *35*, 2304–2319. [CrossRef]

31. Owen, D.R.; Howell, O.W.; Tang, S.P.; Wells, L.A.; Bennacef, I.; Bergstrom, M.; Gunn, R.N.; Rabiner, E.A.; Wilkins, M.R.; Reynolds, R.; et al. Two binding sites for [3H] PBR28 in human brain: Implications for TSPO PET imaging of neuroinflammation. *J. Cereb. Blood Flow Metab.* **2010**, *30*, 1608–1618. [CrossRef]

32. Owen, D.R.; Gunn, R.N.; Rabiner, E.A.; Bennacef, I.; Fujita, M.; Kreisl, W.C.; Innis, R.B.; Pike, V.W.; Reynolds, R.; Matthews, P.M.; et al. Mixed-affinity binding in humans with 18-kDa translocator protein ligands. *J. Nucl. Med.* **2011**, *52*, 24–32. [CrossRef] [PubMed]

33. Mizrahi, R.; Rusjan, P.M.; Kennedy, J.; Pollock, B.; Mulsant, B.; Suridjan, I.; De Luca, V.; Wilson, A.A.; Houle, S. Translocator protein (18 kDa) polymorphism (rs6971) explains in-vivo brain binding affinity of the PET radioligand [18F]-FEPPA. *J. Cereb. Blood Flow Metab.* **2012**, *32*, 968–972. [CrossRef]

34. Guo, Y.; Kalathur, R.C.; Liu, Q.; Kloss, B.; Bruni, R.; Ginter, C.; Kloppmann, E.; Rost, B.; Hendrickson, W.A. Structure and activity of tryptophan-rich TSPOs. *Science* **2015**, *347*, 551–555. [CrossRef] [PubMed]

35. Li, F.; Liu, J.; Zheng, Y.; Garavito, R.M.; Ferguson-Miller, S. Crystal structures of translocator protein (TSPO) and mutant mimic of a human polymorphism. *Science* **2015**, *347*, 555–558. [CrossRef]

36. Jaremko, Ł.; Jaremko, M.; Giller, K.; Becker, S.; Zweckstetter, M. Structure of the mitochondrial translocator protein in complex with a diagnostic ligand. *Science* **2014**, *343*, 1363–1366. [CrossRef]

37. Delavoie, F.; Li, H.; Hardwick, M.; Robert, J.C.; Giatzakis, C.; Péranzi, G.; Yao, Z.X.; Maccario, J.; Lacapere, J.J.; Papadopoulos, V. In vivo and in vitro peripheral-type benzodiazepine receptor polymerization: Functional significance in drug ligand and cholesterol binding. *Biochemistry* **2003**, *42*, 4506–4519. [CrossRef]

38. Yeliseev, A.A.; Kaplan, S. TspO of rhodobacter sphaeroides a structural and functional model for the mammalian peripheral benzodiazepine receptor. *J. Biol. Chem.* **2000**, *275*, 5657–5667. [CrossRef]

39. Hinsen, K.; Vaitinadapoule, A.; Ostuni, M.A.; Etchebest, C.; Lacapere, J.J. Construction and validation of an atomic model for bacterial TSPO from electron microscopy density, evolutionary constraints, and biochemical and biophysical data. *Biochim. Biophys. Acta Biomembr.* **2015**, *1848*, 568–580. [CrossRef]

40. Zeng, J.; Guareschi, R.; Damre, M.; Cao, R.; Kless, A.; Neumaier, B.; Bauer, A.; Giorgetti, A.; Carloni, P.; Rossetti, G. Structural prediction of the dimeric form of the mammalian translocator membrane protein TSPO: A key target for brain diagnostics. *Int. J. Mol. Sci.* **2018**, *19*, 2588. [CrossRef]

41. Si Chaib, Z.; Marchetto, A.; Dishnica, K.; Carloni, P.; Giorgetti, A.; Rossetti, G. Impact of Cholesterol on the Stability of Monomeric and Dimeric Forms of the Translocator Protein TSPO: A Molecular Simulation Study. *Molecules* **2020**, *25*, 4299. [CrossRef]

42. Rao, R.; Diharce, J.; Dugué, B.; Ostuni, M.A.; Cadet, F.; Etchebest, C. Versatile dimerization process of translocator protein (TSPO) revealed by an extensive sampling based on a coarse-grained dynamics study. *J. Chem. Inf. Model.* **2020**, *60*, 3944–3957. [CrossRef]

43. Issop, L.; Ostuni, M.A.; Lee, S.; Laforge, M.; Péranzi, G.; Rustin, P.; Benoist, J.F.; Estaquier, J.; Papadopoulos, V.; Lacapère, J.J. Translocator protein-mediated stabilization of mitochondrial architecture during inflammation stress in colonic cells. *PLoS ONE* **2016**, *11*, e0152919. [CrossRef]

44. Sousounis, K.; Haney, C.E.; Cao, J.; Sunchu, B.; Tsonis, P.A. Conservation of the three-dimensional structure in non-homologous or unrelated proteins. *Hum. Genom.* **2012**, *6*, 10. [CrossRef]

45. Illergård, K.; Ardell, D.H.; Elofsson, A. Structure is three to ten times more conserved than sequence—A study of structural response in protein cores. *Proteins Struct. Funct. Bioinform.* **2009**, *77*, 499–508. [CrossRef]

46. Madeira, F.; Park, Y.M.; Lee, J.; Buso, N.; Gur, T.; Madhusoodanan, N.; Basutkar, P.; Tivey, A.R.; Potter, S.C.; Finn, R.D.; et al. The EMBL-EBI search and sequence analysis tools APIs in 2019. *Nucleic Acids Res.* **2019**, *47*, W636–W641. [CrossRef]

47. Consortium, T.U. UniProt: The universal protein knowledgebase. *Nucleic Acids Res.* **2016**, *45*, D158–D169. [CrossRef]

48. Consortium, U. UniProt: A worldwide hub of protein knowledge. *Nucleic Acids Res.* **2019**, *47*, D506–D515. [CrossRef]

49. Li, F.; Liu, J.; Liu, N.; Kuhn, L.A.; Garavito, R.M.; Ferguson-Miller, S. Translocator protein 18 kDa (TSPO): An old protein with new functions? *Biochemistry* **2016**, *55*, 2821–2831. [CrossRef]

50. Humphrey, W.; Dalke, A.; Schulten, K. VMD: Visual molecular dynamics. *J. Mol. Graph.* **1996**, *14*, 33–38. [CrossRef]

51. Rohmer, M.; Bouvier-Nave, P.; Ourisson, G. Distribution of hopanoid triterpenes in prokaryotes. *Microbiology* **1984**, *130*, 1137–1150. [CrossRef]

52. Russ, W.P.; Engelman, D.M. The GxxxG motif: A framework for transmembrane helix-helix association. *J. Mol. Biol.* **2000**, *296*, 911–919. [CrossRef] [PubMed]

53. Senes, A.; Gerstein, M.; Engelman, D.M. Statistical analysis of amino acid patterns in transmembrane helices: The GxxxG motif occurs frequently and in association with β-branched residues at neighboring positions. *J. Mol. Biol.* **2000**, *296*, 921–936. [CrossRef]

54. Senes, A.; Ubarretxena-Belandia, I.; Engelman, D.M. The Cα–H ... O hydrogen bond: A determinant of stability and specificity in transmembrane helix interactions. *Proc. Natl. Acad. Sci. USA* **2001**, *98*, 9056–9061. [CrossRef]

55. Doura, A.K.; Fleming, K.G. Complex interactions at the helix–helix interface stabilize the glycophorin A transmembrane dimer. *J. Mol. Biol.* **2004**, *343*, 1487–1497. [CrossRef] [PubMed]

56. Brosig, B.; Langosch, D. The dimerization motif of the glycophorin A transmembrane segment in membranes: Importance of glycine residues. *Protein Sci.* **1998**, *7*, 1052–1056. [CrossRef]

57. Li, F.; Xia, Y.; Meiler, J.; Ferguson-Miller, S. Characterization and modeling of the oligomeric state and ligand binding behavior of purified translocator protein 18 kDa from Rhodobacter sphaeroides. *Biochemistry* **2013**, *52*, 5884–5899. [CrossRef]

58. Iatmanen-Harbi, S.; Papadopoulos, V.; Lequin, O.; Lacapere, J.J. Characterization of the high-affinity drug ligand binding site of mouse recombinant TSPO. *Int. J. Mol. Sci.* **2019**, *20*, 1444. [CrossRef]

59. Trott, O.; Olson, A.J. AutoDock Vina: Improving the speed and accuracy of docking with a new scoring function, efficient optimization, and multithreading. *J. Comput. Chem.* **2010**, *31*, 455–461. [CrossRef]

60. Jaremko, M.; Jaremko, Ł.; Giller, K.; Becker, S.; Zweckstetter, M. Structural integrity of the A147T polymorph of mammalian TSPO. *Chembiochem. Eur. J. Chem. Biol.* **2015**, *16*, 1483. [CrossRef]

61. Owen, D.R.; Yeo, A.J.; Gunn, R.N.; Song, K.; Wadsworth, G.; Lewis, A.; Rhodes, C.; Pulford, D.J.; Bennacef, I.; Parker, C.A.; et al. An 18-kDa translocator protein (TSPO) polymorphism explains differences in binding affinity of the PET radioligand PBR28. *J. Cereb. Blood Flow Metab.* **2012**, *32*, 1–5. [CrossRef] [PubMed]

62. Biovia, D.S. Discovery Studio Modeling Environment, Release 2017, 2016. Available online: https://www.3ds.com/products-services/biovia/ (accessed on 1 June 2020).

63. Bruno, A.; Barresi, E.; Simola, N.; Da Pozzo, E.; Costa, B.; Novellino, E.; Da Settimo, F.; Martini, C.; Taliani, S.; Cosconati, S. Unbinding of translocator protein 18 kda (tspo) ligands: From in vitro residence time to in vivo efficacy via in silico simulations. *ACS Chem. Neurosci.* **2019**, *10*, 3805–3814. [CrossRef] [PubMed]

64. Dixon, T.; Uyar, A.; Ferguson-Miller, S.; Dickson, A. Membrane-mediated ligand unbinding of the PK-11195 ligand from the translocator protein (TSPO). *Biophys. J.* **2020**, *120*, 158–167. [CrossRef]

65. Murail, S.; Robert, J.C.; Coïc, Y.M.; Neumann, J.M.; Ostuni, M.A.; Yao, Z.X.; Papadopoulos, V.; Jamin, N.; Lacapère, J.J. Secondary and tertiary structures of the transmembrane domains of the translocator protein TSPO determined by NMR. Stabilization of the TSPO tertiary fold upon ligand binding. *Biochim. Biophys. Acta Biomembr.* **2008**, *1778*, 1375–1381. [CrossRef] [PubMed]

66. Jaremko, Ł.; Jaremko, M.; Giller, K.; Becker, S.; Zweckstetter, M. Conformational flexibility in the transmembrane protein TSPO. *Chem. Eur. J.* **2015**, *21*, 16555–16563. [CrossRef] [PubMed]

67. Dahl, A.C.E.; Chavent, M.; Sansom, M.S. Bendix: Intuitive helix geometry analysis and abstraction. *Bioinformatics* **2012**, *28*, 2193–2194. [CrossRef] [PubMed]

68. von Heijne, G. Proline kinks in transmembrane α-helices. *J. Mol. Biol.* **1991**, *218*, 499–503. [CrossRef]

69. Wilman, H.R.; Shi, J.; Deane, C.M. Helix kinks are equally prevalent in soluble and membrane proteins. *Proteins Struct. Funct. Bioinform.* **2014**, *82*, 1960–1970. [CrossRef]

70. Corpet, F. Multiple sequence alignment with hierarchical clustering. *Nucleic Acids Res.* **1988**, *16*, 10881–10890. [CrossRef]

71. Šali, A.; Blundell, T.L. Comparative protein modeling by satisfaction of spatial restraints. *J. Mol. Biol.* **1993**, *234*, 779–815. [CrossRef]

72. Eramian, D.; Shen, M.Y.; Devos, D.; Melo, F.; Sali, A.; Marti-Renom, M.A. A composite score for predicting errors in protein structure models. *Protein Sci.* **2006**, *15*, 1653–1666. [CrossRef]

73. Studer, G.; Biasini, M.; Schwede, T. Assessing the local structural quality of transmembrane protein models using statistical potentials (QMEANBrane). *Bioinformatics* **2014**, *30*, i505–i511. [CrossRef] [PubMed]

74. Waterhouse, A.; Bertoni, M.; Bienert, S.; Studer, G.; Tauriello, G.; Gumienny, R.; Heer, F.T.; de Beer, T.A.P.; Rempfer, C.; Bordoli, L.; et al. SWISS-MODEL: Homology modeling of protein structures and complexes. *Nucleic Acids Res.* **2018**, *46*, W296–W303. [CrossRef]

75. Bienert, S.; Waterhouse, A.; de Beer, T.A.P.; Tauriello, G.; Studer, G.; Bordoli, L.; Schwede, T. The SWISS-MODEL Repository—New features and functionality. *Nucleic Acids Res.* **2016**, *45*, D313–D319. [CrossRef] [PubMed]

76. Guex, N.; Peitsch, M.C.; Schwede, T. Automated comparative protein structure modeling with SWISS-MODEL and Swiss-PdbViewer: A historical perspective. *Electrophoresis* **2009**, *30*, S162–S173. [CrossRef] [PubMed]

77. Benkert, P.; Biasini, M.; Schwede, T. Toward the estimation of the absolute quality of individual protein structure models. *Bioinformatics* **2010**, *27*, 343–350. [CrossRef]

78. Bertoni, M.; Kiefer, F.; Biasini, M.; Bordoli, L.; Schwede, T. Modeling protein quaternary structure of homo-and hetero-oligomers beyond binary interactions by homology. *Sci. Rep.* **2017**, *7*, 10480. [CrossRef]

79. Karplus, M.; Petsko, G.A. Molecular dynamics simulations in biology. *Nature* **1990**, *347*, 631. [CrossRef]

80. Karplus, M.; McCammon, J.A. Molecular dynamics simulations of biomolecules. *Nat. Struct. Mol. Biol.* **2002**, *9*, 646. [CrossRef]

81. Lomize, M.A.; Pogozheva, I.D.; Joo, H.; Mosberg, H.I.; Lomize, A.L. OPM database and PPM web server: Resources for positioning of proteins in membranes. *Nucleic Acids Res.* **2011**, *40*, D370–D376. [CrossRef]

82. Pettersen, E.F.; Goddard, T.D.; Huang, C.C.; Couch, G.S.; Greenblatt, D.M.; Meng, E.C.; Ferrin, T.E. UCSF Chimera—A visualization system for exploratory research and analysis. *J. Comput. Chem.* **2004**, *25*, 1605–1612. [CrossRef] [PubMed]

83. Gerhard, H.; Walter, N. Lipid composition of mitochondrial outer and inner membranes of Neurospora crassa. *Hoppe Seyler'S Z. Physiol. Chem.* **1974**, *355*, 279–288.

84. Martin, B.; Lily, S. The effect of cholesterol on the viscosity of protein-lipid monolayers. *Chem. Phys. Lipids* **1976**, *17*, 416–422.

85. Osterberg, P.M.; Senturia, S.D. "Membuilder": An Automated 3D Solid Model Construction Program for Microelectromechanical Structures. In Proceedings of the International Solid-State Sensors and Actuators Conference-TRANSDUCERS'95, Stockholm, Sweden, 25–29 June 1995; Volume 2, pp. 21–24.

86. Ghahremanpour, M.M.; Arab, S.S.; Aghazadeh, S.B.; Zhang, J.; van der Spoel, D. MemBuilder: A web-based graphical interface to build heterogeneously mixed membrane bilayers for the GROMACS biomolecular simulation program. *Bioinformatics* **2013**, *30*, 439–441. [CrossRef]

87. Schmidt, T.H.; Kandt, C. LAMBADA and InflateGRO2: Efficient membrane alignment and insertion of membrane proteins for molecular dynamics simulations. *J. Chem. Inf. Model.* **2012**, *52*, 2657–2669. [CrossRef]

88. Berendsen, H.J.; van der Spoel, D.; van Drunen, R. GROMACS: A message-passing parallel molecular dynamics implementation. *Comput. Phys. Commun.* **1995**, *91*, 43–56. [CrossRef]

89. Van Der Spoel, D.; Lindahl, E.; Hess, B.; Groenhof, G.; Mark, A.E.; Berendsen, H.J. GROMACS: Fast, flexible, and free. *J. Comput. Chem.* **2005**, *26*, 1701–1718. [CrossRef]

90. Jaembeck, J.P.; Lyubartsev, A.P. An extension and further validation of an all-atomistic force field for biological membranes. *J. Chem. Theory Comput.* **2012**, *8*, 2938–2948. [CrossRef]

91. Ponder, J.W.; Case, D.A. Force fields for protein simulations. In *Advances in Protein Chemistry*; Elsevier: Amsterdam, The Netherlands, 2003; Volume 66, pp. 27–85.

92. Price, D.J.; Brooks, C.L., III. A modified TIP3P water potential for simulation with Ewald summation. *J. Chem. Phys.* **2004**, *121*, 10096–10103. [CrossRef] [PubMed]

93. Wang, J.; Wolf, R.M.; Caldwell, J.W.; Kollman, P.A.; Case, D.A. Development and testing of a general amber force field. *J. Comput. Chem.* **2004**, *25*, 1157–1174. [CrossRef]

94. Sprenger, K.; Jaeger, V.W.; Pfaendtner, J. The general AMBER force field (GAFF) can accurately predict thermodynamic and transport properties of many ionic liquids. *J. Phys. Chem. B* **2015**, *119*, 5882–5895. [CrossRef] [PubMed]

95. Frisch, M.; Trucks, G.; Schlegel, H.; Scuseria, G.; Robb, M.; Cheeseman, J.; Montgomery, J., Jr.; Vreven, T.; Kudin, K.; Burant, J.; et al. *Gaussian 09, Revision A02*; Pittsburgh, P.A., Pople, J.A., Eds.; Gaussian Inc.: Wallingford, UK, 2009.

96. da Silva, A.W.S.; Vranken, W.F. ACPYPE-Antechamber python parser interface. *BMC Res. Notes* **2012**, *5*, 367. [CrossRef] [PubMed]

97. Parrinello, M.; Rahman, A. Polymorphic transitions in single crystals: A new molecular dynamics method. *J. Appl. Phys.* **1981**, *52*, 7182–7190. [CrossRef]
98. Martyna, G.J.; Tobias, D.J.; Klein, M.L. Constant pressure molecular dynamics algorithms. *J. Chem. Phys.* **1994**, *101*, 4177–4189. [CrossRef]
99. Hoover, W.G. Canonical dynamics: Equilibrium phase-space distributions. *Phys. Rev. A* **1985**, *31*, 1695. [CrossRef]
100. Nosé, S. A molecular dynamics method for simulations in the canonical ensemble. *Mol. Phys.* **1984**, *52*, 255–268. [CrossRef]
101. Hünenberger, P.H. Thermostat algorithms for molecular dynamics simulations. In *Advanced Computer Simulation*; Springer: Berlin/Heidelberg, Germany, 2005; pp. 105–149.
102. Hess, B.; Bekker, H.; Berendsen, H.J.; Fraaije, J.G. LINCS: A linear constraint solver for molecular simulations. *J. Comput. Chem.* **1997**, *18*, 1463–1472. [CrossRef]

Article

Study of Endogen Substrates, Drug Substrates and Inhibitors Binding Conformations on MRP4 and Its Variants by Molecular Docking and Molecular Dynamics

Edgardo Becerra [1,2], Giovanny Aguilera-Durán [1,3], Laura Berumen [2], Antonio Romo-Mancillas [3,*] and Guadalupe García-Alcocer [2,*]

1 Posgrado en Ciencias Químico Biológicas, Facultad de Química, Universidad Autónoma de Querétaro, Cerro de las Campanas S/N, Querétaro 76010, Mexico; ebecerra1989@gmail.com (E.B.); giovanny.aguilera@uaq.mx (G.A.-D.)
2 Centro Universitario, Unidad de Investigación Genética, Facultad de Química, Universidad Autónoma de Querétaro, Querétaro 76010, Mexico; lcbsq@yahoo.com
3 Centro Universitario, Laboratorio de Diseño Asistido por Computadora y Síntesis de Fármacos, Facultad de Química, Universidad Autónoma de Querétaro, Querétaro 76010, Mexico
* Correspondence: ruben.romo@uaq.mx (A.R.-M.); leguga@email.com (G.G.-A.); Tel.: +52-442-192-1200-75032 (A.R.-M.); +52-442-192-1200-5529(G.G.-A.); Fax: +52-442-192-1302 (G.G.-A.)

Abstract: Multidrug resistance protein-4 (MRP4) belongs to the ABC transporter superfamily and promotes the transport of xenobiotics including drugs. A non-synonymous single nucleotide polymorphisms (nsSNPs) in the ABCC4 gene can promote changes in the structure and function of MRP4. In this work, the interaction of certain endogen substrates, drug substrates, and inhibitors with wild type-MRP4 (WT-MRP4) and its variants G187W and Y556C were studied to determine differences in the intermolecular interactions and affinity related to SNPs using protein threading modeling, molecular docking, all-atom, coarse grained, and umbrella sampling molecular dynamics simulations (AA-MDS and CG-MDS, respectively). The results showed that the three MRP4 structures had significantly different conformations at given sites, leading to differences in the docking scores (DS) and binding sites of three different groups of molecules. Folic acid (FA) had the highest variation in DS on G187W concerning WT-MRP4. WT-MRP4, G187W, Y556C, and FA had different conformations through 25 ns AA-MD. Umbrella sampling simulations indicated that the Y556C-FA complex was the most stable one with or without ATP. In Y556C, the cyclic adenosine monophosphate (cAMP) and ceefourin-1 binding sites are located out of the entrance of the inner cavity, which suggests that both cAMP and ceefourin-1 may not be transported. The binding site for cAMP and ceefourin-1 is quite similar and the affinity (binding energy) of ceefourin-1 to WT-MRP4, G187W, and Y556C is greater than the affinity of cAMP, which may suggest that ceefourin-1 works as a competitive inhibitor. In conclusion, the nsSNPs G187W and Y556C lead to changes in protein conformation, which modifies the ligand binding site, DS, and binding energy.

Keywords: MRP4; SNPs; variants; protein threading modeling; molecular docking; molecular dynamics; binding site

Citation: Becerra, E.; Aguilera-Durán, G.; Berumen, L.; Romo-Mancillas, A.; García-Alcocer, G. Study of Endogen Substrates, Drug Substrates and Inhibitors Binding Conformations on MRP4 and Its Variants by Molecular Docking and Molecular Dynamics. *Molecules* **2021**, *26*, 1051. https://doi.org/10.3390/molecules26041051

Academic Editors: Marco Tutone and Anna Maria Almerico

Received: 17 December 2020
Accepted: 8 February 2021
Published: 17 February 2021

Publisher's Note: MDPI stays neutral with regard to jurisdictional claims in published maps and institutional affiliations.

1. Introduction

The transport of xenobiotics out of the cell across membranes is a mechanism used by cells to detoxify. This mechanism is mediated by ATP-binding cassette (ABC) transporters [1]. Multidrug resistance protein-4 (MRP4) is a member of the ABCC subfamily and mediates the transport of xenobiotics such as cardiovascular, antiviral, and anticancer drugs. The substrates for MRP4 are mainly glucuronide conjugates and organic anions [2]. MRP4 can modify drug pharmacokinetics and contributes to the manifestation of side effects or multidrug resistance. In addition, the tumor energy metabolism is related to multidrug resistance due to the high production of ATP to enhance the activity of MRP4

and other ABC transporters [3]. An increase in MRP4 activity or expression leads to a decrease in drug efficacy. In another instance, a decrease in MRP4 activity or expression could enhance toxicity due to drug accumulation. Since MRP4 is expressed in the kidneys, liver, erythrocytes, lymphocytes, adrenal glands, platelets, brain, and pancreas in humans, it can modify cellular exposure to drugs. In addition, the toxicity produced by MRP4 will depend on the type of drug or endogen substrate [4]. The MRP4 dysregulation has been reported in several pathological disorders, especially in cancer [5]; thus, MRP4 represents an attractive therapeutic target. The design of pharmacological agents with the ability to selectively modulate the activity of this ABC transporter or modify its affinity of a given substrate represents a challenge in chemical biology and drug design [6,7].

MRP4 consists of 1325 amino acids and is the shortest member of the ABCC subfamily [8]. The basic MRP4 core structure is comprised of transmembrane domains (TMDs) and intracellular nucleotide binding domains (NBDs). The domain arrangement for MRP4 is TMD-NBD-TMD-NBD [9]. Each TMD consists of six transmembrane helixes (TMHs) that determine the ligand specificity and allow ligand binding. In addition, NDBs bind and hydrolyze ATP to trigger substrate transport [10]. MRP4 is codified by the ABCC4 gen, located on chromosome 13q32.1 [11]. Alternative splicing leads to four isoforms, of which isoform 1 has been the most studied [1]. MRP4 is a highly polymorphic gene [12]; however, limited data are available on the function of MRP4 variants. Recent studies have been focused on the relationship between ABCC4 nsSNPs and drug disposition. In most cases, nsNSPs have little or no effect on the protein structure or function, but sometimes nsSNPs promotes non-functional or highly functional proteins [13]. The nsSNPs that occur in protein coding regions always alter the encoded amino acid, and the effect on the structure or function of the protein depends on the mutated site [14].

Nada-Abla and coworkers in 2008 reported that the MRP4 variants G187W and G487E show a significantly reduced function of azidothymidine and adefovir transport compared to wild-type MRP4 (WT-MRP4). G187W is a non-synonymous ABCC4 variant and the mutation is located at the cytosolic loop 1 in the TMD (Figure 1); it has undergone the greatest structural change in terms of composition, polarity, and molecular volume. G187W also has a 50% reduction in function, and this could be clinically relevant [15]. On the other hand, Y556C is another non-synonymous ABCC4 variant and is located at NBD1 (Figure 1). Mayukh-Banerjee and coworkers in 2016 reported that the Y556C variant exhibited a 1.8-fold increase in dimethylarsinic acid effectiveness relative to WT-MRP4. Experiments on MRP4 transfection into the HEK cell line showed that the Y556C variant had 50% less expression than WT-MRP4. Both G487E and Y556C had appropriate cellular membrane localization [16].

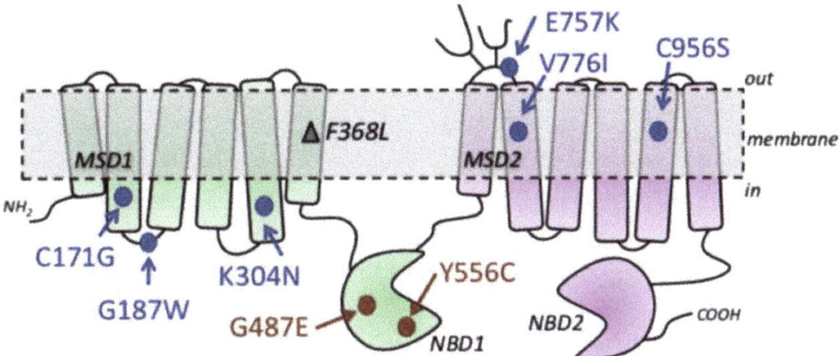

Figure 1. MRP4 structure of the cell membrane and localization of MRP4 variants (adapted from Banerjee at al., 2016) [16].

The crystallographic structure of MRP4 is not available; thus, a protein threading model can be built based on the homology of the template and the construction of loops.

Protein threading by I-TASSER relies in the identification of structural templates from the Protein Data Bank (PDB) using a local meta-threading server, a method for template-based protein structure prediction. Three-dimensional models are generated for a given sequence by collecting high-scoring structural templates from locally installed threading programs [17,18]. After model building, it can be refined by molecular mechanics calculations, such as energy minimization and molecular dynamics simulations. Protein threading models are considered working tools that can be used to generate hypotheses related to protein structure, protein function, and protein–ligand interactions. The molecular docking of drug molecules into their binding sites allows us to identify relevant amino acids for ligand–protein interactions in order to select such amino acids for further site-directed mutagenesis studies [9]. In the present study, three different MRP4 structures (WT and its variants, G187W and Y556C) were built through protein threading to study, by molecular docking, the interactions between endogen substrates, drug substrates, and inhibitors, with MRP4-WT, G187W, and Y556C, allowed to observe changes in the pattern of intermolecular interactions, adapted from Russel and coworkers in 2008, where they report the IC_{50} of several molecules, in vitro, over the MRP4 protein [8], and to calculate the binding energy (ΔG) between FA and cAMP and MRP4 structures.

2. Results and Discussion

2.1. WT-MRP4, G187W, and Y556C Model Building and CG-MD Simulations

MRP4 plays a critical role in the distribution of different xenobiotics and endogen substrates, which can lead to different effects in the organism. Differences in the MRP4 activity depend, among other factors, on the expression or mutational changes of the ABCC4 gene, leading to significantly higher or lower transport activity [4]. The WT-MRP4 and its variants Y556C and G187W were built by protein threading in the I-TASSER server [19]. Protein threading and homology modeling are based on the principle that similar primary sequences will lead to similar 3D protein structures. According to the BLAST server, the template structure MRP1 from *Bos taurus* and human MRP4 had a 36.56% identity sequence similarity. When the primary sequence of a protein has 30% of identity as referred to a template (crystallographic structure), the protein threading and homology models are considered functional because the root mean standard deviation (RMSD) of the positions of their atoms is 2.0 Å or less with regard to the template structure [20–22].

The best model by the I-TASSER of each MRP4 structure was selected for further analysis with coarse-grained molecular dynamics simulations (CG-MDS) of 1 μs. Figure 2 shows the three MRP4 models and the most representative structures (cluster 1) obtained in I-TASSER and by CG-MDS at timesteps 630.40 ns for WT-MRP4, 564.90 ns for G187W, and 674.90 ns for Y556C. The conformations of WT-MRP4 and variants were in an "inward-facing conformation" [23], while, in CG-MDS, the three MRP4 structures were in a closed state. All the loops that connect the alpha helixes of the three MRP4 structures have different conformations and distributions over the protein.

In this work, RMSD values higher than 2.0 Å were considered significant, hence the protein conformations were considered different. Figure 3 shows the different MRP4 sites studied and each region is illustrated with a different color, where the green color represents WT-MRP4; those sites are the nucleotide-binding domains (NBD), the transmembrane domains (TMD), and the residues relevant to substrate interaction (r85-236 and r715-866).

The WT-MRP4, G187W, and Y556C conformations during the first 100 ns of CG-MDS changed significantly, according to the RMSD values (Figure 4a), which indicates a large movement of the protein to further stabilization from 250 to 1000 ns. The RMSD of WT-MRP4 was higher than those of its variants, considering the complete structure. In addition, different regions of the MRP4 structure were studied focusing on the ligand binding sites, nsSNPs, and ATP pocket binding. Figure 4b shows the RMSD values for TMDs of the WT-MRP4 and its variants. According to the RMSD plot, the changes in the TMDs' conformations are quite similar among the three MRP4 structures.

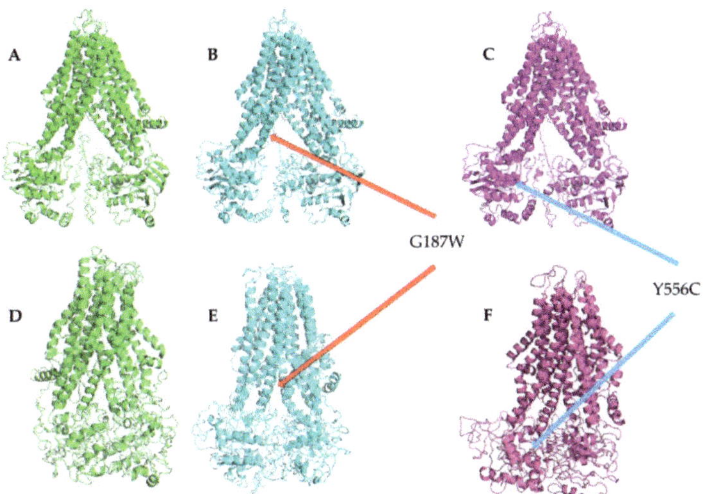

Figure 2. MRP4 models built by homology modeling in I-TASSER and cluster 1 from CG-MDS. Green, (**A**) WT-MRP4. Cyan, (**B**) G187W. Magenta, (**C**) Y556C. (**D–F**) represent cluster 1 obtained from GC-MDS for WT-MRP4, G187W, and Y556C, respectively. The arrows indicate the location of mutations.

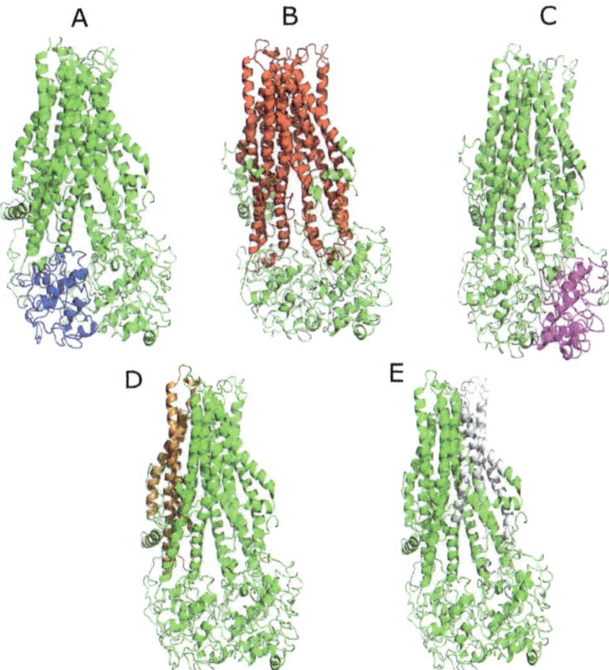

Figure 3. Representation of different sites of the MRP4 protein in the complete structure. (**A**) Blue represents NBD1. (**B**) Red represents TMDs. (**C**) Magenta represents NBD2. (**D**) Orange represents r85-236. (**E**) Gray represents r715-866.

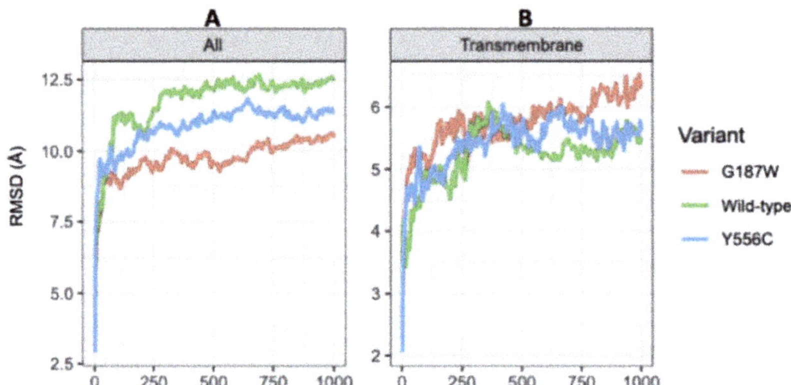

Figure 4. plot for the complete WT-MRP4 structures and their variants (**A**) and the TMDs (**B**) throughout 1000 ns of MDS.

In contrast to the complete structure, the TMDs do not obtain stabilization according to the RMSD values, which increase and decrease over 1000 ns. Moreover, the RMSD values for G187W remain increasing from 750 to 1000 ns, and such behavior could be due to the mutation is in the cytosolic loop 1 that connects the transmembrane helix (TMH) 1 and TMH2. The NBD1 conformation in WT-MRP4 remained unstable and the RMSD values kept increasing throughout the 1000 ns of CG-MDS, while the NBD1 conformation in G187W and Y556C was stable with an RMSD value around 6.0 Å (Figure 5). The RMSD values for NBD2 of the WT-MRP4 and its variants were similar even though there was no stabilization through the simulation. In addition, the RMSD values of G187W and Y556C tended to increase while the RMSD values of WT-MRP4 tended to decrease at the end of the simulation.

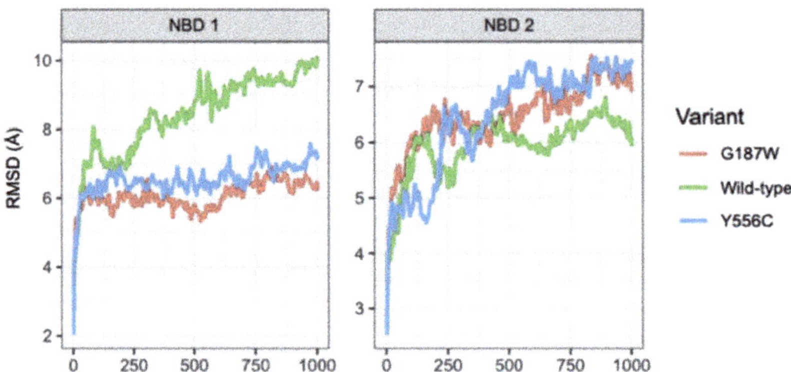

Figure 5. RMSD plot for NBD1 and NBD2 of WT-MRP4 and its variants throughout the 1000 ns of MDS.

According to the RMSD plot in Figure 6, the conformation in residues (r) 85 to 236 (r85-236 site corresponding to TMH1-TMH4) of WT-MRP4 remained stable from 250 to 1000 ns while in G187W it stabilized at the last 250 ns. r85-236 site in Y556C did not stabilize throughout the 1000 ns. In the same Figure 6, the RMSD plot indicates that the conformation in the r715-866 site (TMH7-TMH10) in WT-MRP4 and Y556C kept constant, with a tendency toward increasing motion in Y556C and decreasing motion in WT-MRP4. Besides this, the conformation in r715-866 site in G187W did not stabilize and the RMSD

values suggest high motion throughout the 1000 ns of simulation. In the case of ATP sites 1 and 2 (Figure 7), the conformation remained unstable, suggesting that it is a site with high motion, with exception of ATP site 2 of WT-MRP4, which kept stable through the simulation. Regarding the three different MRP4 structures, alignments on the NBDs, TMDs, and TMH1-TMH4 and TMH7-TMH10 sites were performed to determine differences among the structures. In all the alignments (Supplementary Material S1), it was observed that all the sites studied presented significant structural differences, according to the RMSD values, comparing WT-MRP4 vs. its variants and G187W vs. Y566C. Moreover, the TMHs are responsible for the specificity for the substrate, and the r-85-236 and r715-866 sites in WT-MRP4 were significantly different with respect to G187W and Y556C, which could lead to differences in the ligand affinity, the ligand binding site, and the motion of the protein [23,24].

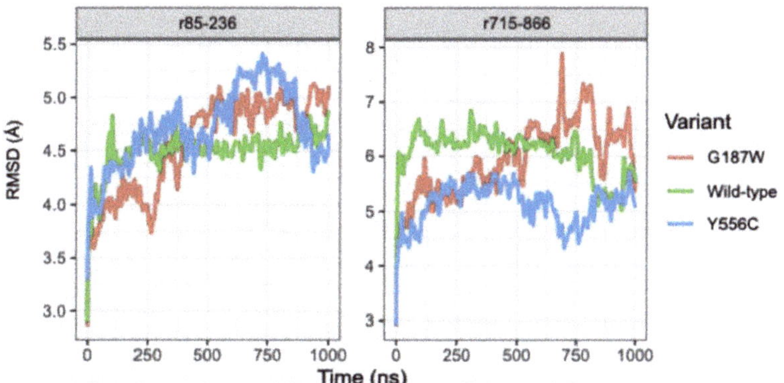

Figure 6. RMSD plot for r85-236 and r715-866 of WT-MRP4 and its variants. r86-236 represents TMH1-TMH4 and r715-866 represents TMH7-TMH10.

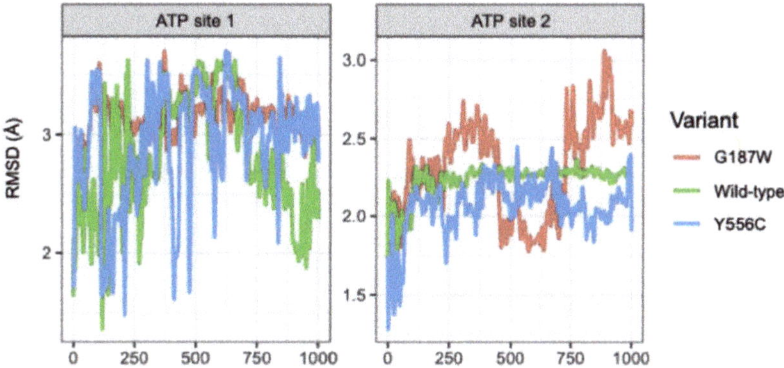

Figure 7. RMSD plot for ATP binding site 1 (ATP site 1) and ATP binding site 2 (ATP site 2) of WT-MRP4 and its variants.

2.2. Molecular Docking in MRP4 and Variants

By using cluster 1 from CG-MDS for each structure of MRP4, molecular docking was performed to explore the effect of the MRP4 variants on the affinity of three different groups of molecules, previously reported as substrates or inhibitors in vitro. Table 1 presents the docking score (DS), expressed as kcal/mol, related to the interaction between endogenous substrates and WT-MRP4, Y556C, and G187W. In this work, significant differences between docking poses were considered when a difference greater than 1 kcal/mol in DS was

present. According to the DS, most endogen substrates significantly changed their DS in G187W and Y556C with respect to WT-MRP4. The Y556C mutation is located at NBD1 and leads to a different conformation with respect to WT-MRP4, which causes the most substantial change in the ligand binding site of all the molecules studied, even more than those molecules with significant changes in DS compared to WT-MRP4.

Table 1. Endogen substrate DS (kcal/mol) in the WT-MRP4 and MRP4 variants.

Endogen Substrate	WT-MRP4	G187W	Y556C
cAMP	−5.88	−6.35	−5.82
Cholic acid	−10.55	−7.2 *	−8.51 *
Folic acid	−8.56	−1.51 *	−2.39 *
Glycolic acid	−7.35	−9.09 *	−5.42 *
Leukotriene B4	−8.19	−7.82	−6.07 *
PGE1	−8.21	−7.31	−7.48
PGE2	−7.36	−7.31	−7.00
Prasterone sulphate	−10.65	−9.24 *	−9.57 *
Taurocholic acid	−4.61	−9.9 *	−6.07 *
Uric acid	−4.77	−5.13	−5.06

cAMP: cyclic adenosine monophosphate. * Represents more than 1 kcal/mol of difference in the DS compared with WT.

Since cAMP is considered to be the main molecule transported by MRP4 [25], it was used as a control to compare the effect of nsSNPs and ATP binding. The cAMP DS was not considered significantly different in WT-MRP4 with respect to Y556C and G187W. Cholic acid DS was significantly different, with more than 3 kcal/mol when comparing WT-MRP4 with respect to G187W, while cholic acid DS in WT-MRP4 with respect to Y556C was significantly different at over 2 kcal/mol. The taurocholic acid DS was significantly different, with more than 4 kcal/mol when comparing WT-MRP4 with respect to G187W. Folic acid (FA) was the molecule with the highest variation in DS when comparing WT-MRP4 to G187W and Y556C, with 7.05 and 6.17 kcal/mol differences.

Table 2 presents the DSs of drug substrates. Cefazoline and olmesartan DSs were significantly different in WT-MRP4 with respect to G187W, while cefazoline, furosemide, leucovorin, methotrexate, tenofovir, and topotecan DSs were significantly different in WT-MRP4 with respect to Y556C. According to DS, most drug substrates could present more affinity for Y556C than for WT-MRP4.

Table 2. Drug substrate DS (kcal/mol) in the WT-MRP4 and MRP4 variants.

Drug Substrate	WT-MRP4	G187W	Y556C
6-Mercaptopurine	−4.34	−5.06	−4.81
Adefovir	−3.54	−2.88	−4.42
Cefazoline	−8.71	−5.65 *	−10.12 *
Cefotaxime	−7.04	−6.08	−6.88
Ceftizoxime	−7.02	−7.16	−6.59
Furosemide	−5.77	−5.52	−7.00 *
Hydrochlorothiazide	−6.70	−6.39	−6.73
Leucovorin	−6.46	−6.60	−3.78 *
Methotrexate	−6.98	−7.22	−3.76 *
Olmesartan	−7.30	−8.75 *	−7.4
Tenofovir	−2.76	−3.15	−4.58 *
Topotecan	−6.09	−6.39	−8.26 *

* Represents more than 1 kcal/mol difference in the DS compared against WT.

The DS of inhibitors is presented in Table 3. Glafenine DS was significantly different in WT-MRP4 with respect to both G187W and Y556C. Ceefourin-1, indomethacin, and sildenafil DSs were significantly different in WT-MRP4 with respect to G187W, while losartan DS was significantly different between WT-MRP4 and Y556C. Endogen substrates

were the group of molecules with more variation in the DS, and thus such mutations on the ABCC4 gene could lead to changes in cell metabolism related to changes in the distribution of endogen substrates across cell membranes. Several mutational studies on MRP1 have demonstrated that amino acids in several TMDs are involved in substrate binding and nsSNPs can modify the ligand binding site or the affinity of ligands in a selective manner [11]. Yet, the DS values are not totally related to function. Regarding drug substrates and inhibitors, a small number of molecules changed their DS significantly to G187W and Y556C as related to WT-MRP4, which could modify the pattern of transport of cefazoline, ceftizoxime, olmesartan, topotecan, and ceefourin-1, which could, possibly, mean that they act as substrates or inhibitors depending on the MRP4 variant.

Table 3. Drug inhibitor DS (kcal/mol) in the WT-MRP4 and MRP4 variants.

Inhibitors	WT-MRP4	G187W	Y556C
ABSF	−5.41	−5.85	−5.97
Artesunate	−6.02	−5.42	−6.59
Ceefourin1	−7.52	−6.19 *	−7.22
Celecoxib	−7.57	−8.16	−7.98
Dipyridamole	−3.86	−4.78	−4.23
Glafenine	−8.27	−5.48 *	−6.32 *
Indomethacin	−8.42	−7.07 *	−8.43
Losartan	−8.82	−8.91	−6.55 *
MK-571	−8.86	−9.65	−8.87
Parthenolide	−7.88	−6.92	−7.67
Prazosin	−6.71	−6.98	−7.25
Probenecid	−6.59	−5.89	−6.34
Quercetin	−6.16	−6.2	−6.63
Sildenafil	−7.67	−8.43 *	−8.25
Sulindac	−8.41	−7.82	−8.41
Tyrphostin	−7.74	−7.32	−8.11

* Represents more than 1 kcal/mol difference in the DS compared against WT.

2.3. Differences in the Interaction Pattern in MRP4 Structures

Ligand interaction diagrams (LIDs) represent the pattern of intermolecular interactions of molecules with MRP4 amino acids. Those molecules with a >2 kcal/mol difference in the three different MRP4 structures appear in LIDs in Supplementary Material S2–S5 and the LIDs for FA appear in Figures 8–10. The interaction sites in the three different MRP4 structures were different, mainly in Y556C, for all the molecules exhibited in the LIDs. Cholic acid interacts mainly with hydrophobic residues on WT-MRP4 and Y556C, while on G187W it interacts with hydrophobic and polar residues. H-bonds are only exerted through hydrophobic residues, except on Y556C, where arginine exerts an H-bond with cholic acid. In addition, cholic acid interacts with positively charged amino acids in the three MRP4 structures, lysine on G187W and WT-MRP4, and arginine on Y556C (Supplementary Material S2). The taurocholic acid binding site was different in WT-MRP4 with respect to G187W and was totally different with respect to Y556C. The intermolecular interactions in Y556C and G187W were H-bonds, polar, hydrophobic, interactions with positively and negatively charged amino acids, while interactions in WT-MRP4 were hydrophobic and polar but did not interact with negatively charged amino acids and did not exert H-bonds (Supplementary Material S3). The cefazoline binding site was quite different on the three different MRP4 structures. Hydrophobic and polar residues on WT-MRP4, Y556C, and G187W interact with cefazoline, but only on G187W do H-bonds with arginine occur. WT-MRP4 and Y556C interact with cefazoline through negatively charged amino acids (glutamate), while the G187W interacts through positively charged amino acids (arginine) (Supplementary Material S4). Ceefourin-1 interacts with hydrophobic, polar, negative, and positively charged residues on WT-MRP4 and Y556C, even though the binding site is different in each MRP4 structure, which leads to only one difference; ceefourin-1 interacts via H-bonds in Y556C. Additionally, ceefourin-1 interacts in G187W

with hydrophobic, polar, and positively charged residues, and via H-bonds and π-π stacking (Supplementary Material S5). The Y556C mutation led to a substantial change in the binding site of most of the molecules analyzed, while G187W mutation did not substantially change the binding site with respect to WT-MRP4. Although the molecule's binding site was different among the MRP4, the pattern of intermolecular interactions could be similar to that observed in the LIDs. The molecules presented in the LIDs had a substantial modification to their binding site and intermolecular interactions in Y556C, which could be related to changes in the transport-rate, IC50, entrance to the inner cavity, and effect on the Y556C conformational movement.

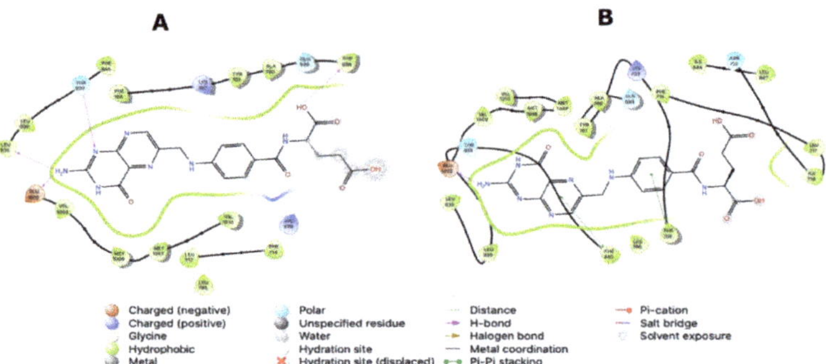

Figure 8. FA binding interaction diagram in WT-MRP4 at T0 (**A**) and 16.20 ns (**B**) in AA-MDS. The nomenclature for the intermolecular interactions is shown in the bottom of the figure.

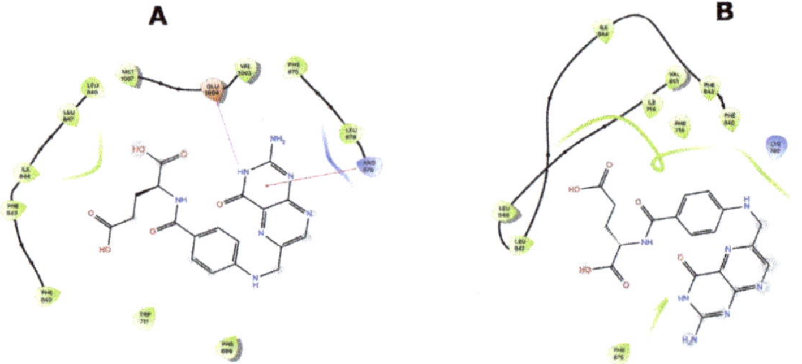

Figure 9. FA binding diagram interactions on G187W at T0 (**A**) and at 8.0 ns (**B**) in AA-MDS.

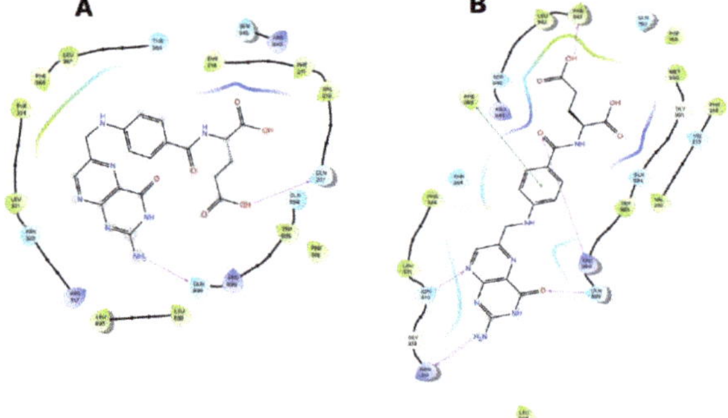

Figure 10. FA binding interaction diagrams in Y556C at T0 (**A**) and 15.0 ns (**B**) in AA-MDS.

Table 4 shows the most important residues required for the interactions of endogen substrates, drug substrates, and inhibitors in each MRP4 structure. It is worth noting that in the inhibitor group, only Lys 329 and Arg 951 appear in both endogen and drug substrates, suggesting that such residues may play an important role in the binding site. In the endogen and drug substrate groups, at least two residues are repeated—Arg 946 and Lys 702.

Table 4. Residues considered as important for ligand–MRP4 interactions.

	WT-MRP4	G187W	Y556C
Endogen substrates	Glu 1002, Lys 702, Thr 839, Lys 106, Lys 329, Glu 374, Gln 251, Phe 698	Arg 946, Arg 998. Gly 991, Lys 702, Arg 946	Arg 951, Trp 947, Glu 7, Gln 6, Val4
Drug substrates	Arg 946, Arg 998, Lys 702, arg 946, Thr 994, arg 312, Lys 329, Phe 698	Ser 945, Lys 329, Leu 987, Arg 946, Arg 951, Arg 312, Ser 306	Arg 946, Asp 15, Phe 329, Lys 32, Phe 939
Inhibitors	Phe 993, Pro 867, Arg 951, Arg 362, Thr 366, Gln 251, Leu 247	Phe 325, Lys 329, Thr 364, Glu 102, Glu 103, Trp 995	Phe 993, Pro 867, Arg 951, Trp 947

2.4. All-Atom Molecular Dinaymics (AA-MD) Simulations and Umbrella Sampling Studies

FA was the molecule with the highest DS variation in G187W and Y556C related to WT-MRP4 in molecular docking studies, considering that ATP was not bound; hence, to study the effect of the presence of ATP and the mutations, changes in the pattern of intermolecular interactions and affinity to MRP4 structures, 25 ns AA-MDS, and 10 ns of Umbrella sampling simulations were carried out on the MRP4-FA complexes and compared with cAMP as a control molecule. The C1 in AA-MDS was at 16.2 ns in WT-MRP4, 8.0 ns in G187W, and 15.0 ns in Y556C. Figure 8 shows the FA LID in WT-MRP4 at T0 and 16.2 ns. In this simulation, the differences in the FA binding site and the pattern of intermolecular interaction can be observed according to the WT-MRP4 conformation at a given time. The FA binding site was the same at T0 and 16.2 ns, suggesting that the protein can be in an inward-facing conformation for 25 ns or even more. Moreover, the pattern of intermolecular interactions between FA and WT-MRP4 is quite similar, consisting of H-bonds; π-π stacking; and interactions with polar, hydrophobic, and negatively and positively charged residues. The differences in the pattern of intermolecular interactions rely on the H-bonds, with 4 H-bonds at T0 with Phe698 and Leu835, Glu1002 and Thr839. Meanwhile, at 16.2 ns the interactions were mainly hydrophobic (π-π stacking) and there was one H-bond with Glu1002. Therefore, it seems that FA from 0 to 16.2 ns interacts with MRP4 to achieve its optimal bonding only. A longer simulation will help to determine all the FA binding sites across the WT-MRP4. Besides this, the AA-MDS studies were performed on each MRP4-FA complex to determine differences in the patterns of interactions and conformations in frames from 0 to 25 ns every 5 ns referred to T0 (Supplementary Material S6–S23). As mentioned above, FA did not change its conformation significantly in WT-MRP4 from 0 to 25 ns, while in G187W the conformation was significantly different at 20 and 25 ns, as the carboxyl group was responsible for the FA conformational changes and multiple intermolecular interactions, such as H-bonds, π-π, and π-cation. FA in G187W did not change its binding site throughout all the AA-MD simulation, but it kept moving throughout 25 ns to obtain a DS greater than that of the most stable conformation at 8 ns (Supplementary Material S12–S17). The G187W mutation is located at cytosolic loop 1, close to the entrance of the inner cavity [26], and leads to changes in the MRP4 conformation which could block or interfere with ligand binding or the entrance to its binding site. Notwithstanding this, it is not possible to determine with this study whether there is a relevant effect on the FA transport by G187W. Figure 9 shows the FA LID in G187W at T0 and 8.0 ns in AA-MDS. The FA binding site was slightly different, with intermolecular

interactions by an H-bond and π-cation at T0, while at 8.0 no intermolecular interactions were observed, but the pocket binding was composed of hydrophobic amino acids in both time steps.

Figure 10 shows the FA LID in Y556C at T0 and 15.00 ns in AA-MDS. In this case, the FA binding site was the same, but the intermolecular interactions by H-bonds at 15.00 ns were higher than at T0, which seems to confer to FA a stronger binding over time. Y556C promoted the greatest change in the FA binding site with respect to WT-MRP4 and G187W, but throughout the 25 ns in AA-MDS the binding site did not change. On the contrary, the pattern of intermolecular interactions as well as the FA conformation was different across the AA-MDS (Supplementary Material S18–S23).

The AA-MDS studies were conducted in cAMP as a substrate control to determine if the nsSNPs affect the pattern of intermolecular interactions in the same manner as FA. First, AA-MDS were performed on the wild type and variants for MRP4-cAMP complex to obtain the C1 conformations. Further, ATP and Mg^{2+} were added into the MRP4 structures to obtain a more complete system, and further AA-MDS were performed to determine a new C1. This procedure was applied in the MRP4-FA complexes as well. Both C1 conformations before ATP and after ATP were used for umbrella sampling studies to obtain the free binding energy (ΔG) of each ligand. Since two ATP molecules bind to MRP4, we referred to them as ATP1 and ATP2 for the NBD1 and NBD2, respectively. In all the MRP4-FA and MRP4-cAMP complexes with ATP bonds, the binding site and the ligand conformation were slightly different concerning the MRP4-FA and MRP4-cAMP complexes without ATP. The presence or the absence of ATP modifies the pattern of intermolecular interactions as well. The H-bonds, π-π stacking, and π-cation interactions are present together or individually in all the complexes, except in G187W-FA without ATP. Given the fact that the ATP influences the ligand conformation, it modifies the intermolecular interactions as well. Figures 11 and 12 show the ATP binding site in NBD1 and NBD2 from the WT-MRP4, G187W, and Y556C structures.

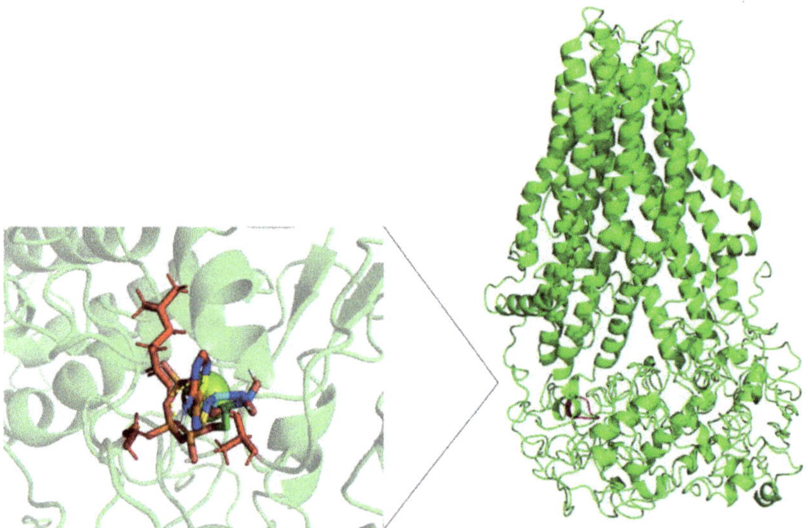

Figure 11. Representation of the ATP binding site 1 in the complete MRP4 structure and an amplification of such site. Color pink in the complete MRP4 structure represents the ATP site 1. Color red and represent the residues interacting with ATP, and rainbow colors represent ATP. The yellow dotted lines indicate polar interactions.

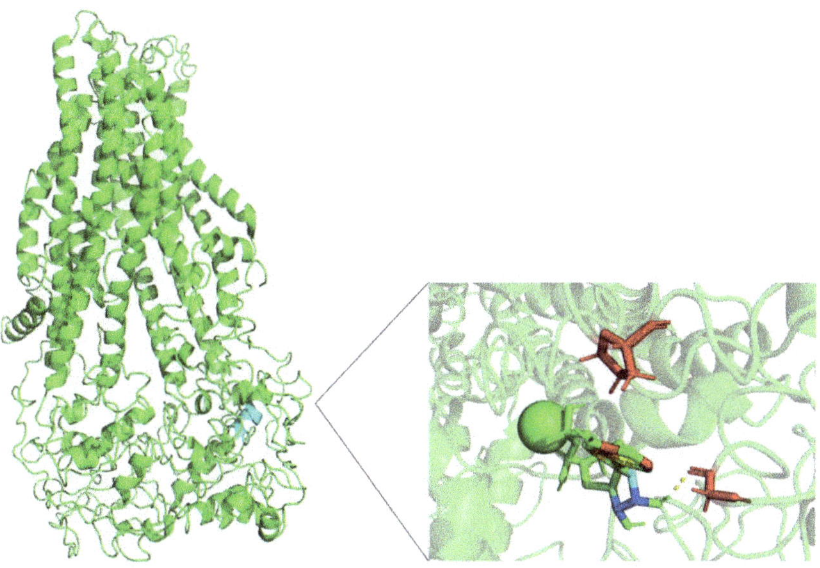

Figure 12. Representation of the ATP binding site 2 in the complete MRP4 structure and an amplification of such site. Color cyan in the complete MRP4 structure represents the ATP site 2. Color red represent the residues interacting with ATP, and rainbow colors represent ATP. The yellow dotted lines indicate polar interactions.

Figures 13–15 show the FA binding sites in WT-MRP4, G187W, and Y556C, respectively. All the afore mentioned figures contain the full MRP4 structure and a close-up of the ligand binding site. As observed in Figures 13 and 14, the FA binding site is almost the same and it is surrounded by the TMH1-TMH3; meanwhile, Figure 15 shows that the FA binding site in Y556C is surrounded by TMH2-TMH7 deep in the inner cavity. This finding is interesting, due to the fact that the Y556C mutation could promote the greater affinity of FA for the Y556C variant by switching into a "high affinity inward-facing orientation", making the binding of FA easier [27]. Nevertheless, this does not mean that the FA can be transported properly, since MRP4 has an altered conformation within the NBD2, which blocks the ATP hydrolysis but not ATP binding. However, MRP4 can function with the energy provided by one ATP hydrolysis, but the Y556C mutation could interfere with such ATP hydrolysis [28].

Figure 16 shows that the cAMP binding site in Y556C is quite different compared to WT-MRP4 and G187W. The cAMP is located out of the normal binding site out of the TMDs and it may not be transported at a proper rate or efficacy. It seems that Y556C is in the "low-affinity outward-facing orientation", where the main characteristic in this stage is that the affinity of the transported entity switches from high affinity (low chemical potential of the substrate) to low affinity (high chemical potential). Such change in affinity is due to the gate to the inside is closed, and the gate to the outside of the membrane is opened [23]. The global effect will depend on the cell type. Focusing on leukemia cells, could be harmful to them, because the intracellular cAMP levels would be increased and thus, leading to apoptosis [25]. On the other hand, the cAMP binding site in Y556C could represent the initial interaction only and further re-location such as the WT-MTP4. Longer AA-MDS are required to test this hypothesis.

To contrast the effect of nsSNPs in MRP4, AA-MDS and umbrella sampling studies were performed in ceefourin-1, which is possesses a high selectivity over MRP4 inhibition. It has been suggested that ceefourin-1 may act as a competitive inhibitor in MRP4 [7], which is consistent to the AA-MDS where the binding site for cAMP and ceefourin-1 is similar

in WT-MRP4, G187W and Y556C with ATP in the complex (Figures 17–19). Interestingly, ceefourin-1 binds at the same abnormal binding site as cAMP in Y556C-ATP. Moreover, Y556C seems to be in the same "low-affinity outward-facing orientation" where the low affinity (high chemical potential) of the ligand leads to switch in the binding site [23]. As mentioned before, such abnormal binding site could be the initial binding site but longer AA-MDS is required to demonstrate it.

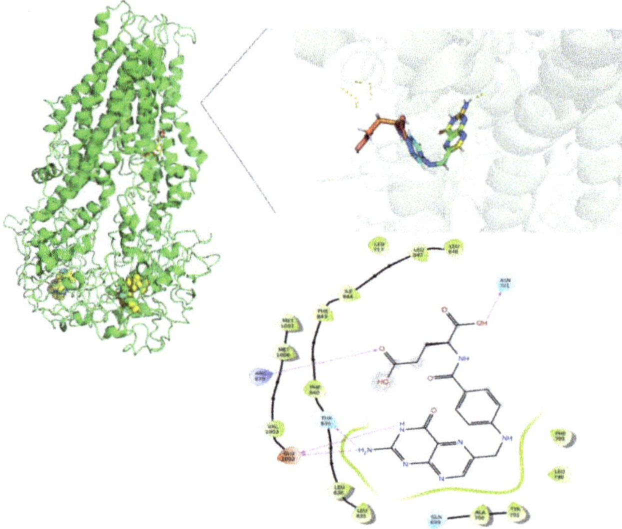

Figure 13. FA binding site in WT-MRP4 and amplification of such site. Rainbow colors in both the complete MRP4 structure and amplification of the binding site represent the FA structure while the yellow spheres and stick represent the ATP. The yellow dotted lines indicate polar interactions. The LID of WT-MRP4-FA is presented at the bottom.

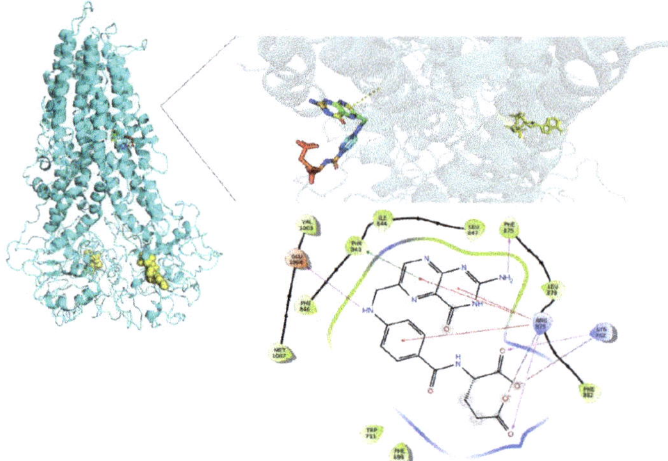

Figure 14. FA binding site in G187W and amplification of such site. Rainbow colors in both the complete WT-MRP4 and amplification of the binding site represent the FA structure while the yellow spheres and stick represent the ATP. The yellow dotted lines indicate polar interactions. The LID of G187W-FA is presented at the bottom.

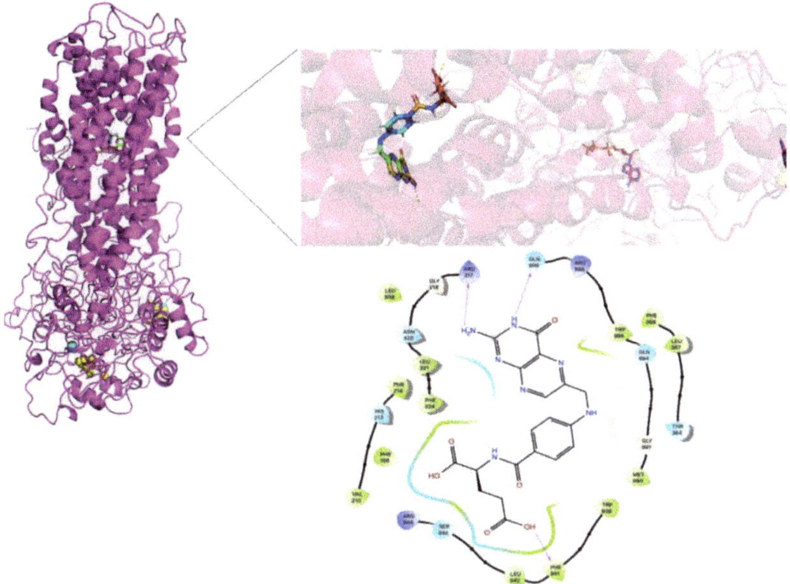

Figure 15. FA binding site in Y556C and amplification of such a site. Rainbow colors in both the complete G187W and amplification of the binding site represent the FA structure while yellow spheres represent the ATP. The yellow dotted lines indicate polar interactions. The LID of Y556C-FA is presented at the bottom.

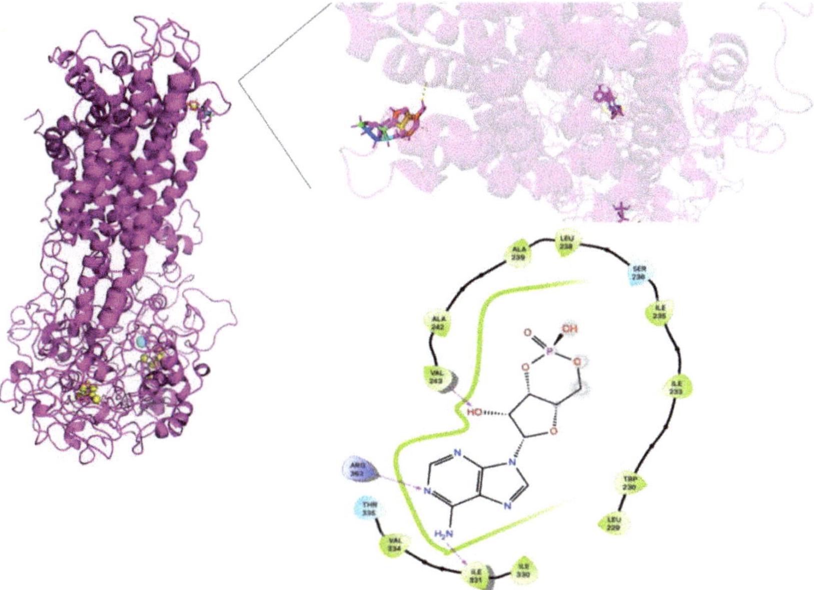

Figure 16. cAMP binding site in Y556C and amplification of such site. Rainbow colors in both the complete Y556C and amplification of the binding site represent the cAMP structure while yellow spheres represent the ATP. The yellow dotted lines indicate polar interactions. The LID of Y556C-cAMP is presented at the bottom.

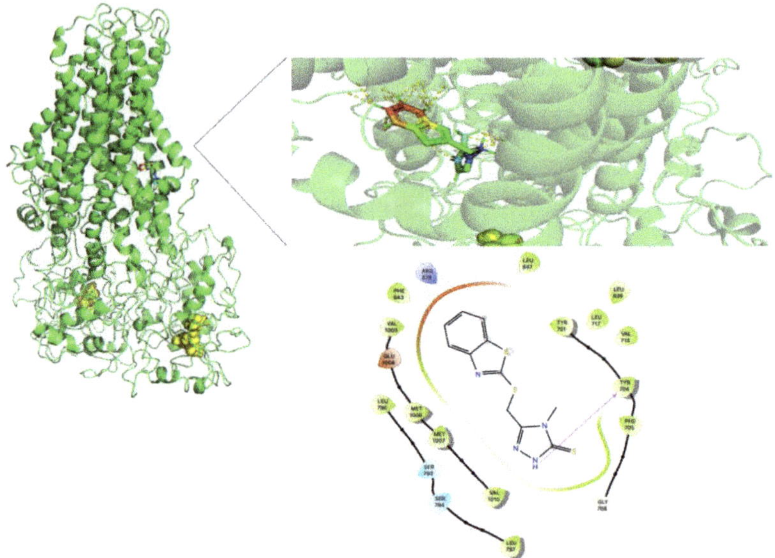

Figure 17. Ceefourin-1 binding site in WT-MRP4 and amplification of such site. Rainbow colors in both the complete WT-MRP4 and amplification of the binding site represent the ceefourin-1 structure while yellow spheres represent the ATP. The yellow dotted lines indicate polar and non-polar interactions. The LID of WT-MRP4-ceefourin-1 is presented at the bottom.

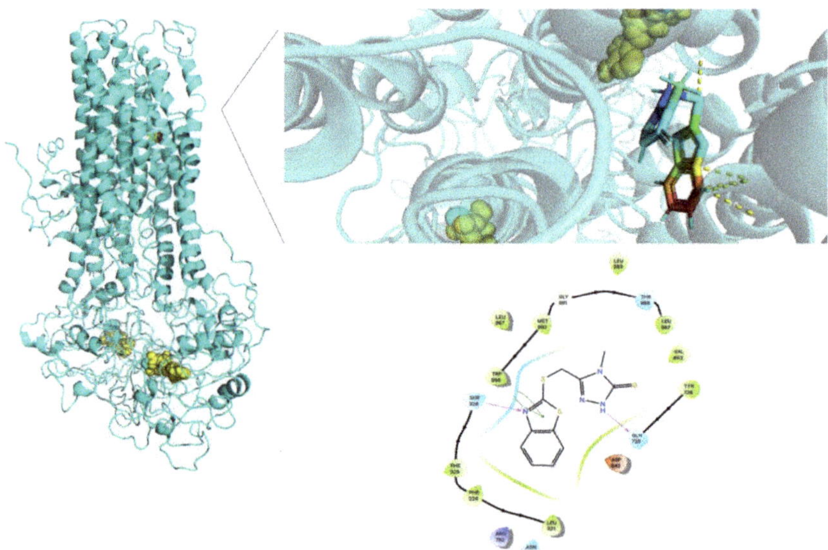

Figure 18. Ceefourin-1 binding site in G187W and the amplification of such a site. Rainbow colors in both the complete G187W and amplification of the binding site represent the ceefourin-1 structure, while yellow spheres represent the ATP. The yellow dotted lines indicate polar and non-polar interactions. The LID of G187W-ceefourin-1 is presented at the bottom.

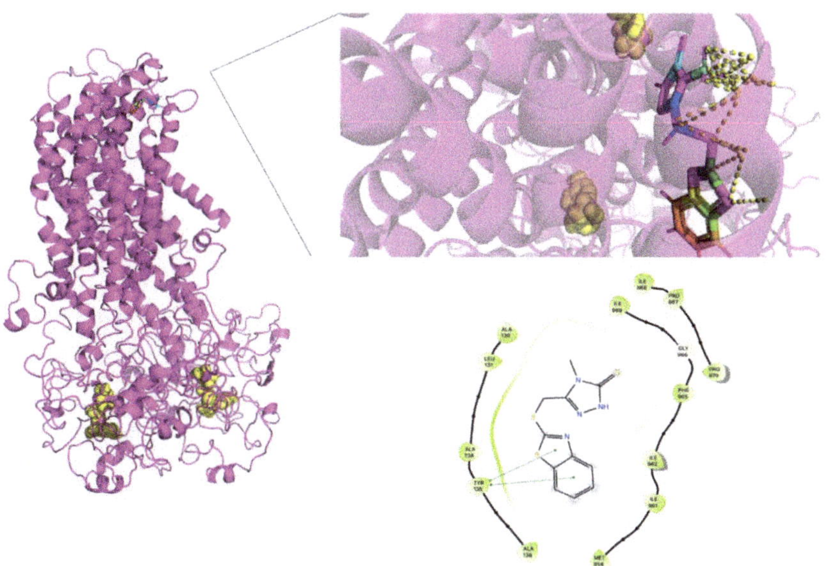

Figure 19. Ceefourin-1 binding site in Y556C and the amplification of such a site. Rainbow colors in both the complete Y556C and amplification of the binding site represent the ceefourin-1 structure, while yellow spheres represent the ATP. The yellow dotted lines indicate polar and non-polar interactions. The LID of Y556C-ceefourin-1 is presented at the bottom.

The cAMP and ceefourin-1 abnormal binding site in Y556C was one of the most remarkable findings in this work and showed the importance of studying the global effect of Y556C in different cell lines.

Umbrella sampling simulations were performed to determine the theoretical affinity, expressed as the free energy of binding (ΔG), of an event related to protein–ligand interactions along a reaction coordinate. Table 5 shows the ΔG obtained by Umbrella sampling through 10 ns of simulation on each complex. According to Umbrella sampling results, the Y556C-FA complex is the most stable with or without ATP bound to MRP4. Interestingly, ΔG for WT-FA and Y556C-FA complexes were lower in those MRP4 without ATP. In addition, ΔGs for G187W-cAMP and Y556C-cAMP complexes were higher without ATP, suggesting that the ATP bound is required for both FA and cAMP binding stabilization. ΔG for WT-MRP4-cAMP with or without ATP was the same. This represents an interesting finding because in WT-MRP4, the binding of ATP may promote the stabilization and conformational changes on MRP4 protein to promote an adequate interaction with FA to be properly transported out of the cell. Regarding the information provided by the LIDs, it is not possible to link the intermolecular interactions to ΔG because of the lack of a pattern in all the complexes. The intermolecular interactions do not define the increase or decrease of ΔG in this case. To confirm these observations, experimentally testing the FA efflux on cells expressing equal levels of WT-MRP4, G187W, and Y556C, as well as MRP4 inhibition by ceefourin-1 to measure the transport rate would afford further information of the differences among mutants and wild-type MRP4, experiments that are currently being carried out and will be reported in future articles.

The ΔG for the WT-MRP4-ATP1 complex was lower than that for FA and cAMP, which is reasonable considering its role in MRP4 functioning and needs to be bound to NBDs for longer time than ligands. In the case of ATP1/2 complexes with MRP4 variants, the most remarkable finding was that the Y556C-ATP1 complex was the least stable, theoretically demonstrated through the Umbrella sampling results, and it may be related to the mutation in the NBD1. The mechanism for the MRP4 function requires the

transmission of the molecular motion from the NBDs to the TMDs. At this point, the ATP-binding can be considered as the power stroke in which the chemical potential of the transported entity changed, and ATP hydrolysis leads to the formation of an extra negative charge, thus opening the closed nucleotide sandwich structure and the opening of the nucleotide sandwich structure facilitates Pi release and ADP dissociation, which in turn allows the TMDs and access gates to reset to the high-affinity orientation on the original side of the membrane to continue the transport cycle [23,29]. If the Y556C mutation promotes a decreased affinity of ATP1 for its binding site and, in turn, a blockade of ATP1 hydrolysis, the Y556C variant activity could be diminished or truncated, also considering that WT-MTP4 and its variants lack the ability to hydrolyze ATP2 in NBD2; thus, the lack of Y556C efficacy in the transport substrates. Additionally, ATP2 had the highest affinity for Y556C, and it would be interesting to test if high affinity could reestablish the ATP2 hydrolysis in NBD2.

Table 5. Binding free energies (ΔGs) obtained by umbrella sampling.

	ΔG (kcal/mol)		
Ligand	WT-MRP4	G187W	Y556C
FA	−25.340	−23.288	−46.4898
FA-ATP	−16.490	−23.429	−38.9452
FA-ATP1	−46.010	−29.343	−18.7461
FA-ATP2	−37.226	−18.690	−61.100
cAMP	−17.700	−31.014	−12.5676
cAMP-ATP	−17.721	−39.300	−31.560
cAMP-ATP1	−43.966	−33.690	−49.7395
cAMP-ATP2	−23.860	−47.230	−38.5368
Ceefourin-1	−25.9938	−33.8589	−18.8326
Ceefourin-1-ATP	−22.9107	−19.7664	−22.5503
Ceefourin-1-ATP1	−40.9182	−43.1199	−43.1179
Ceefourin-1-ATP2	−37.2706	−52.9313	−53.6592

FA-ATP, cAMP-ATP, and ceefourin-1-ATP: the ΔG is referred to FA, cAMP or ceefourin-1 in complex with ATP. FA-ATP1, FA-ATP2, cAMP-ATP1, cAMP-ATP2, ceefourin-1-ATP1, and ceefourin-1-ATP2: ΔG calculation is focused on ATP1 or ATP2

The abnormal cAMP and ceefourin-1 binding in Y556C (Figures 16 and 19) is consistent with the ΔG for Y556C-cAMP and Y556-ceefourin-1, which had the lowest affinity with respect to the complexes with WT-MRP4 and G187W, and it indicates a low affinity at that binding site, suggesting two possibilities: (a) it is an initial cAMP or ceefourin-1 binding site to further binding at the inner cavity, or (b) possible deficiency in cAMP and ceefourin-1 transport. The latter is the most feasible due to cAMP interacts to G187W and WT-MRP4 in its normal binding site, but Y556C seems to be in the "low affinity outward-facing orientation" as mentioned before. To predict the effect of nsSNPs on the efficacy of chemotherapeutics, it is important to determine in a further study, the relation between ΔG in silico and the transport rate in vitro of several substrates. This suggests that ΔGs higher than that for the WT-cAMP-ATP complex may be related to a low rate of cAMP transport. The ΔGs for ceefourin-1 in all the complexes with MRP4 and its variants had the best affinity compared to those complexes of cAMP, supporting the idea that ceefourin-1 may act as a competitive inhibitor at least with cAMP. However, the presence of ATP in WT-MRP4 and its variants promotes a better affinity to cAMP than ceefourin-1 in the complexes with G187W and Y556C. A competitive assay in vitro is required to determine if ceefourin-1 had the best affinity to WT-MRP4 and its variants. The Figures 20 and 21 represent the LIDs for FA and cAMP, respectively. In both figures, the intermolecular interactions of ligands with the absence of ATP are presented in the upper panel while the intermolecular interactions of ligands with the presence of ATP are shown in the bottom panel. The FA binding site is almost the same with or without ATP, but the number and the type of intermolecular interactions is different. It seems that ATP promotes a greater number of H-bonds between FA and all WT-MRP4 and its variants. Regarding the cAMP

binding site, the ATP binding did not modify it; thus, the effect of ATP seems to be related with changes in the intermolecular interactions and affinity that, according to Umbrella sampling results, the ATP decreases the ΔG in the G187W-cAMP-ATP and Y556C-cAMP-ATP complexes. Figure 22 presents the LIDs for ceefourin-1. The binding site is different in all the complexes; interestingly, such a binding site does not change in the presence of ATP. The main difference induced for ATP is the intermolecular interaction pattern, which led to great amount of π-π stacking and H-bonds.

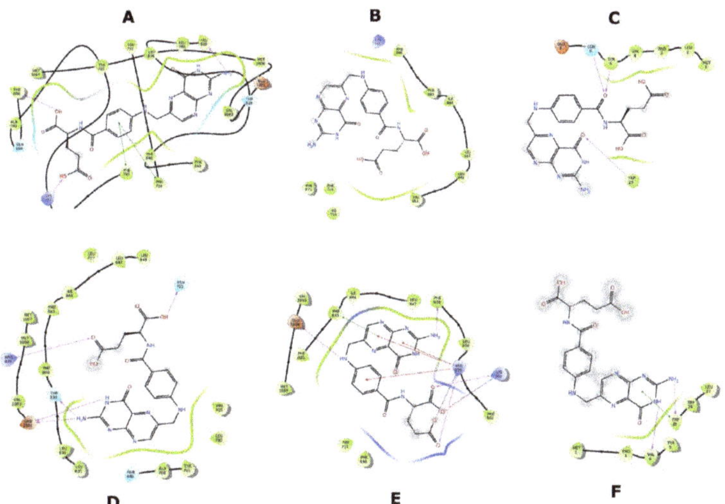

Figure 20. LIDs for FA with the different MRP4 structures with or without ATP. (**A**) WT-MRP4-FA complex. (**B**) G187W-FA complex. (**C**) Y556C-FA complex. (**D**) WT-MRP4-FA-ATP complex. (**E**) G187W-FA-ATP complex. (**F**) Y556C-FA-ATP complex.

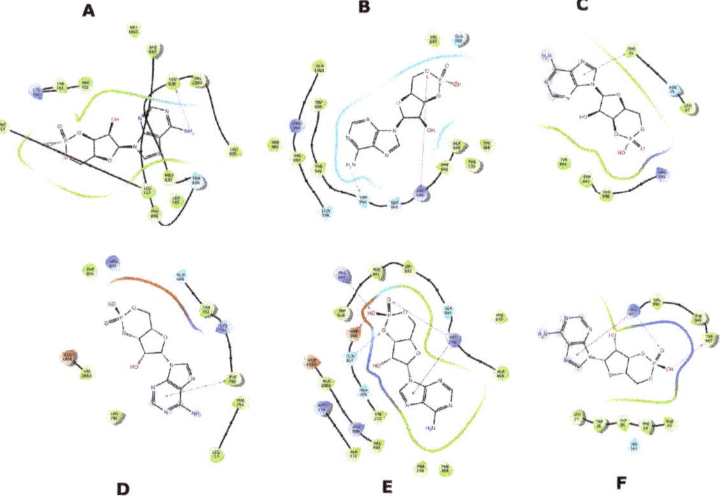

Figure 21. LIDs for cAMP with the different MRP4 structures with or without ATP. (**A**) WT-MRP4-cAMP complex. (**B**) G187W-cAMP complex. (**C**) Y556C-cAMP complex. (**D**) WT-MRP4-cAMP-ATP complex. (**E**) G187W-cAMP-ATP complex. (**F**) Y556C-cAMP-ATP complex.

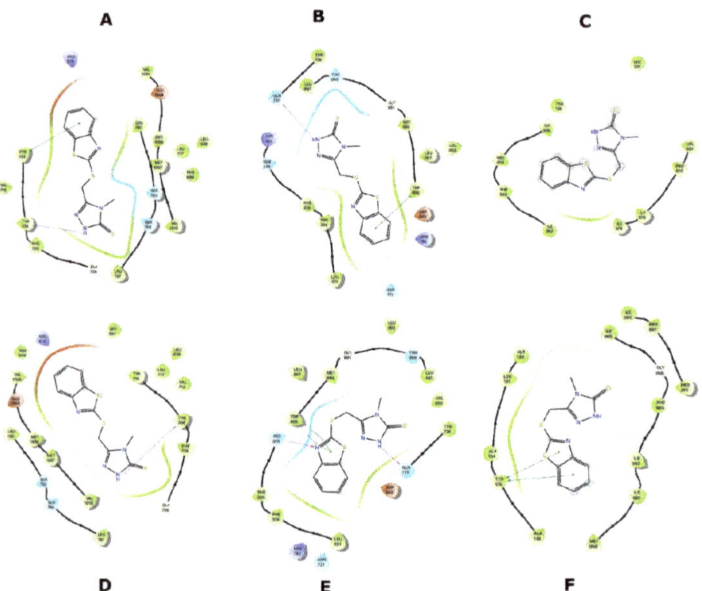

Figure 22. LIDs for ceefourin-1 with the different MRP4 structures with or without ATP. (**A**) WT-MRP4-ceefourin-1 complex. (**B**) G187W-ceefourin-1 complex. (**C**) Y556C-ceefourin-1 complex. (**D**) WT-MRP4-ceefourin-1-ATP complex. (**E**) G187W-ceefourin-1-ATP complex. (**F**) Y556C-ceefourin-1-ATP complex.

The MRP4 variants may predispose the population to a given disease regarding the site of the MRP4 that was affected by the mutation and the change in the affinity of a given substrate. The clinical implications of MRP4 variants have been studied over the past years and it is crucial to describe the relation of the MRP4 variants with diseases [30].

3. Materials and Methods

3.1. Protein Threading for WT-MRP4 and Its Variants G187W and Y556C

Structure prediction by protein threading for MRP4 was performed using its primary sequence (code: O15439) from the UniProt database [31,32]. MRP4 mutant models were made by the substitution of amino acids, G187W, or Y556C into the WT-MRP4 primary sequence. Each primary sequence was uploaded into the I-TASSER [17,18] server for the calculations of the models. The crystallography of MRP1 (PDBD code: 5UJ9) from *Bos taurus* was used as a template in all cases [33,34].

3.2. Coarse-Grained Molecular Dynamics (CG-MD) Simulations

All the systems (WT, G187W, and Y556C) for the simulations were built in the Martini Maker module [35] from CHARMM-GUI [36,37] using the Martini2.2p force field and adding a phosphatidylcholine (POPC) lipid bilayer membrane in an isothermal-isobaric ensemble (NPT) at 310.15 K. The simulations of 1 μs CG-MD were carried out in the Gromacs 2018.7 program [38,39], after of the minimizations and equilibrium protocols suggested by CHARM-GUI server. The module "cluster" of Gromacs 2018.7 was used to find the relevant conformations of the simulation using the "gromos" algorithm and the backbone for alignment. The most representative structure of the largest conformation cluster of the three simulations was converted into all atoms in the Martini Maker/All-Atom converter from CHARMM-GUI for the following calculations.

All the CG-MD simulations were performed in the ADA cluster of the National Laboratory of Advanced Scientific Visualization at Campus Juriquilla of the National Autonomous University of Mexico (LAVIS-UNAM).

3.3. Molecular Docking on WT-MRP4, G187W and Y556C

The 3D structures of the selected ligand groups, substrate drugs, endogenous substrates, and inhibitors were obtained from the PubChem public database [40]. Molecular docking was performed with AutoDock 4.2.6 optimized for graphical-processing units using a total of 50 runs and 25,000,000 evaluations; a grid box of 22.5 × 22.5 × 22.5 Å^3 centering on the relevant amino acids reported by Ravna 2008 and 2009, El-Sheink 2008, and Chen 2018 [9,41–43]; in a Lamarckian genetic algorithm and Solis-Wets local search [44].

3.4. All-Atom Molecular Dynamics (AA-MD) Simulations

For molecular dynamic studies with FA as a ligand, the protein–ligand complexes WT-MRP4-FA, G187W-FA, or Y556C-FA, obtained from the molecular docking, were used, and AA-MD simulations were performed in Desmond 3.6 as an application of the Maestro software [45,46] as graphical interface. The AA-MDS systems were built using the System Setup module with an OPLS force-field (Optimized Potentials for Liquid Simulations), adding a POPC lipid membrane and simple point charge (SPC) water model in an NPT assembly at 310.15 K. Once the system was built, the standard relaxation protocol for system relaxation with increasing temperatures and decreasing restraints was used and an MD production simulation of 25 ns (100 frames) was performed in the Molecular Dynamics module. Clustering was performed in the Desmond Trajectory Clustering module to obtain the most representative conformation of the largest cluster (cluster1, C1). Trajectory analyses were performed in the Simulation Event Analysis module in Maestro.

3.5. Umbrella Sampling

Using the C1 of the 25 ns AA-MD trajectory, the system was built under the previously mentioned conditions. Once the system was obtained and relaxed using the AA-MD relaxation protocol, a 10 ns (100 frames) MD simulation was performed in the Metadynamics module of Desmond using the protein and ligand center-of-mass distance as the collective variable, with 0.3 kcal/mol height and 0.1 kcal/mol width as the Gaussian parameters for the umbrella protocol, on an NPT ensemble at 310.15 K and 1.01325 bar. Finally, the analysis was performed in the Metadynamics Analysis module of Desmond.

3.6. Manipulation of the Complexes and Figures

All the protein figures and alignments presented in this work were made in PyMOL software (The PyMOL Molecular Graphics System, Version 1.2r3pre, Schrödinger) [47]. The manipulation of the WT-MRP4, G187W, and Y556C structures was performed in Maestro (Schrödinger Release 2019-3: Maestro, Schrödinger, LLC, New York, NY, USA) [48].

4. Conclusions

To obtain the 3D structure of MRP4 and its variants, which are not resolved by NMR or X-ray, protein threading was performed in this study and relaxing the structures was performed by CG-MDS to carry out the molecular docking, while MDS and umbrella sampling studies were performed to yield relevant information regarding the residues involved in the binding of the studied molecules groups and changes in the ΔG of FA and cAMP in the presence or the absence of ATP, which also allowed us to observe the relevance of the mutations in the binding and movement of MRP4 and its variants. According to our results, the nsSNPs G187W and Y556C led to changes in the ligand binding site, DS, and binding energy (ΔG). In addition, the ATP binding to MRP4 significantly modifies the intermolecular interactions (at least, for FA and cAMP) and the binding energy compared to the complexes where ATP was not bound to MRP4. The effect of the abnormal binding site of cAMP in Y556C is consistent with the highly selective MRP4 inhibitor ceefourin-1,

which makes it interesting to study such mutations in vitro. The affinity of ceefourin-1 for WT-MRP4 and its variants is higher than the affinity of cAMP. Cofactors such as ATP and Mg^{2+} should be included in the in silico analyses related to MRP4. It is well known that non-synonymous mutations usually affect the protein function or activity and its conformation, but this is the first report that suggests that most endogen substrates change their affinity and binding site in G187W and Y556C, which could modify the cell metabolism. We will report in further works the measure of substrate efflux and the relation between G187W or Y556C expression, location, and cell viability to determine the overall effect of the nsSNPs G187W and Y556C in vitro.

Supplementary Materials: The following are available online: Table S1: RMSD values obtained for the alignment of C1 cluster from AA-MDS for different sites of WT-MRP4, G187W and Y556C, Figure S2: Binding site and intermolecular interaction of cholic acid in WT-MRP4 (A), G187W (B) and Y556C (C). D Nomenclature of LIDs, Figure S3: Binding site and intermolecular interaction of taurocholic acid in WT-MRP4 (A), G187W (B) and Y556C (C), Figure S4: Binding site and intermolecular interactions of cefazoline in WT-MRP4 (A), G187W (B) and Y556C (C), Figure S5: Binding site and intermolecular interaction of ceefourin-1 in WT-MRP4 (A), G187W (B) and Y556C (C), Figure S6: WT-MRP4-FA complex at T0 in AA-MDS, Figure S7: WT-MRP4-FA complex at 5 ns in AA-MDS, Figure S8: WT-MRP4-FA complex at 10 ns in AA-MDS, Figure S9: WT-MRP4-FA complex at 15 ns in AA-MDS, Figure S10: WT-MRP4-FA complex at 20 ns in AA-MDS, Figure S11: WT-MRP4-FA complex at 25 ns in AA-MDS, Figure S12: G187W-FA complex at T0 in AA-MDS, Figure S13: G187W-FA complex at 5 ns in AA-MDS, Figure S14: G187W-FA complex at 10 ns in AA-MDS, Figure S15: G187W-FA complex at 15 ns in AA-MDS, Figure S16: G187W-FA complex at 20 ns in AA-MDS, Figure S17: G187W-FA complex at 25 ns in AA-MDS, Figure S18: Y556C-FA complex at T0 in AA-MDS, Figure S19: Y556C-FA complex at 5 ns in AA-MDS, Figure S20: Y556C-FA complex at 10 ns in AA-MDS, Figure S21: Y556C-FA complex at 15 ns in AA-MDS, Figure S22: Y556C-FA complex at 20 ns in AA-MDS, Figure S23: Y556C-FA complex at 25 ns in AA-MDS.

Author Contributions: Conceptualization: E.B., L.B., and G.G.-A.; calculations: E.B., G.A.-D., and A.R.-M.; data analysis: E.B., G.A.-D., and A.R.-M.; writing—original draft preparation: E.B., G.A.-D.; writing—review and editing: E.B., L.B., G.G.-A., G.A.-D., L.B., and A.R.-M.; project administration and funding acquisition: L.B. and G.G.-A. All authors have read and agreed to the published version of the manuscript.

Funding: This research was funded by the University of Queretaro (UAQ) through the FOFI-UAQ research grant FCQ2018-39 and the National Council of Research and Technology (CONACYT) doctoral scholarship no. 335389 for Edgardo Becerra and no. 335496 for Giovanny Aguilera-Durán.

Institutional Review Board Statement: Not Applicable.

Informed Consent Statement: Not Applicable.

Acknowledgments: The authors gratefully acknowledge the technical support from Luis Aguilar, Alejandro de León, Carlos Flores, and Jair García of the National Laboratory of Advanced Scientific Visualization (LAVIS-UNAM).

Conflicts of Interest: The authors declare no conflict of interest.

References

1. Berthier, J.; Arnion, H.; Saint-Marcoux, F.; Picard, N. Multidrug resistance-associated protein 4 in pharmacology: Overview of its contribution to pharmacokinetics, pharmacodynamics and pharmacogenetics. *Life Sci.* **2019**, *231*, 116540. [CrossRef]
2. Lee, K.; Belinsky, M.G.; Bell, D.W.; Testa, J.R.; Kruh, G.D. Isolation of MOAT-B, a widely expressed multidrug resistance-associated protein/canalicular multispecific organic anion transporter-related transporter. *Cancer Res.* **1998**, *58*, 2741–2747. [PubMed]
3. Fan, T.; Sun, G.; Sun, X.; Zhao, L.; Zhong, R.; Peng, Y. Tumor energy metabolism and potential of 3-bromopyruvate as an inhibitor of aerobic glycolysis: Implications in tumor treatment. *Cancers* **2019**, *11*, 317. [CrossRef]
4. Borst, P.; De Wolf, C.; Van De Wetering, K. Multidrug resistance-associated proteins 3, 4, and 5. *Pflügers Arch. Eur. J. Physiol.* **2006**, *453*, 661–673. [CrossRef]
5. Copsel, S.; Garcia, C.; Diez, F.; Vermeulem, M.; Baldi, A.; Bianciotti, L.G.; Russel, F.G.M.; Shayo, C.; Davio, C. Multidrug resistance protein 4 (MRP4/ABCC4) regulates cAMP cellular levels and controls human leukemia cell proliferation and differentiation. *J. Biol. Chem.* **2011**, *286*, 6979–6988. [CrossRef] [PubMed]

6. Belleville-Rolland, T.; Sassi, Y.; Decouture, B.; Dreano, E.; Hulot, J.-S.; Gaussem, P.; Bachelot-Loza, C. MRP4 (ABCC4) as a potential pharmacologic target for cardiovascular disease. *Pharmacol. Res.* **2016**, *107*, 381–389. [CrossRef] [PubMed]

7. Yaneff, A.; Sahores, A.; Gómez, N.; Carozzo, A.; Shayo, C.; Davio, C. MRP4/ABCC4 as a new therapeutic target: Meta-analysis to determine cAMP binding sites as a tool for drug design. *Curr. Med. Chem.* **2019**, *26*, 1270–1307. [CrossRef]

8. Russel, F.G.M.; Koenderink, J.B.; Masereeuw, R. Multidrug resistance protein 4 (MRP4/ABCC4): A versatile efflux transporter for drugs and signalling molecules. *Trends Pharmacol. Sci.* **2008**, *29*, 200–207. [CrossRef]

9. Ravna, A.W.; Sager, G. Molecular modeling studies of ABC transporters involved in multidrug resistance. *Mini-Rev. Med. Chem.* **2009**, *9*, 186–193. [CrossRef]

10. Wittgen, H.G.; van den Heuvel, J.J.M.W.; Krieger, E.; Schaftenaar, G.; Russel, F.G.; Koenderink, J.B. Phenylalanine 368 of multidrug resistance-associated protein 4 (MRP4/ABCC4) plays a crucial role in substrate-specific transport activity. *Biochem. Pharmacol.* **2012**, *84*, 366–373. [CrossRef]

11. Deeley, R.G.; Westlake, C.; Cole, S.P.C. Transmembrane transport of endo- and xenobiotics by mammalian ATP-binding cassette multidrug resistance proteins. *Physiol. Rev.* **2006**, *86*, 849–899. [CrossRef] [PubMed]

12. Saito, S.; Iida, A.; Sekine, A.; Miura, Y.; Ogawa, C.; Kawauchi, S.; Higuchi, S.; Nakamura, Y. Identification of 779 genetic variations in eight genes encoding members of the ATP-binding cassette, subfamily C (ABCC/MRP/CFTR). *J. Hum. Genet.* **2002**, *47*, 147–171. [CrossRef]

13. Yates, C.M.; Sternberg, M.J. The effects of non-synonymous single nucleotide polymorphisms (nsSNPs) on protein–protein interactions. *J. Mol. Biol.* **2013**, *425*, 3949–3963. [CrossRef] [PubMed]

14. Singh, A.; Thakur, M.; Singh, S.K.; Sharma, L.K.; Chandra, K. Exploring the effect of nsSNPs in human YPEL3 gene in cellular senescence. *Sci. Rep.* **2020**, *10*, 15301. [CrossRef]

15. Abla, N.; Chinn, L.W.; Nakamura, T.; Liu, L.; Huang, C.C.; Johns, S.J.; Kawamoto, M.; Stryke, D.; Taylor, T.R.; Ferrin, T.E.; et al. The human multidrug resistance protein 4 (MRP4, ABCC4): Functional analysis of a highly polymorphic gene. *J. Pharmacol. Exp. Ther.* **2008**, *325*, 859–868. [CrossRef] [PubMed]

16. Banerjee, M.; Marensi, V.; Conseil, G.; Le, X.C.; Cole, S.P.; Leslie, E.M. Polymorphic variants of MRP4/ABCC4 differentially modulate the transport of methylated arsenic metabolites and physiological organic anions. *Biochem. Pharmacol.* **2016**, *120*, 72–82. [CrossRef]

17. Zhang, Y. I-TASSER server for protein 3D structure prediction. *BMC Bioinform.* **2008**, *9*, 40. [CrossRef]

18. Roy, A.; Kucukural, A.; Zhang, Y. I-TASSER: A unified platform for automated protein structure and function prediction. *Nat. Protoc.* **2010**, *5*, 725–738. [CrossRef]

19. Yang, J.; Zhang, Y. Protein structure and function prediction using I-TASSER. *Curr. Protoc. Bioinform.* **2015**, *52*, 5.8.1–5.8.15. [CrossRef]

20. Chothia, C.; Lesk, A.M. The relation between the divergence of sequence and structure in proteins. *EMBO J.* **1986**, *5*, 823–826. [CrossRef]

21. Nikolaev, D.M.; Shtyrov, A.A.; Panov, M.S.; Jamal, A.; Chakchir, O.B.; Kochemirovsky, V.A.; Olivucci, M.; Ryazantsev, M.N. A comparative study of modern homology modeling algorithms for rhodopsin structure prediction. *ACS Omega* **2018**, *3*, 7555–7566. [CrossRef]

22. Xiang, Z. Advances in homology protein structure modeling. *Curr. Protein Pept. Sci.* **2006**, *7*, 217–227. [CrossRef]

23. Liu, X.; Pan, G. (Eds.) Drug Transporters in Drug Disposition, Effects and Toxicity. In *Advances in Experimental Medicine and Biology*; Springer Singapore: Singapore, 2019; Volume 1141, ISBN 978-981-13-7646-7.

24. Beis, K. Structural basis for the mechanism of ABC transporters. *Biochem. Soc. Trans.* **2015**, *43*, 889–893. [CrossRef] [PubMed]

25. Perez, D.R.; Smagley, Y.; Garcia, M.; Carter, M.B.; Evangelisti, A.; Matlawska-wasowska, K.; Winter, S.S.; Sklar, A.; Chigaev, A. Cyclic AMP efflux inhibitors as potential therapeutic agents for leukemia. *Oncotarget* **2016**, *7*, 33960–33982. [CrossRef]

26. Dai, F.; Yoo, W.G.; Lee, J.-Y.; Lu, Y.; Pak, J.H.; Sohn, W.-M.; Hong, S.-J. Multidrug resistance-associated protein 4 is a bile transporter of Clonorchis sinensis simulated by in silico docking. *Parasites Vectors* **2017**, *10*, 578. [CrossRef]

27. Martinez, L.; Falson, P. Multidrug resistance ATP-binding cassette membrane transporters as targets for improving oropharyngeal candidiasis treatment. *Adv. Cell. Mol. Otolaryngol.* **2014**, *2*, 23955. [CrossRef]

28. Procko, E.; O'Mara, M.L.; Bennett, W.F.D.; Tieleman, D.P.; Gaudet, R. The mechanism of ABC transporters: General lessons from structural and functional studies of an antigenic peptide transporter. *FASEB J.* **2009**, *23*, 1287–1302. [CrossRef]

29. Sauna, Z.E.; Nandigama, K.; Ambudkar, S.V. Multidrug resistance protein 4 (ABCC4)-mediated ATP hydrolysis. *J. Biol. Chem.* **2004**, *279*, 48855–48864. [CrossRef] [PubMed]

30. Tanner, C.; Boocock, J.; Stahl, E.A.; Dobbyn, A.; Mandal, A.K.; Cadzow, M.; Phipps-Green, A.J.; Topless, R.K.; Hindmarsh, J.H.; Stamp, L.K.; et al. Population-specific resequencing associates the ATP-binding cassette subfamily C member 4 gene with gout in New Zealand Māori and Pacific men. *Arthritis Rheumatol.* **2017**, *69*, 1461–1469. [CrossRef]

31. Null, N. Activities at the Universal Protein Resource (UniProt). *Nucleic Acids Res.* **2014**, *42*, D191–D198. [CrossRef]

32. The UniProt Consortium. UniProt: The universal protein knowledgebase. *Nucleic Acids Res.* **2017**, *45*, D158–D169. [CrossRef] [PubMed]

33. Altschul, S.F.; Gish, W.; Miller, W.; Myers, E.W.; Lipman, D.J. Basic local alignment search tool. *J. Mol. Biol.* **1990**, *215*, 403–410. [CrossRef]

34. States, D.J.; Gish, W. QGB: Combined use of sequence similarity and codon bias for coding region identification. *J. Comput. Biol.* **1994**, *1*, 39–50. [CrossRef] [PubMed]
35. Zhang, J.Z.H.; Ingólfsson, H.I.; Cheng, X.; Lee, J.; Marrink, S.J.; Im, W. CHARMM-GUI martini maker for coarse-grained simulations with the martini force field. *J. Chem. Theory Comput.* **2015**, *11*, 4486–4494. [CrossRef]
36. Jo, S.; Kim, T.; Iyer, V.G.; Im, W. CHARMM-GUI: A web-based graphical user interface for CHARMM. *J. Comput. Chem.* **2008**, *29*, 1859–1865. [CrossRef]
37. Lee, J.; Cheng, X.; Swails, J.M.; Yeom, M.S.; Eastman, P.K.; Lemkul, J.A.; Wei, S.; Buckner, J.; Jeong, J.C.; Qi, Y.; et al. CHARMM-GUI input generator for NAMD, GROMACS, AMBER, OpenMM, and CHARMM/OpenMM simulations using the CHARMM36 additive force field. *J. Chem. Theory Comput.* **2016**, *12*, 405–413. [CrossRef] [PubMed]
38. Berendsen, H.; Van Der Spoel, D.; Van Drunen, R. GROMACS: A message-passing parallel molecular dynamics implementation. *Comput. Phys. Commun.* **1995**, *91*, 43–56. [CrossRef]
39. Van Der Spoel, D.; Lindahl, E.; Hess, B.; Groenhof, G.; Mark, A.E.; Berendsen, H.J.C. GROMACS: Fast, flexible, and free. *J. Comput. Chem.* **2005**, *26*, 1701–1718. [CrossRef]
40. Kim, S.; Chen, J.; Cheng, T.; Gindulyte, A.; He, J.; He, S.; Li, Q.; Shoemaker, B.A.; Thiessen, P.A.; Yu, B.; et al. PubChem 2019 update: Improved access to chemical data. *Nucleic Acids Res.* **2019**, *47*, D1102–D1109. [CrossRef]
41. Ravna, A.W.; Sager, G. Molecular model of the outward facing state of the human multidrug resistance protein 4 (MRP4/ABCC4). *Bioorganic Med. Chem. Lett.* **2008**, *18*, 3481–3483. [CrossRef]
42. El-Sheikh, A.A.K.; van den Heuvel, J.J.M.W.; Krieger, E.; Russel, F.G.M.; Koenderink, J.B. Functional role of arginine 375 in transmembrane helix 6 of multidrug resistance protein 4 (MRP4/ABCC4). *Mol. Pharmacol.* **2008**, *74*, 964–971. [CrossRef] [PubMed]
43. Chen, Y.; Yuan, X.; Xiao, Z.; Jin, H.; Zhang, L.; Liu, Z. Discovery of novel multidrug resistance protein 4 (MRP4) inhibitors as active agents reducing resistance to anticancer drug 6-Mercaptopurine (6-MP) by structure and ligand-based virtual screening. *PLoS ONE* **2018**, *13*, e0205175. [CrossRef] [PubMed]
44. Santos-Martins, D.; Solis-Vasquez, L.; Koch, A.; Forli, S. Accelerating AutoDock4 with GPUs and gradient-based local search. *ChemRvix* **2019**, preprint. [CrossRef]
45. Bowers, K.J.; Chow, D.E.; Xu, H.; Dror, R.O.; Eastwood, M.P.; Gregersen, B.A.; Klepeis, J.L.; Kolossvary, I.; Moraes, M.A.; Sacerdoti, F.D.; et al. Scalable Algorithms for Molecular Dynamics Simulations on Commodity Clusters. In Proceedings of the ACM/IEEE SC 2006 Conference (SC '06), Tampa, FL, USA, 11–17 November 2006; IEEE: Piscataway, NJ, USA, 2006; p. 43.
46. D.E. Shaw Research. *Schrödinger Release 2020-1: Desmond Molecular Dynamics System*; Maestro-Desmond Interoperability Tools; Schrödinger LLC: New York, NY, USA, 2020.
47. Pymol, L. *The PyMOL Molecular Graphics System, Version 1.2r3pre*; Schrödinger LLC: New York, NY, USA, 2011.
48. *Schrödinger Release 2019-3 Maestro*; Schrödinger LLC: New York, NY, USA, 2019.

Article

Biocomputational Prediction Approach Targeting FimH by Natural SGLT2 Inhibitors: A Possible Way to Overcome the Uropathogenic Effect of SGLT2 Inhibitor Drugs

Mutaib M. Mashraqi [1,†], Navaneet Chaturvedi [2,3,†], Qamre Alam [4], Saleh Alshamrani [1], Mosa M. Bahnass [1,5], Khurshid Ahmad [6], Amany I. Alqosaibi [7], Mashael M. Alnamshan [7], Syed Sayeed Ahmad [6], Mirza Masroor Ali Beg [6], Abha Mishra [2], Sibhghatulla Shaikh [6,*] and Syed Mohd Danish Rizvi [8,*]

Citation: Mashraqi, M.M.; Chaturvedi, N.; Alam, Q.; Alshamrani, S.; Bahnass, M.M.; Ahmad, K.; Alqosaibi, A.I.; Alnamshan, M.M.; Ahmad, S.S.; Beg, M.M.A.; et al. Biocomputational Prediction Approach Targeting FimH by Natural SGLT2 Inhibitors: A Possible Way to Overcome the Uropathogenic Effect of SGLT2 Inhibitor Drugs. *Molecules* **2021**, *26*, 582. https://doi.org/10.3390/molecules26030582

Academic Editors: Marco Tutone and Anna Maria Almerico

Received: 18 December 2020

Accepted: 19 January 2021

Published: 22 January 2021

Publisher's Note: MDPI stays neutral with regard to jurisdictional claims in published maps and institutional affiliations.

[1] Department of Clinical Laboratory Sciences, College of Applied Medical Sciences, Najran University, Najran 61441, Saudi Arabia; mmmashraqi@nu.edu.sa (M.M.M.); saalshamrani@nu.edu.sa (S.A.); mmbohnass@nu.edu.sa (M.M.B.)

[2] Biomolecular Engineering Laboratory, School of Biochemical Engineering, Indian Institute of Technology, Banaras Hindu University, Varanasi 221005, India; 14.navneet@gmail.com (N.C.); abham.bce@itbhu.ac.in (A.M.)

[3] Department of Molecular and Cell Biology, Leicester Institute of Structural and Chemical Biology, University of Leicester Henry Wellcome Building, Lancaster Road Leicester, Leicester LE1 7HB, UK

[4] Medical Genomics Research Department, King Abdullah International Medical Research Center (KAIMRC), King Saud Bin Abdulaziz University for Health Sciences, King Abdulaziz Medical City, Ministry of National Guard Health Affairs, Riyadh 11426, Saudi Arabia; alamqa@ngha.med.sa

[5] Department of Animal Medicine (Infectious Diseases), Faculty of Veterinary Medicine, Zagazig University, Zagazig 44519, Egypt

[6] Department of Medical Biotechnology, Yeungnam University, Gyeongsan 38541, Korea; ahmadkhursheed2008@gmail.com (K.A.); sayeedahmad4@gmail.com (S.S.A.); mirzamasroor1986@gmail.com (M.M.A.B.)

[7] Department of Biology, College of Science, Imam Abdulrahman bin Faisal University, P.O. Box 1982, Dammam 31441, Saudi Arabia; amgosaibi@iau.edu.sa (A.I.A.); malnamshan@iau.edu.sa (M.M.A.)

[8] Department of Pharmaceutics, College of Pharmacy, University of Hail, P.O. Box 2440, Hail 81451, Saudi Arabia

[*] Correspondence: sibhghat.88@gmail.com (S.S.); syedrizvi10@yahoo.com (S.M.D.R.)

[†] These two authors contributed equally to this work.

Abstract: The Food and Drug Administration (FDA) approved a new class of anti-diabetic medication (a sodium–glucose co-transporter 2 (SGLT2) inhibitor) in 2013. However, SGLT2 inhibitor drugs are under evaluation due to their associative side effects, such as urinary tract and genital infection, urinary discomfort, diabetic ketosis, and kidney problems. Even clinicians have difficulty in recommending it to diabetic patients due to the increased probability of urinary tract infection. In our study, we selected natural SGLT2 inhibitors, namely acerogenin B, formononetin, (−)-kurarinone, (+)-pteryxin, and quinidine, to explore their potential against an emerging uropathogenic bacterial therapeutic target, i.e., FimH. FimH plays a critical role in the colonization of uropathogenic bacteria on the urinary tract surface. Thus, FimH antagonists show promising effects against uropathogenic bacterial strains via their targeting of FimH's adherence mechanism with less chance of resistance. The molecular docking results showed that, among natural SGLT2 inhibitors, formononetin, (+)-pteryxin, and quinidine have a strong interaction with FimH proteins, with binding energy (ΔG) and inhibition constant (ki) values of −5.65 kcal/mol and 71.95 μM, −5.50 kcal/mol and 92.97 μM, and −5.70 kcal/mol and 66.40 μM, respectively. These interactions were better than those of the positive control heptyl α-D-mannopyranoside and far better than those of the SGLT2 inhibitor drug canagliflozin. Furthermore, a 50 ns molecular dynamics simulation was conducted to optimize the interaction, and the resulting complexes were found to be stable. Physicochemical property assessments predicted little toxicity and good drug-likeness properties for these three compounds. Therefore, formononetin, (+)-pteryxin, and quinidine can be proposed as promising SGLT2 inhibitors drugs, with add-on FimH inhibition potential that might reduce the probability of uropathogenic side effects.

Keywords: sodium–glucose co-transporters 2; FimH; uropathogenic bacteria; urinary tract infections; diabetes

1. Introduction

Globally, diabetes mellitus is one of the most prevalent metabolic diseases and is estimated to increase to 552 million cases by 2030 [1]. Diabetes has been considered to enhance vulnerability to infectious diseases and has also been linked with an enhanced risk of death from infectious illness in some [2,3], but not all [4], investigations.

Sodium–glucose cotransporter-2 (SGLT2) inhibitors are a novel group of drugs used to treat patients with type 2 diabetes mellitus (T2DM). These drugs exert their effects by preventing glucose reabsorption at the proximal renal tubule and enhancing the excretion of urinary glucose [5]. Owing to the elevated urine glucose levels, SGLT2 inhibitors enhance the risk of urinary tract infections (UTIs) [6]. In addition, pharmacologically-induced urine glucose levels with SGLT2 inhibitors in diabetic patients might further sustain bacterial growth [7]. By themselves, SGLT2 inhibitors can possibly enhance the risk of UTIs and susceptibility to genital infections when used to manage patients [8]. SGLT2 inhibitors might cause serious UTIs, as the FDA warned in December 2015 [9]. Empagliflozin and canagliflozin are the preferred drugs suggested for T2DM patients with established cardiovascular disease [10,11]. The United States-based public safety advisory reported 19 cases of fatal renal or blood infection from March 2013 to October 2014. The origin of these cases was traced to a UTI induced by SGLT2 inhibitor intake [9].

Bacterial pili are proteinaceous projections extending from the bacterial cell surface and are used for attachment and cell motility [12]. Gram-negative bacteria use Type 1 pili to adhere to the host tissue and, therefore, Type 1 pili have been established as an important virulence factor in UTIs. Type 1 pili are made up of repeated subunits of FimA protein. These subunits are arranged to form a helical wound cylinder that composes a thick pilus rod. The distal flexible tip of the pilus rod is comprised of a single copy of proteins—FimF and FimG—while the tip adhesin bears FimH. The distinct binding of FimH (terminal adhesin) to mannosylated host glycoproteins mediates the adhesion of bacterial pathogens to the host tissue. UTI pathogenesis is caused by the FimH and, therefore, is a promising curative target [13].

Acerogenin B is a cyclic diarylheptanoid obtained from the bark of *Acer nikoense*, and it has been found to inhibit both SGLT1 and SGLT2 [14]. Kurarinone and formononetin are flavonoids isolated from the dried root of *Sophora flavescens*. Kurarinone has demonstrated inhibitory activity against both SGLT1 and SGLT2; however, formononetin was reported to inhibit only SGLT2 and not SGLT1 [15]. Quinidine is isolated from the bark of the cinchona tree and (+)-pteryxin is extracted from the plant *Peucedanum* spp. Recently, both of these natural compounds have been found to inhibit both SGLT1 and SGLT2 [16]. Heptyl α-D-mannopyranoside is a FimH antagonist [17] and, in this study, it was used as a reference compound.

In the present study, we explored natural SGLT2 inhibitors (acerogenin B, kurarinone, formononetin, quinidine, and (+)-pteryxin) that might have less severe uropathogenic side effects than does the approved SGLT2 inhibitor canagliflozin, also known as gliflozins. We speculated that formononetin, (+)-pteryxin, and quinidine would be promising SGLT2 inhibitors with less severe uropathogenic side effects.

2. Methodology

2.1. SGLT2 Inhibitors and Target Protein Structure Retrieval

The 3-dimensional structure of FimH was taken from the protein data bank (PDB ID: 4AV5), while the 3-dimensional structure of SGLT2 was made by employing the SWISS-MODEL Workspace after retrieving its amino acid sequence from Uniprot (P31639). The PDB structure 2XQ2 was used as a template and the generated model was validated

using various in silico tools, viz., SAVES v6.0 and VADAR (Volume, Area, Dihedral Angle Reporter). The PDB structure of the compounds canagliflozin (CID: 24812758), acerogenin B (CID: 10913542), formononetin (CID: 5280378), (−)-kurarinone (CID: 10812923), (+)-pteryxin (CID: 511787), quinidine (CID: 441074), and the FimH antagonist heptyl α-D-mannopyranoside (CID: 11300413) were retrieved from the PubChem database.

2.2. Physicochemical Properties and Toxicity Potential Prediction

The physicochemical properties and toxicity potential of SGLT2 inhibitors and FimH antagonists were calculated by applying the Orisis Datawarrior property explorer tool (http://www.openmolecules.org/datawarrior/download.html). At first, molecular weight, the number of rotatable bonds, the number of hydrogen bond acceptors and donors, cLogP value, topological polar surface area (TPSA), and the Lipinski's rule violation [18] were estimated to check physicochemical parameters. Later, the method outlined by Zhao et al. [19] was applied to calculate percentage of absorption; here, the following formula was used: absorption % = 109 − (0.345 × TPSA). Toxicity was also predicted for SGLT2 inhibitors and FimH antagonists by using toxicity features such as irritability, reproductive effects, tumorigenicity, and mutagenicity. Orisis Datawarrior property explorer tool toxicity predictions are based on comparative analysis of our tested compounds with the pre-estimated set of known structural molecules.

2.3. Molecular Docking

The ligands were docked to the SGLT2 and FimH proteins using "AutoDock 4.2" by following the protocol described by Rizvi et al. [20]. To minimize the energy usage of the ligand molecules, a Merck molecular force field (MMFF94) was employed. The ligand atoms were added with Gasteiger partial charges. Docking calculations were done on the target proteins. Essential hydrogen atoms, Kollman united atom type charges and solvation parameters were added by using AutoDock tools. Consequently, the binding pocket was added with conserved water molecules to mimic the *in vivo* environment. An auto grid program was used to generate the affinity (grid) maps sized at 60 × 60 × 60°, the aim of which was to target the grid coordinates in the catalytic site of the target protein (SGLT2 and FimH). The x, y, and z coordinate values for the FimH protein targeting the catalytic site were taken as 3.305, −16.68, and −13.57, respectively. Docking simulations were done using the Lamarckian genetic algorithm and the Solis and Wets local search method. The initial position, orientation, and torsions of the ligands were set randomly. Each docking experiment was derived from 100 different runs that were set to terminate after a maximum of 2,500,000 energy evaluations. The population size was set to 150. The Discovery Studio Visualizer 2.5 (Accelrys, San Diego, CA, USA) was used to generate the final figures.

2.4. LIGPLOT+ ANALYSIS

After the completion of all docking experiments, the best combination of ligand–FimH and ligand–SGLT2 was selected and analyzed using LIGPLOT+ Version v.2.1 (EMBL-EBI, Cambridgeshire, UK). The hydrogen and hydrophobic interactions between the important amino acid residues of FimH/SGLT2 with each ligand were analyzed by LIGPLOT. For each interaction, the 3-D structures generated were converted into 2-D figures by applying the LIGPLOT algorithm.

2.5. Molecular Dynamics Simulation

System building: GROMACS 4.6.7 packages [21,22] were used to prepare the system and perform molecular dynamics (MD) simulations using the gromos53a6 force field [23]. The protein solute was solvated by explicit SPC216 water [24] in a dodecahedral box with a margin of 10 Å between the solute and the box walls. Systems were brought to neutrality by the addition of sodium counter ions.

Simulation detail: A 10 Å cut-off distance was taken under the particle-mesh Ewald method [25] to calculate long-range electrostatic interactions, and a 10 Å cut-off distance was also considered to evaluate van der Waals interactions. The LINCS algorithm of fourth-

order expansion was used to constrain bond lengths [26]. The steepest descent algorithm was applied to optimize the removal of steric clashes between atoms for 10,000 steps. The system was equilibrated for 1 ns with position restraints on all heavy atoms. Berendsen weak coupling schemes were used to maintain the system at 300 K and 1 atom, using separate baths for the system. Initial velocities were generated randomly using a Maxwell–Boltzmann distribution corresponding to 300 K. Finally, the production run was performed for 20 ns. Furthermore, xmgrace (http://plasma-gate.weizmann.ac.il) was used for preparing graphs. Ligand topology preparation was implemented by using the PRODRG server, using the option specifying no chirality, full charge, and no energy minimization [27].

Trajectory analysis: The g_rms tool of the GROMACS package was used to calculate the root-mean-square deviation (RMSD) of each trajectory. The covariance matrix and eigenvectors of the trajectories were calculated through g_covar and g_anaeig programs.

3. Results and Discussion

3.1. Physicochemical Properties and Molecular Docking

Many varieties of plant-based natural compounds have been reported, which have significant anti-diabetic effects. In streptozotocin-stimulated diabetic mice, bakuchiol (a polyphenol compound) decreases glucose levels and enhances serum insulin levels [28]. Caffeine (an alkaloid) lowers glucose, induces insulin secretion in vitro, and improves glucose absorption in skeletal muscle [29]. Vanillic acid isolated from *Fagara tessmannii Engl.* (Family: Rutaceae) exhibits α-glucosidase inhibitory actions in vitro [30]. Christinin-A is a triterpenoidal saponin glycoside that exhibits an anti-hyperglycemic effect in both type 1 and type 2 diabetic rats [31].

Molecular docking has an important role in drug discovery, assisting in digging out the active or lead compounds from a library of natural compounds [32]. It is one of the most widely used virtual screening tools, particularly when the three-dimensional structure of the target protein is available. Docking enables the prediction of both ligand–target binding affinity and the structure of the protein–ligand complex, which are useful for optimizing the lead [33,34].

Prior to the molecular docking study, we checked the physicochemical properties and toxicity potential of SGLT2 inhibitors and FimH antagonists (Tables 1 and 2). The selected SGLT2 inhibitors were known drug canagliflozin and natural SGLT2 inhibitors, namely, acerogenin B, formononetin, (−)-kurarinone, (+)-pteryxin, and quinidine, while heptyl α-D-mannopyranoside was selected as an FimH antagonist. During physicochemical property assessment, we found that out of all the compounds tested, (−)-kurarinone showed one violation of the Lipinski rule [18], i.e., a cLogP value (Logarithm of compound partition coefficient between n-octanol) higher than 5 (Table 1). On the other hand, all the tested compounds showed no toxicity except (+)-pteryxin. (+)-pteryxin was predicted to have a high irritant effect with no mutagenic, tumorigenic, or reproductive toxicity (Table 2).

Table 1. Physicochemical properties of natural sodium–glucose co-transporter 2 (SGLT2) inhibitors and control compounds.

S.No.	Compound Name	Physiochemical Parameters							
		% of Absorption **	Topological Polar Surface Area (Å)^2	Molecular Weight	cLogP ***	Hydrogen Bond Donors	Hydrogen Bond Acceptors	Number of Rotatable Bonds	Lipinski's Violation
	Rule	-	-	<500	$\leq$5	<5	<10	$\leq$10	$\leq$1
1.	Acerogenin B	91.85	49.69	298.38	4.50	2	3	0	0
2.	Formononetin	89.76	55.76	268.26	2.24	1	4	2	0
3.	(−)-kurarinone	75.80	96.22	438.51	6.11	3	6	7	1
4.	(+)-pteryxin	78.59	88.13	386.39	3.34	0	7	5	0
5.	Quinidine	93.27	45.59	324.42	2.61	1	4	4	0
6.	Canagliflozin *	68.15	118.39	444.52	3.27	4	5	5	0
7.	Heptyl α-D-mannopyranoside *	74.71	99.38	278.34	0.485	4	6	8	0

* Control drugs/compounds; ** Percentage of Absorption (% of Absorption) was calculated by the following equation: % of Absorption = 109 − (0.345 × Topological Polar Surface Area); *** Logarithm of the compound partition coefficient between n-octanol and water.

Table 2. Toxicity potential of natural SGLT2 inhibitors and control compounds.

S.No.	Compound Name	Toxicity Risks			
		Mutagenic	Tumorigenic	Reproductive Effect	Irritant
1.	Acerogenin B	None	None	None	None
2.	Formononetin	None	None	None	None
3.	(−)-kurarinone	None	None	None	None
4.	(+)-pteryxin	None	None	None	High
5.	Quinidine	None	None	None	None
6.	Canagliflozin *	None	None	None	None
7.	Heptyl α-D-mannopyranoside *	None	None	None	None

* Control drugs/compounds.

The predicted model of SGLT2 revealed that 81.25 percent of the residues had an average 3D-1D score of ≥ 0.2, resulting in a "pass" in SAVES v6.0.0 by the VERIFY3D tool. In addition, the Ramachandran plot (showing 93% of the residues in the allowed region), fractional accessible surface area, stereo/packing quality index, fractional residue volume, and 3-D profile quality index (produced by the VADAR 1.8 server) showed that the predicted 3-D model was within an appropriate range (Figure 1).

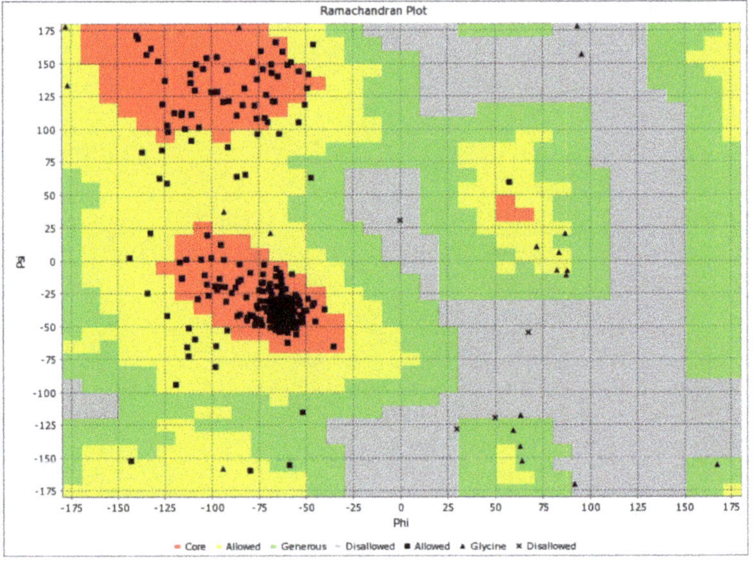

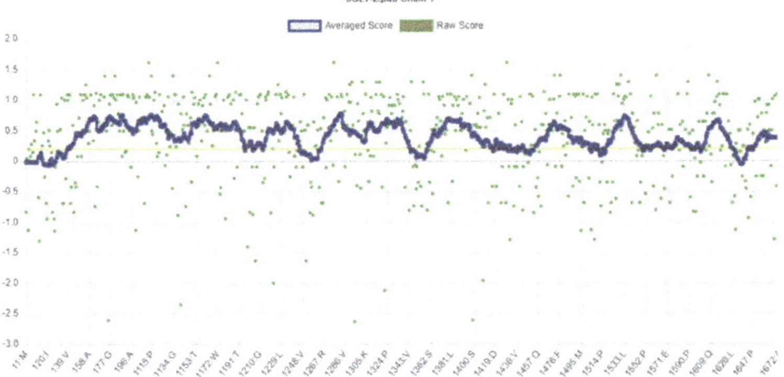

Figure 1. *Cont.*

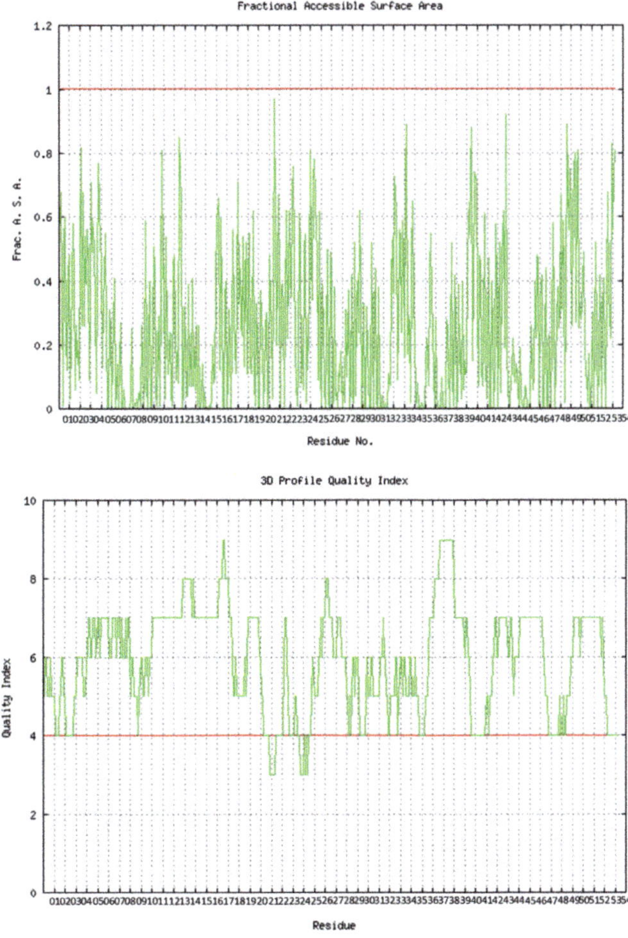

Figure 1. *Cont.*

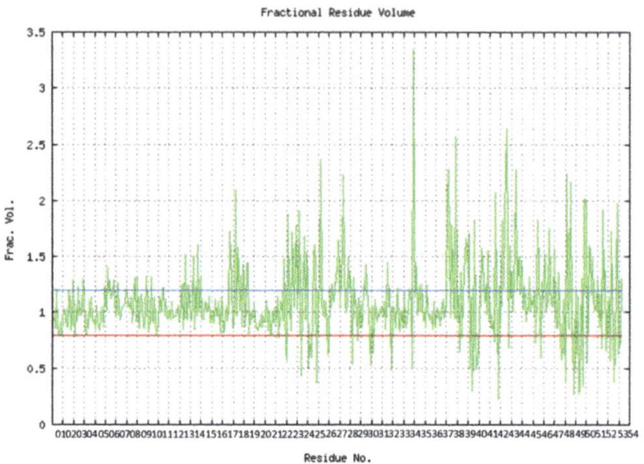

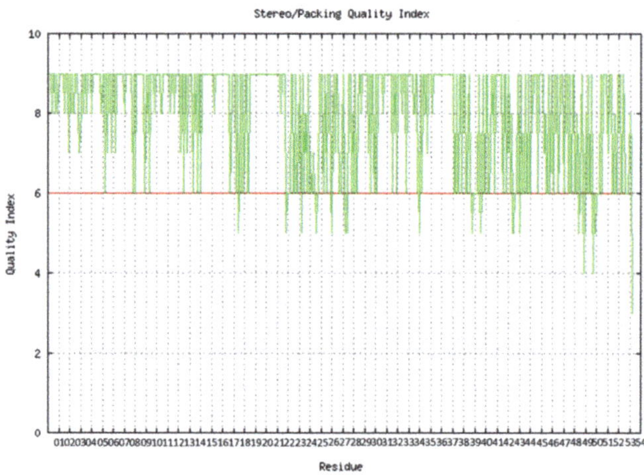

Figure 1. Validation of the predicted 3-D structure of SGLT2.

The inhibition of SGLT2 has been considered a novel pharmacotherapy for T2DM treatment [35]. Accordingly, molecular docking studies have revealed that the natural SGLT2 inhibitors formononetin, (+)-pteryxin, and quinidine were efficiently bound with SGLT2. Formononetin was found to interact with the F98, E99, A102, L149, K152, T153, V286, S287, Y290, W291, I456, Q457, and S460 binding pocket residues of SGLT2 (Figure 2a); while the S74, G79, H80, K154, I155, V157, D158, S161, S393, I397, and I456 residues of SGLT2 were observed to bind with (+)-pteryxin (Figure 2b). Furthermore, quinidine was found to interact with the S74, N75, H80, L84, F98, E99, V286, S287, Y290, W291, F453, I456, and Q457 residues of SGLT2 (Figure 2c). Consistent with this, the amino acid residues H80, F98, T153, K154, V157, D158, V290, I397, F453, I456, Q457, and S460 were shown to be important for the inhibition of SGLT2 [36]. Amino acid residues H80, V286, Y290, W291, and I456 are the main hydrophobic residues of SGLT2, interacting with the dock inhibitors formononetin, (+)-pteryxin and quinidine (Figure 3a–c). This concurs with a previous report wherein the amino acid residues H80, Y290, and I456 of SGLT2 have also been reported have a hydrophobic interaction with the inhibitor [36,37]. The binding energy (BE) for the catalytic domain interactions of formononetin–SGLT2, (+)-pteryxin–SGLT2, and quinidine–SGLT2 were found to be −7.63 kcal/mol, −9.01 kcal/mol, and −8.77 kcal/mol,

respectively, while their inhibition constants (Ki) were 2.57 µM, 0.245 µM, and 0.371 µM, respectively (Table 3).

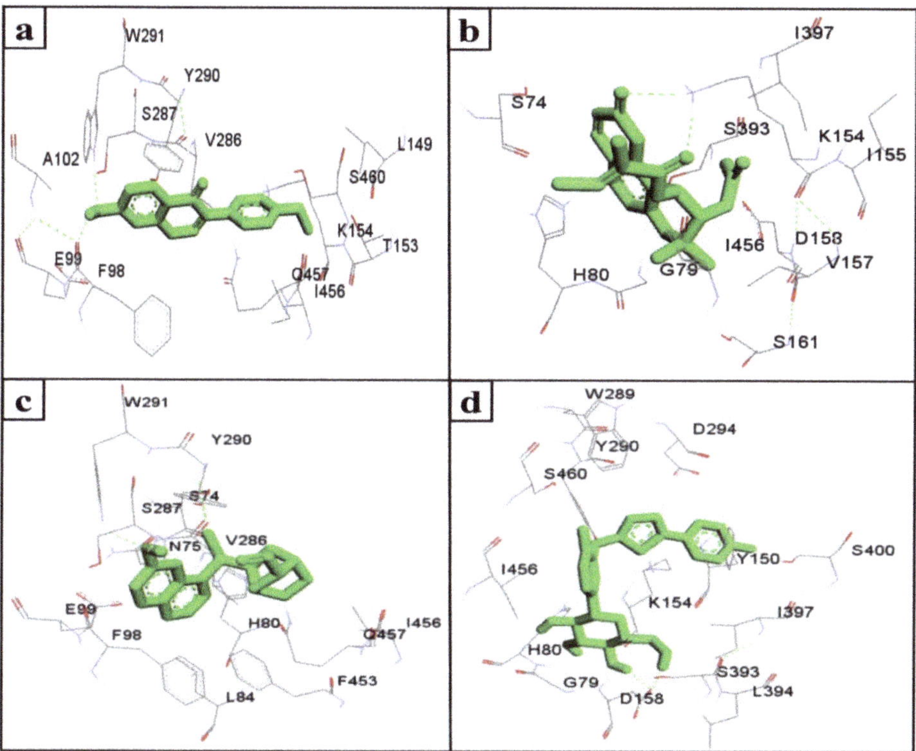

Figure 2. The amino acid residue of SGLT2 interacting with formononetin (**a**), (+)-pteryxin (**b**), quinidine (**c**), and canagliflozin (**d**). The ligands (formononetin, (+)-pteryxin, quinidine, and canagliflozin) are represented as green stick forms and hydrogen bonds are indicated as green dashed lines.

Table 3. The docking results of the molecular interactions of SGLT2 and FimH with natural SGLT2 inhibitors and control compounds.

Compounds	SGLT2		FimH	
	Binding Energy (kcal/mol)	Inhibition Constant (µM)	Binding Energy (kcal/mol)	Inhibition Constant (µM)
Canagliflozin *	−7.23	5.04	−3.56	2450
Acerogenin B	−6.30	24.25	−4.40	598.52
Formononetin	−7.63	2.57	−5.65	71.95
(−)-kurarinone	−7.23	5.03	−3.93	1310
(+)-pteryxin	−9.01	0.248	−5.50	92.97
Quinidine	−8.77	0.371	−5.70	66.40
Heptyl α-D-mannopyranoside **	-	-	−4.46	109.49

* Control compound for SGLT2; ** Control compound for FimH.

The FDA approved canagliflozin, an SGLT2 inhibitor, for use in T2DM treatment in 2013 [38]. In the present study, canagliflozin was used as a positive control against SGLT2. Canagliflozin was observed to bound with G79, H80, Y150, K154, D158, W289, Y290, D294, S393, I397, S460, and I456 residues of SGLT2 (Figures 2d and 3d). Interestingly, these amino acids of SGLT2 have also been found to interact with SGLT2 inhibitors (formononetin,

(+)-pteryxin, and quinidine). I456 was one of the most reactive common hydrophobic residues of SGLT2, interacting with formononetin, (+)-pteryxin, and quinidine, as well as canagliflozin (Figure 3a–d). The Y290 residue of SGLT2 was found to make H-bonds with quinidine, while the same residue was observed to hydrophobically interact with formononetin and canagliflozin (Figure 3a,c,d). In addition, G79, H80, and I397 were the common hydrophobic interacting residues of SGLT2 with (+)-pteryxin and canagliflozin (Figure 3b,d).

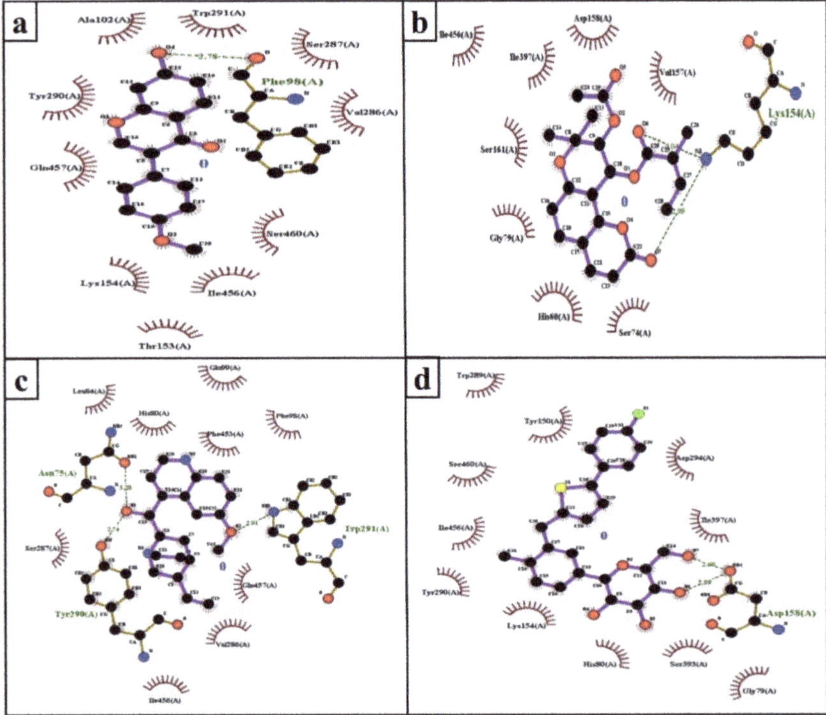

Figure 3. Ligplot analysis of formononetin (**a**), (+)-pteryxin (**b**), quinidine (**c**), and canagliflozin (**d**) in terms of their interaction with SGLT2. The amino acid residues forming hydrophobic interactions are shown as red arcs while the hydrogen bonds are shown as green dashed lines with indicated bond lengths.

The function of human SGLT1 protein is dramatically affected by amino acid at position 457. It has been shown that residue 457 (i.e., Q457) in human SGLT1 directly interacts with sugar for its reabsorption [39,40]. The amino acid sequences of SGLT1 and SGLT2 revealed that both of these proteins have glutamine at the residue 457 position, and glucose–galactose malabsorption occurs due to mutation in the glutamine residue (Q457) [41]. Consistent with this, in the present study, formononetin and quinidine were observed to interact with the Q457 residue that is suggested to impair SGLT2 function.

Since the discovery of the first natural SGLT2 inhibitor (i.e., phlorizin), several other synthetic glucoside analogs have been developed [42]. Tofogliflozin is a selective SGLT2 inhibitor that enhances urinary glucose excretion in a dose-dependent manner [43]. Luseogliflozin is an orally active second-generation SGLT2 inhibitor that has protective effects on pancreatic beta-cell mass and function [44]. Furthermore, efforts have been made to find new active natural ingredients from *S. Flavescents* [15], alkaloids from *A. macrophylla* [45], and *Schisandrae Chinensis Fructus* for the development of SGLT2 inhibition [46]. A 4-O-methyl derivative of sergliflozin-A has been reported to exhibit SGLT2 inhibition activity [47].

SGLT2 inhibitors are some of the most promising anti-diabetic agents introduced into clinical practice over the last decade. The therapeutic benefits of SGLT2 inhibitors include weight loss, a reduction in blood pressure, and an enhancement in high-density lipoprotein level [48,49]. However, the SGLT2 inhibitors dapagliflozin, canagliflozin, and empagliflozin have been found to cause UTIs and genital infections in diabetic patients [50–52]. Additionally, the FDA has issued warnings about the possible UTI-inducing side effects of SGLT2 inhibitors in December 2015 [9].

UTIs pose a severe medical issue worldwide, with more than 85% of UTIs caused by uropathogenic *Escherichia coli* [53]. FimH is a bacterial adhesin that facilitates the colonization of uropathogenic *E. coli* on the cell surface of the human and murine bladder and leads to the formation of biofilm [54]. Therefore, this adhesin has been considered a virulence factor and a promising therapeutic target for the treatment of UTIs [55]. Interestingly, the docking results indicate that the SGLT2 inhibitors formononetin, (+)-pteryxin, and quinidine show strong binding with FimH. FimH was found to interact with formononetin through 11 amino acid residues, namely F1, N46, D47, Y48, I52, D54, Q133, N135, Y137, N138, and D140 (Figure 4a), while (+)-pteryxin was found to interact with 6 amino acid residues, namely Y48, T51, I52, Y137, N138, and D140 (Figure 4b). Similarly, 10 amino acid residues, namely the F1, I13, Y48, I52, D54, Q133, N135, Y137, N138, and D140 residues of FimH, were found to interact with quinidine (Figure 4c). The BE for formononetin–FimH, (+)-pteryxin–FimH, and quinidine–FimH interactions were found to be −5.65 kcal/mol, −5.50 kcal/mol, and −5.70 kcal/mol, respectively. The corresponding estimated inhibition constants (Ki) for the above-mentioned complexes were determined to be 71.95 μM, 92.97 μM, and 66.40 μM, respectively (Table 3).

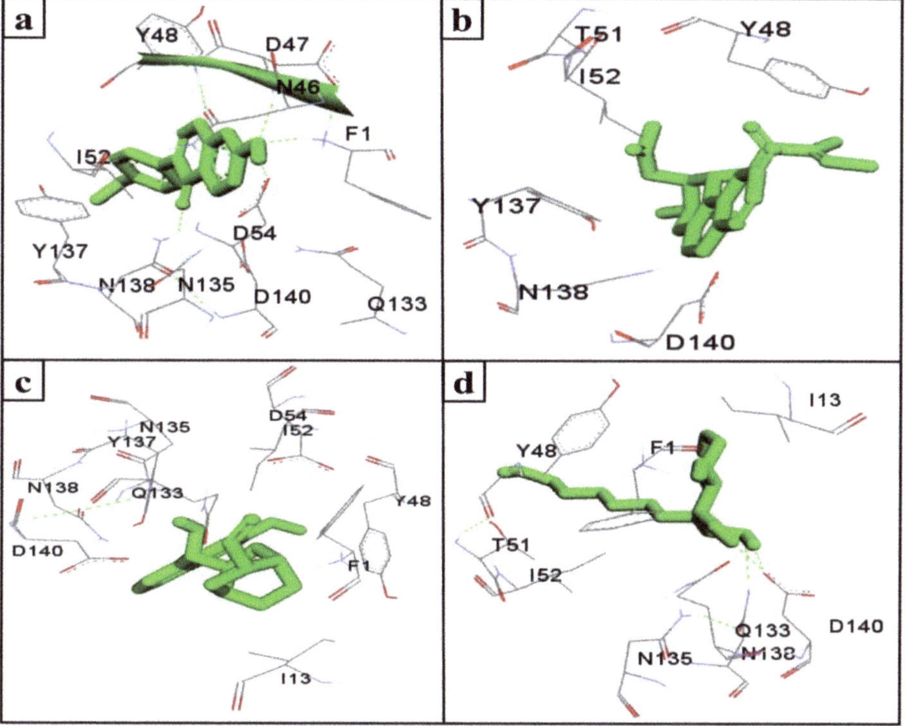

Figure 4. The amino acid residue of FimH's interaction with formononetin (**a**), (+)-pteryxin (**b**), quinidine (**c**), and heptyl α-D-mannopyranoside (**d**). The ligands (formononetin, (+)-pteryxin, quinidine, and heptyl α-D-mannopyranoside) are represented as green stick forms and hydrogen bonds are represented as green dashed lines.

The FimH protein has two domains, the C-terminal pilin domain and the N-terminal lectin domain. The FimH lectin domain has a mannose-binding pocket (N46, D47, D54, Q133, N135, N138, and D140), in which sugar contributes to the formation of various hydrogen bonds with FimH. Hydrophobic regions are present in the surrounding region of the mannose-binding pocket and consist of hydrophobic support (F1, I13, and F142), the tyrosine gate (Y48, I52, and Y137), and the residue T51 [56]. Consistent with this, in the present study, the D47, D54, and N138 of the FimH protein were involved in hydrogen binding with formononetin (Figure 5a), while Y48, I52, Y137, and D140 were the common hydrophobic amino acid residues interacting with formononetin, (+)-pteryxin and quinidine (Figure 5a–c).

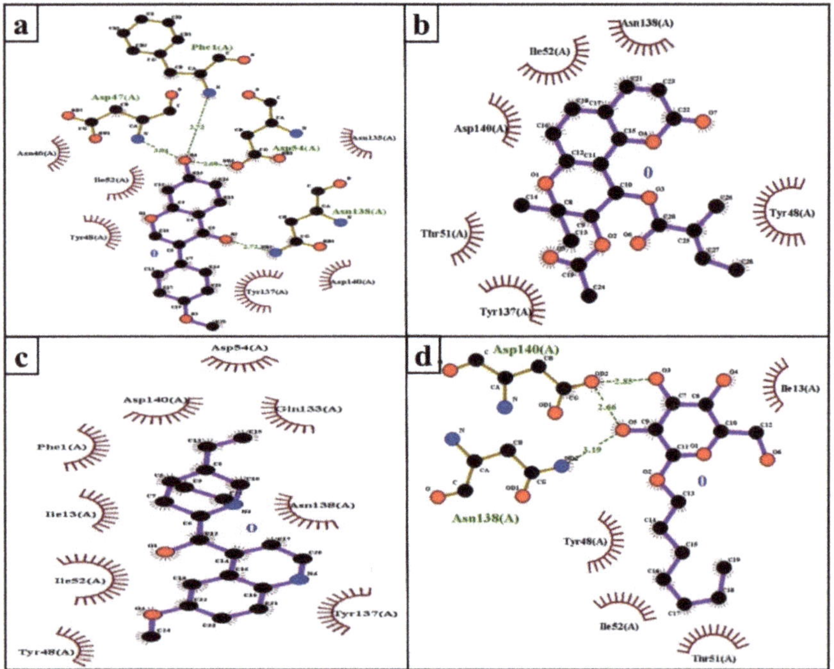

Figure 5. Ligplot analysis of the interactions of formononetin (**a**), (+)-pteryxin (**b**), quinidine (**c**), and heptyl α-D-mannopyranoside (**d**) with SGLT2. The amino acid residues forming hydrophobic interactions are shown as red arcs while the hydrogen bonds are shown as green dashed lines with indicated bond lengths.

In the present study, heptyl α-D-mannopyranoside (a FimH antagonist) was used as a positive control against the FimH protein. Molecular docking analysis revealed that amino acid residues, namely the F1, I13, Y48, T51, Q133, N135, N138, D140, and I52 residues of the FimH protein, have a vital role in binding with heptyl α-D-mannopyranoside (Figures 4d and 5d). The main FimH binding pocket amino acid residues are F1, N46, D47, D54, E133, N135, D140, and F142, and mutation in these individual main residues leads to them affecting FimH function and reducing its virulence [57,58]. Interestingly, we found that these amino acid residues of FimH were also determined to interact with natural SGLT2 inhibitors (formononetin, (+)-pteryxin, and quinidine).

O- and C-linked alpha-D-mannosides with hydrophobic and aryl substituents are effective FimH antagonists. The substitution of biphenyl derivatives may provide additional advantages for the FimH antagonist molecule. The addition of 1,10-biphenyl pharmacophore and various aglycone atoms enhanced the alpha-D-mannose derivatives' suitability as FimH inhibitors [59,60]. Further, glycomimetics based on mannose scaffolding

has also been synthesized and widely studied for their aptitude as FimH antagonists [61,62]. Besides synthetic compounds, natural substances, like cranberry and its derivatives (such as myricetin, proanthocyanidin (PAC)-standardized cranberry extract, and polyphenol metabolites extracted from PAC), have anti-adhesive effects on uropathogenic *E. coli* [63].

In docking experiments, it is usually crucial to look for a ligand that can bind efficiently with the receptor, using Gibbs free energy as a parameter of better binding [64,65]. The strength of an interaction between a ligand and a receptor is measured in terms of the free energy of binding. The lowest BE is the output of the efficient binding of the drug/ligand to the active site of the receptor [66]. A higher (negative) BE is a sign of efficient interaction between the ligand and the receptor [67]. Accordingly, in the present study, formononetin, (+)-pteryxin, and quinidine exhibited strong interaction with the FimH protein, with a high BE relative to that of the positive control heptyl α-D-mannopyranoside, suggesting that these compounds could be promising SGLT2 inhibitors with less severe uropathogenic side effects.

Diabetic ketoacidosis and its associated events have been reported at a low frequency in T2DM patients treated with canagliflozin [68]. Interestingly, in the present study, the BE of canagliflozin with FimH relative to the other SGLT2 inhibitor indicates that canagliflozin has less potency in inhibiting the FimH protein, thereby having a high susceptibility to causing diabetic ketosis and UTIs.

3.2. Root Mean Square Deviation

Root mean square deviation (RMSD) is the most significant dynamic to explore in terms of conformational changes by means of stability in the structure and dynamic behavior of the receptor [69]. The complexes of formononetin, quinidine, and (+)-pteryxin with FimH obtained in molecular docking that interacted the best were further probed by MD simulation. The binding of ligands to their receptor protein can result in large conformational deviations in the resulting complex, specifically at the binding site. In this study, values of RMSD were increased at the beginning, with respect to the native structure of the RMSD. Figure 6A reports the RMSD of the backbone atoms of FimH protein as they interacted with each ligand molecule, where formononetin–FimH, quinidine–FimH and (+)-pteryxin–FimH interactions are represented with a black, red, and green color code, respectively. RMSD changes in all systems were initiated at the same point (~0.125 nm); however, in the formononetin–FimH complex, the conformational changes took place throughout the stabilization process. The change in the RMSD values in the formononetin–FimH complex indicated a small fluctuation until ~30 ns, and, afterwards, RMSD was found to be stable, remaining at around 0.35 nm in value. Due to the initial ups and downs in the RMSD values of the formononetin–FimH complex, the topologies of the structures were observed to significantly change during simulation. The observed initial RMSD change predicted the large conformational changes in regions near the binding pocket. Although the simulation lasted for 50 ns, the discrepancy from the initial structure within the first 30 ns was sufficient to point out the protein structures that were substantially denatured at binding sites and adjacent areas. While the RMSD values of the quinidine–FimH complex showed little turbulence until ~20 ns, the values reached a plateau later on. The value observed was greater than 0.3 nm initially; however, after 35 ns it went steady at 0.37 nm. In general, overall RMSD values fluctuated within 0.5 nm $\pm$ 0.05 nm throughout the simulation.

Like the above complexes, RMSD values showed frequent conformational changes and binding residue incorporation during the simulation analysis of the pteryxin–FimH complex. Conformational changes were observed at a functional domain near the binding site of the complex, and a considerable movement of the pteryxin ligand was also predicted at that site. Moreover, the RMSDs of all three ligand candidates were calculated to evaluate the behavior of the ligands within the binding site of the protein. Figure 6B shows the plot representation of RMSD for each ligand. Ligands (formononetin and quinidine) have generally shown high stability within the close contact residues of protein, whereas, in the

case of pteryxin, this ligand demonstrated several fluctuations within the binding site. Due to the movement of the pteryxin ligand, an initial sudden elevation (~0.13 nm) followed by fluctuations until 0.15 ns was observed during the simulation. Although the RMSDs maintained a plateau later on, an average value of 0.21 nm was calculated throughout the simulation. Moreover, the difference in RMSDs due to the movement of pteryxin from the initial to the final frame at the binding site was observed as 0.15 nm, which revealed that pteryxin shows considerably weaker binding with interacting amino acid residues during the simulation. On the other hand, it was observed that formononetin and quinidine stably interacted during simulation, and the pattern of contact between the binding residues and both ligands (formononetin and quinidine) predicted regulation of the activity of the FimH protein.

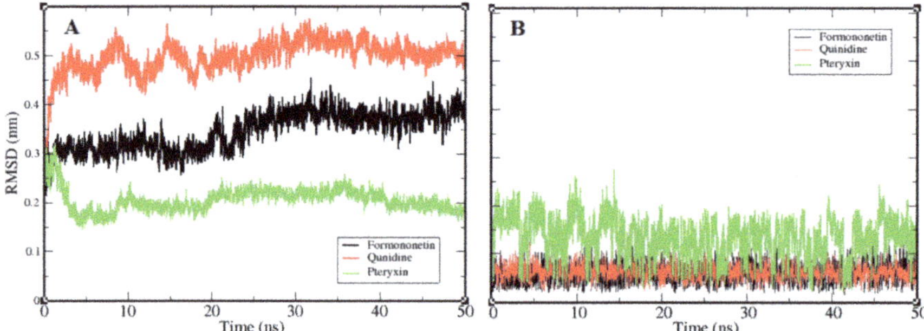

Figure 6. Representation of an RMSD plot of each complex in terms of their relationship between RMSD values and simulation time, shown in three different color codes. Panel (**A**) shows an RMSD plot of the backbone atoms of FimH protein receptor interacting with three ligand candidates, while panel (**B**) represents the RMSD plots of ligand candidates during simulation, which are color coded black (Formononetin), red (Quinidine), and green (Pteryxin). FimH protein complexes with Formononetin and Quinidine possessed high stability during simulation, while the third complex with Pteryxin demonstrated comparatively less stability.

Consequently, the RMSDs of the formononetin–FimH and quinidine–FimH complexes were found to have comparatively more stable trajectories than the (+)–pteryxin–FimH complex during simulation analysis (Figure 6).

3.3. Principle Component Analysis

To identify the overall patterns of motion in the complexes during simulation, we utilized principal components analysis (PCA) for the three complexes. The covariance matrices of the simulations were calculated to produce the eigenvectors, and then a screening of the trajectories was performed around each of the diverse eigenvectors. Furthermore, the dominant motions during a four-vector RMSD for each complex were observed (Figure 7). A large portion of the overall variations in the receptor protein can be explained by a few reduced amplitude eigenvectors with large eigenvalues. The analysis illustrated that the fourth eigenvector accounts for the depletion of the RMSD value, although the rest of the three vectors showed elevation at the initial time of the simulation. It was shown that after ~8 ns the RMSD values reached a plateau and maintained their value until the end of the simulation (Figure 7).

To know the movement of the backbone atoms of the receptor protein, 100 frames of the first principal component were collectively loaded into VMD, which show the motility of the residues (Figure 8), although there might be differences between each complex with respect to the motions sampled. The width of the band is proportional to amplitude; a narrow band signifies those segments that hardly moved, while wide bands signify the sections most affected by the transitions. The middle domain of the protein showed less movement than the adjacent domains that predicted the stability of the middle domain.

The rest of the protein exhibited comparatively wider bands, which might be due to an increase in functionality.

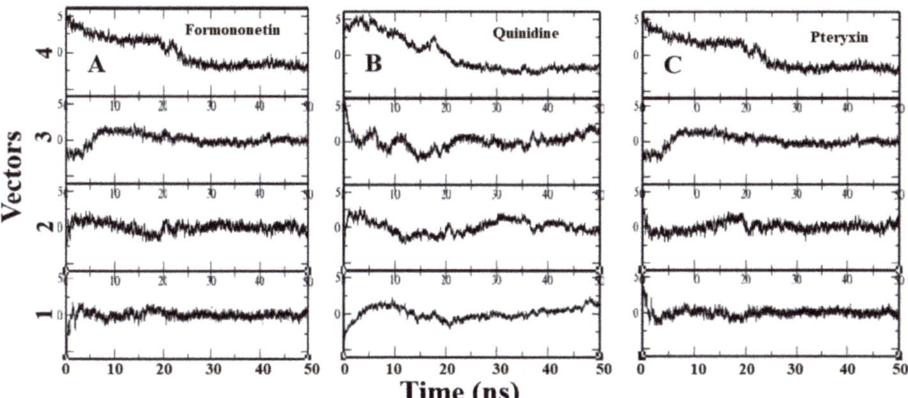

Figure 7. Each panel (**A–C**) reports the calculation of projection of particular trajectory on eigenvectors for each three complexes. The RMSDs for every four eigenvectors are depicted as four subplots in each panel. Plot shows the projections of each vector are fitted to the structure in the eigenvector.

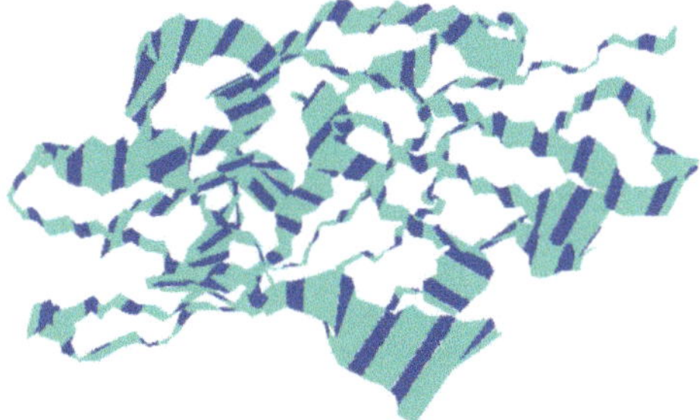

Figure 8. The motion of the backbone atoms of the protein receptor during simulation of the trajectory as calculated by principal component analysis. The width of the bands is proportional to amplitude.

It is worth mentioning that the binding energy values and MD simulations obtained can only demonstrate the binding efficacy and stability of inhibitors with the target protein. Further benchwork experiments are required to optimize these compounds (formononetin, (+)-pteryxin, and quinidine) as promising SGLT2 inhibitors with add-on FimH inhibition potential that might reduce the probability of uropathogenic side effects.

4. Conclusions

SGLT2 inhibitors are a newer class of drugs that have enormous anti-diabetic potential. Unfortunately, the side effects, specifically those that contribute to the development of UTIs, have impeded their success rate. Several clinicians, as well as diabetic patients, are still in a dilemma over the use of this class of anti-diabetic medication. In contrast, FimH is blooming as a potential alternative target for UTI treatment, and FimH inhibitors are currently under development. These inhibitors act against the uropathogenic bacterial strains adherence to the mucosal surface of the urinary tract, resulting in a lower chance of encountering resistance. In the present study, we docked natural SGLT2 inhibitors with FimH target proteins to predict their FimH interaction potential. To the best of our knowledge, this is the first time that SGLT2 inhibitors have been explored in terms of their use against FimH. Our findings suggest that among all SGLT2 inhibitors examined, formononetin,

(+)-pteryxin, and quinidine exhibited strong interactions with the FimH protein with high binding energy, in comparison to the positive control heptyl α-D-mannopyranoside. On the other hand, the FDA-approved SGLT2 inhibitor canagliflozin indicated lower interaction with FimH. Thus, we hypothesize that if explored further, natural SGLT2 inhibitors might reduce the probability of UTIs when used against FimH. The authors anticipate researchers to duly use the findings of this study to design more versatile and potent dual inhibitors against SGLT2 and FimH to cope with the uropathogenic side effects of SGLT2 inhibitor class anti-diabetic medications.

Author Contributions: Conceptualization, S.S. and S.M.D.R.; methodology, S.S., N.C., and S.M.D.R.; writing—Original draft preparation, S.S. and S.M.D.R.; writing—Review and editing, M.M.M., Q.A., S.A., M.M.B., K.A., S.S.A., M.M.A.B., A.M., A.I.A., and M.M.A. All authors have read and agreed to the published version of the manuscript.

Funding: This research was funded by the Deanship of Scientific Research, Najran University, Najran, Saudi Arabia grant number NU/MID/17/091.

Institutional Review Board Statement: Not applicable.

Informed Consent Statement: Not applicable.

Data Availability Statement: Not applicable.

Conflicts of Interest: The authors declare no conflict of interest.

References

1. Leung, M.Y.; Pollack, L.M.; Colditz, G.A.; Chang, S.H. Life years lost and lifetime health care expenditures associated with diabetes in the U.S., national health interview survey, 1997–2000. *Diabetes Care* **2015**, *38*, 460–468. [CrossRef] [PubMed]
2. Shah, B.R.; Hux, J.E. Quantifying the risk of infectious diseases for people with diabetes. *Diabetes Care* **2003**, *26*, 510–513. [CrossRef] [PubMed]
3. Thomsen, R.W.; Hundborg, H.H.; Lervang, H.H.; Johnsen, S.P.; Schonheyder, H.C.; Sorensen, H.T. Diabetes mellitus as a risk and prognostic factor for community-acquired bacteremia due to enterobacteria: A 10-year, population-based study among adults. *Clin. Infect. Dis.* **2005**, *40*, 628–631. [CrossRef] [PubMed]
4. Thomsen, R.W.; Hundborg, H.H.; Lervang, H.H.; Johnsen, S.P.; Sorensen, H.T.; Schonheyder, H.C. Diabetes and outcome of community-acquired pneumococcal bacteremia: A 10-year population based cohort study. *Diabetes Care* **2004**, *27*, 70–76. [CrossRef]
5. Dardi, I.; Kouvatsos, T.; Jabbour, S.A. SGLT2 inhibitors. *Biochem. Pharmacol.* **2016**, *101*, 27–39. [CrossRef]
6. Benfield, T.; Jensen, J.S.; Nordestgaard, B.G. Influence of diabetes and hyperglycaemia on infectious disease hospitalization and outcome. *Diabetologia* **2007**, *50*, 549–554. [CrossRef]
7. Boyko, E.J.; Fihn, S.D.; Scholes, D.; Abraham, L.; Monsey, B. Risk of urinary tract infection and asymptomatic bacteriuria among diabetic and nondiabetic postmenopausal women. *Am. J. Epidemiol.* **2005**, *161*, 557–564. [CrossRef]
8. Geerlings, S.; Fonseca, V.; Castro-Diaz, D.; List, J.; Parikh, S. Genital and urinary tract infections in diabetes: Impact of pharmacologically induced glucosuria. *Diabetes Res. Clin. Pract.* **2014**, *103*, 373–381. [CrossRef]
9. U.S. Food and Drug Administration. FDA Drug Safety Communication: FDA Revises Labels of SGLT2 Inhibitors for Diabetes to Include Warnings About Too Much Acid in the Blood and Serious Urinary Tract Infections. 2015. Available online: http://www.fda.gov/Drugs/DrugSafety/ucm475463.htm (accessed on 10 November 2016).
10. Zinman, B.; Wanner, C.; Lachin, J.M.; Fitchett, D.; Bluhmki, E.; Hantel, S.; Mattheus, M.; Devins, T.; Johansen, O.E.; Woerle, H.J.; et al. Empagliflozin, cardiovascular outcomes, and mortality in type 2 diabetes. *N. Engl. J. Med.* **2015**, *373*, 2117–2128. [CrossRef]
11. Neal, B.; Perkovic, V.; Mahaffey, K.W.; de Zeeuw, D.; Fulcher, G.; Erondu, N.; Shaw, W.; Law, G.; Desai, M.; Matthews, D.R. Canagliflozin and cardiovascular and renal events in type 2 diabetes. *N. Engl. J. Med.* **2017**, *377*, 644–657. [CrossRef]
12. Fronzes, R.; Remaut, H.; Waksman, G. Architectures and biogenesis of non-flagellar protein appendages in Gram-negative bacteria. *EMBO J.* **2008**, *27*, 2271–2280. [CrossRef] [PubMed]
13. Mydock-McGrane, L.K.; Cusumano, Z.T.; Janetka, J.W. Mannose-derived FimH antagonists: A promising anti-virulence therapeutic strategy for urinary tract infections and Crohn's disease. *Expert Opin. Ther. Pat.* **2016**, *26*, 175–197. [CrossRef] [PubMed]
14. Shimokawa, Y.; Akao, Y.; Hirasawa, Y.; Awang, K.; Hadi, A.H.; Sato, S.; Aoyama, C.; Takeo, J.; Shiro, M.; Morita, H. Gneyulins A and B, stilbene trimers, and noidesols A and B, dihydroflavonol-C-glucosides, from the bark of Gnetum gnemonoides. *J. Nat. Prod.* **2010**, *73*, 763–767. [CrossRef] [PubMed]
15. Sato, S.; Takeo, J.; Aoyama, C.; Kawahara, H. Na+-glucose cotransporter (SGLT) inhibitory flavonoids from the roots of Sophora flavescens. *Bioorg. Med. Chem.* **2007**, *15*, 3445–3449. [CrossRef] [PubMed]

16. Oranje, P.; Gouka, R.; Burggraaff, L.; Vermeer, M.; Chalet, C.; Duchateau, G.; van der Pijl, P.; Geldof, M.; de Roo, N.; Clauwaert, F.; et al. Novel natural and synthetic inhibitors of solute carriers SGLT1 and SGLT2. *Pharmacol. Res. Perspect.* **2019**, *7*, e00504. [CrossRef]

17. Abgottspon, D.; Rölli, G.; Hosch, L.; Steinhuber, A.; Jiang, X.; Schwardt, O.; Cutting, B.; Smiesko, M.; Jenal, U.; Ernst, B.; et al. Development of an Aggregation Assay to Screen FimH Antagonists. *J. Microbiol. Methods* **2010**, *82*, 249–255. [CrossRef]

18. Lipinski, C.A.; Lombardo, F.; Dominy, B.W.; Feeney, P.J. Experimental and computational approaches to estimate solubility and permeability in drug discovery and development settings. *Adv. Drug Deliv. Rev.* **2001**, *46*, 3–26. [CrossRef]

19. Zhao, Y.H.; Abraham, M.H.; Le, J.; Hersey, A.; Luscombe, C.N.; Beck, G.; Sherborne, B.; Cooper, I. Rate-limited steps of human oral absorption and QSAR studies. *Pharm. Res.* **2002**, *19*, 1446–1457. [CrossRef]

20. Rizvi, S.M.D.; Shakil, S.; Haneef, M.A. simple click by click protocol to perform docking: AutoDock 4.2 made easy for non-bioinformaticians. *EXCLI J.* **2013**, *12*, 831–857.

21. Spoel, D.V.D.; Lindahl, E.; Hess, B.; Groenhof, G.; Mark, A.E.; Berendsen, H.J. GROMACS: Fast, flexible, and free. *J. Comput. Chem.* **2005**, *26*, 1701–1718. [CrossRef]

22. Pronk, S.; Páll, S.; Schulz, R.; Larsson, P.; Bjelkmar, P.; Apostolov, R.; Shirts, M.R.; Smith, J.C.; Kasson, P.M.; van der Spoel, D.; et al. GROMACS 4.5: A high-throughput and highly parallel open source molecular simulation toolkit. *Bioinformatics* **2013**, *29*, 845–854. [CrossRef] [PubMed]

23. Oostenbrink, C.; Villa, A.; Mark, A.E.; van Gunsteren, W.F. A biomolecular force field based on the free enthalpy of hydration and solvation: The GROMOS force-field parameter sets 53A5 and 53A6. *J. Comput. Chem.* **2004**, *25*, 1656–1676. [CrossRef] [PubMed]

24. Toukan, K.; Rahman, A. Molecular-dynamics study of atomic motions in water. *Phys. Rev. B* **1985**, *31*, 2643. [CrossRef] [PubMed]

25. Essmann, U.; Perera, L.; Berkowitz, M.L. A smooth particle mesh Ewald method. *J. Chem. Phys.* **1995**, *103*, 8577. [CrossRef]

26. Hess, B.; Bekker, H.; Berendsen, H.J.C.; Fraaije, J.G.E.M. LINCS: A linear constraint solver for molecular simulations. *J. Comput. Chem.* **1997**, *18*, 1463–1472. [CrossRef]

27. Schüttelkopf, A.W.; Van Aalten, D.M. PRODRG: A tool for high-throughput crystallography of protein-ligand complexes. *Acta Crystallogr. Sect. D Biol. Crystallogr.* **2004**, *60*, 1355–1363. [CrossRef]

28. Seo, E.; Lee, E.K.; Lee, C.S.; Chun, K.H.; Lee, M.Y.; Jun, H.S. Psoraleacorylifolia L. seed extract ameliorates streptozotocin-induced diabetes in mice by inhibition of oxidative stress. *Oxid. Med. Cell Longev.* **2014**, *2014*, 897296. [CrossRef]

29. Monteiro, M.; Farah, A.; Perrone, D.; Trugo, L.C.; Donangelo, C. Chlorogenic acid compounds from coffee are differentially absorbed and metabolized in humans. *J. Nutr.* **2007**, *137*, 2196–2201. [CrossRef]

30. Mbaze, L.M.; Poumale, H.M.; Wansi, J.D. alpha-Glucosidase inhibitory pentacyclic triterpenes from the stem bark of Fagara tessmannii (Rutaceae). *Phytochemistry* **2007**, *68*, 591–595. [CrossRef]

31. Abdel-Zaher, A.O.; Salim, S.Y.; Assaf, M.H.; Abdel-Hady, R.H. Antidiabetic activity and toxicity of Zizyphus spina-christi leaves. *J. Ethnopharmacol.* **2005**, *101*, 129–138. [CrossRef]

32. Meng, X.Y.; Zhang, H.X.; Mezei, M.; Cui, M. Molecular docking: A powerful approach for structure-based drug discovery. *Curr. Comput. Aided Drug Des.* **2011**, *7*, 146–157. [CrossRef] [PubMed]

33. Udrescu, L.; Bogdan, P.; Chiş, A.; Sîrbu, I.O.; Topîrceanu, A.; Văruţ, R.M.; Udrescu, M. Uncovering New Drug Properties in Target-Based Drug-Drug Similarity Networks. *Pharmaceutics* **2020**, *12*, 879. [CrossRef] [PubMed]

34. Pinzi, L.; Rastelli, G. Molecular Docking: Shifting Paradigms in Drug Discovery. *Int. J. Mol. Sci.* **2019**, *20*, 4331. [CrossRef] [PubMed]

35. Rajesh, R.; Naren, P.; Sudha, V.; Unnikrishnan, M.K.; Pandey, S.; Varghese, M.; Gang, S. Sodium Glucose Co transporter 2 (SGLT2) Inhibitors: A New Sword for the Treatment of Type 2 Diabetes Mellitus. *Int. J. Pharma Sci. Res.* **2010**, *1*, 139–147.

36. Shaikh, S.; Rizvi, S.M.D.; Shakil, S.; Riyaz, S.; Biswas, D.; Jahan, R. Forxiga (Dapagliflozin): Plausible role in the treatment of diabetes associated neurological disorders. *Biotechnol. Appl. Biochem.* **2016**, *63*, 145–150. [CrossRef]

37. Rizvi, S.M.D.; Shakil, S.; Biswas, D.; Shakil, S.; Shaikh, S.; Bagga, P.; Kamal, M.A. Invokana (Canagliflozin) as a Dual Inhibitor of Acetylcholinesterase and Sodium Glucose Co-Transporter 2: Advancement in Alzheimer's Disease-Diabetes Type 2 Linkage via an Enzoinformatics Study. *CNS Neurol. Disord.-Drug Targets* **2014**, *13*, 447–451. [CrossRef]

38. Kaushal, S.; Singh, H.; Thangaraju, P.; Singh, J. Canagliflozin: A novel SGLT2 inhibitor for Type 2 Diabetes Mellitus. *N. Am. J. Med. Sci.* **2014**, *6*, 107–113.

39. Díez-Sampedro, A.; Barcelona, S. Sugar binding residue affects apparent Na affinity and transport stoichiometry in mouse sodium/glucose cotransporter type 3B. *J. Biol. Chem.* **2011**, *286*, 7975–7982. [CrossRef]

40. Liu, T.; Krofchick, D.; Silverman, M. Effects on conformational states of the rabbit sodium/glucose cotransporter through modulation of polarity and charge at glutamine 457. *Biophys. J.* **2009**, *96*, 748–760. [CrossRef]

41. Wright, E.M.; Turk, E.; Martin, M.G. Molecular basis for glucose galactose malabsorption. *Cell Biochem. Biophys.* **2002**, *36*, 115–121. [CrossRef]

42. Choi, C.I. Sodium-Glucose Cotransporter 2 (SGLT2) Inhibitors from Natural Products: Discovery of Next-Generation Antihyperglycemic Agents. *Molecules* **2016**, *21*, 1136. [CrossRef] [PubMed]

43. Ito, S.; Hosaka, T.; Yano, W.; Itou, T.; Yasumura, M.; Shimizu, Y.; Kondo, T. Metabolic effects of tofogliflozin are efficiently enhanced with appropriate dietary carbohydrate ratio and are distinct from carbohydrate restriction. *Physiol. Rep.* **2018**, *6*, e13642. [CrossRef] [PubMed]

44. Okauchi, S.; Shimoda, M.; Obata, A.; Kimura, T.; Hirukawa, H.; Kohara, K.; Kaneto, H. Protective effects of SGLT2 inhibitor Luseogliflozin on pancreatic β-cells in obese type 2 diabetic db/db mice. *Biochem. Biophys. Res. Commun.* **2016**, *470*, 772–782. [CrossRef] [PubMed]

45. Arai, H.; Hirasawa, Y.; Rahman, A.; Kusumawati, I.; Zaini, N.C.; Sato, S.; Aoyama, C.; Takeo, J.; Morita, H. Alstiphyllanines E-H, picraline and ajmaline-type alkaloids from *Alstonia macrophylla* inhibiting sodium glucose cotransporter. *Bioorg. Med. Chem.* **2010**, *18*, 2152–2158. [CrossRef]

46. Qu, Y.; Chan, J.Y.; Wong, C.W.; Cheng, L.; Xu, C.; Leung, A.W.; Lau, C.B. Antidiabetic Effect of Schisandrae Chinensis Fructus Involves Inhibition of the Sodium Glucose Cotransporter. *Drug Dev. Res.* **2015**, *76*, 1–8. [CrossRef]

47. Cao, X.; Zhang, W.; Yan, X.; Huang, Z.; Zhang, Z.; Wang, P.; Shen, J. Modification on the O-glucoside of Sergliflozin-A: A new strategy for SGLT2 inhibitor design. *Bioorg. Med. Chem. Lett.* **2016**, *26*, 2170–2173. [CrossRef]

48. Wilding, J.P.; Charpentier, G.; Hollander, P.; González-Gálvez, G.; Mathieu, C.; Vercruysse, F.; Usiskin, K.; Law, G.; Black, S.; Canovatchel, W.; et al. Efficacy and safety of canagliflozin in patients with type 2 diabetes mellitus inadequately controlled with metformin and sulphonylurea: A randomized trial. *Int. J. Clin. Pract.* **2013**, *67*, 1267–1282. [CrossRef]

49. Rosenstock, J.; Jelaska, A.; Frappin, G.; Salsali, A.; Kim, G.; Woerle, H.J.; Broedl, U.C.; EMPA-REG MDI Trial Investigators. Improved glucose control with weight loss, lower insulin doses, and no increased hypoglycemia with empagliflozin added to titrated multiple daily injections of insulin in obese inadequately controlled type 2 diabetes. *Diabetes Care* **2014**, *37*, 1815–1823. [CrossRef]

50. Jabbour, S.A.; Hardy, E.; Sugg, J.; Parikh, S. Dapagliflozin is effective as add-on therapy to sitagliptin with or without metformin: A 24-Week, multicenter, randomized, double blind, placebo-controlled study. *Diabetes Care* **2014**, *37*, 740–750. [CrossRef]

51. Bode, B.; Stenlof, K.; Harris, S.; Sullivan, D.; Fung, A.; Usiskin, K.; Meininger, G. Long-term efficacy and safety of canagliflozin over 104 weeks in patients aged 55–80 years with type 2 diabetes. *Diabetes Obes. Metab.* **2015**, *17*, 294–303. [CrossRef]

52. Haering, H.U.; Merker, L.; Christiansen, A.V.; Roux, F.; Salsali, A.; Kim, G.; Meinicke, T.; Woerle, H.J.; Broedl, U.C.; EMPA-REG EXTEND™ METSU Investigators. Empagliflozin as add-on to metformin plus sulphonylurea in patients with type 2 diabetes. *Diabetes Res. Clin. Pract.* **2015**, *110*, 82–90. [CrossRef] [PubMed]

53. Ronald, A.R.; Nicolle, L.E.; Stamm, E.; Krieger, J.; Warren, J.; Schaeffer, A.; Naber, K.G.; Hooton, T.M.; Johnson, J.; Chambers, S.; et al. Urinary tract infection in adults: Research priorities and strategies. *Int. J. Antimicrob. Agents* **2001**, *17*, 343–348. [CrossRef]

54. Anderson, G.G.; Palermo, J.J.; Schilling, J.D.; Roth, R.; Heuser, J.; Hultgren, S.J. Intracellular bacterial biofilm-like pods in urinary tract infections. *Science* **2003**, *301*, 105–107. [CrossRef] [PubMed]

55. Mydock-McGrane, L.K.; Hannan, T.J.; Janetka, J.W. Rational Design Strategies for FimH Antagonists: New Drugs on the Horizon for Urinary Tract Infection and Crohn's Disease. *Expert Opin. Drug Discov.* **2017**, *12*, 711–731. [CrossRef]

56. Wellens, A.; Garofalo, C.; Nguyen, H.; Van Gerven, N.; Slättegård, R.; Hernalsteens, J.P.; Wyns, L.; Oscarson, S.; De Greve, H.; Hultgren, S.; et al. Intervening with urinary tract infections using anti-adhesives based on the crystal structure of the FimH-oligomannose-3 complex. *PLoS ONE* **2008**, *3*, e2040. [CrossRef]

57. Hung, C.S.; Bouckaert, J.; Hung, D.; Pinkner, J.; Widberg, C.; DeFusco, A.; Auguste, C.G.; Strouse, R.; Langermann, S.; Waksman, G.; et al. Structural basis of tropism of *Escherichia coli* to the bladder during urinary tract infection. *J. Mol. Microbiol.* **2002**, *44*, 903–915. [CrossRef]

58. Chen, S.L.; Hung, C.S.; Pinkner, J.S.; Walker, J.N.; Cusumano, C.K.; Li, Z.; Bouckaert, J.; Gordon, J.I.; Hultgren, S. Positive selection identifies an *in vivo* role for FimH during urinary tract infection in addition to mannose binding. *Proc. Natl. Acad. Sci. USA* **2009**, *106*, 22439–22444. [CrossRef]

59. Mousavifar, L.; Vergoten, G.; Charron, G.; Roy, R. Comparative study of aryl O-, C-, and S-mannopyranosides as potential adhesion inhibitors toward uropathogenic *E. coli* FimH. *Molecules* **2019**, *24*, 3566. [CrossRef]

60. Mousavifar, L.; Touaibia, M.; Roy, R. Development of mannopyranoside therapeutics against adherent-invasive *Escherichia coli* infections. *Acc. Chem. Res.* **2018**, *51*, 2937–2948. [CrossRef]

61. Sarshar, M.; Behzadi, P.; Ambrosi, C.; Zagaglia, C.; Palamara, A.T.; Scribano, D. FimH and anti-adhesive therapeutics: A disarming strategy against uropathogens. *Antibiotics* **2020**, *9*, 397. [CrossRef]

62. Ribć, R.; Meštrović, T.; Neuberg, M.; Kozina, G. Effective anti-adhesives of uropathogenic *Escherichia coli*. *Acta Pharm.* **2018**, *68*, 1–18. [CrossRef] [PubMed]

63. Rafsanjany, N.; Senker, J.; Brandt, S.; Dobrindt, U.; Hensel, A. In vivo consumption of cranberry exerts ex vivo antiadhesive activity against FimH-Dominated uropathogenic *Escherichia coli*: A combined in vivo, ex vivo, and in vitro study of an extract from vaccinium macrocarpon. *J. Agric. Food Chem.* **2015**, *63*, 8804–8818. [CrossRef] [PubMed]

64. Ansari, M.A.; Shaikh, S.; Shakil, S.; Rizvi, S.M. An enzoinformatics study for prediction of efficacies of three novel penem antibiotics against New Delhi metallo-β-lactamase-1 bacterial enzyme. *Interdiscip. Sci. Comput. Life Sci.* **2014**, *6*, 208–215. [CrossRef] [PubMed]

65. Chaturvedi, N.; Yadav, B.S.; Pandey, P.N.; Tripathi, V. The effect of β-glucan and its potential analog on the structure of Dectin-1 receptor. *J. Mol. Graph. Model* **2017**, *74*, 315–325. [CrossRef] [PubMed]

66. Rizvi, S.M.D.; Shakil, S.; Zeeshan, M.; Khan, M.S.; Shaikh, S.; Biswas, D.; Ahmad, A.; Kamal, M.A. An enzoinformatics study targeting polo-like kinases-1 enzyme: Comparative assessment of anticancer potential of compounds isolated from leaves of Ageratum houstonianum. *Pharmacogn. Mag.* **2014**, *10*, 14–21.

67. Verma, A.; Rizvi, S.M.D.; Shaikh, S.; Ansari, M.A.; Shakil, S.; Ghazal, F.; Siddiqui, M.H.; Haneef, M.; Rehman, A. Compounds isolated from Ageratum houstonianum inhibit the activity of matrix metalloproteinases (MMP-2 and MMP-9): An oncoinformatics study. *Pharmacogn. Mag.* **2014**, *10*, 18–26.
68. Erondu, N.; Desai, M.; Ways, K.; Meininger, G. Diabetic ketoacidosis and related events in the canagliflozin type 2 diabetes clinical program. *Diabetes Care* **2015**, *38*, 1680–1686. [CrossRef]
69. Kuzmanic, A.; Zagrovic, B. Determination of ensemble-average pairwise root mean-square deviation from experimental B-factors. *Biophys. J.* **2010**, *98*, 861–871. [CrossRef]

Article

Doxorubicin Encapsulation in Carbon Nanotubes Having Haeckelite or Stone–Wales Defects as Drug Carriers: A Molecular Dynamics Approach

Leonor Contreras [1,*] , Ignacio Villarroel [2], Camila Torres [2] and Roberto Rozas [1]

[1] Laboratorio de Química Computacional y Propiedad Intelectual, Departamento de Ciencias del Ambiente, Facultad de Química y Biología, Universidad de Santiago de Chile, USACH, Avenida Libertador Bernardo O'Higgins 3363, Casilla 40, Correo 33, Santiago 9170022, Chile; roberto.rozas@usach.cl

[2] Departamento de Computación e Informática, Facultad de Ingeniería, Universidad de Santiago de Chile, USACH, Avenida Ecuador 3659, Santiago 9170022, Chile; gatrevolution@gmail.com (I.V.); ctorres@kpitec.com (C.T.)

* Correspondence: leonor.contreras@usach.cl; Tel.: +56-2-2718-1151

Abstract: Doxorubicin (DOX), a recognized anticancer drug, forms stable associations with carbon nanotubes (CNTs). CNTs when properly functionalized have the ability to anchor directly in cancerous tumors where the release of the drug occurs thanks to the tumor slightly acidic pH. Herein, we study the armchair and zigzag CNTs with Stone–Wales (SW) defects to rank their ability to encapsulate DOX by determining the DOX-CNT binding free energies using the MM/PBSA and MM/GBSA methods implemented in AMBER16. We investigate also the chiral CNTs with haeckelite defects. Each haeckelite defect consists of a pair of square and octagonal rings. The armchair and zigzag CNT with SW defects and chiral nanotubes with haeckelite defects predict DOX-CNT interactions that depend on the length of the nanotube, the number of present defects and nitrogen doping. Chiral nanotubes having two haeckelite defects reveal a clear dependence on the nitrogen content with DOX-CNT interaction forces decreasing in the order 0N > 4N > 8N. These results contribute to a further understanding of drug-nanotube interactions and to the design of new drug delivery systems based on CNTs.

Keywords: carbon nanotubes; Stone–Wales defects; haeckelite defects; doxorubicin encapsulation; drug delivery system; binding free energies; noncovalent interactions; molecular dynamics

Citation: Contreras, L.; Villarroel, I.; Torres, C.; Rozas, R. Doxorubicin Encapsulation in Carbon Nanotubes Having Haeckelite or Stone–Wales Defects as Drug Carriers: A Molecular Dynamics Approach. *Molecules* **2021**, *26*, 1586. https://doi.org/10.3390/molecules26061586

Academic Editors: Marco Tutone and Anna Maria Almerico

Received: 2 March 2021
Accepted: 11 March 2021
Published: 13 March 2021

Publisher's Note: MDPI stays neutral with regard to jurisdictional claims in published maps and institutional affiliations.

1. Introduction

Doxorubicin (DOX), an antineoplasic drug, approved for medical use by the FDA [1,2], has been used for more than 40 years to combat various types of cancers despite the cardiological risks associated with its use. Researchers at the Mayo Clinic [3] have reviewed its mechanism of action and, mainly thanks to the work of Denard et al. and Zhang et al. [4,5], have proposed two alternatives that explain its function. In one of them, DOX would stabilize a complex formed by double-stranded DNA and topoisomerase, which later it would cut both strands of DNA. The alternative is the production of a larger quantity of ceramides which would produce the translocation of a CREB3L-1 protein from the endoplasmic reticulum to the Golgii apparatus. There, some proteases would break the CREB3L-1 protein in such a way that its N-terminal fragment would be translocated to the nucleus where it would direct the DNA transduction, to finally express p21 proteins, which would be those that inhibit tumor growth. Other mechanisms of action of DOX reviewed by Ferreira et al. consider the intercalation of DOX in nuclear DNA and mitochondrial DNA, inhibition of topoisomerase-IIß, and epigenetic factors that involve methylation and deacetylation reactions [6].

In order to increase drug bioavailability and avoid its adverse effects, several types of drug carriers have been used, among which carbon nanotubes (CNTs) have shown

to form stable associations with DOX [7]. However, CNTs also exhibit some toxicity problems [8–11]. Fortunately, CNTs can be functionalized with fragments to increase their water solubility, which prevents them from being deposited as agglomerates in the body. Under certain conditions of concentration and purity CNTs are non-toxic [12]. Additionally, the functionalization of the CNTs facilitates the anchoring of the nanotubes right in the tumor to be attacked. The physicochemical and conductive properties of CNTs give them a versatility of applications in various fields such as electronics, photonics, catalysts, drug carriers, biotechnology, bone tissue engineering and others [13–15].

In the current work we are interested in the ability of CNTs to adsorb drugs and transport them to the target site. It is important to determine the structural parameters that facilitate the DOX-CNT association and to allow the development of strong DOX-CNT intermolecular interaction forces, which help to inhibit the drug from being released before reaching its target. Once there, the acidic pH of the tumor environment causes the release of the drug. Indeed, several studies show experimentally that DOX release is favored at pH of 5 or less [16–19], which is also demonstrated at a theoretical level [20].

Another technologically important characteristic of CNTs is their chirality which significantly modify their conductive properties. For example, armchair (n, n) nanotubes are conductive while zigzag (n, 0) and chiral (n, m) nanotubes are semiconductors except when the value of the difference (n − m) is a multiple of 3, since nanotubes become conductive [21–24]. The diameter of the nanotube has also been shown to be an important structural point since, depending on the diameter, the degree of curvature of the nanotube can be controlled, which seems to be a decisive factor in the stability of the DOX-CNT associations.

As can be deduced, the chirality and the diameter of the nanotubes are properties that determine their behavior and also the presence of structural defects which can change the chirality [25]. The carbon rings of the nanotube that differ from the hexagons are called defects, and depending on their ordering and distribution in the nanotube they confer different properties. For example, the five- and seven-membered rings, when distributed around the perimeter of the nanotube, constitute defects called bumpy. If they are distributed axially in the nanotube they are called zipper defects [26]. Other four-membered rings, along with eight-membered rings, are called haeckelite defects [22,27]. These defects are formed by the addition of two carbon atoms or ad-dimers. However, there are other defects that are formed by rearrangement of their bonds, such as the Stone–Wales defects formed by a pair of rings of five and eight members [28].

The presence of some of these defects has significant technological importance. For example, zigzag nanotubes in the presence of ad-dimers can induce plastic transformations in a material that would otherwise be brittle [29]. Chiral nanotubes stand out, which in the presence of bumpy defects considerably increase their conductivity over armchair and zigzag nanotubes according to DFT studies considering dispersion-corrected B3LYP-D3 functional [30]. Zigzag nanotubes that contain bumpy defects show greater conductive ability and capacity to be reduced [30]; nitrogen doping increases the conductive ability of armchair nanotubes [30]. In addition, armchair nanotubes with bumpy defects notably increase their ability to adsorb hydrogen with very convenient hydrogen adsorption energy values, for their use in the management of clean energy [30].

CNTs form stable associations with doxorubicin [7]. Various theoretical and molecular dynamics studies predict a high capacity of CNTs as carriers [20,31–36] and several molecular dynamics studies of functionalized DOX-CNT systems with various organic groups have contributed to the study of DOX loading and release [37–43]. Although there are several works on the adsorption of DOX in CNT, there are no studies that report on the optimal structure that a nanotube should have to behave as a DOX nanocarrier. The situation is complicated because there are also no experimental data available on the formation and characterization of DOX-CNT complexes and the determination of their DOX-CNT binding energies. For non-functionalized nanotubes, Wang and Xu [20], systematically studied the adsorption and encapsulation of DOX in armchair nanotubes of different diameters,

using the theoretical methods PM6-DH2 and M06-2X in the ONIOM scheme and found that the diameter of the nanotube at which the best DOX encapsulation occurred was 14 Å and corresponded to (10,10) armchair nanotubes. This same behavior was confirmed through a study of molecular dynamics for armchair, zigzag and chiral nanotubes, finding that the strongest DOX-CNT interactions were produced for 14 Å in diameter nanotubes, regardless of chirality [32]. A different situation occurs in the presence of defects in the nanotube. In the case of bumpy defects, a dependence on chirality is observed, since armchair nanotubes with bumpy defects present weaker DOX-CNT interactions than armchair nanotubes without defects. In contrast, bumpy defects in chiral nanotubes favor the DOX-CNT interaction [32].

Several methods of synthesis of CNTs have been reviewed [44]. However, there is a lack of comparative systematic experimental antecedents on this issue, which makes it possible to pose as a valid hypothesis for a theoretical study that the presence of defects in the nanotube, the type and number of defects and their position, also modify the DOX-CNT association properties, along with the chirality and size of the nanotube.

The previous antecedents also lead us to investigate if there is a general trend for some type of nanotube, for example, with chirality or type of defect that accounts for the degree of DOX-CNT association. Our research questions include: (i) how does structural or nitrogen doping defects affect the ability of CNTs as drug carriers, in this case, DOX. (ii) Is the effect produced by the defects the same, regardless of the chirality and the size of the nanotube? (iii) Is there any type of defect that has better characteristics than the others? (iv) How does it affect the number of defects present?

In this work DOX-CNT binding energies are determined for chiral nanotubes with haeckelite defects (with rings of 4 and 8 carbon atoms) and for armchair and zigzag nanotubes with Stone–Wales defects (SW), by means of the MM/PBSA and MM/GBSA methods implemented in the AMBER program of molecular dynamics.

2. Results

Below are the results obtained by molecular dynamics (MD) simulation for DOX encapsulation systems in chiral CNTs with haeckelite (Hk) defects and also in armchair and zigzag CNTs with Stone–Wales (SW) defects. The Hk defects consist of a pair of rings of 4 and 8 members each, while the SW defects are made up of a pair of rings of 5 and 7 members each, as shown in Figure 1.

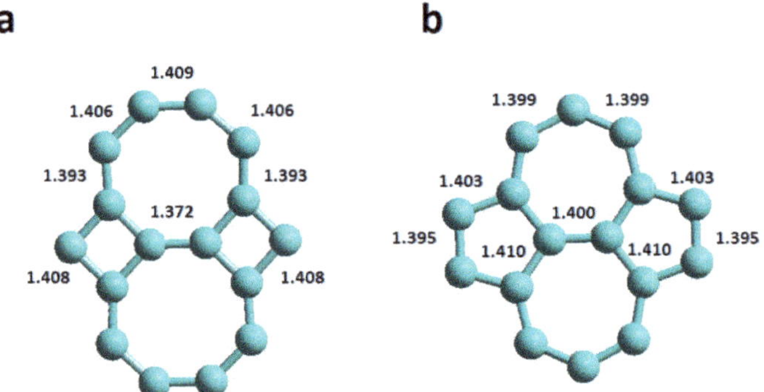

Figure 1. Representation of carbon nanotube structural defects and their C–C bond distances in Å. (**a**) haeckelite defect; (**b**) Stone–Wales defect.

2.1. Chiral Nanotubes with Hk Defects

Chiral nanotubes Ch(13,08) with one Hk defect (named Hk1) and two Hk defects (named Hk2) having 0N, 4N and 8N were studied considering different initial positions of the DOX: in the region of the defect (D), in the regular region of the nanotube (R) (there

are no defects in that area) with the DOX NH_2 group pointing towards the center of the nanotube (v1 orientation) or to the inverse direction (v2 orientation) as shown in Figure 2 for Hk2 chiral nanotubes. Other additional DOX orientations refer to Hk1 nanotubes: when the DOX NH_2 group is oriented in a direction proximal to the defect (p) or is in the direction opposite to the defect (o) as shown in Figure 3.

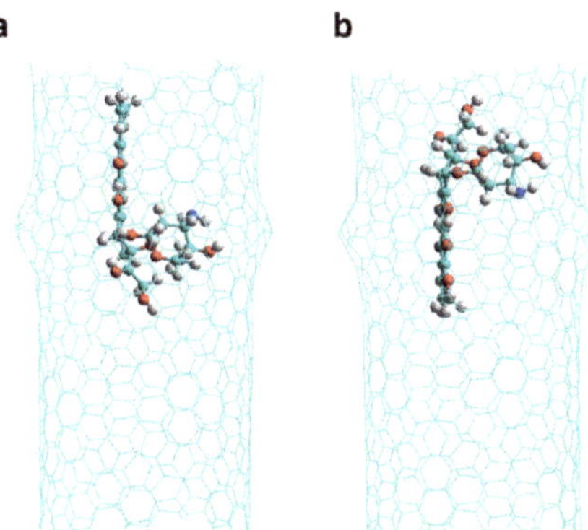

Figure 2. DOX orientation when encapsulated in a chiral nanotube having two haeckelite defects. (**a**) v1 orientation; (**b**) v2 orientation. Lateral views.

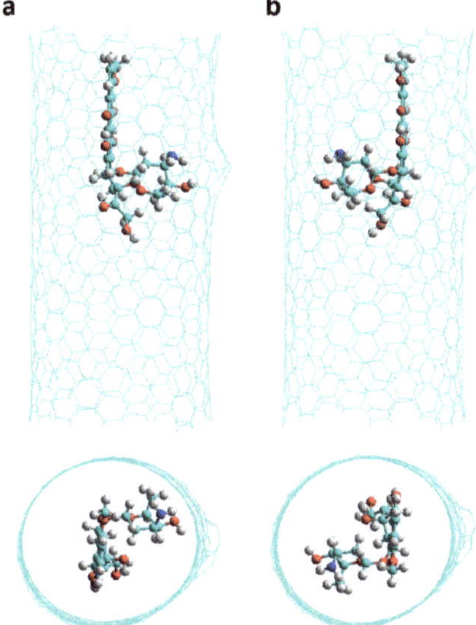

Figure 3. The encapsulated DOX orientations into a chiral nanotube having one haeckelite defect. (**a**) Hk1-DoxDIn.v1p; the DOX NH_2 group is oriented in a direction proximal to the defect; (**b**) Hk1-DoxDIn.v1o; the DOX NH_2 group is oriented in an opposite direction. Lateral and frontal views shown.

2.1.1. Chiral Nanotubes with HK1 Defects

For chiral nanotubes with Hk1 defects (Hk1 chiral nanotubes), the results predict a similar behavior for both undoped, 0N, and nitrogen doped nanotubes having 4N and 8N. In all three cases, the DOX-CNT interaction is favored when the DOX is located in the defect area with the NH_2 group pointing towards the center of the nanotube (v1 orientation)

in the proximal direction close to the defect as shown in Table 1 (runs 2, 10 and 18 for 0N, 4N and 8N, respectively) with Poisson–Boltzman (PB) binding energies of -102, -99 and -102 kcal/mol, respectively. Coherently, most of these systems exhibit equilibrium distances with values between 3.2 and 3.6 Å evidencing stronger DOX-CNT interactions which are favored by the orientation of the DOX that facilitates the NH-π interaction. In Figure 4 the initial conformation of the Ch(13,08)8N-Hk1-DoxDIn.v1p complex is shown together with the final conformations after 2 ns of MD simulation and after 100 ns. It is observed that DOX does not move towards the regular part of the nanotube but interacts with the defect and as a result, in that area, there is a significant deformation of the nanotube. These results obtained for chiral nanotubes show the same behavior as was reported for armchair nanotubes with more favorable DOX-CNT interactions for systems in which DOX is located in the defect region and when it is oriented with its nitrogen atom directed towards the center of the nanotube. However, armchair nanotubes having one haeckelite defect exhibit DOX-CNT binding energies that are more exothermic suggesting stronger DOX-CNT interactions [33].

Table 1. DOX-CNT Poisson–Boltzman (PB) and generalized bond (GB) binding energies (in kcal/mol) for the nitrogen doped and undoped chiral nanotubes (34 Å length) having one haeckelite defect, Hk1, considering encapsulated system. $d_{p\text{-}NT}$ are the distances from the DOX anthraquinonic plane to the nanotube wall; $d_{N\text{-}NT}$ is the distance from the DOX nitrogen atom to the nanotube wall. All distances are expressed in Å.

Run	Type [1]	PB	GB	$d_{p\text{-}NT}$	$d_{N\text{-}NT}$
1	Ch(13,08)0N-HK1-DoxDIn.v1o	-78	-77	3.7	3.7
2	Ch(13,08)0N-HK1-DoxDIn.v1p	-102	-104	3.6	3.5
3	Ch(13,08)0N-HK1-DoxDIn.v2o	-79	-80	3.6	3.9
4	Ch(13,08)0N-HK1-DoxDIn.v2p	-85	-87	3.8	3.6
5	Ch(13,08)0N-HK1-DoxRIn.v1o	-78	-77	3.4	3.1
6	Ch(13,08)0N-HK1-DoxRIn.v1p	-80	-80	3.4	3.6
7	Ch(13,08)0N-HK1-DoxRIn.v2o	-79	-78	3.8	4.9
8	Ch(13,08)0N-HK1-DoxRIn.v2p	-80	-80	3.5	3.2
9	Ch(13,08)4N-HK1-DoxDIn.v1o	-77	-76	3.5	3.4
10	Ch(13,08)4N-HK1-DoxDIn.v1p	-99	-102	3.4	3.3
11	Ch(13,08)4N-HK1-DoxDIn.v2o	-78	-80	3.8	3.9
12	Ch(13,08)4N-HK1-DoxDIn.v2p	-79	-80	3.4	3.0
13	Ch(13,08)4N-HK1-DoxRIn.v1o	-77	-77	3.6	4.0
14	Ch(13,08)4N-HK1-DoxRIn.v1p	-79	-81	3.6	3.5
15	Ch(13,08)4N-HK1-DoxRIn.v2o	-78	-79	4.2	4.6
16	Ch(13,08)4N-HK1-DoxRIn.v2p	-79	-80	3.5	3.2
17	Ch(13,08)8N-HK1-DoxDIn.v1o	-78	-79	3.6	3.5
18	Ch(13,08)8N-HK1-DoxDIn.v1p	-102	-104	3.4	3.2
19	Ch(13,08)8N-HK1-DoxDIn.v2o	-78	-79	3.8	3.6
20	Ch(13,08)8N-HK1-DoxDIn.v2p	-78	-79	3.4	3.4
21	Ch(13,08)8N-HK1-DoxRIn.v1o	-79	-80	3.4	3.4
22	Ch(13,08)8N-HK1-DoxRIn.v1p	-81	-81	3.6	3.4
23	Ch(13,08)8N-HK1-DoxRIn.v2o	-73	-76	3.5	3.2
24	Ch(13,08)8N-HK1-DoxRIn.v2p	-81	-82	3.5	3.1

[1] DoxD means DOX position is in the defect zone; DoxR is for the DOX in the regular part of the nanotube; v1 means the nitrogen atom of the DOX is oriented towards the center of the tube; v2 indicates the inverse orientation; v1p indicates that the DOX nitrogen atom is located in a proximal space regarding the defect meanwhile v1o is used to indicate that the DOX nitrogen atom is located in an opposite space regarding the defect as shown in Figure 3.

DOX-CNT systems, doped with 4N and containing Hk1 defect exhibit quite similar DOX-CNT PB binding energy values between -79 and -77 kcal/mol, probably accounting for an electronic distribution that interacts with the drug in a similar way regardless of DOX position and orientation. This could be due to the arrangement of nitrogen atoms which are part of two pyrimidine rings placed opposite each other on the walls of the nanotube.

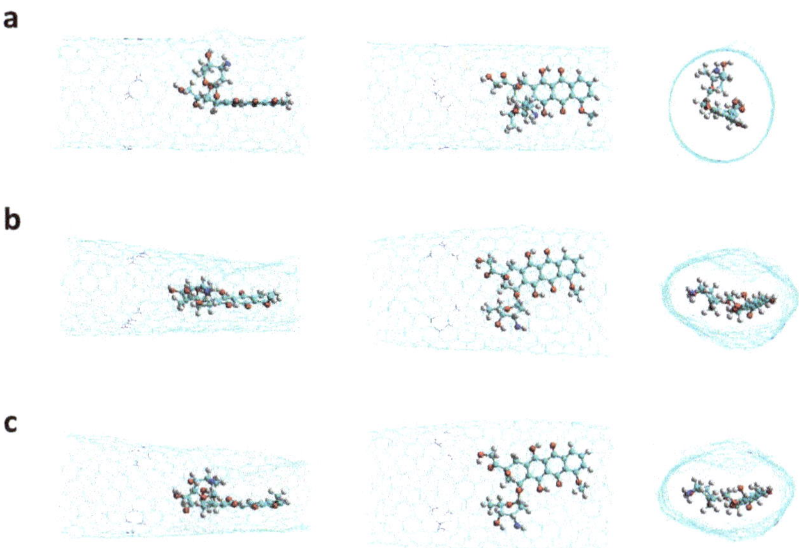

Figure 4. Representation of encapsulated DOX-CNT complex for nitrogen-doped chiral nanotube with one haeckelite defect, Ch(13,08)8N-Hk1-DoxDIn.v1p, at different simulation times. (**a**) 0 ns; (**b**) 2 ns; (**c**) 100 ns. Two laterals and one frontal view are shown.

2.1.2. Chiral Nanotubes with Hk2 Defects

Carbon nanotube diameter effect. The nitrogen doped and undoped Hk2 chiral (13,08) CNTs of 14 Å diameter and 19 Å length showed better DOX-CNT PB and GB binding energies than the corresponding Hk2 chiral (13,10) CNTs of 16 Å diameter (calculated with RESP charges for DOX) as shown in Figure 5. This was an expected result considering PM6-DH2 and M06-2X theoretical calculations in the scheme of ONIOM for the DOX encapsulation in armchair CNTs without defects [20], and also molecular dynamics studies on armchair, zigzag and chiral nanotubes with reported values of 14 Å as an optimal value of the nanotube diameter for encapsulating the DOX [32]. A diameter of 14 Å allows the proper curvature of the nanotube for the formation of different attractive and complementary non-covalent interactions between the nanotube and the DOX that stabilize the entire system which is also fulfilled in this case of nanotubes containing two haeckelite defects in their structure.

Hk2 chiral nanotubes of diameter 14 Å exhibit less favorable DOX-CNT PB binding energies for nitrogen doped nanotubes in the order 0N > 4N > 8N with values of -101, -97 and -74 kcal/mol, respectively (see Figure 5), which predicts stronger DOX-CNT interactions for undoped Hk2 chiral (13,08) nanotubes. These values were calculated using RESP charges for DOX. RESP (restrained electrostatic potential) approach to derive partial charges has been reported as having a lower average error than MM3 and CHARMm in a study considering 55 molecules [45]. Mean SD for the PB binding energy values between 2.6 and 3.2 kcal/mol and between 2.6 and 3.1 kcal/mol for the GB binding energy values were observed, being in all cases less than 4.2%.

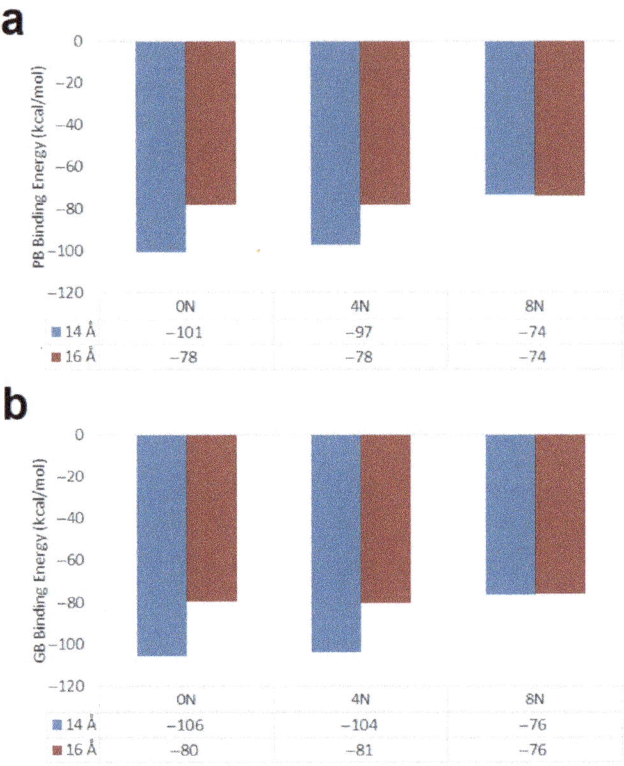

Figure 5. DOX-CNT binding energies for nitrogen-doped and undoped chiral nanotubes of 19 Å length with different diameter, having Hk2 defects and calculated with RESP charges for DOX. (**a**) PB binding energies; (**b**) GB binding energies. Blue is for 14 Å diameter and red is for 16 Å diameter.

Carbon nanotube length and DOX pose effects. Longer Hk2 chiral nanotubes (34 Å length) exhibit more exothermic DOX-CNT PB binding energy values than Hk2 chiral nanotubes of 19 Å length, for both nitrogen doped and undoped nanotubes. For the undoped nanotubes (0N) and those doped with 4N, a clear preference of the DOX for the nanotube defect zone is shown with DOX-CNT PB binding energy values of −109 and −104 kcal/mol, respectively (runs 1 and 5, Table 2). When the DOX is in the regular zone of the nanotube, no significant differences are observed between the v1 or v2 orientations of the DOX. However, in cases where the DOX is initially located in the defect zone, stronger interactions are predicted for v1 DOX orientation for both undoped and 4N doped systems as shown in Table 2 (runs 1 vs. 2 and 5 vs. 6).

Table 2. Poisson–Boltzman (PB) and generalized bond (GB) binding energies (in kcal/mol) for the nitrogen doped and undoped chiral nanotubes (34 Å length) having two haeckelite defects, Hk2, considering encapsulated system. $d_{p\text{-NT}}$ are the distances from the DOX anthraquinonic plane to the nanotube wall; $d_{N\text{-NT}}$ is the distance from the DOX nitrogen atom to the nanotube wall. All distances are expressed in Å.

Run	Type [1]	PB	GB	$d_{p\text{-NT}}$	$d_{N\text{-NT}}$
1	Ch(13,08)0N-HK2-DoxDin.v1	−109	−112	3.4	3.4
2	Ch(13,08)0N-HK2-DoxDin.v2	−82	−83	3.7	3.4
3	Ch(13,08)0N-HK2-DoxRin.v1	−80	−79	3.3	3.4
4	Ch(13,08)0N-HK2-DoxRin.v2	−79	−79	3.2	4.8
5	Ch(13,08)4N-HK2-DoxDin.v1	−104	−108	3.5	4.3
6	Ch(13,08)4N-HK2-DoxDin.v2	−82	−83	3.8	3.5
7	Ch(13,08)4N-HK2-DoxRin.v1	−79	−78	4.1	3.4
8	Ch(13,08)4N-HK2-DoxRin.v2	−80	−80	3.3	3.5
9	Ch(13,08)8N-HK2-DoxDin.v1	−80	−78	3.4	3.3
10	Ch(13,08)8N-HK2-DoxDin.v2	−87	−89	3.6	3.2
11	Ch(13,08)8N-HK2-DoxRin.v1	−80	−80	3.6	3.3
12	Ch(13,08)8N-HK2-DoxRin.v2	−80	−79	3.8	3.7

[1] DoxD means DOX position is in the defect zone; DoxR is for the DOX in the regular part of the nanotube; v1 means the nitrogen atom of the DOX is pointing towards the center of the tube; v2 indicates the inverse orientation.

The best system for DOX encapsulation in Hk2 chiral nanotubes is therefore Ch(13,08)0N-HK2-DoxDIn.v1 (run 1, Table 2) with the DOX in the defect region and v1 orientation (with the nitrogen pointing towards the center of the nanotube). In this conformation, the formation of the non-covalent DOX-CNT interactions is facilitated, which are mainly constituted by π–π stacking interactions complemented by NH-π, CH-π, C=O-π and van der Waals interactions [20,31,46]. Figure 6 shows the non-covalent interactions for the most favorable case with the DOX in the area of the nanotube defect (a large green surface is observed) and as a comparison, the same nanotube with the DOX encapsulated in the regular area (less green regions and more red regions are observed), with PB DOX-CNT binding energy values of −109 and −80 kcal/mol, respectively (runs 1 and 3, Table 2). A program specially developed for the visualization of non-covalent interactions (NCI) was used [47].

Under similar MD simulation conditions but considering RESP charges for DOX, Hk2 chiral (13,08) nanotubes (−101 kcal/mol, Figure 5) predict stronger DOX-CNT interactions than reported Hk2 armchair (10,10) nanotube (−83.4 kcal/mol) with the encapsulated DOX located in the defect zone in v1 orientation [33]. Hk2 chiral nanotubes predict stronger DOX-CNT interactions than reported Hk2 zigzag (18,0) nanotubes which exhibit values of DOX-CNT PB binding energies of −78.7 kcal/mol for undoped nanotubes [33]. The three types of CNTs in comparison have diameters of 14 Å. In this way, in terms of chirality and according to the indicated results, Hk2 nanotubes exhibit the following order of ability to encapsulate the DOX: chiral > armchair > zigzag, despite chiral nanotubes are shorter than zigzag and armchair nanotubes (19 vs. 34 Å length). The enhanced ability of chiral nanotubes with respect to other nanotubes to encapsulate DOX was reported also for perfect CNTs through MD simulation studies considering RESP charges for DOX [32].

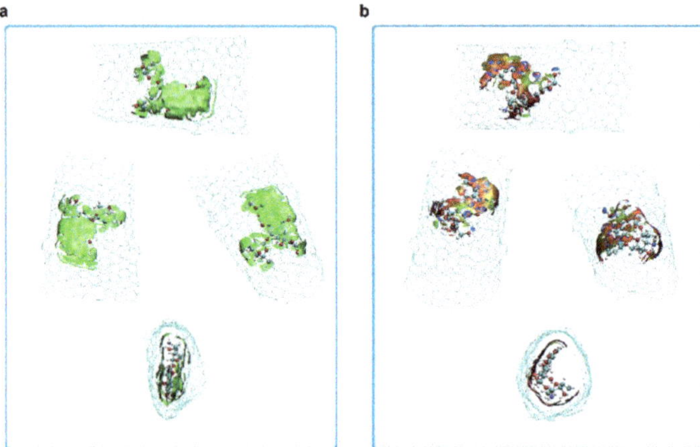

Figure 6. Representation of non-covalent interactions (NCI) for DOX encapsulation inside chiral carbon nanotubes having two haeckelite defects. (**a**) DOX D position; (**b**) DOX R position. Cutplot 0.01 0.1. Blue surfaces indicate strong interactions; green means weak interactions; red means repulsion [47]. Different comparative lateral and frontal views shown.

2.2. Nanotubes with Stone–Wales Defects

The encapsulation of the DOX was studied in armchair and zigzag nanotubes that have one and two Stone–Wales defects (SW1 and SW2, respectively) as shown in Figure 7.

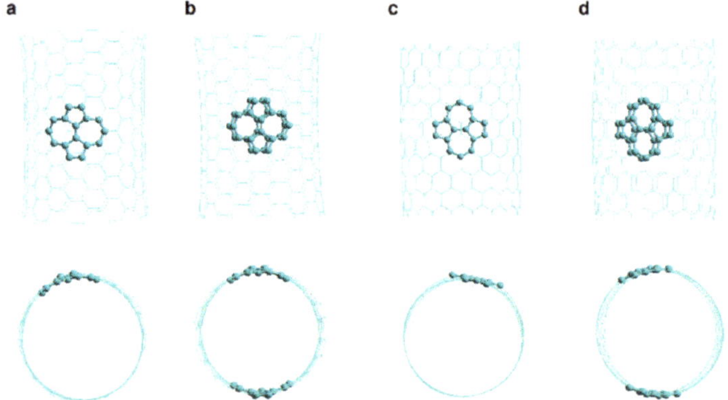

Figure 7. Representation of carbon nanotubes having one and two Stone–Wales defects. (**a**) SW1 armchair; (**b**) SW2 armchair; (**c**) SW1 zigzag; (**d**) SW2 zigzag. Side and front views shown.

2.2.1. SW1 and SW2 Armchair Nanotubes

Armchair (10,10) nanotubes of 20 Å and 34 Å in length having Stone–Wales defects were studied, which showed different behaviors in DOX encapsulation. The shorter nanotubes (20 Å length) exhibit a significant stronger interaction with the DOX in two situations: (i) when they are of the SW2 type (with two defects, doped and undoped) in comparison with SW1 nanotubes and (ii) when they are doped with 4N (SW1 and SW2) as is clearly depicted in Figure 8a. Nitrogen doped SW1 and SW2 armchair (10,10) nanotubes of 20 Å length predict stronger interactions with the DOX than corresponding longer SW1 and SW2 nanotubes of 34 Å in length as shown in Figure 8 particularly for 4N doped nanotubes.

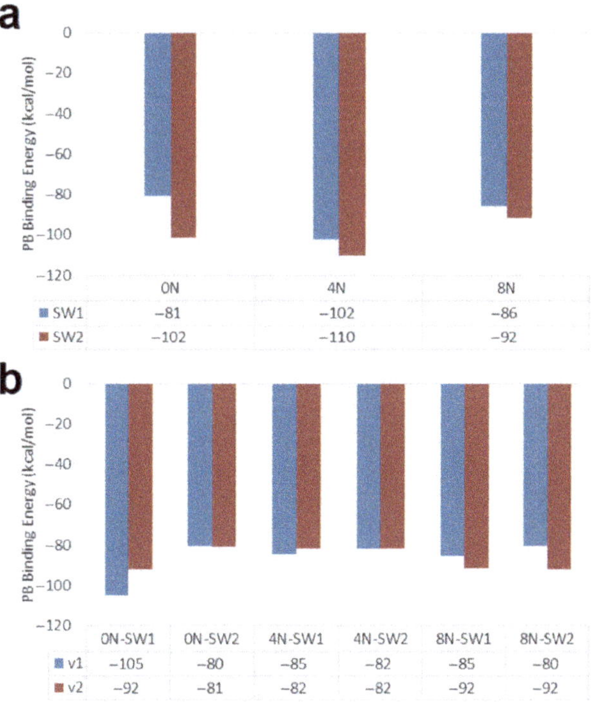

Figure 8. DOX-CNT PB binding energies for nitrogen-doped and undoped armchair (10,10) nanotubes of different length having one and two Stone–Wales defects. (**a**) 20 Å length; (**b**) 34 Å length.

In contrast, longer SW1 armchair nanotubes predict somewhat stronger interactions than longer SW2 armchair nanotubes the difference being more significant for the undoped SW1 and SW2 nanotubes with PB DOX-CNT binding energy values of −105 and −80 kcal/mol for, respectively, with DOX v1 orientation, and for DOX v2 orientation −92 vs. −81 kcal/mol, respectively, as shown in Figure 8b.

The most exothermic PB DOX-CNT binding energy with a value of −110 kcal/mol correspond to the shorter 4N-doped SW2 armchair nanotube which predicts the stronger DOX-CNT interactions. In the shorter SW2 armchair nanotubes the DOX is symmetrically located and can interact with both of the two defects which favors DOX-CNT interactions. Meanwhile for shorter SW1 armchair nanotubes the DOX interacts with just one defect only. In contrast, the less exothermic PB DOX-CNT binding energies correspond to longer SW1 and SW2 armchair nanotubes with values between −92 and −80 kcal/mol. The only exception is the longer SW1 armchair nanotube where the DOX is in the v1 orientation showing a PB binding energy of −105 kcal/mol. Figure S1 (Supplementary Materials) clearly shows the differences in relative DOX-CNT binding energies. For short nanotubes (20 Å long), SW2 exhibits more exothermic DOX-CNT binding energies than SW1, with 4N doping predicting the strongest DOX-CNT interactions. Within the long nanotubes (34 Å long) the most exothermic DOX-CNT binding energies correspond to the undoped SW1 nanotubes with v1 orientation. In Figure 9 its initial structure is shown and also at 2 ns and 74 ns of simulation where a probable double π–π interaction of the DOX with the two opposite walls of the nanotube is appreciated which generates a significant deformation of the nanotube in addition to its NH-π interaction with the DOX amino group that helps stabilize the system. In the longer nanotubes, the DOX interacts with the regular part of the nanotube also. Apparently the interactions DOX-Stone–Wales defects are stronger than the interactions DOX-regular CNTs.

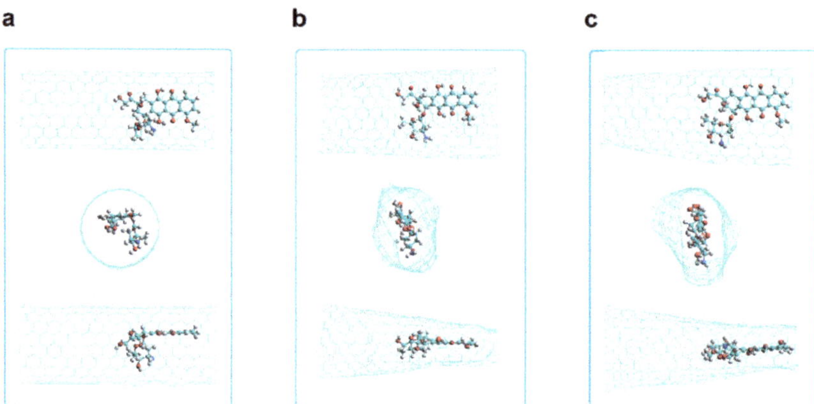

Figure 9. Representation of encapsulated DOX-CNT complex for undoped *armchair* nanotube with one Stone–Wales defect, A(10,10)0N-SW1-DoxDIn.v1 at different molecular dynamics simulation times. (**a**) 0 ns; (**b**) 2 ns; (**c**) 74 ns. Two lateral and frontal views shown.

2.2.2. SW1 and SW2 Zigzag Nanotubes

Zigzag (18,0) nanotubes of 20 Å in length and 14 Å diameter having one and two Stone–Wales defects were studied. In Figure 10 the PB and GB DOX-CNT binding energies are shown for undoped nanotubes and for nanotubes doped with 4 and 8 nitrogen atoms. It is observed that both types of binding energies show the same tendency (as was also observed in Figure 5 for chiral nanotubes) and that undoped zigzag nanotubes predict stronger DOX-CNT interactions than doped ones, both with one or two SW defects.

Zigzag nanotubes having one SW defect (SW1) exhibit stronger DOX-CNT interactions than those having two SW defects (SW2). This behavior is more significant for nanotubes doped with 4 nitrogen atoms. Structurally, the presence of 4N means that there are two pyrimidine rings in the nanotube. Zigzag nanotubes with a SW1 defect doped with 8 nitrogen atoms (have four pyrimidine rings in the nanotube) show weaker DOX-CNT interactions than in the case of nanotubes doped with 4N. Undoped SW1 and SW2 nanotubes show similar DOX-CNT PB binding energies of −102 and −100 kcal/mol, respectively. Apparently DOX accommodates better in a space with SW defects but free from the influence of the nitrogen electron cloud. No great difference is observed between the DOX-CNT binding energies obtained for undoped SW1 and SW2 nanotubes and those for SW1 nanotubes doped with 4N with a PB binding energy of −101 kcal/mol; but for 4N-doped SW2 nanotubes, the DOX-CNT PB binding energy decreases significantly to −78 kcal/mol (probably DOX is prevented from accommodating in a narrower space). The same effect is observed when SW1 zigzag nanotubes doped with 4N are compared with those doped with 8N (−101 kcal/mol vs. −82 kcal/mol). Mean SD for the PB binding energy values ranged between 2.7 and 4.0 kcal/mol and between 2.9 and 4.1 kcal/mol for the PB and GB binding energy values, respectively, being in all cases less than 4%. Figure S2 (Supplementary Materials) clearly shows the DOX-CNT relative binding energies with respect to the non-doped nanotubes. SW1 nanotubes have more exothermic DOX-CNT binding energies than SW2. Furthermore, the non-doped SW1 nanotubes exhibit DOX-CNT binding energy values not very distant from the 4N-doped nanotubes, being the 8N-doped nanotubes the ones with the least favorable binding energies.

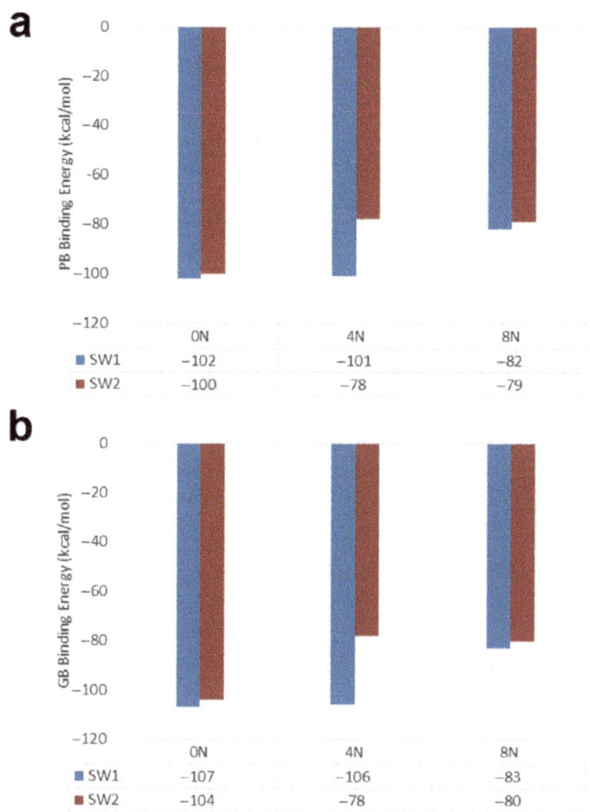

Figure 10. Representation of the PB and GB DOX-CNT binding energies for nitrogen-doped and undoped zigzag (18,0) nanotubes having one and two Stone–Wales defects. (**a**) PB binding energy; (**b**) GB binding energy.

The equilibrium distances $d_{p\text{-}NT}$ between the DOX anthraquinonic planar part and the nanotube wall ranged between 3.3 and 3.7 Å and the equilibrium distances $d'_{p\text{-}NT}$ between the same DOX planar part and the nanotube wall in the opposite direction it ranged between 3.8 and 3.9 Å for cases having DOX-CNT PB binding energy between -100 and -102 kcal/mol. On the other hand, for cases with PB binding energy between -78 and -82 kcal/mol, the distances $d'_{p\text{-}NT}$ ranged between 7.6 and 8.2 Å. So, for binding energies that predict strongest DOX-CNT interactions, a deformation of the nanotube is observed, which can also be seen in Figures 4 and 9 for chiral and armchair nanotubes, respectively, probably due principally to a double $\pi-\pi$ stacking between the DOX anthraquinonic planar rings and both of the opposite CNT walls, and DOX-CNT van der Waals interactions. The equilibrium $d_{N\text{-}NT}$ distances between the DOX nitrogen atom and the nanotube inner wall exhibit values between 3.3 and 4.0 Å regardless of the value of the binding energies.

3. Discussion

It is interesting to find out how the structural parameters of nanotubes affect the relative DOX-CNT binding energies which could allow us to infer about the DOX-CNT interactions for nanotubes of different types. The aim is to predict nanotube structures that can develop more exothermic binding energies; that is, that produce more favorable DOX-CNT interactions, without fear that the desorption of the drug be difficult since the

acidic pH existing in the tumor environment facilitates protonation and release of the drug, which has been verified in different systems [14–17,38].

In the present work, it was found that a diameter of 14 Å favors the DOX-CNT interaction for Hk2 chiral nanotubes, in agreement with works reported for other defect-free nanotubes [20,32]. Furthermore, it was found that the strength of the DOX-CNT interaction decreases as the number of doping nitrogen atoms increases (0N > 4N > 8N). This relative trend was found both for long Hk2 chiral nanotubes (33 Å length) using Mulliken charges for DOX, as well as for short Hk2 chiral nanotubes (19 Å length) using RESP charges for DOX. However, using similar simulation conditions, Hk1 and Hk2 armchair and zigzag nanotubes (33 Å length) showed that nitrogen-doped systems had more favorable binding energies than non-doped ones [33]. Chirality shows to be an important parameter that controls the effect of the presence of nitrogen in the nanotube, modifying the distribution of electron density and therefore the DOX-CNT interactions in such a way that the presence of nitrogen in the chiral nanotubes destabilizes the association DOX-CNT unlike of what happens in armchair and zigzag nanotubes.

The initial poses of the DOX in the CNT proved to be important, finding better interactions for chiral nanotubes when the DOX is located in the defect part, oriented in such a way that its nitrogen atom points towards the center of the nanotube and is close to the defect. Apparently, this conformation favors π–π, van der Waals interactions and also electrostatic interactions of the NH-π, OH-π, C=O-π type as was observed in Figure 6. These DOX-CNT interactions in the cases of more electronegative binding energies translate into a significant deformation of the nanotube, as observed in Figures 4 and 9 with PB binding energies of -102 and -105 kcal/mol, for chiral and armchair nanotubes, respectively, which could be explained by the π–π interaction of the flat anthraquinonic system of the DOX with both opposite walls of the nanotube. The deformation of the nanotube has been observed in other MD simulation works [32,33], and also in works carried out using the PM6-DH2 and M06-2X methods [20].

On the other hand, the short SW2 armchair nanotubes (20 Å length) showed better DOX-CNT binding energies than the long SW2 armchair nanotubes (34 Å length). Although the short SW2 armchair nanotubes showed more exothermic DOX-CNT binding energies than the respective short SW1 armchair nanotubes, this trend did not hold for the long armchair nanotubes.

In contrast, the short SW1 zigzag nanotubes showed more exothermic DOX-CNT binding energies than the short SW2 zigzag nanotubes. For the SW1 and SW2 zigzag nanotubes it was found that the strength of the DOX-CNT interactions decreased in the presence of nitrogen in the order 0N > 4N > 8N, in a similar way to that shown by the Hk2 chiral nanotubes. Nitrogen-doped SW1 armchair and zigzag nanotubes showed stronger DOX-CNT interactions than the respective Hk2 chiral nanotubes.

There is a lack of experimental studies on the determination of DOX-CNT binding energies. Only one estimation is known in an aqueous system, of about 11.5 kcal/mol for DOX-CNT complexes 100 nm long and 2 to 3 nm in diameter [19]. However, the formation of stable DOX-CNT conjugates has been determined experimentally, verified by atomic force microscopy (AFM) and scanning tunneling microscopy (STM) images, using single-walled CNTs (SWCNTs) with diameter 1–1.5 nm and a few hundred nanometers long, purchased from ILJIN Co., Inc., Korea. [7]. In none of these cases is the chirality of the nanotube or the presence of defects specified.

It is not easy to confront theoretical studies that depend on a variety of factors, such as, in the field of molecular dynamics, the assignment of the type of atom involved in the consideration of the bonded and non-bonded parameters, or the definition of dihedrals and other interpretations of the description of the force field used [48].

The method used in the present work is very useful to obtain the DOX-CNT bonding energy values since it is comparatively fast and reproducible. Its calculation is based on the MM/PBSA and MM/GBSA approaches starting from the equilibrium energy values obtained through molecular dynamics simulation. Although the calculation does not

consider the translational entropy (huge computational cost), it has been shown that it does provide adequate values of the relative binding energies, which have been experimentally validated in biological systems [49,50] and it has recently been reported that when the MM/PBSA and MM/GBSA are used together with empirical corrections, they allow better experimental correlations [51].

DOX-CNT binding energies for the encapsulation of DOX on the walls of armchair nanotubes without defects at the PM6-DH2 level, in aqueous solution, showed values between -51.6 and -53.7 kcal/mol as a function of diameter [20].

As a reference, the nitrosamine adsorption free energy on open-ended Stone–Wales defective (5,5) armchair nanotube calculated by DFT methods at the B3LYP/6-31G(d) level of theory showed a value of -137.14 kcal/mol [52]. DOX-CNT binding free energy values obtained by the MM/PBSA and MM/GBSA methods for (10,10) armchair nanotubes without defects were -43 kcal/mol for DOX adsorption and fluctuated between -109 and -90 kcal/mol for DOX encapsulation [32]. For (10,10) armchair nanotubes with bumpy defects they ranged between -96 and -83 kcal/mol. (10,10) armchair nanotubes with haeckelite defects ranged from -104 to -80 kcal/mol depending on the number of defects and the presence of nitrogen [33].

Interestingly, although the DOX-CNT binding free energy values previously reported by us using the current methodology [32], are far from the values calculated by Wang and Xu [20] by means of PM6-DH2 and M06-2X methods, in both works, after a systematic study, it was found that the best nanotube diameter for encapsulate the DOX is 14 Å. This fact, together with the deformation of the nanotube observed with both methods means that the behavior and ability ranking of the nanotubes to associate with DOX coincides in both works, so it could be considered as an indirect validation of the use of the MM/PBSA and MM/GBSA methods to predict the DOX-CNT binding free energies for different nanotube structures.

The binding energy calculations are very useful to study molecular interactions, complex stability, getting information necessary for drug design, carrier design, inhibitor design, etc. The methods that allow obtaining the most accurate values are expensive in computational time and resources. Semi-empirical methods have been found to allow a good level of accuracy when compared to experimental measurements, and to be relatively fast and of low computational cost. The prediction of binding energies through the MM/PBSA approach increases in efficiency and accuracy when using the partial charges of the ligand determined by semi-empirical methods. A study with 50 protein-ligand systems revealed an excellent performance of the PM7 and AM1 systems, comparable to B3LYP which requires a huge computational cost, reaching the conclusion that the semi-empirical methods AM1 or PM7 provide partial charges of the proteins that help to improve the prediction of binding energies through methods such as solvated interaction energy (SIE) and MM/PBSA in a highly efficient and accurate way [53]. These results validate in some way the use of AM1 in this work.

To summarize, the structural parameters that dominate the behavior of defective nanotubes related to DOX encapsulation are interdependent and in order of importance, from what is observed in this work, the following can be mentioned: chirality, diameter and length of the nanotube; presence of nitrogen atoms as dopants; number of defects present and type of defects. Additionally, the initial pose of the DOX in the nanotube also affects the intermolecular attractive forces. We have found that the initial pose of DOX affects the DOX-CNT binding energies and we have been able to clarify that DOX-CNT interactions are favored when the DOX amino group points towards the center of the nanotube with an orientation close to the defects. It would be interesting in a next stage to do a simulation that considers different conformations of DOX.

As future work, we are doing research on chiral nanotubes with Stone–Wales defects. However, based on the known simulation studies, which are not directly comparable, we believe a global systematic approach is needed. This new comparable approach should consider the best values reported for each system (and also some of the values found as

unfavorable, for comparison) to perform the new simulations. To have a valid ranking of CNTs activity, as drug carriers, with or without defects, at least two conditions should be considered for the new simulations: (i) comparable structural parameters (for example, same diameter, length and number of defects, including nitrogen); and (ii) comparable simulation conditions (same version of the package of programs, same force field, type of charges, type of solvation, among others). Due to the complexity and the high number of variables, we believe that the use of Artificial Intelligence tools will be very useful to improve the study of the design of new drug carriers and also for other related applications. Anyway, refinement of the results for the best systems found could be done using simulation methods and conditions that guarantee greater accuracy.

4. Simulation Methods

The nanotubes were prepared as single-walled open nanotubes finished in hydrogen with the help of HyperTube [54] and Hyperchem [55] and optimized to the Austin Model One (AM1) level. Chiral CNTs with one and two haeckelite defects (Hk1 and Hk2, respectively) and also armchair and zigzag CNTs with one and two Stone–Wales defects (SW1 and SW2, respectively) were prepared as shown in Figure 7. Molecular dynamics simulations were performed with AMBER16 [56,57]. A program made at home was use for the preparation of the necessary files and instructions to run the MD program [32]. The combined GAFF and ff99SB force fields were used. The DOX was optimized at the level of HF/6-31G* for some specific cases, just to get the restrained electrostatic potential (RESP) partial charges by using the antechamber AMBER program [45,58]. For the rest of the structures the AM1-Mulliken charges were used. All the DOX-CNT complexes were neutral systems, solvated in an explicit solvent, in a 10 Å octahedron water box using bondi radii under periodic boundary conditions. TIP3P was used as water model.

Simulations were run after some steps of minimization (1000 stages) and heating (from 0 to 300 K), both at constant volume, density balance (50 ps) and equilibrium (500 ps) both of them at constant pressure. Then, the production step consisting of six independent short stages (250 ps each) at constant pressure was carried out. This procedure of using several independent short simulations instead of a single long simulation has being recognized as efficient and accurate [50,59]. In effect, as it is shown in Figure 4, a simulation performed for 100 ns (using several short simulation) revealed similar energy and geometry results than the simulation procedure with short stages as it was indicated above. The drug-nanotube binding free energies were determined through the MM/PBSA and MM/GBSA methods implemented in AMBER [56]. These methods were applied on an ensemble of 200 uncorrelated snapshots collected from the equilibrated molecular dynamics simulation for calculating the Gibbs free energy difference of the solvated bound ($G_{\text{CNT-DOX}}$) and unbound states of the drug (G_{DOX}) and nanotube (G_{CNT}) molecules Equation (1).

$$\Delta G = G_{\text{CNT-DOX}} - G_{\text{CNT}} - G_{\text{DOX}} \tag{1}$$

Each of these terms were obtained according to Equation (2):

$$G = E_{vdW} + E_{bond} + E_{el} + E_{pol} + E_{np} - TS \tag{2}$$

where E_{vdW} (van der Waals), E_{bond} (bond, angle, dihedral) and E_{el} (electrostatic) are the standard molecular mechanics (MM) energy terms; E_{pol} (polar term) is calculated by solving the Poisson–Boltzmann (PB) or the generalized born (GB) equation; E_{np} (non-polar term) is estimated from a linear relation with the solvent accessible surface area (SA); T is the absolute temperature and S is the entropy term (estimated through a normal-mode analysis of the vibrational frequencies). The fact that the binding free energies are calculated without considering the translational entropy must be borne in mind when interpreting the results. To obtain a good ranking of nanotube ability for encapsulating DOX based on the DOX-CNT binding energy values this is a convenient method in terms of computational resources, but not for obtaining absolute DOX-CNT binding energy values.

5. Conclusions

The DOX-CNT binding energies calculated in this work allow us to establish interesting trends such as:

- When using RESP charges, DOX encapsulation inside Hk2 nanotubes of 14 Å diameter, predicts that the DOX-CNT attractive forces decrease in the following order of nanotube chirality: chiral > armchair > zigzag in agreement with that reported in a MD study under similar conditions, for defect-free nanotubes.
- Hk2 chiral nanotubes (short and long) favor DOX-CNT interactions which decrease in the presence of nitrogen as a dopant in the order 0N > 4N > 8N.
- Short armchair nanotubes promote more favorable DOX-CNT interactions if they are SW2; but long armchair nanotubes favor the DOX-CNT interactions if they are SW1. When the DOX is located close to the defect, a better interaction occurs than when it is located in the defect-free zone of the nanotube. The interaction is also improved if the DOX nitrogen atom is close to the defect.
- Short SW1 zigzag nanotubes favor DOX-CNT interactions which decrease in the presence of nitrogen as a dopant in the order 0N > 4N > 8N, like the Hk2 chiral nanotubes. SW1 zigzag nanotubes exhibit more exothermic DOX-CNT binding energies than SW2 zigzag nanotubes. In both cases, the undoped SW defective zigzag nanotubes are predicted to exhibit stronger DOX-CNT interactions than corresponding nitrogen-doped nanotubes.

Supplementary Materials: The following are available online, Figure S1: DOX-CNT PB relative binding energies for nitrogen-doped and undoped (10,10) armchair nanotubes of different length having one and two Stone–Wales defects. (a) 20 Å length; reference: 4N doped SW2 nanotube with −110 kcal/mol PB DOX-CNT binding energy; (b) 34 Å length; reference: 0N SW1 nanotube with −105 kcal/mol PB DOX-CNT binding energy. Figure S2: Representation of the PB and GB DOX-CNT relative binding energies for nitrogen-doped and undoped (18,0) zigzag nanotubes having one and two Stone–Wales defects. (a) PB binding energy. Reference: 0N SW1 nanotube with −102 kcal/mol PB DOX-CNT binding energy; (b) GB binding energy. Reference: 0N SW1 nanotube with −107 kcal/mol GB DOX-CNT binding energy.

Author Contributions: Conceptualization, L.C. and R.R.; funding acquisition, L.C.; investigation, L.C. and R.R.; methodology, L.C. and C.T.; project administration, L.C.; resources, C.T.; software, I.V. and C.T.; validation, L.C. and C.T.; visualization, I.V.; writing—original draft, L.C.; writing—review and editing, L.C., I.V. and R.R. All authors have read and agreed to the published version of the manuscript.

Funding: This research was funded by University of Santiago de Chile, USACH, grant number DICYT 061941CF and CIA SDT 2981.

Institutional Review Board Statement: Does not apply.

Informed Consent Statement: Does not apply.

Data Availability Statement: Does not apply.

Acknowledgments: We are grateful to David A. Case for facilitating access to using AMBER software and Rodrigo Yáñez for computer facilities.

Conflicts of Interest: The authors declare no conflict of interest.

Sample Availability: Not available.

References

1. Etheridge, M.L.; Campbell, S.A.; Erdman, A.G.; Haynes, C.L.; Wolf, S.M.; McCullough, J. The big picture on nanomedicine: The state of investigational and approved nanomedicine products. *Nanomed. Nanotechnol. Biol. Med.* **2013**, *9*, 1–14. [CrossRef] [PubMed]
2. Marketing Bureau. Zydus Cadila Receives US FDA Approval for Generic Doxil Liposome Injection. Available online: http://www.pharmabiz.com/NewsDetails.aspx?aid=131053&sid=2 (accessed on 12 September 2020).
3. Patel, A.G.; Kaufmann, S.H. How does doxorubicin work? *eLife* **2012**, *1*, e00387. [CrossRef]

4. Denard, B.; Lee, C.; Ye, J. Doxorubicin blocks proliferation of cancer cells through proteolytic activation of CREB3L1. *eLife* **2012**, *1*, e00090. [CrossRef] [PubMed]
5. Zhang, S.; Liu, X.; Bawa-Khalfe, T.; Lu, L.-S.; Lyu, Y.L.; Liu, L.F.; Yeh, E.T.H. Identification of the molecular basis of doxorubicin-induced cardiotoxicity. *Nat. Med.* **2012**, *18*, 1639–1642. [CrossRef] [PubMed]
6. Ferreira, L.L.; Oliveira, P.J.; Cunha-Oliveira, T. *Epigenetics in Doxorubicin Cardiotoxicity*; Elsevier BV: Amsterdam, The Netherlands, 2019; pp. 837–846.
7. Rodríguez-Galván, A.; Amelines-Sarria, O.; Rivera, M.; Carreón-Castro, M.D.P.; Basiuk, V.A. Adsorption and Self-Assembly of Anticancer Antibiotic Doxorubicin on Single-Walled Carbon Nanotubes. *Nano* **2016**, *11*, 1650038. [CrossRef]
8. Lalan, M.; Jani, D. Toxicological Aspects of Carbon Nanotubes, Fullerenes and Graphenes. *Curr. Pharm. Des.* **2021**, *27*, 556–564. [CrossRef]
9. Contreras , M.L.; Rozas, R. Carbon Nanotubes: Toxicological Properties, Use as Drug Delivery Material and Computational Contribution to Predict their Properties Including Structures with Topological Defects. *Adv. Clin. Toxicol.* **2020**, *5*, 1–7. [CrossRef]
10. Mamidi, N. Cytotoxicity Evaluation of Carbon Nanotubes for Biomedical and Tissue Engineering Applications. *Perspect. Carbon Nanotub.* **2019**, *12*. [CrossRef]
11. Ali-Boucetta, H.; Kostarelos, K. Pharmacology of carbon nanotubes: Toxicokinetics, excretion and tissue accumulation. *Adv. Drug Deliv. Rev.* **2013**, *65*, 2111–2119. [CrossRef] [PubMed]
12. Nayak, T.; Leow, P.; Ee, P.-L.; Arockiadoss, T.; Ramaprabhu, S.; Pastorin, G. Crucial Parameters Responsible for Carbon Nanotubes Toxicity. *Curr. Nanosci.* **2010**, *6*, 141–154. [CrossRef]
13. Venkataraman, A.; Amadi, E.V.; Chen, Y.; Papadopoulos, C. Carbon Nanotube Assembly and Integration for Applications. *Nanoscale Res. Lett.* **2019**, *14*, 1–47. [CrossRef]
14. Pei, B.; Wang, W.; Dunne, N.; Li, X. Applications of Carbon Nanotubes in Bone Tissue Regeneration and Engineering: Superiority, Concerns, Current Advancements, and Prospects. *Nanomaterials* **2019**, *9*, 1501. [CrossRef] [PubMed]
15. Saliev, T. The Advances in Biomedical Applications of Carbon Nanotubes. *C J. Carbon Res.* **2019**, *5*, 29. [CrossRef]
16. Yan, Y.; Wang, R.; Hu, Y.; Sun, R.; Song, T.; Shi, X.; Yin, S. Stacking of doxorubicin on folic acid-targeted multiwalled carbon nanotubes for in vivo chemotherapy of tumors. *Drug Deliv.* **2018**, *25*, 1607–1616. [CrossRef] [PubMed]
17. Le, C.M.Q.; Cao, X.T.; Kim, D.W.; Ban, U.H.; Lee, S.H.; Lim, K.T. Preparation of poly(styrene-alt-maleic anhydride) grafted multi-walled carbon nanotubes for pH-responsive release of doxorubicin. *Mol. Cryst. Liq. Cryst.* **2017**, *654*, 181–189. [CrossRef]
18. Kumar, D.S.; Hasumura, T.; Nagaoka, Y.; Yoshida, Y.; Maekawa, T.; Jeyamohan, P. Accelerated killing of cancer cells using a multifunctional single-walled carbon nanotube-based system for targeted drug delivery in combination with photothermal therapy. *Int. J. Nanomed.* **2013**, *8*, 2653–2667. [CrossRef] [PubMed]
19. Liu, Z.; Sun, X.; Nakayama-Ratchford, N.; Dai, H. Supramolecular Chemistry on Water-Soluble Carbon Nanotubes for Drug Loading and Delivery. *ACS Nano* **2007**, *1*, 50–56. [CrossRef] [PubMed]
20. Wang, Y.; Xu, Z. Interaction mechanism of doxorubicin and SWCNT: Protonation and diameter effects on drug loading and releasing. *RSC Adv.* **2016**, *6*, 314–322. [CrossRef]
21. Das, D. *Carbon Nanotube and Graphene Nanoribbon Interconnects*; CRC Press: Boca Raton, FL, USA, 2014; pp. 1–196.
22. Charlier, J.-C. Defects in Carbon Nanotubes. *Accounts Chem. Res.* **2002**, *35*, 1063–1069. [CrossRef] [PubMed]
23. Crespi, V.H.; Cohen, M.L.; Rubio, A. In Situ Band Gap Engineering of Carbon Nanotubes. *Phys. Rev. Lett.* **1997**, *79*, 2093–2096. [CrossRef]
24. Saito, R.; Fujita, M.; Dresselhaus, G.; Dresselhaus, M.S. Electronic structure of chiral graphene tubules. *Appl. Phys. Lett.* **1992**, *60*, 2204–2206. [CrossRef]
25. Ebbesen, T.W.; Takada, T. Topological and SP3 defect structures in nanotubes. *Carbon* **1995**, *33*, 973–978. [CrossRef]
26. Sternberg, M.; Gruen, D.M.; Kedziora, G.; Horner, D.A.; Zapol, P.; Curtiss, L.A.; Redfern, P.C. Carbon Ad-Dimer Defects in Carbon Nanotubes. *Phys. Rev. Lett.* **2006**, *96*, 075506. [CrossRef]
27. Terrones, H.; Hernández, E.; Grobert, N.; Charlier, J.-C.; Ajayan, P.M. New Metallic Allotropes of Planar and Tubular Carbon. *Phys. Rev. Lett.* **2000**, *84*, 1716–1719. [CrossRef] [PubMed]
28. Stone, A.J.; Wales, D.J. Theoretical studies of icosahedral C60 and some related species. *Chem. Phys. Lett.* **1986**, *128*, 501–503. [CrossRef]
29. Orlikowski, D.; Bernholc, J.; Roland, C.; Nardelli, M.B. Ad-dimers on Strained Carbon Nanotubes: A New Route for Quantum Dot Formation? *Phys. Rev. Lett.* **1999**, *83*, 4132–4135. [CrossRef]
30. Fuentes, M.L.C.; Soto, R.R. Carbon Nanotubes: Molecular and Electronic Properties of Regular and Defective Structures. *Density Funct. Calc.* **2018**, *9*, 197–218. [CrossRef]
31. Pakdel, M.; Raissi, H.; Shahabi, M. Predicting doxorubicin drug delivery by single-walled carbon nanotube through cell membrane in the absence and presence of nicotine molecules: A molecular dynamics simulation study. *J. Biomol. Struct. Dyn.* **2019**, *38*, 1488–1498. [CrossRef] [PubMed]
32. Contreras, M.L.; Torres, C.; Villarroel, I.; Rozas, R. Molecular dynamics assessment of doxorubicin–carbon nanotubes molecular interactions for the design of drug delivery systems. *Struct. Chem.* **2019**, *30*, 369–384. [CrossRef]
33. Torres, C.; Villarroel, I.; Rozas, R.; Contreras, L. Carbon Nanotubes Having Haeckelite Defects as Potential Drug Carriers. Molecular Dynamics Simulation. *Molecules* **2019**, *24*, 4281. [CrossRef] [PubMed]

34. Zhang, L.; Peng, G.; Li, J.; Liang, L.; Kong, Z.; Wang, H.; Jia, L.; Wang, X.; Zhang, W.; Shen, J.-W. Molecular dynamics study on the configuration and arrangement of doxorubicin in carbon nanotubes. *J. Mol. Liq.* **2018**, *262*, 295–301. [CrossRef]

35. Izadyar, A.S.; Farhadian, N.; Chenarani, N. Molecular dynamics simulation of doxorubicin adsorption on a bundle of functionalized CNT. *J. Biomol. Struct. Dyn.* **2015**, *34*, 1797–1805. [CrossRef]

36. Sornmee, P.; Rungrotmongkol, T.; Saengsawang, O.; Arsawang, U.; Remsungnen, T.; Hannongbua, S. Understanding the Molecular Properties of Doxorubicin Filling Inside and Wrapping Outside Single-Walled Carbon Nanotubes. *J. Comput. Theor. Nanosci.* **2011**, *8*, 1385–1391. [CrossRef]

37. Wolski, P.; Nieszporek, K.; Panczyk, T. Carbon nanotubes and short cytosine-rich telomeric DNA oligomers as platforms for controlled release of doxorubicin-a molecular dynamics study. *Int. J. Mol. Sci.* **2020**, *21*, 3619. [CrossRef]

38. Maleki, R.; Afrouzi, H.H.; Hosseini, M.; Toghraie, D.; Piranfar, A.; Rostami, S. pH-sensitive loading/releasing of doxorubicin using single-walled carbon nanotube and multi-walled carbon nanotube: A molecular dynamics study. *Comput. Methods Programs Biomed.* **2020**, *186*, 105210. [CrossRef] [PubMed]

39. Pennetta, C.; Floresta, G.; Graziano, A.C.E.; Cardile, V.; Rubino, L.; Galimberti, M.; Rescifina, A.; Barbera, V. Functionalization of Single and Multi-Walled Carbon Nanotubes with Polypropylene Glycol Decorated Pyrrole for the Development of Doxorubicin Nano-Conveyors for Cancer Drug Delivery. *Nanomaterials* **2020**, *10*, 1073. [CrossRef]

40. Kordzadeh, A.; Amjad-Iranagh, S.; Zarif, M.; Modarress, H. Adsorption and encapsulation of the drug doxorubicin on covalent functionalized carbon nanotubes: A scrutinized study by using molecular dynamics simulation and quantum mechanics calculation. *J. Mol. Graph. Model.* **2019**, *88*, 11–22. [CrossRef] [PubMed]

41. Karnati, K.R.; Wang, Y. Understanding the co-loading and releasing of doxorubicin and paclitaxel using chitosan functionalized single-walled carbon nanotubes by molecular dynamics simulations. *Phys. Chem. Chem. Phys.* **2018**, *20*, 9389–9400. [CrossRef]

42. Wolski, P.; Nieszporek, K.; Panczyk, T. Pegylated and folic acid functionalized carbon nanotubes as pH controlled carriers of doxorubicin. Molecular dynamics analysis of the stability and drug release mechanism. *Phys. Chem. Chem. Phys.* **2017**, *19*, 9300–9312. [CrossRef] [PubMed]

43. Rungnim, C.; Rungrotmongkol, T.; Poo-Arporn, R.P. pH-controlled doxorubicin anticancer loading and release from carbon nanotube noncovalently modified by chitosan: MD simulations. *J. Mol. Graph. Model.* **2016**, *70*, 70–76. [CrossRef]

44. Janas, D. From Bio to Nano: A Review of Sustainable Methods of Synthesis of Carbon Nanotubes. *Sustain. J. Rec.* **2020**, *12*, 4115. [CrossRef]

45. Wang, J.; Cieplak, P.; Kollman, P.A. How well does a restrained electrostatic potential (RESP) model perform in calculating conformational energies of organic and biological molecules? *J. Comput. Chem.* **2000**, *21*, 1049–1074. [CrossRef]

46. Veclani, D.; Tolazzi, M.; Melchior, A. Molecular Interpretation of Pharmaceuticals' Adsorption on Carbon Nanomaterials: Theory Meets Experiments. *Processes* **2020**, *8*, 642. [CrossRef]

47. Johnson, E.R.; Keinan, S.; Mori-Sánchez, P.; Contreras-García, J.; Cohen, A.J.; Yang, W. Revealing Noncovalent Interactions. *J. Am. Chem. Soc.* **2010**, *132*, 6498–6506. [CrossRef]

48. Khalak, Y.; Tresadern, G.; de Groot, B.L.; Gapsys, V. Non-equilibrium approach for binding free energies in cyclodextrins in SAMPL7: Force fields and software. *J. Comput. Mol. Des.* **2021**, *35*, 49–61. [CrossRef] [PubMed]

49. Westermaier, Y.; Ruiz-Carmona, S.; Theret, I.; Perron-Sierra, F.; Poissonnet, G.; Dacquet, C.; Boutin, J.A.; Ducrot, P.; Barril, X. Binding mode prediction and MD/MMPBSA-based free energy ranking for agonists of REV-ERBα/NCoR. *J. Comput. Mol. Des.* **2017**, *31*, 755–775. [CrossRef] [PubMed]

50. Chéron, N.; Shakhnovich, E.I. Effect of sampling on BACE-1 ligands binding free energy predictions via MM-PBSA calculations. *J. Comput. Chem.* **2017**, *38*, 1941–1951. [CrossRef]

51. Eken, Y.; Almeida, N.M.S.; Wang, C.; Wilson, A.K. SAMPL7: Host–guest binding prediction by molecular dynamics and quantum mechanics. *J. Comput. Mol. Des.* **2021**, *35*, 63–77. [CrossRef]

52. Ruangpornvisuti, V. Molecular modeling of dissociative and non-dissociative chemisorption of nitrosamine on close-ended and open-ended pristine and Stone-Wales defective (5,5) armchair single-walled carbon nanotubes. *J. Mol. Model.* **2009**, *16*, 1127–1138. [CrossRef]

53. Peng, C.; Wang, J.; Yu, Y.; Wang, G.; Chen, Z.; Xu, Z.; Cai, T.; Shao, Q.; Shi, J.; Zhu, W. Improving the accuracy of predicting protein–ligand binding-free energy with semiempirical quantum chemistry charge. *Futur. Med. Chem.* **2019**, *11*, 303–321. [CrossRef] [PubMed]

54. Contreras, M.L.; Ávila, D.; Alvarez, J.; Rozas, R. Computational algorithms for fast-building 3D carbon nanotube models with defects. *J. Mol. Graph. Model.* **2012**, *38*, 389–395. [CrossRef] [PubMed]

55. *HyperChem Release*, Version 7.5; Hypercube Inc.: Gainesville, FL, USA, 2004.

56. Case, D.A.; Betz, R.M.; Cerutti, D.S.; Cheatham, T.E., III; Darden, T.A.; Duke, R.E.; Giese, T.J.; Gohlke, H.; Goetz, A.W.; Homeyer, N.; et al. *AMBER 2016*; University of California: San Francisco, CA, USA, 2016.

57. Case, D.A.; Cheatham, T.E., III; Darden, T.; Gohlke, H.; Luo, R.; Merz, K.M., Jr.; Onufrie, A.; Simmerling, C.; Wang, B.; Woods, R. The Amber biomolecular simulation programs. *J. Comput. Chem.* **2005**, *26*, 1668–1688. [CrossRef] [PubMed]

58. Bayly, C.I.; Cieplak, P.; Cornell, W.; Kollman, P.A. A well-behaved electrostatic potential based method using charge restraints for deriving atomic charges: The RESP model. *J. Phys. Chem.* **1993**, *97*, 10269–10280. [CrossRef]

59. Genheden, S.; Ryde, U. The MM/PBSA and MM/GBSA methods to estimate ligand-binding affinities. *Expert Opin. Drug Discov.* **2015**, *10*, 449–461. [CrossRef] [PubMed]

molecules

Article

Identification of New *Mycobacterium tuberculosis* Proteasome Inhibitors Using a Knowledge-Based Computational Screening Approach

Tahani M. Almeleebia [1], Mesfer Al Shahrani [2], Mohammad Y. Alshahrani [2,3], Irfan Ahmad [2], Abdullah M. Alkahtani [4], Md Jahoor Alam [5], Mohd Adnan Kausar [6], Amir Saeed [7,8], Mohd Saeed [5] and Sana Iram [9,10,*]

[1] Department of Clinical Pharmacy, College of Pharmacy, King Khalid University, Abha 61421, Saudi Arabia; talmelby@kku.edu.sa
[2] Department of Clinical Laboratory Sciences, College of Applied Medical Sciences, King Khalid University, Abha 61421, Saudi Arabia; mesferm@kku.edu.sa (M.A.S.); moyahya@kku.edu.sa (M.Y.A.); irfancsmmu@gmail.com (I.A.)
[3] Research Center for Advanced Materials Science (RCAMS), King Khalid University, Abha 61421, Saudi Arabia
[4] Department of Microbiology and Clinical Parasitology, College of Medicine, King Khalid University, Abha 61421, Saudi Arabia; abdalqahtani@kku.edu.sa
[5] Department of Biology, College of Sciences, University of Hail, Hail 2240, Saudi Arabia; j.alam@uoh.edu.sa (M.J.A.); mo.saeed@uoh.edu.sa (M.S.)
[6] Department of Biochemistry, College of Medicine, University of Hail, Hail 2240, Saudi Arabia; ma.kausar@uoh.edu.sa
[7] Department of Clinical Laboratory Sciences, College of Applied Medical Sciences, University of Hail, Hail 2240, Saudi Arabia; am.saeed@uoh.edu.sa
[8] Faculty of Medical Laboratory Sciences, Department of Medical Microbiology, University of Medical Sciences & Technology, Khartoum 12810, Sudan
[9] Department of Medical Biotechnology, Yeungnam University, Gyeongsan 38541, Korea
[10] Nanomedicine & Nanobiotechnology Laboratory, Department of Biosciences, Integral University, Lucknow 226026, India
* Correspondence: sanairam157@gmail.com or sanairam157@yu.ac.kr

check for **updates**

Citation: Almeleebia, T.M.; Shahrani, M.A.; Alshahrani, M.Y.; Ahmad, I.; Alkahtani, A.M.; Alam, M.J.; Kausar, M.A.; Saeed, A.; Saeed, M.; Iram, S. Identification of New *Mycobacterium tuberculosis* Proteasome Inhibitors Using a Knowledge-Based Computational Screening Approach. *Molecules* **2021**, *26*, 2326. https://doi.org/10.3390/molecules26082326

Academic Editor: Marco Tutone

Received: 8 March 2021
Accepted: 13 April 2021
Published: 16 April 2021

Publisher's Note: MDPI stays neutral with regard to jurisdictional claims in published maps and institutional affiliations.

Abstract: *Mycobacterium tuberculosis* (Mtb) is a deadly tuberculosis (TB)-causing pathogen. The proteasome is vital to the survival of Mtb and is therefore validated as a potential target for anti-TB therapy. Mtb resistance to existing antibacterial agents has enhanced drastically, becoming a worldwide health issue. Therefore, new potential therapeutic agents need to be developed that can overcome the complications of TB. With this purpose, in the present study, 224,205 natural compounds from the ZINC database have been screened against the catalytic site of Mtb proteasome by the computational approach. The best scoring hits, ZINC3875469, ZINC4076131, and ZINC1883067, demonstrated robust interaction with Mtb proteasome with binding energy values of -7.19, -7.95, and -7.21 kcal/mol for the monomer (K-chain) and -8.05, -9.10, and -7.07 kcal/mol for the dimer (both K and L chains) of the beta subunit, which is relatively higher than that of reference compound HT1171 (-5.83 kcal/mol (monomer) and -5.97 kcal/mol (dimer)). In-depth molecular docking of top-scoring compounds with Mtb proteasome reveals that amino acid residues Thr1, Arg19, Ser20, Thr21, Gln22, Gly23, Asn24, Lys33, Gly47, Asp124, Ala126, Trp129, and Ala180 are crucial in binding. Furthermore, a molecular dynamics study showed steady-state interaction of hit compounds with Mtb proteasome. Computational prediction of physicochemical property assessment showed that these hits are non-toxic and possess good drug-likeness properties. This study proposed that these compounds could be utilized as potential inhibitors of Mtb proteasome to combat TB infection. However, there is a need for further bench work experiments for their validation as inhibitors of Mtb proteasome.

Keywords: *Mycobacterium tuberculosis*; tuberculosis; proteasome; natural compounds

1. Introduction

Mycobacterium tuberculosis (Mtb) is a deadly tuberculosis (TB)-causing pathogen. TB is a communicable disease that ranks in the world's top 10 causes of death. Besides, it is the leading cause of single infectious agent fatality (higher than HIV/AIDS), and approximately 10 million people fell ill with TB in 2019 [1]. A person with a weakened immune system is highly susceptible to TB infection; thus, their involvement with HIV is the major cause of fatality for these patients [2]. Mtb resistance to existing antibacterial agents has enhanced drastically as well as multidrug-resistant and extensively drug-resistant Mtb strains. These strains are becoming a worldwide health issue [3,4] and are involved in the host immune system's resistance to nitric oxide stress.

Proteasomes are multi-subunit proteolytic complexes that have a vital role in various cellular functions. Proteasome inhibition has appeared as a prevailing approach for the management of various infectious diseases [5]. Mtb proteasome is vital for the bacterium pathogenesis; hence, it is regarded as an attractive target for the development of new agents that may inhibit Mtb. The Mtb mutans lacking the proteasome (proteasome subunits silencing) are viable in vitro, but the infection cannot be maintained in the TB mouse model [6,7]. Hence, it seems that Mtb proteasomes are vital for their propagation in mammalian hosts and are involved in the host immune system's resistance to nitric oxide stress [8].

The proteasome is a heap-shaped protein made up of four rings of heptamers. Its length and width are 150 and 115 Å, respectively [9,10]. Inner beta rings are formed by seven identical prcB subunits, and outer alpha rings are formed by seven identical prcA subunits, which give a path to inner beta rings with active sites when they are open. As a result, it provides the overall organization of $\alpha7\beta7\beta7\alpha7$. The active site of the bacterial proteasome is identical to that of the archaeal and eukaryotic proteasomes and is found primarily in β subunits [9,11].

In the literature, several effective Mtb proteasome inhibitors have been documented. Among them, HT1171, GL5, MLN273, and fellutamide-B are the most potent Mtb proteasome inhibitors [9,12,13]. The Mtb proteasome was revealed to be inhibited by 15 psoralens from a library of 92 analogs, and compounds 8, 11, 13, and 15 exhibited potent inhibition in a fluorescence-based enzymatic assay [14]. Several plant-derived natural products were discovered to inhibit Mtb proteasome with IC_{50} values ranging from 25 to 120 M using the chymotrypsin substrate Suc-LLVY-AMC [15]. Various bortezomib analogs have been developed, with the phenol- and halogen-substituted analogs being more specific for the Mtb proteasome than the human proteasome [16].

Drug development requires the identification of some potential hits from a huge library of chemical components. Screening such a huge compounds library using wet-lab assays can be a difficult task. Protein-ligand docking is a powerful tool in drug development because it aids in the identification of active or lead compounds from a library of compounds [17,18]. This method is also capable of accurately identifying inhibitor binding modes to the target proteins [17,19]. Computational screening before laboratory testing is a successful approach in decreasing the number of candidate inhibitors for benchwork-based screening [20–22]. The present study aimed to identify new possible hits from the natural compounds databases using in silico, state-of-the-art techniques that could serve as Mtb proteasome inhibitors to combat TB infection.

2. Methodology

2.1. Protein Structure Preparation

Mtb proteasome 3D structure (PDB ID: 5TRG) was retrieved from the protein data bank and prepared in monomer form by Discovery Studio (DS) 2020. Since the Mtb proteasome core particle has 14 chains in the beta subunit, all of which have the same active sites, the present study focused on chain K as a monomer, and the K and L chains as a dimer.

2.2. Database Collection and Refinement

Natural compounds were accessed from the ZINC database (https://zinc.docking.org accessed on 29 January 2021), limiting the outcomes by choosing "natural products" as a subset, resulting in a total of 224,205 compounds and was then downloaded in SDF format. Furthermore, these compounds were processed for minimization and preparation for screening using the "ligand preparation" tool in DS 2020.

2.3. Receptor-Based Virtual Screening

In order to identify possible leads, the prepared librarian was screened against the Mtb proteasome active site using AutoDock vina (version 1.1.2). Then, the top-scored hits were further processed for in-depth molecular docking studies.

2.4. Molecular Docking

Lead hits were docked with Mtb proteasome (monomer; K-chain) by Autodock4.2 to determine the ligand–protein interaction and their binding affinities [23]. All input files were prepared using AutoDock Tools, adding polar hydrogen in protein, and assigning the charges with the Kollman charges method. The grid center points X, Y, and Z were kept as 36.731, -11.255, and 43.313, respectively. Grid points were fixed as $60 \times 60 \times 60$ Å with the spacing of 0.375 Å. Additionally, these hits were also docked with the dimer form (K and L chains) of the beta subunit of the Mtb proteasome, keeping the grid center points X, Y, and Z as 41.22, 0.60, and 32.31, respectively. All docking calculation parameters were kept as a default value. The highest negative binding energy (BE) value was ranked as the most promising binding pose.

2.5. LIGPLOT$^+$ Analysis

The H-bond and hydrophobic interactions between "hit compounds–Mtb proteasome" complexes were determined by the LIGPLOT+ Version v.2.1. The 3-D structures of the "compound–Mtb proteasome" interaction produced were transformed into 2-D figures using the LIGPLOT algorithm.

2.6. Drug-Likeness

Top-scoring molecules (ZINC3875469, ZINC4076131, and ZINC1883067) were further used to estimate drug-likeness, toxicity, and pharmacokinetic properties using the pkCSM and SwissADME tools [24]. SMILE IDs of the molecules retrieved from the ZINC database were entered into the pkCSM tool to evaluate drug-likeness [25].

2.7. Molecular Dynamics (MD) Simulations

To study the dynamic behavior of the protein–ligand complex in simulated physiological conditions, MD simulations were performed using Gromacs Ver. 2020.4. The protein–ligand complexes were solvated in a $10 \times 10 \times 10$ Å orthorhombic periodic box built with TIP3P water molecules. By adding a sufficient number of 9 Na counterions, the entire system was neutralized. This solvated system was energy-minimized and position-restrained with CHARMM 36 as a force-field [26]. Further, 20 ns of MD run was carried out at 1 atm pressure and 300 K temperature implementing NPT ensemble with a recording interval of 100 ps. This resulted in a total of 1000 reading frames. In the end, various parameters of MD simulation study such as ligand binding site analysis, root-mean-square deviation (RMSD), root-mean-square fluctuation (RMSF), radius of gyration, Mindistance, H-bond analysis, etc., were analyzed for the stability, compactness, structural variations, and protein–ligand interactions in the solvated system.

3. Results and Discussion

3.1. Virtual Screening, Molecular Docking, and LIGPLOT

Mtb is the only known bacterial pathogen that has proteasome activity [6]. The increase in drug-resistant TB is a major public health concern and requires the development of new agents that can evade the resistance and effectively control TB. With this purpose, we conducted the computational screening of 224,205 natural compounds from the ZINC database targeting the Mtb proteasome. Among them, the selected hits ZINC3875469, ZINC4076131, and ZINC1883067 showed strong binding with the Mtb proteasome. Two-dimensional structures of hit compounds are shown in Figure 1. ZINC3875469 interacted with proteasome through 10 amino acid residues: Thr1, Ser20, Thr21, Gln22, Gly23, Ala46, Gly47, Thr48, Gly140, and Ser141 (Figure 2a); while Thr1, Arg19, Ser20, Asn24, Thr21, Gln22, Gly23, Asn24, Lys33, Gly47, Ala49, and Ala180 residues of the proteasome were observed to bind with ZINC4076131 (Figure 2b). In a similar way, ZINC1883067 was found to interact with Thr1, Arg19, Ser20, Thr21, Lys33, Ala46, Gly47, Gly140, Ser141, and Ala180 residues of the proteasome (Figure 2c).

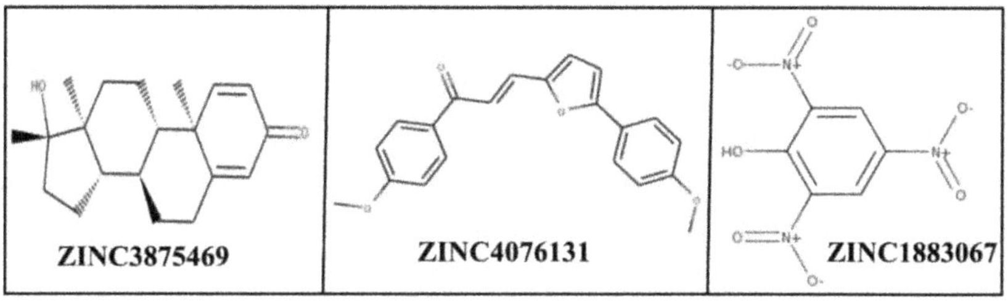

Figure 1. 2D structures of hit compounds.

The active site pocket residues of Mtb proteasome were determined as Ser20, Thr21, Gln22, Val31, Ile45, Ala46, Thr48, Ala49, Val53, Asp124, Asp128, and Asp130 [27]. Interestingly, ZINC3875469, ZINC4076131, and ZINC1883067 were also found to bind with these residues of Mtb proteasome. In a study, small molecules were reported to interact with Thr1, Arg19, Ser20, Thr21, Gln22, Lys33, Gly47, Ala49, Gly140, and Ser141 residues of Mtb proteasome [27]. Consistent with this, in the present study, the selected hits were observed to bind with the similar residues of Mtb proteasome.

Oxathiazol-2-one compounds altered the Mtb proteasome by interacting with the Thr1 residues of the core complex beta-subunit. Consequently, Thr1 is cyclocarbonylated, which greatly alters the active site environment and causes an alternative protein conformation in which the substrate-binding pocket is disrupted. As a result, Mtb proteasome substrates were unable to obtain access to the proteasome, causing toxic proteins and peptides to accumulate within mycobacterial cells. Notably, the oxathiazol-2-one compounds were thought to have no effect on the substrate-binding pocket of human proteasomal beta-subunits. It was proposed that this was due to the fact that the residues involved in preserving the altered conformation were largely non-conserved and thus not susceptible to cyclocarbonylation inactivation. This provided highly selective inhibition of Mtb proteasomes while leaving host proteasomal activity unaffected [13]. Interestingly, in this study, the selected hits ZINC3875469, ZINC4076131, and ZINC1883067 were found to interact with Thr1 residues of the Mtb proteasome.

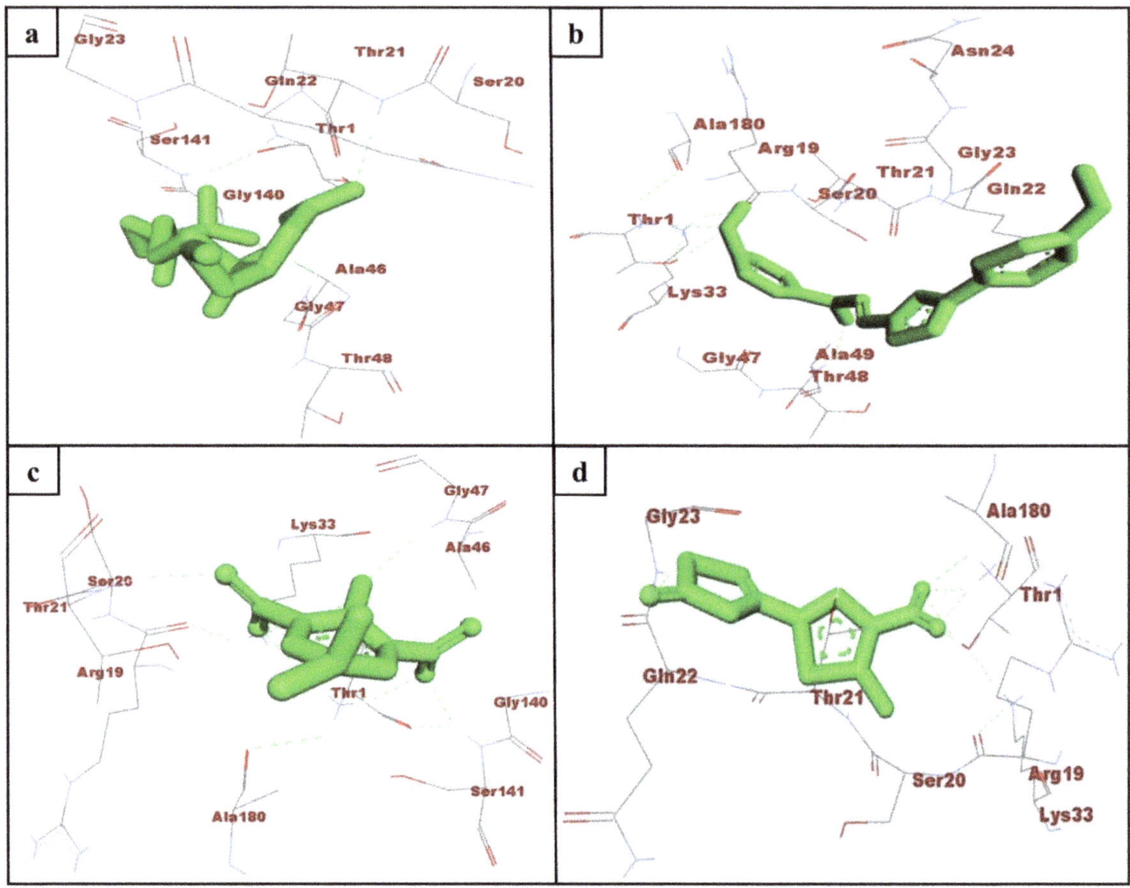

Figure 2. Interacting residues of Mtb proteasome (monomer; K-chain) with ZINC3875469 (**a**), ZINC4076131 (**b**), ZINC1883067 (**c**), and HT1171 (**d**).

We also performed the molecular docking of selected hits with the dimer model (K and L chains) of the Mtb proteasome to determine possible interactions with the adjacent chain (L-chain). Asp124 of the adjacent chain of the Mtb proteasome has been reported as an important residue for inhibitor interaction [28]. Interestingly, in addition to interacting with K-chain residues, ZINC3875469 and ZINC4076131 interacted with several other Mtb proteasome L-chain residues including the Asp124 (Figure 3a,c). Despite the fact that ZINC1883067 did not interact with the Asp124 residue of the L- chain, it did interact with several other residues of both chains of the Mtb proteasome (Figure 3b).

The BE values for hits ZINC3875469, ZINC4076131, and ZINC1883067 with the Mtb proteasome were found to be −7.19, −7.95, and −7.21 kcal/mol, respectively, for the monomer, and −8.05, −9.10, and −7.07 kcal/mol, respectively, for the dimer (Table 1).

HT1171 is a well-characterized inhibitor of the Mtb proteasome [13], which was used as the control compound in this study. HT1171 has been reported to bind with Thr1, Thr21, Arg19, Ser20, Val31, and Ala49 residues of Mtb proteasome [27]. Interestingly, Thr1, Ser20, and Thr21 are the common interacting residues of the Mtb proteasome with HT1171 and the selected hit compounds in this study (Figure 2a–d). BE of HT1171 against the Mtb proteasome was noted as −5.83 kcal/mol for the monomer and -5.97 kcal/mol for the dimer (Table 1).

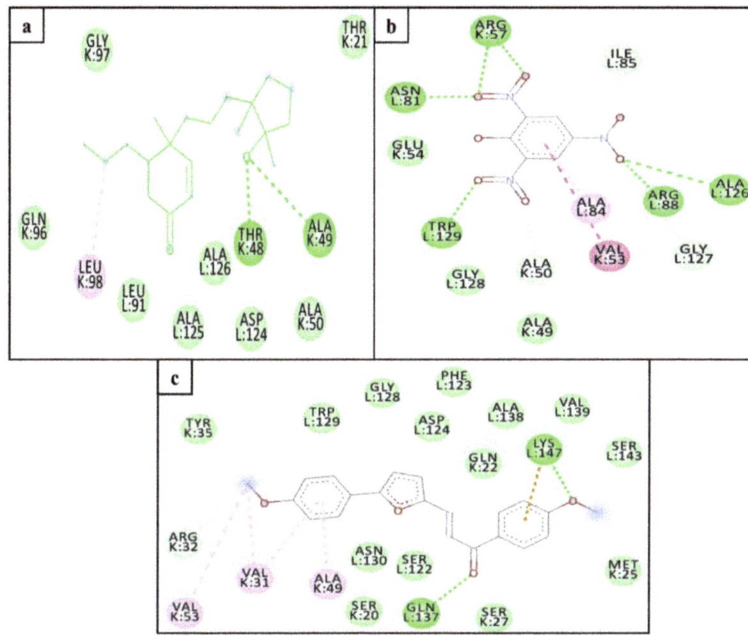

Figure 3. Interacting residues of Mtb proteasome (dimer; K and L chains) with ZINC3875469 (**a**), ZINC1883067 (**b**), and ZINC4076131 (**c**).

Table 1. Binding energy of hit compounds with Mtb proteasome (monomer and dimer).

Compound	Binding Energy (kcal/mol)		Inhibition Constant (µM)	
	Monomer	Dimer	Monomer	Dimer
ZINC3875469	−7.19	−8.05	28.9	26.54
ZINC4076131	−7.95	−9.10	43.24	0.213
ZINC1883067	−7.21	−7.07	27.92	23.96
HT1171 *	−5.83	−5.97	45.36	47.01

* Positive control for Mtb proteasome.

The hydrophobic interaction and H-bond help to elucidate the potency of inhibitors to the target protein and contribute an important role in "inhibitor–protein" complex stability [29,30]. The Mtb proteasome residues involved in H-bond (Table 2) and hydrophobic interaction with selected compounds are shown in Figure 4.

Table 2. H-bond interactions between compounds and the Mtb proteasome (monomer; K-chain).

S.No.	Target	Compound	H-Bond	H-Bond Length (Å)
1.		ZINC1883067	THR1:HN3-UNK0:O8	2.78
			THR1:HG1-UNK0:O14	2.68
			THR21:HN-UNK0:O15	3.06
			SER141:HN-UNK0:O8	2.83
2.	Mtb proteasome	ZINC4076131	THR1:HT3-UNK0:O2	3.17
			ALA49:HN-UNK0:O8	2.95
3.		ZINC3875469	THR21:HN-UNK0:O13	2.15
			GLY47:HN-UNK0:O3	2.49
4.		HT1171	GLY23:HN-UNK0:O6	2.71

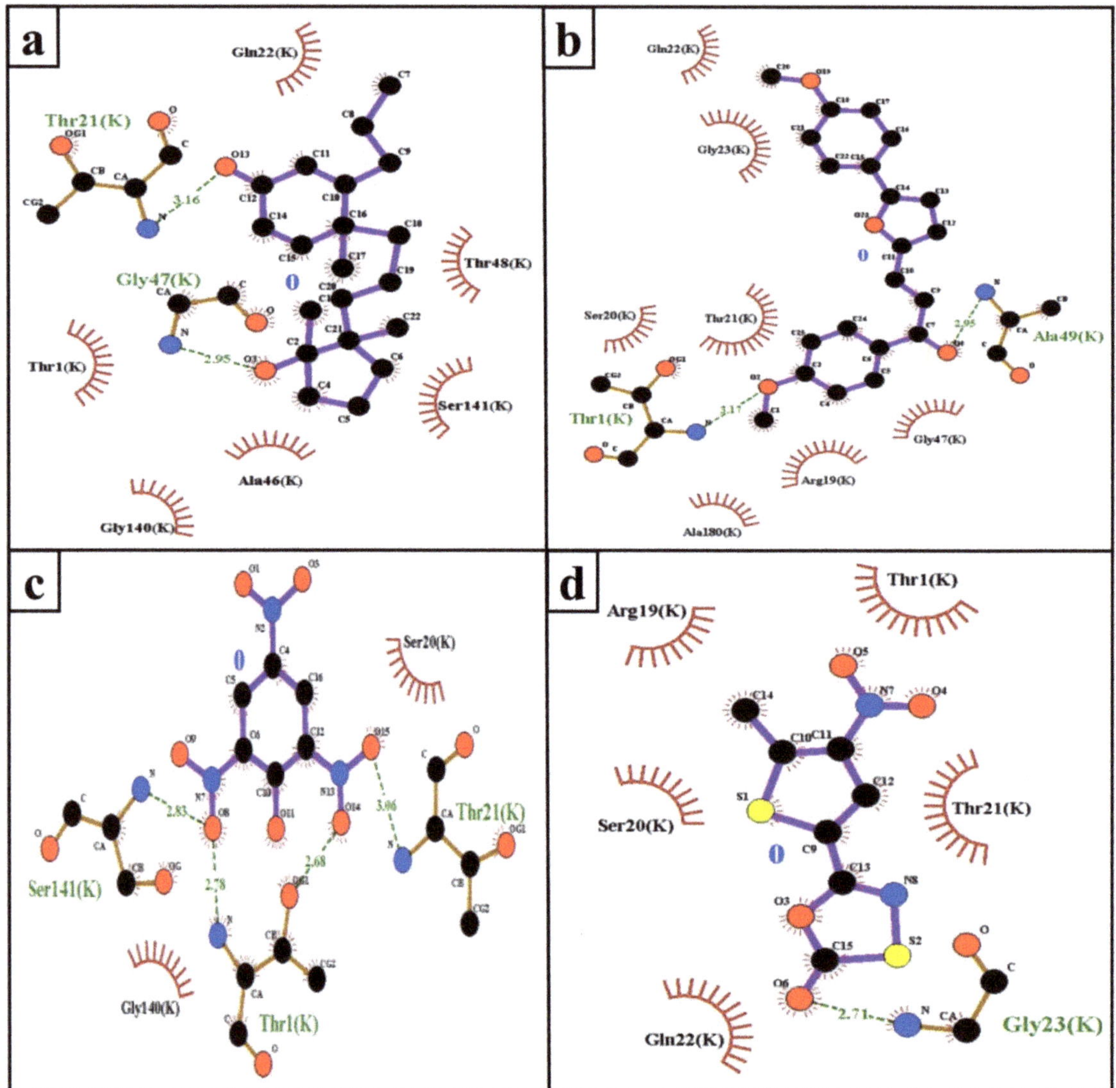

Figure 4. H-bond (green dashed line) and hydrophobic interacting (red arc) residues of the Mtb proteasome (monomer; K-chain) with ZINC3875469 (**a**), ZINC4076131 (**b**), ZINC1883067 (**c**), and HT1171 (**d**).

In the docking study, more negative BE corresponded to the strong binding of hits to the target protein. Furthermore, it is a fact that weaker binding will ultimately have a rapid dissociation rate [31]. Accordingly, in this study, the hits (ZINC3875469, ZINC4076131, and ZINC1883067) exhibited lower BE (strong binding) with the Mtb proteasome than the control compound (HT1171), suggesting that these compounds could be utilized as an inhibitor of the Mtb proteasome to combat TB. The results of the pkCSM and SwissADME tool show that top-scored compounds (ZINC3875469, ZINC4076131, and ZINC1883067) retained an acceptable range of ADMET and drug-likeness (Lipinski) (Table 3).

Table 3. Pharmacokinetic properties of top-scoring ligands.

Property	Model Name		Predicted Value			Unit
			ZINC1883067	ZINC4076131	ZINC3875469	
Absorption	Water solubility		−3.67	−5.213	−4.624	log mol/L
	Caco2 permeability		−0.275	1.432	1.569	log Papp in 10–6 cm/s
	Intestinal absorption (human)		81.583	97.413	96.726	% Absorbed
	Skin Permeability		−2.786	−2.629	−2.985	log Kp
Distribution	VDss (human)		−0.44	0.315	0.397	log L/kg
	Fraction unbound (human)		0.173	0.118	0.105	Fu
	BBB permeability		−1.036	0.099	0.2	log BB
	CNS permeability		−2.744	−1.483	−2.42	log PS
Metabolism	CYP2D6 substrate		No	No	No	Yes/No
	CYP3A4 substrate		No	Yes	Yes	
	inhibitor	CYP1A2	Yes	Yes	No	
		CYP2C19	No	Yes	No	
		CYP2C9	No	Yes	No	
		CYP2D6	No	No	No	
		CYP3A4	No	No	No	
Excretion	Total Clearance		0.587	0.742	0.636	log mL/min/kg
	Renal OCT2 substrate		No	No	Yes	Yes/No
Toxicity	AMES toxicity		Yes	No	No	
	Max. tolerated dose (human)		−0.58	0.547	−0.423	log mg/kg/day
	hERG I inhibitor		No	No	No	Yes/No
	Oral Rat Acute Toxicity (LD50)		2.321	2.347	1.837	mol/kg
	Oral Rat Chronic Toxicity (LOAEL)		1.35	1.913	1.708	log mg/kg_bw/day
	Hepatotoxicity		No	No	No	Yes/No
	Skin Sensitisation		Yes	No	No	
	T. pyriformis toxicity		0.804	0.569	1.054	(log µg/L)
	Minnow toxicity		0.644	−2.307	0.712	
Druglikeness	Lipinski		Yes	Yes	Yes	(Yes/No)

3.2. MD Simulation

MD simulations of the protein–ligand complex were performed using Gromacs 2020.4 on the Linux platform. MD simulation provides information about the receptor–ligand complex with time, so we performed the MD simulation for 20 ns on the three complexes (hit compounds). After the simulation, we analyzed the trajectory files for RMSD, RMSF, protein–ligand interactions, etc.

3.2.1. RMSD

The RMSD value determines the mean deviation of the complex at a specific time. It is an indicator that defines the average change in an atom's displacement in the specific molecular conformation of the reference conformation. In trajectory analysis, the complex RMSD was found within the range of 0.25 Å. The initial RMSD value of the complex was 0.1 Å. The backbone atoms were monitored, and the stability, compactness, structural fluctuations, and protein–ligand interactions in a solvated system were examined. RMSD of the backbone was also noted to 0.2 Å (Figure 5). RMSD was found to be constant at 0.2 Å

after 10 ns. The low RMSD value suggests that complexes are more stable. Moreover, it was noted that among the three complexes, complex ZINC3875469 had the lowest RMSD values. On the other hand, complex ZINC3873067 had a comparatively larger RMSD value. This suggests that complex ZINC3875469 has more stability.

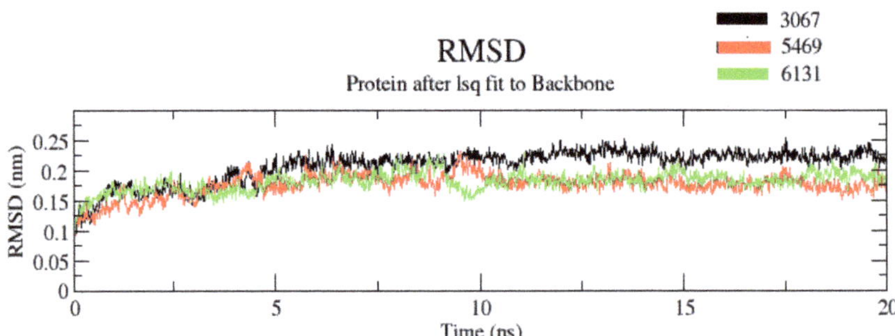

Figure 5. RMSD of protein and ligand after the initial RMSD values were stabilized. The RMSD graph for the backbone is shown in black color (complex ZINC1883067), red color (complex ZINC3875469), and green color (complex ZINC4076131).

3.2.2. RMSF

The RMSF is useful for characterizing local protein mobility in the protein–ligand complex, which is calculated throughout the simulation. It relates to the root-mean-square displacement of each frame conformation residue relative to the average conformation used to determine the flexibility of a protein region. In an RMSF plot, the peak shows the protein area fluctuates more throughout the simulation, while the lower RMSF values reflect the less conformational transition. The atomic profile fluctuations were found to be almost similar in all three complexes. The analysis revealed that the RMSF plot (Figure 6) displayed minimal fluctuations in the protein structures for complex ZINC3875469. The protein–ligand complex displayed lower flexibility, and the RMSF plot revealed variations in certain regions of protein residues. It is suggested that the ligand binding site remained approximately rigid throughout the simulation.

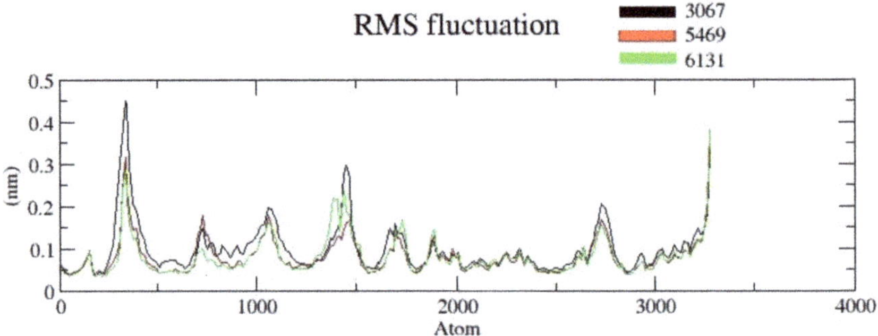

Figure 6. RMSF protein backbone and ligand complex is shown in black color (complex ZINC1883067), red color (complex ZINC3875469), and green color (for complex ZINC4076131).

3.2.3. Radius of Gyration (Rg)

Rg is used to assess a characterization parameter that evaluates changes in protein structures. For the measurement of the transition in Rg of the protein backbone atoms, the gmx gyrate software was used. Figure 7 shows that the Rg values of complexes ZINC1883067, ZINC3875469, and ZINC4076131 did not change significantly throughout the simulation and continued to fluctuate at 1.73 and 1.69 nm, respectively, indicating that the ligand had little influence on protein structures. It was observed that the Rg value of complex ZINC3875469 was lower and had little fluctuation comparatively throughout the 20 ns of simulation. This suggests that the ligand–protein interaction in complex ZINC3875469 is very high, which makes its structure more compact.

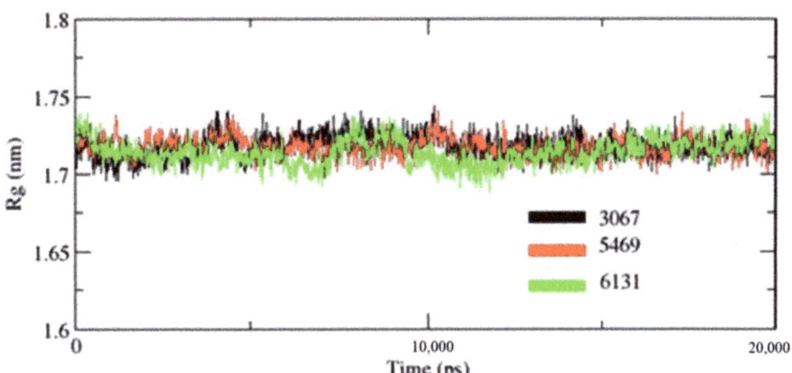

Figure 7. Plot for Rg of backbone atoms of the proteins in the presence of ligands. Black color (complex ZINC1883067), red color (complex ZINC3875469), and green color (complex ZINC4076131).

3.2.4. Minimum Distance

The minimum distance between protein and ligand is given in Figure 8. The average value was found to be 1.5 nm. Interestingly, complex ZINC3875469 had the comparatively lowest minimum distance of 0.25 nm during the entire simulation, indicating more compactness and stability of complex ZINC3875469. This suggests that complex ZINC3875469 is more stable than other complexes comparatively.

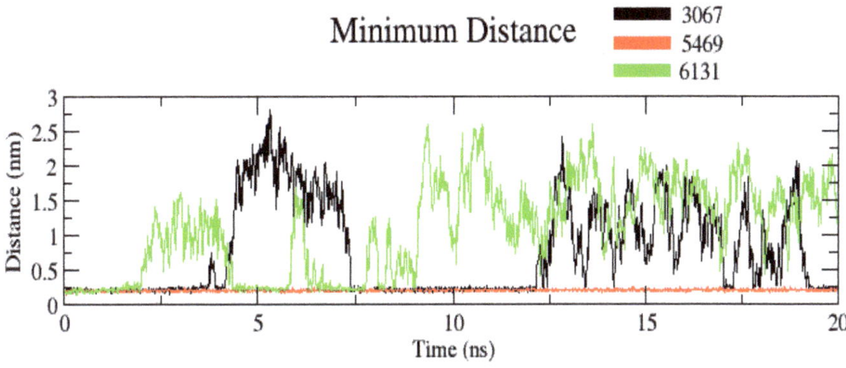

Figure 8. Plot showing minimum distance for complexes. Black color (complex ZINC1883067), red color (complex ZINC3875469), and green color (complex ZINC4076131).

3.2.5. Number of Hydrogen Bonds (H-Bond Number)

The number of H-bonds was measured to find out the robustness of the complex using a cut-off value 0.35 nm. It was noticed that complexes ZINC1883067 and ZINC3875469 had the most H-bonds, but the complex ZINC3875469 H-bonds were more stable over the entire simulation and thus play a significant role in stabilizing protein–ligand interactions (Figure 9).

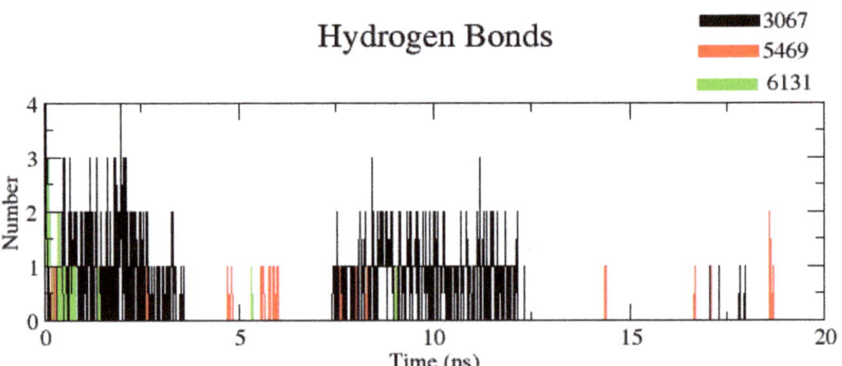

Figure 9. Plot showing number of hydrogen bonds for complexes. Black color (complex ZINC1883067), red color (complex ZINC3875469), and green color (complex ZINC4076131).

It should be noted that the BE values and MD simulations obtained can only show the binding effectiveness and stability of inhibitors with the target protein. However, further bench work studies are required to validate these hits (ZINC3875469, ZINC4076131, and ZINC1883067) as Mtb proteasome inhibitors.

4. Conclusions

In summary, natural compounds from the ZINC database were screened against the Mtb proteasome. The top hit compounds (ZINC3875469, ZINC4076131, and ZINC1883067) demonstrated robust binding with the monomer as well as dimer Mtb proteasome. Molecular docking of these selected hits with the Mtb proteasome dimer model revealed that, in addition to interacting with K-chain residues, they also interacted with many other residues of the L-chain. These results open the way for further experimental confirmation in the quest for a novel Mtb proteasome inhibitor to combat TB.

Author Contributions: Conceptualization, T.M.A., and S.I.; methodology, T.M.A., and S.I.; writing—original draft preparation, T.M.A., and S.I.; writing—review and editing, M.A.S., M.Y.A., I.A., A.M.A., M.J.A., M.A.K., A.S., and M.S. All authors have read and agreed to the published version of the manuscript.

Funding: The authors would like to express their gratitude to the Research Center of Advanced Materials—King Khalid University, Saudi Arabia for support by grant number (RCAMS/KKU/0020/20).

Institutional Review Board Statement: Not applicable.

Informed Consent Statement: Not applicable.

Data Availability Statement: Not applicable.

Conflicts of Interest: The authors declare no conflict of interest.

Sample Availability: Not available.

References

1. MacNeil, A.; Glaziou, P.; Sismanidis, C.; Maloney, S.; Floyd, K. Global Epidemiology of Tuberculosis and Progress Toward Achieving Global Targets—2017. *MMWR Morb. Mortal. Wkly. Rep.* **2019**, *68*, 263–266. [CrossRef]
2. Bloom, B.R.; Atun, R.; Cohen, T.; Dye, C.; Fraser, H.; Gomez, G.B.; Knight, G.; Murray, M.; Nardell, E.; Rubin, E.; et al. Tuberculosis. In *Major Infectious Diseases*, 3rd ed.; Holmes, K.K., Bertozzi, S., Bloom, B.R., Jha, P., Eds.; The International Bank for Reconstruction and Development/The World Bank: Washington, DC, USA, 2017. [CrossRef]
3. Raviglione, M.C.; Smith, I.M. XDR tuberculosis—Implications for global public health. *N. Engl. J. Med.* **2007**, *356*, 656–659. [CrossRef]
4. Saxena, A.K.; Singh, A. Mycobacterial tuberculosis Enzyme Targets and their Inhibitors. *Curr. Top Med. Chem.* **2019**, *19*, 337–355. [CrossRef] [PubMed]
5. Bibo-Verdugo, B.; Jiang, Z.; Caffrey, C.R.; O'Donoghue, A.J. Targeting proteasomes in infectious organisms to combat disease. *FEBS J.* **2017**, *284*, 1503–1517. [CrossRef] [PubMed]
6. Darwin, K.H.; Ehrt, S.; Gutierrez-Ramos, J.C.; Weich, N.; Nathan, C.F. The proteasome of Mycobacterium tuberculosis is required for resistance to nitric oxide. *Science* **2003**, *302*, 1963–1966. [CrossRef] [PubMed]
7. Gandotra, S.; Schnappinger, D.; Monteleone, M.; Hillen, W.; Ehrt, S. In vivo gene silencing identifies the Mycobacterium tuberculosis proteasome as essential for the bacteria to persist in mice. *Nat. Med.* **2007**, *13*, 1515–1520. [CrossRef] [PubMed]
8. Zhai, W.; Wu, F.; Zhang, Y.; Fu, Y.; Liu, Z. The Immune Escape Mechanisms of Mycobacterium Tuberculosis. *Int. J. Mol. Sci.* **2019**, *20*, 340. [CrossRef]
9. Hu, G.; Lin, G.; Wang, M.; Dick, L.; Xu, R.M.; Nathan, C.; Li, H. Structure of the Mycobacterium tuberculosis proteasome and mechanism of inhibition by a peptidyl boronate. *Mol. Microbiol.* **2006**, *59*, 1417–1428. [CrossRef]
10. Groll, M.; Ditzel, L.; Lowe, J.; Stock, D.; Bochtler, M.; Bartunik, H.D.; Huber, R. Structure of 20S proteasome from yeast at 2.4 A resolution. *Nature* **1997**, *386*, 463–471. [CrossRef]
11. Kwon, Y.D.; Nagy, I.; Adams, P.D.; Baumeister, W.; Jap, B.K. Crystal structures of the Rhodococcus proteasome with and without its pro-peptides: Implications for the role of the pro-peptide in proteasome assembly. *J. Mol. Biol.* **2004**, *335*, 233–245. [CrossRef]
12. Lin, G.; Li, D.; Chidawanyika, T.; Nathan, C.; Li, H. Fellutamide B is a potent inhibitor of the Mycobacterium tuberculosis proteasome. *Arch. Biochem. Biophys.* **2010**, *501*, 214–220. [CrossRef] [PubMed]
13. Lin, G.; Li, D.; de Carvalho, L.P.; Deng, H.; Tao, H.; Vogt, G.; Wu, K.; Schneider, J.; Chidawanyika, T.; Warren, J.D.; et al. Inhibitors selective for mycobacterial versus human proteasomes. *Nature* **2009**, *461*, 621–626. [CrossRef] [PubMed]
14. Rozman, K.; Alexander, E.M.; Ogorevc, E.; Bozovicar, K.; Sosic, I.; Aldrich, C.C.; Gobec, S. Psoralen Derivatives as Inhibitors of Mycobacterium tuberculosis Proteasome. *Molecules* **2020**, *25*, 1305. [CrossRef] [PubMed]
15. Zheng, Y.; Jiang, X.; Gao, F.; Song, J.; Sun, J.; Wang, L.; Sun, X.; Lu, Z.; Zhang, H. Identification of plant-derived natural products as potential inhibitors of the Mycobacterium tuberculosis proteasome. *BMC Complement Altern. Med.* **2014**, *14*, 400. [CrossRef] [PubMed]
16. Lin, G.; Tsu, C.; Dick, L.; Zhou, X.K.; Nathan, C. Distinct specificities of Mycobacterium tuberculosis and mammalian proteasomes for N-acetyl tripeptide substrates. *J. Biol. Chem.* **2008**, *283*, 34423–34431. [CrossRef] [PubMed]
17. Pinzi, L.; Rastelli, G. Molecular Docking: Shifting Paradigms in Drug Discovery. *Int. J. Mol. Sci.* **2019**, *20*, 4331. [CrossRef]
18. Ivanova, L.; Karelson, M.; Dobchev, D.A. Multitarget Approach to Drug Candidates against Alzheimer's Disease Related to AChE, SERT, BACE1 and GSK3beta Protein Targets. *Molecules* **2020**, *25*, 1846. [CrossRef]
19. Sethi, A.; Joshi, K.; Sasikala, K.; Alvala, M. Molecular docking in modern drug discovery: Principles and recent applications. In *Drug Discovery and Development—New Advances*; IntechOpen : London, UK, 2019; pp. 27–39. [CrossRef]
20. Langdon, S.R.; Westwood, I.M.; van Montfort, R.L.; Brown, N.; Blagg, J. Scaffold-focused virtual screening: Prospective application to the discovery of TTK inhibitors. *J. Chem. Inf. Model.* **2013**, *53*, 1100–1112. [CrossRef]
21. Yang, S.C.; Chang, S.S.; Chen, H.Y.; Chen, C.Y. Identification of potent EGFR inhibitors from TCM Database@Taiwan. *PLoS Comput. Biol.* **2011**, *7*, e1002189. [CrossRef]
22. Valasani, K.R.; Vangavaragu, J.R.; Day, V.W.; Yan, S.S. Structure based design, synthesis, pharmacophore modeling, virtual screening, and molecular docking studies for identification of novel cyclophilin D inhibitors. *J. Chem. Inf. Model.* **2014**, *54*, 902–912. [CrossRef]
23. Morris, G.M.; Huey, R.; Lindstrom, W.; Sanner, M.F.; Belew, R.K.; Goodsell, D.S.; Olson, A.J. AutoDock4 and AutoDockTools4: Automated docking with selective receptor flexibility. *J. Comput. Chem.* **2009**, *30*, 2785–2791. [CrossRef]
24. Daina, A.; Michielin, O.; Zoete, V. SwissADME: A free web tool to evaluate pharmacokinetics, drug-likeness and medicinal chemistry friendliness of small molecules. *Sci. Rep.* **2017**, *7*, 42717. [CrossRef] [PubMed]
25. Pires, D.E.; Blundell, T.L.; Ascher, D.B. pkCSM: Predicting Small-Molecule Pharmacokinetic and Toxicity Properties Using Graph-Based Signatures. *J. Med. Chem.* **2015**, *58*, 4066–4072. [CrossRef] [PubMed]
26. MacKerell, A.D., Jr. Atomistic models and force fields. *Comput. Biochem. Biophys.* **2001**, *9*, 7–38.
27. Tyagi, R.; Srivastava, M.; Jain, P.; Pandey, R.P.; Asthana, S.; Kumar, D.; Raj, V.S. Development of potential proteasome inhibitors against Mycobacterium tuberculosis. *J. Biomol. Struct. Dyn.* **2020**, *19*, 1–15. [CrossRef] [PubMed]
28. Mehra, R.; Chib, R.; Munagala, G.; Yempalla, K.R.; Khan, I.A.; Singh, P.P.; Khan, F.G.; Nargotra, A. Discovery of new Mycobacterium tuberculosis proteasome inhibitors using a knowledge-based computational screening approach. *Mol. Divers.* **2015**, *19*, 1003–1019. [CrossRef]

29. Shaikh, S.; Zainab, T.; Shakil, S.; Rizvi, S.M. A neuroinformatics study to compare inhibition efficiency of three natural ligands (Fawcettimine, Cernuine and Lycodine) against human brain acetylcholinesterase. *Network* **2015**, *26*, 25–34. [CrossRef] [PubMed]
30. Rizvi, S.M.; Shaikh, S.; Khan, M.; Biswas, D.; Hameed, N.; Shakil, S. Fetzima (levomilnacipran), a drug for major depressive disorder as a dual inhibitor for human serotonin transporters and beta-site amyloid precursor protein cleaving enzyme-1. *CNS Neurol. Disord. Drug Targets* **2014**, *13*, 1427–1431. [CrossRef] [PubMed]
31. Copeland, R.A. Conformational adaptation in drug-target interactions and residence time. *Future Med. Chem.* **2011**, *3*, 1491–1501. [CrossRef] [PubMed]

Article

Identification of Novel Src Inhibitors: Pharmacophore-Based Virtual Screening, Molecular Docking and Molecular Dynamics Simulations

Yi Zhang, Ting-jian Zhang, Shun Tu, Zhen-hao Zhang and Fan-hao Meng *

School of Pharmacy, China Medical University, Shenyang 110122, China; cruckzhang0304@163.com (Y.Z.); todayfy@outlook.com (T.-j.Z.); tushun2018@163.com (S.T.); zhangzhenhao314@163.com (Z.-h.Z.)
* Correspondence: fhmeng@cmu.edu.cn; Tel.: +86-133-8688-7639

Academic Editors: Marco Tutone and Anna Maria Almerico
Received: 25 July 2020; Accepted: 5 September 2020; Published: 8 September 2020

Abstract: Src plays a crucial role in many signaling pathways and contributes to a variety of cancers. Therefore, Src has long been considered an attractive drug target in oncology. However, the development of Src inhibitors with selectivity and novelty has been challenging. In the present study, pharmacophore-based virtual screening and molecular docking were carried out to identify potential Src inhibitors. A total of 891 molecules were obtained after pharmacophore-based virtual screening, and 10 molecules with high docking scores and strong interactions were selected as potential active molecules for further study. Absorption, distribution, metabolism, elimination and toxicity (ADMET) property evaluation was used to ascertain the drug-like properties of the obtained molecules. The proposed inhibitor–protein complexes were further subjected to molecular dynamics (MD) simulations involving root-mean-square deviation and root-mean-square fluctuation to explore the binding mode stability inside active pockets. Finally, two molecules (ZINC3214460 and ZINC1380384) were obtained as potential lead compounds against Src kinase. All these analyses provide a reference for the further development of novel Src inhibitors.

Keywords: Src inhibitors; pharmacophore model; virtual screening; molecular docking; molecular dynamics simulations

1. Introduction

The Src family kinases (SFKs) are a family of non-receptor tyrosine kinases, which are involved in a wide variety of essential functions to sustain cellular homeostasis, where they regulate cell cycle progression, motility, proliferation, differentiation and survival, among other cellular processes [1]. As a prototypical member of the SFKs, Src contains Yes, Fyn, Lyn, Lck, Hck, Fgr, Yrk, Frk and Blk kinases [2]. Src consists of four homology domains (SH1, SH2, SH3 and SH4) and a unique domain (Figure 1). The SH1 domain (also called the catalytic domain) is composed of two subdomains (generally termed N-terminal and C-terminal lobes) separated by a cleft. The N-terminal lobe contains the highly conserved hinge region that is implicated in the interaction with the ATP-adenine ring and to which most of the Src inhibitors anchor through hydrogen bonding. The C-terminal lobe is larger, comprises an activation loop that contains a tyrosine residue that can be autophosphorylated (Tyr419 in human c-Src) and is the positive regulatory site responsible for maximizing kinase activity. The phosphorylation of this residue stabilizes the kinases in an active conformation accessible to ATP and substrates. On the contrary, when another tyrosine residue located in the C-terminal lobe tail (Tyr530 in human c-Src) is phosphorylated, a closed conformation is induced [3]. The SH2 and SH3 domains regulate the Src catalytic activity through both intramolecular and protein–protein interactions.

The SH4 domain is a 15-amino acid sequence whose myristoylation allows the binding of Src members to the inner surface of the plasma membrane. The unique domain is included in the N-terminal segment of the proteins, together with SH4, and is composed of 50–70 residues. Unlike the SH domains, it displays the greatest divergence among the SFKs and thus probably contributes to the differentiation of their biological functions [4]. Src is a central signaling hub that can be activated by many factors, including immune-response receptors, integrins and other adhesion receptors, receptor protein tyrosine kinases, G protein-coupled receptors and cytokine receptors [5]. In normal cells, Src is only transiently activated during the multiple cellular events in which it is involved. Conversely, Src is overexpressed and/or hyperactivated in a large variety of solid tumors and is probably a strong promoting factor for the development of metastatic cancer phenotypes [6]. Src is responsible for many human cancers such as lung [7], neuronal [8], ovarian [9], esophageal [10] and gastric cancers [11], as well as melanoma [12] and Kaposi's sarcoma [13]. Due to its involvement in many cellular processes related to cancer development, Src has long been considered a potential drug target in oncology.

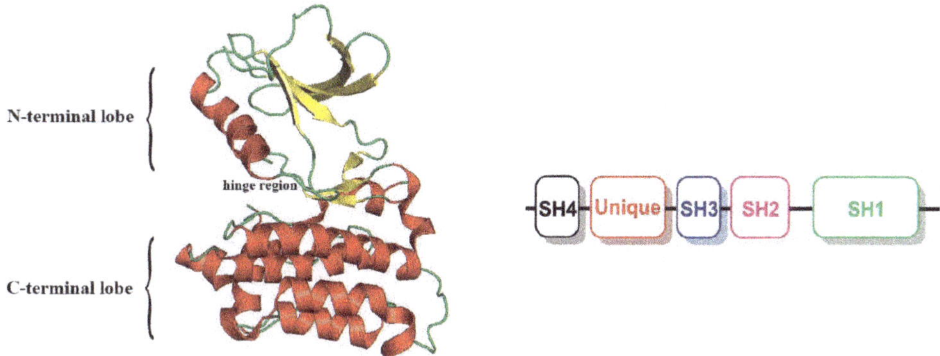

Figure 1. The crystal structure of the Src kinase and schematic domain structure.

The Src inhibitors developed to date are generally categorized into three major classes: (1) tyrosine kinase activity inhibitors (ATP-competitive inhibitors); (2) protein–protein interaction inhibitors (SH2, SH3 or substrate-binding domain); (3) enzyme destabilizers that provide a correlation between Src and its united molecular chaperone, i.e., heat shock protein 90 (Hsp90) [14,15]. The search for small molecules with an inhibitory activity toward Src kinases constitutes a growing field of study. Several compounds have entered clinical trials, with two compounds ultimately approved by the FDA: dasatinib, approved in 2006, and bosutinib, approved in 2012 [16]. However, dasatinib is known to inhibit over 40 kinases, while bosutinib inhibits over 45 kinases, making it impossible to use these compounds as selective mechanistic probes for Src-dependent pharmacology [17,18]. Furthermore, most Src inhibitors reported share similar scaffolds such as pyrazolo [3,4-d] pyrimidine, quinoline and quinazoline (Figure 2). To this end, it is meaningful to find more effective and selective Src inhibitors with new chemical scaffolds.

In this work, we report an integrated screening method containing pharmacophore-based virtual screening; molecular docking; absorption, distribution, metabolism, elimination and toxicity (ADMET) prediction; and molecular dynamics (MD) simulations to find novel Src inhibitors.

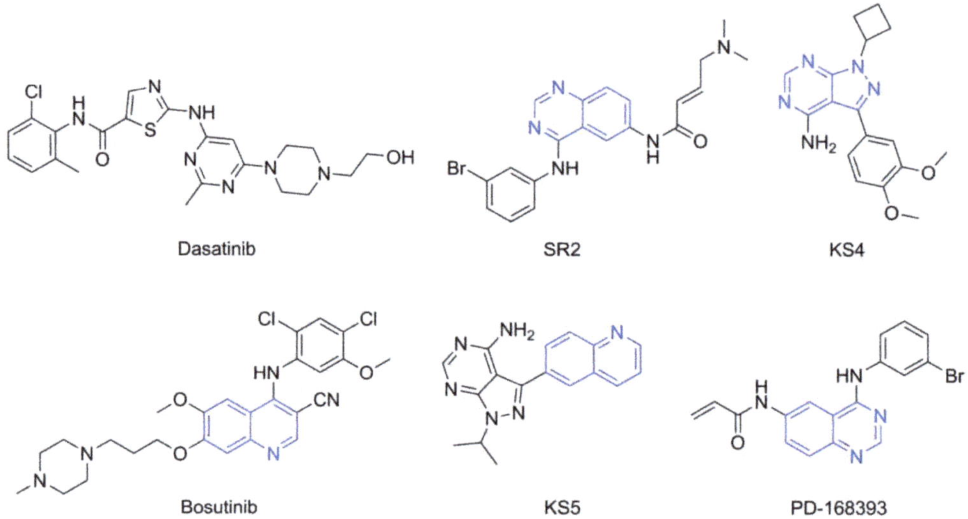

Figure 2. Chemical structures of previously reported Src inhibitors.

2. Results and Discussion

2.1. Preparation of Chemical Database

Prior to performing the virtual screening, the database needed to undergo several filtering and preparation steps to reduce the enormous number of compounds [19]. All the selected ligands were downloaded from the ZINC database (https://zinc.docking.org/) using filters such as "in-stock" and "drug-like", and the selection of ligands was performed based on Lipinski's rule of five (molecular weight limit of 300 to 600 Da, hydrogen-bond acceptor limit of 10, hydrogen-bond donor limit of 5, rotatable-bond limit of 7, and log P limit of 5). To produce a more refined and precise set of chemical data, some built-in functions such as the "partial charges" and "energy minimize" tools of the Molecular Operating Environment software (MOE, Version 2015.10) were applied on the data set. The resulting database comprised 1,033,419 molecules with lowest energy in 3D format.

2.2. Generation and Validation of Pharmacophore Model

The protein–ligand complex serves as the starting template for this modeling, wherein intermolecular interactions are perceived as feature points for subsequent virtual screening. Most often, a single protein–ligand complex is used as the template to align and score the database molecules, from which the best-fitted molecules are prioritized as potential hits [20]. Based on the crystal structure of Src kinase selected (PDB ID: 3F3V), five key pharmacophore features were generated, including one hydrogen-bond donor (Don), two hydrogen-bond acceptors (Acc), one hydrophobic and aromatic center (Hyd/Aro) and one aromatic center (Aro). As shown in Figure 3, the pharmacophore model was designed in consideration of the binding poses of the original ligand (RL45, yellow sticks). Three hydrogen-bond features were present for the ligand–protein complex: the hydrogen-bond donor (F1, purple sphere) of RL45 interacted with Glu310; the hydrogen-bond acceptors (F2 and F3, cyan sphere) of RL45 interacted with Asp404 and Met341, respectively. One hydrophobic and aromatic center (F4, yellow sphere) and one aromatic center (F5, orange sphere) were also present for the ligand.

For the validation of generated pharmacophore model, a test database was built including 18 Src reported inhibitors and 18 collected decoy molecules, which can be seen in the Supplementary Materials. The test database was then subjected to screening against the pharmacophore model to validate its precision. As a result, we obtained 14 active molecules as hits, and none of the inactive molecules was

mapped to the pharmacophore model. The results from the test database revealed the precision of the generated pharmacophore model.

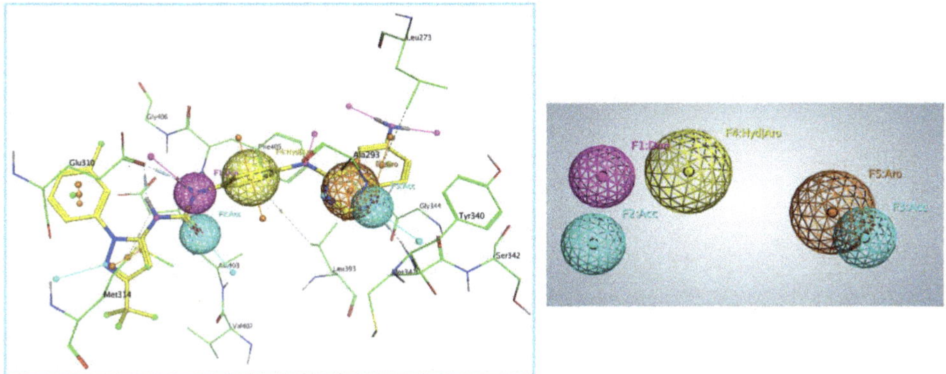

Figure 3. Pharmacophore features generated by Molecular Operating Environment (MOE).

2.3. Pharmacophore-Based Virtual Screening

The pharmacophore-model-based screening of databases has been considered as an important tool for computer-aided drug discovery techniques and provides information about geometric and electronic features that are involved in interaction with receptors [21]. In this part, the chemical database comprised 1,033,419 molecules with lowest energy in 3D format and was generated by applying various filters. We utilized the protein–ligand complex reported for Src for pharmacophore model generation and performed virtual screening of the prepared database to find out the best matches against the model. As a result, 891 molecules were obtained as hits on the basis of pharmacophore features.

2.4. Molecular Docking

The hits obtained from the pharmacophore-based virtual screening were subjected to molecular docking studies; the top 10 molecules with the highest docking scores were selected for the study of binding modes (Table 1). Among these 10 molecules, ZINC23247639 and ZINC10479320 have been reported as broad-spectrum kinase inhibitors that interacted with the highly conserved ATP-binding sites of many human protein kinases [22–24]; thus, we filtered them out in this investigation. As we know, hydrogen bonds established between receptor and ligand play a major role in the functionality and stability of the complex. Hence, we observed the dominant hydrogen-bond interactions between the groups of the other eight hit molecules and the residues of the active site, and then, ZINC3214460 and ZINC1380384 caught our attention.

Using the default GBVI/WSA dG as a docking function in the MOE software, ZINC3214460 and ZINC1380384's docking scores were calculated as −9.6287 and −8.9096 kcal/mol, respectively. The binding interactions of ZINC3214460 and ZINC1380384 were illustrated by the PyMOL [25] software (Figure 4). As we can see, some key amino acid residues were involved in hydrogen-bonding interactions with ZINC3214460; a carbonyl group accepted an H-bond from the Asp404, and an N atom of the isoxazole formed an H-bond with Met341. The H-bond distances were 2.87 and 3.19 Å, respectively, and hydrogen-bonding energy components contributed −4.6 kcal/mol to the binding. Correspondingly, ZINC1380384 formed three H-bond interactions with the kinase. The amide fragment formed two H-bonds with Asp404 and Glu310, respectively, and an N atom of the benzimidazole accepted an H-bond from a Met341 residue. The hydrogen-bonding energy components contributed −6.9 kcal/mol to the binding. On the basis of good binding energies and their pattern of binding interaction with active pocket, two selected molecules showed strong interaction with the key amino acid residues of Src kinase. Furthermore, these two molecules have not been reported as kinase

inhibitors or related with cancer yet, so our attention is focused on ZINC3214460 and ZINC1380384, which were further subjected to ADMET prediction and molecular dynamics simulations.

Table 1. The structures and docking results of molecules.

ZINC ID	Structure	Src Docking Score (kcal/mol)
ZINC3214460	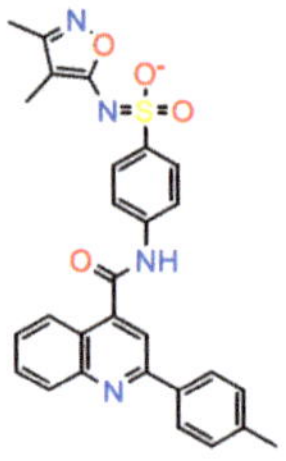	−9.6287
ZINC61925676	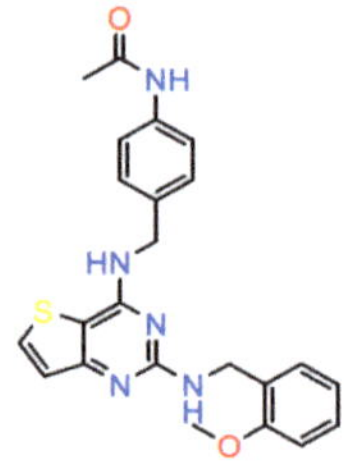	−9.1879
ZINC58158745	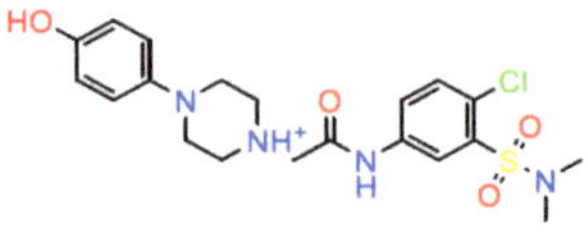	−9.1320
ZINC12075400	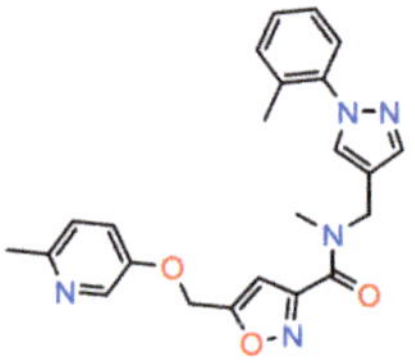	−8.9992
ZINC1380384	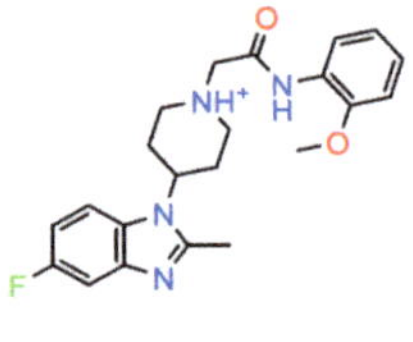	−8.9096
ZINC12853028	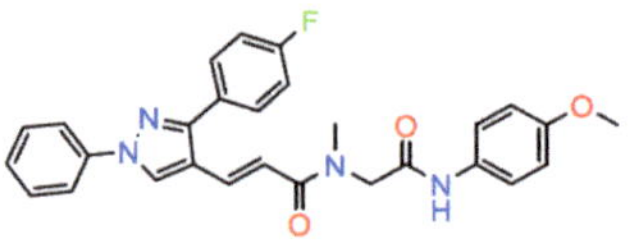	−8.7889

Table 1. *Cont.*

ZINC ID	Structure	Src Docking Score (kcal/mol)
ZINC23247639		−8.7219
ZINC949873		−8.5887
ZINC36389462		−8.5816
ZINC10479320		−8.5090

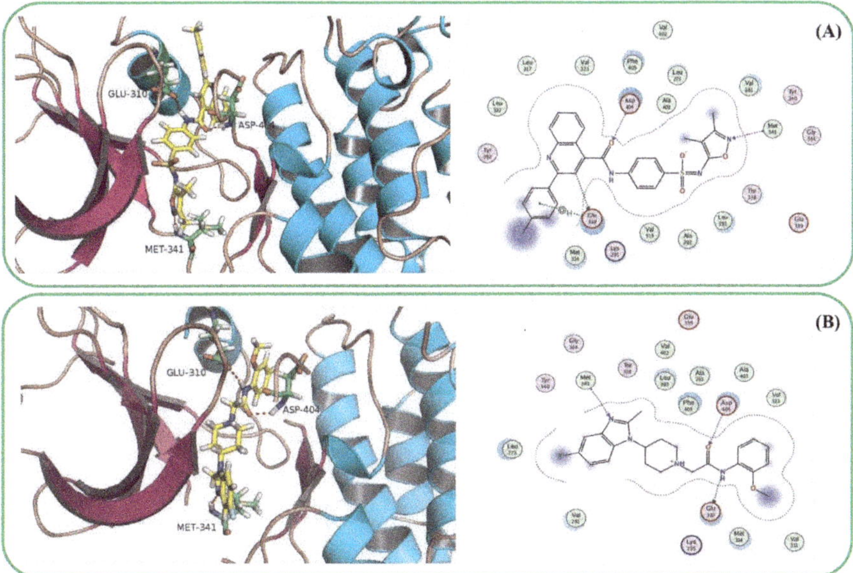

Figure 4. Binding interactions of two hit molecules with active pocket of Src kinase. The hit molecules ZINC3214460 (**A**) and ZINC1380384 (**B**) are displayed in yellow sticks, and catalytic residues are displayed in green sticks. Hydrogen bonds are shown as red dashes.

2.5. ADMET Prediction

The lead compounds in drug development were found to have favorable absorption, distribution, metabolism, elimination and toxicity (ADMET) properties [26]. Pharmacokinetic properties predict the drug-likeness of ligand molecules. Therefore, the ADMET properties of molecules are essential for the development of an effective druggable molecule. In this section, ADMET characteristics such as buffer solubility, blood–brain barrier penetration (BBB), Caco-2 permeability, human intestinal absorption (HIA), plasma protein binding (PPB), cytochrome P450 2D6 (CYP2D6) modulation and hERG inhibition were studied for ZINC3214460, ZINC1380384, dasatinib and bosutinib. The results are summarized in Table 2. Solubility and human intestinal absorption are two key factors that affect oral bioavailability. We found that ZINC3214460 and ZINC1380384 show extremely high values of solubility. Low blood–brain barrier permeability was found, which served in reducing the side effects and toxicity to the brain, and the values of all the compounds are less than 1 (C.brain/C.blood < 1), suggesting that they are inactive in the CNS (central nervous system). Caco-2 permeability was used to evaluate the suitability of compounds for oral dosing, and the proposed compounds have slightly worse human intestinal permeability than two drugs approved by the FDA. In addition, the comparable intestinal absorption (HIA) for dasatinib and bosutinib indicated that two proposed compounds possess good bioavailability. The high plasma-protein binding of ZINC3214460 means a long half-life and stable efficacy, which could maintain a durable potency and adequate stability of the compound. The inhibition of CYP2D6 by a drug constitutes the majority of cases of drug–drug interaction. It was found that none of the compounds may inhibit CYP2D6. The cardiotoxicity may be related with the high risk of hERG inhibition, as shown in Table 2; ZINC3214460 has a low risk of inhibiting hERG, meaning that it has little chance of causing cardiac problems. However, according to the high-risk level of hERG inhibition, the potential cardiotoxicity of ZINC1380384 should be considered in the future. More ADMET prediction data for these two proposed molecules can be found in the Supplementary Materials (Figures S1 and S2).

Table 2. The absorption, distribution, metabolism, elimination and toxicity (ADMET) prediction for the investigated compounds.

Compound	Buffer Solubility [1]	BBB [2]	Caco-2 [3]	HIA [4]	PPB [5]	CYP2D6 Inhibition	hERG Inhibition
ZINC3214460	81.69	0.01036	18.87	97.41	100	None	Low risk
ZINC1380384	3735.39	0.4491	26.62	92.11	34.90	None	High risk
Dasatinib	0.3113	0.03504	32.01	93.59	70.29	None	Medium risk
Bosutinib	5.500	0.06055	50.35	97.23	85.07	None	Medium risk

[1] Buffer solubility: water solubility in buffer system (SK atomic types, mg/L), [2] BBB: blood–brain barrier penetration (C.brain/C.blood), [3] Caco-2: in vitro Caco-2 cell permeability (nm/sec), [4] HIA: human intestinal absorption (%), [5] PPB: plasma protein binding (%).

2.6. Molecular Dynamics Simulations

MD simulations were conducted to check the stability of the complexes predicted by molecular docking. After 50 ns MD simulations, the root-mean-square deviation (RMSD) of the backbone of Src kinase and the ligands at 300 K was plotted against time (ns). As can be seen in Figure 5, the RMSDs of 3F3V-ZINC3214460 and 3F3V-ZINC1380384 were discovered to be relatively stable at about 0.27 and 0.22 nm, respectively. There were some fluctuations in the beginning, and then, the complexes gradually tended to equilibrium until the time reached 25 ns of simulation. The RMSD values of RL45 and ZINC1380384 in the binding site of Src are similar; however, the 3F3V-ZINC3214460 complex has a larger RMSD value at around 0.30 nm. The root-mean-square fluctuation (RMSF) is an important parameter that yields data about the structural adaptability of every residue in the system. The RMSF values for all the residues in the 3F3V-ZINC3214460 and 3F3V-ZINC1380384 complexes were calculated (Figure 6). In general, the Src inhibitors are settled to the highly conserved region of

the protein (Glu339, Asp348 and Asp404 in the affinity pocket and Glu310 and Leu317 in the hinge region) through hydrogen bonds, van der Waals forces and hydrophobic interaction, etc. We found that the RMSF fluctuation values of the conserved amino acids (e.g., Glu310, Leu317 and Glu339) in 3F3V-ZINC3214460 and 3F3V-ZINC1380384 were lower than those in 3F3V-RL45, reflecting that the two proposed molecules could form stronger interactions with these conserved residues. The proteins showed distinct behavior with a varied magnitude of fluctuations at the loop region (e.g., Ser522 and Tyr527), which could be used to discover more selective inhibitors of Src. Based on the above analysis, we can conclude that these two selected molecules matched very well with the Src binding pocket, suggesting reasonable binding modes.

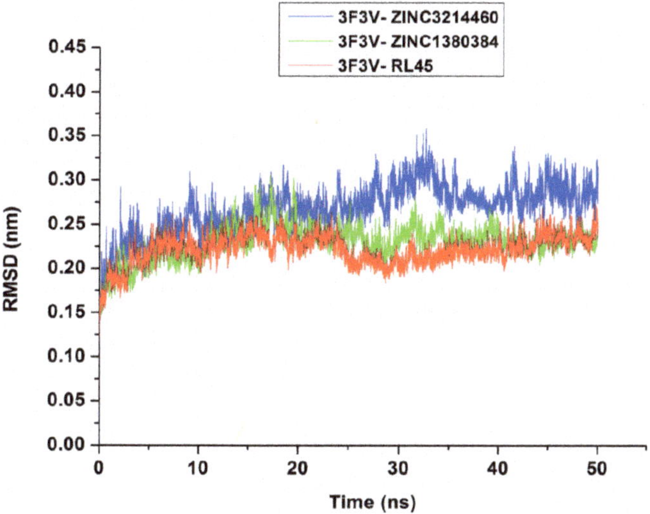

Figure 5. The root-mean-square deviation (RMSD) trajectories of 3F3V–inhibitor complexes during 50 ns simulations.

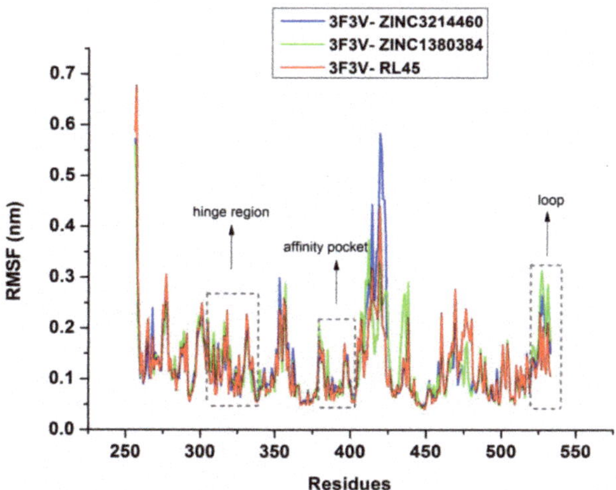

Figure 6. The root-mean-square fluctuation (RMSF) maps of 3F3V–inhibitor complexes during simulations.

3. Materials and Methods

3.1. Generation and Validation of Pharmacophore Model

In this study, a protein–ligand complex-based pharmacophore model was generated by using the pharmacophore query editor protocol of MOE. Binding interactions induce significant chemical features, which were taken into account for the creation of the pharmacophore model [27]. To generate a pharmacophore model with good quality, MOE utilizes an in-built set of pharmacophore features including hydrogen-bond donor (Don), hydrogen-bond acceptor (Acc), aromatic center (Aro), Pi ring center (PiR), aromatic ring or Pi ring normal (PiN), hydrophobic atom (HydA), anionic atom (Ani), cationic atom (Cat) and so on. We summarized all the binding interactions of Src complexes in the DFG-out state from the RCSB Protein Data Bank (PDB, https://www.rcsb.org/), and several common features (e.g., hydrophobic interactions, hydrogen bonding modes and catalytic residues) were utilized for generating the model. Finally, the crystal structure of Src kinase in complex with a substrate-based inhibitor, 1-(4-((6-aminoquinazolin-4-yl) amino) phenyl)-3-(3-(tert-butyl)-1-(m-tolyl)-1H-pyrazol-5-yl) urea (RL45) (PDB ID: 3F3V; resolution, 2.6 Å), was chosen for the creation of a complex-based pharmacophore system. The inhibitor RL45 had strong interaction with key amino acid residues of Src kinase (i.e., Asp404, Glu310 and Met341) and was further used as a reference in molecular docking (Figure 7). The binding interactions of RL45 with the active pocket of Src are illustrated by the PyMOL software (2.3.2).

The generated pharmacophore model was validated via a test database including 18 Src reported inhibitors and 18 collected decoy molecules. The molecules of Src inhibitors were downloaded from DrugBank (https://www.drugbank.ca/). The decoy molecules refer to those compounds with reported activities not related to Src, whereas their physical properties, including molecular weight, number of hydrogen-bond donors and acceptor, number of rotatable bonds, and Log P were similar to those of known Src inhibitors. The test database can be seen in the Supplementary Materials.

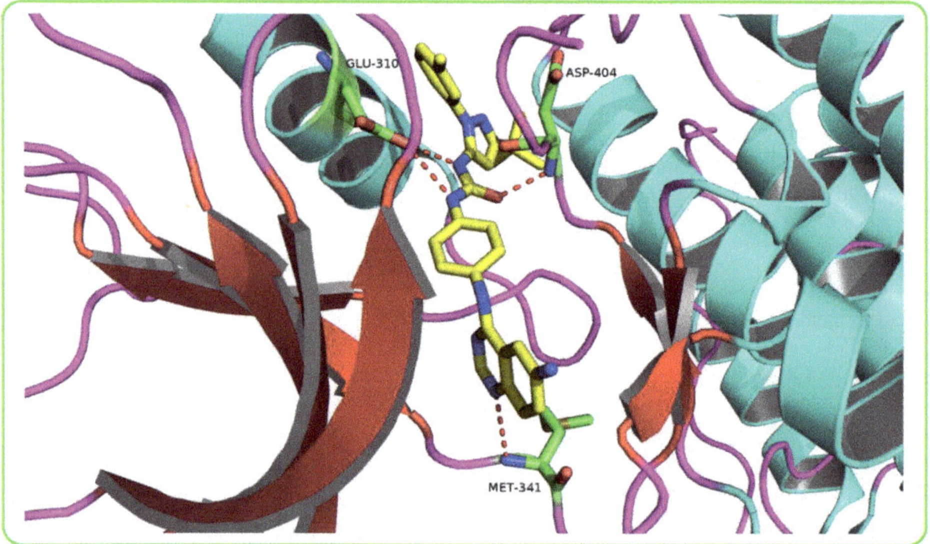

Figure 7. Binding interaction of RL45 with active pocket of Src kinase. The ligand RL45 is displayed in yellow sticks, and catalytic residues are displayed in green sticks. Hydrogen bonds are shown as red dashes.

3.2. Pharmacophore-Based Virtual Screening

In Computer-Aided Drug Designing (CADD), virtual screening is one of the time-saving methods for the discovery of novel, potent and drug-like compounds [28]. Pharmacophore models have the advantage that they can be used not only to identify novel active compounds in virtual screening but also for anti-target modeling to avoid side-effects resulting from off-target activity [29]. Especially, when structural information about the target protein or the ligand's active conformation is available, pharmacophore-based models are superior to docking and quantitative structure–activity relationship (QSAR) methods [30]. Based on the pharmacophore model generated above, virtual screening was conducted by a pharmacophore search protocol in MOE with an EHT scheme. MOE's pharmacophore search is used to apply a query to a database of molecular conformations and report those conformations as hits that satisfy the pharmacophore features. The hit molecules were preferred and kept in a separate database for the further evaluation of interactions.

3.3. Molecular Docking

The crystal structure of Src kinase was obtained from the RCSB Protein Data Bank (PDB ID: 3F3V; resolution, 2.6 Å). The protein was prepared for docking using the quickprep tool of MOE, including correcting structural issues, protonating the structure, deleting unbound water molecules and minimizing the structure to a specified gradient, to make the pocket available for the docking of new molecules. The original ligand (RL45) was used to define the binding site of the Src active pocket. For the docking parameters, we set the force field to MMFF94x and used the triangle matcher placement algorithm [31], which returned thirty poses; we also used the Rigid Receptor refinement method, which returned five poses. The London dG method was applied to score the poses in both steps [32]. By studying the top-scored docking poses, only the molecular poses for which the binding modes satisfied the pharmacophore features were retained. Each molecule with the highest docking score was regarded as a docking result for further analysis.

3.4. ADMET Prediction

Owing to poor pharmacokinetic parameters, many drugs have not passed through clinical trial stages [33]. Thus, it is necessary to predict the pharmacokinetics and toxicity of newly obtained molecules, which were selected from the docking results for further analysis. In this section, the Pre-ADMET server application (https://preadmet.bmdrc.kr) was used. The Pre-ADMET approach is based on different classes of molecular parameters that are considered for generating quantitative structure properties [34].

3.5. Molecular Dynamics Simulations

Based on the docking results, the best-posed complex was subjected to MD simulation studies using the Groningen Machine for Chemicals Simulations (GROMACS) 5.0 package [35] with a CHARMM36 force field [36] under periodic boundary conditions for molecules. Ligand topology files were generated using the CHARMM General Force Field [37]. The charge of the system was neutralized by the addition of the ions. The energy was minimized using a steepest-gradient method to remove any close contacts. The particle mesh Ewald (PME) method was employed for energy calculation and for electrostatic and Van der Waals interactions. The systems were equilibrated in the NVT ensemble for 50,000 steps, followed by equilibration in the NPT ensemble for an additional 50,000 steps. Finally, 50 ns molecular dynamics simulations were performed at 300 K with a 2.0 fs time step, and coordinates were saved every picosecond for analysis [38,39].

4. Conclusions

In this study, in order to find novel Src inhibitors, an integrated screening method was employed; through pharmacophore-based virtual screening and molecular docking, the top 10 molecules with

good binding scores were selected for the study of binding modes. ADMET prediction and molecular dynamics simulations were used to predict the pharmacokinetic properties and stabilities of proposed ligand–protein complexes. Finally, two molecules (ZINC3214460 and ZINC1380384) were selected with excellent properties and stable binding modes. In addition, ZINC1380384, possessing a novel benzo [d] imidazole scaffold, was valuable for further optimization and provides a reference for the development of novel potent Src inhibitors.

Supplementary Materials: Figure S1: The ADMET prediction for ZINC3214460; Figure S2: The ADMET prediction for ZINC1380384.

Author Contributions: Conceptualization, T.-j.Z.; methodology, Y.Z.; software, Y.Z. and T.-j.Z.; validation, Y.Z., S.T. and Z.-h.Z.; writing—original draft preparation, Y.Z.; writing—review and editing, Y.Z.; project administration, F.-h.M. All authors have read and agreed to the published version of the manuscript.

Funding: This research was funded by the National Natural Science Foundation of China (Grant No. 81803355 and 81573687) and the Scientific Research Fund of Liaoning Provincial Education Department (LQNK201739).

Conflicts of Interest: The authors have no conflict of interest to declare.

Abbreviations

Src: sarcoma; SFKs: Src family kinases; MOE: molecular operating environment; ADMET: absorption, distribution, metabolism, elimination and toxicity; MD: molecular dynamics; RMSD: root-mean-square deviation; RMSF: root-mean-square fluctuation

References

1. Roskoski, R., Jr. Src protein-tyrosine kinase structure, mechanism, and small molecule inhibitors. *Pharmacol. Res.* **2015**, *94*, 9–25. [CrossRef] [PubMed]

2. Olgen, S. Design strategies, structures and molecular interactions of small molecule Src inhibitors. *Anti-Cancer Agents Med. Chem.* **2016**, *16*, 992–1002. [CrossRef]

3. Zhang, H.; Forman, H.J. 4-Hydroxynonenal activates Src through a non-canonical pathway that involves EGFR/PTP1B. *Free Radic. Biol. Med.* **2015**, *89*, 701–707. [CrossRef] [PubMed]

4. Le Roux, A.L.; Mohammad, I.L.; Mateos, B.; Arbesú, M.; Gairí, M.; Khan, F.A.; Teixeira, J.M.C.; Pons, M. A myristoyl-binding site in the SH3 domain modulates c-Src membrane anchoring. *iScience* **2019**, *12*, 194–203. [CrossRef] [PubMed]

5. Ge, M.M.; Zhou, Y.Q.; Tian, X.B.; Manyande, A.; Tian, Y.K.; Ye, D.W.; Yang, H. Src-family protein tyrosine kinases: A promising target for treating chronic pain. *Biomed. Pharmacother.* **2020**, *125*, 110017. [CrossRef]

6. Tintori, C.; Fallacara, A.L.; Radi, M.; Zamperini, C.; Dreassi, E.; Crespan, E.; Maga, G.; Schenone, S.; Musumeci, F.; Brullo, C.; et al. Combining X-ray crystallography and molecular modeling toward the optimization of pyrazolo[3,4-d]pyrimidines as potent c-Src inhibitors active in vivo against neuroblastoma. *J. Med. Chem.* **2015**, *58*, 347–361. [CrossRef] [PubMed]

7. Liu, L.; Wang, W.; Gao, S.; Wang, X. MicroRNA-208a directly targets Src kinase signaling inhibitor 1 to facilitate cell proliferation and invasion in non-small cell lung cancer. *Mol. Med. Rep.* **2019**, *20*, 3140–3148. [CrossRef]

8. Poli, G.; Martinelli, A.; Tuccinardi, T. Computational approaches for the identification and optimization of Src family kinases inhibitors. *Curr. Med. Chem.* **2014**, *21*, 3281–3293. [CrossRef]

9. Kadife, E.; Chan, E.; Luwor, R.; Kannourakis, G.; Findlay, J.; Ahmed, N. Paclitaxel-induced Src activation is inhibited by dasatinib treatment, independently of cancer stem cell properties, in a mouse model of ovarian cancer. *Cancers* **2019**, *11*, 243. [CrossRef]

10. Liu, Z.; Chen, Z.; Wang, J.; Zhang, M.; Li, Z.; Wang, S.; Dong, B.; Zhang, C.; Gao, J.; Shen, L. Mouse avatar models of esophageal squamous cell carcinoma proved the potential for EGFR-TKI afatinib and uncovered Src family kinases involved in acquired resistance. *J. Hematol. Oncol.* **2018**, *11*, 109. [CrossRef]

11. Da Costa, J.D.E.F.F.B.; Sant'Anna, C.D.; Muniz, J.A.P.C.; Da Rocha, C.A.M.; Lamarao, L.M.; Nunes, C.D.F.A.M.; De Assumpcao, P.P.; Burbano, R.R. Deregulation of the SRC family tyrosine kinases in gastric carcinogenesis in non-human primates. *Anticancer Res.* **2018**, *38*, 6317–6320. [CrossRef]

12. Feddersen, C.R.; Schillo, J.L.; Varzavand, A.; Vaughn, H.R.; Wadsworth, L.S.; Voigt, A.P.; Zhu, E.Y.; Jennings, B.M.; Mullen, S.A.; Bobera, J.; et al. Src-dependent DBL family members drive resistance to vemurafenib in human melanoma. *Cancer Res.* **2019**, *79*, 5074–5087. [CrossRef] [PubMed]

13. Abere, B.; Samarina, N.; Gramolelli, S.; Rückert, J.; Gerold, G.; Pich, A.; Schulz, T.F. Kaposi's sarcoma-associated herpesvirus nonstructural membrane protein pK15 recruits the class II phosphatidylinositol 3-kinase PI3K-C2 alpha to activate productive viral replication. *J. Virol.* **2018**, *92*, e00544-18. [CrossRef] [PubMed]

14. Wadhawan, A.; Smith, C.; Nicholson, R.I.; Barrett-Lee, P.; Hiscox, S. Src-mediated regulation of homotypic cell adhesion: Implications for cancer progression and opportunities for therapeutic intervention. *Cancer Treat. Rev.* **2011**, *37*, 234–241. [CrossRef]

15. Ungefroren, H.; Sebens, S.; Groth, S.; Gieseler, F.; Fändrich, F. Differential roles of Src in transforming growth factor-ß regulation of growth arrest, epithelial-to-mesenchymal transition and cell migration in pancreatic ductal adenocarcinoma cells. *Int. J. Oncol.* **2011**, *38*, 797–805. [CrossRef]

16. Du, G.; Rao, S.; Gurbani, D.; Henning, N.J.; Jiang, J.; Che, J.; Yang, A.; Ficarro, S.B.; Marto, J.A.; Aguirre, A.J.; et al. Structure-based design of a potent and selective covalent inhibitor for SRC kinase that targets a P-Loop cysteine. *J. Med. Chem.* **2020**, *63*, 1624–1641. [CrossRef]

17. Li, J.; Rix, U.; Fang, B.; Bai, Y.; Edwards, A.; Colinge, J.; Bennett, K.L.; Gao, J.; Song, L.; Eschrich, S.; et al. A chemical and phosphoproteomic characterization of dasatinib action in lung cancer. *Nat. Chem. Biol.* **2010**, *6*, 291–299. [CrossRef]

18. Remsing Rix, L.L.; Rix, U.; Colinge, J.; Hantschel, O.; Bennett, K.L.; Stranzl, T.; Müller, A.; Baumgartner, C.; Valent, P.; Augustin, M.; et al. Global target profile of the kinase inhibitor bosutinib in primary chronic myeloid leukemia cells. *Leukemia* **2009**, *23*, 477–485. [CrossRef]

19. Gentile, F.; Barakat, K.H.; Tuszynski, J.A. Computational characterization of small molecules binding to the human XPF active site and virtual screening to identify potential new DNA repair inhibitors targeting the ERCC1-XPF endonuclease. *Int. J. Mol. Sci.* **2018**, *19*, 1328. [CrossRef]

20. Braga, R.C.; Andrade, C.H. Assessing the performance of 3D pharmacophore models in virtual screening: How good are they? *Curr. Top. Med. Chem.* **2013**, *13*, 1127–1138. [CrossRef]

21. Dhanjal, J.K.; Sharma, S.; Grover, A.; Das, A. Use of ligand-based pharmacophore modeling and docking approach to find novel acetylcholinesterase inhibitors for treating Alzheimer's. *Biomed. Pharmacother.* **2015**, *71*, 146–152. [CrossRef] [PubMed]

22. Deibler, K.K.; Mishra, R.K.; Clutter, M.R.; Antanasijevic, A.; Bergan, R.; Caffrey, M.; Scheidt, K.A. A Chemical probe strategy for interrogating inhibitor selectivity across the MEK kinase family. *Acs Chem. Biol.* **2017**, *12*, 1245–1256. [CrossRef] [PubMed]

23. Gattelli, A.; García Solá, M.E.; Roloff, T.C.; Cardiff, R.D.; Kordon, E.C.; Chodosh, L.A.; Hynes, N.E. Chronic expression of wild-type Ret receptor in the mammary gland induces luminal tumors that are sensitive to Ret inhibition. *Oncogene* **2018**, *37*, 4046–4054. [CrossRef] [PubMed]

24. Nguyen, H.P.; Koutsoukas, A.; Mohd Fauzi, F.; Drakakis, G.; Maciejewski, M.; Glen, R.C.; Bender, A. Diversity selection of compounds based on 'protein affinity fingerprints' improves sampling of bioactive chemical space. *Chem. Biol. Drug Des.* **2013**, *82*, 252–266. [CrossRef] [PubMed]

25. Lill, M.A.; Danielson, M.L. Computer-aided drug design platform using PyMOL. *J. Comput. Aided Mol. Des.* **2011**, *25*, 13–19. [CrossRef]

26. Clark, R.D. Predicting mammalian metabolism and toxicity of pesticides in silico. *Pest Manag. Sci.* **2018**, *74*, 1992–2003. [CrossRef]

27. Takagi, Y.; Matsui, K.; Nobori, H.; Maeda, H.; Sato, A.; Kurosu, T.; Orba, Y.; Sawa, H.; Hattori, K.; Higashino, K.; et al. Discovery of novel cyclic peptide inhibitors of dengue virus NS2B-NS3 protease with antiviral activity. *Bioorg. Med. Chem. Lett.* **2017**, *27*, 3586–3590. [CrossRef]

28. Wadood, A.; Riaz, M.; Uddin, R.; Ul-Haq, Z. In silico identification and evaluation of leads for the simultaneous inhibition of protease and helicase activities of HCV NS3/4A protease using complex based pharmacophore mapping and virtual screening. *PLoS ONE* **2014**, *9*, e89109. [CrossRef]

29. Schuster, D. 3D pharmacophores as tools for activity profiling. *Drug Discov. Today. Technol.* **2010**, *7*, e205–e211. [CrossRef]

30. Drwal, M.N.; Griffith, R. Combination of ligand- and structure-based methods in virtual screening. *Drug Discov. Today. Technol.* **2013**, *10*, e395–e401. [CrossRef]

31. Courcot, B.; Bridgeman, A.J. Modeling the interactions between polyoxometalates and their environment. *J. Comput. Chem.* **2011**, *32*, 3143–3153. [CrossRef] [PubMed]

32. Corbeil, C.R.; Williams, C.I.; Labute, P. Variability in docking success rates due to dataset preparation. *J. Comput. Aided Mol. Des.* **2012**, *26*, 775–786. [CrossRef] [PubMed]

33. Jia, W.Q.; Liu, Y.Y.; Feng, X.Y.; Xu, W.R.; Cheng, X.C. Discovery of novel and highly selective PI3Kδ inhibitors based on the p110δ crystal structure. *J. Biomol. Struct. Dyn.* **2020**, *38*, 2499–2508. [CrossRef] [PubMed]

34. Teng, Y.; Lu, X.; Xiao, M.; Li, Z.; Zou, Y.; Ren, S.; Cheng, Y.; Luo, G.; Xiang, H. Discovery of potent and highly selective covalent inhibitors of Bruton's tyrosine kinase bearing triazine scaffold. *Eur. J. Med. Chem.* **2020**, *199*, 112339. [CrossRef]

35. Abraham, M.J.; Murtola, T.; Schulz, R.; Páll, S.; Smith, J.C.; Hess, B.; Lindahl, E. GROMACS: High performance molecular simulations through multi-level parallelism from laptops to supercomputers. *Softwarex* **2015**, *1*, 19–25. [CrossRef]

36. Huang, J.; MacKerell, A.D., Jr. CHARMM36 all-atom additive protein force field: Validation based on comparison to NMR data. *J. Comput. Chem.* **2013**, *34*, 2135–2145. [CrossRef]

37. Vanommeslaeghe, K.; Hatcher, E.; Acharya, C.; Kundu, S.; Zhong, S.; Shim, J.; Darian, E.; Guvench, O.; Lopes, P.; Vorobyov, I.; et al. CHARMM general force field: A force field for drug-like molecules compatible with the CHARMM all-atom additive biological force fields. *J. Comput. Chem.* **2010**, *31*, 671–690. [CrossRef]

38. Dolezal, R.; Soukup, O.; Malinak, D.; Savedra, R.M.L.; Marek, J.; Dolezalova, M.; Pasdiorova, M.; Salajkova, S.; Korabecny, J.; Honegr, J.; et al. Towards understanding the mechanism of action of antibacterial N-alkyl-3-hydroxypyridinium salts: Biological activities, molecular modeling and QSAR studies. *Eur. J. Med. Chem.* **2016**, *121*, 699–711. [CrossRef]

39. Childers, M.C.; Daggett, V. Validating molecular dynamics simulations against experimental observables in light of underlying conformational ensembles. *J. Phys. Chem. B* **2018**, *122*, 6673–6689. [CrossRef]

Sample Availability: Samples of the compounds are not available from the authors.

Article

In Silico Study Identified Methotrexate Analog as Potential Inhibitor of Drug Resistant Human Dihydrofolate Reductase for Cancer Therapeutics

Rabia Mukhtar Rana [1], Shailima Rampogu [1], Noman Bin Abid [2], Amir Zeb [1], Shraddha Parate [1], Gihwan Lee [1], Sanghwa Yoon [1], Yumi Kim [1], Donghwan Kim [1] and Keun Woo Lee [1,*]

[1] Division of Life Sciences, Division of Applied Life Science (BK21 Plus), Research Institute of Natural Science (RINS), Gyeongsang National University (GNU), 501 Jinju-daero, Jinju 52828, Korea; rabia.mukhtar.rana@gmail.com (R.M.R.); shailima.rampogu@gmail.com (S.R.); zebamir85@gmail.com (A.Z.); parateshraddha@gmail.com (S.P.); pika890131@gmail.com (G.L.); jsyoon0517@gmail.com (S.Y.); yumikim@gnu.ac.kr (Y.K.); donghwanz@naver.com (D.K.)
[2] Division of Life Science and Applied Life Science (BK 21), College of Natural Sciences, Gyeongsang National University, Jinju 52828, Korea; noman_abid@gnu.ac.kr
*   Correspondence: kwlee@gnu.ac.kr; Tel.: +82-55-772-1360

Academic Editors: Marco Tutone and Anna Maria Almerico
Received: 18 June 2020; Accepted: 30 July 2020; Published: 31 July 2020

Abstract: Drug resistance is a core issue in cancer chemotherapy. A known folate antagonist, methotrexate (MTX) inhibits human dihydrofolate reductase (hDHFR), the enzyme responsible for the catalysis of 7,8-dihydrofolate reduction to 5,6,7,8-tetrahydrofolate, in biosynthesis and cell proliferation. Structural change in the DHFR enzyme is a significant cause of resistance and the subsequent loss of MTX. In the current study, wild type hDHFR and double mutant (engineered variant) F31R/Q35E (PDB ID: 3EIG) were subject to computational study. Structure-based pharmacophore modeling was carried out for wild type (WT) and mutant (MT) (variant F31R/Q35E) hDHFR structures by generating ten models for each. Two pharmacophore models, WT-pharma and MT-pharma, were selected for further computations, and showed excellent ROC curve quality. Additionally, the selected pharmacophore models were validated by the Guner-Henry decoy test method, which yielded high goodness of fit for WT-hDHFR and MT-hDHFR. Using a SMILES string of MTX in ZINC[15] with the selections of 'clean', in vitro and in vivo options, 32 MTX-analogs were obtained. Eight analogs were filtered out due to their drug-like properties by applying absorption, distribution, metabolism, excretion, and toxicity (ADMET) assessment tests and Lipinski's Rule of five. WT-pharma and MT-pharma were further employed as a 3D query in virtual screening with drug-like MTX analogs. Subsequently, seven screening hits along with a reference compound (MTX) were subjected to molecular docking in the active site of WT- and MT-hDHFR. Through a clustering analysis and examination of protein-ligand interactions, one compound was found with a ChemPLP fitness score greater than that of MTX (reference compound). Finally, a simulation of molecular dynamics (MD) identified an MTX analog which exhibited strong affinity for WT- and MT-hDHFR, with stable RMSD, hydrogen bonds (H-bonds) in the binding site and the lowest MM/PBSA binding free energy. In conclusion, we report on an MTX analog which is capable of inhibiting hDHFR in wild type form, as well as in cases where the enzyme acquires resistance to drugs during chemotherapy treatment.

Keywords: methotrexate; drug resistance; human dihydrofolate reductase; pharmacophore modeling; virtual screening; molecular docking; molecular dynamics simulation.

1. Introduction

A major complication in cancer treatment with chemotherapy is the development of resistance to previously effective drugs. Clinically, two main types of drug resistance exist: intrinsic resistance, which is not associated with drug exposure, but rather, with an innate ability of tumor cells; and acquired resistance, which occurs after exposure to the drug [1]. Various mechanisms like increased rates of drug efflux, alterations in drug metabolism, variations in drug targets, increased target expression, activation of survival pathways, increased expression of anti-apoptotic proteins and mutation of drug targets are involved in acquiring resistance to chemotherapeutic agents [2].

Human dihydrofolate reductase (hDHFR) catalyzes the reduction of 7,8-dihydrofolate (DHF) to 5,6,7,8-tetrahydrofolate in a nicotinamide adenine dinucleotide phosphate (NADPH) dependent manner. Tetrahydrofolate is an essential cofactor in several metabolic pathways like purine and thymidylate biosynthesis, playing a vital role in cell division and proliferation [3]. Due to the significance of its crucial role in nucleoside biosynthesis, hDHFR has widely been studied and exploited as a drug target [4,5].

Methotrexate (MTX) ($C_{20}H_{22}N_8O_5$) is an antimetabolite, an analogue of folic acid and a derivative of aminopterin antiproliferative drugs that inhibits dihydrofolate reductase [6]. The drug primarily penetrates intracellular targets through an active carrier transport mechanism which is shared by reduced folates and facilitated by the reduced folate carrier (RFC) [7]. This process is carried out by the enzyme folylpolyglutamate synthetase (FPGS) through the accumulation of glutamate residues [8,9]. MTX and polyglutamylated conformations of MTX are tightly bound inhibitors of hDHFR and hinder pyrimidine, and hence thymidylate biosynthesis [10–12]. Decreased MTX affinity has been detected in cell lines exposed to increased dosage causing mutations in hDHFR [13–18]. Mutations in dihydrofolate reductase variants with amino acid substitutions at residues Phe31 [19], Arg70 [20], Leu22 [21,22], Val115 [23] and Phe34 [24] existing in folate binding site have been detected in MTX-resistant cancer cell lines.

MTX-resistant point mutant hDHFR crystal structures have provided an understanding of the details of decreased binding of MTX or other antifolates [25–30]. Volpato et al. reported a combinatorial MTX-resistant hDHFR variant F31R/Q35E which exhibited >650-fold decreased binding to MTX to reveal the structural details of MTX resistance in the F31R/Q35E variant, and obtained the crystal structure of this variant bound with MTX at 1.7 Å resolution (PDB ID: 3EIG) [31]. This highly MTX-resistant variant is an effective selectable marker for several mammalian cell types, along with murine hematopoietic stem cells [32]. Since mutations triggering MTX resistance have not been studied in mammals, and MTX is an approved drug for human treatment, engineered resistant DHFRs provide highly capable ex vivo or in vivo selective markers for human [33].

In recent decades, advances in computational techniques have led to an acceleration of drug discovery [34]. For example, cheminformatics allows us to understand and characterize the molecular properties and chemical activities of specific compounds and produce huge libraries of small molecules to screen against particular therapeutic processes [35]. After candidate identification, other cheminformatics techniques can be utilized to generate libraries of compounds which are structurally and chemically similar to the identified "hits" in order to improve stability, toxicity and kinetics. Additionally, bioinformatics methodologies can be applied to study the therapeutic activity of candidate drugs predicting interactions between drugs and proteins, to analyze the impact on biological pathways and functions and to determine genomic variants that may vary drug response [36]. Accordingly, several approaches have been developed to reduce the research expense and risk of failure for drug discovery, among which computer-aided drug design (CADD) is one of the most effective [37].

Since drug resistance is hinders chemotherapy, there is an urgent need to discover the drugs that inhibit hDHFR in wild type as well as in mutant form, i.e., after acquiring resistance to MTX. Singh et al. developed a small peptide as an anticancer drug targeting hDHFR which was supposed to be effective in MTX-resistant hDHFR because of a larger drug–protein interaction area [38]. Despite the larger interaction area of the peptide, it was specifically designed to inhibit only wild type hDHFR. We carried

out a computational study to identify a candidate molecule capable of inhibiting wild type along with mutant hDHFR. Structure-based pharmacophore modeling was performed exploiting hDHFR wild type and drug-resistant F31R/Q35E variant structures in complex with methotrexate to allocate important chemical features of protein-ligand interactions. Pharmacophore models, WT-pharma and MT-Pharma, with four features comprising key residues were selected from ten models generated for each structure. The selected pharmacophore models exhibited the highest area under the receiver operating characteristics (ROC) curve, verifying the sensitivity of the models to retrieve active compounds. WT- and MT- pharma were further subjected to validation by the Guner-Henry decoy test method.

ZINC was initially developed as an open-access database and toolset to support access to compounds for virtual screening. The upgraded version ZINC[15] makes it possible to carry out similarity searches and to explore the analogs of a given structure or part of a structure according to the input line employed [39]. The MTX structure in the SMILES (Simplified Molecular-Input Line-Entry System) format was used in ZINC[15] to retrieve MTX-analog structures. The obtained analogs were filtered through ADMET and Lipinski's Rule of five to categorize drug-like compounds. The validated pharmacophores WT-pharma and MT-pharma were then used as 3D-query to screen against drug-like MTX-analogs. The analogs mapped with WT- and MT-pharma were carried out for molecular docking where two compounds were found with a higher docking score than the reference (MTX). Further, molecular dynamics simulation confirmed one compound with a stronger affinity for WT and MT hDHFR yielding stable RMSD and strong molecular interactions with catalytic active site residues. Additionally, binding free energy calculation through MM/PBSA (Molecular Mechanics/Poisson-Boltzmann Surface Area) demonstrated robust binding affinity of Hit molecules with WT and MT hDHFR. Accordingly, in this study, we predicted an analog compound of MTX as a potential inhibitor for wild type and drug-resistant hDHFR for cancer therapeutics.

2. Results

2.1. Generation of Structures Based Pharmacophore Models

Crystal structures of wild type and F31R/Q35E variant of hDHFR in complex with methotrexate downloaded from the protein data bank were carried out for structure-based pharmacophore modeling. A total of 10 pharmacophore models for each structure, were generated while producing a ROC curve with each model. All the pharmacophores were attributed in terms of the total number of features, types of features, and selectivity score and ROC curve quality (Table 1). All ten models for wild type and for mutant structures yielded the same selectivity score with the difference in location of pharmacophoric features.

Table 1. Receptor-ligand based pharmacophores characteristics details.

Sr. No.	Number of Features	WT hDHFR Phrmacophore Details			MT hDHFR Phrmacophore Details		
		Features Set	Selectivity Score	ROC Curve Quality	Features Set	Selectivity Score	ROC Curve Quality
Pharmacophore_1	4	HBD, HBD, HYP, NI	11.090	0.832	HBD, HBD, NI, NI	12.455	0.944
Pharmacophore_2	4	HBD, HBD, NI, RA	11.090	0.924	HBA, HBD, NI, NI	12.455	0.929
Pharmacophore_3	4	HBD, HBD, HYP, NI	11.090	0.886	HBA, HBD, NI, NI	12.455	0.913
Pharmacophore_4	4	HBD, HBD, NI, RA	11.090	0.951	HBD, NI, NI, RA	12.455	0.985
Pharmacophore_5	4	HBD, HBD, NI, RA	11.090	0.903	HBD, NI, NI, RA	12.455	0.968
Pharmacophore_6	4	HBA, HBD, HBD, NI	11.090	0.941	HBD, HYP, NI, NI	12.455	0.946
Pharmacophore_7	4	HBA, HBD, NI, RA	11.090	0.937	HBD, HYP, NI, NI	12.455	0.955
Pharmacophore_8	4	HBD, HYP, NI, RA	11.090	0.822	HBD, NI, NI, RA	12.455	0.958
Pharmacophore_9	4	HBD, HYP, NI, RA	11.090	0.907	HBD, NI, NI, RA	12.455	0.962
Pharmacophore_10	4	HBA, HBD, NI, RA	11.090	0.989	HBD, HYP, NI, NI	12.455	0.933

The pharmacophore models comprising of features including key residues Glu30, Asn64, Arg70, and Val115 were selected so-called WT-pharma and Mt-pharma for wild type and mutant structures respectively. WT-pharma consisting of four pharmacophoric features included one hydrogen bond acceptor (HBA), one hydrogen bond donor (HDB), one negative ionizable (NI) and one ring aromatic (RA). MT-pharma comprised of features one hydrogen bond donor (HDB), two negative ionizable (NI) and one ring aromatic (RA) (Figure 1).

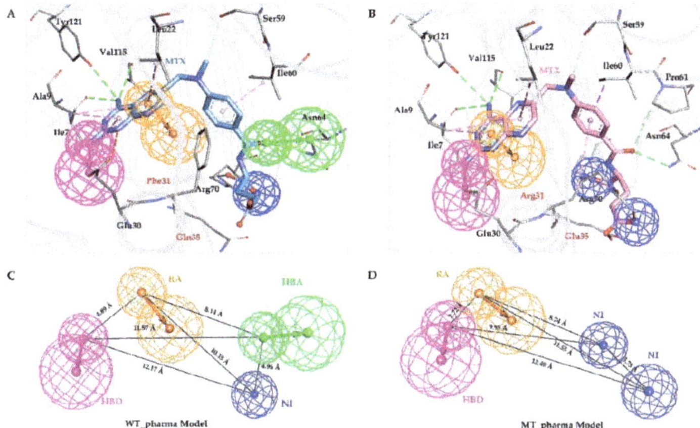

Figure 1. Structure-based pharmacophore generation. (**A**) Residues of wild type hDHFR active site complementing pharmacophoric features are shown as a thin stick. Bound inhibitor (MTX) is shown as a light blue colored thick stick model. HBA, HBD, RA, and NI are colored as green, magenta, orange and blue respectively. (**B**) Residues of mutant hDHFR active site complementing pharmacophoric features are shown as a thin stick. Bound inhibitor (MTX) is shown as a pink-colored thick stick model. HBA, HBD, RA, and NI are colored as green, magenta, orange and blue respectively. (**C**) Interfeature distance illustration of WT-pharma (**D**) Interfeature distance illustration of MT-pharma.

2.2. Pharmacophore Models Validation

Selected pharmacophore models termed as WT-pharma and MT-pharma were assessed for their sensitivity to retrieve the active compounds by receiver operating characteristics (ROC) curve. ROC curves were plotted with the generation of pharmacophore models by utilizing the option *Validation* that is available in the *receptor-ligand Phamacophore* module in DS for structure-based pharmacophore modeling. For this purpose, sets of 46 active and 24 inactive molecules were employed to testify model efficacy by creating the ROC curve. Higher the area under the ROC curve interpreted higher sensitivity of the model. For WT-pharma ROC displayed 0.989 and for MT-pharma 0.985 curve quality indicating 98.9% and 98.5% area under the curve illustrated as highly sensitive pharmacophore models to identify active molecules (Figure 2).

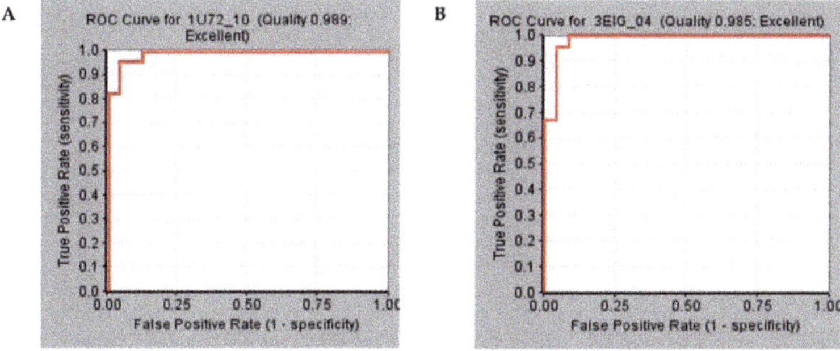

Figure 2. Receiver Operating Characteristics curves for validation of selected pharmacophore models between true positive and false-positive rates. (**A**) ROC curve shown in the red line for the WT-pharma model with 0.989 curve quality depicts 98.9% area under the curve. (**B**) ROC curve shown in the red line for the MT-pharma model with 0.985 curve quality depicts 98.5% area under the curve.

Additionally, Decoy set validation was implemented using a *Ligand Pharmacophore Mapping* module in DS. The accuracy of WT-pharma and Mt-pharma was evaluated by four factors i.e., false positive, false negative, enrichment factor (EF), and goodness of fit (GF). EF and GF were computed by applying the data of various parameters given in Table 2. Other properties of WT-pharma and MT-pharma including a percentage of the number of active yields (%Y), percent ratio of actives in the hit list (%A), false negatives, and false positives were also measured (Table 2).

Table 2. Decoy set validation for WT & MT hDHFR structure-based pharmacophore models. WT-pharma and MT-pharma obtained the highest goodness of fit score suggesting the suitability of the models for virtual screening.

Parameters	Values (WT hDHFR)	Values (MT hDHFR)
Total no. of molecules in the database (D)	90	90
Total no. of actives in the database (A)	20	20
Total no. of hit molecules from the database (H_t)	25	17
Total no. of active molecules in hit list (H_a)	19	17
Percentage Yield of actives [(H_a/H_t) × 100]	76%	100%
Percentage Ratio of actives [(H_a/A) × 100]	95%	85%
Enrichment Factor [EF = (H_a/H_t)/(A/D)]	3.4	4.5
False negatives (A − H_a)	1	13
False positive (H_t − H_a)	6	0
Goodness of fit score [GF = ($H_a/4H_tA$)(3A + H_t) × [{1 − (H_t − H_a)/(D − A)}]]	0.93	0.96

2.3. Obtaining Methotrexate Analog Structures

For the generation of structures analogous to methotrexate (MTX), we exploited MTX structure using SMILES format in ZINC[15] for similarity search. Consequently, 32 compounds were retrieved (Table S1. These compounds were downloaded in SDF format to visualize in DS and to carry out for further computations.

2.4. Drug-Likeness of MTX-analogs and Virtual Screening with Pharmacophore Models

The compounds downloaded were subjected to ADMET and Lipinski's Rule of five assessment tests to filter out drug-like MTX analogs. The ADMET assessment test gauged the pharmacokinetic features of the compounds obtained from ZINC[15]. In the ADMET assessment test, compounds were estimated for noninhibition to CYP2D6 and nonhepatotoxicity. The pharmacokinetic properties of blood brain barrier (BBB), optimal solubility, and good intestinal absorption were evaluated by setting their values to 3, 3, and 0, respectively. Lipinski's rule of five assessment was carried out after ADMET evaluation. Through Lipinski's rule of five filtration, compounds with AlogP value less than 5, number of HBD <5, number of HBA <10, molecular weight <500 Da, and fewer than 10 rotatable bonds were determined [40,41]. Accordingly, 8 compounds were found obeying drug-like properties. The drug-like MTX-analogs were carried out for virtual screening against WT-pharma and MT-pharma. All 8 compounds were aligned with WT-pharma but one compound was not in agreement with MT-pharma. Subsequently, 7 MTX-analogs were recognized as screening Hits for further computations.

2.5. Molecular Docking of Screening Hits in Active Site of hDHFR

To explore the binding mode of 7 drug-like compounds retrieved from virtual screening against WT-pharma and MT-pharma, molecular docking simulations were carried out using GOLD v 5.2.2. The 3D structure of wild type and F31R/Q35E variant of hDHFR in complex with inhibitor MTX were taken from protein data bank (PDB ID: 1U72 and 3EIG respectively). Both the structures have a high resolution of 1.9 Å for wild type and 1.7 Å for the F31R/Q35E variant. The co-crystal bound inhibitor (MTX) was docked in the active site of wild type hDHFR in order to optimize the docking protocol. The docked pose of MTX showed a low RMSD value of 0.58 Å with the crystallographic pose of

MTX in the active site of the wild type hDHFR as shown in Figure S1 of supplementary material. The WT- and MT-pharma retrieved drug-like (candidate) compounds were docked by implementing the same optimized protocol. Docking results showed that ChemPLP fitness scores and ASP fitness scores of MTX as a reference compound were 99.23 and 56.65 for wild type hDHFR while 88.98 and 49.84 for mutant hDHFR, respectively. These docking scores were considered as cut-off values for wild type and mutant hDHFR docking results analysis. The candidate compounds for wild type hDHFR were selected based on ChemPLP and ASP fitness scores greater than 99.23 and 56.65 respectively. For mutant hDHFR, compounds yielding ChemPLP and ASP fitness scores higher than 88.98 and 49.84 were selected (Table 3).

Table 3. Comparison of ChemPLP and ASP dock scores of MTX (reference inhibitor) and Hit compound in the active sites of WT and MT hDHFR.

System	ChemPLP Score	ASP Score
WT hDHFR + MTX	99.23	56.65
WT hDHFR + Hit	103.74	57.70
MT hDHFR + MTX	88.98	49.84
MT hDHFR + Hit	91.07	47.59

Additionally, the compounds were investigated about ligand conformations effectively showing essential interactions in the active site of hDHFR. Finally, one compound which contained pharmacophoric features of wild type and mutant hDHFR structures and fulfilled the above-mentioned criteria of docking scores was selected as docking Hit. The pharmacophore mapping of Hit compound (ID: ZINC000013508844) with WT-pharma and MT-pharma models are shown (Figure 3).

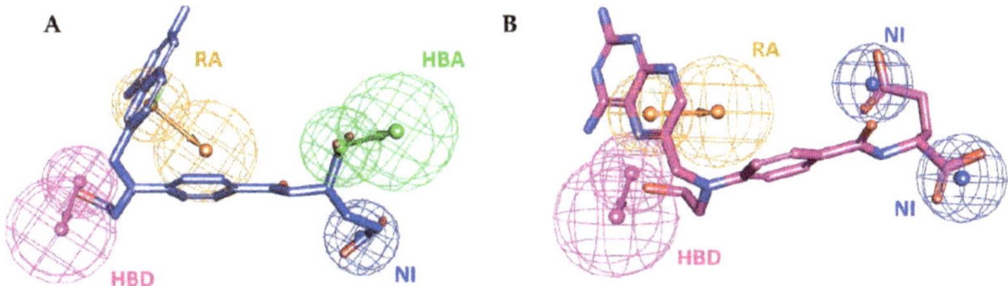

Figure 3. Hit compound (MTX-analog) mapping with pharmacophore models. (**A**) Hit compound represented in dark blue colored thick stick model mapping with WT-pharma. (**B**) Hit compound represented in magenta-colored thick stick model mapping with MT-pharma.

2.6. Molecular Dynamic Simulations for Structures Stability Evaluation

MD simulations were executed to estimate the binding stability of Hit compound after docking in the active site of wild type and mutant hDHFR. Four MD simulation systems were employed as one for each complex i.e., for hit compound and reference compound (MTX) in complex with wild type and mutant hDHFR structures, respectively. The initial details of each system subjected to simulation are given in Table 4.

Table 4. The specifications of four systems used for molecular dynamics simulations.

System	No. of TIP3P Water Molecules	No. of Na$^+$ Ions	System Size (nm)
WT hDHFR + MTX [a]	7726	1	$7.11 \times 7.11 \times 5.03$
WT hDHFR + Hit	7646	1	$7.11 \times 7.11 \times 5.03$
MT hDHFR + MTX	8258	2	$7.11 \times 7.11 \times 5.03$
MT hDHFR + Hit	8181	1	$7.11 \times 7.11 \times 5.03$

MTX [a]: the reference inhibitor.

Root mean square deviation (RMSD) was measured of the protein-ligand complex for each simulation system to assess ligand binding with hDHFR. In the results of 50 ns simulation, protein-ligand RMSD of reference compound (MTX) in complex with wild type hDHFR was recorded at an average of 0.21 nm throughout the simulation (Figure 4A). The average RMSD of MTX with mutant hDHFR was observed 0.21 nm up to 38.9 ns but afterward it significantly increased to an average of 0.62 nm indicating loss of MTX binding with MT-hDHFR. Accordingly, the representative structure of each system was taken from the last 8 ns (30–38 ns) before the loss of MTX binding with MT-hDHFR. The Hit compound obtained from docking results showed stable RMSD in complex with WT- and MT- hDHFR. The average root means square deviation values of Hit compound in complex with WT-hDHFR and MT-hDHFR were observed at an average of 0.21 nm and 0.22 nm respectively, throughout the simulation depicting that both the systems were well converged (Figure 4A). Additionally, per residue RMSF (root mean square fluctuation) calculated for each complex which was noted about 0.3 nm for all residues except for the MTX which showed RMSF about 2.3 nm in complex with MT-hDHFR (Figure 4B).

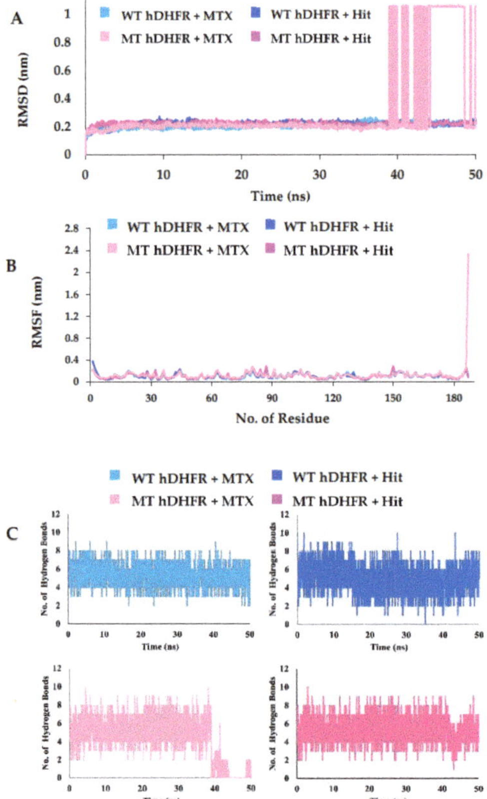

Figure 4. RMSD analysis of the reference (MTX) and hit compound (MTX-analog). (**A**) RMSD of the protein-ligand complex of wild type and mutant hDHFR revealed their stability throughout the simulation, with no abnormal behavior in all systems except for MTX in complex with MT hDHFR. (**B**) RMSF per residue plot for all the systems portrayed their residues RMSD is stable except for MT hDHFR ligand (MTX) which showed a high fluctuation level. (**C**) The number of intermolecular hydrogen bonds between protein and ligand during 50 ns MD simulations. Light blue and pink colors represent MTX in wild type and mutant hDHFR, respectively, while dark blue and magenta represent the Hit compound in wild type and mutant hDHFR, respectively.

The superimposition of representative structures demonstrated that the binding pattern and conformational alignment of Hit in the active site of hDHFR was similar to that of MTX (Figure 5).

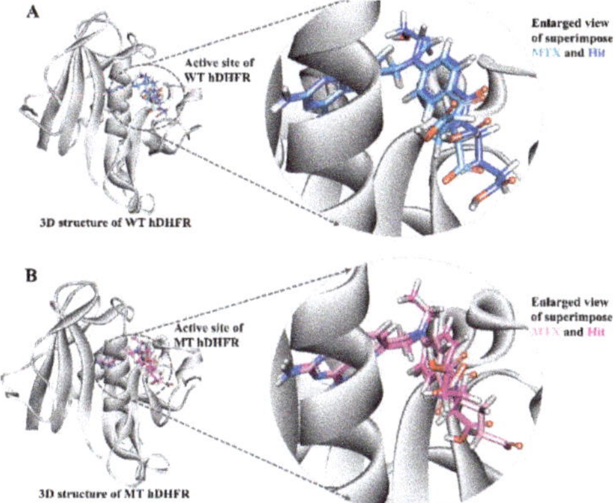

Figure 5. The binding patterns of the reference inhibitor (MTX) and hit compound in the active site of wild type and mutant hDHFR. Compounds are displayed by their representative structures superimposed (left) and enlarged (right). The protein is shown in white color. (**A**) Light blue and dark blue colors represent MTX and Hit compound in wild type hDHFR. (**B**) Pink and magenta colors represent MTX and Hit compound respectively in mutant hDHFR.

The substrate-binding site of hDHFR is mainly comprised of Ile7, Glu30, Phe31, Gln35, Ser59, Pro61, Asn64, Arg70 and Val115 [42]. Our results suggested that the reference compound (MTX) could bind with substrate binding residues of WT-hDHFR but lost its binding with MT-hDHFR, in agreement with Volpato et al. [32]. In contrast with MTX, the Hit compound exhibited strong binding with the active site of both WT- and MT-hDHFR. The MTX formed H-bonds with Ile7, Glu30, and Val115, Phe31, Asn64 and Arg70, as well as one carbon–hydrogen with Pro61 of WT-hDHFR (Figure 6A, Table 5). Furthermore, MTX established π-interactions with Ile7, Ala9, Leu22, Phe34 and Ile60 and showed van der Waals contacts with Val8, Asp21, Phe31, Arg32, Tyr33, Gln35, Thr56, Ser59, Leu67, Lys68, Tyr121 and Thr136 (Table 5).

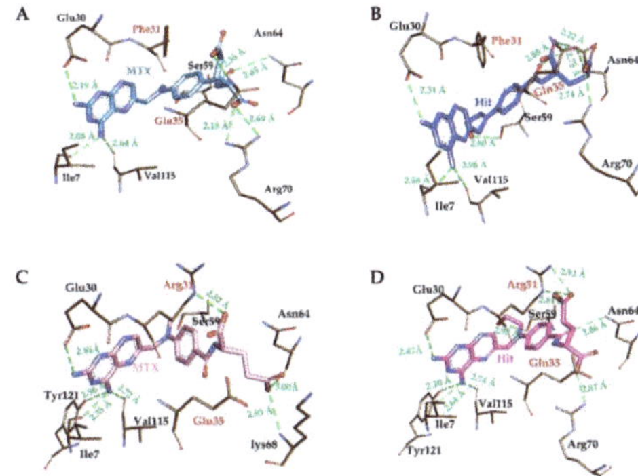

Figure 6. Molecular interactions analyses. The reference inhibitor MTX and Hit compound interacted with essential residues in the active site of hDHFR. MTX in WT hDHFR (**A**), Hit in WT hDHFR (**B**), MTX in MT hDHFR (**C**) and Hit in MT hDHFR (**D**) are depicted as light blue, dark blue, pink, and magenta-colored stick representation. The H-bond forming residues of hDHFR are displayed as a brown stick model. H-bonding and bond distances are represented as green dashed lines and measured in angstrom (Å), respectively.

Table 5. Molecular interactions between the ligands (MTX and hit compound) and the active site residues of WT and MT hDHFR.

Compound	Hydrogen Bond Residues (<3Å)	van der Waals Residues	Carbon Hydrogen Bond Residues	π-Interaction Residues
MTX (with WT hDHFR)	Ile7, Glu30, Asn64, Arg70(2), Gln35, Val115	Val8, Asp21, Phe31, Arg32, Tyr33,, Thr56, Ser59, Leu67, Lys68, Tyr121, Thr136	Pro61	Ile7, Ala9, Leu22, Phe34, Ile60
Hit (with WT hDHFR)	Ile7, Glu30, Gln35, Ser59, Asn64(2), Arg70, Val115	Val8, Asp21, Phe31, Tyr33, Phe34, Thr56, Leu67, Thr136	Pro61, Lys68	Ile7, Ala9, Leu22, Ile60
MTX (with MT hDHFR)	Ile7, Glu30, Arg31, Asn64, Lys68, Val115, Tyr121	Asp21, Phe34, Tyr33, Glu35, Thr56, Pro61, Arg70, Phe134, Thr136	Val8, Leu67, Ser59, Lys68	Ile7, Ala9, Leu22, Arg31, Ile60
Hit (with MT hDHFR)	Ile7, Glu30, Arg31 (2), Ser59, Asn64, Arg70, Val115, Tyr121	Val8, Asp21, Arg28, Arg32, Phe34, Glu35, Thr56, Pro61,Leu67, Thr136	Ser59	Ile7, Ala9, Leu22, Arg31, Ile60

In the representative structure of MT-hDHFR which was obtained before the disruption of MTX binding, molecular interactions were observed to analyze the difference of MTX binding with wild type hDHFR, leading us to speculate about the segment of resistance to MTX in mutant hDHFR. Accordingly, MTX was shown to form H-bond interactions with Ile7, Glu30, Arg31, Asn64, Lys68, Val115, Tyr121 and van der Waals interaction with Asp21, Phe34, Tyr33, Glu35, Thr56, Pro61, Arg70, Phe134 and Thr136 (Figure 6C, Table 5). Our findings also indicate that MTX formed carbon–hydrogen bonds with Val8, Leu67, Ser59, Lys68 and π-interactions with Asp21, Phe34, Tyr33, Glu35, Thr56, Pro61, Arg70, Phe134 and Thr136. The Hit compound in complex with WT-hDHFR formed H-bonds with Ile7, Glu30, Gln35, Asn64 (2), Arg70 and Val115, as well as carbon–hydrogen bonds with Pro61 and Lys68 (Figure 6B, Table 5). Additionally, Hit showed van der Waals interactions with the hydrophobic pocket residues of WT-hDHFR such as Val8, Asp21, Phe31, Tyr33, Phe34, Thr56, Ser59, Leu67 and Thr136, as well as π-interactions with Ile7, Ala9, Leu22 and Ile60 (Table 5). In the case of MT-hDHFR, the Hit compound established H-bonds with Ile7, Glu30, Arg31, Ser59, Asn64, Arg70, Val115 and Tyr121 (Figure 6D, Table 5). The Hit compound showed hydrophobic van der Waals interactions with Val8, Asp21, Arg28, Arg32, Phe34, Glu35, Thr56, Pro61, Leu67 and Thr136 residues of the WT-hDHFR while π-interactions were formed with Ile7, Ala9, Leu22, Arg31 and Ile60. The residue Ser59 also exhibited carbon–hydrogen bonds with C11 atoms in addition to conventional H-bonds with O13 atoms in Hit molecules. The conventional H-bond was formed only by Hit in the MT-hDHFR binding site. Throughout the simulation period, the total number of intermolecular H-bonds of WT- and MT-hDHFR in complex with MTX and Hit were calculated. Our results showed that the Hit compound formed an average number of H-bonds with WT- and MT-hDHFR comparable to that of MTX (reference) in WT-hDHFR. Since MTX has a very weak binding with MT-hDHFR, it could not maintain average number of H-bonds after 38.9 ns (Figure 4C), which enhanced the results obtained from the RMSD plots. Our results suggest that Hit (MTX analog) is capable of binding tightly with wild type, as well as MTX-resistant, F31R/Q35E hDHFR variants.

2.7. Binding Free Energy Calculations for MTX and Hit Compound

MM/PBSA binding free energies were calculated for MTX and Hit in complex with WT- and MT-hDHFR. The free energies of MTX and Hit in complex with WT-hDHFR were observed as −646.76 kJ/mol and −642.12 kJ/mol, whereas MT-hDHFR MTX could yield only −49 kJ/mol, while the Hit compound attained −571.38 kJ/mol. The binding free energy evaluations underscore our findings that the Hit molecule is tightly bound with WT- and MT-hDHFR, displaying comparable free energy of MTX in complex with WT-hDHFR. The decomposition analysis of the binding free energy indicated that electrostatic and van der Waals forces are significant characteristics in hDHFR inhibition (Figure 7, Table 6).

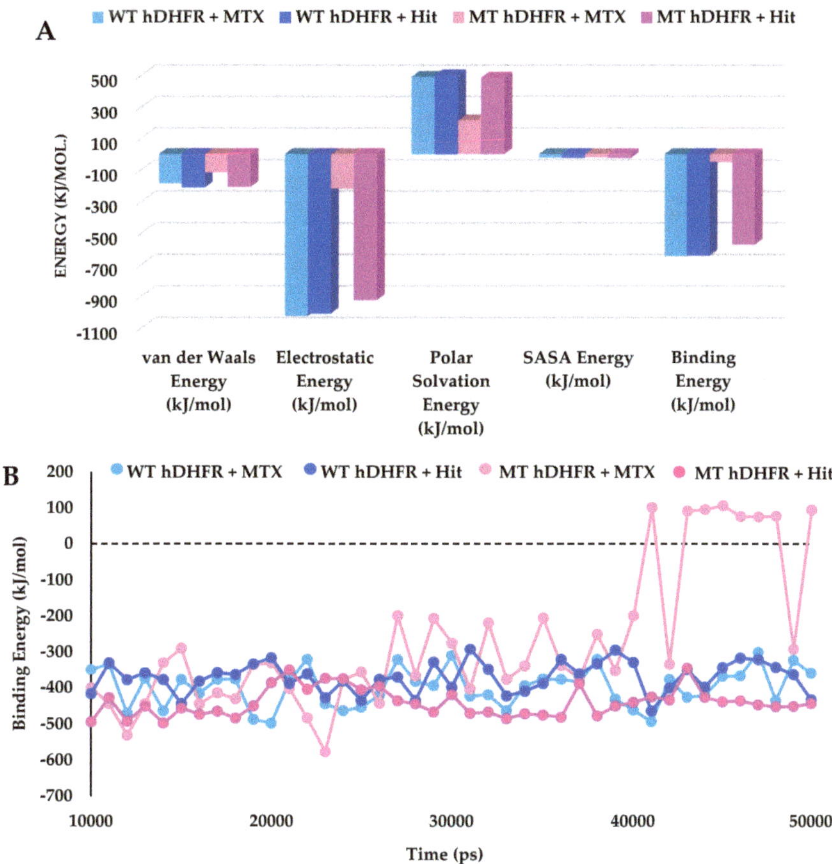

Figure 7. Binding free energy analyses. (**A**) Graphical representation of MM/PBSA estimated binding free energy of wild type and mutant hDHFR in complex with MTX (reference) and Hit compound throughout the simulation time. The reference compound is depicted as light blue and dark blue for wild type and mutant hDHFR, respectively. The Hit compound is shown in pink and magenta colors for wild type and mutant hDHFR, respectively. (**B**) The binding free energy decomposition analysis of the final hits in the active site of hDHFR infers that the Hit compound was comparably strongly bound with WT and MT hDHFR, while MTX lost its binding with the mutant structure.

Table 6. Decomposition of binding free energy.

Complex	Van der Waals Energy (kJ/mol)	Electrostatic Energy (kJ/mol)	Polar Solvation Energy (kJ/mol)	SASA [b] Energy (kJ/mol)	Binding Energy (kJ/mol)
WT hDHFR + [a]MTX	−184.057	−1023.945	489.982	−22.594	−646.767
WT hDHFR + Hit	−210.358	−1007.98	499.622	−22.622	−642.123
MT hDHFR + MTX	−116.884	−217.191	212.294	−18.862	−49.299
MT hDHFR + Hit	−207.152	−923.188	483.648	−23.977	−571.381

[a] MTX: methotrexate as reference inhibitor. SASA [b]: Solvent accessible surface area.

Altogether, our results verified that the newly identified MTX-analog favorably adapted the active site of wild type and double mutant hDHFR and acquired polar and nonpolar interactions with the catalytic active residues.

The structure of the Hit compound, which was modified by adding a carbon-oxygen group (C11-C12-O13) with a *p*-ABA moiety, is illustrated in its 2D structure in Figure 8.

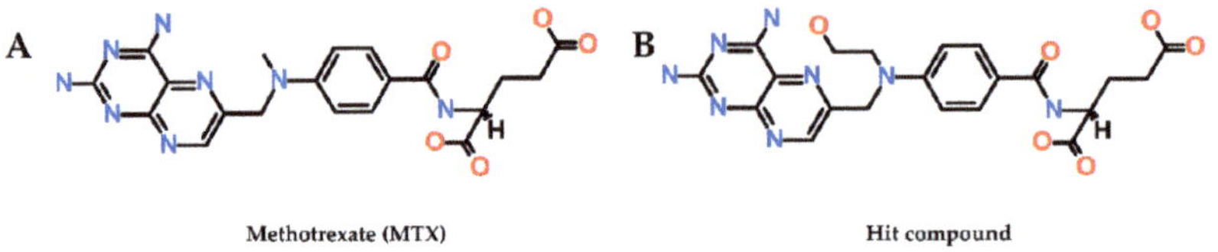

Figure 8. (**A**) 2D structure of MTX (**B**) 2D structure of Hit compound (MTX analog, ZINC ID: ZINC000013508844).

3. Discussion

Chemotherapeutics are very effective in the treatment of cancers, but drug resistance is often a limiting factor. Acquired resistance is the type of drug resistance that can develop through various adaptive responses such as mutations, increased expression of the therapeutic target and activation of alternative compensatory signaling pathways arising over the course of the treatment of tumors [43] Human dihydrofolate reductase (hDHFR) is an enzyme that is responsible for the catalysis of the reduction of 7,8-dihydrofolate (DHF) to 5,6,7,8-tetrahydrofolate, which is crucial for DNA synthesis and cell proliferation [44]. Therefore, hDHFR has been widely used as a target for cancer therapeutics for several decades [45]. Methotrexate is a well-known inhibitor that displays a high affinity with hDHFR, but mutation in the active site of hDHFR results in the loss of MTX binding [31].

The present study aimed to identify an analog of methotrexate that was capable of binding tightly, and hence inhibiting, wild type and doubly mutant hDHFR (F31R/Q/35E) by employing several computational methods including structure-based pharmacophore modeling, virtual screening, molecular docking and molecular dynamics simulations studies [46]. Structure-based pharmacophore models of crystal structures of wild type (PDB ID: 1U72) and variant (PDB ID: 3EIG) hDHFR in complex with methotrexate were obtained, with four features in each model. The best pharmacophore models of each the structure were selected by analyzing the inclusion of key residues in pharmacophoric features and sensitivity of the models to retrieve true positive compounds depicted by the highest area under the ROC curve. The selected pharmacophore models, termed as WT-pharma and MT-pharma for wild type and mutant hDHFR structures, respectively, were rationally assessed for the inclusion of conserved hydrogen bond residue Glu30 and other key residues, such as Asn64, Arg70 and Val115 [44,47]. Further, with each pharmacophore model, the ROC curve was formed between the number of false positive (FP) and true positive (TP) compounds retrieved by that model from the datasets of 46 active and 24 inactive compounds. Higher AUC values in the ROC curves infer greater sensitivity of WT-pharma and MT-Pharma in retrieving actives, and specificity for ignoring inactives [48,49]. Using ZINC[15], Mayorga et al. found a high number of compounds when they utilized a small fragment of the original structure [50]. In our study, to explore analogs of MTX, we used the full structure of MTX and selected 'in vitro', 'in vivo 'and 'clean' options in the section of 'subsets', which resulted in the generation of only 32 analogous compounds (Table S1. An ADMET assessment test and Lipinski's Rule of five scrutinized the downloaded compounds from ZINC[15] for their drug-like properties, and found eight compounds satisfying the required criteria to qualify as lead compounds [51,52]. The validated pharmacophore models of wild type and mutant structures of hDHFR were applied as 3D query for virtual screening with the drug like compounds capable of binding with wild type and mutant hDHFR as well [53]. A molecular docking study was employed to inspect the most suitable and consistent binding mode of the molecules in the binding sites of receptor proteins. Consequently, the best binding modes obtained from docking based on scoring functions and key interactions with the active site residues of wild type and mutant hDHFR were used in MD simulations to assess their stability [54]. The RMSD plots inferred that the Hit compound showed similar modes of interaction in wild type and mutant hDHFR active sites as MTX in the active site of wild type hDHFR. Specifically, the average RMSD profiles (<0.25 nm) obtained for protein-ligand complexes of Hit with wild type and mutant hDHFR exhibited that the systems were uniform and compact, as the stability of the system can be inferred by an RMSD

value of less than 0.3 nm [40,41]. The RMSD plot for MTX in complex with mutant hDHFR showed abrupt fluctuation after 38.9 ns, which indicated the loss of MTX binding with the active site of hDHFR. Furthermore, a high RMSF value (2.34 nm) of MTX (residue187) indicated a loss of ligand binding with only mutant hDHFR protein. Our results showed that Hit compound established stable H-bonds with the active site residues of wild type and mutant hDHFR. Similar to the molecular interactions of MTX (reference compound), most H-bonds were formed by pterin moiety and α-glutamate moiety with hDHFR active site residues, while *p*-aminobenzoic acid (*p*-ABA) moiety formed mainly hydrophobic interactions [31]. The conserved hydrogen bond with OE1 atom of catalytic residue Glu30 was formed with the pterin moiety of Hit molecule [42,55]. The additional oxygen atom in the structure of Hit compound formed a hydrogen bond with Ser59 in both wild type and mutant hDHFR, while Ser59 belonged to the coenzyme NADPH binding site [56]. It was speculated that the hydrogen bond between Ser59 and the modified *p*-ABA moiety of Hit compound contributed to the strong binding of the Hit compound with the mutant structure of hDHFR. Furthermore, the conserved hydrogen bonds formed by an α-carboxylate group of MTX with the side chains of Arg70 and Gln35 while *p*-aminobenzoyl keto group with Asn64 were also observed in Hit compound's interactions with wild type hDHFR [26].

In the mutant structure, due to the substitution of Glu35, a hydrogen bond was not formed because of unfavorable close electrostatic contact of two negative charges between Glu35 side chain and glutamate moiety of MTX and Hit compound. In contrast, Arg31, which was substituted at the position of Phe31, was observed to form hydrogen bonds through the guanidinium group with an α-glutamate moiety of MTX. Hit compound displayed a double hydrogen bond with Arg31; double hydrogen bonds are considered to be crucial for strong binding in protein-ligand interactions [57]. Furthermore, the hydrogen bond of the Arg70 side-chain with an α-carboxylate group of MTX was lost in Mt hDHFR, while Lys68 was formed a hydrogen bond with the α-carboxylate group. On the other hand, the Hit compound retained the H-bond of the Arg70 side chain with the α-carboxylate group, as it did in the wild type. Moreover, the only hydrogen bond formed by a *p*-ABA group of MTX with Asn64 in wild type hDHFR was also shifted to a H-bond with an α-carboxylate group of MTX in mutant hDHFR. While the *p*-ABA group of Hit compound formed a hydrogen bond with Asn64, as in the wild type, an additional H-bond was formed between Ser59 and modified oxygen atom added to *p*-ABA group. Accordingly, a comparative analysis of protein-ligand interactions of MTX and Hit compound suggested that Hit (MTX-analog) may be capable of retaining its strong binding with WT and MT hDHFR. Additionally, the binding free energy evaluations performed by the MM/PBSA method also inferred that the complexes of WT and MT hDHFR with Hit compound were comparably stable, like MTX in WT hDHFR; meanwhile, the binding free energy profile noticeably depicts the loss of MTX binding in MT hDHFR.

4. Materials and Methods

4.1. Structure Based Pharmacophore Modeling

Ligand binding features were assessed by the structures of wild type (PDB ID: 1U72) and drug-resistant (PDB ID: 3EIG) human DHFR in complex with methotrexate taken from protein data bank [26,31]. Pharmacophore models were generated using the *Receptor-ligand Pharmacophore Generation* module in Discovery Studio (DS) v.4.5 (Dassault System, BIOVIA Corp, San Diego, CA, USA). FAST (Features from Accelerated Segment Test) algorithm was applied for Conformation Generation, while the Fitting Method was set to Flexible. The Validation option was set to *True*, for which a set of 46 active and 24 inactive compounds, downloaded from BindingDB (https://www.bindingdb.org/bind/index.jsp) were exploited to generate a ROC curve for each pharmacophore model.

4.2. Decoy Test Validation

The ability of pharmacophore to identify hDHFR inhibitors was assessed by the Guner−Henry method (Decoy test method) [44]. A test set was prepared by collecting hDHFR inhibitors whose

experimental activities (IC_{50} values) were measured by the same biological assays. The test set was composed of active and inactive molecules of hDHFR. The selected pharmacophore models of wild type and mutant structures were employed as a 3D query to obtain the best-fitted molecules from the test set. Screening of the test set was executed by the Ligand Pharmacophore Mapping protocol embedded in DS. Accordingly, several parameters, like Guner–Henry (GH) score, enrichment factor (EF), the percentage ratio of actives (%A), percentage yield of actives (%Y), false negative and false-positive, were calculated, which determined the efficacy of WT-pharma and MT-pharma

$$
\begin{aligned}
&EF = (H_a/H_t)/(A/D) \\
&GF = (H_a/4H_tA)\,(3A + H_t) \times [\{1 - (H_t - H_a)/(D - A)\}]
\end{aligned}
\tag{1}
$$

where D is the total molecules in the data set, A specifies the total number of active compounds in the data set, H_t indicates the total number of Hits retrieved and H_a refers to the number of actives present in the retrieved Hits.

4.3. Methotrexate Analogs Generation

Methotrexate structure was subjected to a similarity search in $ZINC^{15}$ using SMILES string of MTX. For the generation of clean analogs, in vivo and in vitro options were selected in the available range of Subsets to Check. Subsequently, the structures were downloaded in the SDF (Spatial Data File) format, generated by the webserver, to carry out for further computations in DS.

4.4. Drug-Likeness Prediction and Virtual Screening

The molecules retrieved from $ZINC^{15}$ were tested through ADMET and Lipinski's Rule of five embedded assessment techniques in DS to identify drug-like compounds. Subsequently, the compounds exhibiting such properties were carried out for virtual screening with WT-pharma and MT-pharma. The compounds which fitted with both pharmacophores were considered as screening compounds in our molecular docking study.

4.5. Molecular Docking Simulation

A docking study was employed through the Genetic Optimization of Ligand Docking (GOLD) package v5.2.2 (The Cambridge Crystallographic Data Centre, Cambridge, United Kingdom). GOLD software provides full flexibility of ligands and limited flexibility of protein; hence, it delivers more reliable results in computational biology the crystal structures of wild type (PDB ID: 1U72) and variant (PDB ID: 3EIG) hDHFR in complex with Methotrexate were taken from protein data bank. The wild type and variant structures of hDHFR were prepared for docking by eliminating water molecules in DS. Chemistry at Harvard macromolecular mechanisms (CHARMm) force field was applied to add hydrogen atoms to the structures of hDHFR. The binding sites of wild type and mutant hDHFR were identified within the radius of 9Å of bound inhibitor (MTX) using the *Define and Edit Binding Site* module, planted in DS. During docking, MTX-analogs retrieved from virtual screening along with methotrexate as reference were treated as ligand molecules. The ChemPLP (Piecewise Linear Potential) score and ASP (Astex Statistical Potential) score were used as the default scoring and rescoring functions, respectively. The ChemPLP is the default scoring function in GOLD software which is empirically optimized for pose prediction. It is implemented to establish the steric complementarity between protein and ligand, distance- and angle-dependent hydrogen and metal bonding terms as well as the heavy atoms clash- and torsional potential. The ASP scoring function measures the atom–atom potential and has similar precision to Chemscore and Goldscore fitness functions [58,59]. During docking, GA (genetic algorithm) was run to produce 100 poses for each drug-like molecule. The bound ligand (MTX) was employed as a reference compound throughout the analyses. Cluster analysis was performed to scrutinize hit compounds exhibiting a higher dock score than cut off (dock score of reference molecule).

4.6. Molecular Dynamics (MD) Simulation

Molecular dynamics simulations were performed using CHARMm36 all-atom force field [60] in Groningen Machine for Chemical Simulation (GROMACS) v5.0.6 package [61]. For every protein-ligand complex, an independent simulation system was generated. The topology and coordinates files for MTX and docking hits were generated using SwissParam [62]. Transferable intermolecular potential with three points (TIP3P) water model in a cubic box was used for solvation of each system. Solvent molecules were substituted with sodium ions (Na^+) to nullify the total charge of simulation systems. The energies of the systems were minimized by applying steepest decent algorithwhere the maximum force was kept less than 10 kJ/mol I order to avoid any bad contacts likely to be occurred in the production run. Initially, the systems were equilibrated in two steps. First, the number of particles at constant volume and temperature (NVT) equilibration was carried out for 100 ps at 300 K. The temperature of the system was kept constant using V-rescale thermostat [63]. In second phase, 100 ps equilibration was performed with number of particles at constant pressure of 1 bar (NPT) and temperature 300 K [64]. Accordingly following the protocol mentioned earlier, all the systems were carried out for production run. In short, bond lengths of heavy atoms were sustained using Linear Constraint Solver (LINCS) algorithm [65]. Particle Mesh Ewald (PME) method was employed to calculate the long-range electrostatic interactions [66]. Short-range interactions length was kept to 12 Å. All simulations were performed with the periodic boundary conditions to make infinite systems. Time interval was kept of 10 ps to save coordinates data. Finally, result's visualization and analysis were performed using GROMACS and DS.

4.7. Binding Free Energy Calculations

The binding free energy was calculated by employing the Molecular Mechanics/Poisson-Boltzmann Surface Area (MM/PBSA) method [67]. Following the MM/PBSA protocol to compute free energies of the protein-ligand complex, equidistant snapshots of the hDHFR-ligand complex were extracted. The binding free energy of the protein-ligand complex is stated as:

$$\Delta G_{binding} = G_{complex} - (G_{protein} + G_{ligand}) \tag{2}$$

where $G_{complex}$ denotes the sum of the free energy of the complex, and $G_{protein}$ and G_{ligand} specify the free energies of portion and ligand in their unbound states.

Free energy can be defined as:

$$G_X = E_{MM} + G_{solvation} \tag{3}$$

where X can be a protein, a ligand, or their complex. E_{MM} signifies the average molecular mechanics potential energy in vacuum, while $G_{solvation}$ indicates the free energy of solvation.

Accordingly, molecular mechanics potential energy in vacuum can be calculated by implementing the equation:

$$E_{MM} = E_{bonded} + E_{nonbonded} = E_{bonded} + (E_{vdw} + E_{elec}) \tag{4}$$

E_{bonded} denotes the bonded interactions, while $E_{nonbonded}$ terms the nonbonded interactions. The value of ΔE_{bonded} is generally treated as zero.

The combined energetic terms of electrostatic (G_{polar}) and apolar ($G_{nonpolar}$) give the solvation free energy which is measured as:

$$G_{solvation} = G_{polar} + G_{nonpolar} \tag{5}$$

Here Poisson-Boltzmann (PB) equation is implemented to compute G_{polar}, while the $G_{nonpolar}$ is calculated from the solvent-accessible surface area (SASA) as:

$$G_{nonpolar} = \gamma^{SASA} + b \tag{6}$$

where, γ represents the coefficient of solvent surface tension, while b is its fitting parameter, whose values are 0.02267 kJ/mol/Å^2 and 3.849 kJ/mol, respectively.

5. Conclusions

In the current study, structure-based pharmacophore modeling, virtual screening, molecular docking and molecular dynamics simulation methods were utilized to identify a potential inhibitor that was capable of strong binding with wild type as well as drug-resistant (mutant) hDHFR. Structure-based pharmacophore models for WT and MT hDHFR in complex with MTX were generated and validated by the decoy test and ROC curve. Methotrexate analogs were generated by exploiting the MTX structure in ZINC^{15}, and carried out for ADMET and Lipinski's Rule of five assessment tests to evaluate drug-likeness of compounds obtained from ZINC. The drug-like compounds were used in virtual screening with validated WT and MT pharmacophore models as a 3D query to identify potential hits of wild type and mutant hDHFR. The compounds obtained from virtual screening were docked in the active site sites of WT and MT hDHFR. Subsequently, through docking results analysis, one compound was found to have a higher dock score than the reference compound (MTX), displaying essential molecular interactions with key residues of the hDHFR active site. Furthermore, MD simulation and binding free energy calculations for the Hit compound and MTX in complex with WT and MT hDHFR were also used to evaluate the stability of the Hit compound with WT and MT hDHFR. Taken together, our findings indicate MTX analog (ZINC000013508844) to be a potential inhibitor of wild type hDHFR and drug-resistant F31R/Q35E variant of hDHFR. In future work, we will try to synthesize the Hit compound to verify our findings through bioassay by collaborating with an experimental lab. These findings can also be extended to assess other drug resistant hDHFR variants for cancer therapeutics.

Supplementary Materials: Available online. Table S1 2D structures and ZINC IDs of Methotrexate analogs, Figure S1: Superimposition of X-ray structure pose and docked pose of MTX.

Author Contributions: Conceptualization, K.W.L.; Data curation, R.M.R., A.Z., S.P.; Formal analysis, N.B.A., S.R., A.Z. and S.P.; Investigation, R.M.R. and Y.K.; Methodology, R.M.R., S.R., G.L., and S.Y.; Project administration, K.W.L.; Supervision, K.W.L.; Validation, S.Y., G.L., D.K.; Visualization, R.M.R.; Writing and editing, R.M.R., N.B.A., S.R. All authors have read and agreed to the published version of the manuscript.

Funding: This research was supported by the Bio and Medical Technology Development Program of the National Research Foundation (NRF) and funded by the Korean government (MSIT) (No. NRF-2018M3A9A7057263). Also supported by Basic Science Research Program (2017R1D1A3B03035738) through the National Research Foundation of Korea (NRF) funded by the Ministry of Education of Republic of Korea.

Conflicts of Interest: The authors declare no competing interests. The funders had no role in the design of the study; in the collection, analyses, or interpretation of data; in the writing of the manuscript, or in the decision to publish the results.

References

1. Bertino, J.R. Ode to methotrexate. *J. Clin. Oncol.* **1993**, *11*, 5–14. [CrossRef] [PubMed]
2. Holohan, C.; Van Schaeybroeck, S.; Longley, D.B.; Johnston, P.G. Cancer drug resistance: An evolving paradigm. *Nat. Rev. Cancer* **2013**, *13*, 714–726. [CrossRef] [PubMed]
3. Tran, P.N.; Tate, C.J.; Ridgway, M.C.; Saliba, K.J.; Kirk, K.; Maier, A.G. Human dihydrofolate reductase influences the sensitivity of the malaria parasite Plasmodium falciparum to ketotifen - A cautionary tale in screening transgenic parasites. *Int. J. Parasitol. Drugs Drug Resist.* **2016**, *6*, 179–183. [CrossRef] [PubMed]
4. Schweitzer, B.I.; Dicker, A.P.; Bertino, J.R. Dihydrofolate reductase as a therapeutic target. *Faseb J.* **1990**, *4*, 2441–2452. [CrossRef]

5. Beierlein, J.M.; Frey, K.M.; Bolstad, D.B.; Pelphrey, P.M.; Joska, T.M.; Smith, A.E.; Priestley, N.D.; Wright, D.L.; Anderson, A.C. Synthetic and crystallographic studies of a new inhibitor series targeting Bacillus anthracis dihydrofolate reductase. *J. Med. Chem.* **2008**, *51*, 7532–7540.

6. Czarnecka-Operacz, M.; Sadowska-Przytocka, A. Review paper The possibilities and principles of methotrexate treatment of psoriasis-the updated knowledge. *Magdal. Czarnecka-Oper. Postep Derm Alergol* **2014**, *6*, 392–400.

7. Goldman, I.D.; Matherly, L.H. The cellular pharmacology of methotrexate. *Pharmacol. Ther.* **1985**, *28*, 77–102. [CrossRef]

8. Whitehead, V.; Rosenblatt, D.; Vuchich, M.; Shuster, J.; Witte, A.; Beaulieu, D. Accumulation of methotrexate and methotrexate polyglutamates in lymphoblasts at diagnosis of childhood acute lymphoblastic leukemia: A pilot prognostic factor analysis. *Blood* **1990**, *76*, 44–49. [CrossRef]

9. McGuire, J.J.; Hsieh, P.; Coward, J.K.; Bertino, J.R. Enzymatic synthesis of folylpolyglutamates. Characterization of the reaction and its products. *J. Biol. Chem.* **1980**, *255*, 5776–5788.

10. Allegra, C.J.; Chabner, B.A.; Drake, J.C.; Lutz, R.; Rodbard, D.; Jolivet, J. Enhanced inhibition of thymidylate synthase by methotrexate polyglutamates. *J. Biol. Chem.* **1985**, *260*, 9720–9726.

11. Allegra, C.J.; Drake, J.C.; Jolivet, J.; Chabner, B.A. Inhibition of phosphoribosylaminoimidazolecarboxamide transformylase by methotrexate and dihydrofolic acid polyglutamates. *Proc. Natl. Acad. Sci. USA* **1985**, *82*, 4881–4885. [CrossRef] [PubMed]

12. Jolivet, J.; Cowan, K.H.; Curt, G.A.; Clendeninn, N.J.; Chabner, B.A. The pharmacology and clinical use of methotrexate. *N. Engl. J. Med.* **1983**, *309*, 1094–1104. [CrossRef] [PubMed]

13. Simonsen, C.C.; Levinson, A.D. Isolation and expression of an altered mouse dihydrofolate reductase cDNA. *Proc. Natl. Acad. Sci. USA* **1983**, *80*, 2495–2499. [CrossRef] [PubMed]

14. Haber, D.A.; Beverley, S.M.; Kiely, M.L.; Schimke, R.T. Properties of an altered dihydrofolate reductase encoded by amplified genes in cultured mouse fibroblasts. *J. Biol. Chem.* **1981**, *256*, 9501–9510. [PubMed]

15. Srimatkandada, S.; Schweitzer, B.I.; Moroson, B.A.; Dube, S.; Bertino, J.R. Amplification of a polymorphic dihydrofolate reductase gene expressing an enzyme with decreased binding to methotrexate in a human colon carcinoma cell line, HCT-8R4, resistant to this drug. *J. Biol. Chem.* **1989**, *264*, 3524–3528. [PubMed]

16. Melera, P.W.; Davide, J.P.; Hession, C.A.; Scotto, K.W. Phenotypic expression in Escherichia coli and nucleotide sequence of two Chinese hamster lung cell cDNAs encoding different dihydrofolate reductases. *Mol. Cell. Biol.* **1984**, *4*, 38–48. [CrossRef] [PubMed]

17. Dicker, A.P.; Volkenandt, M.; Schweitzer, B.I.; Banerjee, D.; Bertino, J.R. Identification and characterization of a mutation in the dihydrofolate reductase gene from the methotrexate-resistant Chinese hamster ovary cell line Pro-3 MtxRIII. *J. Biol. Chem.* **1990**, *265*, 8317–8321.

18. Mclvor, R.S.; Simonsen, C.C. Isolation and characterization of a variant dihydrofolate reductase cDNA from methotrexate-resistant murine L5178Y cells. *Nucleic Acids Res.* **1990**, *18*, 7025–7032. [CrossRef]

19. Chundurus, S.K.; Cody, V.; Luftg, J.R.; Pangborno, W.; Applemansll, J.R.; Blakleysii, R.L. Methotrexate-resistant variants of human dihydrofolate reductase. Effects of Phe31 substitutions. *J. Biol. Chem.* **1994**, *269*, 9547–9555.

20. Thompson, P.D.; Freisheim, J.H. Conversion of Arginine to Lysine at Position 70 of Human Dihydrofolate Reductase: Generation of a Methotrexate-Insensitive Mutant Enzyme. *Biochemistry* **1991**, *30*, 8124–8130. [CrossRef]

21. Ercikan-Abali, E.A.; Waltham, M.C.; Dicker, A.P.; Schweitzer, B.I.; Gritsman, H.; Banerjee, D.; Bertino, J.R. Variants of human dihydrofolate reductase with substitutions at Leucine-22: Effect on catalytic and inhibitor binding properties. *Mol. Pharm.* **1996**, *49*, 430–437.

22. Lewis, W.S.; Cody, V.; Galitsky, N.; Luft, J.R.; Pangborn, W.; Chunduru, S.K.; Spencer, H.T.; Appleman, J.R.; Blakley, R.L. Methotrexate-resistant variants of human dihydrofolate reductase with substitutions of leucine 22. Kinetics, crystallography, and potential as selectable markers. *J. Biol. Chem.* **1995**, *270*, 5057–5064. [CrossRef]

23. Fossati, E.; Volpato, J.P.; Poulin, L.; Guerrero, V.; Dugas, D.-A.; Pelletier, J.N. 2-tier bacterial and in vitro selection of active and methotrexate-resistant variants of human dihydrofolate reductase. *J. Biomol. Screen.* **2008**, *13*, 504–514. [CrossRef] [PubMed]

24. Nakano, T.; Spencer, H.T.; Appleman, J.R.; Blakley, R.L. Critical role of phenylalanine 34 of human dihydrofolate reductase in substrate and inhibitor binding and in catalysis. *Biochemistry* **1994**, *33*, 9945–9952. [CrossRef] [PubMed]

25. Gangjee, A.; Vidwans, A.P.; Vasudevan, A.; Queener, S.F.; Kisliuk, R.L.; Cody, V.; Li, R.; Galitsky, N.; Luft, J.R.; Pangborn, W. Structure-based design and synthesis of lipophilic 2,4-diamino-6-substituted quinazolines and their evaluation as inhibitors of dihydrofolate reductases and potential antitumor agents. *J Med Chem* **1998**, *41*, 3426–3434.

26. Cody, V.; Luft, J.R.; Pangborn, W. Understanding the role of Leu22 variants in methotrexate resistance: Comparison of wild-type and Leu22Arg variant mouse and human dihydrofolate reductase ternary crystal complexes with methotrexate and NADPH. *Acta Cryst. D. Biol. Cryst.* **2005**, *61*, 147–155. [CrossRef]

27. Cody, V.; Galitsky, N.; Luft, J.R.; Pangborn, W.; Rosowsky, A.; Blakley, R.L. Comparison of two independent crystal structures of human dihydrofolate reductase ternary complexes reduced with nicotinamide adenine dinucleotide phosphate and the very tight-binding inhibitor PT523. *Biochemistry* **1997**, *36*, 13897–13903. [CrossRef]

28. Cody, V.; Galitsky, N.; Luft, J.R.; Pangborn, W.; Gangjee, A. Analysis of two polymorphic forms of a pyrido[2,3-d]pyrimidine N9-C10 reversed-bridge antifolate binary complex with human dihydrofolate reductase. *Acta Cryst. D. Biol. Cryst.* **2003**, *59*, 654–661. [CrossRef]

29. Cody, V.; Luft, J.R.; Pangborn, W.; Gangjee, A. Analysis of three crystal structure determinations of a 5-methyl-6-N-methylanilino pyridopyrimidine antifolate complex with human dihydrofolate reductase. *Acta Cryst. D. Biol. Cryst.* **2003**, *59*, 1603–1609. [CrossRef]

30. Cody, V.; Luft, J.R.; Pangborn, W.; Gangjee, A.; Queener, S.F. Structure determination of tetrahydroquinazoline antifolates in complex with human and Pneumocystis carinii dihydrofolate reductase: Correlations between enzyme selectivity and stereochemistry. *Acta Cryst.. D. Biol. Cryst.* **2004**, *60*, 646–655. [CrossRef]

31. Volpato, J.P.; Yachnin, B.J.; Blanchet, J.; Guerrero, V.; Poulin, L.; Fossati, E.; Berghuis, A.M.; Pelletier, J.N. Multiple Conformers in Active Site of Human Dihydrofolate Reductase F31R/Q35E Double Mutant Suggest Structural Basis for Methotrexate Resistance. *J. Biol. Chem.* **2009**, *284*, 20079–20089. [CrossRef]

32. Volpato, J.P.; Fossati, E.; Pelletier, J.N. Increasing methotrexate resistance by combination of active-site mutations in human dihydrofolate reductase. *J. Mol. Biol.* **2007**, *373*, 599–611. [CrossRef]

33. Patel, M.; Sleep, S.E.H.; Lewis, W.S.; Spencer, H.T.; Mareya, S.M.; Sorrentino, B.P.; Blakley, R.L. Comparison of the protection of cells from antifolates by transduced human dihydrofolate reductase mutants. *Hum. Gene.* **1997**, *8*, 2069–2077. [CrossRef]

34. Sliwoski, G.; Kothiwale, S.; Meiler, J.; Lowe, E.W. Computational methods in drug discovery. *Pharmacol. Rev.* **2014**, *66*, 334–395. [CrossRef] [PubMed]

35. Blaney, J.M.; Martin, E.J. Computational approaches for combinatorial library design and molecular diversity analysis. *Curr. Opin. Chem. Biol.* **1997**, *1*, 54–59. [CrossRef]

36. Drews, J. Drug discovery: A historical perspective. *Science (80-.).* **2000**, *287*, 1960–1964. [CrossRef] [PubMed]

37. Ou-yang, S.; Kong, X.; Liang, Z.; Luo, C.; Jiang, H. Computational drug discovery. *Acta Pharm. Sin.* **2012**, *33*, 1131–1140. [CrossRef] [PubMed]

38. Singh, A.; Deshpande, N.; Pramanik, N.; Jhunjhunwala, S.; Rangarajan, A.; Atreya, H.S. Optimized peptide based inhibitors targeting the dihydrofolate reductase pathway in cancer OPEN. *Sci. Rep.* **2018**, *8*, 1–8.

39. Sterling, T.; Irwin, J.J. ZINC 15—Ligand Discovery for Everyone. *J. Chem. Inf. Model.* **2015**, *55*, 2324–2337. [CrossRef]

40. Brigo, A.; Lee, K.W.; Mustata, G.I.; Briggs, J.M. Comparison of Multiple Molecular Dynamics Trajectories Calculated for the Drug-Resistant HIV-1 Integrase T66I/M154I Catalytic Domain. *Biophys. J.* **2005**, *88*, 3072–3082. [CrossRef]

41. Vinay Kumar, C.; Swetha, R.G.; Anbarasu, A.; Ramaiah, S. Computational Analysis Reveals the Association of Threonine 118 Methionine Mutation in PMP22 Resulting in CMT-1A. *Adv. Bioinform.* **2014**, *2014*, 1–10. [CrossRef] [PubMed]

42. Rao, A.S.; Road, K. A study on dihydrofolate reductase and its inhibitors: A review. *Int. J* **2013**, *4*, 2535–2547.

43. Longley, D.B.; Johnston, P.G. Molecular mechanisms of drug resistance. *J. Pathol.* **2005**, *205*, 275–292. [CrossRef] [PubMed]

44. Schnell, J.R.; Dyson, H.J.; Wright, P.E. Structure, dynamics, and catalytic function of dihydrofolate reductase. *Annu. Rev. Biophys. Biomol. Struct* **2004**, *33*, 119–159. [CrossRef] [PubMed]

45. Raimondi, M.V.; Randazzo, O.; Franca, M.L.; Barone, G.; Vignoni, E.; Rossi, D.; Collina, S. DHFR Inhibitors: Reading the Past for Discovering Novel Anticancer Agents. *Molecules* **2019**, *24*, 1140. [CrossRef]

46. Lu, X.; Yang, H.; Chen, Y.; Li, Q.; He, S.; Jiang, X.; Feng, F.; Qu, W.; Sun, H. The development of pharmacophore modeling: Generation and recent applications in drug discovery. *Curr. Pharm. Des.* **2018**, *24*, 3424–3439. [CrossRef]

47. Oefner, C.; D'Arcy, A.; Winkler, F.K. Crystal structure of human dihydrofolate reductase complexed with folate. *Eur. J. Biochem.* **1988**, *174*, 377–385. [CrossRef]

48. Lasko, T.A.; Bhagwat, J.G.; Zou, K.H.; Ohno-Machado, L. The use of receiver operating characteristic curves in biomedical informatics. *J. Biomed. Inform.* **2005**, *38*, 404–415. [CrossRef]

49. Triballeau, N.; Acher, F.; Brabet, I.; Pin, J.P.; Bertrand, H.O. Virtual screening workflow development guided by the "receiver operating characteristic" curve approach. Application to high-throughput docking on metabotropic glutamate receptor subtype 4. *J. Med. Chem.* **2005**, *48*, 2534–2547.

50. Herrera-Mayorga, V.; Lara-Ramírez, E.E.; Chacón-Vargas, K.F.; Aguirre-Alvarado, C.; Rodríguez-Páez, L.; Alcántara-Farfán, V.; Cordero-Martínez, J.; Nogueda-Torres, B.; Reyes-Espinosa, F.; Bocanegra-García, V.; et al. Structure-based virtual screening and in vitro evaluation of new trypanosoma cruzi cruzain inhibitors. *Int. J. Mol. Sci.* **2019**, *20*, 1742. [CrossRef]

51. Lipinski, C.A.; Lombardo, F.; Dominy, B.W.; Feeney, P.J. Experimental and computational approaches to estimate solubility and permeability in drug discovery and development settings. *Adv. Drug Deliv. Rev.* **2012**, *64*, 4–17. [CrossRef]

52. Ramasamy, S.; Chin, S.P.; Sukumaran, S.D.; Buckle, M.J.C.; Kiew, L.V.; Chung, L.Y. In Silico and in vitro analysis of bacoside A aglycones and its derivatives as the constituents responsible for the cognitive effects of Bacopa monnieri. *PLoS ONE* **2015**, *10*, e0126565. [CrossRef] [PubMed]

53. Fradera, X.; Babaoglu, K. Overview of methods and strategies for conducting virtual small molecule screening. In *Current Protocols in Chemical Biology*; John Wiley & Sons, Inc.: Hoboken, NJ, USA, 2017; Volume 9, pp. 196–212.

54. Pikkemaat, M.G.; Linssen, A.B.M.; Berendsen, H.J.C.; Janssen, D.B. Molecular dynamics simulations as a tool for improving protein stability. *Protein Eng. Des. Sel.* **2002**, *15*, 185–192. [CrossRef] [PubMed]

55. El-Hamamsy, M.H.R.I.; Smith, A.W.; Thompson, A.S.; Threadgill, M.D. Structure-based design, synthesis and preliminary evaluation of selective inhibitors of dihydrofolate reductase from Mycobacterium tuberculosis. *Bioorg. Med. Chem.* **2007**, *15*, 4552–4576. [CrossRef]

56. Bhosle, A.; Chandra, N. Structural analysis of dihydrofolate reductases enables rationalization of antifolate binding affinities and suggests repurposing possibilities. *Febs J.* **2016**, *283*, 1139–1167. [CrossRef] [PubMed]

57. Chen, D.; Oezguen, N.; Urvil, P.; Ferguson, C.; Dann, S.M.; Savidge, T.C. Regulation of protein-ligand binding affinity by hydrogen bond pairing. *Sci. Adv.* **2016**, *2*, e1501240. [CrossRef]

58. Liu, J.; Su, M.; Liu, Z.; Li, J.; Li, Y.; Wang, R. Enhance the performance of current scoring functions with the aid of 3D protein-ligand interaction fingerprints. *BMC Bioinform.* **2017**, *18*, 343. [CrossRef]

59. Li, Y.; Han, L.; Liu, Z.; Wang, R. Comparative assessment of scoring functions on an updated benchmark: 2. evaluation methods and general results. *J. Chem. Inf. Model.* **2014**, *54*, 1717–1736. [CrossRef]

60. Huang, J.; Mackerell, A.D. CHARMM36 all-atom additive protein force field: Validation based on comparison to NMR data. *J. Comput. Chem.* **2013**, *34*, 2135–2145. [CrossRef]

61. Rosta, E.; Buchete, N.-V.; Hummer, G. Thermostat artifacts in replica exchange molecular dynamics simulations. *J. Chem. Theory Comput.* **2009**, *5*, 1393–1399. [CrossRef]

62. Zoete, V.; Cuendet, M.A.; Grosdidier, A.; Michielin, O. SwissParam: A fast force field generation tool for small organic molecules. *J. Comput. Chem.* **2011**, *32*, 2359–2368. [CrossRef] [PubMed]

63. Bussi, G.; Donadio, D.; Parrinello, M. Canonical sampling through velocity-rescaling. *J. Chem. Phys.* **2007**, *126*, 014101. [CrossRef] [PubMed]

64. Haigis, V.; Belkhodja, Y.; Vuilleumier, R.; Boutin, A. Challenges in first-principles NPT molecular dynamics of Soft Porous Crystals: A case study on MIL-53(Ga). *J. Chem. Phys.* **2014**, *141*, 064703. [CrossRef] [PubMed]

65. Hess, B.; Bekker, H.; Berendsen, H.J.C.; Fraaije, J.G.E.M. LINCS: A linear constraint solver for molecular simulations. *J. Comput. Chem.* **1997**, *18*, 1463–1472. [CrossRef]

66. Abraham, M.J.; Gready, J.E. Optimization of parameters for molecular dynamics simulation using smooth particle-mesh Ewald in GROMACS 4.5. *J. Comput. Chem.* **2011**, *32*, 2031–2040. [CrossRef]

67. Kumari, R.; Kumar, R.; Lynn, A.; Lynn, A. *g_mmpbsa* —A GROMACS Tool for High-Throughput MM-PBSA Calculations. *J. Chem. Inf. Model.* **2014**, *54*, 1951–1962. [CrossRef]

Sample Availability: Samples of the compounds are not available from the authors.

 *molecules*

Article

Structure-Based Discovery of Dual-Target Hits for Acetylcholinesterase and the α7 Nicotinic Acetylcholine Receptors: In Silico Studies and In Vitro Confirmation

Sebastian Oddsson [1,2,3] , Natalia M. Kowal [1,2,3], Philip K. Ahring [2,3], Elin S. Olafsdottir [1] and Thomas Balle [2,3,*]

[1] Faculty of Pharmaceutical Sciences, School of Health Sciences, University of Iceland, 107 Reykjavík, Iceland; seb20@hi.is (S.O.); nmp@hi.is (N.M.K.); elinsol@hi.is (E.S.O.)
[2] Sydney School of Pharmacy, Faculty of Medicine and Health, The University of Sydney, Sydney, NSW 2006, Australia; philip.ahring@sydney.edu.au
[3] Brain and Mind Centre, The University of Sydney, Camperdown, NSW 2050, Australia
* Correspondence: thomas.balle@sydney.edu.au; Tel.: +61-290-367-035

Academic Editors: Marco Tutone and Anna Maria Almerico
Received: 3 June 2020; Accepted: 15 June 2020; Published: 22 June 2020

Abstract: Despite extensive efforts in the development of drugs for complex neurodegenerative diseases, treatment often remains challenging or ineffective, and hence new treatment strategies are necessary. One approach is the design of multi-target drugs, which can potentially address the complex nature of disorders such as Alzheimer's disease. We report a method for high throughput virtual screening aimed at identifying new dual target hit molecules. One of the identified hits, N,N-dimethyl-1-(4-(3-methyl-[1,2,4]triazolo[4,3-a]pyrimidin-6-yl)phenyl)ethan-1-amine (Ýmir-2), has dual-activity as an acetylcholinesterase (AChE) inhibitor and as an α7 nicotinic acetylcholine receptor (α7 nAChR) agonist. Using computational chemistry methods, parallel and independent screening of a virtual compound library consisting of 3,848,234 drug-like and commercially available molecules from the ZINC15 database, resulted in an intersecting set of 57 compounds, that potentially possess activity at both of the two protein targets. Based on ligand efficiency as well as scaffold and molecular diversity, 16 of these compounds were purchased for in vitro validation by Ellman's method and two-electrode voltage-clamp electrophysiology. Ýmir-2 was shown to exhibit the desired activity profile (AChE IC_{50} = 2.58 ± 0.96 μM; α7 nAChR activation = 7.0 ± 0.9% at 200 μM) making it the first reported compound with this particular profile and providing further evidence of the feasibility of in silico methods for the identification of novel multi-target hit molecules.

Keywords: high-throughput virtual screening; dual-target lead discovery; neurodegenerative disorders; Alzheimer's disease; dual mode of action; multi-modal; nicotinic acetylcholine receptor; acetylcholinesterase; molecular docking

---

1. Introduction

The development of treatments for neurodegenerative disorders such as Alzheimer's disease (AD), the most prevalent type of dementia, is a pressing matter due to the incurable, progressive, and debilitating nature of the disease [1]. The situation is further aggravated by the ageing of populations worldwide and reflected by the estimated triplication of the AD-affected from 50 million in 2018 to 152 million in 2050 [2]. Neuropathologically, AD is characterized by its accompanying lesions, most notably but not exclusively by amyloid plaques, neurofibrillary tangles, and neuronal and synaptic loss [3,4], that lead to a variety of neurochemical changes, prominently in the cholinergic

system [5]. Taken together, these alterations are the cause of cognitive symptoms such as memory loss, language impairment, visuospatial dysfunction, and executive functioning issues [6] which are often accompanied by other behavioral and psychological symptoms (BPSD) [7,8].

Despite extensive research efforts to understand AD, its causes and the underlying disease mechanisms remain poorly understood. This renders the search for effective drugs difficult. AD treatments have yet to succeed in slowing down disease progression and AD medications currently approved in the US are palliative in nature but nevertheless continue to provide the biggest benefits for patients [9]. The drugs fall into two categories, acetylcholinesterase (AChE) inhibitors (galantamine (**1**), donepezil (**2**), rivastigmine (**3**); Figure 1) that raise synaptic acetylcholine (ACh) levels and memantine (**4**), an NMDA receptor antagonist that regulates glutamatergic activity [10,11].

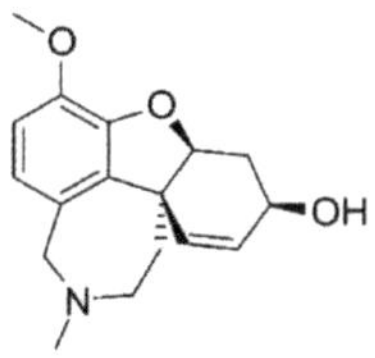

Galantamine (**1**)

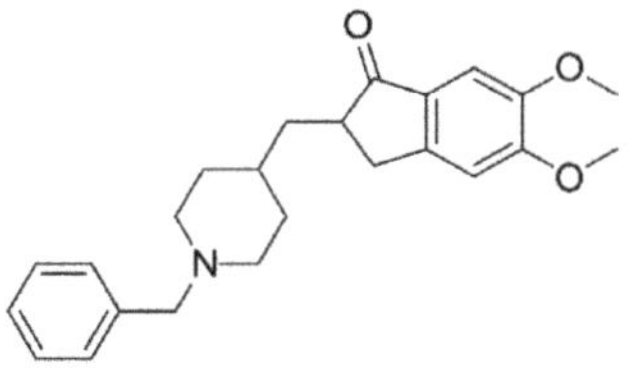

Donepezil (**2**)

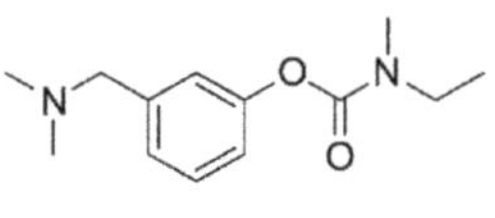

Rivastigmine (**3**)

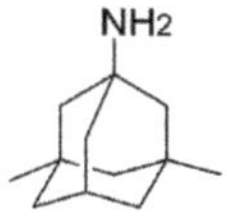

Memantine (**4**)

Figure 1. Structures of currently FDA-approved drugs for the treatment of AD, galantamine (**1**), donepezil (**2**), rivastigmine (**3**) and memantine (**4**). Donepezil and memantine are also approved as a combination therapy.

Investigations into the effects of administering memantine in conjunction with AChE inhibitors started about two decades ago [12,13] and led to the approval of Namzaric, which is a combination therapy relying on fixed doses of donepezil and memantine [14]. Other combination therapies are also of interest, e.g., drugs with pro-cognitive effects combined with anti-inflammatory drugs [15]. In fact, treatments that target more than one disease mechanism have become an emergent therapeutic strategy, especially for the treatment of complex, multifactorial diseases [16]. Besides addressing their intricate nature, the combination of two molecules with different activities in one drug formulation can result in drug synergism, as well as prevent unwanted compensatory mechanisms and drug tolerance, for instance [17].

An alternative to combination therapies is multi-modal compounds [18,19]. These compounds have essentially the same advantages as drug combinations but have, analogously to single molecule therapies, an easier therapeutic regimen, fewer drug-drug interactions and no differences in pharmacological properties to be considered [17,19]. However, the search for multi-target ligands also poses new challenges [20]. This conundrum is commonly tackled by the synthesis of bivalent hybrid ligands [21], but computational approaches can also be used to screen for novel ligands, i.e., by exploiting pharmacophore models underlying a set of known reference ligands or models of the corresponding binding sites [22]. Virtual screening (VS) methods allow for the screening of large compound databases against these models within a reasonable timeframe at comparably low cost. A prerequisite is the availability of enough data of adequate quality in order to construct predictive models. With the number of published X-ray structures, ligand datasets, and the ever-increasing size of screening compound databases, these strategies are becoming increasingly more feasible [23].

Nevertheless, VS for multi-target-directed ligands is not as commonly used as might be expected due to the complexity of the task and a lack of established protocols [23].

In previous work, we embarked on the search for AChE inhibitors that concurrently increase receptor activation at pro-cognitive nicotinic acetylcholine receptors (nAChRs) in a subtype-specific manner [24]. In addition to the already established benefits of increasing synaptic ACh levels by AChE inhibitors, the activity profile would thus extend to direct modulation of the cholinergic system. Since the dosage of AChE inhibitors is often limited due to adverse effects, boosting the activity of a pro-cognitive receptor such as the $\alpha7$ nAChR [25] could increase the overall treatment efficiency. Importantly, we established the feasibility of VS for the identification of compounds targeting AChE as well as the $\alpha7$ nAChR [24]. Top-scoring VS hits obtained from one target were subsequently screened against the second target, which allowed for the identification of dual-activity compounds, however, the disadvantage of this approach was introduction of bias towards the first target screened and that none of these screening hits showed agonistic activity at the $\alpha7$. Therefore, in the present study, the size of the screening database was increased ~45 fold by shifting the focus from natural products and their derivatives to drug-like molecules. Furthermore, the entirety of the screening database was screened against both targets individually. Donepezil and a recently published $\alpha7$ nAChR agonist were selected as reference compounds and structure-based, parallel and independent VS was performed. Here, we present the new protocol along with in silico and in vitro results, which show that it is feasible to identify multi-target compounds with the desired activity profile.

2. Results

Our goal was to find new compounds with dual activity: (i) inhibitors at the AChE and (ii) agonists of the $\alpha7$ nAChR as illustrated in Figure 2. The approach taken was to screen a pre-filtered compound database against an AChE X-ray structure and an $\alpha7$ nAChR homology model in parallel using identical parameters. The intersecting set between the two independently obtained sets of VS hits was considered to contain potential bimodal compounds as illustrated by the Venn diagram in Figure 2. A subset of these compounds were selected for in vitro testing at human $\alpha7$ nAChR expressed in *Xenopus laevis* oocytes by two-electrode voltage-clamp electrophysiology and at the human recombinant AChE using Ellman's colorimetric method [26] to validate the in silico predictions.

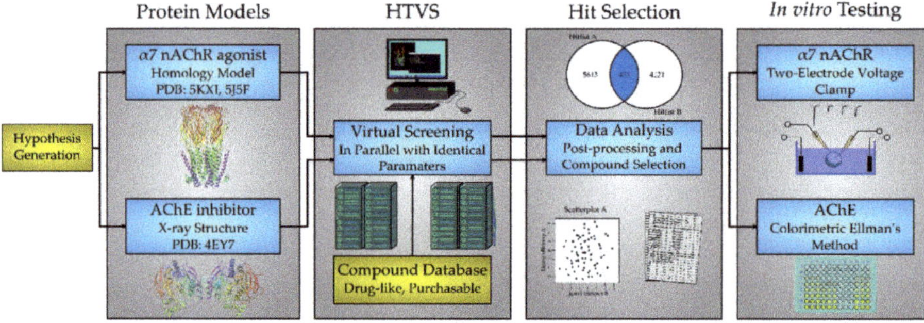

Figure 2. Workflow of the parallel HTVS, hit selection and in vitro evaluation. Two protein targets were selected for in silico studies and protein models suitable for docking prepared. The same database of compounds was then individually screened against each model using identical parameters. After post-processing, common compounds from the two independent screening hit lists were used to identify compounds destined for in vitro testing.

2.1. Protein Structures and Homology Modeling

Based on structure-activity considerations for AChE inhibitors and $\alpha7$ nAChR agonists, an X-ray structure of AChE co-crystallized with donepezil (**2**) determined to a resolution of

2.35 Å (PDB 4EY7) was selected for the high-throughput virtual screening (HTVS) [27]. Since the structure of α7 nAChR has not been determined to date, a homology model was constructed using an $(\alpha 4)_2(\beta 2)_3$ nAChR structure with a resolution of 3.94 Å (PDB 5KXI) [28] as primary template augmented with an additional α4 subunit to facilitate modelling of an α7 homopentamer. In addition, an acetylcholine binding protein (AChBP) from *Lymnaea stagnalis* co-crystallized with *N4,N4*-bis[(pyridin-2-yl)methyl]-6-(thiophen-3-yl)pyrimidine-2,4-diamine (Compound 44; α7-pEC$_{50}$ = 6.3) from a recently published ligand series (PDB 5J5F) [29] served to define and bias the binding site. This structure was chosen because donepezil is an extended, linear structure like Compound 44 (**5**), as illustrated in Figure 3. We hypothesized that including the AChBP structure with its co-crystallized ligand as an additional template in the model building process would result in an α7 nAChR model, in which the binding pocket would be opened up in a way that would facilitate the docking of extended molecules resembling donepezil in size and shape, thus enhancing the probability of identifying dual active molecules.

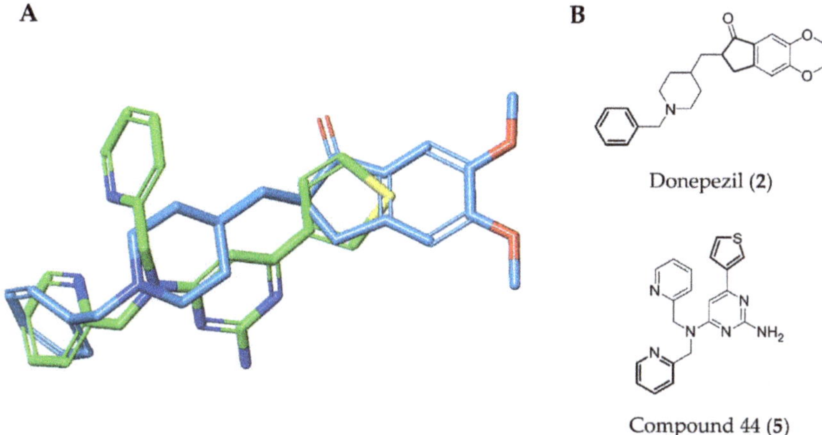

Figure 3. (**A**) Superposition of reference ligands in their bioactive, co-crystallized conformations. AChE inhibitor Donepezil from PDB 4EY7 is shown in blue and α7 nAChR agonist Compound 44 from PDB 5J5F in green. Atoms of the rings shown in bold in Panel B were used for superimposition of the ligands. (**B**) Chemical structures of Donepezil and Compound 44.

We selected the homology model with the lowest Discrete Optimized Protein Energy (DOPE) score [30] and inspected its Φ, Ψ angle distributions (Figure 4). A small fraction of the modelled amino acid residues occupies energetically unfavorable regions of the plot. Since all these residues are in non-conserved loop regions of the protein distant from the binding site, the model was deemed suitable for docking. Both the α7 nAChR model and AChE structure were validated by docking the respective reference ligands (RMSD Compound 44 = 0.87 Å, Donepezil = 0.62 Å).

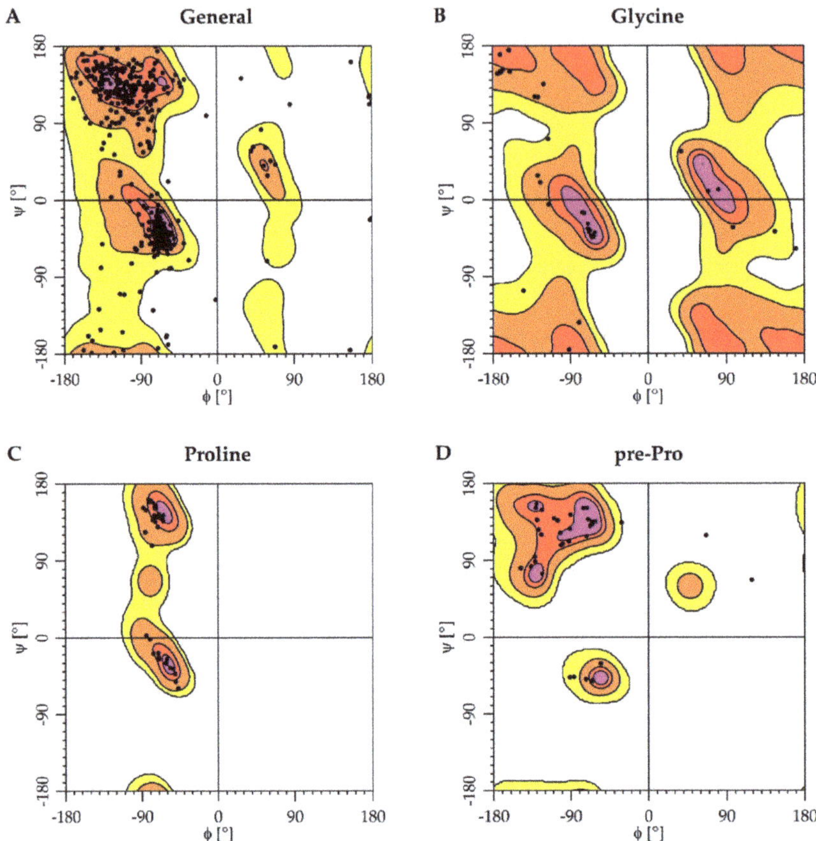

Figure 4. Detailed Ramachandran diagram of chains A and B of the α7 nAChR homology model by residue type. (**A**) The general case of non-glycine, non-proline and non-pre-proline residues is depicted. The special cases having either significantly less (**B**; Glycine) or more conformational restraints (**C,D**; Proline and preproline). The contours of Glycine are twofold symmetrized based on the dataset from Lovell et al. [31].

2.2. Screening Database

To optimize the chances of success for the VS, we considered the importance of various factors that concern compounds making up the screening database. The ZINC15 database [32], which is freely available and suited for VS, provides compound annotations, such as different degrees of commercial availability and the ability to download specific subsets of the database based thereon, and was chosen as starting point. In a first step, a subset of 7,679,852 compounds was downloaded from ZINC15 (Figure 5) that corresponds to compounds which are "anodyne" (lowest reactivity), "in-stock" (highest commercial availability), have molecular weights (MWs) between 250 and 500 Da and a logP within the range of −1 to 5. This puts an emphasis on drug-likeliness and on the ability to verify the results in the laboratory. In addition to the property filters built into ZINC15, another set of property and structure filters was applied. The property filters were based on Lipinski's rule of five [33] with the intention of enhancing the drug-likeliness of the compounds while the structural filters were more focused on reducing the probability of identifying reactive, assay-interfering or otherwise problematic ligands as exemplified by REOS (rapid elimination of swill) [34] or PAINS (pan-assay interference compounds) [35] filters. Roughly half of the initial set was discarded during this extensive filtering

process. To finalize the screening database, the remaining 3,848,234 compounds were prepared for docking resulting in 5,213,053 entities. At this stage, the total charge per molecule was limited to 0 and 1 predicated on the knowledge that nicotinic receptor ligands in most instances contain a basic nitrogen atom crucial for binding.

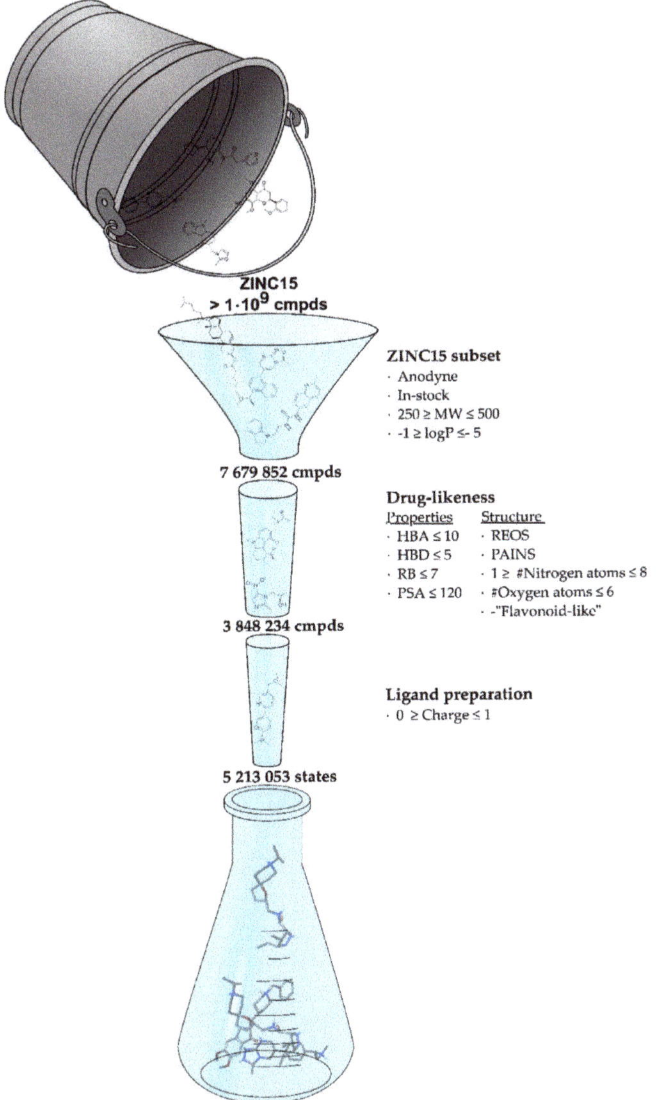

Figure 5. Schematic overview depicting the number of molecules during database preparation, starting from the compound database ZINC15 (bucket) through various filtering stages (funnel) to the final database used for VS (flask). States refers to the number of different entities for which 3D coordinates were generated and includes protonation states and tautomers.

2.3. Virtual Screening

The database was docked against both protein targets in parallel using the glide-based Schrodinger virtual screening workflow and identical parameters. Following three rounds of docking exercises, the number of compounds was narrowed down to approximately 15,000 for each target. After post-processing, the compounds below the 0.5 quantile of the normalized ligand efficiencies were considered reasonable VS hits for each target and the intersect between the two VS screening results was then investigated in more detail (Figure 6). The median set of the putative ligands consisted of 4734 and 3824 compounds for AChE and α7 nAChR, respectively, with 57 compounds shared between the two sets. The docking poses and the structural diversity of these compounds were then analyzed in order to select 15 compounds for in vitro testing. Notably, all compounds contained a basic nitrogen and at least one aromatic ring. Several hits were analogous structures, hence compounds were prioritized for in vitro testing based on the ligand efficiency values from docking and the observed molecular interactions in the docked ligand poses such as hydrogen bonding to TRP-149 of the α7 nAChR, while also ensuring that no close derivatives were present in the final set.

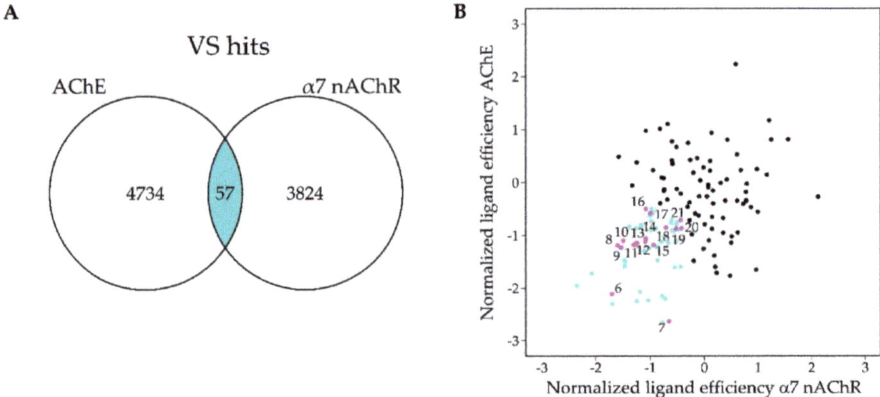

Figure 6. Venn diagram of AChE and α7 nAChR screening hits in the 0.5 quantile of normalized ligand efficiencies. (**B**) Scatterplot of normalized, strain-corrected ligand efficiencies. Compounds which are in the 0.5 quantile are colored blue and tested compounds are labeled as indicated Table 1 and shown in magenta.

2.4. AChE and nAChR α7 Activity Testing

The selected compounds (**6–21**, Table 1) were screened for AChE inhibitory activity at 200 μM by employing Ellman's colorimetric analysis [26]. For five poorly soluble compounds, the methanol supernatants were tested. Ellman's colorimetric analysis is based on the breakdown of acetylthiocholine (ATCI) by AChE to thiocholine and acetic acid. Ellman's reagent in turn reacts with the thiol group of thiocholine resulting in a yellow color indicating the extent of enzymatic activity. Of the eleven compounds soluble in MeOH, ten were shown to inhibit AChE significantly at 200 μM varying from 66.6–100% while one compound (**17**), showed low inhibition (22.9 ± 1.1%, Table 1), which is in agreement with it having the lowest docking score for AChE of all the compounds tested. The five poorly soluble compounds also exhibited significant inhibitory activities between 45.9 and 100%. Overall, fourteen compounds inhibited AChE enzymatic activity more than 60%, eleven more than 80% and eight more than 90%. Compounds that inhibited AChE more than 80% in the initial screening at 200 μM had IC$_{50}$ values in the low micromolar range between 1.10 and 33.47 μM, which is two to three orders of magnitude lower than donepezil, the reference ligand, which showed an IC$_{50}$ value of 0.06. μM. The IC$_{50}$ of compounds **6** and **15**, the least potent AChE inhibitors, were determined to be 58.47 and 114.70 μM.

Table 1. Docking and in vitro results for tested compounds (**6–21**). For AChE, compounds were initially screened at 200 μM followed by IC_{50} measurements for soluble compounds with ≥ 80% inhibition. For $\alpha 7$ nAChR agonism, activities were measured at 200 μM concentrations. For determination of modulation of ACh responses at the $\alpha 7$ nAChR, 100 μM compound was co-applied with 30 μM ACh. Measured data represent the mean ± S.E.M of at least three AChE replicates and independent oocyte experiments.

Compd. ID	Structure	AChE G-Score [a] Norm. LE [b] %Inhibition [c,d] IC_{50} (μM)(pIC_{50} ± SEM)	$\alpha 7$ nAChR G-Score [a] Norm. LE [b] %Activation [c] %Inhibition [c]	Compd. ID	Structure	AChE G-Score [a] Norm. LE [b] %Inhibition [c,d] IC_{50} (μM) (pIC_{50} ± SEM)	$\alpha 7$ nAChR G-Score [a] Norm. LE [b] %Activation [c] %Inhibition [c]
6	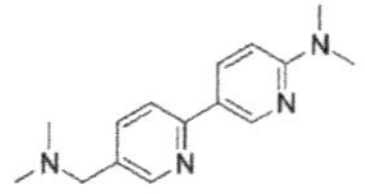	−12.30 −2.11 76.1 ± 0.9% IC_{50} = 58.47 (4.23 ± 0.02)	−13.85 −1.71 0.4 ± 0.4% 47.2 ± 3%	14	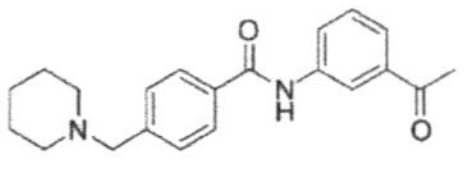	−13.04 −1.06 100% IC_{50} = 1.10 (5.96 ± 0.02)	−15.56 −1.09 0.3 ± 0.3% 60.0 ± 1%
Ýmir-2 (7)[f]	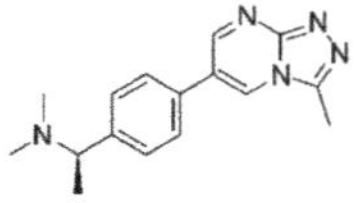	−14.90 −2.63 98.4 ± 0.9% IC_{50} = 2.58 (5.59 ± 0.02)	−11.50 −0.65 7.0 ± 0.9% -	Ýmir-10 (15)	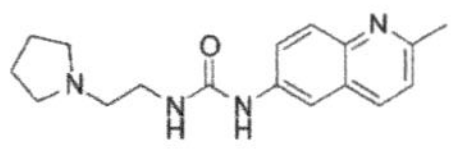	−11.79 −1.18 66.6 ± 1.7% IC_{50} = 114.70 (3.94 ± 0.02)	−13.13 −0.94 2.3 ± 0.4% -
8	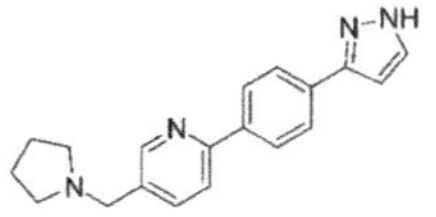	−12.34 −1.19 100%[d] -	−16.35 −1.60 0.3 ± 0.3% 59.0 ± 5%	16	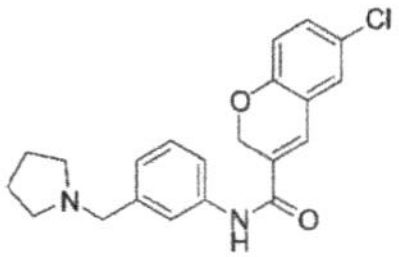	−11.81 −0.50 100%[d] -	−16.16 −1.08 1.0 ± 0.0% 90.7 ± 1%
9	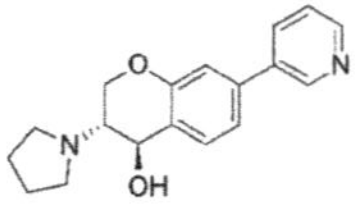	−11.91 −1.23 45.9 ± 3.8%[d] - -	−15.40 −1.54 1.0 ± 0.6% 66.7 ± 4%	17	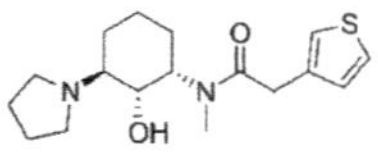	−10.22 −0.59 22.9 ± 1.1% - -	−13.35 −1.00 - -
10	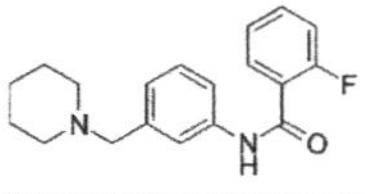	−12.11 −1.10 90.8 ± 0.2% IC_{50} = 11.28 (4.95 ± 0.02)	−15.93 −1.50 0.3 ± 0.3% 85.0 ± 2%	18	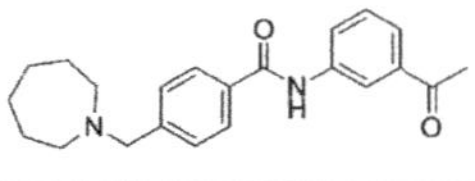	−12.92 −0.86 91.6 ± 2.2% IC_{50} = 17.88 4.75 ± 0.02	−14.49 −0.71 0.7 ± 0.3% 68.7 ± 1%

Table 1. *Cont.*

11	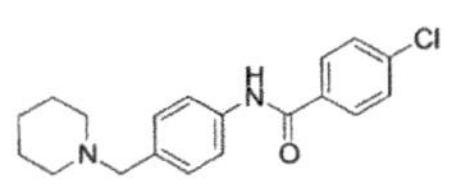	−12.32 −1.18 100%[d] - -	−15.16 −1.30 0.7 ± 0.3% 73.3 ± 6%	Ÿmir-14 (**19**)	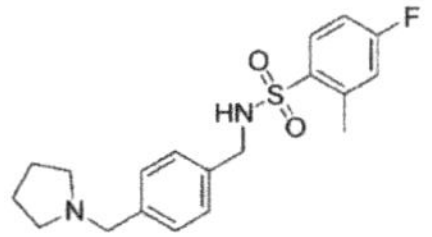	−12.48 −0.88 100% IC$_{50}$ = 1.32 (5.88 ± 0.01)	−13.13 −0.52 0.7 ± 0.3% 96.0 ± 1%
12	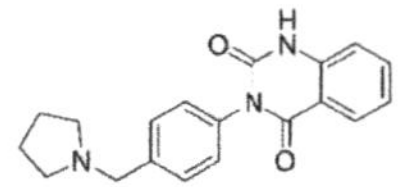	−12.75 −1.14 63.7 ± 4.4%[d] - -	−15.61 −1.25 0.3 ± 0.3% 81.7 ± 1%	**20**[f]	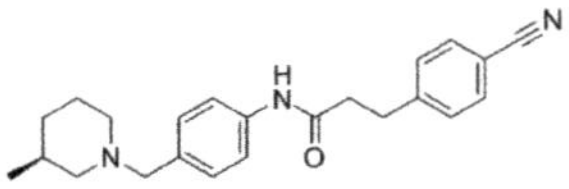	−13.45 −0.87 86.0 ± 0.1% IC$_{50}$ = 33.47 (4.48 ± 0.02)	−13.71 −0.42 0.3 ± 0.3% 94.3 ± 1%
13	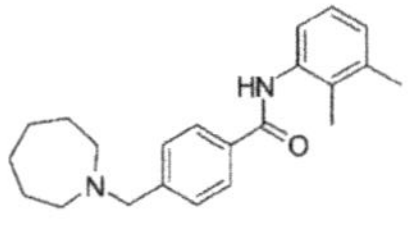	−13.20 −1.12 82.5 ± 1.1% IC$_{50}$ = 30.24 (4.52 ± 0.02)	−15.60 −1.10 0.7 ± 0.7% 68.0 ± 2%	Ÿmir-16 (**21**)	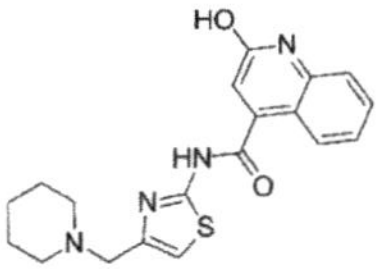	−12.48 −0.71 87.7 ± 0.6% IC$_{50}$ = 7.76 (5.10 ± 0.02)	−13.27 −0.44 0.3 ± 0.3% 97.3 ± 1%

[a] Glide G-score (kcal/mol); [b] Normalized ligand efficiency as defined in Formula 1; [c] "-" indicates "value not determined"; [d] % inhibition of supernatant (MeOH); [e] galantamine pIC$_{50}$ = 6.64 ± 0.02%, donepezil pIC$_{50}$ = 7.22 ± 0.02; [f] tested as racemic mixture.

Subsequently, all compounds except compound **17** were assessed at the α7 nAChR expressed in *Xenopus* oocytes using two-electrode voltage-clamp electrophysiology. Compounds were tested for direct activation of the α7 nAChR in a 0.2–200 μM concentration range. Compounds displaying less than 1% direct activation were further evaluated at 100 μM for their ability to alter currents evoked by 30 μM ACh. Compound **7** (Ýmir-2) and **15** (Ýmir-10) exhibited activation of α7 nAChR with 7.0 ± 0.9% and 2.3 ± 0.4% at 200 μM, respectively (Figure 7). Attempts to establish their potency were unsuccessful due to limited solubility. However, application of 2, 20, and 200 μM, as evident from Figure 7, established a concentration dependent effect. The remaining thirteen compounds exhibited less than 1% agonism indicating that they were either inefficient at mediating receptor activation or inactive at the tested concentrations. When tested as antagonists for their ability to inhibit ACh-evoked currents, all compounds showed inhibition in a range of 47.2–97.3% at 100 μM, with compounds **19** (Ýmir-14) and **21** (Ýmir-16) displaying almost full inhibition of 96% and 97%, respectively, at 100 μM (Figure 8).

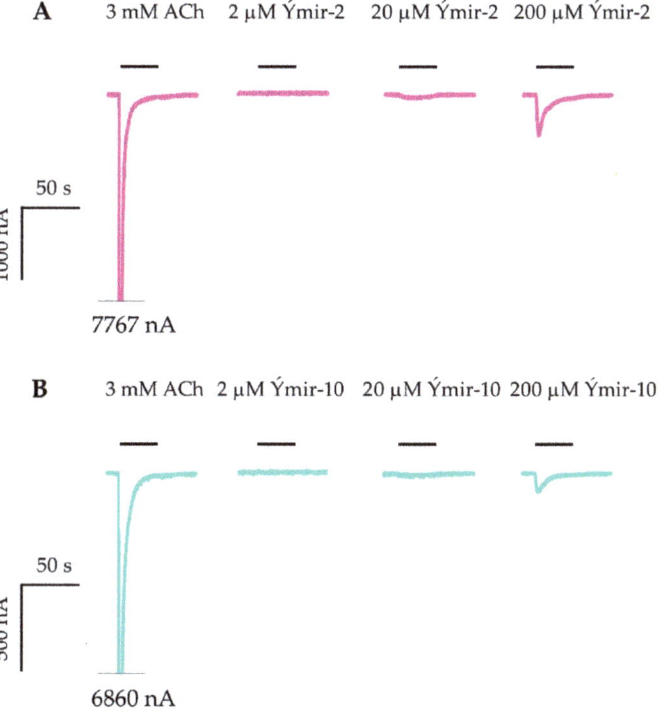

Figure 7. Evaluation of compounds as agonists. Representative current traces for ACh, Ýmir-2 (**7**) (**A**) and Ýmir-10 (**15**) (**B**) at α7 nAChRs expressed in *Xenopus* oocytes. Cells were subjected to two-electrode voltage-clamp electrophysiology experiments where the oocyte membrane potential was clamped at −60 mV. The representative traces were baseline subtracted. Bars above the traces represent the application periods and the respective test solution concentrations are indicated above them. Note that the majority of the washing periods (3 min) between each trace are omitted in the figure.

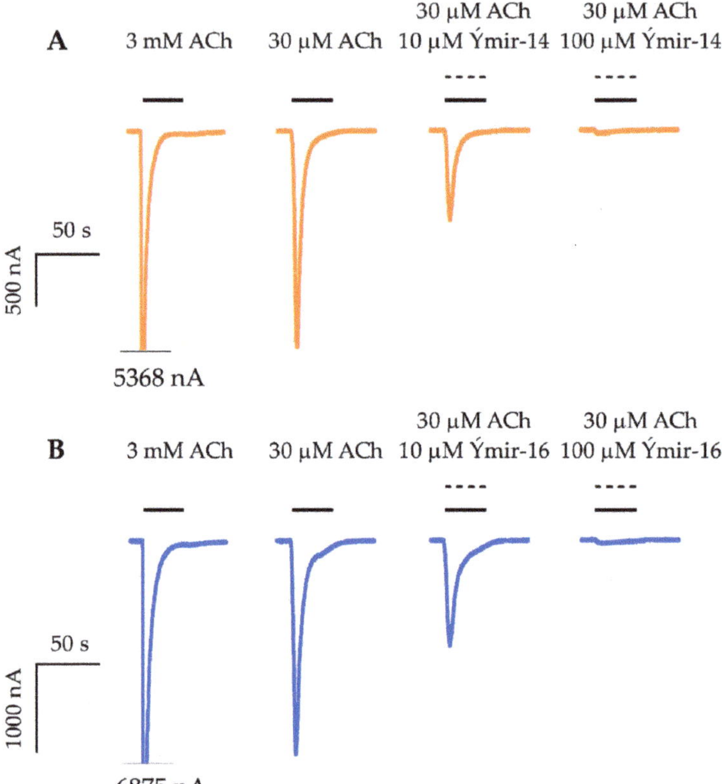

Figure 8. Evaluation of compounds as antagonists. Representative current traces for ACh and 10 and 100 µM Ýmir-14 (**19**) (**A**) and Ýmir-16 (**21**) (**B**) co-applied with 30 µM ACh at α7 nAChRs expressed in *Xenopus* oocytes. Cells were subjected to two-electrode voltage-clamp electrophysiology experiments where the oocyte membrane potential was clamped at −60 mV. The representative traces were baseline subtracted. Bars above the traces represent the application periods and the respective test solution concentrations are indicated above them. Note that the majority of the washing periods (3 min) between each trace is omitted in the figure.

3. Discussion

We embarked on the search for bimodal compounds with the help of computational methods. In accordance with the hypothesis from our previous study [24], we searched for hit molecules that target α7 nAChR as agonists and AChE as inhibitors. A drug based on this new activity profile could provide a new strategy for treating AD by the dual modulation of cholinergic signaling.

Despite the requirements of VS for high quality models of binding sites and screening databases, it has proven useful for the identification of new ligands for single targets and many methodological improvements have been made over the past decades [36,37]. Adding a second biological target adds another significant constraint to the problem, which is often addressed by pre-filtering or pre-screening the compound database based on one target before testing the second target [23].

In the current study, we conducted a VS without pre-screening our ligand database and docked the entire dataset to both targets. AChE and α7 nAChR are structurally and functionally distinct proteins but both evolved to accommodate ACh in their respective binding pockets. Sharing the same endogenous ligand and hence pharmacophoric elements should increase the probability of finding a

molecule that fits both pockets. In addition, we constrained the search to ligands that are extended and linear based on two reference ligands.

We successfully employed this HTVS approach and identified two compounds (Ýmir-2, Ýmir-10) that showed AChE inhibition and activation of the α7 nAChR, confirming the feasibility of VS for the search of bimodal compounds at these targets as reported previously [24]. We observed a remarkably high hit rate for AChE inhibitors, where all but one (**17**) of the tested compounds showed activity at 200 μM. Moreover, 10 out of the 11 compounds soluble in MeOH showed significant inhibition of the enzymatic activity of AChE, as did all the compounds where only the supernatants were tested due to low solubility, indicating that these compounds also interact with the active site of AChE. However, further analysis to determine IC_{50} values could not be performed for these compounds.

The hit rate at the α7 nAChR was likewise high, with 2/15 compounds displaying direct agonism. Ýmir-2 (**7**) and Ýmir-10 (**15**) showed partial activation of 7.0% and 2.3%, respectively, at 200 μM. Due to solubility issues, the maximal receptor activation could not be determined. The remaining compounds exhibited α7 nAChR antagonism, in a range between 47.2% and 97.3% at 100 μM when co-applied with 30 μM ACh. Two compounds, Ýmir-14 (**19**) and Ýmir-16 (**21**) inhibited the response of ACh almost fully and in a concentration dependent manner (Figure 8). This, and the fact that the agonist-based VS project yielded compounds that are structurally similar to each other as well as to known α7 nAChR agonists, suggests that the investigated compounds interact with the binding pocket where they likely act as antagonists although weak partial agonism cannot be ruled out. Hence, as all compounds but one displayed some activity at the nAChR, and as the difference between an antagonist and agonist is often subtle, the VS in terms of binding to the α7 nAChR orthosteric binding site could be considered as high as 14/15. However, it cannot be excluded that receptor inhibition for some compounds is mediated by a different mechanism such as direct blockage of the ion channel pore. Further experiments would be required to elucidate the exact mechanism of inhibitory interaction.

All 57 compounds identified in both screenings were extended, linear molecules. Other common properties such as the presence of a basic nitrogen at one end of the molecule and an aromatic moiety at the other were also observed. As the set included derivatives of the same compound but also compounds that differed only in their basic amine (e.g., piperidine and pyrrolidine) we identified these two regions as the main source of variability in the set of putative bimodal compounds. The low diversity of the intersecting set is likely a consequence of the protocol and the constraints applied during docking, but the structural patterns are also in accordance with known nicotinic ligands indicating that this is a general feature of nAChR ligands. The docked poses of Ýmir-2 (**7**) and Ýmir-10 (**15**) display characteristic interactions of AChE inhibitors and nicotinic receptor ligands around the basic amine, i.e., the positively charged amines are coordinated in the aromatic cage of the α7 nAChR and the anionic site of AChE (Figure 9). The lack of hydrogen bonds and strong hydrophobic interactions distal to the basic amines in the α7 nAChR suggests that the activity of these ligands could be further improved in this region. It is also noteworthy that the basic amine of Ýmir-2 (**7**) does not display the hydrogen bond to the backbone carbonyl of TRP-149 which is often considered crucial [38] and observed in multiple co-crystallized complexes between ligands and nicotinic receptors as well as acetylcholine binding proteins.

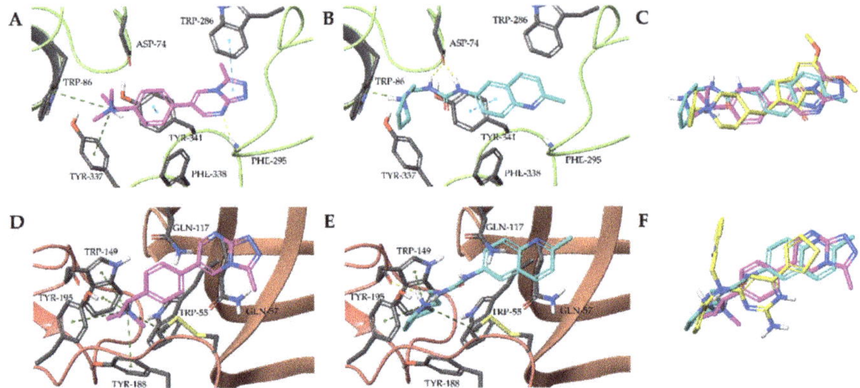

Figure 9. Interactions of Ýmir-2 (**7**) (magenta) and Ýmir-10 (**15**) (blue) in AChE (**A,B**) and α7 nAChR (**D,E**) and superposition of docking poses to reference ligands donepezil (**C**) and Compound 44 (**5**) (**F**). Hydrogen-bonding and is indicated by yellow, dotted lines, cation-π interactions by green, dotted lines and π–π interactions by blue dotted lines.

4. Materials and Methods

4.1. Materials

Plant origin galantamine hydrobromide analytical standard was purchased from PhytoLab GmbH & Co. KG (Vestenbergsgreuth, Germany). Screened compounds (**6–16**, **18**, **20** and **21**) were purchased from Molport (Riga, Latvia), compound **17** from Asinex (Winston Salem, NC, USA) and **19** from AsisChem (Waltham, MA, USA). Restriction enzymes were from New England Bio Labs Inc. (Ipswich, MA, USA) and DNA and RNA purification kits were from QIAGEN N.V. (Venlo, The Netherlands). mMessage mMachine T7 transcription kit was from ThermoFisher Scientific (Waltham, MA, USA). Human recombinant acetylcholinesterase (C1682), acetylcholine chloride, acetylthiocholine iodide, 5,5-dithiobis-(2-nitro-benzoic acid), bovine serum albumin, kanamycin, theophylline, tricaine, collagenase, HEPES, Trizma, salts and other chemicals not mentioned specifically were purchased from Sigma-Aldrich Co. LLC (St. Louis, MO, USA) and were of analytical grade.

4.2. Protein Models

4.2.1. nAChR α7 Homology Model

The amino acid sequence of the human α7 nAChR was obtained from the UniProt protein knowledgebase (entry P36544 [39]) and aligned to the sequences of the α4- and β2-subunits of PDB 5KXI [28] (α- and β-subunits or chains A and B respectively) and 5J5F [29] (chain A) with the T-Coffee Expresso structural alignment tool [40]. Residues used to build the homology model were then manually selected ensuring that the binding site was mainly built on the 5J5F template, an AChBP structure in complex with Compound 44 (**5**) (N4,N4-bis[(pyridin-2-yl)methyl]-6-(thiophen-3-yl)pyrimidine-2,4-diamine), whereas the tertiary and quaternary structure was largely modelled after 5KXI (Figure 10). Since the α7 receptor consists of α-subunits only, a single α-subunit from the $(α4)_2(β2)_3$ template was also provided separately as template where there is a β-subunits in the 5KXI complex. The α7 nAChR homology models were built with Modeller (Version 9.16, Automodel class, Salilab, UCSF, San Francisco, CA, USA) [41]. Protein secondary structures as well as the pentameral symmetry were supplied as constraints to Modeller. Out of 100 generated models, the model with the highest scoring DOPE potential [30] was selected for further studies.

Figure 10. Sequence alignment of the templates used in the homology model of α7 nAChR. The sequence of one α- and one β-subunit of the (α4)₂(β2)₃ receptor co-crystallized with nicotine (PDB 5KXI) and of an AChBP co-crystallized with Compound 44 (**5**) were aligned using T-Coffee and modified for homology model building. Residues used as templates in homology modelling are shown in bold and conserved amino acids highlighted green. The overall sequence identity between the α7 nAChR and the (α4)₂(β2)₃ nAChR subunits used as primary templates are 45% (62% sequence similarity) and 37% (57% sequence similarity), respectively. The sequence identity between α7 nAChR and AChBP is 28% (46% sequence similarity) overall and 39% (64% sequence similarity) for the part that was used as template.

The resulting receptor model was then prepared with the Schrodinger Protein preparation wizard [42] (hydrogens added and H-bonds optimized, protonation according to pH 7.4, restrained minimization (RMSD 0.3 Å)). The quality of the selected model was assessed based on a Ramachandran plot drawn in R [43] using the data published by Lovell et al. [31]. Subsequently, the protein-ligand complex was refined with Schrodinger Prime [44] for the ligand its surrounding protein residues within

5 Å using Monte Carlo sampling [45], VSGB solvation model [46], OPLS3 force field [47], and otherwise default parameters.

4.2.2. AChE Model

The protein structure of AChE (PDB 4EY7 [27]) was downloaded from the Protein Data Bank [48] and all waters, ions, and small molecules except for donepezil removed. It was then prepared with the Schrodinger Protein preparation wizard [42] as described above for the α7 nAChR model.

4.3. Screening Library

The ZINC15 database [32] was downloaded in smiles format after application of the following filters: MW 250–500; logP −1–5; in-stock; anodyne. The smiles strings were then converted to canonical smiles and duplicates removed using OpenBabel 2.4.0 [49]. REOS [34] and PAINS [35] structure filters were applied using Schrodinger utilities. A set of additional filters were defined and applied to remove "flavonoid-like" structures, restricting the amount and position of halogen atoms and limiting the number of oxygen and nitrogen atoms to 6 and 1–8, respectively. Finally, physicochemical properties were calculated using Schrodinger software (Release 2018-1, Schrödinger, LLC, New York, NY, USA) and filtered using a custom perl script (HBD 5; HBA 10; PSA ≤ 120, RB ≤ 7).

4.4. Virtual Screening

With respect to VS, 14 Å^3 Glide Receptor Grids [50,51] were centered around donepezil and Compound **44** in the AChE and α7 nAChR protein models respectively and generated with the receptor scaling factor set to 0.9, the partial charge cut off to 0.3 and three positional restraints along the reference ligands in each of the binding sites. For the α7 nAChR, the areas concerned were in the hydrophobic pocket as well as the area corresponding to the position of the thiophen moiety of the *N4,N4*-bis[(pyridin-2-yl)methyl]-6-(thiophen-3-yl)pyrimidine-2,4-diamine ligand in the 5J5F structure and for AChE they were set at the anionic site as well as the hydrophobic gorge occupied by the indanone moiety of donepezil. These positional restraints were applied after each round of docking, thereby ensuring that only ligand poses occupying crucial areas in the binding site were considered without artificially enriching them. To perform the Ligand Docking, the Virtual Screening Workflow available in the Schrodinger suite based on Glide docking [50,51] was used. Ligands were prepared with OPLS_2005 forcefield [52] at pH 7.4 and a maximum of 8 different stereoisomers per ligand generated. Only neutral and mono charged states were kept (See Section 4.3 also). For docking, the ligand scaling factor set to 0.85 and the partial charge cut-off to 0.2. Moreover, 300,000 compounds were kept after the initial HTVS, 30,000 after the SP (Standard Precision), and none were removed after the XP (extra precision) docking stage. Docking scores were strain corrected with the help of the Strain Rescore script (OPLS3 [47], aqueous solvation model, Energy Offset 0 kcal/mol and default cartesian constraint settings). Subsequently, the dataset was further processed in R [43], where glide docking scores were used to calculate ligand efficiencies as

$$\text{Ligand efficiency} = (\text{Glide docking score} - \text{strain penalty}) \times (N_{\text{Heavy atoms}})^{-1} \qquad (1)$$

and the values obtained normalized. Afterwards, compounds with positive docking scores as well as compounds with strain energy larger than 4 kcal/mol were removed from the dataset. The intersecting set of both obtained lists of docking hits was then considered to contain the putative bimodal compounds. Subsequently, compounds in the lower 0.5 quantile of both normalized ligand efficiencies were excluded. From the resulting list, compounds were selected based on diversity, docking pose, and commercial availability.

4.5. AChE Activity Testing

In vitro AChE inhibitory activity was studied using the colorimetric method of Ellman [26] using human recombinant AChE enzyme. To a 96-well microplate, test solutions were applied along with 1 mg/mL bovine serum albumin, 0.5 mM 5,5-dithiobis-(2-nitro-benzoic acid), and 0.5 mM acetylthiocholine iodide. The reaction was initiated by addition of 0.20 units/mL AChE enzyme and followed by colorimetric detection at 405 nm. Experiments were conducted in triplicate. All compounds were dissolved in methanol (maximum 2% methanol at assay conditions that did not affect the enzyme activity) and screened at 200 μM with galantamine as a positive control.

Compounds that fully solubilized in MeOH and exhibited more than 70% inhibition of ACh degradation were further analyzed and their IC_{50} values determined using between five and eight concentrations.

4.6. nAChR α7 Activity Testing

4.6.1. Molecular Biology

Human α7 nAChR receptor subunits were cloned and inserted into expression vectors as described previously [53]. Plasmid cDNAs were linearized using a downstream Not I restriction site and purified. cRNA was prepared and capped from the linearized cDNA using the mMessage mMachine T7 transcription kit according to the manufactures protocol. Purified cRNA was aliquoted and stored at a concentration of 0.5 μg·μL^{-1} at $-80\,^{\circ}$C until further use.

4.6.2. Expression of α7 nAChR in *Xenopus laevis* Oocytes

Xenopus laevis oocytes were obtained as described previously [54], briefly, ovary lobes were removed by surgical incision, sliced into small pieces and defolliculated by collagenase treatment. The protocol for this specific study was approved by the Animal Ethics Committee of the University of Sydney (Protocol number: 2013/5915) and carried out according to these guidelines. Stage V and VI oocytes were injected with a total of ~25 ng of cRNA encoding human α7 nAChR with RIC3 (in 5:1 ratio), a protein enhancing the expression of the receptor. Injected oocytes were incubated for 3–5 days at 18 $^{\circ}$C in a saline solution (96 mM NaCl, 2 mM KCl, 1 mM $MgCl_2$, 1.8 mM $CaCl_2$, 5 mM HEPES (hemisodium, pH 7.4)) supplemented with 2.5 mM sodium pyruvate, 0.5 mM theophylline, and 50 μM gentamycin.

4.6.3. Oocyte Electrophysiology

Electrophysiological recordings from *Xenopus laevis* oocytes were performed using the two-electrode voltage-clamp technique as described previously [54,55]. Briefly, oocytes were placed in a custom-built recording chamber and continuously perfused with a saline solution. The saline solution contained 96 mM NaCl, 2 mM KCl, 1 mM $MgCl_2$, 1.8 mM $CaCl_2$, 5 mM HEPES (hemisodium, pH 7.4). Pipettes were backfilled with 3 M KCl and open-pipette resistances ranged from 0.3 to 1.5 MΩ when submerged in the saline solution. Oocytes were voltage clamped at a holding potential of -60 mV using an Axon Geneclamp 500B amplifier (Molecular Devices, LLC, San Jose, CA, USA). Rapid solution exchange in the oocyte vicinity (order of a few seconds) was ensured by application through a 1.5 mm diameter capillary tube placed approximately 2 mm from the oocyte as described previously [54]. The solution flow rate through the capillary was 2.0 mL/min. Experiments were performed as follows: nAChR currents were initially evoked with three $ACh_{control}$ applications (~EC_{20}, 30 μM), a maximum efficacious concentration of ACh_{max} (EC_{100}, 3 mM) followed by three additional $ACh_{control}$ applications. Thereafter, test compounds in increasing concentration were applied (25 s), the maximal tested concentration was 200 μM. A wash period of at least 2 min was kept between each application. After the agonist test, new ACh controls were applied (ACh_{max} followed by three additional $ACh_{control}$ applications) and compounds that displayed <1% activation were tested for their ability to modulate the effect of ACh. In these experiments, 10 and 100 μM concentration of

a test compound was co-applied with 30 μM ACh. Peak current amplitudes were normalized with respect to the amplitude of current elicited by 3 mM or 30 μM ACh for the agonist and antagonist test, respectively. All experiments were conducted at least in triplicate.

ACh was initially dissolved in milliQ water as 10 mM stock solution. Screened compounds were dissolved as a 50 mM stock solution in DMSO, except for Ýmir-2 and compound **12** which were kept as 10 mM stock solutions. The maximal DMSO concentration in the final dilution did not exceed 2%. This DMSO concentration did not evoke any current from the receptors. Compound dilutions were prepared in a saline solution on the day of the experiment.

4.7. Data Analysis

Data analysis was performed as reported previously [24]. Electrophysiological data were analyzed using pClamp 10.2 (Molecular Devices, LLC, San Jose, CA, USA). During analysis, traces were baseline subtracted and responses to individual applications quantified as peak-current amplitudes. Statistical calculations were performed using GraphPad Prism 7 (GraphPad Software, GraphPad, San Diego, CA, USA). Activation of the $\alpha 7$ nAChR was calculated as a percentage of E_{max} response to ACh. For evaluation of the inhibitory activity, the percentage of remaining peak-current amplitudes relative to that of the $ACh_{control}$ (EC_{20}) application was calculated. AChE inhibition data were analyzed using GraphPad Prism 7. Absorption readings from the AChE inhibition assay were plotted versus time and linear regression was performed. From the obtained slopes the percentages of inhibition were calculated normalized to the control (Galantamine) and IC_{50} values were determined from non-linear regression analysis. Data were fitted with the slope set to 1 and remaining current amplitude at infinitely high compound concentrations set to 0.

5. Conclusions

It was confirmed that HTVS approaches can be applied in the search for novel bimodal drug hits active at AChE and $\alpha 7$ nAChR with good hit rates. Ýmir-2 and Ýmir-10 represent novel compounds with a dual activity profile, i.e., inhibition of AChE and activation of the $\alpha 7$ nAChR, reported for the first time. The successful identification of two bimodal compounds is an encouraging outcome for VS for AD drug hits. Derivatives of Ýmir-2 could be developed into compounds with improved physicochemical properties and activity profiles of clinical interest.

Author Contributions: S.O., T.B., and E.S.O. designed the studies. S.O. designed and performed the in silico and AChE experiments. N.M.K. and P.K.A. designed and performed the $\alpha 7$ nAChR experiments. S.O. and T.B. drafted the manuscript, and all authors contributed to the final version of the manuscript. E.S.O. and T.B. acquired funding and performed supervision of the project. All authors approved the final version of the manuscript. All authors have read and agreed to the published version of the manuscript.

Funding: This research was supported by the Icelandic Centre for Research [grant number: 152604], doctoral grant and financial support from the University of Iceland.

Conflicts of Interest: The authors declare no conflict of interest.

References

1.   Gitler, A.D.; Dhillon, P.; Shorter, J. Neurodegenerative disease: Models, mechanisms, and a new hope. *Dis. Model. Mech.* **2017**, *10*, 499–502. [CrossRef] [PubMed]

2.   Patterson, C.; World Alzheimer. Report 2018: The state of the art of dementia research: New frontiers. *ADI* **2018**, *15*, 1473.

3.   Perl, D.P. Neuropathology of Alzheimer's disease. *Mt. Sinai J. Med.* **2010**, *77*, 32–42. [CrossRef]

4.   Serrano-Pozo, A.; Frosch, M.P.; Masliah, E.; Hyman, B.T. Neuropathological alterations in Alzheimer disease. *Cold Spring Harb. Perspect. Med.* **2011**, *1*, a006189. [CrossRef]

5.   Hampel, H.; Mesulam, M.M.; Cuello, A.C.; Farlow, M.R. Giacobini, E.; Grossberg, G.T.; Khachaturian, A.S.; Vergallo, A.; Cavedo, E.; Snyder, P.J.; et al. The cholinergic system in the pathophysiology and treatment of Alzheimer's disease. *Brain* **2018**, *141*, 1917–1933. [CrossRef] [PubMed]

6. Nelson, P.T.; Alafuzoff, I.; Bigio, E.H.; Bouras, C.; Braak, H.; Cairns, N.J.; Castellani, R.J.; Crain, B.J.; Davies, P.; Del Tredici, K.; et al. Correlation of Alzheimer disease neuropathologic changes with cognitive status: A review of the literature. *J. Neuropathol. Exp. Neurol.* **2012**, *71*, 362–381. [CrossRef] [PubMed]

7. Cerejeira, J.; Lagarto, L.; Mukaetova-Ladinska, E.B. Behavioral and psychological symptoms of dementia. *Front. Neurol.* **2012**, *3*, 73. [CrossRef] [PubMed]

8. van der Linde, R.M.; Dening, T.; Stephan, B.C.; Prina, A.M.; Evans, E.; Brayne, C. Longitudinal course of behavioural and psychological symptoms of dementia: Systematic review. *Br. J. Psychiatry* **2016**, *209*, 366–377. [CrossRef]

9. Zemek, F.; Drtinova, L.; Nepovimova, E.; Sepsova, V.; Korabecny, J.; Klimes, J. Kuca, K. Outcomes of Alzheimer's disease therapy with acetylcholinesterase inhibitors and memantine. *Expert Opin. Drug Saf.* **2014**, *13*, 759–774.

10. Allain, H.; Bentue-Ferrer, D.; Tribut, O.; Gauthier, S.; Michel, B.F.; Drieu-La Rochelle, C. Alzheimer's disease: The pharmacological pathway. *Fundam. Clin. Pharmacol.* **2003**, *17*, 419–428. [CrossRef]

11. Tan, C.C.; Yu, J.T.; Wang, H.F.; Tan, M.S.; Meng, X.F.; Wang, C.; Jiang, T.; Zhu, X.C.; Tan, L. Efficacy and safety of donepezil, galantamine, rivastigmine, and memantine for the treatment of Alzheimer's disease: A systematic review and meta-analysis. *J. Alzheimers Dis.* **2014**, *41*, 615–631. [CrossRef]

12. Schmitt, B.; Bernhardt, T.; Moeller, H.J.; Heuser, I.; Frolich, L. Combination therapy in Alzheimer's disease: A review of current evidence. *CNS Drugs* **2004**, *18*, 827–844. [CrossRef] [PubMed]

13. Matsunaga, S.; Kishi, T.; Iwata, N. Combination Therapy with Cholinesterase Inhibitors and Memantine for Alzheimer's Disease: A Systematic Review and Meta-Analysis. *Int. J. Neuropsychopharmacol.* **2015**, *18*, P859–P860. [CrossRef] [PubMed]

14. *Namzaric (Memantine Hydrochloride Extended-Release/Donepezil Hydrochloride) Capsules*; FDA: Silver Spring, MD, USA, 2014.

15. Kumar, A.; Singh, A.; Ekavali. A review on Alzheimer's disease pathophysiology and its management: An update. *Pharmacol. Rep.* **2015**, *67*, 195–203. [CrossRef] [PubMed]

16. Anighoro, A.; Bajorath, J.; Rastelli, G. Polypharmacology: Challenges and opportunities in drug discovery. *J. Med. Chem.* **2014**, *57*, 7874–7887. [CrossRef] [PubMed]

17. Rosini, M. Polypharmacology: The rise of multitarget drugs over combination therapies. *Future Med. Chem.* **2014**, *6*, 485–487. [CrossRef]

18. Van der Schyf, C.J.; Geldenhuys, W.J. Multimodal drugs and their future for Alzheimer's and Parkinson's disease. *Int. Rev. Neurobiol.* **2011**, *100*, 107–125. [PubMed]

19. Rosini, M.; Simoni, E.; Caporaso, R.; Minarini, A. Multitarget strategies in Alzheimer's disease: Benefits and challenges on the road to therapeutics. *Future Med. Chem.* **2016**, *8*, 697–711. [CrossRef]

20. Talevi, A. Multi-target pharmacology: Possibilities and limitations of the "skeleton key approach" from a medicinal chemist perspective. *Front. Pharmacol.* **2015**, *6*, 205. [CrossRef]

21. Michalska, P.; Buendia, I.; Barrio, L.D.; Leon, R. Novel Multitarget Hybrid Compounds for the Treatment of Alzheimer's Disease. *Curr. Top. Med. Chem.* **2017**, *17*, 1027–1043. [CrossRef]

22. Zhou, S.; Li, Y.; Hou, T. Feasibility of using molecular docking-based virtual screening for searching dual target kinase inhibitors. *J. Chem. Inf. Model.* **2013**, *53*, 982–996. [CrossRef] [PubMed]

23. McKie, S.A. Polypharmacology: In silico methods of ligand design and development. *Future Med. Chem.* **2016**, *8*, 579–602. [CrossRef] [PubMed]

24. Kowal, N.M.; Indurthi, D.C.; Ahring, P.K.; Chebib, M.; Olafsdottir, E.S.; Balle, T. Novel approach for the search for chemical scaffolds with activity at both acetylcholinesterase and the alpha 7 nicotinic acetylcholine receptor: A perspective on scaffolds with dual activity for the treatment of neurodegenerative disorders. *Molecules* **2019**, *24*, 446. [CrossRef]

25. Zoli, M.; Pistillo, F.; Gotti, C. Diversity of native nicotinic receptor subtypes in mammalian brain. *Neuropharmacology* **2015**, *96*, 302–311. [CrossRef] [PubMed]

26. Ellman, G.L.; Courtney, K.D.; Andres, V., Jr.; Feather-Stone, R.M. A new and rapid colorimetric determination of acetylcholinesterase activity. *Biochem. Pharmacol.* **1961**, *7*, 88–95. [CrossRef]

27. Cheung, J.; Rudolph, M.J.; Burshteyn, F.; Cassidy, M.S.; Gary, E.N.; Love, J.; Franklin, M.C.; Height, J.J. Structures of human acetylcholinesterase in complex with pharmacologically important ligands. *J. Med. Chem.* **2012**, *55*, 10282–10286. [CrossRef]

28. Morales-Perez, C.L.; Noviello, C.M.; Hibbs, R.E. X-ray structure of the human alpha4beta2 nicotinic receptor. *Nature* **2016**, *538*, 411–415. [CrossRef]

29. Kaczanowska, K.; Camacho Hernandez, G.A.; Bendiks, L.; Kohs, L.; Cornejo-Bravo, J.M.; Harel, M.; Finn, M.G.; Taylor, P. Substituted 2-aminopyrimidines selective for alpha7-nicotinic acetylcholine receptor activation and association with acetylcholine Binding Proteins. *J. Am. Chem. Soc.* **2017**, *139*, 3676–3684. [CrossRef]

30. Shen, M.Y.; Sali, A. Statistical potential for assessment and prediction of protein structures. *Protein Sci.* **2006**, *15*, 2507–2524. [CrossRef]

31. Lovell, S.C.; Davis, I.W.; Arendall, W.B., 3rd; de Bakker, P.I.; Word, J.M.; Prisant, M.G.; Richardson, J.S.; Richardson, D.C. Structure validation by Calpha geometry: Phi, psi and Cbeta deviation. *Proteins* **2003**, *50*, 437–450. [CrossRef]

32. Sterling, T.; Irwin, J.J. ZINC 15-Ligand Discovery for Everyone. *J. Chem. Inf. Model.* **2015**, *55*, 2324–2337. [CrossRef] [PubMed]

33. Lipinski, C.A.; Lombardo, F.; Dominy, B.W.; Feeney, P.J. Experimental and computational approaches to estimate solubility and permeability in drug discovery and development settings. *Adv. Drug Deliv Rev.* **2001**, *46*, 3–26. [CrossRef]

34. Walters, W.P.; Murcko, A.A.; Murcko, M.A. Recognizing molecules with drug-like properties. *Curr. Opin. Chem. Biol.* **1999**, *3*, 384–387. [CrossRef]

35. Baell, J.B.; Holloway, G.A. New substructure filters for removal of pan assay interference compounds (PAINS) from screening libraries and for their exclusion in bioassays. *J. Med. Chem.* **2010**, *53*, 2719–2740. [CrossRef] [PubMed]

36. Pagadala, N.S.; Syed, K.; Tuszynski, J. Software for molecular docking: A review. *Biophys. Rev.* **2017**, *9*, 91–102. [CrossRef]

37. da Silva Rocha, S.F.L.; Olanda, C.G.; Fokoue, H.H.; Sant'Anna, C.M.R. Virtual Screening Techniques in Drug Discovery: Review and Recent Applications. *Curr. Top. Med. Chem.* **2019**, *19*, 1751–1767. [CrossRef]

38. Celie, P.H.; van Rossum-Fikkert, S.E.; van Dijk, W.J.; Brejc, K.; Smit, A.B.; Sixma, T.K. Nicotine and carbamylcholine binding to nicotinic acetylcholine receptors as studied in AChBP crystal structures. *Neuron* **2004**, *41*, 907–914. [CrossRef]

39. UniProt, C. UniProt: A worldwide hub of protein knowledge. *Nucleic Acids Res.* **2019**, *47*, D506–D515.

40. Notredame, C.; Higgins, D.G.; Heringa, J. T-Coffee: A novel method for fast and accurate multiple sequence alignment. *J. Mol. Biol.* **2000**, *302*, 205–217. [CrossRef]

41. Sali, A.; Blundell, T.L. Comparative protein modelling by satisfaction of spatial restraints. *J. Mol. Biol.* **1993**, *234*, 779–815. [CrossRef]

42. Sastry, G.M.; Adzhigirey, M.; Day, T.; Annabhimoju, R.; Sherman, W. Protein and ligand preparation: Parameters, protocols, and influence on virtual screening enrichments. *J. Comput. Aided Mol. Des.* **2013**, *27*, 221–234. [CrossRef] [PubMed]

43. R Core Team. R: A Language and Environment for Statistical Computing. Available online: https://www.r-project.org/ (accessed on 10 January 2018).

44. Jacobson, M.P.; Friesner, R.A.; Xiang, Z.; Honig, B. On the role of the crystal environment in determining protein side-chain conformations. *J. Mol. Biol.* **2002**, *320*, 597–608. [CrossRef]

45. Metropolis, N.; Ulam, S. The Monte Carlo method. *J. Am. Stat. Assoc* **1949**, *44*, 335–341. [CrossRef] [PubMed]

46. Li, J.; Abel, R.; Zhu, K.; Cao, Y.; Zhao, S.; Friesner, R.A. The VSGB 2.0 model: A next generation energy model for high resolution protein structure modeling. *Proteins* **2011**, *79*, 2794–2812. [CrossRef]

47. Harder, E.; Damm, W.; Maple, J.; Wu, C.; Reboul, M.; Xiang, J.Y.; Wang, L.; Lupyan, D.; Dahlgren, M.K.; Knight, J.L.; et al. OPLS3: A force field providing broad coverage of drug-like small molecules and proteins. *J. Chem. Theory Comput.* **2016**, *12*, 281–296. [CrossRef] [PubMed]

48. Berman, H.M.; Westbrook, J.; Feng, Z.; Gilliland, G.; Bhat, T.N.; Weissig, H.; Shindyalov, I.N.; Bourne, P.E. The Protein Data Bank. *Nucleic Acids Res.* **2000**, *28*, 235–242. [CrossRef]

49. O'Boyle, N.M.; Banck, M.; James, C.A.; Morley, C.; Vandermeersch, T.; Hutchison, G.R. Open Babel: An open chemical toolbox. *J. Cheminform.* **2011**, *3*, 33. [CrossRef]

50. Friesner, R.A.; Banks, J.L.; Murphy, R.B.; Halgren, T.A.; Klicic, J.J.; Mainz, D.T.; Repasky, M.P.; Knoll, E.H.; Shelley, M.; Perry, J.K.; et al. Glide: A new approach for rapid, accurate docking and scoring. 1. Method and assessment of docking accuracy. *J. Med. Chem.* **2004**, *47*, 1739–1749. [CrossRef]

51. Halgren, T.A.; Murphy, R.B.; Friesner, R.A.; Beard, H.S.; Frye, L.L.; Pollard, W.T.; Banks, J.L. Glide: A new approach for rapid, accurate docking and scoring. 2. Enrichment factors in database screening. *J. Med. Chem.* **2004**, *47*, 1750–1759. [CrossRef]

52. Banks, J.L.; Beard, H.S.; Cao, Y.; Cho, A.E.; Damm, W.; Farid, R.; Felts, A.K.; Halgren, T.A.; Mainz, D.T.; Maple, J.R.; et al. Integrated Modeling Program, Applied Chemical Theory (IMPACT). *J. Comput. Chem.* **2005**, *26*, 1752–1780. [CrossRef]

53. Timmermann, D.B.; Gronlien, J.H.; Kohlhaas, K.L.; Nielsen, E.O.; Dam, E.; Jorgensen, T.D.; Ahring, P.K.; Peters, D.; Holst, D.; Christensen, J.K.; et al. An allosteric modulator of the α7 nicotinic acetylcholine receptor possessing cognition-enhancing properties in vivo. *J. Pharmacol. Exp. Ther.* **2007**, *323*, 294–307. [CrossRef] [PubMed]

54. Mirza, N.R.; Larsen, J.S.; Mathiasen, C.; Jacobsen, T.A.; Munro, G.; Erichsen, H.K.; Nielsen, A.N.; Troelsen, K.B.; Nielsen, E.O.; Ahring, P.K. NS11394 [3′-[5-(1-hydroxy-1-methyl-ethyl)-benzoimidazol-1-yl]-biphenyl-2-carbonitrile], a unique subtype-selective GABAA receptor positive allosteric modulator: In vitro actions, pharmacokinetic properties and in vivo anxiolytic efficacy. *J. Pharmacol. Exp. Ther.* **2008**, *327*, 954–968. [CrossRef] [PubMed]

55. Kowal, N.M.; Ahring, P.K.; Liao, V.W.Y.; Indurti, D.C.; Harvey, B.S.; O'Connor, S.M.; Chebib, M.; Olafsdottir, E.S.; Balle, T. Galantamine is not a positive allosteric modulator of human α4β2 or α7 nicotinic acetylcholine receptors. *Br. J. Pharmacol.* **2018**, *175*, 2911–2925. [CrossRef] [PubMed]

Sample Availability: Not available.

 *molecules*

Article

Characterizing Epitope Binding Regions of Entire Antibody Panels by Combining Experimental and Computational Analysis of Antibody: Antigen Binding Competition

Benjamin D. Brooks [1,2,3,*], Adam Closmore [4], Juechen Yang [5], Michael Holland [5], Tina Cairns [3], Gary H. Cohen [3] and Chris Bailey-Kellogg [6]

[1] Department of Biomedical Sciences, Rocky Vista University, Ivins, UT 84738, USA
[2] Inovan Inc., Fargo, ND 58102, USA
[3] Department of Microbiology, School of Dental Medicine, University of Pennsylvania, Philadelphia, PA 19104, USA; tmcairns@upenn.edu (T.C.); ghc@upenn.edu (G.H.C.)
[4] Department of Pharmacy, North Dakota State University, Fargo, ND 58102, USA; Adam.Closmore@gmail.com
[5] Department of Biomedical Engineering, North Dakota State University, Fargo, ND 58102, USA; juechen.yang@gmail.com (J.Y.); mchollandred@gmail.com (M.H.)
[6] Computer Science Department, Dartmouth, Hanover, NH 03755, USA; cbk@cs.dartmouth.edu
*   Correspondence: bbrooks@rvu.edu; Tel.: +1-435-222-1403

Academic Editors: Marco Tutone and Anna Maria Almerico
Received: 23 April 2020; Accepted: 28 July 2020; Published: 11 August 2020

Abstract: Vaccines and immunotherapies depend on the ability of antibodies to sensitively and specifically recognize particular antigens and specific epitopes on those antigens. As such, detailed characterization of antibody–antigen binding provides important information to guide development. Due to the time and expense required, high-resolution structural characterization techniques are typically used sparingly and late in a development process. Here, we show that antibody–antigen binding can be characterized early in a process for whole panels of antibodies by combining experimental and computational analyses of competition between monoclonal antibodies for binding to an antigen. Experimental "epitope binning" of monoclonal antibodies uses high-throughput surface plasmon resonance to reveal which antibodies compete, while a new complementary computational analysis that we call "dock binning" evaluates antibody–antigen docking models to identify why and where they might compete, in terms of possible binding sites on the antigen. Experimental and computational characterization of the identified antigenic hotspots then enables the refinement of the competitors and their associated epitope binding regions on the antigen. While not performed at atomic resolution, this approach allows for the group-level identification of functionally related monoclonal antibodies (i.e., communities) and identification of their general binding regions on the antigen. By leveraging extensive epitope characterization data that can be readily generated both experimentally and computationally, researchers can gain broad insights into the basis for antibody–antigen recognition in wide-ranging vaccine and immunotherapy discovery and development programs.

Keywords: epitope binning; epitope mapping; epitope prediction; antibody:antigen interactions; protein docking; glycoprotein D (gD); herpes simplex virus fusion proteins

---

1. Introduction

The utility of antibodies (Abs) for treatment of disease has been recognized for over a century and the clinical realization of this vision in a host of broad therapeutic indications has begun to

come to fruition [1,2]. The efficacy of an Ab, whether from vaccination or as a therapeutic, is functionally determined by the specific epitope that it recognizes on its cognate antigen (Ag). Thus, even Abs targeting the same Ag have demonstrated variable efficacy depending on their epitope specificities [3,4]. Recent advances in Ab engineering have enabled researchers to generate large panels of Abs with broad epitope coverage allowing for the selection of Ab–Ag pairings with improved safety and efficacy [5–8]. Abs have demonstrated differential effects in both vaccines and therapies depending on their epitopes [9–11]. Moreover, combination immunotherapies with diverse epitopes have demonstrated synergistic efficacy and have reduced the ability of cancer and infectious disease to develop resistance [12,13]. Furthermore, the use of immune repertoire B cell sequencing is supporting expanded clinical applications not only in immunotherapy, but also in the development of vaccines for emergent infectious diseases, both viral and bacterial [14]. The specificity of Abs in vaccine applications reveals various levels of protection depending on which epitopes are targeted by the immune response [12,13,15–19].

1.1. Understanding the Structures of Epitope:Paratope Interactions Guides the Design of Superior Immunotherapies and Vaccines

One of the areas for improvement in Ab engineering is the characterization of the binding interactions between an Ab and its specific Ag, as defined by the epitope–paratope interface [5,20–22]. Due to practical limitations, structural data is only available for a relatively small number of epitope–paratope interactions [23,24], since commonly used analytical techniques for obtaining these data, such as X-ray crystallography, NMR spectroscopy, cryo-electron microscopy, and H-D exchange mass spectrometry, are highly resource-intensive and often require artisan skillsets [25–28]. As such, these techniques are feasible only for late-stage lead molecules where they provide confirmation data rather than early-stage predictive tools that influence candidate selection [5]. These limitations highlight the realization that a more sophisticated strategy is required to characterize a panel of antibodies either to screen immunotherapies or to characterize immune repertoires from natural or vaccinated responses [23]. The localization of these interactions would provide information to help understand biochemical mechanisms of action, which is at the core of advancing the discovery and development of new immunotherapeutics and vaccines [20,29,30].

1.2. High-Throughput Epitope Characterization Assays Provide Valuable Epitopic Information

Epitope characterization has made significant progress in recent years as the throughput of biosensors has improved [5,31]. Foremost amongst emerging techniques for high-throughput epitope characterization is epitope binning, as it merges the speed and functional-site identification capabilities demanded by the biopharmaceutical industry [3–5,32]. Epitope binning is a competitive immunoassay where Abs are tested in a pairwise manner for their simultaneous binding to their specific Ag, thereby generating a blocking profile for each Ab showing how it blocks or does not block the others in the panel [32–35]. Abs with similar blocking profiles can be clustered together into a "bin", or represented as a "community" in a network plot that illustrates the blocking relationships among the Abs. In an oversimplified generalization, Abs in the same community are assumed to recognize the same epitope region and generally block the binding of others in the community. Advances in throughput offered by emerging array-based, label-free methods are allowing epitope binning to be performed early in drug discovery [32–35]. This approach can enable the pipeline to be populated with Abs that are diverse epitopically and, consequently, functionally. A key limitation of epitope binning remains its inability to provide insights into the locations of the epitopes on the target [3–5,32].

1.3. Integrating High-Throughput Experiment and Computation Enables Characterization of Ab-Specific Epitopes

An emerging approach for localizing Ab recognition and characterizing Ab-specific epitopes involves coupling experimental data and computational modeling [9,36]. Purely computational

approaches to epitope prediction are quick and inexpensive but are not yet of sufficient accuracy to be relied upon on their own [36–40]. However, the incorporation of even limited experimental data can help close the gap. For example, the EpiScope approach first constructs a homology model of an Ab based on its sequence, then computationally docks that model onto a structure or high-quality homology model of the Ag, and finally, designs focused mutagenesis experiments to test the docking models [9]. While it is generally not clear from computational scoring alone which docking model is the most accurate, it was shown in both retrospective and prospective studies that a small number (generally 3–5) of binding assays for computationally designed Ag variants can reliably enable the various docking models to be confirmed or rejected and thereby identify the general epitope region [9]. Complementarily, experimental data can be used to focus docking and energy minimization to better define binding mode or epitope [41]. In general, combined computational–experimental approaches balance cost and accuracy in characterizing epitope sites.

Here, we take this general idea and scale it up from individual Abs to sets of Abs, presenting a new integrated experimental-computational approach (Figure 1) to characterize epitopes for an entire panel of Abs against an Ag by combining experimental binning with "dock binning," a new computational counterpart based on analysis of docking models for all the Abs. With the application to a model system, glycoprotein D from herpes simplex virus, we show that this combination of powerful experimental and computational methods can help rapidly identify antigenic regions and localize Ab-specific epitopes. The approach promises to enable better understanding of Ab–Ag interactions at a larger scale, and ultimately improve to the design of vaccines and therapeutics.

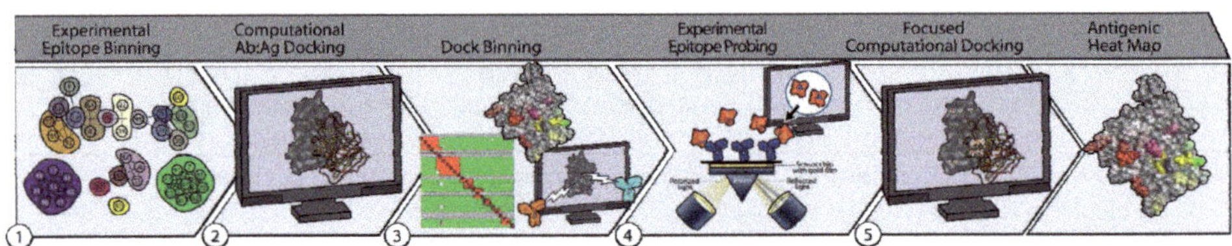

Figure 1. Schematic overview of our method for localizing Ab-specific epitopes by integrated experimental and computational analysis of binding competition. (Step 1) Experimental binning identifies communities of Abs that compete with each other for Ag binding. This grouping allows subsequent analyses to be focused on one or a few representatives from each community, reducing the effort required. (Step 2) For each representative Ab, a homology model of its Fv structure is constructed from its sequence, and Ab:Ag docking models are generated from the Ab model and the Ag structure or homology model. (Step 3) The docking models are computationally clustered, with this dock binning process analogous to experimental epitope binning in identifying patterns of competition. Since the competition is in terms of structural models of Ab:Ag binding, the identified communities correspond to general antigenic regions, and thereby map out potential binding regions on the Ag, summarized across the whole Ab panel as an antigenic heatmap. (Step 4) Experimental data is collected to probe the hypothesized epitope regions, e.g., using site-directed mutagenesis, chimeragenesis, peptide binding, or selection of alternative natural variants, in order to alter a putative epitope and evaluate effects on experimental competition or binding. (Step 5) The experimental data is used to focus docking, redefining bins, better characterizing competition, and better localizing epitope binding regions. This ultimately results in an antigenic heatmap localizing putative binding regions of the different Abs (illustrated by colored patches on the Ag surface).

2. Results

We demonstrate the power of combined experimental and computational binning in application to an important target with a wealth of data available for our use: herpes simplex virus (HSV) glycoprotein D (gD). GD is a fusion protein found in HSV that has served as the standard by which all other HSV-2 vaccines are evaluated for safety and efficacy [42]. GD subunit vaccines have conferred some measure of protection against viral challenge [42–44], but gD subunit vaccines fail to prevent infection or latent

infection [45]. GD serves as a particularly good target for demonstration of our approach due to the availability of a large panel of available Abs as well as variant Ag constructs that may be leveraged for subsequent analyses. The following sections describe our process (Figure 1) to characterize the epitopes of a set of anti-gD Abs.

2.1. Experimental Binning Identified Four Communities of Anti-gD Abs

Step #1 of the workflow (following the schematic in Figure 1) is to perform experimental epitope binning on a panel of Abs, identifying communities of cross-competing Abs and selecting representatives from each community for further investigation. Figure 2 further overviews the general workflow for such epitope binning experiments. In previous studies, this general approach was applied specifically to gD. In particular, high-throughput SPRi technology in a classical sandwich assay format was used to assess competition between pairs of Abs, from a panel of 46 Abs, against four soluble Ag variants, gD from type 1 HSV truncated to the first 285 or 306 residues, and that from type 2 HSV likewise truncated to 285 or 306 residues [46–48]. Subsequent analyses presented here are all based on the 285 residue truncation of gD2.

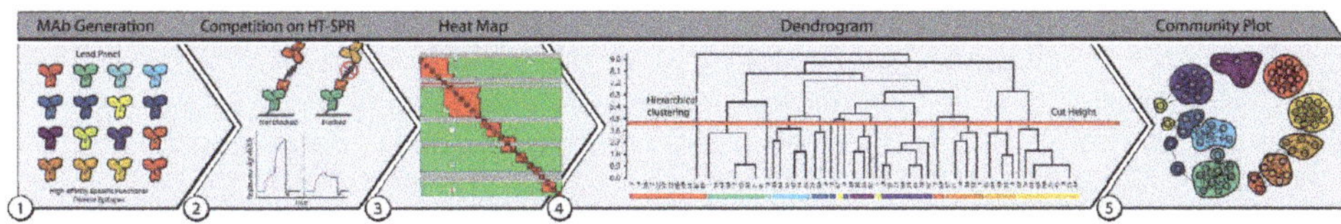

Figure 2. Schematic overview of experimental epitope binning, identifying cross-competition between Abs for binding an Ag. Sensorgrams are assessed for blocking, inferring whether or not two Abs block each other from binding in a competition assay and thus are assumed to bind the same epitopic region on the Ag. The blocking data is collected in a heatmap that indicates whether or not each Ab pair blocks (red blocked, green sandwiched). The heatmap is processed with hierarchal clustering algorithms based on similarity in blocking profiles, thereby generating a dendrogram. Finally, cutting the dendrogram separates Abs into clusters represented in a community network, which has nodes for the Abs and edges indicating which pairs block each other.

The studies showed that the Abs covered much of the surface of gD [4], with numerous Abs in each distinct community of cross-blocking Abs [46–48]. Six "sentinel" Abs, namely DL11, MC23, MC2, MC5, MC14, and 1D3, were selected as representatives (generally the most highly connected within each community) [49,50]. Of particular note, there exists a crystal structure of an additional Ab, E317 (PDB ID 3W9E) [51]. Thus, even though E317 is in the same community as the previously selected DL11, we included it here to serve as a structural "control". These seven Abs represent the communities in serving as the subjects of the following computational and experimental analyses.

2.2. Computational Epitope Prediction Characterized the Putative Epitope Binding Regions

With the representative Abs selected, Step #2 (Figure 1) is to construct Ab homology models and dock them to the structure/model of the Ag. Ab homology modeling is generally very high quality (<1–2 Å level RMSD to native) for everything except the heavy chain CDR 3, which is more variable (more typically 3–6 Å, though it can be better) [52,53]. For the gD study, we used representative state-of-the-art methods, within Schrodinger BioLuminate (BioLuminate, Schrödinger, LLC, New York, NY, USA, 2020.) to perform this modeling, but note that many alternative approaches are available and may yield somewhat different Ab and Ab:Ag models. A homology model was constructed for each of the seven Abs; for control purposes, E317 was homology modeled on a different scaffold from its crystal structure (RMSD 1.43 Å). The crystal structure of the Ag gD2 was taken from PDB id 2C36 [54–56], with missing residues homology modeled. Docking models were then generated for

each Ab model against the gD2 model using the Piper algorithm within Schrodinger BioLuminate. This yielded roughly thirty representative low-energy docking models per Ab [57].

As is common [9], the docking models were spread over much of the gD2 surface. In general, the quality of Ab:Ag complex models produced by docking has steadily improved [58], e.g., for 95% of the cases in one benchmark, a near-native model was within the top 30 models [40]. Thus, while we could not be confident in the accuracy of any particular model, we could hypothesize that, in aggregate, they included the antigenic sites, setting up the next stage in our analysis.

2.3. Dock Binning Grouped Models and Identified Broadly Antigenic Regions

In order to identify the "hottest" putative antigenic sites on the protein, worth experimentally probing across the Ab panel, Step #3 (Figure 1) performs dock binning and constructs an antigenic heatmap. In analogy to experimental binning, this step first characterizes competition among Abs for sites on the Ag, here according to the docking models. At this point in the analysis, since there are many (roughly 30 in our gD results) docking models for each Ab, and they may be widely dispersed over the Ag, competition between a pair of docking models from different Abs does not necessarily imply that the Abs themselves compete. However, it does imply that it would be informative to evaluate the binding of those Abs against the Ag sites for which the docking models compete (e.g., mutate such a site and experimentally assess changes in binding/competition). Such a test would provide, with a single experiment, information regarding both Abs' binding sites. More generally, the more evidence there is for interaction with an Ag residue (the "hotter" that residue), the more generally and experimentally informative it should be to probe that residue for binding across the entire Ab panel. This insight is the basis for performing dock binning and constructing an associated antigenic heatmap.

The dock binning workflow (schematically illustrated in Figure 3) is relatively straightforward, mostly following that of experimental epitope binning (see again Figure 2). Now the competition heatmap is based on overlap in docking models, rather than experimental competition. A variety of different methods are possible for assessing this overlap; the results presented here are based on one that we call the "common interaction metric", which considers two Ab docks that contact the same residue(s) to be competing. This score drives the clustering of the docks based on their patterns of competition. Here, the hierarchical clustering method from experimental epitope binning was used [34,35]. A community network map is then generated from the dendrogram.

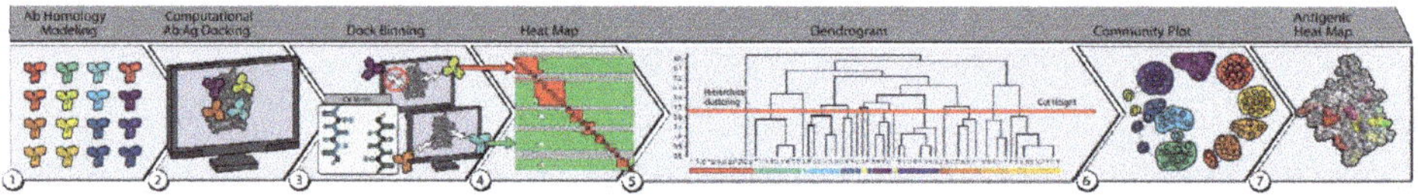

Figure 3. Schematic overview of "dock binning", computational analysis of cross-competition of docking models of Abs against an Ag. Homology models of the Abs are computationally docked against a crystal structure or homology model of the Ag. Docking models are evaluated for "competition" and the extent of competition represented in a heatmap. The competition profiles are subsequently used to cluster docking models, with a community network representing the models (nodes), their competition (edges), and the identified clusters (groups of nodes). In contrast to experimental binning, dock binning is based on structural analysis and thus provides insights into where on the Ag the Abs might be interacting. The antigenic heatmap highlights the most popular Ag residues across the docking models; these are thus most generally informative for subsequent experimentally probing.

Figures 4 and 5 illustrate the results from applying this process general process to the Ab:gD2 docking models. Figure 4 shows the docking model clusters generated for the seven Abs against gD2 and Figure 5 their community relationships. Note that docks from each Ab are found across all communities.

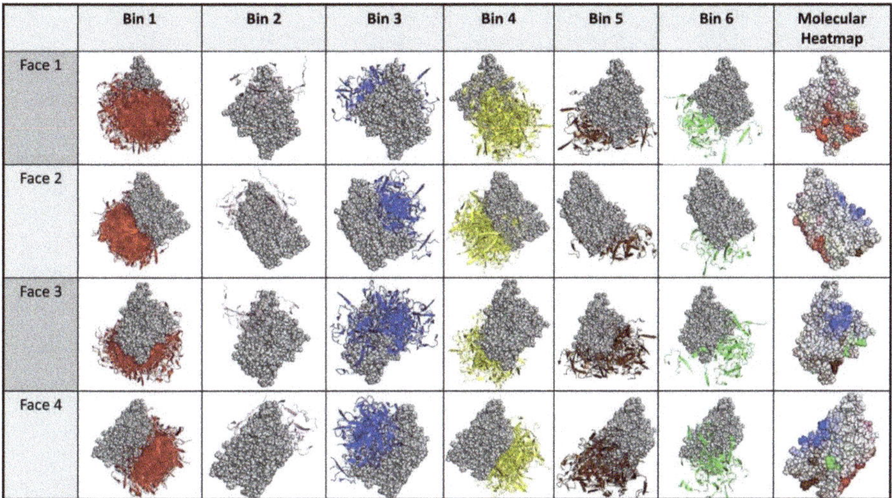

Figure 4. Binned docking models for Abs vs. gD2. Clusters of docking models are arranged in a table format, with each column a different cluster and each row a different perspective. Face #1 is the nectin binding face. Faces #2–4 are rotated by 90, 180, and 270 degrees around the *y*-axis. The rightmost column is an antigenic heat map (see also Figure 6) where residues are colored by community and shaded so that residues with more Ab interactions are colored more darkly.

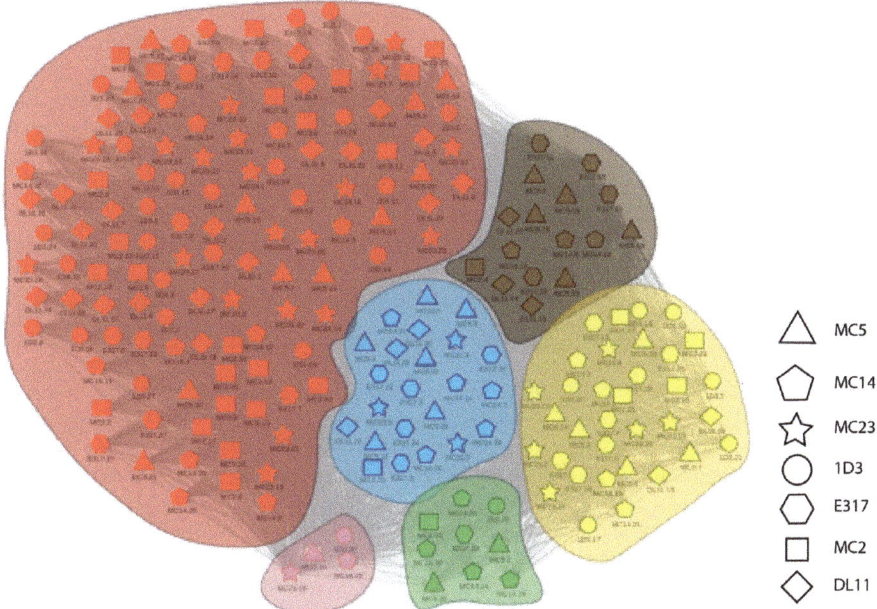

Figure 5. Community map of docking models for Abs vs. gD2. Nodes represent docking models, with different symbols for the different Abs. Edges indicates structural competition between pairs of docking models. Colors and background shading indicate community membership according to the partitioning of a dendrogram. The shape of each marker indicates the antibody in the dock (i.e. each shape is a single antibody).

The last step in dock binning, where it can produce insights beyond those that possible with experimental binning alone, leverages the intuition introduced above: analyze the dock bins to identify common putative antigenic sites which may then be subjected to experimental probing. Here computationally identified noncovalent Ab–Ag interactions (hydrogen, electrostatic, pi, and van der Waals) were aggregated across all docks for all Abs to determine a score for each Ag residue in terms of its total number of interactions. Figure 6 (also the last column of Figure 4) shows the resulting "antigenic heatmap," with residues of the protein structure colored according to the dock binning community (Figures 4 and 5) and with darkness proportional to the aggregated scores. This antigenic heatmap highlights the "hot" antigenic residues whose experimental probing is likely to be most informative and can enable the efficient localization of epitopes of the whole Ab panel.

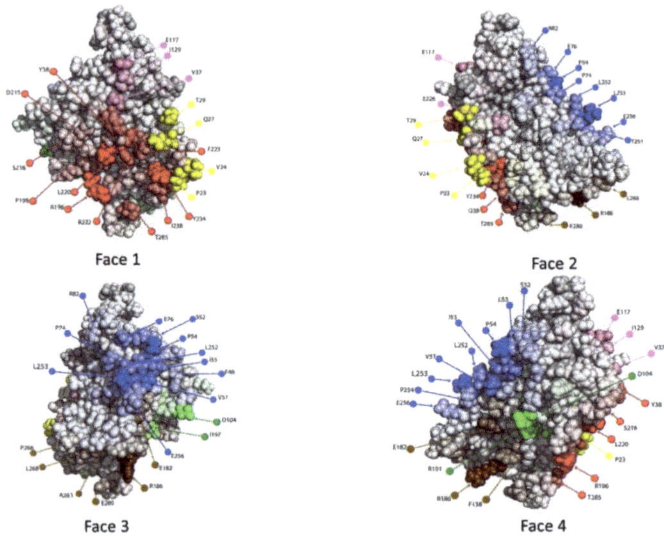

Figure 6. Antigenic heat map for Abs vs. gD2. Each residue is colored according to the Ab community (Figure 5) with which it has the most interactions and shaded so that residues with more interactions are colored more darkly. Face #1 is the nectin binding face. Faces #2–4 are rotated by 90, 180, and 270 degrees around the *y*-axis.

2.4. Dock Binning Enabled Selection of Experimental Assays to Evaluate Antigenic Regions

At this point in the process (Step #4), critical Ag residues according to the antigenic heatmap are probed for Ab recognition (e.g., via point mutagenesis followed by a binding assay, to assess if a mutation away from native disrupts binding). In this study, we were fortunate that there already exists a wealth of existing data available regarding Ab–gD2 binding. We cross-referenced the "hot" residues from the antigenic heatmap against the available data (including Ab:gD2 variant binding, peptide binding, and known "monoclonal Ab resistant", or MAR, mutations), considering the sentinel Abs as well as others in their communities (Supplementary Figure S1) [46,48]. Figure 7 highlights the residues from the antigenic heatmap for which such binding data was available [59,60]. We consider these as the experimental evaluation of the hot residues from the antigenic heatmap. We note that, in other settings, mutations (individual or combination) could be computationally designed to evaluate the disruption of binding while preserving antigenic stability [9,61]. Thus, while we used a large number of experimental measurements in this study, we expect that a much smaller number of tests would suffice in practice, e.g., 3–5 variants sufficed to localize individual Ab epitopes in previous computationally directed studies [9].

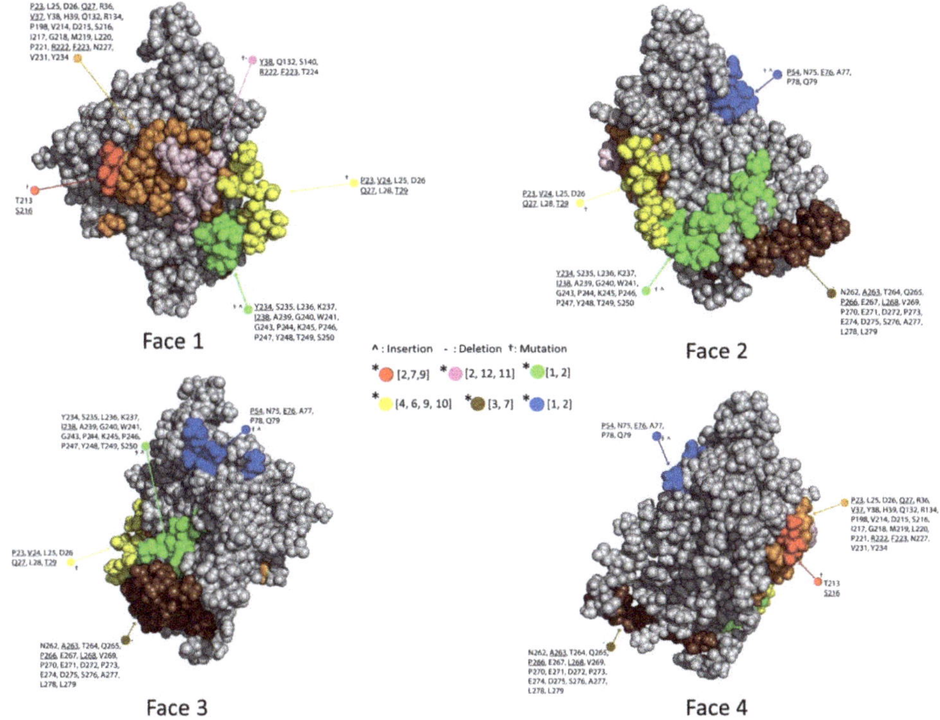

Figure 7. Experimentally probed residues. Residues with associated experimental binding data are labeled; bracketed numbers in the legend indicate primary references in the bibliography. The surface is colored by associated Abs (Supplementary Figure S1), rather than dock binning community. The residues that were also highlighted by dock binning are underlined; this data serves as probes of those particular predicted antigenic hotspots.

2.5. Experimental Data Allowed Re-Docking to Focus Ab:Ag Models Based on Experimental Binding Data, thereby Localizing Each Ab's Epitope Region

Finally (Step #5 from Figure 1), the experimental data is used to focus docking of each Ab against the Ag. Here, the experimental data (Figure 7/Supplementary Figure S1) was used to focus docking toward (with "affinity") residues confirmed to be important for an Ab's binding and away from (with "repulsion") those determined not to be. For example, DL11 was docked with an affinity towards the residues 213 and 218 and with repulsion away from the residues associated with Abs from other communities, since these Abs were determined from the initial experimental binning not to compete with DL11, and thus it is assumed the epitopes don't overlap.

The focused docks are then subjected to dock binning and antigenic heatmap construction as described for Step #3. Figures 8–10 show the focused-docking results for the anti-gD Abs. In contrast to the initial unconstrained docking, docking models for an Ab are now more concentrated, focused on the experimentally important residues for the Ab. Consequently, the communities are now fairly homogeneous for each Ab, and the localization of each Ab on gD can be fairly well inferred from an associated hot region in the antigenic heatmap. For example, when compared to the crystal structure for our "structural control Ab" E317, 10 antigenic heatmap residues agree with those in the crystal structure, while 12 extend further out and four are missed (Supplementary Table S1).

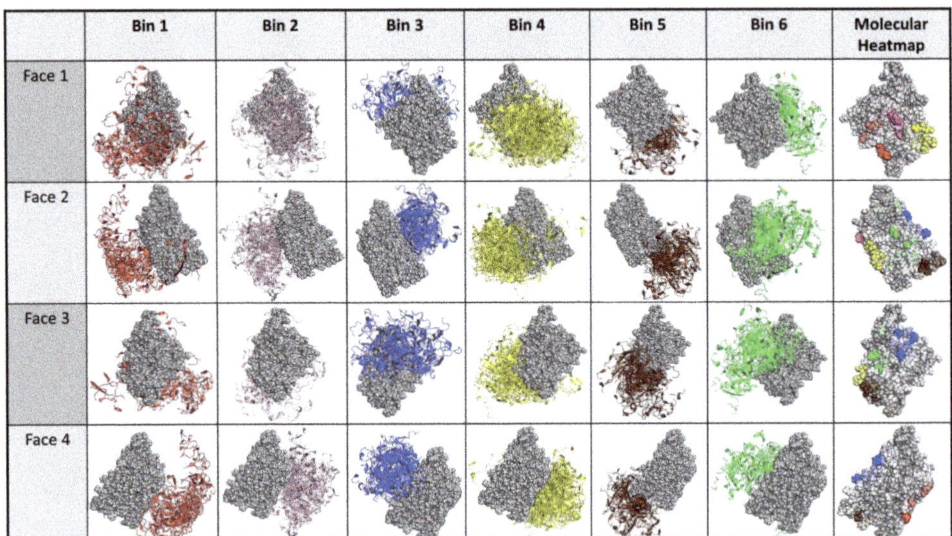

Figure 8. Binned experimentally-focused docking models for Abs vs. gD2. The same representation as Figure 4, but based on docking models that were focused according to experimental data.

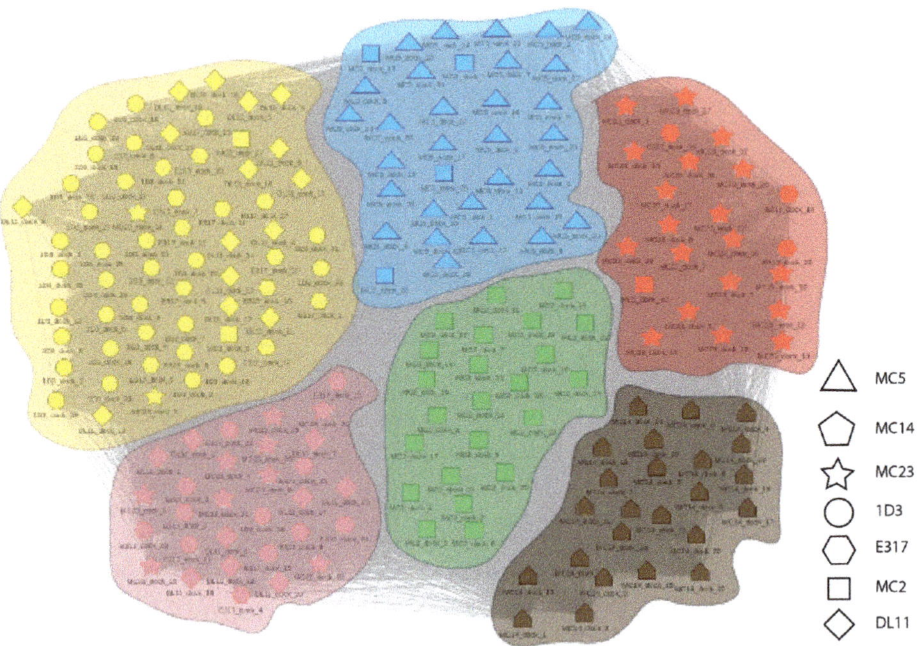

Figure 9. Community map of experimentally-focused docking models for Abs vs. gD2. The same representation as Figure 5, but based on docking models that were focused according to experimental data. Note the relative homogeneity of Abs (different symbols) within each community, contrasting with Figure 5. The shape of each marker indicates the antibody in the dock (i.e., each shape is a single antibody).

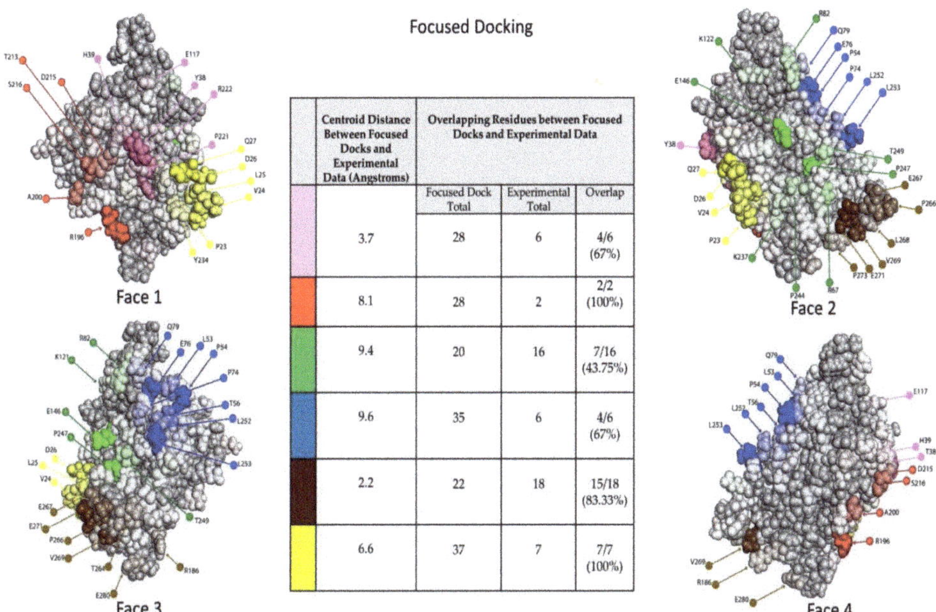

Centroid Distance Between Focused Docks and Experimental Data (Angstroms)	Overlapping Residues between Focused Docks and Experimental Data		
	Focused Dock Total	Experimental Total	Overlap
3.7	28	6	4/6 (67%)
8.1	28	2	2/2 (100%)
9.4	20	16	7/16 (43.75%)
9.6	35	6	4/6 (67%)
2.2	22	18	15/18 (83.33%)
6.6	37	7	7/7 (100%)

Figure 10. Antigenic heat map for Abs vs. gD based on eperimentally-focused docking models. The same representation as Figure 6, but based on docking models that were focused according to experimental data. The resulting hot spots can be interpreted as likely epitope regions of the associated antibody/-ies in the bin. Relative.

While there is no ground truth for the actual epitopes of the Abs other than E317, the agreement between the antigenic heatmap residues and those previously identified by various experiments (Figure 7/Supplementary Figure S1) was quantified in terms of both centroid distances and common residues (table in the middle of Figure 10). Many of the distances are quite small, indicating that the dock binning region was centered in the same general region as the experimentally identified residues used to focus it. However, some of the regions have fairly few experimentally identified residues (e.g., red and blue communities), and the docking models typically expanded to cover a significantly larger region capturing more Ab–Ag surface complementarity. The quantification of agreement between residues in the antigenic heatmap and those experimentally used to focus the docking further illustrates that while docking largely stays in the focused region as intended, it does include some additional residues and omit some others. The disparity could indicate, for example, either that the dock binning missed some important residues contributing to recognition or that the experiments overestimated the importance of some residues. Likewise, either that dock binning found some additional residues that had not been discovered by previous experiments or that it was somewhat off-target. Such differences can then be the subject of further experimental investigation.

3. Discussion

3.1. The Integration of Experimental and Computational Binning Provides Important Epitope Information That Can Inform Discovery

Here, we demonstrate the utility of combining computational modeling and high-throughput experimental data to characterize epitopes early in discovery and thereby enable more effective drug and vaccine development. Epitope binning can facilitate the identification of functional epitopes and is thus increasingly used as a primary or early second screen [5]. However, while experimental

epitope binning can inform competition among Abs for an epitope, it cannot localize the epitope on the Ag [5]. In contrast, while computational methods can identify potential epitopes, they are currently limited in their accuracy [40]. Even in this study, we were confronted with a wide range of seemingly equally reasonable docking models, leading us to consider: (1) how to identify the most accurate dock, (2) how to identify highly immunogenic sites on the target, and (3) how to group the docks into epitope regions in order to compare with experimental binning experiments. We determined that identifying the most accurate dock (question 1) was not realistic based on the current accuracy of the computational methods. However, we realized that docks could help us identify putative highly antigenic sites (question 2). Furthermore, this information would be the next logical step in assessing experimental binning communities and would limit the possibilities to consider when attempting to map communities onto their epitopes (question 3). Thus, this combination of computational and experimental binning (Figure 1) can leverage the advantages of each method, complementarily, in order to localize Ab epitopes across an entire panel.

The initial dock binning step provides not only putative antigenically hot regions but also important information toward the generation of mutants that can be used in epitope mapping and cross-antigen binning to test hypotheses regarding Ab–Ag binding and thereby localize epitopes. Dock binning identifies communities of related Ab docking models, while the antigenic heatmap summarizes possible epitope binding regions on the Ag surface. These two analyses together provide elegant, biologically-centered visualizations characterizing potential binding patterns for an entire Ab panel. Taking the epitope binding regions as hypothesized regions that are generally antigenic allows the design of targeted experiments to evaluate them. The underlying intuition for these experiments is that if an Ab truly recognizes a particular epitope, then a mutation to substantially "re-surface" that epitope should disrupt Ab binding, and thus evaluation of changes to binding can allow us to confirm or reject the epitope for the Ab. By testing a set of variants based on the epitope binding regions, we can thereby simultaneously test all the various hypotheses spanning the Ab panel and potential epitopes.

The results that can be obtained from the dock binning approach are limited by how well the Ab models and the Ab:Ag docking models reflect reality, along with how effectively the selected mutational variants can reveal that reality. We used representative computational approaches here, and overall observed reasonable agreement between the results from our approach and the one available ground truth crystal structure. While the previous purely experimental studies reflect an extensive amount of effort, they also may not fully reflect reality to the same extent as a crystal structure does. Thus, we are limited in our ability to characterize general trends regarding how well docking-based approaches perform and why, and what their particular weaknesses may be. From some previous gD studies, however, we do know that two of the Abs (MC2 and MC5) recognize epitopes that are at least partially obscured until receptor binding allosterically drives a conformational change [46,59]. We thus hypothesize that the epitope region identified for MC2, and presumably that for all members of the community it represents, is not particularly accurate. In general, epitopes, as well as paratopes, in conformationally dynamic or poorly modeled regions may present challenges for this model-driven approach.

3.2. While Binning Is Still Limited, Further Computational Advances Will Improve Its Accuracy and Informativeness

Significant limitations still exist with the prediction of protein-protein interactions, in particular with Ab-Ag epitope identification. Numerous opportunities exist to improve the process and better support vaccine and drug candidate screening. We here summarize some interesting avenues for computational development, following along the path of our workflow (Figure 1):

1. Experimental epitope binning. The fundamental question here is how to cluster the Abs into communities. and new and emerging techniques from network/community analysis, especially in genomics [62–64], may prove beneficial. These metrics can be used to identify communities that require further refinement and/or changes to the clustering. Furthermore, the clustering need

not be discrete, i.e., the partition into communities can be overlapping, and this information can potentially be leveraged throughout the rest of the process. Such analysis could be particularly important in cases of partial or aberrant competition. While the "sentinel" antibodies used here were selected as most representative of clear-cut communities, future approaches could include algorithmic selection of representatives based on properties of the communities (e.g., ensuring adequate coverage of ambiguous clusters) and of the individual antibodies (e.g., based on sequence and structural analysis, favorability for development, etc.).

2. Ab modeling and Ab–Ag docking. We used representative high-quality Ab modeling and Ab–Ag docking methods, but others could be employed; given sufficient data points for different targets, the impacts of these choices could be quantitatively assessed. The complete set of thousands of docking models, instead of cluster representatives, could be considered, thereby potentially leveraging redundancy in weighting region importance. As discussed above regarding gD, accounting for protein flexibility (both Ag and Ab) may change the models substantially in some cases. Docking scores/ranks could be taken into account in order to determine the most important regions. In addition to physically-based docking methods, additional data-driven epitope prediction methods could be used to identify putative antigenic regions [65–67].

3. Dock binning. The same issues with clustering for experimental binning hold here and the same potential solutions can be explored. Machine learning methods could be incorporated in order to train models to integrate the upstream predictions and experimental data, e.g., in a consensus or weighted fashion. New data-driven models that directly seek to predict competition/bins (instead of general-purpose epitope prediction) could be developed.

4. Experimental epitope probing. As presented, mutations are selected to assess the predicted antigenic hot spots and deconvolve which hot spots are associated with which Abs. While this has been done before based simply on maximizing binding disruption according to the models [9,61], more refined metrics and optimization techniques could be developed, e.g., using an information-theoretic approach to maximize what is learned about the communities and hotspots for a given experimental "budget" (e.g., maximum-relevance, minimum-redundancy) [68].

5. Revised dock bins and antigenic hotspots. We employed the approach of docking with affinity/repulsion in order to focus docking according to the experimental data. Docking models could be subsequently refined (e.g., via energy minimization and conformational sampling) and scored not only for physical modeling, but also for the consistency of the experimental data with the predicted effects of the mutations [69].

4. Materials and Methods

All reagents were previously described [46,49,50,70,71].

4.1. Antibodies

Monoclonal Abs (mAbs) were used throughout. The monoclonal antibodies have been created by the Cohen/Eisenberg group at the University of Pennsylvania. The following anti-gD mAbs were previously published: 1D3 [49,72–74]; DL6, DL11, DL15 [51,71,74–78]; E317 [51]; MC1, MC2, MC4, MC5, MC8, MC9, MC10, MC14, MC15, MC23 [71,79]; A18 [80]; AP7, LP2 [81]; HD1, HD2, HD3, H162, H170, H193 [82,83]; 11S, 12S, 45S, 77S, 97S, 106S, 108S, 110S [73,84]; BD78, BD80 [71,85]; and the human mAb VID [86–88], DL15, A18, HD3, H162, H193, 77S, 97S, 106S, 108S and 3D5, 4E3E, 4G4D, and 11B3AG [46,70].

Sentinel Abs were selected based on past studies as most representative of their respective communities. Sentinel Abs were sequenced (Genscript, Piscataway, NJ, U.S.) from mAb clone for computational studies.

4.2. Proteins

Proteins were previously described [70]. Again, these constructs existed before this study and were generated by the Cohen/Eisenberg group. HSV type-1 and type-2 gD, truncated to 285 or 306 residues, were harvested from Sf9 cells infected with baculovirus. Protein was then purified using a DL6 immunosorbent column [70,71,89]. Additional proteins used in this study include: C-terminal truncations 250t, 260t, 275t, and 316t [50,54,70]; deletion mutant Δ(222–224) [50]; point mutants Y38A, V231W, and W294A [54,90,91]; and insertion mutants ins34, ins126, and ins243 [46,77].

4.3. Epitope Binning

Binning experiments were previously described [3,46]. Briefly, a 48-spot microarray of amine-coupled Abs on a CMD200M sensor prism (Xantec GmbH, Kevelaer, Germany) was printed using Continuous Flow Microspotter (CFM, Carterra, Salt Lake City, UT, U.S.). The array was loaded into the MX96 SPR instrument (Ibis). Data were processed using software from Ibis and Carterra as described previously [3,46].

4.4. Protein Modeling and Docking Prediction

The crystal structure of gD2 (285 truncation) was taken from PDB id 2C36 [54–56] and Schrodinger BioLuminate was used to homology model the unstructured regions for missing residues including 257–267. Homology models of the Ab Fvs were also constructed using Schrodinger BioLuminate. For E317, the homology modeling was prevented from using the available crystal structure (PDB id 3W9D); instead, the model was based on the anti-HIV neutralizing Ab 4E10 (PDB id 4M62) as the framework template, yielding a model RMSD value of 1.43 Å to the crystal structure.

Docking models were generated for each Ab model against the gD2 model using the Piper algorithm within Schrodinger BioLuminate [57]. This generated 15–30 different docking models for each sentinel Ab. We note that this method performs rigid docking, not accounting for potential Ab and Ag flexibility.

4.5. Dock Binning

Dock binning takes as input a set of Abs and a set of docking models for each Ab against the same Ag. It assesses docking models for the extent to which they structurally overlap or "compete", and then clusters them based on their profiles of competition against each other in a manner analogous to experimental epitope binning. Finally, it constructs an antigenic heatmap on the Ag surface, highlighting Ag residues according to the frequency with which they contact Ab residues in the docking models comprising a cluster. We now detail these steps.

4.5.1. Competition

Competition between a pair of docking models was assessed with three different scores [21] based on residue-level distances and biophysical interactions across the Ab–Ag interface:

Common interaction metric: the number of Ab–Ag atomic interactions such that one Ag atom has a common interaction with an Ab atom on each Ab. This was computed based on the Interaction Table in BioLuminate, which includes hydrogen bonds, salt bridges, pi stacking, disulfide bonds, and van der Waal interactions.

Cα metric: the number of Cα residues in one Ab within a fixed distance (here 10 Å) from those in the other.

Centroid metric: the distance between the heavy-atom centroids of the two Abs.

Results shown were based on the common interaction metric; results from the other metrics differed only minimally in terms of clustering results.

4.5.2. Clustering

Competition scores were collected into a symmetric matrix, with each row and column representing a docking model and each cell containing the score for the associated pair of models. In this matrix, the row/column for a docking model collects its competition score against each docking model; we call this its "competition score profile". A docking model heatmap was generated based on this matrix (schematically illustrated in Figure 3, with binary scores). The models were hierarchically clustered based on their competition score profiles, here using NBclust. The resulting dendrogram was partitioned to define clusters, also known as communities [92–94]. Carterra Epitope Binning software was used for dendrogram and community plot generation [46].

4.5.3. Antigenic Heatmap

The notion of an antigenic heatmap was developed here to visually represent where the different docking model communities were located on the Ag, i.e., with which Ag residues the Abs in a community interacted the most. Each Ag residue was assigned to the docking model community for which the associated docking models had the most interactions according to the interaction tables generated by BioLuminate (see "common interaction metric" above). The "hotness" of the Ag residue was computed as the number of such interactions, normalized by the size of the community, and the heatmap shade was set accordingly.

In order to compare an antigenic heatmap "footprint" with an experimental epitope region, corresponding sets of residues were identified—those Ag residues comprising a community in the antigenic heatmap and those from the experimental epitope. The centroid of each set of residues was computed and the centroid distances measured, both using Schrodinger PyMOL. In addition, the sets of residues were directly compared for membership to identify how many residues were in common.

Supplementary Materials: The following are available online. Figure S1: Sequenced mAbs and associated experimental data. Table S1: Agreement between epitope residues targeted by our control antibody, E317, according to the crystal structure vs. those identified from our dock binning based approach.

Author Contributions: Conceptualization, B.D.B., G.H.C. and C.B.-K.; Methodology, A.C.; Software, J.Y., A.C. and M.H.; Validation, B.D.B., G.H.C. and C.B.-K.; Formal Analysis, B.D.B., G.H.C. and C.B.-K.; Investigation, T.C.; Resources, G.H.C.; Data Curation, M.H. and R.F.; Writing-Original Draft Preparation, B.D.B.; Writing-Review & Editing, B.D.B., G.H.C. and C.B.-K.; Visualization, A.C. and M.H.; Project Administration, B.D.B.; Funding Acquisition, B.D.B., G.H.C. and C.B.-K. All authors have read and agreed to the published version of the manuscript.

Funding: This research was funded in part by NIH grant numbers 1R43AI132075-01, 1R43AI132075-01, and 2R01GM098977. In addition, the research was supported by AI-18289, AI-142940, AI-139618, and a grant from 549 BIONTECH, Inc. (to G.H.C.) and NSF grants RUI-1904797/ACI-1429467 and XSEDE MCB 550 and The Wellcome Trust, UK.

Acknowledgments: We thank Amanda Brooks for her critical reading of the manuscript, and Wan Ting Saw for helpful discussions. We would like to thank Ron Weed for illustration of the figures.

Conflicts of Interest: B.D.B. has financial interest in a commercial company, Carterra, Inc.

References

1. Gura, T. Therapeutic antibodies: Magic bullets hit the target. *Nature* **2002**, *417*, 584–586. [CrossRef] [PubMed]
2. Chames, P.; Van Regenmortel, M.; Weiss, E.; Baty, D. Therapeutic antibodies: Successes, limitations and hopes for the future. *Br. J. Pharmacol.* **2009**, *157*, 220–233. [CrossRef] [PubMed]
3. Abdiche, Y.N.; Miles, A.; Eckman, J.; Foletti, D.; Van Blarcom, T.J.; Yeung, Y.A.; Pons, J.; Rajpal, A. High-Throughput Epitope Binning Assays on Label-Free Array-Based Biosensors Can Yield Exquisite Epitope Discrimination That Facilitates the Selection of Monoclonal Antibodies with Functional Activity. *PLoS ONE* **2014**, *9*, e92451. [CrossRef] [PubMed]
4. Abdiche, Y.N.; Harriman, R.; Deng, X.; Yeung, Y.A.; Miles, A.; Morishige, W.; Boustany, L.; Zhu, L.; Izquierdo, S.M.; Harriman, W. Assessing kinetic and epitopic diversity across orthogonal monoclonal antibody generation platforms. *Mabs* **2016**, *8*, 264–277. [CrossRef] [PubMed]

5. Ditto, N.T.; Brooks, B.D. The emerging role of biosensor-based epitope binning and mapping in antibody-based drug discovery. *Expert Opin. Drug Discov.* **2016**, *11*, 925–937. [CrossRef]

6. Lonberg, N. Fully human antibodies from transgenic mouse and phage display platforms. *Curr. Opin. Immunol.* **2008**, *20*, 450–459. [CrossRef]

7. Clementi, N.; Mancini, N.; Solforosi, L.; Castelli, M.; Clementi, M.; Burioni, R. Phage display-based strategies for cloning and optimization of monoclonal antibodies directed against human pathogens. *Int. J. Mol. Sci.* **2012**, *13*, 8273–8292. [CrossRef]

8. Murphy, A.J.; Macdonald, L.E.; Stevens, S.; Karow, M.; Dore, A.T.; Pobursky, K.; Huang, T.T.; Poueymirou, W.T.; Esau, L.; Meola, M.; et al. Mice with megabase humanization of their immunoglobulin genes generate antibodies as efficiently as normal mice. *Proc. Natl. Acad. Sci. USA* **2014**, *111*, 5153–5158. [CrossRef]

9. Hua, C.K.; Gacerez, A.T.; Sentman, C.L.; Ackerman, M.E.; Choi, Y.; Bailey-Kellogg, C. Computationally-driven identification of antibody epitopes. *Elife* **2017**, *6*, e29023. [CrossRef]

10. Gan, H.K.; Burgess, A.W.; Clayton, A.H.; Scott, A.M. Targeting of a conformationally exposed, tumor-specific epitope of EGFR as a strategy for cancer therapy. *Cancer Res.* **2012**, *72*, 2924–2930. [CrossRef]

11. Garrett, T.P.; Burgess, A.W.; Gan, H.K.; Luwor, R.B.; Cartwright, G.; Walker, F.; Orchard, S.G.; Clayton, A.H.; Nice, E.C.; Rothacker, J.; et al. Antibodies specifically targeting a locally misfolded region of tumor associated EGFR. *Proc. Natl. Acad. Sci. USA* **2009**, *106*, 5082–5087. [CrossRef] [PubMed]

12. Zolla-Pazner, S.; deCamp, A.; Gilbert, P.B.; Williams, C.; Yates, N.L.; Williams, W.T.; Howington, R.; Fong, Y.; Morris, D.E.; Soderberg, K.A.; et al. Vaccine-Induced IgG Antibodies to V1V2 Regions of Multiple HIV-1 Subtypes Correlate with Decreased Risk of HIV-1 Infection. *PLoS ONE* **2014**, *9*, e87572. [CrossRef] [PubMed]

13. Gottardo, R.; Bailer, R.T.; Korber, B.T.; Gnanakaran, S.; Phillips, J.; Shen, X.; Tomaras, G.D.; Turk, E.; Imholte, G.; Eckler, L.; et al. Plasma IgG to Linear Epitopes in the V2 and V3 Regions of HIV-1 gp120 Correlate with a Reduced Risk of Infection in the RV144 Vaccine Efficacy Trial. *PLoS ONE* **2013**, *8*, e75665. [CrossRef] [PubMed]

14. Wang, X.; Coljee, V.W.; Maynard, J.A. Back to the future: Recombinant polyclonal antibody therapeutics. *Curr. Opin. Chem. Eng.* **2013**, *2*, 405–415. [CrossRef]

15. Steel, J.; Lowen, A.C.; Wang, T.T.; Yondola, M.; Gao, Q.; Haye, K.; García-Sastre, A.; Palese, P. Influenza Virus Vaccine Based on the Conserved Hemagglutinin Stalk Domain. *Mbio* **2010**, *1*, e00018-10. [CrossRef]

16. Gocnik, M.; Fislova, T.; Sladkova, T.; Mucha, V.; Kostolansky, F.; Vareckova, E. Antibodies specific to the HA2 glycopolypeptide of influenza A virus haemagglutinin with fusion-inhibition activity contribute to the protection of mice against lethal infection. *J. Gen. Virol.* **2007**, *88*, 951–955. [CrossRef]

17. Liu, W.-C.; Jan, J.-T.; Huang, Y.-J.; Chen, T.-H.; Wu, S.-C. Unmasking Stem-Specific Neutralizing Epitopes by Abolishing N-Linked Glycosylation Sites of Influenza Virus Hemagglutinin Proteins for Vaccine Design. *J. Virol.* **2016**, *90*, 8496–8508. [CrossRef]

18. Eggink, D.; Goff, P.H.; Palese, P. Guiding the Immune Response against Influenza Virus Hemagglutinin toward the Conserved Stalk Domain by Hyperglycosylation of the Globular Head Domain. *J. Virol.* **2014**, *88*, 699–704. [CrossRef]

19. Margine, I.; Krammer, F.; Hai, R.; Heaton, N.S.; Tan, G.S.; Andrews, S.A.; Runstadler, J.A.; Wilson, P.C.; Albrecht, R.A.; García-Sastre, A.; et al. Hemagglutinin Stalk-Based Universal Vaccine Constructs Protect against Group 2 Influenza A Viruses. *J. Virol.* **2013**, *87*, 10435–10446. [CrossRef]

20. Brooks, B.D.; Miles, A.; Abdiche, Y. High-throughput epitope binning of therapeutic monoclonal antibodies: Why you need to bin the fridge. *Drug Discov. Today* **2014**, *19*, 1040–1044. [CrossRef]

21. Abdiche, Y.N.; Yeung, A.Y.; Ni, I.; Stone, D.; Miles, A.; Morishige, W.; Rossi, A.; Strop, P. Antibodies Targeting Closely Adjacent or Minimally Overlapping Epitopes Can Displace One Another. *PLoS ONE* **2017**, *12*, e0169535. [CrossRef] [PubMed]

22. Estep, P.; Reid, F.; Nauman, C.; Liu, Y.; Sun, T.; Sun, J.; Xu, Y. High throughput solution-based measurement of antibody-antigen affinity and epitope binning. *Mabs* **2013**, *5*, 270–278. [CrossRef] [PubMed]

23. Dimitrov, D.S. Therapeutic antibodies, vaccines and antibodyomes. *Mabs* **2010**, *2*, 347–356. [CrossRef] [PubMed]

24. Renaud, J.-P.; Chung, C.; Danielson, U.H.; Egner, U.; Hennig, M.; Hubbard, R.E.; Nar, H. Biophysics in drug discovery: Impact, challenges and opportunities. *Nat. Rev. Drug Discov.* **2016**, *15*, 679–698. [CrossRef] [PubMed]

25. Zheng, H.; Handing, K.B.; Zimmerman, M.D.; Shabalin, I.G.; Almo, S.C.; Minor, W. X-ray crystallography over the past decade for novel drug discovery–where are we heading next? *Expert Opin. Drug Discov.* **2015**, *10*, 975–989. [CrossRef]

26. Abbott, W.M.; Damschroder, M.M.; Lowe, D.C. Current approaches to fine mapping of antigen–antibody interactions. *Immunology* **2014**, *142*, 526–535. [CrossRef]

27. Weiss, G.A.; Watanabe, C.K.; Zhong, A.; Goddard, A.; Sidhu, S.S. Rapid mapping of protein functional epitopes by combinatorial alanine scanning. *Proc. Natl. Acad. Sci. USA* **2000**, *97*, 8950–8954. [CrossRef]

28. Greenspan, N.S.; Di Cera, E. Defining epitopes: It's not as easy as it seems. *Nat. Biotechnol.* **1999**, *17*, 936–937. [CrossRef]

29. Arkin, M.R.; Randal, M.; DeLano, W.L.; Hyde, J.; Luong, T.N.; Oslob, J.D.; Raphael, D.R.; Taylor, L.; Wang, J.; McDowell, R.S.; et al. Binding of small molecules to an adaptive protein-protein interface. *Proc. Natl. Acad. Sci. USA* **2003**, *100*, 1603–1608. [CrossRef]

30. Novel Inhibitors of DNA Gyrase: 3D Structure Based Biased Needle Screening, Hit Validation by Biophysical Methods, and 3D Guided Optimization. A Promising Alternative to Random Screening-Journal of Medicinal Chemistry (ACS Publications). Available online: https://pubs-acs-org.ezproxy.lib.utah.edu/doi/abs/10.1021/jm000017s (accessed on 28 February 2019).

31. Davidoff, S.N.; Ditto, N.T.; Brooks, A.E.; Eckman, J.; Brooks, B.D. Surface Plasmon Resonance for Therapeutic Antibody Characterization. In *Label-Free Biosensor Methods in Drug Discovery*; Humana Press: New York, NY, USA, 2015; pp. 35–76.

32. Brooks, B. The Importance of Epitope Binning for Biological Drug Discovery. *Curr. Drug Discov. Technol.* **2013**, *11*, 109–112. [CrossRef]

33. Abdiche, Y.N.; Lindquist, K.C.; Pinkerton, A.; Pons, J.; Rajpal, A. Expanding the ProteOn XPR36 biosensor into a 36-ligand array expedites protein interaction analysis. *Anal. Biochem.* **2011**, *411*, 139–151. [CrossRef]

34. Abdiche, Y.N.; Malashock, D.S.; Pons, J. Probing the binding mechanism and affinity of tanezumab, a recombinant humanized anti-NGF monoclonal antibody, using a repertoire of biosensors. *Protein Sci.* **2008**, *17*, 1326–1335. [CrossRef] [PubMed]

35. Abdiche, Y.N.; Malashock, D.S.; Pinkerton, A.; Pons, J. Exploring blocking assays using Octet, ProteOn, and Biacore biosensors. *Anal. Biochem.* **2009**, *386*, 172–180. [CrossRef] [PubMed]

36. Sela-Culang, I.; Kunik, V.; Ofran, Y. The Structural Basis of Antibody-Antigen Recognition. *Front. Immunol.* **2013**, *4*, 302. [CrossRef] [PubMed]

37. Sela-Culang, I.; Ofran, Y.; Peters, B. Antibody specific epitope prediction—Emergence of a new paradigm. *Curr. Opin. Virol.* **2015**, *11*, 98–102. [CrossRef]

38. Zhao, L.; Li, J. Mining for the antibody-antigen interacting associations that predict the B cell epitopes. *BMC Struct. Biol.* **2010**, *10*, S6. [CrossRef]

39. Sircar, A.; Gray, J.J. SnugDock: Paratope Structural Optimization during Antibody-Antigen Docking Compensates for Errors in Antibody Homology Models. *PLoS Comput. Biol.* **2010**, *6*, e1000644. [CrossRef]

40. Brenke, R.; Hall, D.R.; Chuang, G.-Y.; Comeau, S.R.; Bohnuud, T.; Beglov, D.; Schueler-Furman, O.; Vajda, S.; Kozakov, D. Application of asymmetric statistical potentials to antibody–protein docking. *Bioinformatics* **2012**, *28*, 2608–2614. [CrossRef]

41. Wollacott, A.M.; Robinson, L.N.; Ramakrishnan, B.; Tissire, H.; Viswanathan, K.; Shriver, Z.; Babcock, G.J. Structural prediction of antibody-APRIL complexes by computational docking constrained by antigen saturation mutagenesis library data. *J. Mol. Recognit.* **2019**, *32*, e2778. [CrossRef]

42. Long, D.; Madara, T.J.; De Leon, M.P.; Cohen, G.H.; Montgomery, P.C.; Eisenberg, R.J. Glycoprotein D protects mice against lethal challenge with herpes simplex virus types 1 and 2. *Infect. Immun.* **1984**, *43*, 761–764. [CrossRef]

43. Schrier, R.D.; Pizer, L.I.; Moorhead, J.W. Type-specific delayed hypersensitivity and protective immunity induced by isolated herpes simplex virus glycoprotein. *J. Immunol.* **1983**, *130*, 1413–1418. [PubMed]

44. Berman, P.W.; Gregory, T.; Crase, D.; Lasky, L.A. Protection from genital herpes simplex virus type 2 infection by vaccination with cloned type 1 glycoprotein D. *Science* **1985**, *227*, 1490–1492. [CrossRef] [PubMed]

45. Rajčáni, J.; Bánáti, F.; Szenthe, K.; Szathmary, S. The potential of currently unavailable herpes virus vaccines. *Expert Rev. Vaccines* **2018**, *17*, 239–248. [CrossRef] [PubMed]

46. Cairns, T.M.; Ditto, N.T.; Lou, H.; Brooks, B.D.; Atanasiu, D.; Eisenberg, R.J.; Cohen, G.H. Global sensing of the antigenic structure of herpes simplex virus gD using high-throughput array-based SPR imaging. *PLoS Pathog.* **2017**, *13*, e1006430. [CrossRef] [PubMed]

47. Ditto, N.T.; Cairns, T.M.; Lou, H.; Closmore, A.; Eisenberg, R.J.; Cohen, G.C.; Brooks, B.D. Understanding antibody: Antigen Relationships using Antigenic Variants with Array-Based SPRi Epitope Mapping. *J. Drug Res. Dev.* **2016**, *2*, 2470-1009.

48. Hook, L.M.; Cairns, T.M.; Awasthi, S.; Brooks, B.D.; Ditto, N.T.; Eisenberg, R.J.; Cohen, G.H.; Friedman, H.M. Vaccine-induced antibodies to herpes simplex virus glycoprotein D epitopes involved in virus entry and cell-to-cell spread correlate with protection against genital disease in guinea pigs. *PLoS Pathog.* **2018**, *14*, e1007095. [CrossRef]

49. Whitbeck, J.C.; Huang, Z.-Y.; Cairns, T.M.; Gallagher, J.R.; Lou, H.; Ponce-de-Leon, M.; Belshe, R.B.; Eisenberg, R.J.; Cohen, G.H. Repertoire of epitopes recognized by serum IgG from humans vaccinated with herpes simplex virus 2 glycoprotein D. *J. Virol.* **2014**, *88*, 7786–7795. [CrossRef]

50. Whitbeck, J.C.; Muggeridge, M.I.; Rux, A.H.; Hou, W.; Krummenacher, C.; Lou, H.; van Geelen, A.; Eisenberg, R.J.; Cohen, G.H. The major neutralizing antigenic site on herpes simplex virus glycoprotein D overlaps a receptor-binding domain. *J. Virol.* **1999**, *73*, 9879–9890. [CrossRef]

51. Lee, C.-C.; Lin, L.-L.; Chan, W.-E.; Ko, T.-P.; Lai, J.-S.; Wang, A.H.-J. Structural basis for the antibody neutralization of Herpes simplex virus. *Acta Crystallogr. D Biol. Crystallogr.* **2013**, *69*, 1935–1945. [CrossRef]

52. Finn, J.A.; Koehler Leman, J.; Willis, J.R.; Cisneros, A.; Crowe, J.E.; Meiler, J. Improving Loop Modeling of the Antibody Complementarity-Determining Region 3 Using Knowledge-Based Restraints. *PLoS ONE* **2016**, *11*, e0154811. [CrossRef]

53. Messih, M.A.; Lepore, R.; Marcatili, P.; Tramontano, A. Improving the accuracy of the structure prediction of the third hypervariable loop of the heavy chains of antibodies. *Bioinformatics* **2014**, *30*, 2733–2740. [CrossRef] [PubMed]

54. Krummenacher, C.; Supekar, V.M.; Whitbeck, J.C.; Lazear, E.; Connolly, S.A.; Eisenberg, R.J.; Cohen, G.H.; Wiley, D.C.; Carfí, A. Structure of unliganded HSV gD reveals a mechanism for receptor-mediated activation of virus entry. *EMBO J.* **2005**, *24*, 4144–4153. [CrossRef] [PubMed]

55. Kozakov, D.; Brenke, R.; Comeau, S.; Vajda, S. PIPER: An FFT-based protein docking program with pairwise potentials. *Proteins Struct. Funct. Bioinform.* **2006**, *65*, 392–406. [CrossRef] [PubMed]

56. Eddings, M.A.; Johnson, M.A.; Gale, B.K. Determining the optimal PDMS–PDMS bonding technique for microfluidic devices. *J. Micromech. Microeng.* **2008**, *18*, 067001. [CrossRef]

57. Ponomarenko, J.V.; Bourne, P.E. Antibody-protein interactions: Benchmark datasets and prediction tools evaluation. *BMC Struct. Biol.* **2007**, *7*, 64. [CrossRef]

58. Lensink, M.F.; Wodak, S.J. Docking, scoring, and affinity prediction in CAPRI. *Proteins* **2013**, *81*, 2082–2095. [CrossRef]

59. Atanasiu, D.; Saw, W.T.; Lazear, E.; Whitbeck, J.C.; Cairns, T.M.; Lou, H.; Eisenberg, R.J.; Cohen, G.H. Using antibodies and mutants to localize the presumptive gH/gL binding site on HSV gD. *J. Virol.* **2018**, *92*. [CrossRef]

60. Connolly, S.A.; Jackson, J.O.; Jardetzky, T.S.; Longnecker, R. Fusing structure and function: A structural view of the herpesvirus entry machinery. *Nat. Rev. Microbiol.* **2011**, *9*, 369–381. [CrossRef]

61. Choi, Y.; Furlon, J.M.; Amos, R.B.; Griswold, K.E.; Bailey-Kellogg, C. DisruPPI: Structure-based computational redesign algorithm for protein binding disruption. *Bioinformatics* **2018**, *34*, i245–i253. [CrossRef]

62. Eisen, M.B.; Spellman, P.T.; Brown, P.O.; Botstein, D. Cluster analysis and display of genome-wide expression patterns. *Proc. Natl. Acad. Sci. USA* **1998**, *95*, 14863–14868. [CrossRef]

63. Jiang, D.; Tang, C.; Zhang, A. Cluster analysis for gene expression data: A survey. *IEEE Trans. Knowl. Data Eng.* **2004**, *16*, 1370–1386. [CrossRef]

64. Hartuv, E.; Shamir, R. A clustering algorithm based on graph connectivity. *Inf. Process. Lett.* **2000**, *76*, 175–181. [CrossRef]

65. Pittala, S.; Bailey-Kellogg, C. Learning Context-aware Structural Representations to Predict Antigen and Antibody Binding Interfaces. *Bioinformatics* **2020**, *36*, 3996–4003. [CrossRef]

66. Krawczyk, K.; Liu, X.; Baker, T.; Shi, J.; Deane, C.M. Improving B-cell epitope prediction and its application to global antibody-antigen docking. *Bioinformatics* **2014**, *30*, 2288–2294. [CrossRef]

67. Kringelum, J.V.; Lundegaard, C.; Lund, O.; Nielsen, M. Reliable B Cell Epitope Predictions: Impacts of Method Development and Improved Benchmarking. *PLoS Comput. Biol.* **2012**, *8*, e1002829. [CrossRef] [PubMed]

68. Feature Selection Based on Mutual Information: Criteria of Max-Dependency, Max-Relevance, and Min-Redundancy. Available online: https://www.computer.org/csdl/journal/tp/2005/08/i1226/13rRUy3xY96 (accessed on 16 March 2020).

69. Moretti, R.; Lyskov, S.; Das, R.; Meiler, J.; Gray, J.J. Web-accessible molecular modeling with Rosetta: The Rosetta Online Server that Includes Everyone (ROSIE). *Protein Sci.* **2018**, *27*, 259–268. [CrossRef]

70. Rux, A.H.; Willis, S.H.; Nicola, A.V.; Hou, W.; Peng, C.; Lou, H.; Cohen, G.H.; Eisenberg, R.J. Functional Region IV of Glycoprotein D from Herpes Simplex Virus Modulates Glycoprotein Binding to the Herpesvirus Entry Mediator. *J. Virol.* **1998**, *72*, 7091–7098. [CrossRef]

71. Lazear, E.; Whitbeck, J.C.; Ponce-de-Leon, M.; Cairns, T.M.; Willis, S.H.; Zuo, Y.; Krummenacher, C.; Cohen, G.H.; Eisenberg, R.J. Antibody-Induced Conformational Changes in Herpes Simplex Virus Glycoprotein gD Reveal New Targets for Virus Neutralization. *J. Virol.* **2012**, *86*, 1563–1576. [CrossRef]

72. Friedman, H.M.; Cohen, G.H.; Eisenberg, R.J.; Seidel, C.A.; Cines, D.B. Glycoprotein C of herpes simplex virus 1 acts as a receptor for the C3b complement component on infected cells. *Nature* **1984**, *309*, 633–635. [CrossRef]

73. Eisenberg, R.J.; Long, D.; Pereira, L.; Hampar, B.; Zweig, M.; Cohen, G.H. Effect of monoclonal antibodies on limited proteolysis of native glycoprotein gD of herpes simplex virus type 1. *J. Virol.* **1982**, *41*, 478–488. [CrossRef]

74. Cohen, G.H.; Isola, V.J.; Kuhns, J.; Berman, P.W.; Eisenberg, R.J. Localization of discontinuous epitopes of herpes simplex virus glycoprotein D: Use of a nondenaturing ("native" gel) system of polyacrylamide gel electrophoresis coupled with Western blotting. *J. Virol.* **1986**, *60*, 157–166. [CrossRef] [PubMed]

75. Muggeridge, M.I.; Wu, T.T.; Johnson, D.C.; Glorioso, J.C.; Eisenberg, R.J.; Cohen, G.H. Antigenic and functional analysis of a neutralization site of HSV-1 glycoprotein D. *Virology* **1990**, *174*, 375–387. [CrossRef]

76. Carfí, A.; Willis, S.H.; Whitbeck, J.C.; Krummenacher, C.; Cohen, G.H.; Eisenberg, R.J.; Wiley, D.C. Herpes Simplex Virus Glycoprotein D Bound to the Human Receptor HveA. *Mol. Cell* **2001**, *8*, 169–179. [CrossRef]

77. Di Giovine, P.; Settembre, E.C.; Bhargava, A.K.; Luftig, M.A.; Lou, H.; Cohen, G.H.; Eisenberg, R.J.; Krummenacher, C.; Carfi, A. Structure of Herpes Simplex Virus Glycoprotein D Bound to the Human Receptor Nectin-1. *PLoS Pathog.* **2011**, *7*, e1002277. [CrossRef]

78. Lu, G.; Zhang, N.; Qi, J.; Li, Y.; Chen, Z.; Zheng, C.; Gao, G.F.; Yan, J. Crystal Structure of Herpes Simplex Virus 2 gD Bound to Nectin-1 Reveals a Conserved Mode of Receptor Recognition. *J. Virol.* **2014**, *88*, 13678–13688. [CrossRef] [PubMed]

79. Lazear, E.; Whitbeck, J.C.; Zuo, Y.; Carfí, A.; Cohen, G.H.; Eisenberg, R.J.; Krummenacher, C. Induction of conformational changes at the N-terminus of herpes simplex virus glycoprotein D upon binding to HVEM and nectin-1. *Virology* **2014**, *448*, 185–195. [CrossRef]

80. van Kooij, A.; Middel, J.; Jakab, F.; Elfferich, P.; Koedijk, D.G.; Feijlbrief, M.; Scheffer, A.J.; Degener, J.E.; The, T.H.; Scheek, R.M.; et al. High level expression and secretion of truncated forms of herpes simplex virus type 1 and type 2 glycoprotein D by the methylotrophic yeast Pichia pastoris. *Protein Expr. Purif.* **2002**, *25*, 400–408. [CrossRef]

81. Minson, A.C.; Hodgman, T.C.; Digard, P.; Hancock, D.C.; Bell, S.E.; Buckmaster, E.A. An Analysis of the Biological Properties of Monoclonal Antibodies against Glycoprotein D of Herpes Simplex Virus and Identification of Amino Acid Substitutions that Confer Resistance to Neutralization. *J. Gen. Virol.* **1986**, *67*, 1001–1013. [CrossRef]

82. Pereira, L.; Klassen, T.; Baringer, J.R. Type-common and type-specific monoclonal antibody to herpes simplex virus type 1. *Infect. Immun.* **1980**, *29*, 724–732.

83. Pereira, L.; Dondero, D.V.; Gallo, D.; Devlin, V.; Woodie, J.D. Serological analysis of herpes simplex virus types 1 and 2 with monoclonal antibodies. *Infect. Immun.* **1982**, *35*, 363–367. [CrossRef]

84. Showalter, S.D.; Zweig, M.; Hampar, B. Monoclonal antibodies to herpes simplex virus type 1 proteins, including the immediate-early protein ICP 4. *Infect. Immun.* **1981**, *34*, 684–692. [CrossRef] [PubMed]

85. Isola, V.J.; Eisenberg, R.J.; Siebert, G.R.; Heilman, C.J.; Wilcox, W.C.; Cohen, G.H. Fine mapping of antigenic site II of herpes simplex virus glycoprotein D. *J. Virol.* **1989**, *63*, 2325–2334. [CrossRef] [PubMed]

86. Muggeridge, M.I.; Roberts, S.R.; Isola, V.J.; Cohen, G.H.; Eisenberg, R.J. Herpes simplex virus. *Immunochem. Viruses* **1990**, *2*, 459–481.

87. Chiang, H.Y.; Cohen, G.H.; Eisenberg, R.J. Identification of functional regions of herpes simplex virus glycoprotein gD by using linker-insertion mutagenesis. *J. Virol.* **1994**, *68*, 2529–2543. [CrossRef]

88. Seigneurin, J.; Desgranges, C.; Seigneurin, D.; Paire, J.; Renversez, J.; Jacquemont, B.; Micouin, C. Herpes simplex virus glycoprotein D: Human monoclonal antibody produced by bone marrow cell line. *Science* **1983**, *221*, 173–175. [CrossRef]

89. Nicola, A.V.; Willis, S.H.; Naidoo, N.N.; Eisenberg, R.J.; Cohen, G.H. Structure-function analysis of soluble forms of herpes simplex virus glycoprotein D. *J. Virol.* **1996**, *70*, 3815–3822. [CrossRef]

90. Connolly, S.A.; Landsburg, D.J.; Carfi, A.; Whitbeck, J.C.; Zuo, Y.; Wiley, D.C.; Cohen, G.H.; Eisenberg, R.J. Potential Nectin-1 Binding Site on Herpes Simplex Virus Glycoprotein D. *J. Virol.* **2005**, *79*, 1282–1295. [CrossRef]

91. Gallagher, J.R.; Saw, W.T.; Atanasiu, D.; Lou, H.; Eisenberg, R.J.; Cohen, G.H. Displacement of the C Terminus of Herpes Simplex Virus gD Is Sufficient To Expose the Fusion-Activating Interfaces on gD. *J. Virol.* **2013**, *87*, 12656–12666. [CrossRef]

92. Charrad, M.; Ghazzali, N.; Boiteau, V.; Niknafs, A. NbClust Package for Determining the Best Number of Clusters. *R Package Version* **2014**, *2*.

93. Charrad, M.; Ghazzali, N.; Boiteau, V.; Niknafs, A.; Charrad, M.M. Package 'NbClust'. *J. Stat. Softw.* **2014**, *61*, 1–36.

94. Malika, C.; Ghazzali, N.; Boiteau, V.; Niknafs, A. NbClust: An R Package for Determining the Relevant Number of Clusters in a Data Set. *J. Stat. Softw.* **2014**, *61*, 1–36. [CrossRef]

Sample Availability: Samples of the compounds are available from the authors.

Article

Multiscale Virtual Screening Optimization for Shotgun Drug Repurposing Using the CANDO Platform

Matthew L. Hudson and Ram Samudrala *

Department of Biomedical Informatics, Jacobs School of Medicine and Biomedical Sciences, University at Buffalo, Buffalo, NY 14203, USA; mlhudson@buffalo.edu
* Correspondence: ram@compbio.org; Tel.: +1-716-888-4858

Citation: Hudson, M.L.;
Samudrala, R. Multiscale Virtual
Screening Optimization
for Shotgun Drug Repurposing
Using the CANDO Platform.
Molecules **2021**, 26, 2581.
https://doi.org/10.3390/
molecules26092581

Academic Editors: Marco Tutone and
Anna Maria Almerico

Received: 25 February 2021
Accepted: 19 April 2021
Published: 28 April 2021

Publisher's Note: MDPI stays neutral
with regard to jurisdictional claims in
published maps and institutional affiliations.

Abstract: Drug repurposing, the practice of utilizing existing drugs for novel clinical indications, has tremendous potential for improving human health outcomes and increasing therapeutic development efficiency. The goal of multi-disease multitarget drug repurposing, also known as shotgun drug repurposing, is to develop platforms that assess the therapeutic potential of each existing drug for every clinical indication. Our Computational Analysis of Novel Drug Opportunities (CANDO) plat-form for shotgun multitarget repurposing implements several pipelines for the large-scale modeling and simulation of interactions between comprehensive libraries of drugs/compounds and protein structures. In these pipelines, each drug is described by an interaction signature that is compared to all other signatures that are subsequently sorted and ranked based on similarity. Pipelines within the platform are benchmarked based on their ability to recover known drugs for all indications in our library, and predictions are generated based on the hypothesis that (novel) drugs with similar signatures may be repurposed for the same indication(s). The drug-protein interactions used to create the drug-proteome signatures may be determined by any screening or docking method, but the primary approach used thus far has been BANDOCK, our in-house bioanalytical or similarity docking protocol. In this study, we calculated drug-proteome interaction signatures using the pub-licly available molecular docking method Autodock Vina and created hybrid decision tree pipelines that combined our original bio- and chem-informatic approach with the goal of assessing and bench-marking their drug repurposing capabilities and performance. The hybrid decision tree pipeline outperformed the two docking-based pipelines from which it was synthesized, yielding an average indication accuracy of 13.3% at the top10 cutoff (the most stringent), relative to 10.9% and 7.1% for its constituent pipelines, and a random control accuracy of 2.2%. We demonstrate that docking-based virtual screening pipelines have unique performance characteristics and that the CANDO shotgun repurposing paradigm is not dependent on a specific docking method. Our results also provide further evidence that multiple CANDO pipelines can be synthesized to enhance drug repurposing predictive capability relative to their constituent pipelines. Overall, this study indicates that pipelines consisting of varied docking-based signature generation methods can capture unique and useful signals for accurate comparison of drug-proteome interaction signatures, leading to improvements in the benchmarking and predictive performance of the CANDO shotgun drug repurposing platform.

Keywords: drug repurposing; virtual screening; multiscale; multitargeting; polypharmacology; compu-tational biology; drug repositioning; structural bioinformatics; molecular docking; proteomic signature

1. Introduction

1.1. Drug Repurposing

Pharmacological innovation reduces human mortality rates and provides substantial improvements to the quality of life [1]. Therapeutic compounds that have been discovered, lab tested preclinically, and evaluated for risks and efficacy in clinical trials are approved by regulatory bodies such as the United States FDA for specific indications [2]. Potential failures impose high opportunity costs, and the realities of market forces and investment

distort the types of ailments for which treatments are pursued [3–5]. The rate of novel drug discovery has been slowing as costs have been increasing, illustrating the need for more efficient paradigms [6].

Drug discovery traditionally relies on screening a compound or set of compounds against a biological target, typically a protein for a specific indication. Generally, these approaches incorporate high-throughput in vitro compound screens [7] and/or cell-based assays [8] of candidates drawn from wet laboratory studies or computational screens of virtual representations of compounds and biological targets [9–11]. If promising in vitro leads are found, they undergo in vivo testing, eventually leading to approval for clinical use if they continue to demonstrate relative efficacy and safety [2]. Traditional drug discovery methods tend to be focused on a single target and indication [11,12]. However, drugs and other human-ingested compounds interact promiscuously with many proteins in the body [13,14]. These off-target interactions are responsible for side effects and the fact that one drug may be useful for treating multiple indications [15–19]. Single-target approaches may miss promising leads and potentially beneficial off-target side effects, while drugs that have already been discovered and vetted for safety may be used in novel treatment contexts.

Drug repurposing is the practice of finding new uses for existing drugs, taking advantage of prior safety, efficacy, and pharmacological knowledge and data [15]. Drug repurposing has the potential to arbitrarily increase the utility of the FDA-approved drug library [17,20,21], particularly via innovations such as multitarget drug repurposing [22,23]. Drug repurposing has yielded new uses for multiple drugs [15,17] and has demonstrated potential for the treatment of viral [24,25], bacterial [26], and complex indications such as cancer [17].

1.2. Computational Drug Repurposing Using Molecular Docking

Computational models that improve drug discovery and repurposing leverage rapidly increasing computer processing power and vast collections of preclinical (in vitro, in vivo) and clinical data [22]. Although there are a variety of computational approaches, the most relevant ones to this study are structure-based. Structure-based approaches focus on modeling/simulating the effects the three-dimensional (3D) structure of a compound may have on one or more macromolecules, typically protein structures [27]. Structure representations are based on data obtained from X-ray diffraction, NMR spectroscopy, cryogenic electron microscopy, and biochemical and biophysical simulation studies. These models may incorporate other features such as predicted protein-compound binding sites, simulations of the surrounding chemical environments, and the functional characteristics of protein structures.

Molecular docking models the three-dimensional (3D) interaction between small molecule compounds and macromolecular protein structures [28–32]. Typically, these simulations algorithmically calculate the optimal position and orientation of a compound structure that interacts with (or binds to) a particular region of a protein structure and its corresponding interaction strength, using physics-based [33] or knowledge-based [34,35] force fields or scoring functions. The characteristics of a correctly modeled compound-protein structure provide researchers insight into the biological implications of the interaction: for example, a researcher may infer that a signaling pathway may be interrupted if a particular protein were to be inhibited by the compound based on the strength of its binding energy [36]. Molecular docking is also useful when researching large sets of compounds and proteins [37]. By comparing the relative differences in interactions between protein-compound pairs, the researcher can rank and organize pairs according to the strength of their interaction score and/or their similarity to identify patterns that are apparent only when examining large sets with many possible combinations, which is difficult and expensive to do in in vitro or in vivo experiments [22,38]. Molecular docking techniques have varying performance advantages and limitations [39,40]; however, provided that docking

approaches are used wisely in concert with other experimental techniques, they have the potential to be useful for drug repurposing, particularly in a large-scale context [38].

1.3. Shotgun Multitarget Multi-disease Drug Repurposing Using the CANDO Platform

The Computational Analysis of Novel Drug Opportunities (CANDO) platform was developed to mitigate endemic problems in drug discovery and enable multitarget approaches to drug repurposing [22,38,41–45]. The CANDO platform is designed to provide insights about the holistic behavior of compounds interacting within complex biological systems, including how a compound behaves relative to other compounds, and is an extensible standardized framework for building and combining drug repurposing, discovery, and design simulation pipelines. The similarity of drug-protein interaction behavior between a small molecule drug/compound and its macromolecular environment is hypothesized to indicate the similarity of drug therapeutic function [22]. In traditional structure-based and ligand-based drug discovery, therapeutic similarity inferences are often based on molecular target similarity and compound similarity [46]. CANDO extends the similarity assumption principle to include holistic multiscale interaction similarity, that characterizes compounds by the nature of their interaction with entire proteomes and (eventually) interactomes [22,38]. Extending the interaction similarity frontier enables CANDO to account for the promiscuous nature of compound interaction within biological systems and characterize previously unconsidered therapeutic functions of existing approved drugs. CANDO pipelines are evaluated by a benchmarking protocol that examines the relative ranking of every drug for every indication with two or more approved drugs. Analyzing the relative ranking of approved drugs for each indication enables the evaluation of the effectiveness of the platform for recovering known information, comparing relative pipeline performance for particular indications, calculating the accuracy and precision for ranking approved drugs, and determining which components of the platform need improvement.

In this study, we set out to extend the CANDO platform with an additional molecular docking pipeline using the popular software AutoDock Vina [30] to determine whether the prior CANDO performance was dependent on a specific molecular docking protocol, how different molecular docking protocols affect CANDO's performance, and whether hybridizing molecular docking pipelines yields improved performance, as we have previously observed combining structure- and ligand-based CANDO pipelines [43].

2. Results

2.1. Benchmarking Performance of the Different Pipelines

The performance of two new primary pipelines and a hybrid one in the CANDO platform was investigated and compared to those previously created, including a random control (see Figure 1 and Methods). The first is the Vina pipeline that uses the eponymous molecular docking program to screen the CANDO v1.5 3733 drug/compound library against a 134-protein subset of the full proteome library (Vina-134). Multiple binding sites for each protein were predicted and targeted for docking, and the strongest interaction scores were used to construct the drug-proteome signatures.

The second pipeline used was the default CANDO v1.5 pipeline restricted to the same 134 protein subset (v1.5-134). We generated a hybrid decision tree pipeline drawn from a combination of the Vina-134 and the v1.5-134 pipelines. For comparison, we examined the performance of these pipelines with respect to a random control and the v1.5 pipeline implemented with the full CANDO proteome library consisting of 46,784 protein structures (v1.5-full).

Figure 2 illustrates the relative performance of these different pipelines. At the top10 threshold, the hybrid decision tree yielded 13.3% accuracy, v1.5-134 10.9%, and Vina-134 7.11%. The v1.5-134 and v1.5-full pipelines outperformed the Vina pipeline, but the latter was able to substantially contribute to the superior performance of the hybrid pipeline. Notably, the hybrid decision tree pipeline outperformed the v1.5-full pipeline with a top10

accuracy of 12.8% with two orders of magnitude difference in the number of proteins used in the implementation of the pipeline (134 vs. 46,784).

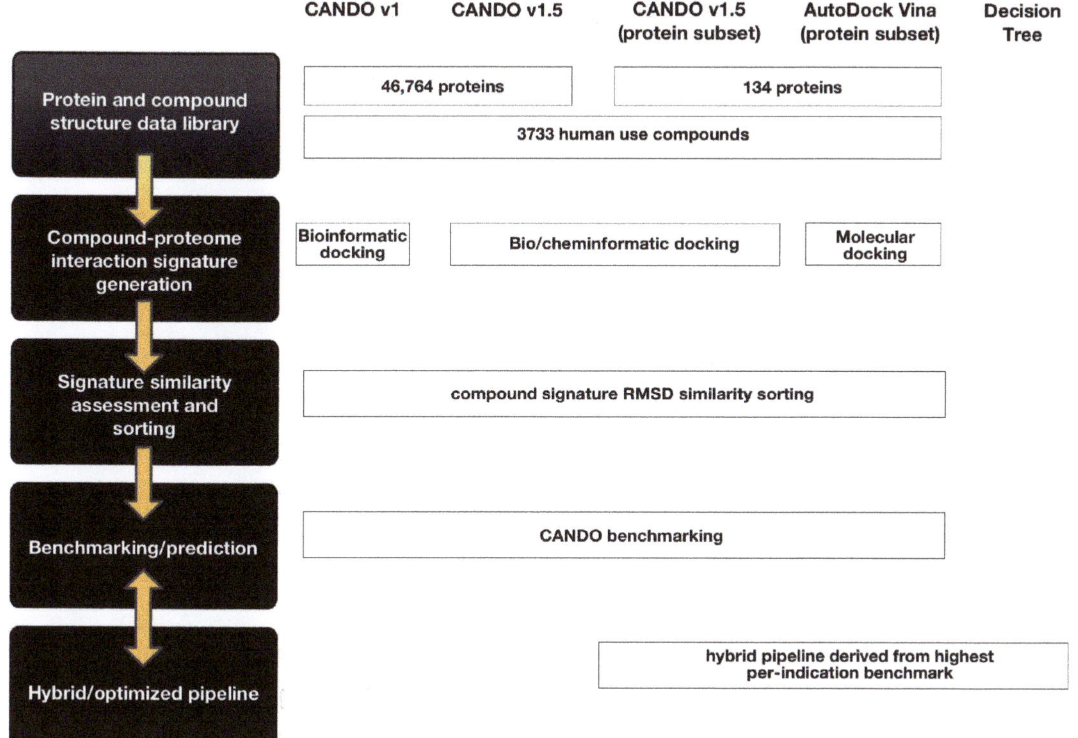

Figure 1. CANDO shotgun drug repurposing platform and pipeline overview. On the left side of the figure is a flow diagram, which indicates the general protocol for implementing a CANDO drug-proteome pipeline. To the right of the flow diagram, each pipeline relevant to this investigation is displayed along with implementation details for each phase of the CANDO protocol. Data curation: The drug-proteome pipelines utilize libraries of protein structure and drug structure representations. Interaction scoring protocol: These pipelines use bioinformatic, cheminformatic, and molecular docking methods to predict the scores between each protein and drug interaction. The set of protein interaction scores for each drug is considered its interaction signature. Each interaction signature can be compared with one another by assessing the root mean squared deviation of their interaction signatures. Drug comparison protocol: Every drug signature is compared with every other drug signature. After every comparison is made, each drug has a list containing the ordered set of every other drug, from most similar signature to least similar signature. Benchmarking protocol: The CANDO benchmarking procedure assesses, for every drug, how many other drugs with the same indication association are found within certain ranking cutoffs. An indication-specific accuracy score is produced by averaging the recovery rate of co-associated drugs for every drug associated with the particular indication for particular ranking cutoffs. The overall pipeline average indication accuracy is the mean of all the individual ones for a particular cutoff. Three pipelines were generated during this investigation: v1.5 (implemented with a subset of the CANDO proteome library), AutoDock Vina (using the same proteome sublibrary), and a hybrid decision tree pipeline derived from the former two pipelines. Each of the subset pipelines utilized a small sublibrary (134 proteins) of the original CANDO v1 and v1.5 pipelines. Although the pipelines used different signature generation approaches (BANDOCK bioanalytical or similarity docking and AutoDock Vina molecular docking), their signatures underwent the same similarity assessment and benchmarking protocol. However, there is room for variation via the use of alternate docking, similarity assessment, and benchmarking approaches.

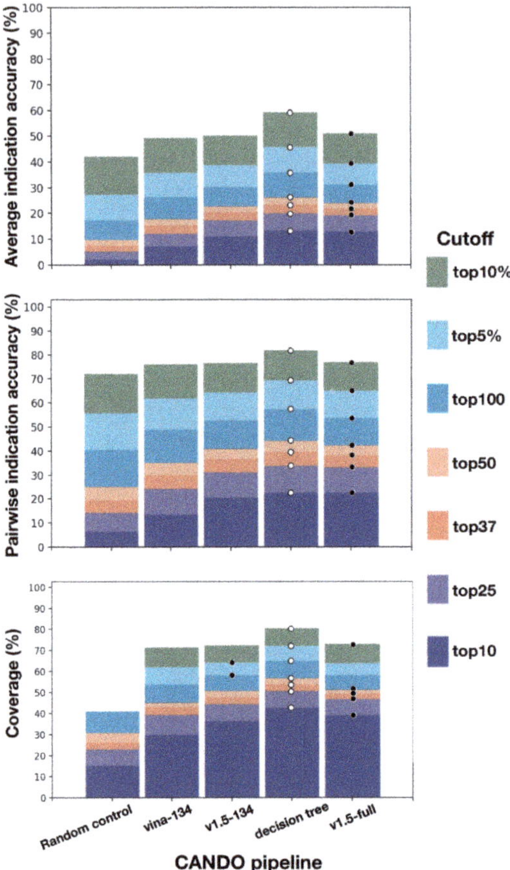

Figure 2. Benchmarking performance of CANDO pipelines used in this study. Three docking-derived pipelines implemented in CANDO: v1.5-full, using the interaction scores generated by the default CANDO v1.5 bioanalytical or similarity docking protocol (BANDOCK) for the full (46,784) proteome library, v1.5-134, using the same interaction scoring protocol for a 134 protein sublibrary, and Vina-134, based on interaction scores generated using AutoDock Vina for the 134 protein sublibrary are compared with (d) a hybrid decision tree pipeline derived from combining Pipelines (b) and (c), as well as (a) the random control reference pipeline calculated numerically from a hypergeometric distribution [43]. The pipelines were assessed by three CANDO platform benchmarking metrics: average indication accuracy (%), pairwise accuracy (%), and coverage (%). Performance cutoffs are denoted by colored bars from most to least stringent: top10 (dark purple), top25 (light purple), top37/top1% (dark pink), top50 (light)pink, top100 (dark blue), top5% (light blue), top10% (dark green), and top50% (light green) for 1439 indications with at least two approved drugs using a leave-one-out benchmarking protocol (see the Methods Section). White dots denote the highest overall accuracy at each threshold. The hybrid decision tree pipeline, which incorporates the highest indication accuracies from the Vina-134 and v1.5-134 pipelines, performed the best at all cutoffs (white dots). Black dots denote high performance in individual pipelines, which was obtained using the two v1.5 pipelines, one based on the 134 proteome sublibrary and the other on the full proteome library. v1.5-134 yielded the highest top50 percent average indication accuracy (85.3%), top100 coverage (52.742%), top5% coverage (64.382%), and top50% coverage (96.064%). Individually, the Vina-134 pipeline significantly outperformed the random control and yielded a significant fraction of the performance of the v1.5 pipelines. The hybrid decision tree pipeline performed the best, indicating that diversity in pipeline simulation implementation can be leveraged to increase drug repurposing performance.

2.2. Divergence in Indication Accuracy at Various Thresholds

Figure 3 illustrates the similarity and divergence of indication accuracy performance at various thresholds: i.e., instances where the Vina-134 pipeline outperforms the v1.5-134 pipeline, instances where the v1.5-134 pipeline outperforms the Vina-134 pipeline, instances where each pipeline yields the same indication accuracy, and instances where each pipeline yields zero percent accuracy. At the top10 threshold, the Vina-134 pipeline had 191 indications (about 13% of all indications) that outperformed the v1.5 pipeline, which had 363 indications outperform Vina-134 (about 25% of all indications). There were 885 equivalently performing indications (with 828 of them at zero percent accuracy) at the top10 cutoff. Overall, the divergence in relative performance increased as the thresholds became less stringent (the CANDO pipeline outperformance share began to decline slightly after the top5% threshold). v1.5 had a higher number of indications in which it outperformed Vina-134. After the top50 cutoff, the proportion of equivalent indication accuracies that were both zero relative to the total equivalent indication accuracies began to decline rapidly.

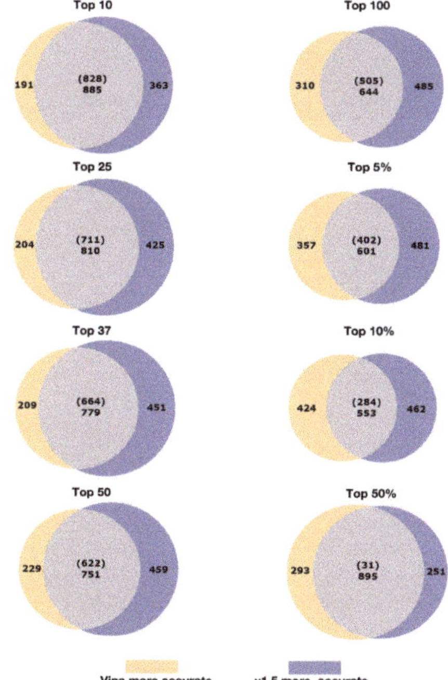

Figure 3. Comparing and contrasting indication accuracies for two CANDO platform pipelines at different cutoffs. Each Venn diagram represents the set of indication accuracies (1439 total) for the Vina-134 and v1.5-134 pipelines at different cutoffs (top10, top25, top37, top50, top100, top5%, top10%, and top50%). Indications that scored higher for the Vina pipeline are in yellow. Indications that scored higher for the v1.5 pipeline are in purple. Indications that scored the same for each pipeline are in gray. The number of indications where both pipelines yield 0% accuracy are provided in parentheses and the total number of equal indication accuracies provided below. At the top10 cutoff, the Vina-134 pipeline yields higher accuracies for 191 indications and the v1.5 pipeline for 363 indications, and each pipeline yields the same accuracy for 885 indications. Although the Vina-134 pipeline produces a substantial number of indication accuracies that are higher or equivalent to v1.5 indication accuracies, the v1.5 pipeline outperforms the Vina-134 pipeline at every cutoff except for top50%. Nonetheless, the orthogonality in the above diagrams indicates that individual pipelines can be synergistically combined into a hybrid pipeline that yields considerable performance improvements.

2.3. Net Differences in Indication Accuracy

Figure 4 elucidates the net differences in top10 accuracies between two pipelines (and the proportion of approved drugs recovered per indication in the top10) for 700 indications. With some notable exceptions, the v1.5-134 pipeline outperformed the Vina-134 pipeline in frequency and magnitude. On a per indication basis, as the total number of approved drugs decreased, the Vina-134 pipeline had a higher number of the outperforming indications in terms of frequency and magnitude.

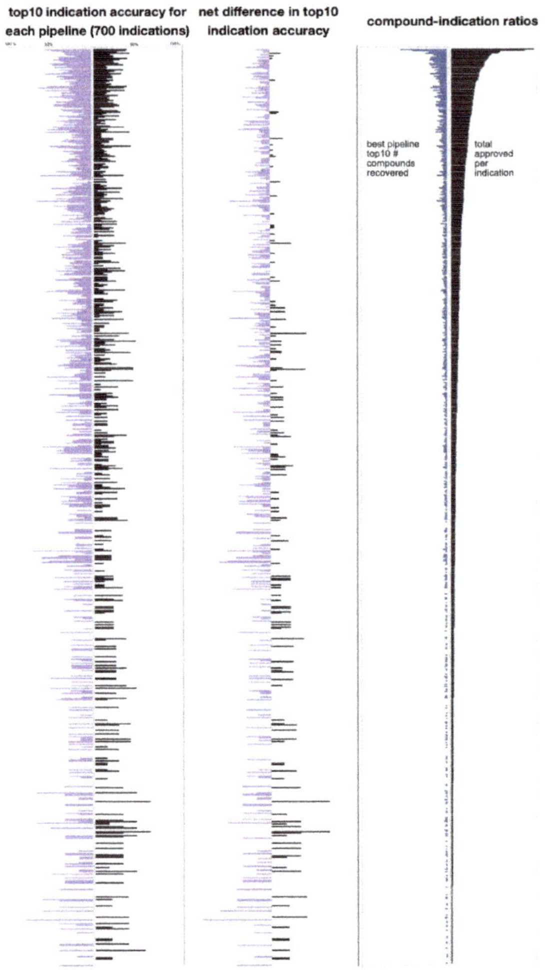

Figure 4. Comparison of 700 indication accuracies for two CANDO platform pipelines at the top10 cutoff. The top10 indication accuracies for 700 indications produced by the Vina-134 and v1.5-134 pipelines are shown in the left panel, with the the v1.5-134 pipeline per indication accuracies in purple on the left side and the Vina-134 pipeline accuracies in black on the right. The net difference in pipeline accuracy for the same indication is shown in the center panel, using the same percentage scale as the left. The number of drugs recovered by the best performing pipeline (in blue on the left side) and the total number of drugs approved per indication (in black on the right side) are shown in the right panel. The number of approved drugs for all three panels ranges from 158 drugs at the top to two drugs at the bottom. Generally, the v1.5-134 pipeline outperforms the Vina-134 pipeline, both by the number of indications and net difference in accuracy per indication.

2.4. Relative Pipeline Indication Accuracy

The pipelines differed at average indication accuracy thresholds and on a per indication basis. In some cases, a pipeline that performed worse overall may do better for a specific indication. Figures 3–5 illustrate the overall divergence, the magnitude of divergence, and the threshold frequency distributions. On a per-indication basis, there was divergence in the relative indication accuracy at various cutoffs, both in terms of net difference in accuracy, recovery at a particular threshold, and the frequency of a particular indication being recovered at a particular interval. The divergence between the per-indication performance of each pipeline elucidated by Figures 2–4 suggests that each pipeline should be used in conjunction with one another for maximum indication inclusivity and accuracy.

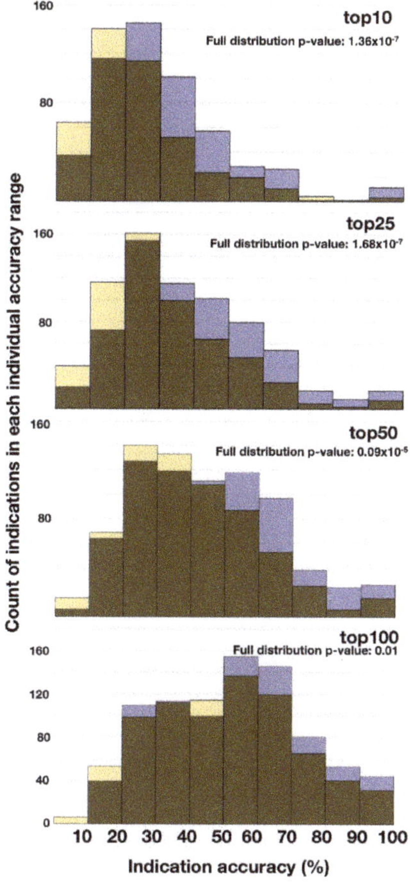

Figure 5. Frequency comparison of indication accuracies for two CANDO platform pipelines at different cutoffs. Shown are four histograms denoting the frequency with which indications fall within particular accuracy ranges (Vina-134 pipeline accuracies are in light yellow and v1.5 accuracies in black). The similarity of each distribution is assessed by the *p*-value using the Kolmogorov–Smirnoff test (*p*-values less than 0.05 are considered to be significant). The v1.5 pipeline outperforms Vina-134 overall, but the *p*-values indicate that the accuracy distributions are different for the two pipelines, indicating the utility of combining pipelines to produce synergistic performance.

2.5. Comparison of the Pipeline Distribution of Per Indication Accuracies

Figure 5 illustrates the distribution of indication accuracies by counting the frequency of each indication that falls within a certain accuracy range. The dissimilarity of pipeline distributions at each cutoff was assessed by applying the Kolmogorov–Smirnoff test. The v1.5 pipeline outperformed Vina-134 overall (which yielded a higher frequency of indications exceeding 50% accuracy).

2.6. Indication Accuracy Distribution

Figure 6 examines the distribution of indications to illustrate their relative performance within each pipeline. Pipelines can also be compared with symmetrical accuracy distribution charts, where individual pipeline accuracy is denoted along the x and y axes. Each point can represent a particular indication (e.g., one of the 1439 indications in the CANDO platform), a defined indication class (e.g., all 39 indications with the string "neoplasm"), or some other way of denoting indications (e.g., indications that occupy a particular branch of the Medical Subjects Heading (MeSH) classification [47] or those that are ontologically similar [48]). When pipelines reach accuracy consensus (or near consensus) for a particular indication (or indication grouping), the point falls on or close to the 45 degree symmetry line. These figures suggest that different pipelines had varying success in benchmarking performance on a per-indication basis. More rigorous clustering analysis, indication classification, and indication definition will yield deeper insight into the relative strengths of each pipeline.

2.7. Distribution of Individual Drug-Indication Pair Rankings

Supplementary Figure S1 plots every drug-indication pair and its corresponding rank within each pipeline. These suggest that there is some substantial ranking consensus between each pipeline, as well as substantial divergence. The distribution was plotted at linear and logarithmic scales to illustrate the density of approved drug-indication pair ranking consensus and divergence. There is a high density of drug-indication pairs that have relatively high ranking in each pipeline. There is also a high density of drug-indication pairs that have a significantly higher ranking in the v1.5-134 pipeline than the Vina-134 pipeline. As with pipeline per-indication accuracy divergence, further investigation into drug-indication pair divergence may help improve the performance of individual and hybrid pipelines, particularly in cases where one pipeline ranked a drug-indication pair substantially higher than the other one (e.g., top100 in one and bottom 50% in the other).

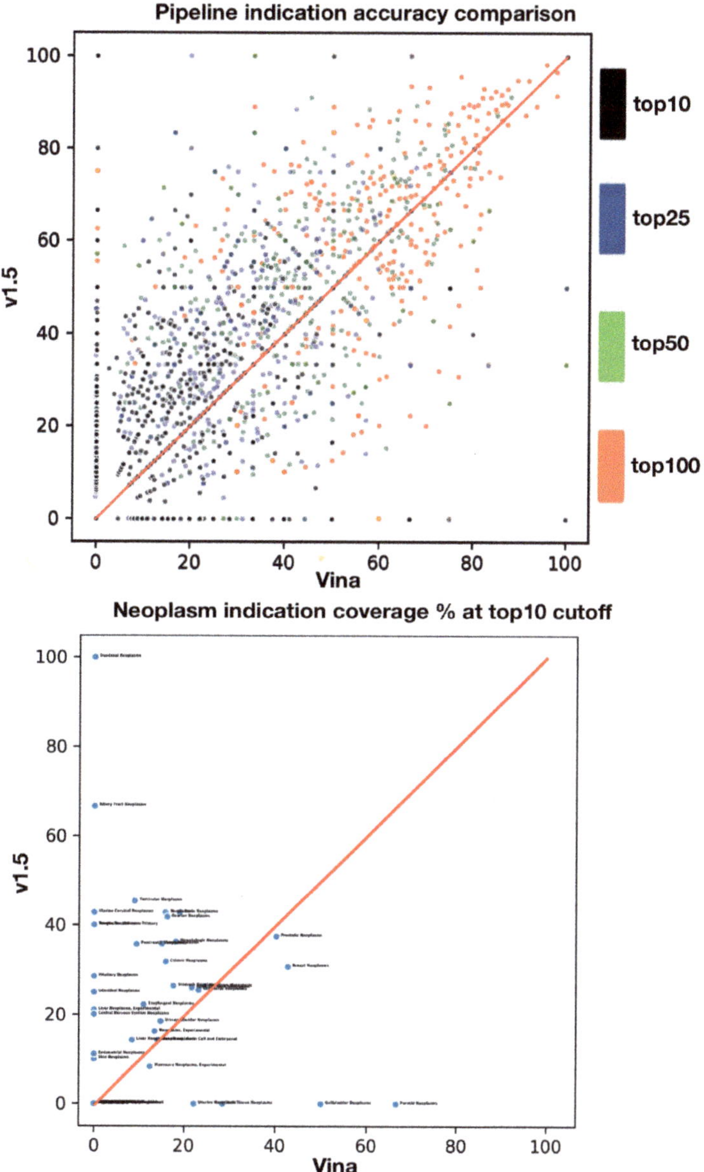

Figure 6. Comparison of the indication accuracies at various cutoffs and for a defined indication class for two CANDO platform pipelines. The top panel denotes a symmetrical accuracy chart. Each axis measures the indication accuracy for each pipeline, and indications are plotted according to their corresponding accuracies (different cutoffs are distributed in alternate colors). Points that land on the 45 degree red line are indications where the pipelines reached consensus; points that fall closer to a particular axis achieved a relatively higher score with the corresponding pipeline. The bottom panel isolates indications for the defined class "neoplasm" comprising 39 indications with the corresponding string. The asymmetrical distribution of the accuracy plot suggests pipeline accuracy differentiation, i.e., different pipelines have differing performance strengths and weaknesses, on a per-indication and indication class level.

3. Discussion

3.1. Multiple Large-Scale Virtual Screening Pipelines

In this investigation, we hypothesized that distinct docking methods would yield distinct drug-proteome interaction signatures due to differing simulation implementation, and correspondingly differing performance for shotgun drug repurposing: BANDOCK, the default bioanalytical or similarity docking in CANDO is a knowledge-based template/comparative modeling protocol [22,38,41,44,45], and AutoDock is a more traditional molecular docking approach with physics-based force fields [30]. Including other molecular docking pipelines beyond the default pipeline implemented in the platform enabled us to evaluate whether or not CANDO as a platform was specifically dependent on the drug-proteome signature generation methodology implemented in the default pipeline.

Our results demonstrated that the CANDO platform was not dependent on a single pipeline implementation, and that combining different virtual screening pipelines can yield better performance relative to using the individual ones. On a platform level, the drug-proteome signature ranking and indication recovery paradigm was viable using more than one means of signature generation. On a pipeline level, the pipelines (the two large-scale virtual screening pipelines and the combined decision tree pipeline) each demonstrated varying degrees of performance and instances of unique signal capture.

The Vina-134 pipeline implemented in this study was viable in that it performed substantially better than the random control and performed at a significant fraction of the performance of the original default pipeline that utilized a bigger protein set. However, small protein libraries have been shown to perform relatively well, and some subsets of protein libraries performed better than others [44]. As previously demonstrated [43], hybrid pipelines can draw from the strengths of each constituent pipeline. As is the case here, the absolute performance of the Vina-134 pipeline was not the best, yet it substantially contributed to the higher performing hybrid pipeline.

3.2. Limitations and Future Work

Although some pipelines yielded superior signal over others in specific circumstances, precisely identifying why this occurred warrants further investigation. On a per-indication basis, it is possible to identify the superior performance of one pipeline over another (Figures 3 and 4), but the MeSH indication classes were not precisely defined or had varying levels of specificity to one another. This issue will be addressed in the future through the use of more precisely defined indication mapping, for instance by using a realism-based ontology [48,49]. We are also using mathematical, statistical, and machine learning techniques to rigorously evaluate and enhance CANDO's pipeline performance, as well as to identify clusters of drug-indication pair rankings when comparing different pipelines and methods [43,50], to yield insight into the ability of each pipeline to accurately recover known per-indication association information and make useful predictions for downstream prospective preclinical and clinical validation [42,51].

4. Materials and Methods

4.1. CANDO Platform and Pipeline Implementation

Figure 1 provides an overview of the CANDO platform and the particular pipeline implementations relevant to this study. The platform uses drug/compound and protein structure libraries curated from public sources and implements protocols for drug- and compound-proteome interaction signature generation, signature similarity calculation and sorting, assessing whether known drugs are ranked highly for the correct indications for single or hybrid pipelines (benchmarking), and generating novel putative drug candidates for specific indications (prediction). CANDO's drug ranking pipelines have utility in many repurposing research contexts. For example, these pipelines can be used for lead generation for subsequent in vitro/in vivo testing and eventual off-label clinical use by physicians. By assessing the top ranking subset of drugs, a researcher or clinician can efficiently infer promising experimental or clinical drug candidates based on drugs ranked

relatively to FDA-approved drug treatments and prior experimental evidence. For many clinical indications, CANDO pipelines are able to identify and highly rank FDA-approved drug treatments along with drugs that are FDA approved for other indications. Researchers can also infer associations between clinical indication classes, diseases, and biological pathways through the examination of indication-indication association networks connected by highly ranked drugs they have in common or other features of their respective compound-proteome signature. As illustrative examples of the broad uses of CANDO, Supplementary Figure S2 and Supplementary Table S1 describe the indication-indication associations for a selection of MeSH neoplasm indications based on shared drugs ranked in the top10 in the Vina pipeline.

4.1.1. Drug/Compound, Protein Structure, and Indication Library Curation

The default CANDO pipelines were implemented using bio-/chem-informatic docking protocols, where interactions were predicted from curated drug and protein libraries. The specific implementations and evolution of the libraries were reported extensively in several publications [22,38,41,44,45]. Briefly, the initial versions of CANDO (v1 and v1.5) incorporated 46,784 proteins and 2030 indication associations for 1439 drugs (out of 3733 compounds in total). Much of the data were drawn from the Protein Data Bank [52], the Food and Drug Administration, PubChem [53], the Comparative Toxicogenomics Database [54], DrugBank [55], protein structure modeling [56], and other sources.

The pipelines used in this study relied on curated sublibraries of the structures of 3733 drugs/compounds and 134 proteins and 13,746 drug-indication mappings, obtained from the same sources as above. We used the sublibraries to rapidly evaluate the utility of multiple molecular docking pipelines.

4.1.2. Drug- and Compound-Proteome Interaction Signature Generation

A CANDO virtual screening pipeline simulates the interactions between all of its proteins and drugs/compounds, usually 3D structures, and is not dependent on any particular approach to accomplish this. These simulations generate proteomic similarity signatures (the vector of drug-protein interaction scores). The default CANDO platform pipelines generate drug-proteome interaction signatures using bioinformatic and cheminformatic docking protocols also described elsewhere extensively [22,38,41,44,45]. These signatures were compared for similarity and ranked. CANDO pipeline Version 1.5 [45] is a refinement of the original default pipeline [22,38,41,44] that uses near identical libraries, but improved interaction scoring [45]. We extended the drug- and compound-proteome interaction signature protocols to include the calculated binding energies generated by the program AutoDock Vina [30], as well as created hybrid pipelines combining molecular docking with bioanalytical/similarity docking (further details below).

4.1.3. Drug- and Compound-Proteome Signature Similarity Calculation and Sorting

Broadly, the CANDO platform works by sorting every drug/compound relative to every other one based on their similarity and then uses known drug-indication associations to assess performance (Figure 2). Various pipelines implemented in CANDO generate drug-proteome interaction signatures for similarity sorting [22,38,43–45]. Underlying this platform is the core assumption that the similarity of drug interaction behavior across a proteome may be used to infer similarity in therapeutic function. The similarity between each drug and every other drug/compound is calculated using the root mean squared deviation of the individual interaction scores across a pair of drug-proteome interaction signatures [38].

Combined drug-proteome interaction signatures form an interaction matrix, with drugs along one axis and proteins on the other. These signatures are compared with one another and then ranked on a per-drug basis, and the quality of the resulting ranking is evaluated using the leave-one-out benchmarking protocol described below.

4.2. Benchmarking CANDO Platform Pipelines

Our benchmarking protocol calculates the performance for every indication with at least two approved drugs (1439 out of 3733 total) at various cutoffs, considering only the the top10 (abbreviated as "top10"), top25, top37 or 1%, top50, top100, top5%, top10%, and top50% of similarly ranked drugs. For each indication, the accuracy was derived from calculating how many known drugs mapped to that indication were "recovered" and highly ranked at various cutoffs.

We utilized three metrics to benchmark pipeline performance: average indication accuracy, pairwise accuracy, and coverage [22,38,41,44,45], all assessed at the different cutoffs. The average (mean) indication accuracy (%) is the average of all individual indication accuracies. The individual indication accuracy metric was calculated using the formula $c/d \times 100$, where c is a count of the number of times at least one approved drug for the indication was recovered within a particular cutoff and d is the total number of drugs approved for that indication. The other two benchmarking metrics were pairwise accuracy (weighted average of indication accuracies using the total number of approved drugs per indication) and coverage (number of indications that have an accuracy greater than zero).

4.3. New and Hybrid Pipelines

The pipelines examined in this study were derived from similarity ranking and benchmarking drug-proteome interaction signatures generated by large-scale bioinformatic and molecular docking. The CANDO platform is not limited to using docking-based virtual screening pipelines and has the potential to incorporate many different approaches to pipeline implementation and data sets (for example, ligand centric approaches have proven quite effective [43]).

4.3.1. Virtual Screening Pipeline Using Autodock Vina

We used a small sublibrary (134 proteins) of the full CANDO proteome library to create the new molecular docking virtual screening pipeline due to computational constraints and also because we previously have shown that appropriately selected sublibraries of a similar size from the full library yield similar or better benchmarking performance [44]. We used the popular software AutoDock Vina Version 1.1.2 [30] for molecular docking of each protein structure against 3733 drugs/compounds from the CANDO v1.5 libraries. As with BANDOCK, we used COFACTOR [57] to predict binding sites, for binding search space size optimization [58], and used the strongest interaction score (lowest calculated binding energy) for each simulation from multiple sites. The best interaction score values for a drug-protein pair were used to generate the drug-proteome signatures.

4.3.2. Decision Tree Pipeline

Prior CANDO platform investigations have demonstrated that multiple pipelines can be combined into a hybrid decision tree to maximize indication accuracy by drawing from pipelines that produce the best performance on a per-indication basis [43]. We used a similar approach in this investigation, using the pipeline that had the highest performance at the top10 cutoff.

4.4. Controls

We also compared the benchmarking performance of the pipelines to values obtained using a hypergeometric distribution that estimates the numerical probability of making a correct prediction by chance. This is one of the random control reference benchmarks used in the CANDO platform, the implementation of which was covered in detail in prior publications [43,45]. Benchmarking performance was also compared to the default pipeline implementations in CANDO Version 1 and Version 1.5 using the complete libraries.

5. Conclusions

Our results indicated that the utilization of multiple diverse docking-based virtual screening approaches in drug repurposing contexts such as the CANDO platform improves benchmarking performance. The Vina-134 pipeline performance indicated that the CANDO platform hypothesis of drug behavior similarity is not limited to the original bionalytical or similarity docking protocol BANDOCK for interaction signature generation. The hybrid decision tree pipeline performance provided further evidence that multiple signature generation pipelines may be combined to yield improved performance. Ongoing and future platform enhancement will incorporate multiple signature generation protocols and pipeline synthesis using AI/machine learning approaches to optimize performance. These improvements in turn will lead to greater predictive power and higher confidence in novel drug candidates generated for specific indications, which will be verified via prospective preclinical and clinical studies.

Supplementary Materials: The following are available online at http://compbio.org/data/mc_cando_multiscale_optimization/ accessed on 24 April 2021, Supplementary Figure S1: Plot of every drug-indication pair and their corresponding rank within each pipeline. Supplementary Figure S2: Indication-indication associations between MeSH neoplasm-associated classes based on the number of compounds predicted in the top10 cutoff by the Vina pipeline. Supplementary Table S1: Raw indication-indication association counts.

Author Contributions: M.L.H. conceived the research design, approach and methods, conducted all experiments and analysis, and drafted the manuscript; R.S. conceived the research design, approach and methods, supervised the activities of M.L.H., and edited the manuscript. All authors have read and agreed to the published version of the manuscript.

Funding: This work was supported in part by a National Institutes of Health Director's Pioneer Award 279 (DP1OD006779), a National Institutes of Health Clinical and Translational Sciences Award (UL1TR001412), a NCATS ASPIRE Design Challenge Award, and startup funds from the Department of Biomedical Informatics at the University at Buffalo.

Institutional Review Board Statement: Not applicable.

Informed Consent Statement: Not applicable.

Data Availability Statement: Publicly available datasets were analyzed in this study. This data can be found here: https://github.com/ram-compbio/CANDO and http://compbio.org/data/ both accessed on April 27, 2021.

Acknowledgments: Additional support provided by the Center for Computational Research at the University at Buffalo. The authors thank James Schuler, Will Mangione, Zackary Falls, Liana Bruggemann, and Manoj Mammen for their valuable input during the development of this manuscript.

Conflicts of Interest: The authors declare no conflicts of interest.

Abbreviations

The following abbreviations are used in this manuscript:

CANDO	Computational Analysis of Novel Drug Opportunities
FDA	Food and Drug Administration
NMR	Nuclear Magnetic Resonance

References

1. Lichtenberg, F.R. *Pharmaceutical Innovation, Mortality Reduction, and Economic Growth*; Technical Report; National Bureau of Economic Research: Cambridge, MA, USA, 1998.
2. FDA, U.S. New Drug Development & Approval Process. fda.gov/drugs accessed April 24, 2021.
3. DiMasi, J.A. New drug development in the United States from 1963 to 1999. *Clin. Pharmacol. Ther.* **2001**, *69*, 286–296.
4. DiMasi, J.A.; Hansen, R.W.; Grabowski, H.G. The price of innovation: New estimates of drug development costs. *J. Health Econ.* **2003**, *22*, 151–185.

5. DiMasi, J.A.; Grabowski, H.G.; Hansen, R.W. Innovation in the pharmaceutical industry: New estimates of R&D costs. *J. Health Econ.* **2016**, *47*, 20–33.
6. Scannell, J.W.; Blanckley, A.; Boldon, H.; Warrington, B. Diagnosing the decline in pharmaceutical R&D efficiency. *Nat. Rev. Drug Discov.* **2012**, *11*, 191.
7. Broach, J.R.; Thorner, J. High-throughput screening for drug discovery. *Nature* **1996**, *384*, 14–16.
8. Michelini, E.; Cevenini, L.; Mezzanotte, L.; Coppa, A.; Roda, A. Cell-based assays: Fuelling drug discovery. *Anal. Bioanal. Chem.* **2010**, *398*, 227–238.
9. Macalino, S.J.Y.; Gosu, V.; Hong, S.; Choi, S. Role of computer-aided drug design in modern drug discovery. *Arch. Pharmacal Res.* **2015**, *38*, 1686–1701.
10. Lionta, E.; Spyrou, G.; K Vassilatis, D.; Cournia, Z. Structure-based virtual screening for drug discovery: Principles, applications and recent advances. *Curr. Top. Med. Chem.* **2014**, *14*, 1923–1938.
11. Patel, C.N.; Georrge, J.J.; Modi, K.M.; Narechania, M.B.; Patel, D.P.; Gonzalez, F.J.; Pandya, H.A. Pharmacophore-based virtual screening of catechol-o-methyltransferase (COMT) inhibitors to combat Alzheimer's disease. *J. Biomol. Struct. Dyn.* **2018**, *36*, 3938–3957.
12. Medina-Franco, J.L.; Giulianotti, M.A.; Welmaker, G.S.; Houghten, R.A. Shifting from the single to the multitarget paradigm in drug discovery. *Drug Discov. Today* **2013**, *18*, 495–501.
13. L Bolognesi, M. Polypharmacology in a single drug: Multitarget drugs. *Curr. Med. Chem.* **2013**, *20*, 1639–1645.
14. Hu, Y.; Bajorath, J. Monitoring drug promiscuity over time. *F1000Research* **2014**, *3*, 218.
15. Ashburn, T.T.; Thor, K.B. Drug repositioning: Identifying and developing new uses for existing drugs. *Nat. Rev. Drug Discov.* **2004**, *3*, 673–683.
16. Langedijk, J.; Mantel-Teeuwisse, A.K.; Slijkerman, D.S.; Schutjens, M.H.D. Drug repositioning and repurposing: Terminology and definitions in literature. *Drug Discov. Today* **2015**, *20*, 1027–1034.
17. Palumbo, A.; Facon, T.; Sonneveld, P.; Blade, J.; Offidani, M.; Gay, F.; Moreau, P.; Waage, A.; Spencer, A.; Ludwig, H.; et al. Thalidomide for treatment of multiple myeloma: 10 years later. *Blood J. Am. Soc. Hematol.* **2008**, *111*, 3968–3977.
18. Roth, B.L.; Sheffler, D.J.; Kroeze, W.K. Magic shotguns versus magic bullets: Selectively non-selective drugs for mood disorders and schizophrenia. *Nat. Rev. Drug Discov.* **2004**, *3*, 353–359.
19. de Lera, A.R.; Ganesan, A. Epigenetic polypharmacology: From combination therapy to multitargeted drugs. *Clin. Epigenetics* **2016**, *8*, 105.
20. Arts, E.J.; Hazuda, D.J. HIV-1 antiretroviral drug therapy. *Cold Spring Harb. Perspect. Med.* **2012**, *2*, a007161.
21. Sardana, D.; Zhu, C.; Zhang, M.; Gudivada, R.C.; Yang, L.; Jegga, A.G. Drug repositioning for orphan diseases. *Brief. Bioinform.* **2011**, *12*, 346–356.
22. Minie, M.; Chopra, G.; Sethi, G.; Horst, J.; White, G.; Roy, A.; Hatti, K.; Samudrala, R. CANDO and the infinite drug discovery frontier. *Drug Discov. Today* **2014**, *19*, 1353–1363.
23. Mangione, W.; Falls, Z.; Chopra, G.; Samudrala, R. Cando. py: Open source software for analyzing large scale drug-protein-disease data. *J. Chem. Inf. Model.* **2020**, *60*, 4131–4136.
24. Xu, M.; Lee, E.M.; Wen, Z.; Cheng, Y.; Huang, W.K.; Qian, X.; Julia, T.; Kouznetsova, J.; Ogden, S.C.; Hammack, C.; et al. Identification of small-molecule inhibitors of Zika virus infection and induced neural cell death via a drug repurposing screen. *Nat. Med.* **2016**, *22*, 1101.
25. Schuler, J.; Hudson, M.L.; Schwartz, D.; Samudrala, R. A systematic review of computational drug discovery, development, and repurposing for Ebola virus disease treatment. *Molecules* **2017**, *22*, 1777.
26. Roder, C.; Thomson, M.J. Auranofin: Repurposing an old drug for a golden new age. *Drugs R D* **2015**, *15*, 13–20.
27. Ou-Yang, S.S.; Lu, J.Y.; Kong, X.Q.; Liang, Z.J.; Luo, C.; Jiang, H. Computational drug discovery. *Acta Pharmacol. Sin.* **2012**, *33*, 1131–1140.
28. Taylor, R.D.; Jewsbury, P.J.; Essex, J.W. A review of protein-small molecule docking methods. *J. Comput. Aided Mol. Des.* **2002**, *16*, 151–166.
29. Pagadala, N.S.; Syed, K.; Tuszynski, J. Software for molecular docking: A review. *Biophys. Rev.* **2017**, *9*, 91–102.
30. Trott, O.; Olson, A.J. AutoDock Vina: Improving the speed and accuracy of docking with a new scoring function, efficient optimization, and multithreading. *J. Comput. Chem.* **2010**, *31*, 455–461.
31. Fine, J.; Konc, J.; Samudrala, R.; Chopra, G. CANDOCK: Chemical atomic network based hierarchical flexible docking algorithm using generalized statistical potentials. *J. Chem. Inf. Model.* **2020**, *60*, 1509–1527.
32. Yuriev, E.; Ramsland, P.A. Latest developments in molecular docking: 2010–2011 in review. *J. Mol. Recognit.* **2013**, *26*, 215–239.
33. Huang, N.; Jacobson, M.P. Physics-based methods for studying protein-ligand interactions. *Curr. Opin. Drug Discov. Dev.* **2007**, *10*, 325.
34. Gohlke, H.; Klebe, G. Statistical potentials and scoring functions applied to protein–ligand binding. *Curr. Opin. Struct. Biol.* **2001**, *11*, 231–235.
35. Muegge, I. A knowledge-based scoring function for protein-ligand interactions: Probing the reference state. *Perspect. Drug Discov. Des.* **2000**, *20*, 99–114.
36. Lokhande, K.B.; Nagar, S.; Swamy, K.V. Molecular interaction studies of Deguelin and its derivatives with Cyclin D1 and Cyclin E in cancer cell signaling pathway: The computational approach. *Sci. Rep.* **2019**, *9*, 1–13.

37. Ma, D.L.; Chan, D.S.H.; Leung, C.H. Molecular docking for virtual screening of natural product databases. *Chem. Sci.* **2011**, *2*, 1656–1665.

38. Sethi, G.; Chopra, G.; Samudrala, R. Multiscale modeling of relationships between protein classes and drug behavior across all diseases using the CANDO platform. *Mini Rev. Med. Chem.* **2015**, *15*, 705–717.

39. Rodrigues, J.; Melquiond, A.; Karaca, E.; Trellet, M.; Van Dijk, M.; Van Zundert, G.; Schmitz, C.; De Vries, S.; Bordogna, A.; Bonati, L.; et al. Defining the limits of homology modeling in information-driven protein docking. *Proteins Struct. Funct. Bioinform.* **2013**, *81*, 2119–2128.

40. Yuriev, E.; Agostino, M.; Ramsland, P.A. Challenges and advances in computational docking: 2009 in review. *J. Mol. Recognit.* **2011**, *24*, 149–164.

41. Chopra, G.; Samudrala, R. Exploring polypharmacology in drug discovery and repurposing using the CANDO platform. *Curr. Pharm. Des.* **2016**, *22*, 3109–3123.

42. Chopra, G.; Kaushik, S.; Elkin, P.L.; Samudrala, R. Combating ebola with repurposed therapeutics using the CANDO platform. *Molecules* **2016**, *21*, 1537.

43. Schuler, J.; Samudrala, R. Fingerprinting CANDO: Increased Accuracy with Structure-and Ligand-Based Shotgun Drug Repurposing. *ACS Omega* **2019**, *4*, 17393–17403.

44. Mangione, W.; Samudrala, R. Identifying protein features responsible for improved drug repurposing accuracies using the CANDO platform: Implications for drug design. *Molecules* **2019**, *24*, 167.

45. Falls, Z.; Mangione, W.; Schuler, J.; Samudrala, R. Exploration of interaction scoring criteria in the CANDO platform. *BMC Res. Notes* **2019**, *12*, 318.

46. Cavasotto, C.N.; Phatak, S.S. Homology modeling in drug discovery: Current trends and applications. *Drug Discov. Today* **2009**, *14*, 676–683.

47. Lipscomb, C.E. Medical subject headings (MeSH). *Bull. Med Libr. Assoc.* **2000**, *88*, 265.

48. Arp, R.; Smith, B.; Spear, A.D. *Building Ontologies with Basic Formal Ontology*; Mit Press: Cambridge, MA, USA, 2015.

49. Schuler, J.; Mangione, W.; Samudrala, R.; Ceusters, W. Foundations for a Realism-based Drug Repurposing Ontology. In Proceedings of the 10th Annual International Conference on Biomedical Ontology, 17 September 2020.

50. Schuler, J.; Falls, Z.; Mangione, W.; Hudson, M.; Bruggemann, L.; Samdurala, R. Evaluating performance of drug repurposing technologies. *Drug Discov. Today* **2020**, *in press.*

51. Mangione, W.; Falls, Z.; Melendy, T.; Chopra, G.; Samudrala, R. Shotgun drug repurposing biotechnology to tackle epidemics and pandemics. *Drug Discov. Today* **2020**, *25*, 1126–1128.

52. Berman, H.M.; Westbrook, J.; Feng, Z.; Gilliland, G.; Bhat, T.N.; Weissig, H.; Shindyalov, I.N.; Bourne, P.E. The protein data bank. *Nucleic Acids Res.* **2000**, *28*, 235–242.

53. Kim, S.; Thiessen, P.A.; Bolton, E.E.; Chen, J.; Fu, G.; Gindulyte, A.; Han, L.; He, J.; He, S.; Shoemaker, B.A.; et al. PubChem substance and compound databases. *Nucleic Acids Res.* **2016**, *44*, D1202–D1213.

54. Davis, A.P.; Murphy, C.G.; Johnson, R.; Lay, J.M.; Lennon-Hopkins, K.; Saraceni-Richards, C.; Sciaky, D.; King, B.L.; Rosenstein, M.C.; Wiegers, T.C.; et al. The comparative toxicogenomics database: Update 2013. *Nucleic Acids Res.* **2012**, *41*, D1104–D1114.

55. Knox, C.; Law, V.; Jewison, T.; Liu, P.; Ly, S.; Frolkis, A.; Pon, A.; Banco, K.; Mak, C.; Neveu, V.; et al. DrugBank 3.0: A comprehensive resource for 'omics' research on drugs. *Nucleic Acids Res.* **2010**, *39*, D1035–D1041.

56. Zhang, Y. I-TASSER server for protein 3D structure prediction. *BMC Bioinform.* **2008**, *9*, 40.

57. Roy, A.; Yang, J.; Zhang, Y. COFACTOR: An accurate comparative algorithm for structure-based protein function annotation. *Nucleic Acids Res.* **2012**, *40*, W471–W477.

58. Feinstein, W.P.; Brylinski, M. Calculating an optimal box size for ligand docking and virtual screening against experimental and predicted binding pockets. *J. Cheminformatics* **2015**, *7*, 1–10.

 *molecules*

Article

The Performance of Gene Expression Signature-Guided Drug–Disease Association in Different Categories of Drugs and Diseases

Xiguang Qi [1], Mingzhe Shen [1], Peihao Fan [1], Xiaojiang Guo [1], Tianqi Wang [2], Ning Feng [3], Manling Zhang [3], Robert A. Sweet [4,5,*], Levent Kirisci [1,*] and Lirong Wang [1,*]

[1] Department of Pharmaceutical Sciences, Computational Chemical Genomics Screening Center, University of Pittsburgh School of Pharmacy, 3501 Terrace St Pittsburgh, PA 15261, USA; Xiq24@pitt.edu (X.Q.); MIS216@pitt.edu (M.S.); pef14@pitt.edu (P.F.); xig53@pitt.edu (X.G.)

[2] Department of Biological Sciences, University of Pittsburgh School of Arts & Sciences, Pittsburgh, PA 15260, USA; Lu_Wang@pitt.edu

[3] Division of Cardiology, Vascular Medicine Institute, University of Pittsburgh School of Medicine, Pittsburgh, PA 15213, USA; FENGN@pitt.edu (N.F.); zhangm5@upmc.edu (M.Z.)

[4] Department of Neurology, University of Pittsburgh School of Medicine, Pittsburgh, PA 15213, USA

[5] Department of Psychiatry, University of Pittsburgh School of Medicine, Pittsburgh, PA 15213, USA

* Correspondence: sweetra@upmc.edu (R.A.S.); levent@pitt.edu (L.K.); Liw30@pitt.edu (L.W.); Tel.: +1-412-383-6089 (R.A.S.); +1-412-624-8118 (L.K.)

Academic Editors: Marco Tutone and Anna Maria Almerico
Received: 21 April 2020; Accepted: 5 June 2020; Published: 16 June 2020

Abstract: A gene expression signature (GES) is a group of genes that shows a unique expression profile as a result of perturbations by drugs, genetic modification or diseases on the transcriptional machinery. The comparisons between GES profiles have been used to investigate the relationships between drugs, their targets and diseases with quite a few successful cases reported. Especially in the study of GES-guided drugs–disease associations, researchers believe that if a GES induced by a drug is opposite to a GES induced by a disease, the drug may have potential as a treatment of that disease. In this study, we data-mined the crowd extracted expression of differential signatures (CREEDS) database to evaluate the similarity between GES profiles from drugs and their indicated diseases. Our study aims to explore the application domains of GES-guided drug–disease associations through the analysis of the similarity of GES profiles on known pairs of drug–disease associations, thereby identifying subgroups of drugs/diseases that are suitable for GES-guided drug repositioning approaches. Our results supported our hypothesis that the GES-guided drug–disease association method is better suited for some subgroups or pathways such as drugs and diseases associated with the immune system, diseases of the nervous system, non-chemotherapy drugs or the mTOR signaling pathway.

Keywords: gene expression signature; drug repositioning approaches; RNA expression regulation

---

1. Introduction

A gene expression signature (GES) is a set of comprehensive gene expression profiles that can reveal the difference between stimulated and normal cell states [1]. Current applications of GES analysis are fruitful in cancer-related areas for disease genotype classification and outcome predictions. For example, Ramaswamy, S. et al. created a GES database for diagnosing and categorizing the tumour type with an accuracy rate of 78% [2]. Wright, G. et al. developed a Bayesian rule-based algorithm to classify diffuse large B cell lymphoma into two subgroups which have a significant difference in the five-year survival rate [3]. Although the GES method is more commonly used in diagnosing

cancer [2–5] and predicting the outcome of certain medical interventions [6,7], some successful cases of application on drug development have also reported [8–10].

Generally, there are two major strategies for applying GES analysis on drug development: drug–drug-based and drug–disease-based. The drug–drug-based method determines the mechanistic actions of drugs by comparing the similarity between the GES induced by a drug of interest to those of drugs with known mechanisms. If two different drugs have similar GES profiles, then they are considered to have "functional similarity", meaning they work in a similar manner. In contrast, the drug–disease-based method compares the similarity between the GES of a drug to that of a disease in order to determine its potential as a new therapeutic agent. If the GES profile of the drug is opposite to opposite to the expression pattern of the disease, then the drug is considered to have a therapeutic effect for the disease. However, if they have similar patterns, then this drug may exacerbate the disease. Studies aimed at drug repurposing or repositioning based on GES analysis usually use one or both of these strategies [8–10]. In addition, there are studies trying to combine the GES method with other methods, such as machine learning, to increase the accuracy of compound indication prediction [11]. However, as those kinds of GES-guided drug repurposing studies usually just reported the successful predicted cases, therefore, the true accuracy of these methods needs to be assessed.

Due to the different and complex mechanisms of disease processes, the idea of an "inverse pattern of a GES between drugs and diseases for therapeutic effect" may not hold, or at least may not be suitable for all categories of drugs and diseases. In other words, a GES may be useful for certain diseases, but not for others. To our knowledge, the application domains of GES-guided drug–disease associations have not been reported. Herein, we conducted a study to validate the power of the GES-guided drug repositioning method and to further explore which specific subgroups of drug–disease pairs are more suitable for this method. Moreover, the most significant subgroup was selected as a case report of detailingwhich genes and/or pathways were more sensitive to the GES-guided drug repositioning method.

2. Results

2.1. GES Profiles Enrollment and Drug–Disease Pairs

After removing signatures from non-human assays and signatures of non-FDA (the U.S. Food and Drug Administration)-approved drugs, we found that GSE10432, GSE7036, GSE6264, GSE38713, GSE31773, GSE11393, GSE8157, GSE13887 and GSE11223 were signatures of both drugs and diseases from the same assays. We kept their disease labels except the CREEDS (crowd extracted expression of differential signatures [12]) ID of dz:297 because this case had information mis-specified (wrong disease information with its original experiment). Two GES profiles from mouse (drug:3288 and dz:724) were mis-specified as human and were also excluded from analysis. The relationship between these Gene Expression Omnibus (GEO) series (GSE) and CREEDS IDs is shown in Table 1. The proportion of data that meets the inclusion criteria is shown in Figure 1.

Table 1. The Gene Expression Omnibus (GEO) series with crowd extracted expression of differential signatures (CREEDS) IDs excluded.

GEO Series	CREEDS IDs	Excluded CREEDS IDs
GSE10432	drug:2772, dz:297	dz:297
GSE7036	drug:3292, dz:181	drug:3292
GSE6264	drug:3064, dz:582	drug:3064
GSE38713	drug:3289, drug:3194, drug:3195, dz:810	drug:3289, drug:3194, drug:3195
GSE31773	drug:2485, dz:712, dz:713, dz:714, dz:715	drug:2485
GSE11393	drug:3401, drug:3196, dz:773, dz:267	drug:3401, drug:3196
GSE8157	drug:2796, dz:880	drug:2796
GSE13887	drug:3181, dz:450	drug:3181,
GSE11223	drug:3294, drug:3287, dz:590, dz:591, dz:593, dz:589, dz:588, dz:587, dz:586, dz:585	drug:3294, drug:3287
GSE7762	drug:3288	drug:3288
GSE3248	dz:724	dz:724

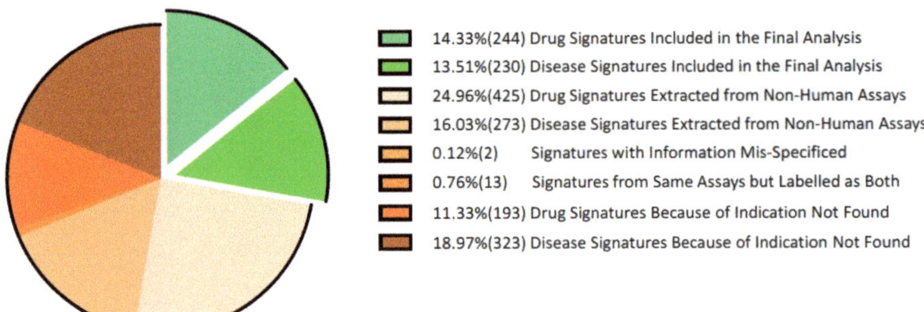

Figure 1. The proportion of data sourced from the crowd extracted expression of differential signatures (CREEDS) database. Numbers of gene signatures are shown in parentheses. "Drug and Disease Signatures Included in the Final Analysis": The proportion of drug or disease gene signatures enrolled in the final analysis. "Drug and Disease Signatures Extracted from Non-Human Assays": The proportion of drug or disease gene signatures extracted from non-human assays. "Signatures with Information Mis-Specified": The proportion of gene signatures with information errors. "Signatures from Same Assays but Labelled as Both": The proportion of gene signatures excluded because of both drug and disease sourcing from the same assay. "Drug and Disease Signatures Because of Indication Not Found": The proportion of gene signatures excluded because no FDA-labelled indication of a relationship was found for the drug or disease (including drugs not approved by FDA).

When the inclusion criteria were applied, and the signatures with no indication relationship were excluded, 230 manual disease signatures and 244 manual drug perturbation signatures from 71 unique diseases and 56 unique drugs, respectively, were enrolled in the final analysis. The average signed Jaccard indexes [12] (SJI) of 3976 unique drug–disease pairs were calculated. Among them, there were 167 pairs with a drug–disease indication from the drug labels. The remaining 3809 unique drug–disease pairs were used as the control group.

2.2. Subgroups Distribution

Among the 56 unique drugs analysed, 32 unique protein targets with 22 categories of Anatomical Therapeutic Chemical (ATC) classification were assigned. Thirteen drugs are classified as chemotherapy drugs, and 44 drugs are not (Methotrexate is both a chemotherapy and a non-chemotherapy drug due to its different main therapeutic targets when against different diseases). For transcription factor (TF) level, 12 drugs are labelled as "directly", 39 drugs are labelled as "not-directly" and 5 drugs

were labelled as "non-Human" (see section *4.4. subgroup classification* for the detailed meanings of labels). Further, 71 diseases are divided into 11 ICD-11 (International Classification of Diseases 11th Revision) categories. In total, 70 subgroups belonging to five categories were assigned (Figure 2, detailed information in Table S1 and Table S2).

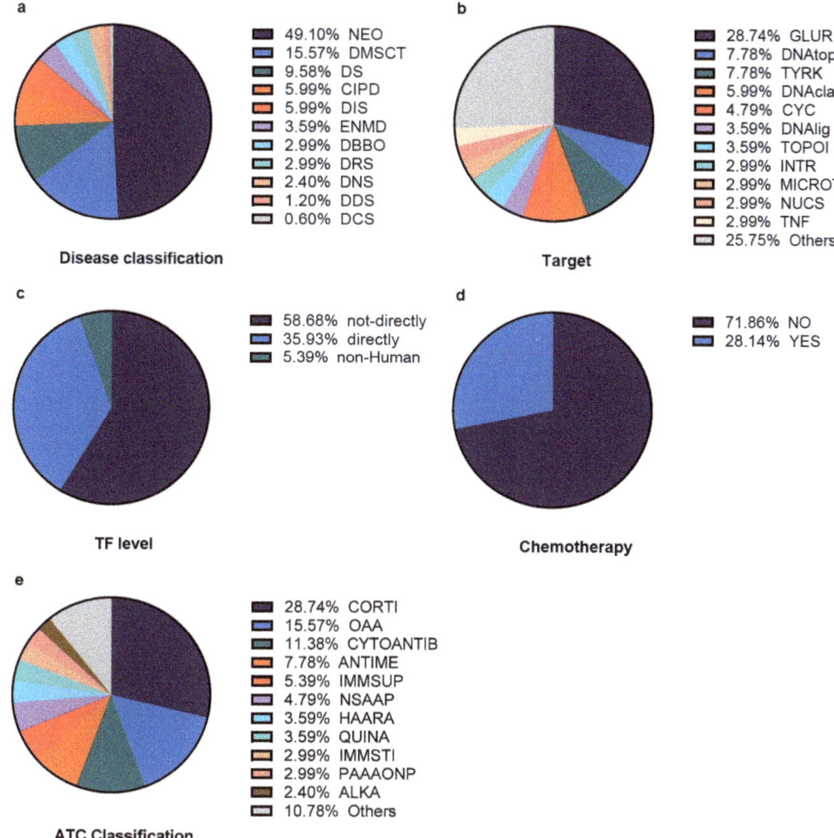

Figure 2. The subgroups proportion of unique 167 indicated drug–disease pairs of different categories. (**a**) Disease classification. NEO: neoplasms, DMSCT: diseases of the musculoskeletal system or connective tissue, DS: diseases of the skin, CIPD: certain infectious or parasitic diseases, DIS: diseases of the immune system, ENMD: endocrine, nutritional or metabolic diseases, DBBO: diseases of the blood or blood-forming organs, DRS: diseases of the respiratory system, DNS: diseases of the nervous system, DDS: diseases of the digestive system, DCS: diseases of the circulatory system. (**b**) Drug target. GLUR: glucocorticoid receptor, DNAtopo: DNA/topoisomerase-human, TYRK: tyrosine kinase, DNAclak: DNA cross-linking/alkylation, CYC: cyclooxygenase, DNAlig: DNA/ligase, TOPOI: topoisomerase-non-human, INTR: interferon receptor, MICROT: microtubules, NUCS: nucleotide synthesis, TNF: tumor necrosis factor. (**c**) TF (transcription factor) level. "Directly": drugs with TFs as its main therapeutic targets. "Not-directly" indicates drugs with main therapeutic targets which are human DNA structures or human proteins but not TFs. "Non-Human" represent drugs interacting

with protein or structures of non-human (for example, from virus or bacterial) as main therapeutic targets. (**d**) Chemotherapy. "YES" or "NO" indicates the drug is a chemotherapy drug or not. (**e**) ATC classification. CORTI: corticosteroids for systemic use, plain, OAA: other antineoplastic agents, CYTOANTIB: cytotoxic antibiotics and related substances, ANTIME: antimetabolites, IMMSUP: immunosuppressants, NSAAP: anti-inflammatory and antirheumatic products, non-steroids, HAARA: hormone antagonists and related agents, QUINA: quinolone antibacterial, IMMSTI: immunostimulants, PAAAONP: plant alkaloids and other natural products, ALKA: alkylating agents.

2.3. Overall Score of GES Similarity of Drug-Indicted Disease Pairs Against Random Drug–Disease Pairs

We observed significantly lower SJI similarity scores of drug–disease indication pairs than those of random drug–disease pairs (p-value of two-side t-test [13] equals to 0.02324). The average similarity score of indicated pairs is −0.00386 with a standard deviation of 0.01794 and that of random control pairs is −0.00072 with a standard deviation of 0.01750, indicating that the GES method can reflect the therapeutic effects of the drugs (The distributions of SJI in both the indication group and the control group are shown in Figure 3).

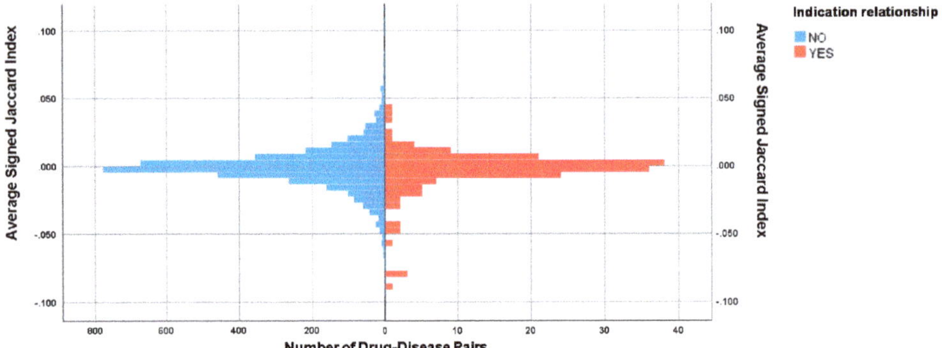

Figure 3. The distribution of signed Jaccard index in the indication group and the control group.

2.4. Subgroup Scores of GES Similarity of Drug-Indicated Disease Pairs Against Random Drug–Disease Pairs

We compared drugs from five different categories of subgroups: (1) disease classifications; (2) drug target; (3) TF level; (4) chemotherapy; and (5) ATC classification. The results are shown in Figure 4, detail information is listed in Table S3 and Table S4. Subgroups with important or significant (q-value according to false discover rate (FDR) lower than 0.05) results according to least squares mean partitions F tests of a generalized linear model (GLM) [14] are listed in Table 2.

Table 2. Subgroups of generalized linear model (GLM) least squares mean partitions F tests results.

Classification Category	Subgroups	Average SJI of Indicated Pairs ± SD	N	Average SJI of Control Pairs ± SD	N	Q value
Disease classification	Diseases of the blood or blood-forming organs	−0.02368 ± 0.03746	6	0.00075 ± 0.02470	138	0.01322
	Diseases of the nervous system	−0.03264 ± 0.03648	4	−0.00054 ± 0.01528	92	0.00704
Drug target classification	Interferon receptor	−0.02314 ± 0.03866	5	0.00916 ± 0.02849	115	0.00110
	Kinase mTOR	−0.05846 ± ———	1	0.00353 ± 0.01580	23	0.01755
Chemotherapy classification	Chemotherapy drugs	0.00048 ± 0.00894	47	−0.00022 ± 0.01221	1049	0.99509
	Non-chemotherapy drugs	−0.00556 ± 0.02026	120	−0.00086 ± 0.01872	2760	0.03937
ATC classification	Immunostimulants	−0.02314 ± 0.03866	5	0.00916 ± 0.02849	115	0.00110
	Other dermatological preparations	−0.05846 ± ———	1	−0.00353 ± 0.01580	23	0.01755
Transcription factor level	Directly	−0.00433 ± 0.02310	60	0.00070 ± 0.01671	1378	0.22309
	Not-directly	−0.00344 ± 0.01443	98	−0.00116 ± 0.01785	2224	0.99509
	Non-Human	−0.00533±0.01574	9	−0.00057 ± 0.01627	207	0.79080

Important subgroups or subgroups with false discover rate (FDR) q-value lower than 0.05 from GLM least squares mean partitions F tests for signed Jaccard index differences between drug-indicted disease pairs and random drug–disease pairs. "———" indicates that subgroups only have one unique drug–disease pair sample with no standard deviation.

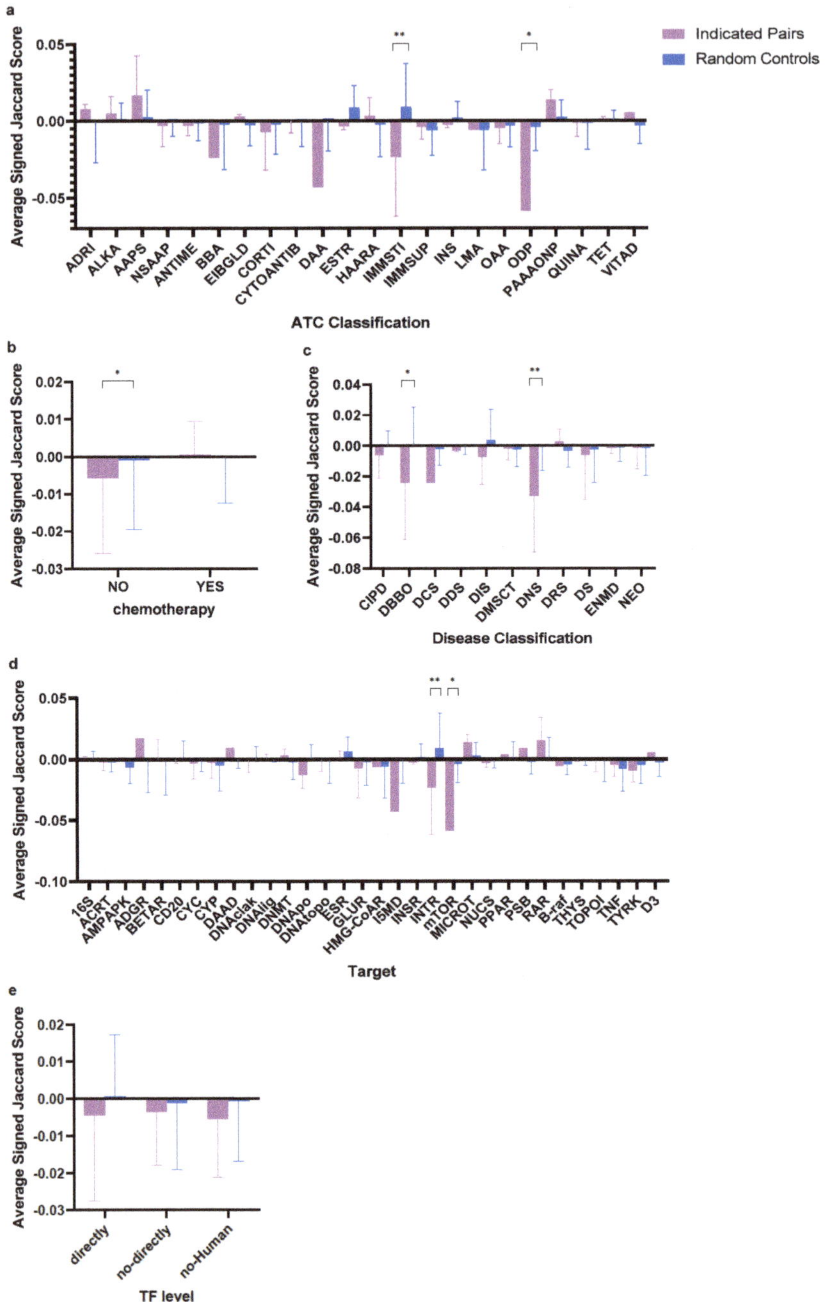

Figure 4. The average signed Jaccard index score of unique indicated drug–disease pairs split by different categories of subgroups. ** indicates FDR Q < 0.01, * indicates FDR Q < 0.05. (**a**) ATC classification. ADRI: adrenergics, inhalants, AAPS: anti-acne preparations for systemic use, EIBGLD:

blood glucose-lowering drugs, excluding insulins, DAA: direct acting antivirals, ESTR: estrogens, INS: insulins and analogues, LMA: lipid modifying agents, plain, ODP: other dermatological preparations, TET: tetracyclines, VITAD: vitamins A and D, including combinations of the two. CORTI, OAA, CYTOANTIB, ANTIME, IMMSUP, NSAAP, HAARA, QUINA, IMMSTI, PAAAONP, ALKA, see Figure 1 legend. (**b**) Chemotherapy. "YES" or "NO" indicates the drug is a chemotherapy drug or not. (**c**) Disease classification. See Figure 1 for abbreviations. (**d**) Target. 16S: 16S ribosomal RNA, ACRT: aminoimidazole caboxamide ribonucleotide transformylase, AMPAPK: AMP-activated protein kinase, ADGR: androgen receptor, BETAR: beta adrenergic receptor, CD20: CD20 antigen, CYP: cytochromes P450, DAAD: delta-aminolevulinic acid dehydratase, DNMT: DNA/methyltransferase, DNApo: DNA/polymerase, ESR: estrogen receptor, HMG-CoAR: HMG-CoA reductase, I5MD: inosine-5′-monophosphate dehydrogenase, INSR: insulin receptor, mTOR: kinase mTOR, PPAR: peroxisome proliferator-activated receptors, PSB: proteasome subunit beta, RAR: retinoic acid receptor, B-raf: serine/threonine-protein kinase B-raf, THYS: thymidylate synthase, D3: vitamin D3 receptor; GLUR, DNAtopo, TYRK, DNAclak, CYC, DNAlig, TOPOI, INTR, MICROT, NUCS, TNF see Figure 1 legend. (**e**) TF (transcription factor) level. "Directly": drugs with TFs as their main therapeutic targets. "Not-directly" indicates drugs with main therapeutic targets which are human DNA structures or human proteins but not TFs. "Non-Human" represents drugs interacting with non-human proteins or structures (for example, from viruses or bacteria) as main therapeutic targets.

2.5. Gene and Pathway Analysis on an Example Drug–Disease GES Pair

Interferon receptor (with the same drug–disease pair content as the immunostimulants subgroup), the subgroup with the lowest q-value, was chosen as a case report for the pathway analysis. The top 5% (93/1898) genes with a relatively reversed expression probability according to the relatively expression probability of a gene's ($G^{I-R\%}$, an indicator of the relative possibility difference of gene expression between the indicated group and the random control group, see below 4.5) scores are shown in Table 3. The top 10 significant biological pathways identified by the ingenuity pathway analysis are shown in Table 4.

Table 3. Top 5% genes with relatively expression probability ($G^{I-R\%}$).

Gene	$G^{I-R\%}$	Gene	$G^{I-R\%}$	Gene	$G^{I-R\%}$	Gene	$G^{I-R\%}$
MX1	−46.87%	FTL	−25.22%	USP18	−19.56%	DUSP6	−16.90%
IFIT3	−41.45%	RPL24	−25.18%	CERS2	−19.38%	TPT1	−16.66%
NME1	−40.50%	ERP29	−23.86%	RPLP0	−19.36%	RSAD2	−16.59%
RPL3	−39.19%	RSL24D1	−23.86%	KLRB1	−19.28%	ADAR	−16.48%
RPS5	−37.61%	PTMA	−23.65%	ADM	−19.23%	DDX58	−16.44%
RPL6	−36.57%	HLA-DRA	−22.88%	PLSCR1	−19.23%	APOBEC3A	−16.40%
MT1HL1	−35.52%	IFIT1	−22.22%	RPLP0P6	−19.14%	PPIB	−16.17%
MT2A	−34.80%	MX2	−22.22%	RPS3A	−19.07%	RGS2	−16.09%
RPSA	−33.55%	LDHB	−22.12%	TRIM22	−19.00%	IRF7	−16.08%
TGFBI	−33.47%	DYNLT1	−21.90%	DDX21	−18.66%	PSMA6	−16.00%
MT1X	−32.30%	ALDH1A1	−21.64%	GCH1	−18.64%	RPL9	−15.94%
HERC5	−32.15%	HSPA1A	−21.53%	GAPDH	−18.55%	OAS1	−15.91%
FAU	−31.82%	SLC25A5	−21.53%	OAS3	−18.48%	RPL31	−15.74%
PLS3	−29.66%	IFIT2	−21.38%	RPS25	−18.40%	PTTG1IP	−15.74%
HLA-A	−29.15%	RPS4X	−21.28%	NDUFB11	−18.40%	BIRC2	−15.74%
RPL22	−28.88%	EIF3E	−20.88%	SNHG6	−18.15%	MYD88	−15.67%
FBL	−28.52%	HMGN2	−20.88%	PSAT1	−18.06%	RPS14P3	−15.64%
RPS8	−27.57%	FTH1P5	−20.80%	IER2	−18.02%	FTH1	−15.62%
ISG15	−26.91%	YWHAZ	−20.72%	UXT	−17.65%	C4orf46	−15.45%
EEF1B2	−26.88%	PFDN5	−20.57%	PARP12	−17.58%	PPT1	−15.42%
PHB2	−26.48%	TMA7	−20.20%	MAFB	−17.40%	YBX1	−15.33%
MT1H	−26.29%	CCT7	−20.12%	LYZ	−17.25%		
RPL8	−26.11%	OASL	−19.89%	NARS	−17.15%		
ATF4	−25.36%	SNHG5	−19.64%	AKR1B1	−17.02%		

Table 4. Top 10 significant biological pathways according to high relatively expression probability genes.

Ingenuity Canonical Pathways	-log(p-value)	Ratio	Genes Overlapped with Datasets
EIF2 Signaling	16.50	8.02% (17/212)	ATF4, EIF3E, FAU, RPL22, RPL24, RPL3, RPL31, RPL6, RPL8, RPL9, RPLP0, RPS25, RPS3A, RPS4X, RPS5, RPS8, RPSA
Activation of IRF by Cytosolic Pattern Recognition Receptors	6.60	9.84% (6/61)	ADAR, DDX58, IFIT2, IRF7, ISG15, PPIB
Regulation of eIF4 and p70S6K Signaling	6.48	5.23% (8/153)	EIF3E, FAU, RPS25, RPS3A, RPS4X, RPS5, RPS8, RPSA
Interferon Signaling	6.34	13.90% (5/36)	IFIT1, IFIT3, ISG15, MX1, OAS1
mTOR Signaling	5.57	3.96% (8/202)	EIF3E, FAU, RPS25, RPS3A, RPS4X, RPS5, RPS8, RPSA
NRF2-mediated Oxidative Stress Response	3.80	3.23% (6/186)	ATF4, CCT7, ERP29, FTH1, FTL, PPIB
Role of Pattern Recognition Receptors in Recognition of Bacteria and Viruses	3.39	3.47% (5/144)	DDX58, IRF7, MYD88, OAS1, OAS3
Neuroinflammation Signaling Pathway	2.78	2.06% (6/291)	ATF4, BIRC2, HLA-A, HLA-DRA, IRF7, MYD88
SPINK1 General Cancer Pathway	2.63	4.92% (3/61)	MT1H, MT1X, MT2A
Systemic Lupus Erythematosus in B Cell Signaling Pathway	2.23	1.89% (5/265)	IFIT2, IFIT3, IRF7, ISG15, MYD88

These 10 pathways are reported to be involved with interferon regulation [15–27]. within inflammatory and immune responses (see Table 5).

Table 5. Top 10 pathways and their function labels.

Ingenuity Canonical Pathways	Function	Reference
EIF2 Signaling	Immune Responses	[28]
Activation of IRF by Cytosolic Pattern Recognition Receptors	Regulate Interferon	[17]
Regulation of eIF4 and p70S6K Signaling	Inflammatory	[18,29]
Interferon Signaling	Immune Responses	[30,31]
mTOR Signaling	Immune Responses	[19]
NRF2-mediated Oxidative Stress Response	Antioxidant Response	[21]
Role of Pattern Recognition Receptors in Recognition of Bacteria and Viruses	Regulate Interferon	[22]
Neuroinflammation Signaling Pathway	Inflammatory	[23]
SPINK1 General Cancer Pathway	Cancer Diagnose	[32]
Systemic Lupus Erythematosus in B Cell Signaling Pathway	Inflammatory	[33]

3. Discussion

It is well-recognized that genes with similar gene expression patterns have a similar function [34]. From the overall score, we can see that FDA-approved drugs listed in the CREEDS database and their indicated diseases generally have inverse GES patterns compared with the random controls. However, the absolute difference between the indicated group and random control group is not

very obvious. For example, in a recent study [35], a significant relationship was found between drug–disease GES similarity and drug therapeutic effect using Cmap [36], with a relatively low overall area under curve (AUC) of 0.57, indicating a real, albeit weak, inverse relationship. The treatment effectors of the drugs identified in this study likely work via the interaction of the genes' protein products, with only a moderate correlation between gene expression and levels of the corresponding protein(s) [37]. Thus, an association study between drugs/diseases and gene expression/pharmaceutical effect is necessary. Also, other mechanisms, for instance, microRNA-based therapeutics, might directly orchestrate the activation/deactivation of the gene expression. However, due to the limitation of available sources, we were unable to investigate other mechanisms of action. Besides, the drugs' TF-levels were not a significant factor that reflect the indication relationship (although drugs directly interacting with TF perform slightly better with q-value of 0.22309 vs. q-value of 0.99509). In our analyses, some subgroups of drugs–diseases pairs with indication associations have positive similarity scores (which means that the drug may exacerbate the disease according to the assumption of gene expression signature similarity) or a score higher than random drug–disease pairs, but these findings were not statistically significant. On the other hand, 7 of 70 subgroups had a significantly lower similarity score when a drug–disease association is indicated.

This study may provide some hints to other future studies utilizing the GES method strategies of comparing drug–disease GES similarity for drug repositioning. That is, certain types of drugs may have a stronger ability to reverse the GES of the diseases they treat, and the disease type may also influence this ability. As such, in specific kinds of subgroups, the drug–disease pairs with higher similarities of reversed GES patterns may have greater therapeutic relationships, which means that focusing on certain kinds of diseases or drugs can increase the true positive rate of the GES-guided drug repositioning method For example, over half (4/7) of the significant subgroups (immunostimulants, interferon receptor, other dermatological preparations, and diseases of the blood or blood-forming organs) are related to diseases associated with the immune system (the disease includes in "other dermatological preparations" atopic dermatitis). This indicates that a drug with drug–disease pairs associated with the immune system tends to have lower similarity scores when compared with the diseases it indicated than random diseases. This means in a GES-guided drug repositioning analysis, an immune-associated drug is more likely to have a potential therapeutic effect on diseases that have a higher inverse similarity with it.

Chemotherapy drugs may not be as good as non-chemotherapy drugs for the GES-guided drug repositioning method (q-values: 0.99509 vs. 0.03937). Unarguably, the high diversity of chemotherapy responses to heterogenetic tumor tissues or even histologically similar tumors has been a challenging problem for a long time [37,38]. The failure of controlling the process of programmed cell death in tissue, one of the major causes of tumors, can be rectified or even overturned by activating/deactivating different pathways under various conditions [39]. This may be the reason that chemotherapy drugs are not good for the GES-guided repositioning approach. On the other hand, non-chemotherapy drugs show a significant result as they interact with cancer cells through more specific mechanisms, such as hormone regulation or mono-target therapy.

For the biological pathway analysis of the interferon receptor subgroup, we found that the genes involved in pathways directly regulated by drugs have the lowest GI-R% scores. It is reasonable that GES-guided drug repositioning methods are more sensitive to drugs directly targeting pathways related to diseases. Furthermore, the significance of mTOR signaling is in accordance with the result in which the subgroup kinase mTOR had a significant indicated-random drug–disease pairs' SJI difference. This result confirmed the high sensitivity of the GES-guided drug repositioning method to this pathway on the other side.

There are some limitations in this study. First, the tissues used for testing the drug effects may not match with the body parts/organs affected by the diseases.. Second, some bias may be caused by the limited number of the CREEDS bio-assay collection which may not have the ability to fully present the pattern of all kinds of drugs and diseases. Additionally, it is important to differentiate

the types of "treatment effect". Some drugs may cure a particular disease while others may just provide symptomatic relief thereby resulting in different patterns of GES for the same disease. Also, some indicated subgroups ("kinase mTOR", and "other dermatological preparations") have too few unique drug–disease pairs (n = 1), which may weaken the analyses' power.

In this study, we systematically analyzed the similarity of gene expression profiles from known drug–disease associations, and we found that indicated pairs have a greater inverse similarity score. We found seven subgroups in which their drugs or diseases may have a greater reversed GES pattern when there is a clear therapeutic effect. These findings suggest that a GES-guided drug repositioning method should be used based on the drug or disease type differences. For example, drugs or diseases associated with the immune system, diseases of the nervous system or non-chemotherapy drugs may be a better choice for drug repositioning. Moreover, our biological pathway enrichment analysis showed that some pathways may be more sensitive to this method, such as the mTOR signaling pathway.

4. Materials and Methods

4.1. Gene Signature Data Collection and Filtering

In this study, all gene signature information was collected from a well-calibrated GES repository, the crowd extracted expression of differential signatures (CREEDS) [12] database. The CREEDS database is maintained by the Ma'ayan Lab of Icahn School of Medicine at Mount Sinai. CREEDS utilized GEO2Enrichr [40] to extract GES profiles from the National Center for Biotechnology Information (NCBI) Gene Expression Omnibus (GEO) and applied a characteristic direction (CD) model [41] to identify differentially expressed genes. This database V1.0 includes 10,797 single-gene perturbations, 2258 disease signatures and 5516 drug perturbation gene signatures. Among these signatures, 2176 manual single-gene perturbations, 828 manual disease signatures and 875 manual drug perturbation signatures were considered to be more accurate compared with the automatically generated GES by the machine learning method. The CREEDS database allows users to compare the similarity between the user-specified GES and the GESs processed and stored in the CREEDS.

We first selected the CREEDS manual GES profiles if the assays were from human tissues and/or human cell linesand if the drugs had FDA approval.

Each GES profile includes a list of up- and down-regulated genes. The SJI [12] (see below), a measurement for the similarity between two GES profiles from the paired drug–disease, was calculated. When a drug or a disease had multiple GES profiles, we calculated the SJIs of all the possible combinations, and an overall score for each unique drug–disease pair was calculated from the average of all scores from pairs sharing the same drug–disease combination. All the disease signatures and drug perturbation signatures were requested through the application program interface (API) provided by CREEDS. GES profiles were removed if they labelled for both a drug treatment and for a disease, because this may cause biased similarity. Under the criteria that (a) the GES profiles must come from assays of human cells/tissues, and (b) drugs must be approved by FDA, the remaining signatures were paired within drugs and diseases according to the indication associations. Signatures without any indicated drug–disease relationship were also excluded from further analysis. For example, cocaine was removed because its indication, local anesthesia, was not in the data of disease signatures and could not be paired. The overall data process is shown in Figure 5.

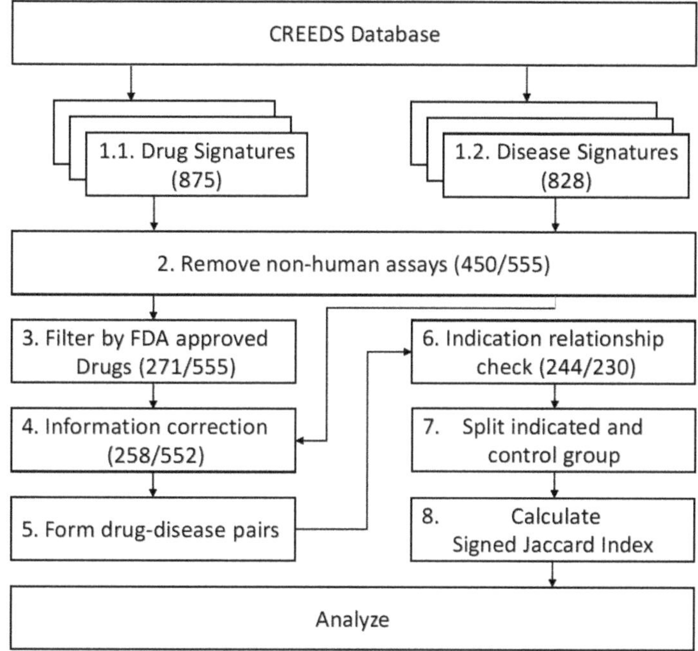

Figure 5. The flow chart of drug and disease gene signature data inclusion process. Numbers of gene signatures left in each step are shown in parentheses: (Number of drug signatures/Number of disease signatures) 1.1. and 1.2. All manual gene signatures retrieved from the CREEDS database. 2. Remove all signatures with assays not labelled as human. 3. Remove all drug signatures not from FDA-approved drugs. 4. Remove signatures with information errors or signatures labelled as both for a drug treatment and for a disease. 5. Remaining drug signatures were paired with each disease signature. 6. Remove signatures with no FDA-labelled indication relationships of drug or disease. 7. Indicated group and control group were divided according to the indication relationship from the FDA drug label. 8. Calculate the signed Jaccard index for each remaining drug–disease pair.

4.2. Similarity Calculation

In our analysis, SJI, which is based on the Jaccard similarity coefficient [42], was used to compute the similarity between GES profiles from a drug and a disease. The Jaccard similarity coefficient is a statistic used to gauge the similarity between different sample sets. It is defined as the size of the intersection divided by the size of the union of two sample sets. It is calculated as follows:

$$Jaccard\ Similarity\ Coefficient(G_1, G_2) = \frac{SAME}{ALL}$$

where G_1 and G_2 stand for two lists of differential expressed gene sets, "SAME" represents the number of same genes between two given gene sets, and "ALL" stands for all the unique genes that appeared in the two gene sets.

SJI, which combines the Jaccard similarity coefficient with the gene regulation direction is calculated as follows:

$$\textbf{Signed Jaccard index}(G_1, G_2) = \frac{J\left(G_1^{up}, G_2^{up}\right) + J\left(G_1^{down}, G_2^{down}\right) - J\left(G_1^{up}, G_2^{down}\right) - J\left(G_1^{down}, G_2^{up}\right)}{2}$$

where J means Jaccard similarity coefficient, and G^{up} and G^{down} are up- or down-regulated genes in the given gene set G, respectively. The index ranges from +1 to −1, where +1 and −1 indicate a same pattern and inverse pattern of two gene sets, respectively. Zero indicates that the two sets have no associations, or the same part is cancelled out by the inverse part. The reason to use an unranked score calculation method (SJI) is to keep in accordance with the same scoring method used in the CREEDS database. The CREEDS API (application programming interface) offers the function to calculate the SJI automatically. However, we found the API could not calculate the SJI correctly when two GES profiles are highly overlapped., therefore, all the SJIs in this study were re-calculated.

4.3. Drug-Related Information Collection

In our analysis, the source of drug-related information is listed as follows:

1. Drug target information was collected from DrugBank [43,44] Release Version 5.1.4 [45] (https://www.drugbank.ca/releases/latest#external-links). Only the targets with the main therapeutic effect in the mechanism of action section were included;

2. The human TF list was collected from the paper published by Samuel A. Lambert et al. [46];

3. ATC classifications on level 3 were collected from the WHO official website (https://www.whocc.no/atc_ddd_index/);

4. The drug indication was from section "indications and usage" of FDA label on FDA website (https://labels.fda.gov/);

5. (Drug-indicated) Disease classification was assigned to each disease based on the International Classification of Diseases 11th Revision (ICD-11), level 1.

4.4. Subgroup Classification

In our analysis, we assessed the following factors that might influence the power of the GES-guided drug repositioning method:

1. Disease classifications: A subgroup was assigned to a disease in a drug–disease pair according to the ICD-11-level 1 code of the disease;

2. Drug target subfamilies: Subgroups were divided by the main therapeutic target of each drug. To avoid group splits being too small, some same subfamilies of targets are grouped as one, such as "Beta-1 adrenergic receptor", "Beta-2 adrenergic receptor" and "Beta-3 adrenergic receptor" are grouped in the same subgroup "Beta adrenergic receptors";

3. The relationship between the drug's main therapeutic targets and human transcription factors: A TF level was assigned according to the relationship between the drugs' main therapeutic targets and human TF. Drugs with main therapeutic targets that can directly interact with at least one TF were labelled as "directly". Drugs with main therapeutic targets which are human DNA structures or human proteins but not TFs were labelled as "not-directly". Drugs interacting with non-human proteins or structures (for example, from viruses or bacteria) as main therapeutic targets were labelled as "non-Human";

4. The drug is a chemotherapy drug or not: Drugs with main therapeutic targets as "DNA cross-linking/alkylation", "DNA/ligase", "DNA/methyltransferase", "DNA/polymerase", "DNA/topoisomerase-human", "micro-tubules", "nucleotide synthesis" or "Thymidylate synthase" were defined as chemotherapy drugs

5. The drug's ATC classification: Subgroups were divided according to the Anatomical Therapeutic Chemical classification system, level 3. Drugs with multiple classifications caused by different administration routes were unified to systematic use.

4.5. Statistical Analysis and Pathway Analysis

The random control group was generated by calculating the average SJI of all possible drug–disease pairs without indicated associations to imitate a GES-guided drug repositioning screening. A t-test [13]

was applied to quantify the mean differences of the SJI between drug-indicated disease pairs and random controls.

For subgroup analysis, GLM [14] least squares mean partitions F tests function was applied to estimate the mean difference between the indicated and control group since the data was unbalanced with multiple factors. A significant result of a certain subgroup indicated that the average SJI of this subgroup was significant between two indication levels (Yes/No). False discovery rate (FDR) q-value of the Benjamini–Hochberg procedure [47] was controlled to 0.05 to avoid an inflated experiment-wise type I error rate caused by multiple comparisons among all subgroups.

Data processing and statistical analysis (student t-tests, GLM, FDR calculation) were conducted using R studio 3.6.1 [48] and SAS software version 9.4. Copyright © 2019 SAS Institute Inc. Cary, NC, USA.

Differentially reversed expression genes (top 5% negative score according to the relatively reverse percentage) from the most significant subgroup will be chosen as examples to conduct biological pathway enrichment analysis.

The relatively reverse percentage is calculated as

$$Relatively\ expression\ probability\ of\ a\ gene\big(G^{I-R\%}\big) = D^{I}\% - D^{R}\%$$

where $D^{I}\%$ **and** $D^{R}\%$ stand for the percentage of the gene which is differentially expressed in all assays of indicated/random drug–diseases pairs. It is calculated as

$$D\% = \frac{NS - NR}{Total\ assays\ pairs}$$

where **NS** and **NR** represent the number of times a gene showed a same or reverse regulation direction between assays of drugs and diseases among all drug–disease assays pairs.

The $G^{I\text{-}R\%}$ ranges from 100% to -100%. A higher positive score indicates that this gene is more likely to be expressed in the same direction in indicated drug–disease assays compared with random drug–disease assays. Likewise, a lower negative score indicates that this gene has a higher probability to express reversely between indicated drug–disease assays compared with random drug–disease assays.

Biological pathway enrichment analysis was conducted by ingenuity pathway analysis (IPA, QIAGEN Inc., https://www.qiagenbioinformatics.com/products/ingenuitypathway-analysis).

Supplementary Materials: The following are available online at http://www.mdpi.com/1420-3049/25/12/2776/s1, Table S1: 70 subgroups with four drug categories, Table S2: 70 subgroups with disease category, Table S3: Indicated drug–disease pair results, Table S4: Random drug–disease pair results.

Author Contributions: Conceptualization, X.Q., M.S. and L.W.; Data Curation, X.Q. and M.S.; Formal Analysis, X.Q. and P.F.; Funding Acquisition, R.A.S. and L.W.; Investigation, X.Q.; Methodology, X.Q. and L.K.; Project Administration, L.W.; Resources, L.W.; Software, X.Q.; Supervision, L.W.; Validation, X.G. and T.W.; Visualization, X.Q; Writing—original draft, X.Q.; Writing—review and editing, N.F., M.Z., R.A.S. and L.W. All authors have read and agreed to the published version of the manuscript.

Funding: This research was funded by the National Institutes of Health [grants AG027224, MH116046, AG005133, PDA035778A]. The APC was funded by the National Institutes of Health [grant MH116046].

Acknowledgments: The authors would like to acknowledge Zefei Li at School of Pharmacy in Sun Yat-sen University for revising the calculation process.

Conflicts of Interest: Dr. Sweet 's work has been funded by NIH [grants AG027224, MH116046, AG005133]. Dr. Wang's work has been funded by NIH [grants MH116046]. Other authors declare no competing interests.

References

1. Alizadeh, A.A.; Eisen, M.B.; Davis, R.E.; Ma, C.; Lossos, I.S.; Rosenwald, A.; Boldrick, J.C.; Sabet, H.; Tran, T.; Yu, X.; et al. Distinct Types of Diffuse Large B-Cell Lymphoma Identified by Gene Expression Profiling. *Nature* **2000**, *403*, 503–511. [CrossRef]

2. Ramaswamy, S.P.; Tamayo, R.; Rifkin, S.; Mukherjee, C.H.; Yeang, M.; Angelo, C.; Ladd, M.; Reich, E.; Latulippe, J.P.; Mesirov, T.; et al. Multiclass Cancer Diagnosis Using Tumor Gene Expression Signatures. *Proc. Natl. Acad. Sci. USA* **2001**, *98*, 15149–15154. [CrossRef]

3. Wright, G.; Tan, B.; Rosenwald, A.; Hurt, E.H.; Wiestner, A.; Staudt, L.M. A Gene Expression-Based Method to Diagnose Clinically Distinct Subgroups of Diffuse Large B Cell Lymphoma. *Proc. Natl. Acad. Sci. USA* **2003**, *100*, 9991–9996. [CrossRef]

4. Yap, Y.L.; Zhang, X.W.; Smith, D.; Soong, R.; Hill, J. Molecular Gene Expression Signature Patterns for Gastric Cancer Diagnosis. *Comput. Biol. Chem.* **2007**, *31*, 275–287. [CrossRef]

5. Ziober, A.F.; Patel, K.R.; Alawi, F.; Gimotty, P.; Weber, R.S.; Feldman, M.M.; Chalian, A.A.; Weinstein, G.S.; Hunt, J.; Ziober, B.L. Identification of a Gene Signature for Rapid Screening of Oral Squamous Cell Carcinoma. *Clin. Cancer Res.* **2006**, *12 Pt 1*, 5960–5971. [CrossRef]

6. Chibon, F. Cancer Gene Expression Signatures—The Rise and Fall? *Eur. J. Cancer* **2000**, *49*, 2000–2009. [CrossRef] [PubMed]

7. Chen, H.Y.; Yu, S.L.; Chen, C.H.; Chang, G.C.; Chen, C.Y.; Yuan, A.; Cheng, C.L.; Wang, C.H.; Terng, H.J.; Kao, S.F.; et al. A Five-Gene Signature and Clinical Outcome in Non-Small-Cell Lung Cancer. *N. Engl. J. Med.* **2007**, *356*, 11–20. [CrossRef] [PubMed]

8. Iorio, F.; Bosotti, R.; Scacheri, E.; Belcastro, V.; Mithbaokar, P.; Ferriero, R.; Murino, L.; Tagliaferri, R.; Brunetti-Pierri, N.; Isacchi, A.; et al. Discovery of Drug Mode of Action and Drug Repositioning from Transcriptional Responses. *Proc. Natl. Acad. Sci. USA* **2010**, *107*, 14621–14626. [CrossRef] [PubMed]

9. Sirota, M.; Dudley, J.T.; Kim, J.; Chiang, A.P.; Morgan, A.A.; Sweet-Cordero, A.; Sage, J.; Butte, A.J. Discovery and Preclinical Validation of Drug Indications Using Compendia of Public Gene Expression Data. *Sci. Transl. Med.* **2011**, *3*, 96ra77. [CrossRef]

10. Hu, G.; Agarwal, P. Human Disease-Drug Network Based on Genomic Expression Profiles. *PLoS ONE* **2009**, *4*, e6536. [CrossRef]

11. Donner, Y.; Kazmierczak, S.; Fortney, K. Drug Repurposing Using Deep Embeddings of Gene Expression Profiles. *Mol. Pharm.* **2018**, *15*, 4314–4325. [CrossRef] [PubMed]

12. Wang, Z.; Monteiro, C.D.; Jagodnik, K.M.; Fernandez, N.F.; Gundersen, G.W.; Rouillard, A.D.; Jenkins, S.L.; Feldmann, A.S.; Hu, K.S.; McDermott, M.G.; et al. Extraction and Analysis of Signatures from the Gene Expression Omnibus by the Crowd. *Nat. Commun.* **2016**, *7*, 12846. [CrossRef] [PubMed]

13. Student. The Probable Error of a Mean. *Biometrika* **1908**, *6*, 1–25. [CrossRef]

14. Nelder, J.A.; Wedderburn, R.W. Generalized Linear Models. *J. R. Stat. Soc. Ser. A-Gen.* **1972**, *135*, 370. [CrossRef]

15. Bustamante, M.F.; Nurtdinov, R.N.; Río, J.; Montalban, X.; Comabella, M. Baseline Gene Expression Signatures in Monocytes from Multiple Sclerosis Patients Treated with Interferon-Beta. *PLoS ONE* **2013**, *8*, e60994. [CrossRef] [PubMed]

16. Bibeau-Poirier, A.; Servant, M.J. Roles of Ubiquitination in Pattern-Recognition Receptors and Type I Interferon Receptor Signaling. *Cytokine Servant* **2008**, *43*, 359–367. [CrossRef]

17. Honda, K.; Taniguchi, T. Irfs: Master Regulators of Signalling by Toll-Like Receptors and Cytosolic Pattern-Recognition Receptors. *Nat. Rev. Immunol. Taniguchi* **2006**, *6*, 644–658. [CrossRef]

18. Su, X.; Yu, Y.; Zhong, Y.; Giannopoulou, E.G.; Hu, X.; Liu, H.; Cross, J.R.; Rätsch, G.; Rice, C.M.; Ivashkiv, L.B. Interferon-Γ Regulates Cellular Metabolism and Mrna Translation to Potentiate Macrophage Activation. *Nat. Immunol. Ivashkiv* **2015**, *16*, 838. [CrossRef]

19. Cao, W.; Manicassamy, S.; Tang, H.; Kasturi, S.P.; Pirani, A.; Murthy, N.; Pulendran, B. Toll-Like Receptor–Mediated Induction of Type I Interferon in Plasmacytoid Dendritic Cells Requires the Rapamycin-Sensitive Pi (3) K-Mtor-P70s6k Pathway. *Nat. Immunol. Pulendran* **2008**, *9*, 1157. [CrossRef]

20. Weinstein, S.L.; Finn, A.J.; Davé, S.H.; Meng, F.; Lowell, C.A.; Sanghera, J.S.; DeFranco, A.L. Phosphatidylinositol 3-Kinase and Mtor Mediate Lipopolysaccharide-Stimulated Nitric Oxide Production in Macrophages Via Interferon-B. *J. Leukoc. Biol. DeFranco* **2000**, *67*, 405–414. [CrossRef]

21. Staitieh, B.S.; Egea, E.E.; Fan, X.; Azih, N.; Neveu, W.; Guidot, D.M. Activation of Alveolar Macrophages with Interferon-Γ Promotes Antioxidant Defenses Via the Nrf2-Are Pathway. *J. Clin. Guidot Cellular Immunol.* **2015**, *6*, 365.

22. Perry, A.K.; Gang, C.H.; Zheng, D.; Hong, T.A.; Cheng, G. Cell research Cheng. The Host Type I Interferon Response to Viral and Bacterial Infections. *Cell Res.* **2005**, *15*, 407–422. [CrossRef]

23. Vieweg, W.V.; Hasnain, M.; Howland, R.H.; Hettema, J.M.; Kogut, C.; Wood, M.A.; Pandurangi, A.K. Citalopram, Qtc Interval Prolongation, and Torsade De Pointes. How Should We Apply the Recent Fda Ruling? *Am. J. Med.* **2012**, *125*, 859–868. [CrossRef] [PubMed]

24. Honda, K.; Yanai, H.; Negishi, H.; Asagiri, M.; Sato, M.; Mizutani, T.; Shimada, N.; Ohba, Y.; Takaoka, A.; Yoshida, N.; et al. Irf-7 Is the Master Regulator of Type-I Interferon-Dependent Immune Responses. *Nature* **2005**, *434*, 772–777. [CrossRef] [PubMed]

25. Hilpert, J.; Beekman, J.M.; Schwenke, S.; Kowal, K.; Bauer, D.; Lampe, J.; Sandbrink, R.; Heubach, J.F.; Stürzebecher, S.; Reischl, J. Biological Response Genes after Single Dose Administration of Interferon B-1b to Healthy Male Volunteers. *J. Neuroimmunol.* **2008**, *199*, 115–125. [CrossRef]

26. Becker, A.M.; Dao, K.H.; Han, B.K.; Kornu, R.; Lakhanpal, S.; Mobley, A.B.; Li, Q.Z.; Lian, Y.; Wu, T.; Reimold, A.M.; et al. Sle Peripheral Blood B Cell, T Cell and Myeloid Cell Transcriptomes Display Unique Profiles and Each Subset Contributes to the Interferon Signature. *PLoS ONE* **2013**, *8*, e67003. [CrossRef]

27. Crow, M.K.; Kirou, K.A.; Wohlgemuth, J. Microarray Analysis of Interferon-Regulated Genes in Sle. *Autoimmunity* **2003**, *36*, 481–490. [CrossRef]

28. Shrestha, N.; Bahnan, W.; Wiley, D.J.; Barber, G.; Fields, K.A.; Schesser, K. Eukaryotic Initiation Factor 2 (Eif2) Signaling Regulates Proinflammatory Cytokine Expression and Bacterial Invasion. *J. Biol. Chem.* **2012**, *287*, 28738–28744. [CrossRef]

29. Flynn, A.; Proud, C.G. The Role of Eif4 in Cell Proliferation. *Cancer Surv.* **1996**, *27*, 293–310.

30. Isaacs, A. Interferon. In *Advances in Virus Research*; Elsevier: Cambridge, MA, USA, 1964; pp. 1–38.

31. Sarasin-Filipowicz, M.; Oakeley, E.J.; Duong, F.H.; Christen, V.; Terracciano, L.; Filipowicz, W.; Heim, M.H. Interferon Signaling and Treatment Outcome in Chronic Hepatitis C. *Proc. Natl. Acad. Sci. USA* **2008**, *105*, 7034–7039. [CrossRef]

32. Flavin, R.; Pettersson, A.; Hendrickson, W.K.; Fiorentino, M.; Finn, S.; Kunz, L.; Judson, G.L.; Lis, R.; Bailey, D.; Fiore, C.; et al. Fiore Spink1 Protein Expression and Prostate Cancer Progression. *Clin. Cancer Res.* **2014**, *20*, 4904–4911. [CrossRef] [PubMed]

33. Smoller, J.W. The Genetics of Stress-Related Disorders: Ptsd, Depression, and Anxiety Disorders. *Neuropsychopharmacology* **2016**, *41*, 297. [CrossRef] [PubMed]

34. Le Roch, K.G.; Zhou, Y.; Blair, P.L.; Grainger, M.; Moch, J.K.; Haynes, J.D.; De la Vega, P.; Holder, A.A.; Batalov, S.; Carucci, D.J.; et al. Carucci, and Elizabeth, A. Winzeler. Discovery of Gene Function by Expression Profiling of the Malaria Parasite Life Cycle. *Science* **2003**, *301*, 1503. [CrossRef] [PubMed]

35. Cheng, J.; Yang, L.; Kumar, V.; Agarwal, P. Systematic Evaluation of Connectivity Map for Disease Indications. *Genome Med.* **2014**, *6*, 540. [CrossRef]

36. Lamb, J.; Crawford, E.D.; Peck, D.; Modell, J.W.; Blat, I.C.; Wrobel, M.J.; Lerner, J.; Brunet, J.P.; Subramanian, A.; Ross, K.N.; et al. The Connectivity Map: Using Gene-Expression Signatures to Connect Small Molecules, Genes, and Disease. *Science* **2006**, *313*, 1929–1935. [CrossRef] [PubMed]

37. Talmadge, J.E.; Benedict, K.; Madsen, J.; Fidler, I.J. Development of Biological Diversity and Susceptibility to Chemotherapy in Murine Cancer Metastases. *Cancer Res.* **1984**, *44*, 3801–3805. [PubMed]

38. Simpson-Herren, L.; Noker, P.E.; Wagoner, S.D. Variability of Tumor Response to Chemotherapy Ii. Contribution of Tumor Heterogeneity. *Cancer Chemother. Pharmacol.* **1988**, *22*, 131–136. [CrossRef] [PubMed]

39. Zimmermann, K.C.; Bonzon, C.; Green, D.R. The Machinery of Programmed Cell Death. *Pharmacol. Ther.* **2001**, *92*, 57–70. [CrossRef]

40. Gundersen, G.W.; Jones, M.R.; Rouillard, A.D.; Kou, Y.; Monteiro, C.D.; Feldmann, A.S.; Hu, K.S.; Ma'ayan, A. Geo2enrichr: Browser Extension and Server App to Extract Gene Sets from Geo and Analyze Them for Biological Functions. *Bioinformatics* **2015**, *31*, 3060–3062. [CrossRef]

41. Clark, N.R.; Hu, K.S.; Feldmann, A.S.; Kou, Y.; Chen, E.Y.; Duan, Q.; Ma'ayan, A. The Characteristic Direction: A Geometrical Approach to Identify Differentially Expressed Genes. *BMC Bioinformatics* **2014**, *15*, 79. [CrossRef]

42. Jaccard, P. Nouvelles Recherches Sur La Distribution Florale. *Bull. Soc. Vaud. Sci. Nat.* **1908**, *44*, 223–270.

43. Wishart, D.S.; Knox, C.; Guo, A.C.; Shrivastava, S.; Hassanali, M.; Stothard, P.; Chang, Z.; Woolsey, J. Drugbank: A Comprehensive Resource for in Silico Drug Discovery and Exploration. *Nucleic Acids Res.* **2006**, *34*, D668–D672. [CrossRef] [PubMed]

44. Wishart, D.S.; Knox, C.; Guo, A.C.; Cheng, D.; Shrivastava, S.; Tzur, D.; Gautam, B.; Hassanali, M. Drugbank: A Knowledgebase for Drugs, Drug Actions and Drug Targets. *Nucleic Acids Res.* **2008**, *36*, D901–D906. [CrossRef] [PubMed]

45. Wishart, D.S.; Feunang, Y.D.; Guo, A.C.; Lo, E.J.; Marcu, A.; Grant, J.R.; Sajed, T.; Johnson, D.; Li, C.; Sayeeda, Z.; et al. Drugbank 5.0: A Major Update to the Drugbank Database for 2018. *Nucleic Acids Res.* **2017**, *46*, D1074–D1082. [CrossRef] [PubMed]

46. Lambert, S.A.; Jolma, A.; Campitelli, L.F.; Das, P.K.; Yin, Y.; Albu, M.; Chen, X.; Taipale, J.; Hughes, T.R.; Weirauch, M.T. The Human Transcription Factors. *Cell* **2018**, *172*, 650–665. [CrossRef]

47. Benjamini, Y.; Hochberg, Y. Controlling the False Discovery Rate: A Practical and Powerful Approach to Multiple Testing. *J. R. Stat. Soc.* **1995**, *57*, 289–300. [CrossRef]

48. *Team, R Core R: A Language and Environment for Statistical Computing*; R Foundation for Statistical Computing: Vienna, Austria, 2013.

Sample Availability: Samples of the compounds ...... are available from the authors.

Article

Is PF-00835231 a Pan-SARS-CoV-2 Mpro Inhibitor? A Comparative Study

Mohammad Hassan Baig [1], Tanuj Sharma [1], Irfan Ahmad [2], Mohammed Abohashrh [3], Mohammad Mahtab Alam [3] and Jae-June Dong [1,*]

1. Department of Family Medicine, Yonsei University College of Medicine, 50-1 Yonsei-ro, Seodaemun-gu, Seoul 120-752, Korea; mohdhassanbaig@gmail.com (M.H.B.); tanush84@gmail.com (T.S.)
2. Department of Clinical Laboratory Sciences, College of Applied Medical Sciences, King Khalid University, Abha 61421, Saudi Arabia; irfancsmmu@gmail.com
3. Department of Basic Medical Sciences, College of Applied Medical Sciences, King Khalid University, Abha 61421, Saudi Arabia; mabuhashra@kku.edu.sa (M.A.); mmalam@kku.edu.sa (M.M.A.)
* Correspondence: s82tonight@yuhs.ac

Abstract: The COVID-19 outbreak continues to spread worldwide at a rapid rate. Currently, the absence of any effective antiviral treatment is the major concern for the global population. The reports of the occurrence of various point mutations within the important therapeutic target protein of SARS-CoV-2 has elevated the problem. The SARS-CoV-2 main protease (Mpro) is a major therapeutic target for new antiviral designs. In this study, the efficacy of PF-00835231 was investigated (a Mpro inhibitor under clinical trials) against the Mpro and their reported mutants. Various in silico approaches were used to investigate and compare the efficacy of PF-00835231 and five drugs previously documented to inhibit the Mpro. Our study shows that PF-00835231 is not only effective against the wild type but demonstrates a high affinity against the studied mutants as well.

Keywords: SARS-CoV-2; main protease; mutants; inhibitors; PF-00835231

Citation: Baig, M.H.; Sharma, T.; Ahmad, I.; Abohashrh, M.; Alam, M.M.; Dong, J.-J. Is PF-00835231 a Pan-SARS-CoV-2 Mpro Inhibitor? A Comparative Study. *Molecules* **2021**, 26, 1678. https://doi.org/10.3390/molecules26061678

Academic Editors: Marco Tutone and Anna Maria Almerico

Received: 3 February 2021
Accepted: 4 March 2021
Published: 17 March 2021

Publisher's Note: MDPI stays neutral with regard to jurisdictional claims in published maps and institutional affiliations.

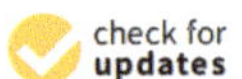

1. Introduction

SARS-CoV-2, the etiological agent of COVID-19, is a pandemic responsible for claiming over a million human lives [1]. More than a thousand drugs are currently in the COVID-19 treatment pipeline; most of which are in the discovery stage and many of these are existing treatments for other conditions currently being evaluated for SARS-CoV-2 [2]. The latest statistics available from ClinicalTrials.gov, a directory of clinical trials funded by the US National Library of Medicine, reveal that 4371 studies are registered worldwide and are growing day by day. Twenty-five trials, including using the most common drug, Hydroxychloroquine, were discontinued [3]. To date only two studies are found in the category of 'Approved for marketing', which are expanded access to convalescent plasma and the drug molecule Remdesivir [4,5]. As an urgent consequence of the novelty of SARS-CoV-2 infections and the lack of appropriate drugs, a wide range of techniques and methods are being used to tackle the emerging worldwide COVID-19 pandemic. To date, the therapies suggested are mainly the repurposing of existing drugs chosen for the similarity of their initial indication such as antivirals/antiretrovirals or for the similarity of their mode of action [6,7].

Pp1a and pp1ab polyproteins are formed by SARS-CoV-2 and are processed by two virally encoded cysteine proteases; the main protease (Mpro) and the papain-like protease [8]. In the viral replication process, it is apparent that the action of the Mpro is vital as it processes the p1a/p1ab polyprotein virus proteolytically at more than 10 junctions to produce a set of non-structural proteins (NSPs) essential for virus replication and transcription including RdRp, helicase and the Mpro itself [9]. Of all recognized forms of coronaviruses, the Mpro is the most explored target for drug development because it has almost the same

mechanism and active site as MERS-CoV (Middle East Respiratory Syndrome Coronavirus) and SARS-CoV [10,11].

Despite the lower mutation rate of the virus, studies have revealed more than 12,000 SARS-CoV-2 genome mutations. Many mutations would not impact the capacity of the virus to transmit or trigger illness because they do not alter the structure of a protein and certain mutations that change proteins are often more likely to damage the virus than to strengthen it [12]. A few vaccines have recently been approved and more than 50 are in different phases of trials. Structural protein mutations that are attacked by the host immune system can impede vaccine efficacy and non-structural protein mutations can develop strains that are resistant to antivirals. Different strains of the virus that are more transmittable than the wild type SARS-CoV-2 have recently been identified in South Africa. In the United Kingdom and several other nations including Europe and Brazil, the extensive spreading of coronavirus variants has placed the world on alert and sparked a new lockdown [13,14]. It picks up minor changes to the genetic code every time the SARS-CoV-2 moves from person to person but researchers are beginning to find variations of how the virus mutates. A major research investment is being made towards the development of new therapeutics or repurposing old drugs as weapons against COVID-19.

Pfizer has started clinical trials (phase I) with a small molecule PF-07304814 that targets the Mpro of SARS-CoV-2 [15]. It may prove to be the first antiviral drug to target this protein (Mpro) to combat COVID-19. PF-07304814 comprises a phosphate group that renders the compound soluble and cleaves the active antiviral PF-00835231 by alkaline phosphatase enzymes in the tissue [15].

Here in this study, we evaluate the binding efficacy of six known inhibitors of the Mpro (Figure 1). The binding efficacy of these inhibitors was measured against the WT and the mutant Mpro (Figure 2). Five drugs (bedaquiline, boceprevir, efonidipine, manidipine and lercanidipine) have been earlier reported to inhibit Mpro activity to below 40 μM [11]. The sixth inhibitor, PF-00835231, is a powerful inhibitor of the SARS-CoV-2 Mpro with sufficient medicinal properties to merit further research as an intravenous COVID-19 therapy [16]. During the 2002–2003 SARS epidemic, PF-00835231 was first discovered by Pfizer chemists to target the SARS-CoV Mpro [16]. Infections petered out, however, and the compound was put on hold along with a collection of other possible coronavirus antivirals. In addition to demonstrating action against two strains of SARS-CoV-2, PF-00835231 was able to kill other coronaviruses in cells as well [15]. PF-00835231 is the active form of PF-07304814 currently being tested (phase 1) in patients with a SARS-CoV-2 infection and mild to moderate symptoms [3,15].

This study involves a state of the art computational evaluation to assess the comparative efficacy of PF-00835231 and other reported inhibitors against the wild type and four reported Mpro mutants. We hypothesize here that PF-00835231 might be less competitive against various SARS-CoV-2 virus mutants; even the results of experimental and clinical studies are still to offer clearer results.

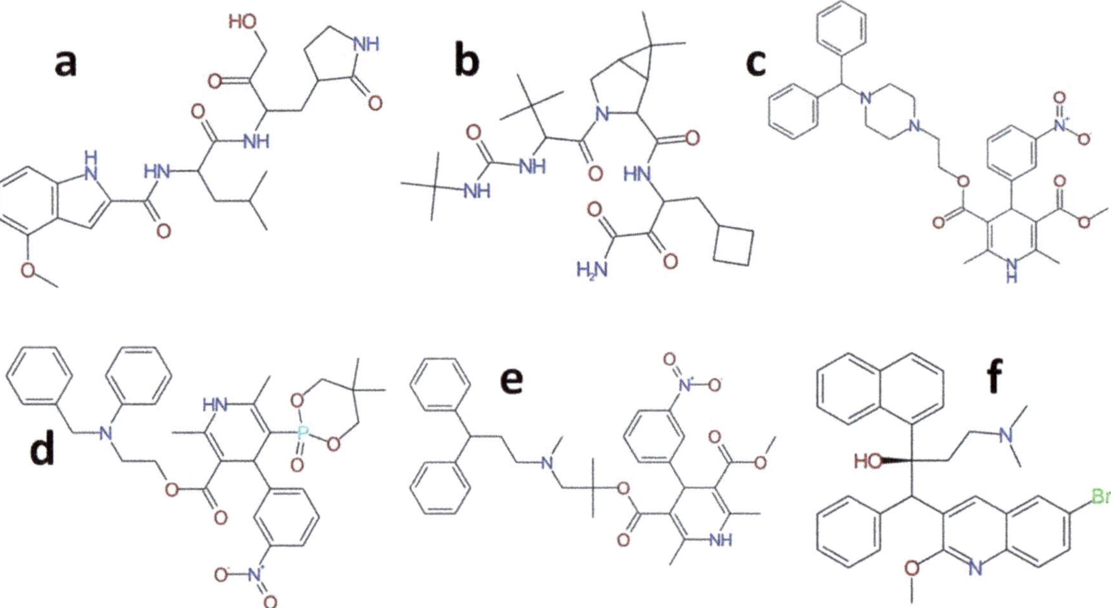

Figure 1. Structure of all of the compounds investigated in this study. (**a**) PF-00835231; (**b**) Boceprevir; (**c**) Manidipine; (**d**) Efonidipine; (**e**) Lercanidipine; (**f**) Bedaquiline.

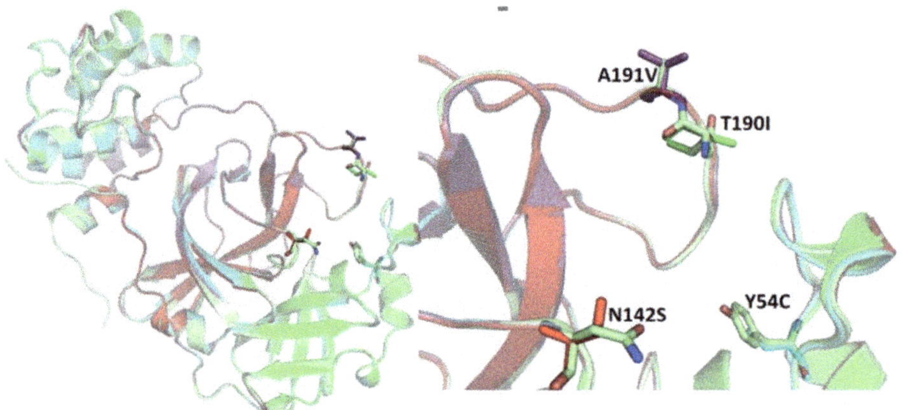

Figure 2. The selected mutant investigated in this study.

2. Result and Discussion

In this study, a comparative analysis of the efficacy of PF-00835231 and five drugs previously documented to inhibit the Mpro (bedaquiline, boceprevir, efonidipine, manidipine and lercanidipine) was performed with the wild type and four reported Mpro mutants (Mutant 1 (Y54C), Mutant 2 (N142S), Mutant 3 (T190I) and Mutant 4 (A191V)).

It was found that in all of the modeled structures, no residues lay in the disallowed region, confirming the significant quality of the structures (Supplementary Figure S1). An ERRAT analysis of all of the structures was also investigated [17]. It was found that the overall structural quality of the modeled structure was very good. A VERIFY_3D [18,19] analysis was also performed and it was found that in all of the modelled structures more than 94% of the residues had an average 3D–1D score > 0.2, proving the great compatibility between the primary sequence to the tertiary structure. Before conducting the molecular docking experiments, the validation of the molecular docking protocol was performed. Different crystal structures of the inhibitor bound SARS-CoV-2 Mpro were retrieved. The binding orientation of the redocked poses was found to be similar to the crystal confirmation of the inhibitor (Supplementary Figure S2 and Table S1). Most of the redocked poses of the inhibitors were found to share a root mean square deviation (RMSD) less than 1 Å (for small molecule inhibitors) than its crystal counterpart (Supplementary Figure S2 and Table S1).

In order to determine the predictive binding effectiveness of small molecules with receptors, a molecular docking evaluation is usually carried out [20]. Six compounds (PF-00835231, bedaquiline, boceprevir, efonidipine, manidipine and lercanidipine) were minimized and prepared for screening within the active site of the Mpro (WT) and modeled Mpro mutants. In terms of the PLP fitness score using GOLD tools, the binding efficacy score of all six selected compounds was calculated.

PF-00835231 was found to be the most effective against the WT Mpro (PLP Fitness score 83.13 (Table 1). This compound was found to be very effective against other selected mutants as well (Tables 2–5). Compared with other selected compounds, PF-00835231 was found to be the most effective inhibitor against Y54C and A191V (Tables 2 and 5) (Figure 3) whereas efonidipine was found to be most effective against N142S and T190I (Tables 3 and 4) (Figure 4). Figures 3 and 4 show the binding of PF-00835231 and efonidipine against the Mpro and the selected mutants. The study also highlighted the important residues playing a crucial role in accommodating the selected compounds within the active site of the Mpro and the mutants. It was also found that the large number of active site residues of the Mpro (WT and mutant) were actively participating in the positioning of all of the molecules. Tables 1–5 represent the details of the interacting residues (amino acid) of the Mpro and mutants interacting with all of the selected molecules. L141, S144, H164, E166, Q189 and Q192 were found to be very prominently involved in making hydrogen bonds with all of the selected compounds (Tables 1–5) (Figures 3 and 4). The crucial role of these residues has been discussed earlier as well [21–23]. Other residues found to be playing an important role in the binding were T25, T26, L27, H41, M49, C145, M165, L167, P168, D187 and R188 (Tables 1–5) (Figures 3 and 4 and Supplementary Figures S3–S7).

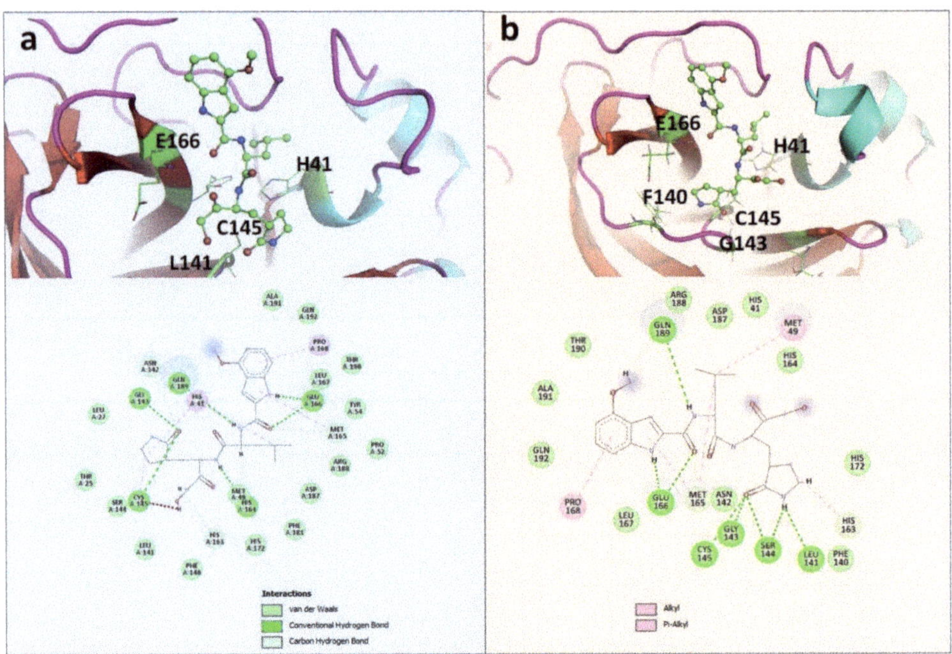

Figure 3. Complex of PF-00835231 within the active site of (**a**) WT; (**b**) Y54C.

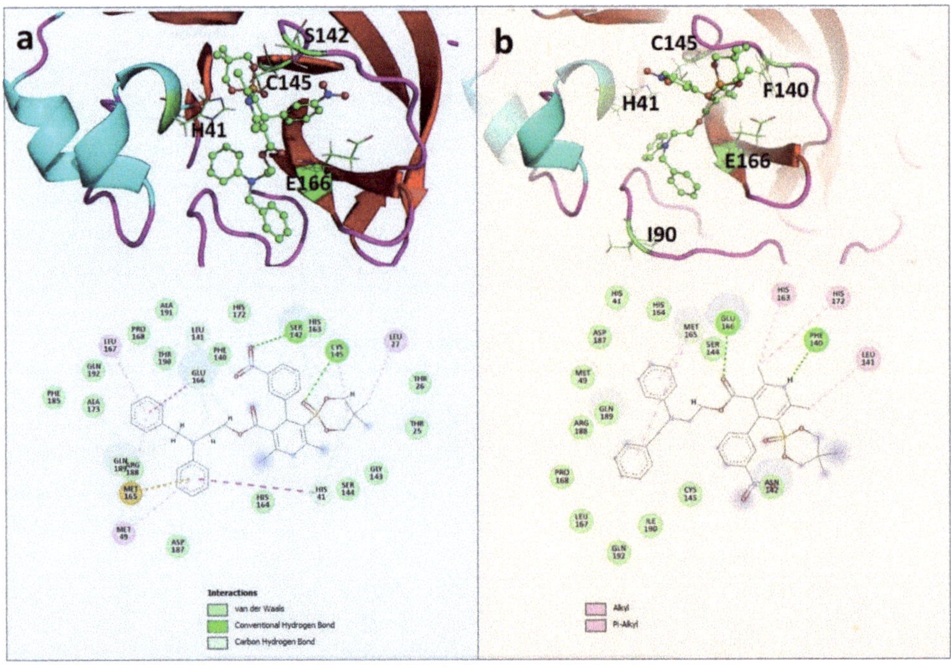

Figure 4. Complex of Efonidipine within the active site of (**a**) N142S; (**b**) T190I.

Table 1. The binding details of all compounds against the WT Mpro.

Protein	Inhibitor	Score	Residues	
			Hydrogen Bond	**Pi Interactions**
WT	PF-00835231	83.13	G143, C145, H164, E166, Q189	H41, P168
	Manidipine	72.93	F140, N142, E166	H41, L141, M165
	Boceprevir	67.42	N142, E166, Q189	H41, M49, C145
	Lercanidipine	69.73	Q189, Q192	C145, M165, L167, P168
	Efonidipine	71.23		H41, C145, M165, P168, A191
	Bedaquiline	66.92	N142	H41

Table 2. The binding details of all compounds against the Mutant 1 (Y54C) Mpro.

Protein	Inhibitor	Score	Residues	
			Hydrogen Bond	**Pi Interactions**
Mutant 1 (Y54C)	PF-00835231	73.17	L141, G143, S144, C145, E166, Q189	M49, P168
	Manidipine	66.68	T26, S46	C145, M165
	Boceprevir	65.96	T26, G143, C145, E166, Q189	H41, M49, H163, M165
	Lercanidipine	65.28	T26, G143, C145	L27, H41, M165
	Efonidipine	68.23	L141, G143, S144, Q189	L27, C145, H163, P168, H172
	Bedaquiline	63.27	H164	L27, H41, M49, C145

Table 3. The binding details of all compounds against the Mutant 2 (N142S) Mpro.

Protein	Inhibitor	Score	Residues	
			Hydrogen Bond	**Pi Interaction**
Mutant 2 N142S	PF-00835231	76.53	L141, S142, G143, E166, Q189	S144, C145, L167, P168
	Manidipine	68.64	F140, E166	L141, M165
	Boceprevir	65.33	Q189	C44, M49, C145, H163, M165
	Lercanidipine	69.05		L27, M49, C145, H163
	Efonidipine	87.05	S142, C145	L27, M49, M165, L167
	Bedaquiline	68.83	H164	H41, M49, C145, M165

Table 4. The binding details of all compounds against the Mutant 3 (T190I) Mpro.

Protein	Inhibitor	Score	Residues	
			Hydrogen Bond	**Pi Interactions**
Mutant 3 T190I	PF-00835231	70.66	E166, D187, Q189	C44, M49, C145, H163, M165
	Manidipine	63.42	F140, N142	M49, L141, M165
	Boceprevir	60.36	N142	C44, T45, M165
	Lercanidipine	66.96	T26	L27, H41, C44, M49, C145, M165
	Efonidipine	72.25	F140, E166	L141, H163, M165, H172
	Bedaquiline	67.87	E166	H41, C145

Table 5. The binding details of all compounds against the Mutant 4 (A191V) Mpro.

Protein	Inhibitor	Score	Residues	
			Hydrogen Bond	**Pi Interaction**
Mutant 4 A191V	PF-00835231	73.61	L141, G143, S144, C145, Q189	M165, P168, H172, V191
	Manidipine	65.95	T25, S46, N142	M165
	Boceprevir	62.08	S46, Q189	C44, M49, M165
	Lercanidipine	67.93	C44, S46, N142	H41, M49, C145, M165
	Efonidipine	71.01	F140	L141, H163, M165, P168
	Bedaquiline	59.99	N142	H41, C145, M165

Recent studies have shown that all of the compounds (bedaquiline, boceprevir, efonidipine, manidipine and lercanidipine) specified in this study carry the potential to inhibit the Mpro with IC50 values below 40 μM [11]. The binding affinity for PF-00835231 has been reported to be in the nano molar range [15]. Our in depth in silico analysis also found PF-00835231 to be carrying a high affinity against the Mpro (WT) compared with other selected compounds. We also hypothesized that these compounds, including PF-00835231, might prove to be effective against the mutants as well. Boceprevir, which is a protease inhibitor and was originally used to treat hepatitis, has been well studied to carry an inhibitory potential against the Mpro [24–26]. This compound was found to carry a very low affinity against the T190I whilst being the most active against the WT. The high binding affinity of this compound against the Mpro (WT) has been reported in several studies [25]. This compound was found to be moderately effective against the WT and the selected mutants. Likewise, other compounds considered in this study showed a moderate affinity against all of the selected mutants. Manidipine [27], which is a calcium channel blocker and is an approved antihypertensive drug, was found to be very effective against the WT. Several studies have reported the potential of manidipine against SARS-CoV-2 [11,28,29]. This compound showed moderate activity against other selected mutants. Lercanidipine [30], another calcium channel blocker and an approved antihypertensive drug, was also found to show moderate activity against all of the selected targets and was most active against the WT and N142S. Bedaquiline [31], another compound considered in this study, is an approved drug for the treatment of active tuberculosis. This compound was found to be moderately active against all of the selected proteins with a maximum binding affinity against N142S and a least binding affinity against A191V.

Considering the high efficacy of PF-00835231 and Efonidipine against all of the selected proteins, we further studied the structure dynamics of WT and the Mpro mutants in complex with these two inhibitors (Figure 5). Root mean square deviation (RMSD) is a very significant parameter to explore the protein dynamics in terms of conformational changes within the protein structure. The backbone RMSD plot revealed that in the presence of PF-00835231, the structures of the WT, Y54C and T190I mutants were stable throughout the 100 ns simulation while the structures of N142S and A191V indicated fluctuations after 50 ns (Figure 5a). The backbone of WT and N142S was found to be stable in the Efonidipine bound structure (Figure 5b) while other mutants, namely Y54C, A191V and T190I, showed fluctuations in the backbone. Our overall investigation found that Efonidipine caused fewer structural variations in the backbone of WT and N142S while PF-00835231 caused fewer structural variations in the WT, Y54C and T190I mutants. This suggested that the association of PF-00835231 within the active site of the Mpro and its mutant was comparatively more stable. Further, the ligand RMSD plot was also analyzed and it was observed that PF-07304814 was more stable with WT, Y54C and N142S throughout the simulation time period compared with Efonidipine (Figure 5c,d). These analyses further provide a strong support that the PF-07304814 bound complexes were very stable. The Hbond analysis also showed that the PF-00835231 bound complexes comparatively made more hydrogen bonds than the Efonidipine bound complexes (Figure 5e,f). This

finding well supports the theory that the stability of PF-07304814 within the active site of the Mpro (WT) and its mutants may be because of the greater number of Hbonds providing the stability to PF-07304814. The overall outcome of this study showed that PF-07304814 could be a very potent inhibitor against the Mpro and its other reported mutants.

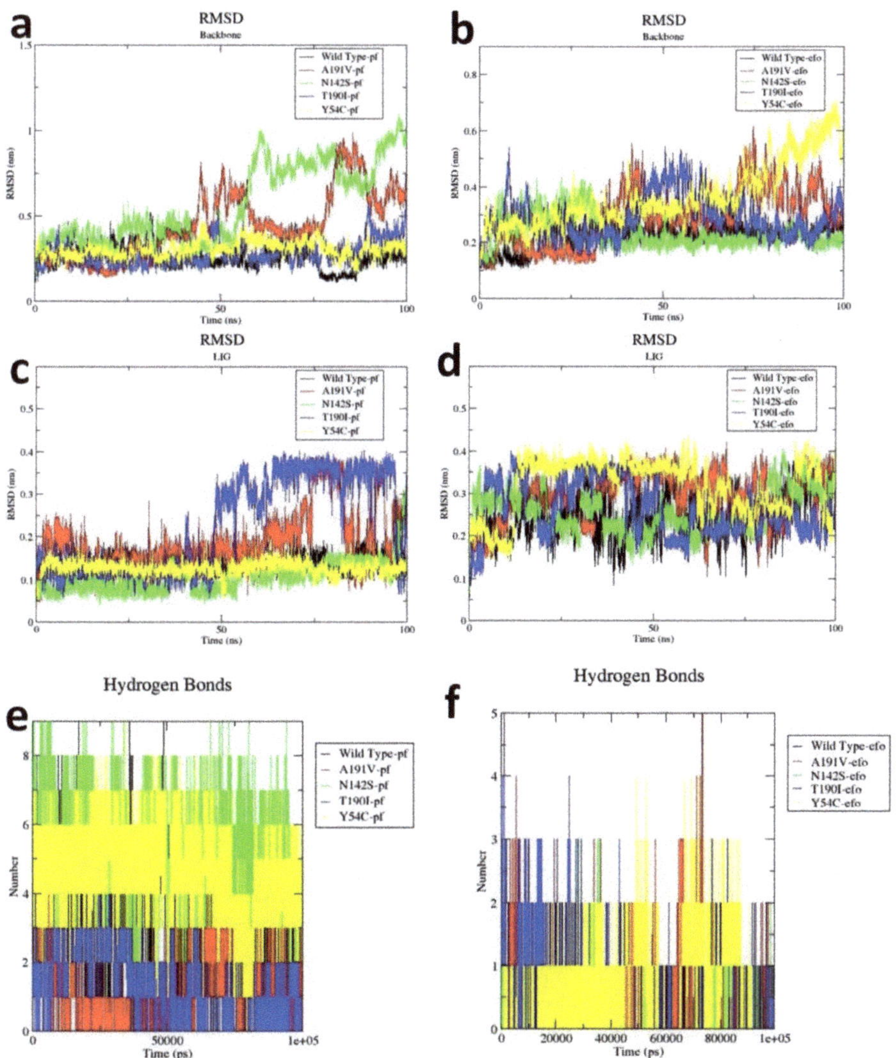

Figure 5. Molecular dynamics results of the PF-07304814 and Efonidipine bound complexes of the WT Mpro (black), Y54C (yellow), N142S (green), T190I (blue) and A1901V (red). The backbone RMSD of the Mpro (WT) and mutants in complex with (**a**) PF-07304814; (**b**) Efonidipine. The ligand RMSD of (**c**) PF-07304814 and (**d**) Efonidipine during the 100 ns. The intermolecular hydrogen bond formations of the (**e**) PF-07304814 and (**f**) Efonidipine bound complex.

3. Materials and Methods

3.1. Protein Structure Preparation

Here in this study, the crystal structure of the SARS-CoV-2 (COVID-19) Mpro in complex with inhibitor UAW248 was retrieved from the RCSB protein databank (pdb id:

6xbi) [32,33]. The crystal bound inhibitor and other heteroatoms were removed. The structure of the mutants was prepared using the molecular modeling technique. The amino acid sequence of all of the selected mutants (Y54C, N142S, T190I and A191V) were retrieved from the NCBI protein database (GenBank: QJD23268.1, QJC19621.1, QJA16866.1 and QIZ14843.1) [34]. To model the structure of the mutants, the 6xbi was taken as a template. The structures were modelled using the modeler 9.23 [35]. All of the modeled structures were validated using various in silico tools [36,37]. The structure of all of the Mpro mutants was modeled and validated as well. The Ramachandran plot was computed for all of the modeled structures using the PROCHECK module of SAVES [38]. All of the structures were subjected to energy minimization using the steepest descent method for 1000 steps.

3.2. Ligand Structure Preparation

The 3D structure of bedaquiline (CID: 5388906), boceprevir (CID: 10324367), efonidipine (CID: 119171), lercanidipine (CID: 65866), manidipine (CID: 4008) and PF-00835231 (CID: 11561899) were retrieved from the PubChem structure database [39]. All of the compounds were energy minimized using the conjugate gradient method for 1000 steps in UCSF Chimera [40].

3.3. Redocking: Co-Crystallized Ligand Pose Validation Study

Several crystal structures of the inhibitor bound SARS-CoV-2 Mpro were retrieved from the RCSB protein databank. The inhibitors were separated and were subjected to a redock within the structure of the Mpro using CCDC GOLD [41]. The docked confirmation of the inhibitors within the active site of the Mpro was compared with the crystal orientations.

3.4. Virtual Screening

All of the selected compounds were subjected to docking within the active site of WT and selected Mpro mutants using Gold 2.2 (CCDC, Cambridge, UK) [41]. The selection was made based on their PLP fitness score. The complexes were visualized using PyMol [42] and a discovery studio visualizer.

3.5. Molecular Dynamics Simulation

The selected complexes were subjected to a molecular dynamics simulation to investigate the stability of these molecules (in complex with WT and mutant Mpro). The molecular dynamics simulation was performed with GROMACS [43,44]. The complexes of PF-00835231 and efonidipine complexed with the Mpro (WT) and mutants prepared using the molecular docking were considered as a starting point for MD study. Here we used a GROMACS 2020.4 package with a Charmm36 force field to perform the MD simulation [45]. GROMACS is a widely used tool for performing MD simulation studies and its utilization in protein-ligand simulation has been reported in a large number of studies [46,47]. The parameter files for all the ligands were generated using SwissParam (https://www.swissparam.ch/, accessed on 15 January 2021), which is an online tool for generating parameters for the Charmm force field. The complexes were solvated within the dodecahedron box of an explicit TIP3P water model with a 0.1 nm margin between the box walls and solute. Na^+ or Cl^- counterions were added to neutralize the system charge [48,49]. The particle mesh Ewald method (cutoff distance of 0.1 nm) was employed for calculating the long-range electrostatic interactions [50]. The Lennard-Jones 6–12 potential was used for evaluating the van der Waals interactions; for this calculation, the cutoff distance was set to 0.1 nm. The LINCS algorithm was used to constrain the bond lengths while setting the time step to 0.002 pico second [51,52]. Further energy minimization was performed using the steepest descent method for 10,000 steps in order to remove the steric clashes between atoms. The whole system was further subjected for equilibration for 1 nano second (ns). To maintain the system at 300 K and 1 atm, Berendsen weak coupling systems were utilized [53,54]. A Maxwell Boltzmann distribution was used

for randomly generating the initial velocities. The final 100 ns production run was performed at 300 K in an NPT ensemble. Furthermore, xmgrace was used to generated graphs (http://plasmagate.weizmann.ac.il, accessed on 15 January 2021); PyMol and VMD were used for further graphical inspections and analysis.

4. Conclusions

In conclusion, this study predicted that PF-00835231, which is already being tested to target the SARS-CoV-2 Mpro, may also be potent against the specified Mpro mutants. Notably, PF-00835231 and five other reported antivirals were investigated for comparative inhibitory efficacy in terms of binding potency against WT and the Mpro mutants. PF-00835231 was found to be the most efficient inhibitor of the Y54C and A191V Mpro mutants with a fitness score of 73.17 and 73.61, respectively, relative to the other listed drugs. Based on our research, it is early to determine but hopefully this potential drug PF-00835231 would most certainly be highly effective against the mutant Mpro and could prove to be a sharp weapon in the fight against the COVID-19 pandemic.

Supplementary Materials: The following are available online: Figure S1: Validation of the modeled structures of mutants (a) Y54C (b) N142S (c) T190I (d) A191V, Figure S2: The superimposed structure of crystal pose (blue) of inhibitor and redocked pose of (a) X47 (green) (pdb id: 6wco), (b) X77 (yellow) (pdb id: 6w63), and (c) ADRAFINIL (red) (pdb id: 7ans) within the binding site of Mpro, Figure S3: Complex of (a) PF-00835231 (b) Boceprevir (c) Manidipine (d) Efonidipine (e) Lercanidipine (f) Bedaquiline within the active site of WT, Figure S4: Complex of all (a) PF-00835231 (b) Boceprevir (c) Manidipine (d) Efonidipine (e) Lercanidipine (f) Bedaquiline within the active site of Y54C, Figure S5: Complex of all (a) PF-00835231 (b) Boceprevir (c) Manidipine (d) Efonidipine (e) Lercanidipine (f) Bedaquiline within the active site of N142S, Figure S6: Complex of all (a) PF-00835231 (b) Boceprevir (c) Manidipine (d) Efonidipine (e) Lercanidipine (f) Bedaquiline within the active site of T190I, Table S1: The inhibitor bound crystal structure of SARS-CoV-2 Mpro considered for the validation of docking protocol.

Author Contributions: Conceptualization, J.-J.D., M.H.B. and I.A.; methodology, T.S., M.H.B., M.A.; software, J.-J.D., M.H.B.; writing—original draft, J.-J.D., M.H.B., I.A. and M.M.A., M.A.; formal analysis, J.-J.D., M.H.B.; investigation, T.S., I.A., M.M.A.; supervision, J.-J.D. All authors have read and agreed to the published version of the manuscript.

Funding: The authors are thankful to the Institute of Research and Consulting Studies at King Khalid University for supporting this research through grant number 26-14-S-2020 and the National Research Foundation of Korea, grant: [NRF-2018R1C1B6009531].

Institutional Review Board Statement: Not applicable.

Informed Consent Statement: Not applicable.

Data Availability Statement: Data is available within the article.

Acknowledgments: The authors are thankful to the National Research Foundation of Korea and the Institute of Research and Consulting Studies at King Khalid University for supporting this research.

Conflicts of Interest: The authors declare no conflict of interest.

Sample Availability: Samples of the compounds are not available from the authors.

References

1. Hu, B.; Guo, H.; Zhou, P.; Shi, Z.L. Characteristics of SARS-CoV-2 and COVID-19. *Nat. Rev. Microbiol.* **2021**, *19*, 141–154. [CrossRef] [PubMed]
2. Robinson, J. Everything you need to know about the COVID-19 therapy trials. *Pharm. J.* **2021**. [CrossRef]
3. Clinicaltrials. gov. Available online: https://www.clinicaltrials.gov/ (accessed on 15 January 2021).
4. Rubin, D.; Chan-Tack, K.; Farley, J.; Sherwat, A. FDA Approval of Remdesivir—A Step in the Right Direction. *N. Engl. J. Med.* **2020**, *383*, 2598–2600. [CrossRef]
5. Pau, A.K.; Aberg, J.; Baker, J.; Belperio, P.S.; Coopersmith, C.; Crew, P.; Grund, B.; Gulick, R.M.; Harrison, C.; Kim, A.; et al Convalescent Plasma for the Treatment of COVID-19: Perspectives of the National Institutes of Health COVID-19 Treatmer Guidelines Panel. *Ann. Intern. Med.* **2021**, *174*, 93–95. [CrossRef]

6. Artigas, L.; Coma, M.; Matos-Filipe, P.; Aguirre-Plans, J.; Farrés, J.; Valls, R.; Fernandez-Fuentes, N.; De La Haba-Rodriguez, J.; Olvera, A.; Barbera, J.; et al. In-silico drug repurposing study predicts the combination of pirfenidone and melatonin as a promising candidate therapy to reduce SARS-CoV-2 infection progression and respiratory distress caused by cytokine storm. *PLoS ONE* **2020**, *15*, e0240149. [CrossRef]

7. Riva, L.; Yuan, S.; Yin, X.; Martin-Sancho, L.; Matsunaga, N.; Pache, L.; Burgstaller-Muehlbacher, S.; De Jesus, P.D.; Teriete, P.; Hull, M.V.; et al. Discovery of SARS-CoV-2 antiviral drugs through large-scale compound repurposing. *Nature* **2020**, *586*, 113–119. [CrossRef] [PubMed]

8. Krichel, B.; Falke, S.; Hilgenfeld, R.; Redecke, L.; Uetrecht, C. Processing of the SARS-CoV pp1a/ab nsp7–10 region. *Biochem. J.* **2020**, *477*, 1009–1019. [CrossRef]

9. Vlachakis, D.; Papakonstantinou, E.; Mitsis, T.; Pierouli, K.; Diakou, I.; Chrousos, G.; Bacopoulou, F. Molecular mechanisms of the novel coronavirus SARS-CoV-2 and potential anti-COVID19 pharmacological targets since the outbreak of the pandemic. *Food Chem. Toxicol.* **2020**, *146*, 111805. [CrossRef] [PubMed]

10. Xue, X.; Yu, H.; Yang, H.; Xue, F.; Wu, Z.; Shen, W.; Li, J.; Zhou, Z.; Ding, Y.; Zhao, Q.; et al. Structures of Two Coronavirus Main Proteases: Implications for Substrate Binding and Antiviral Drug Design. *J. Virol.* **2007**, *82*, 2515–2527. [CrossRef] [PubMed]

11. Ghahremanpour, M.M.; Tirado-Rives, J.; Deshmukh, M.; Ippolito, J.A.; Zhang, C.-H.; De Vaca, I.C.; Liosi, M.-E.; Anderson, K.S.; Jorgensen, W.L. Identification of 14 Known Drugs as Inhibitors of the Main Protease of SARS-CoV-2. *ACS Med. Chem. Lett.* **2020**, *11*, 2526–2533. [CrossRef]

12. Callaway, E. The coronavirus is mutating—Does it matter? *Nat. Cell Biol.* **2020**, *585*, 174–177. [CrossRef]

13. Analytica, O. New COVID-19 Variants Could Prolong Pandemic: Expert Briefings. 2021. Available online: https://www.emerald.com/insight/content/doi/10.1108/OXAN-ES259153/full/html (accessed on 15 January 2021).

14. Law, S.; Leung, A.W.; Xu, C. COVID-19 mutation in the United Kingdom. *Microbes Infect. Dis.* **2021**. [CrossRef]

15. Boras, B.; Jones, R.M.; Anson, B.J.; Arenson, D.; Aschenbrenner, L.; Bakowski, M.A.; Beutler, N.; Binder, J.; Chen, E.; Eng, H.; et al. Discovery of a Novel Inhibitor of Coronavirus 3CL Protease as a Clinical Candidate for the Potential Treatment of COVID-19. *BioRxiv* **2020**. [CrossRef]

16. Hoffman, R.L.; Kania, R.S.; Brothers, M.A.; Davies, J.F.; Ferre, R.A.; Gajiwala, K.S.; He, M.; Hogan, R.J.; Kozminski, K.; Li, L.Y.; et al. Discovery of Ketone-Based Covalent Inhibitors of Coronavirus 3CL Proteases for the Potential Therapeutic Treatment of COVID-19. *J. Med. Chem.* **2020**, *63*, 12725–12747. [CrossRef]

17. Colovos, C.; Yeates, T.O. Verification of protein structures: Patterns of nonbonded atomic interactions. *Protein Sci.* **1993**, *2*, 1511–1519. [CrossRef]

18. Bowie, J.U.; Luthy, R.; Eisenberg, D. A method to identify protein sequences that fold into a known three-dimensional structure. *Science* **1991**, *253*, 164–170. [CrossRef]

19. Lüthy, R.; Bowie, J.U.; Eisenberg, D. Assessment of protein models with three-dimensional profiles. *Nature* **1992**, *356*, 83–85. [CrossRef]

20. Seeliger, D.; De Groot, B.L. Ligand docking and binding site analysis with PyMOL and Autodock/Vina. *J. Comput. Mol. Des.* **2010**, *24*, 417–422. [CrossRef] [PubMed]

21. Jin, Z.; Du, X.; Xu, Y.; Deng, Y.; Liu, M.; Zhao, Y.; Zhang, B.; Li, X.; Zhang, L.; Peng, C.; et al. Structure of Mpro from SARS-CoV-2 and discovery of its inhibitors. *Nature* **2020**, *582*, 289–293. [CrossRef] [PubMed]

22. Bello, M.; Martínez-Muñoz, A.; Balbuena-Rebolledo, I. Identification of saquinavir as a potent inhibitor of dimeric SARS-CoV2 main protease through MM/GBSA. *J. Mol. Model.* **2020**, *26*, 340. [CrossRef] [PubMed]

23. Tejera, E.; Munteanu, C.; López-Cortés, A.; Cabrera-Andrade, A.; Pérez-Castillo, Y. Drugs Repurposing Using QSAR, Docking and Molecular Dynamics for Possible Inhibitors of the SARS-CoV-2 M^{pro} Protease. *Molecules* **2020**, *25*, 5172. [CrossRef]

24. Oerlemans, R.; Ruiz-Moreno, A.J.; Cong, Y.; Kumar, N.D.; Velasco-Velazquez, M.A.; Neochoritis, C.G.; Smith, J.; Reggiori, F.; Groves, M.R.; Dömling, A. Repurposing the HCV NS3–4A protease drug boceprevir as COVID-19 therapeutics. *RSC Med. Chem.* **2020**. [CrossRef]

25. Ma, C.; Sacco, M.D.; Hurst, B.; Townsend, J.A.; Hu, Y.; Szeto, T.; Zhang, X.; Tarbet, B.; Marty, M.T.; Chen, Y.; et al. Boceprevir, GC-376, and calpain inhibitors II, XII inhibit SARS-CoV-2 viral replication by targeting the viral main protease. *Cell Res.* **2020**, *30*, 678–692. [CrossRef]

26. Fu, L.; Ye, F.; Feng, Y.; Yu, F.; Wang, Q.; Wu, Y.; Zhao, C.; Sun, H.; Huang, B.; Niu, P.; et al. Both Boceprevir and GC376 efficaciously inhibit SARS-CoV-2 by targeting its main protease. *Nat. Commun.* **2020**, *11*, 1–8. [CrossRef] [PubMed]

27. Ogihara, T.; Nakagawa, M.; Ishikawa, H.; Mikami, H.; Takeda, K.; Nonaka, H.; Nagano, M.; Sasaki, S.; Kagoshima, T.; Higashimori, K. Effect of manidipine, a novel calcium channel blocker, on quality of life in hypertensive patients. *Blood Press. Suppl.* **1992**, *3*, 135–139.

28. De Wilde, A.H.; Jochmans, D.; Posthuma, C.C.; Zevenhoven-Dobbe, J.C.; Van Nieuwkoop, S.; Bestebroer, T.M.; van den Hoogen, B.G.; Neyts, J.; Snijder, E.J. Screening of an FDA-Approved Compound Library Identifies Four Small-Molecule Inhibitors of Middle East Respiratory Syndrome Coronavirus Replication in Cell Culture. *Antimicrob. Agents Chemother.* **2014**, *58*, 4875–4884. [CrossRef] [PubMed]

29. Maffucci, I.; Contini, A. In Silico Drug Repurposing for SARS-CoV-2 Main Proteinase and Spike Proteins. *J. Proteome Res.* **2020**, *19*, 4637–4648. [CrossRef] [PubMed]

30. McClellan, K.J.; Jarvis, B. Lercanidipine: A review of its use in hypertension. *Drugs* **2000**, *60*, 1123–1140. [CrossRef]

31. Mahajan, R. Bedaquiline: First FDA-approved tuberculosis drug in 40 years. *Int. J. Appl. Basic Med. Res.* **2013**, *3*, 1–2. [CrossRef]
32. Rose, P.W.; Bi, C.; Bluhm, W.F.; Christie, C.H.; Dimitropoulos, D.; Dutta, S.; Green, R.K.; Goodsell, D.S.; Prlić, A.; Quesada, M.; et al. The RCSB Protein Data Bank: New resources for research and education. *Nucleic Acids Res.* **2012**, *41*, D475–D482. [CrossRef] [PubMed]
33. Sacco, M.D.; Ma, C.; Lagarias, P.; Gao, A.; Townsend, J.A.; Meng, X.; Dube, P.; Zhang, X.; Hu, Y.; Kitamura, N.; et al. Structure and inhibition of the SARS-CoV-2 main protease reveal strategy for developing dual inhibitors against Mpro and cathepsin L. *Sci. Adv.* **2020**, *6*, eabe0751. [CrossRef]
34. Wheeler, D.L.; Barrett, T.; Benson, D.A.; Bryant, S.H.; Canese, K.; Chetvernin, V.; Church, D.M.; DiCuccio, M.; Edgar, R.; Federhen, S.; et al. Database resources of the National Center for Biotechnology Information. *Nucleic Acids Res.* **2006**, *35*, D5–D12. [CrossRef] [PubMed]
35. Webb, B.; Sali, A. Comparative Protein Structure Modeling Using MODELLER. *Curr. Protoc. Bioinform.* **2016**, *54*, 5–6. [CrossRef] [PubMed]
36. Pontius, J.; Richelle, J.; Wodak, S.J. Deviations from Standard Atomic Volumes as a Quality Measure for Protein Crystal Structures. *J. Mol. Biol.* **1996**, *264*, 121–136. [CrossRef] [PubMed]
37. Eisenberg, D.; Lüthy, R.; Bowie, J.U. [20] VERIFY3D: Assessment of protein models with three-dimensional profiles. *Meth. Enzymol.* **1997**, *277*, 396–404. [CrossRef]
38. Laskowski, R.A.; MacArthur, M.W.; Moss, D.S.; Thornton, J.M. PROCHECK: A program to check the stereochemical quality of protein structures. *J. Appl. Crystallogr.* **1993**, *26*, 283–291. [CrossRef]
39. Kim, S.; Thiessen, P.A.; Bolton, E.E.; Chen, J.; Fu, G.; Gindulyte, A.; Han, L.; He, J.; He, S.; Shoemaker, B.A.; et al. PubChem substance and compound databases. *Nucleic Acids Res.* **2016**, *44*, D1202–D1213. [CrossRef]
40. Pettersen, E.F.; Goddard, T.D.; Huang, C.C.; Couch, G.S.; Greenblatt, D.M.; Meng, E.C.; Ferrin, T.E. UCSF Chimera—A visualization system for exploratory research and analysis. *J. Comput. Chem.* **2004**, *25*, 1605–1612. [CrossRef]
41. Jones, G.; Willett, P.; Glen, R.C.; Leach, A.R.; Taylor, R. Development and validation of a genetic algorithm for flexible docking. *J. Mol. Biol.* **1997**, *267*, 727–748. [CrossRef]
42. Schrodinger Inc. *PyMOL, Molecular Visualization System, Version 2.4*; Schrodinger Inc.: New York, NY, USA, 2002.
43. Pronk, S.; Páll, S.; Schulz, R.; Larsson, P.; Bjelkmar, P.; Apostolov, R.; Shirts, M.R.; Smith, J.C.; Kasson, P.M.; Van Der Spoel, D.; et al. GROMACS 4.5: A high-throughput and highly parallel open source molecular simulation toolkit. *Bioinformatics* **2013**, *29*, 845–854. [CrossRef]
44. Hess, B.; Kutzner, C.; Van Der Spoel, D.; Lindahl, E. GROMACS 4: Algorithms for Highly Efficient, Load-Balanced, and Scalable Molecular Simulation. *J. Chem. Theory Comput.* **2008**, *4*, 435–447. [CrossRef] [PubMed]
45. Christen, M.; Hünenberger, P.H.; Bakowies, D.; Baron, R.; Bürgi, R.; Geerke, D.P.; Heinz, T.N.; Kastenholz, M.A.; Kräutler, V.; Oostenbrink, C.; et al. The GROMOS software for biomolecular simulation: GROMOS05. *J. Comput. Chem.* **2005**, *26*, 1719–1751. [CrossRef]
46. Liu, X.; Shi, D.; Zhou, S.; Liu, H.; Liu, H.; Yao, X. Molecular dynamics simulations and novel drug discovery. *Expert Opin. Drug Discov.* **2018**, *13*, 23–37. [CrossRef]
47. Bao, Y.; Zhou, L.; Dai, D.; Zhu, X.; Hu, Y.; Qiu, Y. Discover potential inhibitors for PFKFB3 using 3D-QSAR, virtual screening, molecular docking and molecular dynamics simulation. *J. Recept. Signal Transduct.* **2018**, *38*, 413–431. [CrossRef] [PubMed]
48. Mark, P.; Nilsson, L. A Molecular Dynamics Study of Tryptophan in Water. *J. Phys. Chem. B* **2002**, *106*, 9440–9445. [CrossRef]
49. Mark, P.; Nilsson, L. Structure and Dynamics of the TIP3P, SPC, and SPC/E Water Models at 298 K. *J. Phys. Chem. A* **2001**, *105*, 9954–9960. [CrossRef]
50. Darden, T.; York, D.; Pedersen, L. Particle mesh Ewald: An $N \cdot \log(N)$ method for Ewald sums in large systems. *J. Chem. Phys.* **1993**, *98*, 10089–10092. [CrossRef]
51. Hess, B.; Bekker, H.; Berendsen, H.J.; Fraaije, J.G. LINCS: A linear constraint solver for molecular simulations. *J. Comput. Chem.* **1997**, *18*, 1463–1472. [CrossRef]
52. Hess, B. P-LINCS: A Parallel Linear Constraint Solver for Molecular Simulation. *J. Chem. Theory Comput.* **2007**, *4*, 116–122. [CrossRef]
53. Izaguirre, J.A.; Catarello, D.P.; Wozniak, J.M.; Skeel, R.D. Langevin stabilization of molecular dynamics. *J. Chem. Phys.* **2001**, *114*, 2090–2098. [CrossRef]
54. Berendsen, H.J.C.; Postma, J.P.M.; Van Gunsteren, W.F.; DiNola, A.; Haak, J.R. Molecular dynamics with coupling to an external bath. *J. Chem. Phys.* **1984**, *81*, 3684–3690. [CrossRef]

Article

Computational Determination of Potential Multiprotein Targeting Natural Compounds for Rational Drug Design Against SARS-COV-2

Ziyad Tariq Muhseen [1,2], Alaa R. Hameed [3], Halah M. H. Al-Hasani [4], Sajjad Ahmad [5] and Guanglin Li [1,2,*]

[1] Key Laboratory of Ministry of Education for Medicinal Plant Resource and Natural Pharmaceutical Chemistry, Shaanxi Normal University, Xi'an 710062, China; ziyad.tariq82@gmail.com
[2] School of Life Sciences, Shaanxi Normal University, Xi'an 710062, China
[3] Department of Medical Laboratory Techniques, School of Life Sciences, Dijlah University College, Baghdad 00964, Iraq; alaa.raad@duc.edu.iq
[4] Department of Biotechnology, College of Science, University of Diyala, Baqubah 32001, Iraq; halahalhasani@sciences.uodiyala.edu.iq
[5] Foundation University Medical College, Foundation University Islamabad, Islamabad 44000, Pakistan; sajjademaan8@gmail.com
* Correspondence: glli@snnu.edu.cn; Tel.: +86-139-9285-6645

Abstract: SARS-CoV-2 caused the current COVID-19 pandemic and there is an urgent need to explore effective therapeutics that can inhibit enzymes that are imperative in virus reproduction. To this end, we computationally investigated the MPD3 phytochemical database along with the pool of reported natural antiviral compounds with potential to be used as anti-SARS-CoV-2. The docking results demonstrated glycyrrhizin followed by azadirachtanin, mycophenolic acid, kushenol-w and 6-azauridine, as potential candidates. Glycyrrhizin depicted very stable binding mode to the active pocket of the Mpro (binding energy, −8.7 kcal/mol), PLpro (binding energy, −7.9 kcal/mol), and Nucleocapsid (binding energy, −7.9 kcal/mol) enzymes. This compound showed binding with several key residues that are critical to natural substrate binding and functionality to all the receptors. To test docking prediction, the compound with each receptor was subjected to molecular dynamics simulation to characterize the molecule stability and decipher its possible mechanism of binding. Each complex concludes that the receptor dynamics are stable (Mpro (mean RMSD, 0.93 Å), PLpro (mean RMSD, 0.96 Å), and Nucleocapsid (mean RMSD, 3.48 Å)). Moreover, binding free energy analyses such as MMGB/PBSA and WaterSwap were run over selected trajectory snapshots to affirm intermolecular affinity in the complexes. Glycyrrhizin was rescored to form strong affinity complexes with the virus enzymes: Mpro (MMGBSA, −24.42 kcal/mol and MMPBSA, −10.80 kcal/mol), PLpro (MMGBSA, −48.69 kcal/mol and MMPBSA, −38.17 kcal/mol) and Nucleocapsid (MMGBSA, −30.05 kcal/mol and MMPBSA, −25.95 kcal/mol), were dominated mainly by vigorous van der Waals energy. Further affirmation was achieved by WaterSwap absolute binding free energy that concluded all the complexes in good equilibrium and stability (Mpro (mean, −22.44 kcal/mol), PLpro (mean, −25.46 kcal/mol), and Nucleocapsid (mean, −23.30 kcal/mol)). These promising findings substantially advance our understanding of how natural compounds could be shaped to counter SARS-CoV-2 infection.

Keywords: SARS-CoV-2; COVID-19; multiprotein inhibiting natural compounds; virtual screening; MD simulation

Citation: Muhseen, Z.T.; Hameed, A.R.; Al-Hasani, H.M.H.; Ahmad, S.; Li, G. Computational Determination of Potential Multiprotein Targeting Natural Compounds for Rational Drug Design Against SARS-COV-2. *Molecules* **2021**, *26*, 674. https://doi.org/10.3390/molecules26030674

Academic Editor: Marco Tutone
Received: 6 December 2020
Accepted: 21 January 2021
Published: 28 January 2021

Publisher's Note: MDPI stays neutral with regard to jurisdictional claims in published maps and institutional affiliations.

1. Introduction

Coronaviruses (CoVs) cause infection of the upper respiratory tract in higher mammals and humans [1], and several outbreaks have been associated in the recent past with CoVs reported first time in the year 2002 as SARS, in 2012 as MERS, and in late 2019 as COVID-19 [2–5]. The recent pandemic of COVID-19 is caused by a relatively new strain

named SARS-CoV-2 [6–8]. The virus origin is thought to be zoonotic, with potential of transmissibility between person-to-person, resulting in an exponential rise in the number of confirmed cases worldwide [9,10]. Through December 2020, more than 220 countries reported the virus, with more than 64 million individuals infected, and thousands are still getting infected each day. Approximately, the virus has a mortality rate between 5% to 10% [11,12]. Additionally, due to mandatory lockdowns, isolation, and quarantines, millions of lives have been disturbed. The pandemic also badly affected global health, society, and the economy, and these sectors are facing significant challenges [13]. Three vaccines (by Pfizer, Moderna, and AstraZeneca) are authorized by WHO for emergency use and are available to very limited populations. No specific anti-SARS-CoV-2 drugs are currently recommended for SARS-CoV-2 treatment, making the situation difficult to handle. Supportive therapeutics and preventative measures are being taken and are productive in managing the virus [14,15]. Various efforts to target critical proteins of SARS-CoV-2 pathogenesis, including Spike receptor-binding domain (RBD) [16–18], main protease (Mpro) [19], Nucleocapsid N terminal domain (NTD) [20], RNA-dependent RNA polymerase (RdRp) [21], papainlike protease (PLpro) [22], 2′-O-RiboseMethyltransferase [23], viral ion channel (E protein) [24], and angiotensin-converting-enzyme 2 receptor (ACE2) [25], are on the way. Targeting multiple pathogenesis specific proteins within a close network of interaction or dependent functionality would effectively propose effective drugs against the SARS-CoV-2 [26].

SARS-COV-2 Spike protein is key to the host cell infection pathway as it mediates ACE2 recognition, attachment, and fusion to the host cell [16]. The RBD of S1 subunit of the Spike trimer binds explicitly to the ACE2 receptor [27]. This RBD region is an attractive target for therapeutics as it contains conserved residues that are essential in binding to ACE2 [27]. The Mpro of coronaviruses has been studied thoroughly for drug making purposes. These are papainlike proteases involved in processing replicase enzymes [28]. It has 11 cleavage sites in 790 kD-long replicase lab polypeptide, demonstrating its prominent role in proteolytic processing [19,29]. High structural similarity and sequence identity are seen in Mpro from SARS-CoV-2 to that of the SARS-CoV Mpro. It comprises two catalytic domains: chymotrypsin and picornavirus 3C protease like domain. Each contains β-barrel that are six in number and are antiparallelly containing active diad H41 and C145 [30]. These proteases have emerged as essential drug targets as they have a crucial role in replication. Furthermore, inhibitors of Mpro are found to be significantly less cytotoxic as the protein share less similarity with human proteases [31]. Preliminary studies have suggested that HIV protease inhibitors, lopinavir/ritonavir, could be potentially used against SARS-CoV-2 [32]. Additionally, HIV protease inhibitor, Darunavir, and HCV protease inhibitor, Danoprevir, are under clinical studies and in vivo trials for the treatment of SARS-CoV-2 infection [33]. The PLpro enzyme is vital in processing the polypeptide to produce a functional replicase complex and aids in viral spreading [22]. PLpro also plays a role in evading host antiviral immune responses by cleaving proteinaceous modification on the host protein after the post-translation phase [34]. Thus, targeting this enzyme is useful in highlighting therapeutic strategies that can suppress the virus infection and prompt antiviral immunity. The N protein is significant in viral RNA replication and its packing into new virions, making this protein a good candidate for newer drug identification that is specific and biological active [20].

In silico screening of drugs using different computer-aided drug designing applications greatly accelerate the rational drug design process. Ultimately, this saves time, and extra cost goes into the experimentation of leads that fail in the drug discovery process [35–40]. In this investigation, we performed a blind docking approach, followed by molecular dynamics (MD) simulation coupled with binding free energy techniques that dissect the structural dynamics and energy basis of molecular recognition [41,42]. The MPD3 phytochemical database [43] along with a pool of natural antiviral compounds were used against multiple SARS-CoV-2 protein targets to understand their binding mechanism and put forward a hypothesis on how to further optimize these structures to enhance selec-

tivity and maximize anti-SARS-CoV-2 biological potency [44–47]. A schematic summary of the methodology used in this work is provided in Figure 1. The study results might have potential applications in designing new leads against SARS-CoV-2, which can target its multiple proteins as depicted in this study.

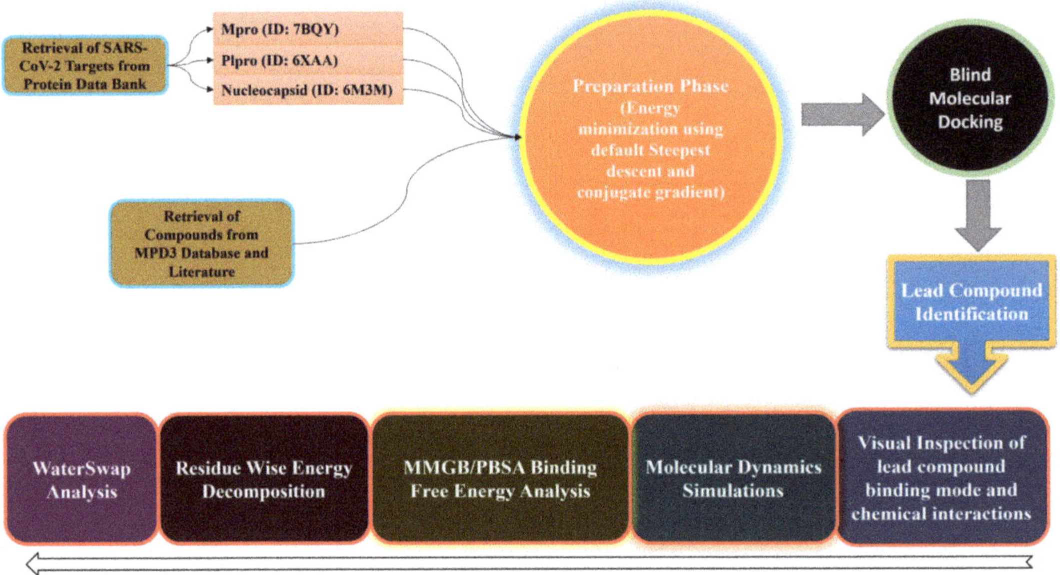

Figure 1. Schematic presentation of the methodology used in this current study.

2. Results and Discussion

2.1. Molecular Docking

Molecular docking is a modeling approach investigating how the receptors and ligands fit together and how the enzymes interact with the ligands [48–52]. Docking calculations were performed in triplicate, and the compound conformations were ranked according to the binding energy in kcal/mol. We used remdesivir as control in docking. The compounds ranked consistently on top with the each receptor and showed a stronger binding score compared to remdesivir were selected for the downward analysis. A general overview of the binding energy of the compounds against the receptors used is presented in Figure 2. The top compound complex with each receptor was generated and subjected first to visual inspections to decipher atomic level interaction and determine the binding conformation. The docking analysis demonstrated glycyrrhizin followed by azadirachtanin, mycophenolic acid, kushenol-w, and 6-azauridine as the best binders among the ~5000 compounds used in this study. The 2D structures of these compounds are presented in Figure 3. Glycyrrhizin also showed stable interactions with the hotspot residues of SARS-CoV-2 spike protein receptor binding domain (RBD) in our previous study [53]. Glycyrrhizin-docked complex of each SARS-CoV-2 protein can be explained separately.

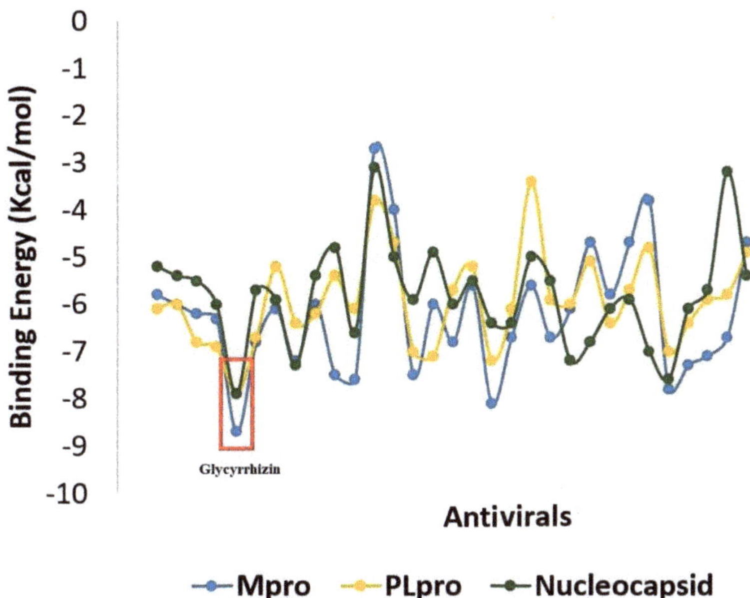

Figure 2. AutoDock binding affinity score of the compounds to the SARS-CoV-2 enzymes.

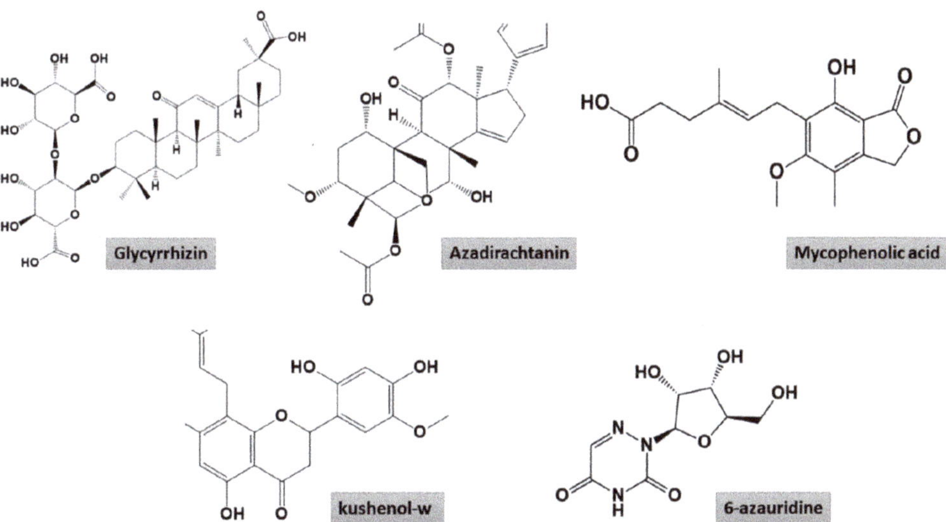

Figure 3. Two-dimensional presentation of high affinity binders to the SARS-CoV-2 proteins.

2.1.1. Mpro–Glycyrrhizin Complex

The Mpro of SARS-CoV-2 is a crucial enzyme and attractive drug target because of its central role in virus transcription and replication [54]. The docking study reported glycyrrhizin again as the best binder among the compounds used to the substrate-binding site of the Mpro (Figure 4). As seen in the binding with other receptors, the compound (2S,3S,4S,5R,6R)-6-(((2S,3R,4S,5S,6S)-6-carboxy-2,4,5-trihydroxytetrahydro-2H-pyran-3-yl)oxy)-3,4,5-trihydroxytetrahydro-2H-pyran-2-carboxylic acid was revealed to contribute

in significant hydrogen bonding and other weak interaction at the active pocket of Mpro. At the binding cavity, the compound engages Asn238 through multiple hydrogen interactions, as well as Asp289. The rest of the compound structure makes a network of hydrophobic interactions mainly dominated by van der Waals contacts. To elucidate further the binding specificity and affinity of the glycyrrhizin for the active pocket residues of Mpro, the interaction profile was compared and contrasted with that for the reported cocrystallized N3 inhibitor [55]. Very low similarity in the binding interaction profile between the compounds was noticed; however, because of the difference in the compound structure, size, and preferred binding site, the pocket residues in contact with glycyrrhizin are close to the N3. This difference in the binding interaction points to the different glycyrrhizin-binding mechanism, where the active moiety favors binding with the P5 binding pocket that is absent in the case of the Mpro–N3 complex. The residues, particularly Asp197 and Thr198, flanked the active site, and any molecule involved in binding with these residues interfere with the natural substrate-binding, thus affecting the enzyme functionality [56]. Additionally, the bulk of the glycyrrhizin structure favors interactions with Domain II and Domain III of the Mpro, in addition to flanking residues of the substrate-binding pocket, thus possibly affecting the dimerization of Domain I and Domain II and rendering the enzyme noncatalytic [57]. Similarly, Zhang et al. reported Mpro complex with an α-ketoamide inhibitor. The cocrystalized lead identified binds to the same substrate binding site reported in this study [28]. Morever, calpain inhibitors and GC-376 analogs are also confirmed to accommodate in the same functional pocket [58]. Beside these, many in silico studies have demonstrated the binding affinity of drug molecules to this active side of Mpro [33,59–61].

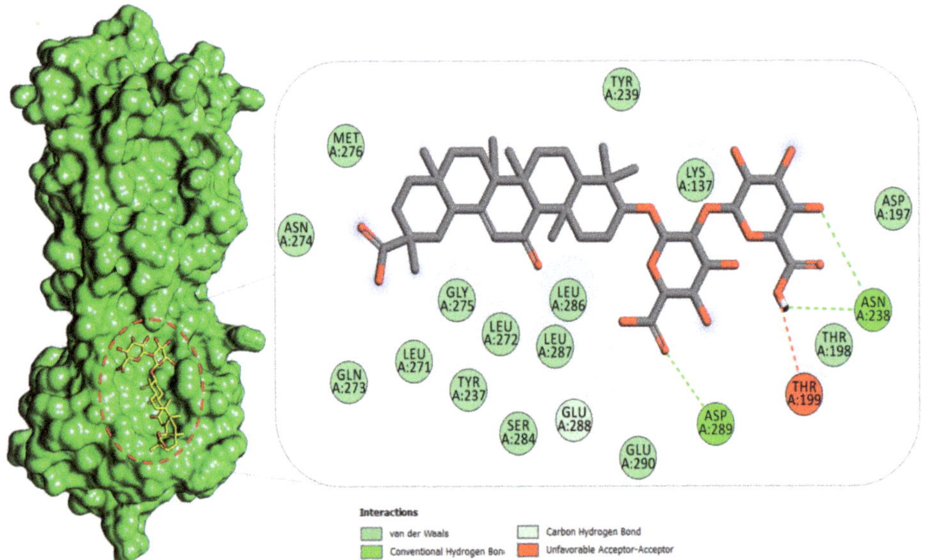

Figure 4. Binding pose of glycyrrhizin (in yellow stick) at the substrate binding pocket of Mpro (in the green surface). A 2D illustration of the glycyrrhizin chemical interactions at the docked site is also provided.

2.1.2. PLpro–Glycyrrhizin Complex

The PLpro enzyme of SARS-CoV-2 is implicated in viral polyproteins processing that generate a replicase complex and assist in virus spreading. The enzyme also plays a fundamental role in cleaving post-translational proteinaceous modifications present on the host protein as a mechanism to avoid antiviral host immune responses [22]. The docked complex between PLpro and glycyrrhizin highlighted the compound binding at the central

palm catalytic cavity (Figure 5). Good binding of the compound-rich electronegative oxygen in the (2*S*,3*S*,4*S*,5*R*,6*R*)-6-(((2*S*,3*R*,4*S*,5*S*,6*S*)-6-carboxy-2,4,5-trihydroxytetrahydro-2*H*-pyran-3-yl)oxy)-3,4,5-trihydroxytetrahydro-2*H*-pyran-2-carboxylic acid at the docked site is the output of several strong hydrogen bond interactions: Gln174, Asp179, and Asn128. Besides these residues, the compounds moiety also formed van der Waals interaction, critical from a stability perspective. The remainder of the compound structure produced van der Waals contacts at this central cavity. The preferred binding of glycyrrhizin is at the central palm, sandwiching the finger and thumb domains, adjacent to the active substrate-binding pocket, which makes a strong bond with many vital catalytic residues. In contrast to the cocrystallized peptide inhibitor VIR251, which has a different conformation and binds to a different substrate cavity site, the glycyrrhizin-binding site is close to the VIR251 site [62]. In terms of interacting binding residues, the glycyrrhizin correlates more with the GRL0617 inhibitor of SAR-CoV-2 PLpro [63]. Further, the effect of conformational change of the BL2 loop upon glycyrrhizin binding is important to evaluate in future studies to disclose the glycyrrhizin recognition mechanism.

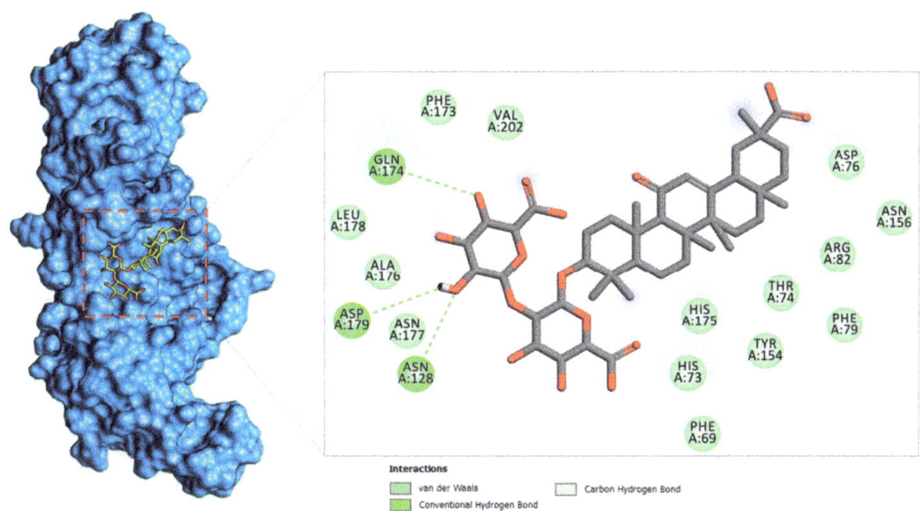

Figure 5. Binding pose of glycyrrhizin (in yellow stick) at the junction binding pocket of PLpro (in blue surface). A 2D illustration of the glycyrrhizin chemical interactions at the docked site is also provided.

In literature, many inhibitors of coronaviruses PLpro are documented that include zinc conjugate inhibitors, naphthalene, and thiopurine derivatives, and natural products [64]. These molecules are known to interact with the active site residues reported in this study. Tanshinones are reported to show inhibition of deubiquitinase and proteolytic activitiy of SARS-CoV PLpro [65]; 8-(Trifluoromethyl)-9*H*-purin-6-amine is a reversible noncovalent inhibitor, whereas N-Ethylmaleimide (NEM) modifies SARS-CoV PLpro Cys [63]. Moreover, 6-mercaptopurine (6MP) and 6-thioguanine (6TG) are slow and competitive inhibitors that form hydrogen bonds with catalytic residues of the SAR-CoV PLpro [66]. Several in silico studies also demonstrated a range of compounds that interfere with the functional site of SARS-CoV-2 PLpro [67–70].

2.1.3. Nucleocapsid–Glycyrrhizin Complex

The SARS-CoV-2 N protein is an RNA binding protein and offers several functions of viral transcription and replication [20]. It particularly plays a pivotal role in helical ribonucleoprotein packing during RNA genome packing, regulating RNA replication, and modulating infected cell metabolism. Blocking of this protein could lead to blocking viral

replication, and thus an attractive target for drug development. The compound glycyrrhizin was found to prefer docking at the loop region 1 at the junction between the β-sheet core and β-hairpin (Figure 6). The molecule is aligned perfectly along the cavity volume where its (*2S,3S,4S,5R,6R*)-6-(((*2S,3R,4S,5S,6S*)-6-carboxy-2,4,5-trihydroxytetrahydro-2*H*-pyran-3-yl)oxy)-3,4,5-trihydroxytetrahydro-2*H*-pyran-2-carboxylic acid part is connected to the β3 and β4 sheets of the β-hairpin. Here, this chemical moiety is involved in hydrogen bonding with Thr92, Arg94, and Arg89, and van der Waals contact with Arg90 and Ala91. The (2S,4aS,6aS,6bR,8aS,12aS,12bR,14bR)-2,4a,6a,6b,9,9,12a-heptamethyl-13-oxo-1,2,3,4,4a,5,6,6a,6b,7,8,8a,9,10,11,12,12a,12b,13,14b-icosahydropicene-2-carboxylic acid region of the compound produced hydrogen bonding with residues (Tyr110 and Arg150) and van der Waal contacts with residues (Asn49,Thr50, Als51, Phe54, Thr55, Tyr112, Pro118, Pro152, and Ala157) of β1, β2, β4, β5, β6, and β7 of the β-sheet core of the protein. Bhowmik et al. reported strong binding of Rutin, Doxycycline, Caffeic acid, Ferulic acid, Simeprevir, and Grazoprevir with several functional residues of the SARS-CoV-2 nucleocapsid protein reported in this study [71].

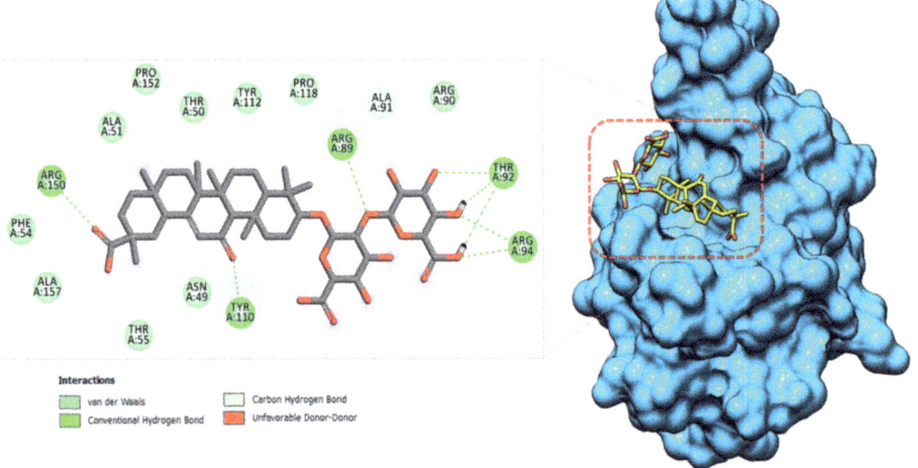

Figure 6. Binding pose of glycyrrhizin (in yellow stick) at the junction binding pocket of PLpro (in deep sky blue surface). A 2D illustration of the glycyrrhizin chemical interactions at the docked site is also provided.

2.2. MD Simulation Analysis

In computer-aided drug design, MD simulations are essential in providing detailed biomolecule dynamical structural information and surface wealth of protein–ligand interactions, energetic data that are foremost to understanding the structural–functionality relationship of target protein principle in ligand recognition/interactions [37,72,73]. This set of information has tremendous applications in guiding novel drug design, thereby making MD simulation a successful tool in the modern drug discovery framework.

2.2.1. Root-Mean-Square Deviation (RMSD) Analysis

MD simulation of 50 ns was performed for each receptor with bound glycyrrhizin to elucidate the compound binding stability and extract receptors/compound structural information that is key in the binding that may be altered to iMprove binding conformation and, ultimately, compound affinity for the target biomolecules. First, RMSD of receptors in each complex was estimated as carbon alpha deviations by superimposing 50,000 snapshots over the initial reference structure versus time (Figure 7A). RMSDs of all three complexes were found: Mpro (maximum, 3.14 Å; mean, 1.97 Å), PLpro (maximum, 2.59 Å; mean,

1.64 Å), and Nucleocapsid (maximum, 2.34 Å; mean, 1.32 Å). All of the receptors are relatively stable in terms of 3D structure, and no flexibility in secondary structures was noticed. As a consequence, glycyrrhizin binding pose was not altered, thus reflecting strong and stable complex formation.

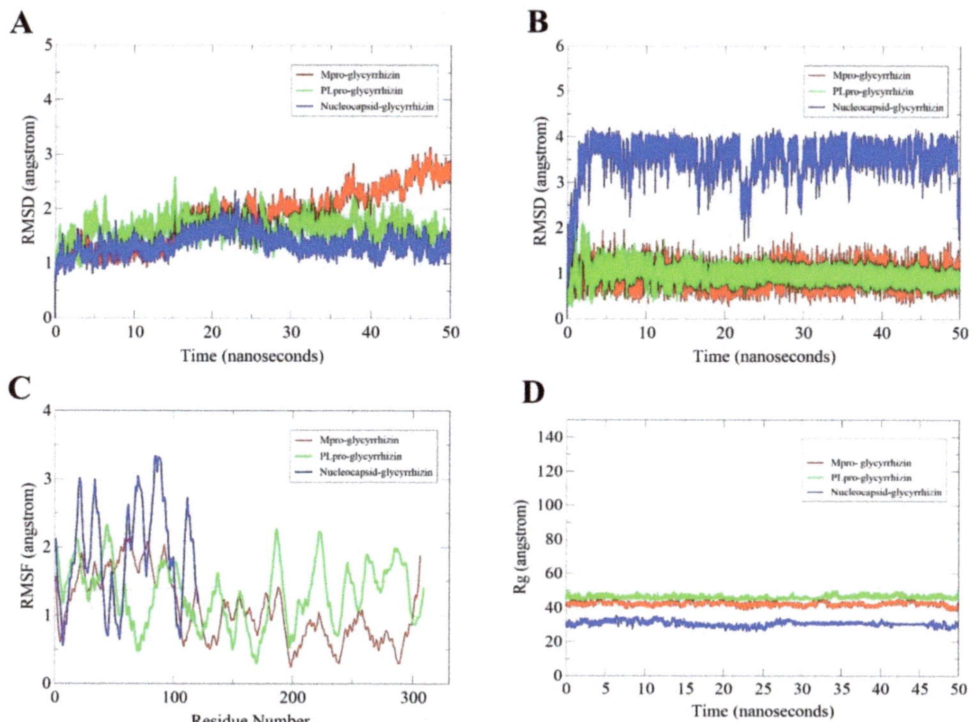

Figure 7. Statistical parameters calculated based on simulation trajectories. Receptor RMSD plots (**A**), glycyrrhizin RMSD plots (**B**), receptor RMSF (**C**), and receptor Rg (**D**).

2.2.2. Glycyrrhizin Conformation Stability

In addition, the MD simulation trajectories were examined to disclose information about the glycyrrhizin conformation stability with the receptors (Figure 7B). The glycyrrhizin RMSD with the receptors is Mpro (maximum, 2.56 Å; mean, 0.93 Å), PL-pro (maximum, 2.14 Å; mean, 0.96 Å), and Nucleocapsid (maximum, 4.20 Å; mean, 3.48 Å). The molecules disclosed high stable, except for some deviations in the gly-cyrrhizin binding mode with the Nucleocapsid protein; therefore, the end MD simulation snapshot over the initial was superimposed to understand the compound dynamics. The (2*S*,3*S*,4*S*,5*R*,6*R*)-6-(((2*S*,3*R*,4*S*,5*S*,6*S*)-6-carboxy-2,4,5-trihydroxytetrahydro-2*H*-pyran-3-yl)oxy)-3,4,5-trihydroxytetrahydro-2*H*-pyran-2-carboxylic acid fragment of the glycyrrhizin is flexible in an attempt to establish a more stable conformation. This moiety left its original site of interaction and moved more towards the β-core sheet for binding (Figure 8).

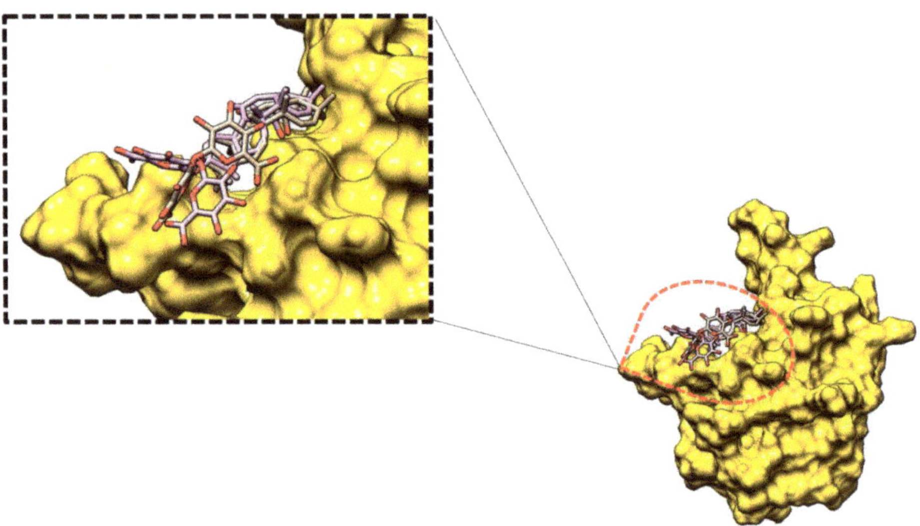

Figure 8. Binding mode dynamics of glycyrrhizin in the MD simulation at the initial time (in tan stick) versus at the end time (in plum stick).

2.2.3. Root-Mean-Square Fluctuation (RMSF) Analysis

The residual flexibility and stability of the receptors in the presence of glycyrrhizin were further elucidated (Figure 7C). Mean RMSF for Mpro is 1.4 Å, PLpro is 1.57 Å, and Nucleocapsid is 1.9 Å. These values suggest good agreement on intermolecular stability.

2.2.4. Radius of Gyration (Rg) Analysis

Additionally, Rg analysis was performed to evaluate protein compactness and structural equilibrium over the simulation time (Figure 7D). The Rg of the systems follows: Mpro–glycyrrhizin (45.62 Å and 42.28 Å), PLpro–glycyrrhizin (50.29 Å and 46.23 Å), and Nucleocapsid–glycyrrhizin (35.71 Å and 30.70 Å). All three systems are quite stable and remain compact.

2.3. MMGB/PBSA Analysis

To get a deeper insight into the compounds binding potential with the SARS-CoV-2 enzymes used, binding free energies were estimated using MMGBSA and MMPBSA techniques. Additionally, per residue decomposition assay was accomplished to highlight residues that contribute majorly to the compound's stability at the docked position and, ultimately, to the strong intermolecular interactions. To this objective, 100 frames were picked at time intervals of 50 ps from the simulation trajectories, discarding the water molecules and counterions. Detailed binding energies of the complexes are listed in Table 1 All of the binding interactions are energetically favorable, resulting in the formation of stable complexes. In all of the complexes, gas-phase energy dominates the system energy with significant contribution from van der Waals compared to electrostatic energy's minor role. The polar solvation energy is illustrated to play a nonfavorable part in binding, whereas the nonpolar energy seems to be vital in complex equilibration. The MMGBSA net binding-energy-ranked stability of the complexes follows: PLpro–glycyrrhizin > Spike–glycyrrhizin > Nucleocapsid–glycyrrhizin > Mpro–glycyrrhizin. The MMPBSA ranking follows: PLpro–glycyrrhizin > Spike–glycyrrhizin > Mpro–glycyrrhizin > N–glycyrrhizin.

Table 1. Binding free energy components of SARS-CoV-2 enzyme complexes with glycyrrhizin. The energy values are provided in units of kcal/mol.

Method	Energy Component	Mpro–Glycyrrhizin	PLpro–Glycyrrhizin	Nucleocapsid–Glycyrrhizin
MMGBSA	Van der Waals Energy	−36.50	−61.10	−37.97
	Electrostatic Energy	−13.92	−8.53	8.75
	Polar Solvation Energy	30.19	26.79	3.15
	Nonpolar Solvation Energy	−4.19	−5.85	−3.97
	Gas Phase Energy	−50.42	−69.63	−29.22
	Solvation Energy	25.99	20.93	−0.82
	Total Binding Energy	−24.42	−48.69	−30.05
MMPBSA	Van der Waals Energy	−36.50	−61.10	−37.97
	Electrostatic Energy	−13.92	−8.53	8.75
	Polar Solvation Energy	42.56	35.77	6.65
	Nonpolar Solvation Energy	−2.94	−4.31	−3.38
	Gas Phase Energy	−50.42	−69.63	−29.22
	Solvation Energy	39.62	31.46	3.27
	Total Binding Energy	−10.80	−38.17	−25.95

2.4. Per-Residue Decomposition

The atomic-level contribution of each residue from the enzymes to the compound binding was elucidated further. Those with an average binding energy of <1 kcal/mol were categorized as hotspot residues because of their significant overall complex stability contribution [74,75]. In the case of Mpro–glycyrrhizin interaction, Asn238 and Asp289 are vital in holding the compound at the docked site. Phe69, Asn128, Gln174, and Asp179 residues are critical in bridging PLpro enzyme with glycyrrhizin compound. The primary hotspot residues in Nucleocapsid–glycyrrhizin complex are Thr92, Arg94, Tyr110, and Arg150. It was further noticed that the van der Waals energy, as noted earlier, dominates the overall binding interaction energy. Hotspot residues of each receptor that are in direct contact and key in the stabilization of glycyrrhizin are presented in Table 2.

2.5. WaterSwap Binding Energy

WaterSwap uses an explicit solvation system that considers interaction details of protein–water, protein–water–ligand, and ligand–water. Such information is not provided in the MMGB/PBSA; therefore, it is not reliable for predicting the role of water molecules in biomolecule–ligand interactions [76]. Specifically, this holds great importance in an instance where the ligand is bridged to the receptor through water molecules. The WaterSwap method has been successfully applied to various biological systems and proved critical in determining absolute binding free energy. For each complex, the WaterSwap energies converged significantly after running 1000 frames. All the values also concluded good stability of intermolecular docked conformation. WaterSwap energies for each complex are shown in Table 3.

Table 2. Hotspot residues identified that played a significant role in interaction with the glycyrrhizin.

Complex	Residues	MMGBSA	MMPBSA
Mpro–Glycyrrhizin	Lys137	−1.74	−1.51
	Asp197	−1.76	−0.45
	Thr198	−1.50	−1.76
	Thr199	−1.18	−2.84
	Tyr237	−1.46	−1.89
	Asn238	−2.98	−3.45
	Tyr239	−1.54	−1.48
	Leu271	−1.24	−1.69
	Leu272	−1.65	−3.48
	Gln273	−1.14	−1.24
	Asn274	−1.56	−1.42
	Met276	−1.73	−1.98
	Ser284	−1.89	−1.51
	Leu286	−1.98	−1.47
	Leu287	−1.48	−2.84
	Glu288	−1.44	−1.84
	Asp289	−3.74	−3.54
	Glu290	−1.88	−5.45
PLpro–Glycyrrhizin	Phe69	−2.54	−3.54
	His73	−2.11	−2.45
	Thr74	−1.82	−1.45
	Asp76	−1.99	−1.68
	Phe79	−1.47	−1.46
	Arg82	−1.82	−1.12
	Asn128	−4.41	−1.39
	Tyr154	−1.61	−1.48
	Asn156	−1.61	−5.24
	Phe173	−1.11	−1.58
	Gln174	−5.48	−3.61
	His175	−1.94	−1.48
	Ala176	−1.69	−1.12
	Asn177	−1.81	−1.62
	Leu178	−1.64	−1.11
	Asp179	−2.47	−1.83
	Val202	−1.43	−1.19
Nucleocapsid–Glycyrrhizin	Asn49	−1.99	−2.54
	Thr50	−1.81	−1.42
	Ala51	−1.25	−2.45
	Phe54	−1.66	−3.15
	Thr55	−1.65	−1.12
	Arg89	−2.74	−1.84
	Thr92	−2.45	−3.65
	Arg94	−4.66	−2.48
	Tyr112	−1.45	−1.24
	Tyr110	−3.74	−3.51
	Pro118	−1.89	−1.48
	Arg150	−2.78	−3.58

Table 3. WaterSwap absolute binding energy estimation for all four complexes.

Algorithm	Mpro–Glycyrrhizin	PLpro–Glycyrrhizin	Nucleocapsid–Glycyrrhizin
Bennett's	−22.39	−25.84	−22.34
Free energy perturbation	−22.48	−25.94	−23.83
Thermodynamic integration	−22.47	−24.61	−23.45
Mean	−22.44	−25.46	−23.30

3. Materials and Methods

3.1. Target Proteins Preparation

The anti-SARS-CoV-2 targets (Mpro PDB code: 7BQY, PLpro PDB code: 6XAA, and Nucleocapsid PDB code: 6M3M) were retrieved and prepared using the AMBER18 program [77]. Ff14SB force field [78] was used for amino acid parameterization. To add complementary hydrogen atoms missed by the crystallography, the tleap module of AmberTools18 was employed. Energy minimization of the targeted proteins was done first for 1000 steepest descent steps, and then by 500 conjugate gradient steps, allowing the step size to be 0.02 Å. Charge addition was done through the Gasteiger method.

3.2. Compound Preparation

The MPD3 phytochemical database (https://www.bioinformation.info/), in addition to reported natural antiviral compounds, were used in this study to filter molecules that show best binding affinity to the selected SARS-CoV-2 multiple targets. The library containing ~5000 natural compounds was imported to PyRx 0.8 software [79], where they were minimized for optimal energy and followed by conversion to pdbqt format for use in virtual screening against the mentioned targets.

3.3. Structure-Based Virtual Screening

Virtual screening of the compounds against of the targets used was done using the AutoDock Vina in PyRx [80] on Windows 10-supported Dell system (processor: Intel(R) Core(TM) i7-8550U CPU @ 1.80 GHz with a 64-bit operating system, ×64-based processor, a memory of 8.00 GB). First, the docking protocol was validated by docking cocrystallized ligands to the protein keeping the docking parameters default except for the sphere around the binding site, which was set to 15 Å. Validation was also done by comparing the best-ranked compounds conformation relative to the crystallized ligand by root-mean-square deviation (RMSD) [81]. Docking of the compound to the targets was accomplished by using the same set of parameters described for the validation procedure and run in triplicates to absolute consistency of the results. The docked solutions were clustered, considering an RMSD value of 1 Å. The binding mode of compounds with the lowest binding energy in kcal/mol was refined in MD simulations.

3.4. MD Simulations

MD simulations of the docked solutions were performed using AMBER18 [77]. Each top complex was explicitly solvated with water molecules, and then to get a neutral system, counter ions were added. Afterward, using the TIP3P solvent model, a water box of thickness 12 Å was created to surround the complex [82]. Simulation of the complex was done through periodic boundary conditions where electrostatic interactions were modeled with the particle–mesh Ewald procedure [74]. In the process, a threshold value of 8 Å was defined for nonbounded interactions. Water molecules were minimized for 500 cycles, followed by complete system minimization for 1000 rounds. Then, each system temperature was gradually scaled to 300 K. Equilibration of the systems was achieved under the NPT ensemble for 100 ps. This involves equilibration of both counter ions and water molecules while considering restraint on solutes in the first phase for 50 ps; subsequent protein side chains were relaxed. MD simulation of 50 ns was performed

at 300 K and 1 atm for two fs under the NPT ensemble. Hydrogen and covalent bonds were constrained using the SHAKE algorithm [83], whereas system temperature was controlled through Langevin dynamics [84]. The initial structure was used as a reference, and CPPTRAJ [85] of AMBER was run to generate a root-mean-square deviation (RMSD) plot to check the system MD simulation convergence [81]. Ligand structural flexibilities were calculated by ligand RMSD. Furthermore, hydrogen bond analysis was performed to investigate hydrogen bonds formed between the compounds and amino acids present within the docked site vicinity.

3.5. MMGB/PBSA Analysis

The binding free energy (ΔG binding) of the complexes was estimated using the AMBER18 MM/PBSA method [42,86]. One hundred snapshots were considered from simulation trajectories at a regular time interval to calculate the free energy difference.

$$\Delta G_{binding} = G_{complex} - (G_{protein} + G_{ligand})$$

$$\Delta G = \Delta G_{gas} + \Delta G_{solv} - T\Delta S$$

$$\Delta G_{gas} = \Delta ele + \Delta Gvdw$$

$$\Delta Gsolv = \Delta G_{GB} + \Delta G_{SA}$$

$$\Delta G_{SA} = \gamma \times SASA \times b$$

In these equations, $G_{complex}$ is delta free energy of the complex, $G_{protein}$ is delta free energy of the protein, and G_{ligand} is delta free energy of the ligand; ΔG_{gas} represents gas-phase energy and can be split into delta electrostatic (ΔE_{ele}), and delta van der Waals (ΔE_{vdw}) energy; and the ΔG_{solv} term stands for solvation free energy, which comprises polar (ΔG_{GB}) and nonpolar (ΔG_{SA}) energy. In the ΔG_{GB}, the εw value is set to 80, and εp is selected as 1.0. Linear combinations of the pairwise overlap method are used to estimate the solvent-accessible surface area (SASA).

3.6. WaterSwap Analysis

WaterSwap [76,87] was additionally done over the last 10 ns of MD simulation for a total of default 1000 iterations, keeping the sample size of Monte Carlo simulation to 1.6×10^9. The absolute binding energy of each complex was estimated using three useful algorithms: thermodynamics integration, free energy perturbation, and Bennett's. The energy value <1 kcal/mol represents a good convergence of the system [75].

4. Conclusions

In this study, we found glycyrrhizin as the most significant natural compound that can act as a double-edged sword and inhibit multiple proteins of SARS-CoV-2. This compound has a high binding affinity for all of the SARS-CoV-2 receptors used in this study and had a stable binding mode in the MD simulation time. The compound revealed important interactions with all receptors, and thus requires further consideration in future anti-SARS-CoV-2 therapeutic studies. Glycyrrhizin has been previously documented to have therapeutic applications against SARS-CoV, chronic hepatitis C, and HIV-1 [88]. The molecule is clinically useful and had few toxic reactions. One way to overcome toxicity is by allowing low concentration of the drug in the cells (<100 µg/mL) [89]. Glycyrrhizin has been reported to inhibit viral penetration and effective both during the viral infection and postinfection [90]. It was previously demonstrated that the glycyrrhizin binds with good affinity to the human ACE2 and interacts with Asp30, Gln288, Arg393, and Arg559 residues, hence also underlines its potential to target the SARS-CoV-2 Spike protein RBD attachment to the human ACE2 receptor [90]. It also was shown that glycyrrhizin can be employed in synergism along with other plant-based molecules to treat SARS-CoVs [91]. From a pharmacological perspective, the glycyrrhizin prevents the production of intracellular reactive oxygen species, activates interferon production, downregulates proinflammatory

cytokines, lowers airway exudate production, and inhibits thrombin [45,92]. The compound was also computationally characterized previously to bind with good affinity to SARS-CoV-2 main protease [93]. Therefore, additional structural modification to lower the side effects and enhance the clinical efficacy of this compound is of high interest to treat SARS-related infections.

Author Contributions: Conceptualization, G.L.; data curation, Z.T.M., A.R.H., and H.M.H.A.-H.; funding acquisition, G.L.; investigation, Z.T.M.; project administration, G.L.; software, S.A.; supervision, G.L.; validation, A.R.H., H.M.H.A.-H., and S.A.; visualization, Z.T.M.; writing—original draft, Z.T.M.; writing—review and editing, A.R.H., H.M.H.A.-H., S.A., and G.L. All authors have read and agreed to the published version of the manuscript.

Funding: This research was funded by the National Natural Science Foundation of China (grant numbers: 31770333, 31370329, and 11631012).

Data Availability Statement: The data presented in this study are available within the article.

Acknowledgments: Authors would like to acknowledge Shaanxi Normal University, Xi'an, China for providing facilities for this study.

Conflicts of Interest: The authors declare that the research was conducted in the absence of any commercial or finan-cial relationships that could be construed as a potential conflict of interest.

Sample Availability: Not available.

References

1. Yang, Y.; Peng, F.; Wang, R.; Guan, K.; Jiang, T.; Xu, G.; Sun, J.; Chang, C. The Deadly Coronaviruses: The 2003 SARS Pandemic and the 2020 Novel Coronavirus Epidemic in China. *J. Autoimmun.* **2020**, *109*, 102434. [CrossRef] [PubMed]
2. Wu, F.; Zhao, S.; Yu, B.; Chen, Y.-M.; Wang, W.; Song, Z.-G.; Hu, Y.; Tao, Z.-W.; Tian, J.-H.; Pei, Y.-Y.; et al. A New Coronavirus Associated with Human Respiratory Disease in China. *Nature* **2020**, *579*, 265–269. [CrossRef] [PubMed]
3. Hui, D.S.; Azhar, E.I.; Madani, T.A.; Ntoumi, F.; Kock, R.; Dar, O.; Ippolito, G.; Mchugh, T.D.; Memish, Z.A.; Drosten, C.; et al. The Continuing 2019-nCoV Epidemic Threat of Novel Coronaviruses to Global Health—The latest 2019 Novel Coronavirus Outbreak in Wuhan, China. *Int. J. Infect. Dis.* **2020**, *91*, 264–266. [CrossRef]
4. Ye, Z.-W.; Yuan, S.; Yuen, K.-S.; Fung, S.-Y.; Chan, C.-P.; Jin, D.-Y. Zoonotic Origins of Human Coronaviruses. *Int J. Biol. Sci.* **2020**, *16*, 1686–1697. [CrossRef]
5. Tahir ul Qamar, M.; Saleem, S.; Ashfaq, U.A.; Bari, A.; Anwar, F.; Alqahtani, S. Epitope-Based Peptide Vaccine Design and Target Site Depiction against Middle East Respiratory Syndrome Coronavirus: An Immune-Informatics Study. *J. Transl. Med.* **2019**, *17*, 362. [CrossRef]
6. Kim, D.; Lee, J.-Y.; Yang, J.-S.; Kim, J.W.; Kim, V.N.; Chang, H. The Architecture of SARS-CoV-2 Transcriptome. *Cell* **2020**, *181*, 914–921.e10. [CrossRef]
7. Zhu, N.; Zhang, D.; Wang, W.; Li, X.; Yang, B.; Song, J.; Zhao, X.; Huang, B.; Shi, W.; Lu, R.; et al. A Novel Coronavirus from Patients with Pneumonia in China, 2019. *N. Engl. J. Med.* **2020**, *382*, 727–733. [CrossRef]
8. Tahir ul Qamar, M.; Alqahtani, S.M.; Alamri, M.A.; Chen, L.-L. Structural Basis of SARS-CoV-2 3CLpro and Anti-COVID-19 Drug Discovery from Medicinal Plants. *J. Pharm. Anal.* **2020**, *10*, 313–319. [CrossRef] [PubMed]
9. Zhou, P.; Yang, X.-L.; Wang, X.-G.; Hu, B.; Zhang, L.; Zhang, W.; Si, H.-R.; Zhu, Y.; Li, B.; Huang, C.-L.; et al. A Pneumonia Outbreak Associated with a New Coronavirus of Probable Bat Origin. *Nature* **2020**, *579*, 270–273. [CrossRef] [PubMed]
10. Zhang, Y.-Z.; Holmes, E.C. A Genomic Perspective on the Origin and Emergence of SARS-CoV-2. *Cell* **2020**, *181*, 223–227. [CrossRef]
11. Fahmi, M.; Kubota, Y.; Ito, M. Nonstructural Proteins NS7b and NS8 are Likely to be Phylogenetically Associated with Evolution of 2019-nCoV. *Infect. Genet. Evol.* **2020**, *81*, 104272. [CrossRef]
12. Tahir ul Qamar, M.; Rehman, A.; Tusleem, K.; Ashfaq, U.A.; Qasim, M.; Zhu, X.; Fatima, I.; Shahid, F.; Chen, L.-L. Designing of a Next Generation Multiepitope Based Vaccine (MEV) against SARS-COV-2: Immunoinformatics and in Silico Approaches. *PLoS ONE* **2020**, *15*, e0244176. [CrossRef]
13. Lenzen, M.; Li, M.; Malik, A.; Pomponi, F.; Sun, Y.-Y.; Wiedmann, T.; Faturay, F.; Fry, J.; Gallego, B.; Geschke, A.; et al. Global Socio-Economic Losses and Environmental Gains from the Coronavirus Pandemic. *PLoS ONE* **2020**, *15*, e0235654. [CrossRef]
14. Boopathi, S.; Poma, A.B.; Kolandaivel, P. Novel 2019 Coronavirus Structure, Mechanism of Action, Antiviral drug Promises and Rule Out against Its Treatment. *J. Biomol. Struct. Dyn.* **2020**, 1–14. [CrossRef]
15. Tahir ul Qamar, M.; Shahid, F.; Aslam, S.; Ashfaq, U.A.; Aslam, S.; Fatima, I.; Fareed, M.M.; Zohaib, A.; Chen, L.-L. Reverse Vaccinology Assisted Designing of Multiepitope-Based Subunit Vaccine Against SARS-CoV-2. *Infect. Dis. Poverty* **2020**, *9*, 1–14. [CrossRef] [PubMed]

16. Yi, C.; Sun, X.; Ye, J.; Ding, L.; Liu, M.; Yang, Z.; Lu, X.; Zhang, Y.; Ma, L.; Gu, W.; et al. Key Residues of the Receptor Binding Motif in the Spike Protein of SARS-CoV-2 That Interact with ACE2 and Neutralizing Antibodies. *Cell. Mol. Immunol.* **2020**, *17*, 621–630. [CrossRef]

17. Chen, R.; Fu, J.; Hu, J.; Li, C.; Zhao, Y.; Qu, H.; Wen, X.; Cao, S.; Wen, Y.; Wu, R.; et al. Identification of the Immunodominant Neutralizing Regions in the Spike Glycoprotein of Porcine Deltacorona-Virus. *Virus Res.* **2020**, *276*, 197834. [CrossRef] [PubMed]

18. Gui, M.; Song, W.; Zhou, H.; Xu, J.; Chen, S.; Xiang, Y.; Wang, X. Cryo-Electron Microscopy Structures of the SARS-CoV Spike Glycoprotein Reveal a Prerequisite Conformational State for Receptor Binding. *Cell Res.* **2017**, *27*, 119–129. [CrossRef]

19. Jin, Z.; Zhao, Y.; Sun, Y.; Zhang, B.; Wang, H.; Wu, Y.; Zhu, Y.; Zhu, C.; Hu, T.; Du, X.; et al. Structural Basis for the Inhibition of SARS-CoV-2 Main Protease by Antineoplastic Drug Carmofur. *Nat. Struct. Mol. Biol.* **2020**, *27*, 529–532. [CrossRef] [PubMed]

20. Kang, S.; Yang, M.; Hong, Z.; Zhang, L.; Huang, Z.; Chen, X.; He, S.; Zhou, Z.; Zhou, Z.; Chen, Q.; et al. Crystal Structure of SARS-CoV-2 Nucleocapsid Protein RNA Binding Domain Reveals Potential Unique Drug Tar-Geting Sites. *Acta Pharm. Sin. B* **2020**, *19*, 1228–1238. [CrossRef]

21. Elfiky, A.A. SARS-CoV-2 RNA Dependent RNA Polymerase (RdRp) Targeting: An in Silico Perspective. *J. Biomol. Struct. Dyn.* **2020**, 1–15. [CrossRef]

22. Shin, D.; Mukherjee, R.; Grewe, D.; Bojkova, D.; Baek, K.; Bhattacharya, A.; Schulz, L.; Widera, M.; Mehdipour, A.R.; Tascher, G.; et al. Papain-Like Protease Regulates SARS-CoV-2 Viral Spread and Innate Immunity. *Nature* **2020**, *587*, 657–662. [CrossRef] [PubMed]

23. Gyebi, G.A.; Ogunro, O.B.; Adegunloye, A.P.; Ogunyemi, O.M.; Afolabi, S.O. Potential Inhibitors of Coronavirus 3-Chymotrypsin-Like Protease (3CLpro): An in Silico Screening of Alkaloids and Terpenoids from African Medicinal Plants. *J. Biomol. Struct. Dyn.* **2020**, 1–19. [CrossRef] [PubMed]

24. Tomar, P.P.S.; Arkin, I.T. SARS-CoV-2 E Protein Is a Potential Ion Channel That Can Be Inhibited by Gliclazide and Memantine. *Biochem. Biophys. Res. Commun.* **2020**, *530*, 10–14. [CrossRef] [PubMed]

25. Yan, R.; Zhang, Y.; Li, Y.; Xia, L.; Guo, Y.; Zhou, Q. Structural Basis for the Recognition of SARS-CoV-2 by full-Length Human ACE2. *Science* **2020**, *367*, 1444–1448. [CrossRef]

26. Hirano, T.; Murakami, M. COVID-19: A New Virus, but a Familiar Receptor and Cytokine Release Syndrome. *Immunity* **2020**, *52*, 731–733. [CrossRef]

27. Du, L.; Zhao, G.; Lin, Y.; Chan, C.; He, Y.; Jiang, S.; Wu, C.; Jin, D.-Y.; Yuen, K.-Y.; Zhou, Y.; et al. Priming with rAAV Encoding RBD of SARS-CoV S Protein and Boosting with RBD-Specific Peptides for T Cell Epitopes Elevated Humoral and Cellular Immune Responses Against SARS-CoV Infection. *Vaccine* **2008**, *26*, 1644–1651. [CrossRef] [PubMed]

28. Zhang, L.; Lin, D.; Sun, X.; Curth, U.; Drosten, C.; Sauerhering, L.; Becker, S.; Rox, K.; Hilgenfeld, R. Crystal Structure of SARS-CoV-2 Main Protease Provides a Basis for Design of Improved α-Ketoamide Inhibitors. *Science* **2020**, *368*, 409–412. [CrossRef]

29. Kumar, Y.; Singh, H.; Patel, C.N. In Silico Prediction of Potential Inhibitors for the Main Protease of SARS-CoV-2 Using Molecular Docking and Dynamics Simulation Based Drug-Repurposing. *J. Infect. Public Health* **2020**, *13*, 1210–1223. [CrossRef]

30. Alamri, M.A.; Tahir ul Qamar, M.; Mirza, M.U.; Bhadane, R.; Alqahtani, S.M.; Muneer, I.; Froeyen, M.; Salo-Ahen, O.M.H. Pharmacoinformatics and Molecular Dynamics Simulation Studies Reveal Potential Covalent and FDA-Approved Inhibitors of SARS-CoV-2 Main Protease 3CLpro. *J. Biomol. Struct. Dyn.* **2020**, 1–13. [CrossRef]

31. Dai, W.; Zhang, B.; Jiang, X.-M.; Su, H.; Li, J.; Zhao, Y.; Xie, X.; Jin, Z.; Peng, J.; Liu, F.; et al. Structure-Based Design of Antiviral Drug Candidates Targeting the SARS-CoV-2 Main Protease. *Science* **2020**, *368*, 1331–1335. [CrossRef] [PubMed]

32. Joshi, R.S.; Jagdale, S.S.; Bansode, S.B.; Shankar, S.S.; Tellis, M.B.; Pandya, V.K.; Chugh, A.; Giri, A.P.; Kulkarni, M.J. Discovery of Potential Multi-Target-Directed Ligands by Targeting Host-Specific SARS-CoV-2 Structurally Conserved Main Proteases. *J. Biomol. Struct. Dyn.* **2020**, 1–16. [CrossRef] [PubMed]

33. Havranek, B.; Islam, S.M. An in Silico Approach for Identification of Novel Inhibitors as Potential Therapeutics Targeting COVID-19 Main Protease. *J. Biomol. Struct. Dyn.* **2020**, 1–12. [CrossRef] [PubMed]

34. Alamri, M.A.; Tahir ul Qamar, M.; Mirza, M.U.; Alqahtani, S.M.; Froeyen, M.; Chen, L.-L. Discovery of Human Coronaviruses Pan-Papain-like Protease Inhibitors Using Computational Approaches. *J. Pharm. Anal.* **2020**, *10*, 546–559. [CrossRef] [PubMed]

35. Riaz, M.; Ashfaq, U.A.; Qasim, M.; Yasmeen, E.; Tahir ul Qamar, M.; Anwar, F. Screening of Medicinal Plant Phytochemicals as Natural Antagonists of p53-MDM2 Interaction to Reactivate p53 Functioning. *Anticancer Drugs* **2017**, *28*, 1032–1038. [CrossRef] [PubMed]

36. Rehan Khalid, R.; Tahir ul Qamar, M.; Maryam, A.; Ashique, A.; Anwar, F.H.; Geesi, M.; Siddiqi, A.R. Comparative Studies of the Dynamics Effects of BAY60-2770 and BAY58-2667 Binding with Human and Bacterial H-NOX Domains. *Molecules* **2018**, *23*, 2141. [CrossRef] [PubMed]

37. Durdagi, S.; Tahir ul Qamar, M.; Salmas, R.E.; Tariq, Q.; Anwar, F.; Ashfaq, U.A. Investigating the Molecular Mechanism of Staphylococcal DNA Gyrase Inhibitors: A Combined Ligand-Based and Structure-Based Resources Pipeline. *J. Mol. Graph. Model.* **2018**, *85*, 122–129. [CrossRef]

38. Muneer, I.; Tusleem, K.; Abdul Rauf, S.; Hussain, H.M.J.; Siddiqi, A.R. Others Discovery of Selective Inhibitors for Cyclic AMP Response Element-Binding Protein: A Combined Ligand and Structure-Based Resources Pipeline. *Anti-Cancer Drugs* **2019**, *30*, 363–373.

39. Yu, W.; MacKerell, A.D. Computer-Aided Drug Design Methods. In *Antibiotics*; Springer: Berlin/Heidelberg, Germany, 2017; pp. 85–106.

40. Tahir ul Qamar, M.; Ashfaq, U.A.; Tusleem, K.; Mumtaz, A.; Tariq, Q.; Goheer, A.; Ahmed, B. In-Silico Identification and Evaluation of Plant Flavonoids as Dengue NS2B/NS3 Protease Inhibitors Using Molecular Docking and Simulation Approach. *Pak. J. Pharm. Sci.* **2017**, *30*, 2119–2137.

41. Durrant, J.D.; McCammon, J.A. Molecular Dynamics Simulations and Drug Discovery. *BMC Biol.* **2011**, *9*, 71. [CrossRef]

42. Genheden, S.; Ryde, U. The MM/PBSA and MM/GBSA Methods to Estimate Ligand-Binding Affinities. *Expert Opin. Drug Discov.* **2015**, *10*, 449–461. [CrossRef] [PubMed]

43. Mumtaz, A.; Ashfaq, U.A.; Qamar, U.M.T.; Anwar, F.; Gulzar, F.; Ali, M.A.; Saari, N.; Pervez, M.T. MPD3: A Useful Medicinal Plants Database for Drug Designing. *Nat. Prod. Res.* **2017**, *31*, 1228–1236. [CrossRef] [PubMed]

44. Santos, I.d.A.; Grosche, V.R.; Bergamini, F.R.G.; Sabino-Silva, R.; Jardim, A.C.G. Antivirals against Coronaviruses: Candidate Drugs for SARS-CoV-2 Treatment? *Front. Microbiol.* **2020**, *11*, 1818. [CrossRef] [PubMed]

45. LuoLiu, P.; Li, J. Pharmacologic Perspective: Glycyrrhizin May Be an Efficacious Therapeutic Agent for COVID-19. *Int. J. Antimicrob. Agents* **2020**, 105995.

46. Elshabrawy, H.A. SARS-CoV-2: An Update on Potential Antivirals in Light of SARS-CoV Antiviral Drug Discoveries. *Vaccines* **2020**, *8*, 335. [CrossRef]

47. Kato, F.; Matsuyama, S.; Kawase, M.; Hishiki, T.; Katoh, H.; Takeda, M. Antiviral Activities of Mycophenolic Acid and IMD-0354 Against SARS-CoV-2. *Microbiol. Immunol.* **2020**, *64*, 635–639. [CrossRef]

48. Case, D.; Ben-Shalom, I.; Brozell, S.; Cerutti, D.; Cheatham III, T.; Cruzeiro, V.; Darden, T.; Duke, R.; Ghoreishi, D.; Gilson, M.; et al. *AMBER 18*; University of California: San Francisco, CA, USA, 2018.

49. Maier, J.A.; Martinez, C.; Kasavajhala, K.; Wickstrom, L.; Hauser, K.E.; Simmerling, C. ff14SB: Improving the Accuracy of Protein Side Chain and Backbone Parameters From ff99SB. *J. Chem. Theory Comput.* **2015**, *11*, 3696–3713. [CrossRef]

50. Dallakyan, S.; Olson, A.J. Small-Molecule Library Screening by Docking with PyRx. In *Chemical Biology*; Springer: Berlin/Heidelberg, Germany, 2015; pp. 243–250.

51. Trott, O.; Olson, A.J. Auto Dock Vina: Improving the Speed and Accuracy of Docking with a New Scoring Function, Efficient Optimization, and Multi-Threading. *J. Comput. Chem.* **2010**, *31*, 455–461.

52. Maiorov, V.N.; Crippen, G.M. Significance of Root-Mean-Square Deviation in Comparing Three-Dimensional Structures of Globular Proteins. *J. Mol. Biol.* **1994**, *235*, 625–634. [CrossRef]

53. Andleeb, S.; Imtiaz-Ud-Din; Rauf, M.K.; Azam, S.S.; Badshah, A.; Sadaf, H.; Raheel, A.; Tahir, M.N.; Raza, S. A One-Pot Multicomponent Facile Synthesis of Dihydropyrimidin-2(1: H)-Thione Derivatives Using Triphenyl-Germane as a Catalyst and Its Binding Pattern Validation. *RSC Adv.* **2016**, *6*, 79651–79661. [CrossRef]

54. Abro, A.; Azam, S.S. Binding Free Energy Based Analysis of Arsenic (+3 Oxidation State) Methyltransferase with S-Adenosylmethionine. *J. Mol. Liq.* **2016**, *220*, 375–382. [CrossRef]

55. Kräutler, V.; Van Gunsteren, W.F.; Hünenberger, P.H. A fast SHAKE Algorithm to Solve Distance Constraint Equations for Small Molecules in Molecular Dynamics Simulations. *J. Comput. Chem.* **2001**, *22*, 501–508. [CrossRef]

56. Izaguirre, J.A.; Catarello, D.P.; Wozniak, J.M.; Skeel, R.D. Langevin Stabilization of Molecular Dynamics. *J. Chem. Phys.* **2001**, *114*, 2090–2098. [CrossRef]

57. Roe, D.R.; Cheatham III, T.E. PTRAJ and CPPTRAJ: Software for Processing and Analysis of Molecular Dynamics Trajectory Data. *J. Chem. Theory Comput.* **2013**, *9*, 3084–3095. [CrossRef] [PubMed]

58. Miller, B.R.; McGee, T.D.; Swails, J.M.; Homeyer, N.; Gohlke, H.; Roitberg, A.E. MMPBSA.py: An Efficient Program for End-State Free Energy Calculations. *J. Chem. Theory Comput.* **2012**, *8*, 3314–3321. [CrossRef]

59. Woods, C.J.; Malaisree, M.; Michel, J.; Long, B.; McIntosh-Smith, S.; Mulholland, A.J. Rapid Decomposition and Visualisation of Protein-Ligand Binding Free Energies by Residue and by Water. *Faraday Discuss.* **2014**, *169*, 477–499. [CrossRef]

60. Woods, C.J.; Malaisree, M.; Hannongbua, S.; Mulholland, A.J. A Water-Swap Reaction Coordinate for the Calculation of Absolute Protein-Ligand Binding Free Energies. *J. Chem. Phys.* **2011**, *134*. [CrossRef]

61. Kiani, Y.S.; Ranaghan, K.E.; Jabeen, I.; Mulholland, A.J. Molecular Dynamics Simulation Framework to Probe the Binding Hypothesis of CYP3A4 Inhibitors. *Int. J. Mol. Sci.* **2019**, *20*, 4468. [CrossRef]

62. Ahmed, B.; Ashfaq, U.A.; Tahir ul Qamar, M.; Ahmad, M. Anticancer Potential of Phytochemicals against Breast Cancer: Molecular Docking and Simulation Approach. *Bangladesh J. Pharmacol.* **2014**, *9*, 545–550. [CrossRef]

63. Tahir ul Qamar, M.; Mumtaz, A.; Ashfaq, U.A.; Adeel, M.M.; Fatima, T. Potential of Plant Alkaloids as Dengue ns3 Protease Inhibitors: Molecular Docking and Simulation Approach. *Bangladesh J. Pharmacol.* **2014**, *9*, 262–267. [CrossRef]

64. Ashfaq, U.A.; Jalil, A.; Tahir ul Qamar, M. Antiviral Phytochemicals Identification from Azadirachta Indica Leaves against HCV NS3 Protease: An in Sili-Co Approach. *Nat. Prod. Res.* **2016**, *30*, 1866–1869. [CrossRef] [PubMed]

65. Tahir ul Qamar, M.; Kiran, S.; Ashfaq, U.A.; Javed, M.R.; Anwar, F.; Ali, M.A.; Gilani, A. ul H. Discovery of Novel Dengue NS2B/NS3 Protease Inhibitors Using Pharmacophore Modeling and Molecular Docking Based Virtual Screening of the Zinc Database. *Int. J. Pharmacol.* **2016**, *12*, 621–632. [CrossRef]

66. Morris, G.M.; Lim-Wilby, M. Molecular Docking. In *Molecular Modeling of Proteins*; Springer: Berlin/Heidelberg, Germany, 2008; pp. 365–382.

67. Muhseen, Z.T.; Hameed, A.R.; Al-Hasani, H.M.H.; Tahir ul Qamar, M.; Li, G. Promising Terpenes as SARS-CoV-2 Spike Receptor-Binding Domain (RBD) Attachment Inhibitors to the Human ACE2 Receptor: Integrated Computational Approach. *J. Mol. Liq.* **2020**, *320*, 114493. [CrossRef] [PubMed]
68. Ton, A.-T.; Gentile, F.; Hsing, M.; Ban, F.; Cherkasov, A. Rapid Identification of Potential Inhibitors of SARS-CoV-2 Main Protease by Deep Docking of 1.3 Billion Compounds. *Mol. Inform.* **2020**, *39*, 2000028. [CrossRef]
69. Jin, Z.; Du, X.; Xu, Y.; Deng, Y.; Liu, M.; Zhao, Y.; Zhang, B.; Li, X.; Zhang, L.; Peng, C.; et al. Structure of M Pro from SARS-CoV-2 and Discovery of its Inhibitors. *Nature* **2020**, *582*, 289–293. [CrossRef] [PubMed]
70. Xue, X.; Yu, H.; Yang, H.; Xue, F.; Wu, Z.; Shen, W.; Li, J.; Zhou, Z.; Ding, Y.; Zhao, Q.; et al. Structures of Two Coronavirus Main Proteases: Implications for Substrate Binding and Antiviral Drug Design. *J. Virol.* **2008**, *82*, 2515–2527. [CrossRef]
71. Kneller, D.W.; Phillips, G.; O'Neill, H.M.; Jedrzejczak, R.; Stols, L.; Langan, P.; Joachimiak, A.; Coates, L.; Kovalevsky, A. Structural Plasticity of the SARS-CoV-2 3CL Mpro Active Site Cavity Revealed by Room Temperature X-ray Crystallography. *Nat. Commun.* **2020**, *11*, 3202. [CrossRef]
72. Sacco, M.D.; Ma, C.; Lagarias, P.; Gao, A.; Townsend, J.A.; Meng, X.; Dube, P.; Zhang, X.; Hu, Y.; Kitamura, N.; et al. Structure and Inhibition of the SARS-CoV-2 Main Protease Reveal Strategy for Developing Dual Inhibitors against Mpro and Cathepsin L. *Sci. Adv.* **2020**, *6*, eabe0751. [CrossRef]
73. Alberto, J.-A.; Ribas-Aparicio, R.M.; Ozores, A.G.; Vega, C.J.A. Virtual Screening of Approved Drugs as Potential SARS-CoV-2 Main Protease Inhibitors. *Comput. Biol. Chem.* **2020**, *88*, 107325. [CrossRef]
74. Ngo, S.T.; Pham, Q.A.N.; Le, T.L.; Pham, D.-H.; Vu, V.V. Computational Determination of Potential Inhibitors of SARS-CoV-2 Main Protease. *J. Chem. Inf. Model.* **2020**, *60*, 5771–5780. [CrossRef]
75. Wang, J. Fast Identification of Possible Drug Treatment of Coronavirus Disease-19 (COVID-19) through Computational Drug Repurposing Study. *J. Chem. Inf. Model.* **2020**, *60*, 3277–3286. [CrossRef] [PubMed]
76. Gao, X.; Qin, B.; Chen, P.; Zhu, K.; Hou, P.; Wojdyla, J.A.; Wang, M.; Cui, S. Crystal structure of SARS-CoV-2 papain-like protease. *Acta Pharm. Sin. B* **2020**, *11*, 237–245. [CrossRef] [PubMed]
77. Fu, Z.; Huang, B.; Tang, J.; Liu, S.; Liu, M.; Ye, Y.; Liu, Z.; Xiong, Y.; Cao, D.; Li, J.; et al. Structural Basis for the Inhibition of the Papain-Like Protease of SARS-CoV-2 by Small Molecules. *Biorxiv* **2020**. [CrossRef]
78. Petushkova, A.I.; Zamyatnin, A. A Papain-Like Proteases as Coronaviral Drug Targets: Current Inhibitors, Opportunities, and Limitations. *Pharmaceuticals* **2020**, *13*, 277. [CrossRef] [PubMed]
79. Park, J.-Y.; Kim, J.H.; Kim, Y.M.; Jeong, H.J.; Kim, D.W.; Park, K.H.; Kwon, H.-J.; Park, S.-J.; Lee, W.S.; Ryu, Y.B. Tanshinones as Selective and Slow-Binding Inhibitors for SARS-CoV Cysteine Proteases. *Bioorg. Med. Chem.* **2012**, *20*, 5928–5935. [CrossRef] [PubMed]
80. Chu, H.-F.; Chen, C.-C.; Moses, D.C.; Chen, Y.-H.; Lin, C.-H.; Tsai, Y.-C.; Chou, C.-Y. Porcine Epidemic Diarrhea Virus Papain-Like Protease 2 Can Be Noncompetitively Inhibited by 6-Thioguanine. *Antiviral Res.* **2018**, *158*, 199–205. [CrossRef]
81. Kouznetsova, V.L.; Zhang, A.; Tatineni, M.; Miller, M.A.; Tsigelny, I.F. Potential COVID-19 Papain-like Protease PLpro Inhibitors: Repurposing FDA-Approved Drugs. *PeerJ* **2020**, *8*, e9965. [CrossRef]
82. Amin, S.A.; Ghosh, K.; Gayen, S.; Jha, T. Chemical-Informatics Approach to COVID-19 Drug Discovery: Monte Carlo based QSAR, vIrtual Screening and Molecular Docking Study of Some in-House Molecules as Papain-Like Protease (PLpro) Inhibitors. *J. Biomol. Struct. Dyn.* **2020**, 1–10. [CrossRef]
83. Mirza, M.U.; Ahmad, S.; Abdullah, I.; Froeyen, M. Identification of Novel Human USP2 Inhibitor and its Putative Role in Treatment of COVID-19 by Inhibiting SARS-CoV-2 Papain-Like (PLpro) Protease. *Comput. Biol. Chem.* **2020**, *89*, 107376. [CrossRef]
84. Bhati, S. Structure-Based Drug Designing of Naphthalene Based SARS-CoV PLpro Inhibitors for the Treatment of COVID-19. *Heliyon* **2020**, *6*, e05558. [CrossRef]
85. Bhowmik, D.; Nandi, R.; Jagadeesan, R.; Kumar, N.; Prakash, A.; Kumar, D. Identification of Potential Inhibitors against SARS-CoV-2 by Targeting Proteins Responsible for Envelope for-Mation and Virion Assembly Using Docking Based Virtual Screening, and Pharmacokinetics Approaches. *Infect. Genet. Evol.* **2020**, *84*, 104451. [CrossRef] [PubMed]
86. Tahir ul Qamar, M.; Maryam, A.; Muneer, I.; Xing, F.; Ashfaq, U.A.; Khan, F.A.; Anwar, F.; Geesi, M.H.; Khalid, R.R.; Rauf, S.A.; et al. Computational Screening of Medicinal Plant Phytochemicals to Discover Potent Pan-Serotype Inhibitors against Dengue Virus. *Sci. Rep.* **2019**, *9*, 1–16. [CrossRef] [PubMed]
87. Karplus, M.; McCammon, J.A. Molecular Dynamics Simulations of Biomolecules. *Nat. Struct. Mol. Biol.* **2002**, *9*, 646. [CrossRef] [PubMed]
88. Cinatl, J.; Morgenstern, B.; Bauer, G.; Chandra, P.; Rabenau, H.; Doerr, H.W. Glycyrrhizin, an Active Component of Liquorice Roots, and Replication of SARS-Associated Coronavirus. *Lancet* **2003**, *361*, 2045–2046. [CrossRef]
89. Ashfaq, U.A.; Masoud, M.S.; Nawaz, Z.; Riazuddin, S. Glycyrrhizin as Antiviral Agent against Hepatitis C Virus. *J. Transl. Med.* **2011**, *9*, 1–7. [CrossRef] [PubMed]
90. Chen, H.; Du, Q. Potential Natural Compounds for Preventing SARS-CoV-2 (2019-nCoV) Infection. *Preprints* **2020**. [CrossRef]
91. Prasad, A.; Muthamilarasan, M.; Prasad, M. Synergistic Antiviral Effects against SARS-CoV-2 by Plant-Based Molecules. *Plant Cell Rep.* **2020**, *39*, 1109–1114. [CrossRef]
92. Chrzanowski, J.; Chrzanowska, A.; Graboń, W. Glycyrrhizin: An Old Weapon against a Novel Coronavirus. *Phyther. Res.* **2020**. [CrossRef]
93. Pham, M.Q.; Vu, K.B.; Pham, T.N.H.; Tran, L.H.; Tung, N.T.; Vu, V.V.; Nguyen, T.H.; Ngo, S.T. Rapid Prediction of Possible Inhibitors for SARS-CoV-2 Main Protease Using Docking and FPL Simulations. *RSC Adv.* **2020**, *10*, 31991–31996. [CrossRef]

molecules

Article

Combining Different Docking Engines and Consensus Strategies to Design and Validate Optimized Virtual Screening Protocols for the SARS-CoV-2 3CL Protease

Candida Manelfi [1], Jonas Gossen [2,3], Silvia Gervasoni [4], Carmine Talarico [1], Simone Albani [2,3], Benjamin Joseph Philipp [2,3], Francesco Musiani [5], Giulio Vistoli [4], Giulia Rossetti [2,6,7], Andrea Rosario Beccari [1] and Alessandro Pedretti [4,*]

1   Dompé Farmaceutici SpA, Via Campo di Pile, 67100 L'Aquila, Italy; candida.manelfi@dompe.com (C.M.); carmine.talarico@dompe.com (C.T.); andrea.beccari@dompe.com (A.R.B.)
2   Computational Biomedicine, Institute for Neuroscience and Medicine (INM-9) and Institute for Advanced Simulations (IAS-5), Forschungszentrum Jülich, 52425 Jülich, Germany; j.gossen@fz-juelich.de (J.G.); s.albani@fz-juelich.de (S.A.); benjamin.joseph@rwth-aachen.de (B.J.P.); g.rossetti@fz-juelich.de (G.R.)
3   Faculty of Mathematics, Computer Science and Natural Sciences, RWTH Aachen, 52062 Aachen, Germany
4   Dipartimento di Scienze Farmaceutiche, Università degli Studi di Milano, Via Mangiagalli, 25, I-20133 Milano, Italy; silvia.gervasoni@unimi.it (S.G.); giulio.vistoli@unimi.it (G.V.)
5   Laboratory of Bioinorganic Chemistry, Department of Pharmacy and Biotechnology, University of Bologna, 40127 Bologna, Italy; francesco.musiani@unibo.it
6   Jülich Supercomputing Center (JSC), Forschungszentrum Jülich, 52425 Jülich, Germany
7   Department of Hematology, Oncology, Hemostaseology and Stem Cell Transplantation University Hospital Aachen, RWTH Aachen University, Pauwelsstraße 30, 52074 Aachen, Germany
*   Correspondence: alessandro.pedretti@unimi.it; Tel.: +39-02-5031-9332

check for **updates**

Citation: Manelfi, C.; Gossen, J.; Gervasoni, S.; Talarico, C.; Albani, S.; Philipp, B.J.; Musiani, F.; Vistoli, G.; Rossetti, G.; Beccari, A.R.; et al. Combining Different Docking Engines and Consensus Strategies to Design and Validate Optimized Virtual Screening Protocols for the SARS-CoV-2 3CL Protease. *Molecules* **2021**, *26*, 797. https://doi.org/10.3390/molecules26040797

Academic Editors: Marco Tutone and Anna Maria Almerico
Received: 13 November 2020
Accepted: 26 January 2021
Published: 4 February 2021

Publisher's Note: MDPI stays neutral with regard to jurisdictional claims in published maps and institutional affiliations.

Abstract: The 3CL-Protease appears to be a very promising medicinal target to develop anti-SARS-CoV-2 agents. The availability of resolved structures allows structure-based computational approaches to be carried out even though the lack of known inhibitors prevents a proper validation of the performed simulations. The innovative idea of the study is to exploit known inhibitors of SARS-CoV 3CL-Pro as a training set to perform and validate multiple virtual screening campaigns. Docking simulations using four different programs (Fred, Glide, LiGen, and PLANTS) were performed investigating the role of both multiple binding modes (by binding space) and multiple isomers/states (by developing the corresponding isomeric space). The computed docking scores were used to develop consensus models, which allow an in-depth comparison of the resulting performances. On average, the reached performances revealed the different sensitivity to isomeric differences and multiple binding modes between the four docking engines. In detail, Glide and LiGen are the tools that best benefit from isomeric and binding space, respectively, while Fred is the most insensitive program. The obtained results emphasize the fruitful role of combining various docking tools to optimize the predictive performances. Taken together, the performed simulations allowed the rational development of highly performing virtual screening workflows, which could be further optimized by considering different 3CL-Pro structures and, more importantly, by including true SARS-CoV-2 3CL-Pro inhibitors (as learning set) when available.

Keywords: SARS-CoV-2; 3CL-Pro; antivirals; virtual screening; docking simulations; drug repurposing; consensus models; binding space; isomeric space

1. Introduction

Coronaviruses (CoVs, subfamily Coronavirinae, and family Coronaviridae) are enveloped viruses consisting of a single positive-strand RNA that can infect humans where they may cause respiratory, gastro-intestinal, and neurological disorders. A recently identified new coronavirus appeared in Wuhan, China [1], at the end of 2019 to cause a world-wide pandemic crisis in the present times [2]. This is mainly responsible for a

pneumonia-like illness that shows severe threats and often requires patient hospitalization with a global lethality equal to 2.1% (data updated to 16 January 2021). By considering the caused illness and its similarity with the SARS coronavirus (SARS-CoV, genome equal to 82%), the virus was called SARS-CoV-2 (severe acute respiratory syndrome-coronavirus-2), while the induced disease was termed Covid-19 [3].

The SARS-CoV-2 genome [4] encodes for structural proteins that are required for viral entry (such as the Spike glycoprotein); non-structural proteins, which comprise enzymes endowed with protease, methyltransferase, helicase, and polymerase activities; and accessory proteins [5], the role of which should be yet fully clarified [6,7]. The non-structural proteins as well as the proteins required for the viral interaction with the host cells represent potential drug targets [8]. In detail, the SARS-CoV-2 replicase gene encodes for two overlapping polyprotein structures (i.e., ppa1a and ppa1ab), which are cleaved in the functional proteins required for viral replication and transcription by a 33.8 kDa protease called M^{Pro} or 3-chymotrypsin like protein, 3CL-Pro [9]. Interestingly, 3CL-Pro itself is comprised within the ppa1a and ppa1ab polyproteins, and indeed, the first enzymatic step involves the autolytic cleavage to liberate 3CL-Pro. This protease is active as a homodimer and shows a peculiar specificity, which differs from the close proteases of the host cells, since it shows a unique substrate preference for glutamine at the P1 site [10].

Such a substrate specificity combined with its key role in the viral life cycle renders such a protein an attractive target to develop anti-SARS-CoV-2 drugs. Thus, a relevant number of publications focused on the identification of potential SARS-CoV-2 3CL-Pro inhibitors appeared in the literature during the last few months [11]. Several studies comprised structure-based virtual screening (VS) campaigns especially targeted for drug repurposing. For example, Gimeno et al. [12] reported a consensus repurposing study which combines three docking programs (Glide, FRED, and AutoDock/Vina). Only the ligands showing favorable and equivalent binding modes in all three programs were considered as actives. Elmezayen et al. combined docking calculations with molecular dynamics (MD) simulations to better investigate the stability of the retrieved hits by evaluating the corresponding free energy using the MM-PBSA method [13]. Again, Meyer-Almes described a method in which molecular docking, ΔG energy calculation, and analysis of the protein ligand interaction fingerprints (PLIF) are combined to increase the predictive performances [14]. Other studies involved computational protocols based on a single docking tool and extended the virtual screenings also to drug-like molecules included within the ZINC database [15] as well as to databases of natural compounds [16]. However, and due to the lack of known SARS-CoV-2 3CL-Pro inhibitors, all the reported VS strategies cannot be optimized and validated a priori, and thus the reliability of the retrieved hits has to be corroborated by combining different computational approaches.

The novelty of the present study is that the known SARS-CoV 3CL-Pro inhibitors can be added to the screened databases and used as a training set of active molecules to evaluate the performances of the applied computational strategies. Although the efficacy of the here simulated SARS-CoV 3CL-Pro inhibitors was not experimentally confirmed against SARS-CoV-2 3CL-Pro, such a choice is justified by the very high conservation degree between the two proteases (SARS-CoV and SARS-CoV-2 3CL-Pro differ only for 12 residues out of 306), which is suggestive of a very similar binding mode within the two enzymatic pockets, as also confirmed by a recent computational analysis [17].

The employment of the known SARS-CoV 3CL-Pro inhibitors as active molecules in the here reported simulations is also confirmed by the recently resolved structures of SARS-CoV-2 3CL-Pro enzyme in complex with known SARS-CoV 3CL-Pro inhibitors, which show ligand arrangements superimposable to those already seen within SARS-CoV 3CL-Pro. A compelling example is offered by the SARS-CoV-2 3CL-Pro structure bound to the potent TG-0205221 inhibitor (PDB Id: 7C8T), which is completely comparable with the corresponding SARS-CoV 3CL-Pro complex (PDB Id: 2GX4, see below). On one hand, such a resolved structure indirectly confirms the potential activity on SARS-CoV-2 3CL-Pro of the most active known inhibitors of SARS-CoV 3CL-Pro. On the other hand, this structure

emphasizes the possibility to employ known SARS-CoV 3CL-Pro inhibitors to guide the design and/or selection of promising hits for SARS-CoV-2 3CL-Pro enzyme.

Hence, the study involves extended docking simulations based on a purposely collected database in which a set of 478 molecules active against SARS-CoV 3CL-Pro with a pIC50 values greater than 5.0 were dispersed within a database of about 13,000 safe in man molecules. In this way, the performed simulations can be exploited to design, optimize, and validate targeted VS protocols by comparing and combining different docking tools (i.e., Glide, LiGen, Plants, and Fred) and different post-docking analyses. Moreover, the best performing approaches can also be used to identify potential hits against SARS-CoV-2 3CL-Pro.

Furthermore, the analysis of the performed VS campaigns involves the combination of the already proposed binding space with the here developed isomeric space. The concept of binding space was recently proposed by some of us to account for the multiple binding modes a ligand can assume within the binding pocket by simultaneously considering more than one pose for each ligand [18]. Similar to property space [19], the binding space can be defined by parameters that encode for both the average and the spread of the scores values, the latter being described by ranges or standard deviations. Previous studies involving both correlative analyses and VS campaigns revealed that the binding space concept can elicit significant improvements in the predictive power compared to the best score values [20].

In this study, the same rationale inspiring the definition of the binding space was applied to define the corresponding isomeric space. In detail, the ligands existing in different tautomers, stereoisomers, or protonation forms are simulated by docking all possible states, and the resulting docking scores are utilized for calculating the corresponding space parameters (i.e., score averages and ranges). As detailed under Methods, the study investigates both the binding and isomeric spaces by considering four possible combinations: (1) without including the parameters of both the binding and isomeric spaces but considering only the best scores for each compound (namely, the canonical conditions); (2) by including the parameters of the sole binding space to confirm the encouraging results already pursued in previous studies; (3) by including the parameters of the sole isomeric space to assess its specific relevance; and (4) by variously combining the parameters of both spaces to investigate their synergistic role. All so-computed score parameters were then utilized to develop consensus models by using the enrichment factor optimization (EFO) approach [21], which generates linear combinations of the considered descriptors and proved particularly efficient in previous VS benchmarking studies [22,23].

2. Results

2.1. Simulated Dataset

Docking simulations involved a database of 13,535 unique molecules including 478 known SARS-CoV inhibitors with pIC50 > 5, taken from literature, which were dispersed within a dataset of 13,057 safe in man compounds. Among the 478 known SARS-CoV inhibitors, 193 compounds are endowed with a pIC50 > 6. These two inhibition thresholds (i.e., pIC50 > 5 and pIC50 > 6) allow the definition of two different (despite partly overlapped) sets of active molecules, and the generation of all analyzed consensus models was repeated by considering the two resulting training sets of SARS-CoV inhibitors. Furthermore, 4515 molecules (i.e., a third of the complete database) required the generation of 14,013 multiple states/isomers with an average of 3.1 isomers per molecule. In this way, the screened database overall included 23,054 ligands.

Due to the different docking protocols and the criteria by which the generated poses were considered as acceptable by each docking tool (as discussed under Methods), the number of docked molecules varies among the docking simulations as summarized in Table 1. In detail, Table 1 shows that the number of simulated molecules by the four tested programs decreases with the following trend: PLANTS > LiGen > Fred > Glide. The number of docked active molecules parallels the same trend except for Glide, which considers a higher number of active molecules while simulating a lower number of ligands (compared to Fred).

The same trend is exhibited by the number of molecules simulated by considering multiple states by the four docking engines.

Table 1. General characteristics of the databases simulated by each docking program and relative generated poses.

Tool	Docked Molecules	pIC50 > 6	pIC50 > 5	Molecules with Multiple States	Monitored Poses
Full Database	13,535	193	478	4548	—
Glide	11,583	164	415	3470	174,071
Fred	11,731	138	411	3379	183,653
Ligen	12,791	190	472	3907	128,133
Plants	13,460	193	478	4475	224,670
Common	10,110	114	353	—[a]	—[a]

[a] The number of both molecules existing in multiple states and generated poses varies among the docking programs.

When considering the number of monitored poses, it should be noted that PLANTS, Fred, and Glide generated 10 poses per ligand, while LiGen provided 10 poses per unique molecule. This implies that for molecules existing in multiple states, LiGen selected the 10 best poses regardless of the involved isomers. Hence, the number of analyzed poses for LiGen is significantly lower than that of the other docking engines.

Despite the described differences, a common dataset composed of those molecules that are simulated by all the utilized docking tools can be extracted. This comprises 10110 ligands (about 75% of the full database) and includes 353 active compounds (i.e., with pIC50 > 5). Such a common database will be used to develop mixed consensus models by combining the docking results from different tools and to compare the best ranked molecules by the four docking tools.

2.2. Results by PLANTS

Table 2 compiles the best enrichment factors in the top 1% (EF 1%) based on the PLANTS simulations as well as the resulting enhancements (as percentage values) induced by the inclusion of the explored space descriptors and by the generation of linear consensus equations. In all reported analyses (Tables 2–5), the consensus models are developed by linearly combining from two to five docking scores using the EFO algorithm (as detailed under Methods). As mentioned above, the analyses are repeated by considering the two sets of active molecules as collected considering two different inhibition thresholds (i.e., pIC50 > 5 and pIC50 > 6).

A preliminary consideration, which affects all these analyses, concerns the unavoidably biased comparison between binding and isomeric spaces. Indeed, the former involves all the screened molecules for which different poses were generated, while the latter affects only those molecules existing in multiple states, which represent about 1/3 of the entire dataset (see above). Specifically, the PLANTS simulations involved 31 compounds existing in multiple states out of 193 for the actives with pIC50 > 6 and 136 out of 478 for the molecules with pIC50 > 5.

Thus, one may understand the different effects exerted by the inclusion of the isomeric space parameters for the two considered pIC50 thresholds. Indeed, the number of examined isomers with pIC50 > 5 is high enough and the isomeric space affords an appreciable enhancing effect, which is greater than that exerted by the binding space. In contrast, the limited number of compounds existing in multiple states with pIC50 > 6 reduces the enhancing effect exerted by the isomeric space. The inclusion of the binding space

parameters induces significant improvements, especially with pIC50 > 6, where it is more effective than the isomeric space, especially when considering the best obtained models.

Table 2. Best EF1% values reached by the various consensus models developed using the PLANTS results plus the relative average values and the corresponding performance enhancements in percentage values. As described under Methods, the consensus equations were generated by linearly combining from two to five docking scores. The EF1% values referring to one variable correspond to the performances reached by single scoring functions.

Number of Variables	Without Spaces	Isomeric Space	Binding Space	Both Spaces Merged	Both Spaces Joint	Mean
			pIC50 > 5			
1	7.1	9.0 (27%) [a]	9.0 (27%) [a)]	10.1 (41%) [a]	9.0 (27%) [a]	8.9
2	10.3	10.3 (0%)	10.3 (0%)	11.3 (10%)	10.3 (0%)	10.5 (18%) [b]
3	10.3	11.8 (14%)	10.3 (0%)	11.3 (10%)	11.3 (15%)	11.1 (24%) [b]
4	10.7	11.8 (10%)	10.7 (0%)	12.2 (14%)	11.6 (10%)	11.4 (28%) [b]
5	10.7	11.8 (10%)	11.6 (8%)	13.0 (22%)	12.0 (12%)	11.8 (33%) [b]
Mean	9.8	10.9 (11%)	10.4 (5%)	11.6 (18%)	11.0 (12%)	10.7 (20%) [b]
			pIC50 > 6			
1	9.4	10.9 (16%)	9.4 (0%)	9.4 (0%)	10.9 (16%)	10.0
2	13.4	14.1 (5%)	13.4 (0%)	14.6 (8%)	14.1 (5%)	13.8 (38%)
3	13.4	15.1 (12%)	13.4 (0%)	15.6 (16%)	16.7 (24%)	14.8 (48%)
4	14.1	15.1 (7%)	18.2 (29%)	17.2 (21%)	18.2 (29%)	16.6 (66%)
5	14.1	15.1 (7%)	18.2 (29%)	18.2 (29%)	18.2 (29%)	16.8 (68%)
Mean	12.9	14.6 (9%)	15.3 (13%)	15.9 (16%)	16.0 (20%)	14.4 (43%)

[a] Here and in the following rows, the performance enhancements are computed in respect to the first column reporting the EF1% values obtained without including space parameters to assess the beneficial role of the inclusion of these parameters and their combinations. The same holds for pIC50 > 6 and Tables 4, 6, and 8. [b] In the last column, the enhancements are computed with respect to the average value obtained by the single scoring functions to evaluate the average effect of the consensus models. The same holds for pIC50 > 6 and Tables 4, 6, and 8.

Table 3. Best EF1% values reached by the various consensus models developed using the LiGen results plus the relative average values and the corresponding performance enhancements in percentage values. As described under Methods, the consensus equations were generated by linearly combining from two to five docking scores. The EF1% values referring to one variable correspond to the performances of single scoring functions.

Number of Variables	Without Spaces	Isomeric Space	Binding Space	Both Spaces Merged	Both Spaces Joint	Mean
			pIC50 > 5			
1	11.3	11.3 (0%)	11.3 (0%)	11.3 (0%)	11.3 (0%)	11.3
2	11.3	12.8 (13%)	12.4 (9%)	11.3 (0%)	12.8 (9%)	12.0 (7%)
3	13.9	13.9 (0%)	14.5 (5%)	13.9 (0%)	14.5 (7%)	14.1 (25%)
4	14.5	14.5 (0%)	16.2 (12%)	14.9 (3 %)	16.2 (12%)	15.3 (34%)
5	14.5	15.1 (4%)	16.2 (12%)	15.6 (7%)	16.2 (12%)	15.5 (36%)
Mean	13.1	13.5 (3%)	14.1 (8%)	13.4 (2%)	14.2 (8%)	13.8 (20%)
5 no PH [a] Distances	9.4	9.4	9.4	9.4	11.11	9.7
5 only PH Distances [a]	14.5	15.1	13.0	8.5	15.6	13.3

Table 3. *Cont.*

Number of Variables	Without Spaces	Isomeric Space	Binding Space	Both Spaces Merged	Both Spaces Joint	Mean
			pIC50 > 6			
1	12.7	12.7 (0%)	12.7 (0%)	12.7 (0%)	12.7 (0%)	12.7
2	13.2	13.2 (0%)	14.8 (12%)	14.3 (8%)	14.8 (12%)	13.8 (8%)
3	13.2	13.8 (4%)	21.2 (60%)	21.2 (60%)	21.2 (60%)	16.6 (31%)
4	13.2	13.8 (4%)	24.9 (88%)	25.4 (92%)	24.9 (88%)	18.2 (43%)
5	13.8	13.8 (0%)	25.9 (89%)	25.4 (84%)	25.9 (89%)	18.5 (46%)
Mean	13.2	13.4 (2%)	19.9 (50%)	19.8 (49%)	19.9 (50%)	16.0 (26%)
5 no PH Distances [a]	10.6	11.6	12.2	21.3	20.1	15.2
5 only PH Distances [a]	9.0	11.1	10.6	12.2	11.1	10.8

[a] In both analyses, the last two rows report the performances reached by the consensus equations generated by combining five variables and excluding the pharmacophoric distances ("five no PH distances") or considering only these parameters ("five only PH distances").

Table 4. Best EF1% values reached by the various consensus models developed using the Fred results plus the relative average values and the corresponding performance enhancements in percentage values. As described under Methods, the consensus equations were generated by linearly combining from two to five docking scores. The EF1% values referring to one variable correspond to the performances of single scoring functions.

Number of Variables	Without Spaces	Isomeric Space	Binding Space	Both Spaces Merged	Both Space Joint	Mean
			pIC50 > 5			
1	13.7	14.9 (9%)	15.1 (11%)	16.6 (21%)	15.1 (11%)	15.1
2	14.9	15.1 (2%)	16.3 (10%)	16.6 (11%)	17.1 (15%)	16.0 (6%)
3	16.1	16.3 (1%)	16.6 (3%)	17.3 (8%)	17.6 (9%)	16.8 (11%)
4	16.1	16.3 (1%)	16.6 (3%)	17.3 (8%)	17.6 (9%)	16.8 (11%)
5	16.1	16.3 (1%)	16.6 (5%)	17.3 (8%)	17.6 (9%)	16.8 (11%)
Mean	15.4	15.8 (3%)	16.3 (6%)	17.0 (11%)	17.0 (11%)	16.3 (8%)
			pIC50 > 6			
1	24.7	24.7 (0%)	24.7 (0%)	24.7 (0%)	24.7 (0%)	24.7
2	25.4	25.4 (0%)	25.4 (0%)	25.4 (0%)	25.4 (0%)	25.4 (3%)
3	25.4	25.4 (0%)	25.4 (0%)	26.9 (6%)	27.6 (9%)	26.2 (6%)
4	25.4	25.4 (0%)	26.9 (6%)	26.9 (6%)	27.6 (9%)	26.4 (7%)
5	25.4	25.4 (0%)	27.6 (9%)	27.6 (9%)	27.6 (9%)	26.7 (8%)
Mean	25.3	25.3 (0%)	26.0 (3%)	26.3 (4%)	26.6 (5%)	25.9 (5%)

Table 5. Best EF1% values reached by the various consensus models developed using the Glide results plus the relative average values and the corresponding performance enhancements in percentage values. As described under Methods, the consensus equations were generated by linearly combining from two to five docking scores. The EF1% values referring to one variable correspond to the performances reached by single scoring functions.

Number of Variables	Without Spaces	Isomeric Space	Binding Space	Both Spaces Merged	Both Spaces Joint	Mean
			pIC50 > 5			
1	12.9	13.8 (8%)	13.1 (2%)	12.9 (0%)	13.8 (8%)	13.3
2	13.8	15.3 (11%)	14.3 (4%)	14.3 (4%)	16.5 (19%)	14.9 (12%)
3	14.3	16.0 (12%)	16.0 (12%)	16.3 (14%)	16.5 (15%)	15.8 (19%)
4	14.3	16.8 (17%)	16.3 (14%)	16.3 (14%)	17.2 (20%)	16.2 (22%)
5	14.3	16.8 (17%)	16.3 (14%)	16.3 (14%)	17.5 (22%)	16.2 (22%)
Mean	13.9	15.7 (13%)	15.2 (9%)	15.2 (9%)	16.3 (17%)	15.3 (15%)
			pIC50 > 6			
1	19.7	22.7 (16%)	20.3 (3%)	19.7 (0%)	22.7 (16%)	21
2	22.1	25.8 (17%)	24.0 (8%)	23.3 (6%)	25.8 (17%)	24.2 (15%)
3	22.7	27.6 (22%)	25.8 (14%)	25.2 (11%)	28.3 (24%)	25.9 (23%)
4	24.0	27.6 (15%)	27.0 (13%)	25.8 (8%)	28.3 (18%)	26.5 (26%)
5	24.0	27.6 (15%)	27.0 (13%)	25.8 (8%)	28.3 (18%)	26.5 (26%)
Mean	22.5	26.3 (17%)	24.8 (10%)	24.0 (8%)	26.7 (19%)	24.8 (18%)

Regardless of the adopted strategy, the combination of the two monitored spaces appears to be productive only with pIC50 > 5, where it affords average improvements greater than those reached by the individual spaces. In contrast, the space combination exerts very modest effects with pIC50 > 6, since the resulting best models do not exceed the performances reached by the sole binding space. Overall, PLANTS affords encouraging performances that benefit from both the inclusion of the space descriptors and the generation of consensus linear equations. The latter induces an average improvement of around 39% with the best improvements around 70%.

While avoiding the systematic analysis of the generated consensus equations, attention will be here focused on the occurrence of the various scores in all generated models. The interested reader can find all computed consensus equations with the resulting performances for all docking programs and all performed analyses at http://www.exscalate4cov.network/. Table S1 reports the occurrence of the diverse scores in the 20 best consensus models generated by the EFO algorithm in all the 10 analyses included in Table 2 (for a total of 200 equations). In detail, Table S1 reveals the remarkable role played by the XScore [24] and PLANTS [25] scoring functions (the latter here comprising also the primary scores). Table S1 also evidences the interesting role played by the scores encoding for non-polar interactions as exemplified by molecular lipophilic potential (MLP) [26] and the VEGA-based scores [27], which here correspond to the Lennard–Jones interaction energies as computed by using the CHARMM and CVFF force fields. Concerning the type of score values (in the models including the space descriptors), the best values play a prevailing role, while spread and mean values have a lesser and rather similar incidence.

2.3. Results by LiGen

As described under Methods, the analyses based on the LiGen simulations involved the primary scores, the scoring functions derived by rescoring analyses, and 30 pharmacophoric distances (PH) as generated by the LiGen software. Table 3 reports the performances reached by the LiGen results as well as the performance enhancements due

to both the inclusion of the space parameters and the generation of consensus models (as percentage values).

The first key observation concerns the very limited role played by the isomeric space for both pIC50 thresholds. Such a result cannot be ascribed to a too low number of active molecules existing in multiple states, since they roughly correspond to those already considered by PLANTS (i.e., 132 out of 472 with pIC50 > 5 and 29 out of 190 with pIC50 > 6). Instead, this finding can be explained by considering that limiting the analysis to 10 poses per unique molecule does not allow an extensive exploration of the isomeric space. Moreover, the scoring functions and the search algorithm implemented by LiGen might be not so sensitive to the isomeric differences among the simulated ligands. When considering the overall satisfactory performances reached by LiGen (see below), the poor results of the isomeric space can be positively evaluated, since they suggest that this docking program can conveniently perform VS campaigns without requiring excessive isomeric expansions of the screened databases with a beneficial reduction of computational time and complexity.

In contrast, the inclusion of the binding space descriptors markedly improves the EF1% values with an enhancement effect, which is particularly evident with pIC50 > 6, where the binding space leads to almost doubled EF1% values. When considering that the LiGen program produced the lowest number of poses, the remarkable results yielded by the binding space parameters emphasize that this reduction is yet able to extract the significant poses, thus minimizing redundant results and reducing the computational costs. Based on the different effects induced by the two analyzed spaces, one may understand why their combinations are unproductive since they yield EF1% values, which are comparable with those obtained by the sole binding space.

The LiGen results seem to particularly benefit from the linear combination of diverse scoring functions: this effect is noticeable in all the experiments, as confirmed by the average increases around 30%, but is particularly remarkable with pIC50 > 6, where the consensus models lead to the doubling of the corresponding EF1% value. Taken together, LiGen provides satisfactory performances, reaching EF1% values around 25 for pIC50 > 6 and with the inclusion of the binding space descriptors.

Since the LiGen analyses also comprise the pharmacophoric distances, a specific experiment was performed to investigate their specific role. In detail, the analysis involved the generation of two additional five-variable consensus equations for each pIC50 threshold. The first model was based on the pharmacophoric distances only and the second on all computed scoring functions except for the pharmacophoric distances. The resulting EF1% values (see Table 3) highlight an overall synergistic effect between pharmacophoric distances and scoring functions. Nevertheless, Table 3 reveals different behaviors depending on the considered inhibition threshold. The pharmacophoric distances play indeed a prevailing role with pIC50 > 5, while the score values assume a more marked role with pIC50 > 6.

Table S2 compiles the incidence of the various docking scores in all the generated consensus equations and reveals the prevailing role played by the pharmacophoric distances as computed by LiGen. Apart from these distances, only PLANTS scores show a significant relevance, while the scores encoding for specific interactions exhibit negligible roles. As seen above, the best values confirm their prevailing role, but here the spread values reveal an incidence that is almost double compared to the score means.

2.4. Results by Fred

Table 4 compiles the EF1% values and the corresponding enhancements as derived by the Fred docking results. The first key consideration is that the single scoring functions provide satisfactory results even without space parameters and consensus combinations as seen with pIC50 > 6. The notable performances elicited by simple docking scores can explain why the inclusion of the descriptors for both explored spaces exerts only limited effects. Similar to what was seen for LiGen, the isomeric space also plays here an almost negligible role, while the binding space parameters yield modest improvements

by increasing the corresponding EF1% values of about 10% in the best consensus models. This result can be explained by considering that the beneficial effects of such post-docking procedures (such as rescoring, the inclusion of binding and isomeric spaces, and consensus approaches) depend on the room for improvement that the simple docking simulations show, and they inevitably decrease when the standard docking results already reveal notable predictive powers.

The docking simulations by Fred involved several active molecules existing in multiple states roughly comparable to those simulated by Plants and LiGen (despite the lowest number of total simulated isomers: i.e., 118 out of 411 with pIC50 > 5 and 27 out of 138 with pIC50 > 6). Hence, the modest relevance of the isomeric space suggests that this docking program does not require significant isomeric expansion of the screened databases to perform reliable VS campaigns. Similarly, the modest effect also exerted by the binding space suggests that the poses generated by Fred show a limited variability and tend to be focused around the best (and reasonably reliable) pose. In this way, the score averages roughly correspond to the best values, and the score spreads lose most of their relevance. As discussed above for the LiGen results, these insensitivities can be seen as positive features, which allow docking simulations to be performed without significant isomeric expansions of the screened datasets and reducing the number of poses computed per ligand thus minimizing the computational costs. Despite the modest effects exerted by the single spaces alone, their combinations yield encouraging results, especially with pIC50 > 5.

Table S3 shows the relative incidence of the computed scoring functions, as seen in all the consensus models generated by Fred. The key observation is that here PLANTS and XScore functions show an almost exclusive role, which minimizes the relevance of all other docking scores. While considering the unsatisfactory results afforded by space parameters and consensus strategies (see above), these results indicate that at least the rescoring calculations played a key role in enhancing the predictive power of the Fred simulations. About the type of score values, the best scores represent the most abundant values, while the mean values play a more relevant role here than the score ranges.

2.5. Results by Glide

Table 5 compiles the EF1% values (and the corresponding enhancements) as derived by using Glide. Despite the lowest number of active compounds simulated in multiple states (i.e., 105 out of 415 with pIC50 > 5 and 24 out of 164 with pIC50 > 6), Table 5 shows the markedly beneficial effect exerted by the isomeric space with both pIC50 thresholds. Although the binding space also elicits encouraging EF1% improvements for both thresholds, the isomeric space affords better results in terms of both reached EF1% values and relative performance enhancements. As already evidenced by previous studies [28], these results suggest that the search algorithms and the scoring functions implemented by Glide are strongly dependent on isomeric differences and invite the exhaustive expansion of the screened databases by considering as many as reasonable isomers/states, even when the number of simulated molecules existing in multiple states is relatively low.

While in the previous analyses the space combinations provided limited enhancements without significant differences between the two applied strategies; here, the joint combination of the two spaces yields remarkable performance enhancements compared to the single spaces with both pIC50 thresholds. In contrast, the merging combination approach appears to be constantly unproductive. These results suggest that the Glide-based docking scores are able to properly account for both multiple states (by isomeric space) and multiple binding modes (by binding space). Such a sensitivity implies that the two explored spaces encode for different information, and thus their descriptors can be synergistically combined, but cannot be fused into a unique space, the descriptors of which would detrimentally confuse the specific roles of the two spaces (as seen with the merging combination). Taken globally, the Glide simulations provide satisfactory predictive performances, which are, on average, comparable with those offered by LiGen with best EF1% values ranging between 25 and 30. As discussed above, the obtained performances

markedly benefit from both explored spaces (and their joint combination) as well as from the development of consensus linear equations as assessed by an overall improvement of about 20% (with best results around 25%).

Table S4 reports the incidence of the various scoring functions in all the equations developed by using the Glide results and emphasizes the relevant role played by the 23 different primary Glide scores [29]. Notably, they appear to be particularly abundant in those experiments, which afforded the best performances. The equations seem to benefit from the inclusion of scoring functions encoding for non-polar interactions. Indeed, the VEGA-based scores, which here correspond to the CHARMM- and CVFF-based Lennard–Jones interaction energies, and the MLP values show an overall incidence of about 20%. Concerning the score types, the best values represent about 50%, and mean and spread values are equally abundant.

2.6. Overall Comparison

Although the differences in the protocols adopted and in the successfully docked molecules by the four tested docking programs prevent a precise comparison of the reached performances, an overall assessment of the previously discussed performances is here reported to compare the specific relevance of the computed space parameters. To do this, Figure 1 compares the reached AUC values as derived from the ROC curves corresponding to the best developed consensus models in all performed docking experiments and for both inhibition thresholds.

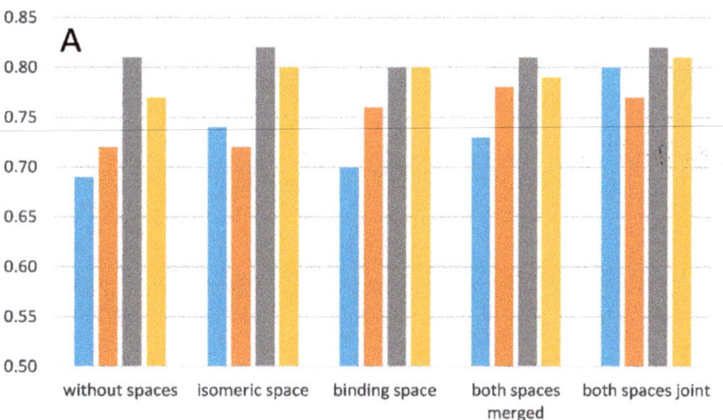

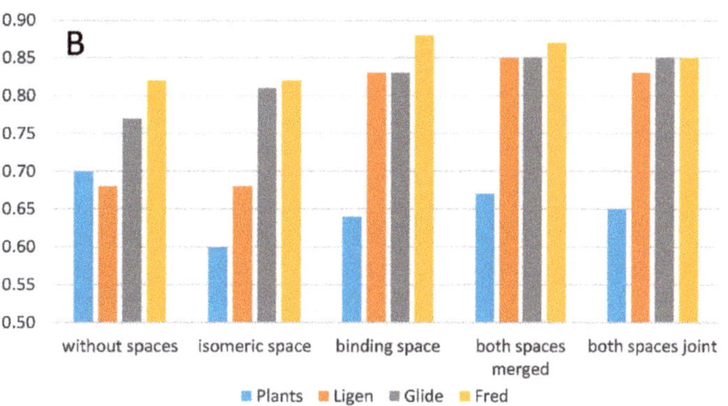

Figure 1. Comparison of the AUC values from the ROC curves for the best performing models generated by the four utilized docking programs in the five tested conditions with (**A**) pIC50 > 5 and (**B**) pIC50 > 6.

The analysis of the reported AUC values reveals results in substantial agreement with those previously discussed for EF1% values and allows for some considerations, which can be summarized as follows. The two explored spaces induce similar overall enhancing effects with the isomeric space, which appears to be more relevant for analyses with pIC50 > 5, reasonably due to the higher number of involved molecules existing in multiple states. The combinations of the two space parameters provide comparable performances, and rarely do they surpass those reached by the individual spaces. The LiGen program is that best benefitting from the inclusion of space parameters, while Fred is the most insensitive tool. When considering the best AUC values, Figure 1 reveals that the performances of the four docking programs are overall comparable for the screening campaigns with pIC50 > 5, while PLANTS yield lower AUC values with pIC50 > 6 compared to the other three pieces of software, which in turn afford rather similar performances.

Comparative analyses were also performed by calculating the corresponding consensus models using the common database. The obtained results (Figure S1 and Table S5, Supplementary Materials) are in clear agreement with those previously discussed. The only difference involves the more limited enhancing role played by the isomeric space, which is ascribable to the reduced number of considered compounds existing in multiple states. The best consensus models developed using the common database will be used to compare the resulting rankings from the four tested docking programs (see below).

Figure 2 focuses on the enhancing role played by the development of the consensus models by showing the progressive effect exerted when including from two to five variables (in respect to the single scores). As already seen, the best improvements are reached by PLANTS and LiGen, while Glide and especially Fred show more limited effects. The beneficial role on the LiGen results might be also ascribed to the fact that their analyses involve the largest number of computed score values due to the inclusion of 30 pharmacophoric distances. However, a relation between performances and the number of variables cannot be evidenced for the other three docking programs.

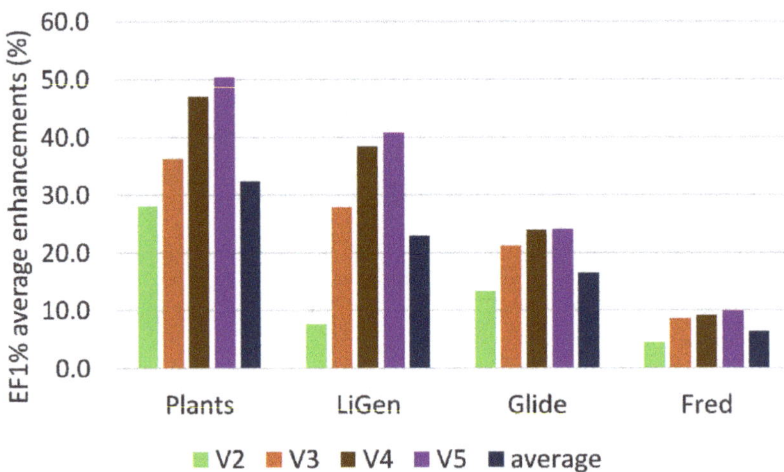

Figure 2. Average EF1% enhancements (in percentage values) due to the generation of the consensus models when including two (V2), three (V3), four (V4), or five (V5) variables plus overall averages. The reported values are computed by averaging the EF1% enhancements for both inhibition thresholds.

When considering the progressive contribution when including from two to five variables in the consensus models, Figure 2 shows that the largest improvements are observed when shifting from one to two (around 14%) and from two to three variables (around 10%), while the inclusion of additional variables induces more limited EF1% increases (6% and 2% for four and five variables, respectively). On one hand, these results

justify the choice made here of avoiding the calculation of consensus models with more than five variables. On the other hand, Figure 2 suggests that a simpler and faster analysis might be focused on consensus equations, including at most three variables that represent an optimal balance between performances, reliability, and computational costs.

To conclude this general analysis, Table S6 compiles the occurrence of the various scores as obtained by all the computed consensus models and reveals the major role played by the primary scores. Notably, primary scores also include the LiGen pharmacophoric distances, which alone represent more than 50% of all primary score values. Table S6 highlights the overall relevant role of both PLANTS and XScore scoring functions and consequently emphasizes the crucial role of rescoring procedures for enhancing the predictive power of all performed VS campaigns. The scores encoding for specific interactions play a minor role, even though Table S5 underlines the appreciable effect played by scoring functions encoding for non-polar interactions as seen by summing the relevance of both MLP and VEGA-based scores. Concerning score types, Table S5 confirms that best values represent about half of all included values and spread and mean scores show a more limited and similar incidence (around 25%).

Even though the analysis of the computed poses goes beyond the primary objective of the study, which was designed to evaluate the effects of the monitored space parameters in enhancing the performances of the performed VS campaigns, Figure 3 compares the best computed poses for a close TG-0205221 analogue included in the screened database with the recently resolved SARS-CoV-2 3CL-Pro complex [30]. While considering that the performed VS campaigns cannot simulate the formation of the covalent bond between Cys145 and the bound ligand, as seen in the reference structure (Figure 3A), the computed poses are in encouraging agreement with that of TG-0205221. Indeed, in all four shown structures, the 2-oxopyrrolidin ring approaches Asn142, the leucine side chain, which replaces the cyclohexyl alanine of TG-0205221 contacts His41, and the benzoyloxy moiety approaches Pro168 and Gln192. The four complexes slightly differ for the arrangement of the electrophilic warhead, even though it appears to be always close enough to Cys145 to yield the Michael adduct.

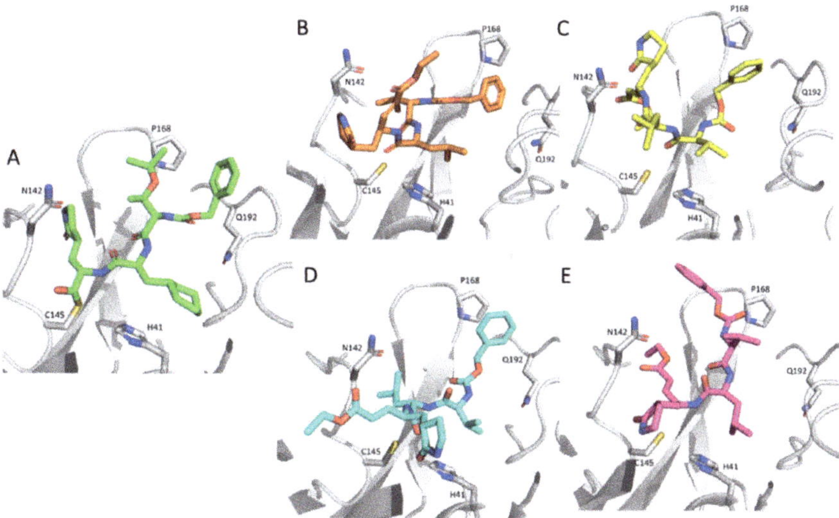

Figure 3. Comparison of the resolved complex of TG-0205221 with the SARS-CoV-2 3CL Pro enzyme (**A**) with the best poses as computed for its close analogue by PLANTS (**B**), LiGen (**C**), Fred (**D**), and Glide (**E**).

2.7. Mixed Consensus Models

The analyses on the common database was primarily carried out to develop mixed consensus equations by linearly combining scores coming from different docking simulations. For simplicity's sake and considering the observed differences among the four docking runs even for the common database, the development of the mixed models was focused on the docking scores as such avoiding the space parameters.

Table 6 reports the best EF1% values as obtained by generating consensus equations that linearly combine scores of pairs of docking engines and the corresponding performance improvements (in percentage values). Overall, the reported EF1% values reveal interesting synergistic effects that affect most of the tested combinations with only four out of 12 cases showing no enhancement. On average, the synergistic enhancements are rather similar for both inhibition thresholds (i.e., 12% for pIC50 > 5 and 19% for pIC50 > 6). The best results are afforded by the combination of Glide plus Fred, which yield for both pIC50 thresholds EF1% better than those of the single docking program, while the best synergistic enhancement is seen when combining LiGen and PLANTS scores with pIC50 > 6 (40%).

Table 6. Best EF1% values as obtained by combining the simple score values (without space parameters) of two or three different docking programs and the corresponding synergistic effect (as enhancements in percentage values). For easy comparison and concerning the results for pairs of docking tools, the diagonal cells reports the best EF1 value obtained by the single docking tool (in italics).

Program	PLANTS	LiGen	Fred	Glide
Results for Pairs of Programs				
pIC50 > 5				
PLANTS	*10.7*	15.3 (6%)	16.1 (0%)	16.2 (13%)
LiGen		*14.5*	16.1 (0%)	16.7 (15%)
Fred			*16.1*	18.1 (12%)
Glide				*14.3*
pIC50 > 6				
PLANTS	*14.1*	19.8 (40%)	28.1 (11%)	24.0 (0%)
LiGen		*13.8*	27.7 (9%)	24.0 (0%)
Fred			*25.4*	29.0 (16%)
Glide				*24*
Results for triplet of programs				
	Fred Glide PLANTS	LiGen Glide PLANTS	LiGen Glide Fred	LiGen Fred PLANTS
pIC50 > 5	18.1 (12%)	17.1 (18%)	18.7 (16%)	16.4 (2%)
pIC50 > 6	29.9 (18%)	24 (0%)	27.2 (7%)	28.1 (11%)

Given these encouraging results, the next analysis involved the combination of triplets of programs. Table 6 also includes the EF1% values reached by these analyses and reveals an appreciable synergistic effect for these combinations with only one case being ineffective and most cases with an EF1% increase greater than 10%. Notably, these consensus models allow reaching EF1% around 20 for pIC50 > 5 and very close to 30 for pIC50 > 6. Unfortunately, the full combination of the scores coming from all tested docking programs did not yield further improvements (results not shown).

2.8. Analysis of the Best Rankings

The last section of this study analyzes the rankings obtained by applying the best consensus models for pIC50 > 6 using the common database (see Table S7). The first part

of the analysis investigates how the frequency of the molecules shared at the same time by two, three, or four rankings varies when browsing the first half of the ranking positions (Figure S1 and Table S7, from 1 to 5000). Figure S1A shows the computed trends and reveals that the frequency of common molecules found in three rankings increases with a linear trend, while the frequencies of molecules included in all the four rankings or only in two rankings show symmetric and parabolic trends. The former grows when increasing the number of monitored ranking positions, and the latter symmetrically decreases.

Figure S1B illustrates the parabolic trends as computed by considering specific pairs of rankings. This allows a graphical evaluation of the increasing overlapping between the results of two docking programs. Figure S1B reveals that the highest frequency of shared compounds is provided by combining the LiGen and Fred rankings, while the other five pairs of programs yield rather similar profiles with the pairs LiGen–Plants and Plants–Fred showing the lowest frequencies of shared molecules. While being detectable even within the best top 100 ranking positions, the differences between the frequencies become appreciable when considering at least the first 500 ranking positions. Hence, the following analyses will focus on the molecules included in the top 500 ranking positions.

The first analysis of the top 500 molecules of each ranking deals with their physicochemical profiling. Table 7 reports the corresponding averages values and standard deviations for some key geometrical and physicochemical descriptors and allows for some relevant considerations. Firstly, limited differences are seen for the average values of the number of rotors and H-bonding groups, while molecular size (as encoded by M.W. and SAS averages) and polarity (as parameterized by PSA and log P averages) reveal a more marked variability. In detail, Table 7 shows that Glide and PLANTS select the bulkiest and the smallest set of ligands, respectively, while Fred and LiGen unravel intermediate averages. Additionally, there is an expected relation between size and lipophilicity for PLANTS, LiGen, and Fred, while Glide selects the ligands with a peculiar profile, since they comprise the bulkiest and the most polar molecules. Finally, concerning the property variability, Table 7 reports modest differences among the four monitored sets of ligands and for all computed descriptors, even though the Glide set shows, on average, the highest standard deviations, thus suggesting a conceivable relation between molecular size and property variability.

Table 7. Average values plus standard deviations for some key geometrical and physicochemical properties as computed by considering the Top 500 molecules of each ranking.

Tool	Rotors	HB_Tot	M.W.	PSA	SAS	Log P
Plants	9.6 ± 3.8	9.6 ± 2.2	510.9 ± 98.2	131.6 ± 45.2	541.8 ± 111.3	2.3 ± 2.7
LiGen	10.2 ± 4.2	9.2 ± 2.9	554.2 ± 95.6	120.8 ± 54.3	593.0 ± 98.3	3.6 ± 2.2
Fred	9.9 ± 4.4	9.2 ± 2.9	545.0 ± 97.8	120.3 ± 56.4	579.9 ± 104.6	2.9 ± 2.4
Glide	10.9 ± 4.2	11.1 ± 3.1	566.0 ± 103.1	148.3 ± 58.9	590.6 ± 109.4	2.2 ± 2.9

The next analysis on the top 500 ligands concerned the overlapping between the four sets of selected ligands. Figure 4 shows the resulting Venn diagram with the corresponding frequency values. As seen in Figure S1B, Figure 4 confirms that the pairs Fred–LiGen and, to a minor extent, Glide–PLANTS show the highest frequencies of common molecules, while the pairs PLANTS–LiGen as well as PLANTS–Fred show the lowest degree of overlapping. Accordingly, the two triplets including Ligen and Fred (LiGen–Glide–Fred and PLANTS–LiGen–Fred) show the highest overlapping degree, while PLANTS–LiGen–Glide reveals the lowest number of shared ligands. Consequently, LiGen and Fred show the lowest numbers of unshared molecules, while PLANTS has the highest number of unique ligands,

which roughly correspond to one half of the analyzed set. Finally, the molecules common to all the analyzed sets are 62, and this is a remarkable result, since they represent about 12.5% of the monitored ranking positions.

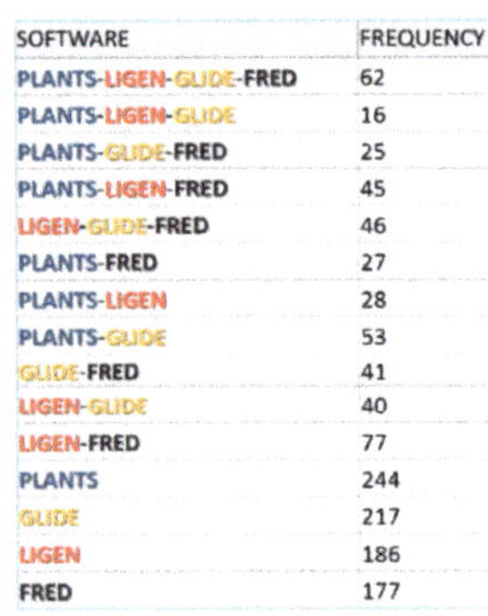

SOFTWARE	FREQUENCY
PLANTS-LIGEN-GLIDE-FRED	62
PLANTS-LIGEN-GLIDE	16
PLANTS-GLIDE-FRED	25
PLANTS-LIGEN-FRED	45
LIGEN-GLIDE-FRED	46
PLANTS-FRED	27
PLANTS-LIGEN	28
PLANTS-GLIDE	53
GLIDE-FRED	41
LIGEN-GLIDE	40
LIGEN-FRED	77
PLANTS	244
GLIDE	217
LIGEN	186
FRED	177

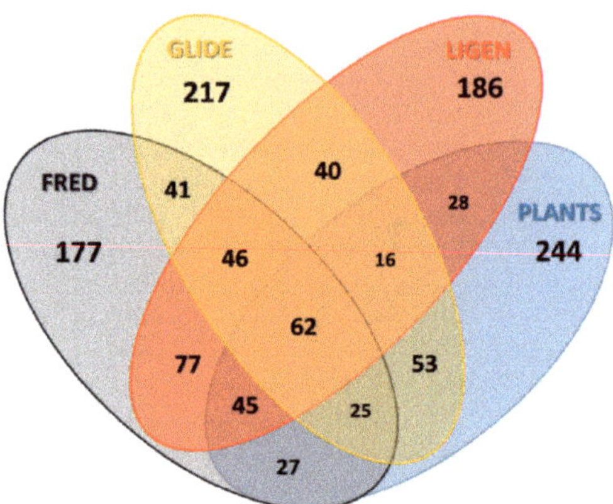

Figure 4. Venn diagram representing the overlapping degree between the Top 500 molecules as derived by applying the best equations developed for each docking software. A table detailing the resulting frequencies is also included (the color code for the four docking tools is the same adopted in Figure 1).

A similar diagram was also obtained when analyzing the common scaffolds, as detected within the screened Top 500 molecules (Figure S2). The scaffold frequencies are in excellent agreement with the ligand frequencies seen in Figure 4 ($r^2 = 0.97$) and the total number of considered ligand frequencies (i.e., 1284) is slightly higher than that of scaffolds (i.e., 828). This means that each detected scaffold is shared on average by $\cong 1.5$ ligands. Stated differently, these findings emphasize that the selected Top 500 molecules do not include congeneric series, and the chemical spaces covered by the top-ranked molecules of each docking software are rather similar (as also suggested by Table 7) with PLANTS and Glide showing the highest number of unshared scaffolds.

Finally, Table S8 compiles the common molecules shared by at least three rankings. In detail, the so collected common molecules are 194 (62 and 132 molecules shared by four and three rankings, respectively), among which 92 belong to the set of active compounds (with pIC50 > 5, i.e., 47%). This finding affords a further validation of the overall predictive power of the reported VS strategies, especially considering that the inhibition activity of several SARS-CoV 3CL-Pro inhibitors against the SARS-CoV-2 3CL-Pro enzyme was experimentally confirmed (as discussed above for TG-0205221 analogues. Among the other 102 molecules, there are 37 compounds that are known inhibitors of SARS-CoV 3CL-Pro but with pIC50 < 5, and 17 known inhibitors of the main proteases of other viruses (such as norovirus and HIV), and these compounds represent a further confirmation of the efficacy of the performed VS campaigns, since some of the retrieved hits (rupintrivir [31], saquinavir [32], and lopinavir [33]) were experimentally confirmed as promising inhibitors of SARS-CoV-2 3CL-Pro. Finally, among the other common molecules, cobicistat [34] and galloyl analogues [35] were identified as SARS-CoV-2 3CL-Pro inhibitors.

3. Methods

3.1. Library and Protein Structure Preparation

Virtual screening studies were performed on a repurposing library, containing a unique list of 13,057 drugs. They comprise the set of safe in man drugs, commercialized or under active development in clinical phases and retrieved from the Integrity database, plus the Fraunhofer's BROAD Repurposing Library provided by Fraunhofer IME. The screened database also includes a set of 478 molecules, in particular preclinical compounds, identi-

fied as "CoV Inhibitors", which were considered as the active training set in the reported optimization/validation analyses. Hence, the screened dataset overall included 13,535 unique molecules and the set of presumably active compounds was composed of 478 inhibitors with pIC50 > 5 of which 193 molecules show a pIC50 > 6.

Even though the database was collected by selecting safe in man molecules so that the obtained results could be used also for repurposing purposes, Figure S4 compares some representative physicochemical properties of active and inactive, which show rather similar distributions. To better assess the reliability of the screened database and to appreciate the role of docking simulations, a ligand-based VS campaign involving about 100 structural and physicochemical descriptors was performed. The best EF1% values as obtained by the consensus models with five variables are very low (EF1% = 5.5 and 3.7 with pIC50 > 5 and pIC50 > 6, respectively). On one hand, these poor results confirm that the simulated database does not show significant differences between active and inactive ligands, which can bias the here presented docking results. On the other hand, the very modest performances reached by the ligand descriptors further emphasize the relevance of the here described structure-based VS workflows.

The reported docking simulations were performed by applying procedures as homogeneous as possible to render the obtained results as comparable as possible. Nevertheless, minor differences remain concerning the preparation of the 3D structures of both protein and ligands, primarily due to specific requirements of the docking software. Thus, the common procedures applied to prepare the input structures are here described, while the specific tasks required by each software will be reported in the following sections. All compounds were converted to 3D structures and prepared by using Schrödinger's LigPrep tool. This process generated multiple states for stereoisomers, tautomers, ring conformations (one stable ring conformer by default), and protonation states. In particular, another Schrödinger package, Epik, was used to assign tautomers and protonation states that would be dominant at a selected pH range (pH = 7 ± 1) [36]. Ambiguous chiral centers were enumerated, allowing a maximum of 32 isomers to be produced from each input structure. Then, energy minimization was performed with the OPLS3 force [37]. In this way, 4515 compounds were characterized by multiple states (with an average of 3.1 states per compound), and a total of 23,654 ligands were generated. Docking simulations involved the monomer A of the first resolved SARS-CoV-2 3CL-Pro structure (PDB Id: 6LU7) in a covalent complex with the N3 inhibitor [9]. The protein structure preparation and the binding site characterization were performed as previously described [38]. Briefly, the protein structure was prepared by removing water solvents, crystallization additives, and the covalently bound N3 ligand. The hydrogen atoms were added by using the VEGA program [27] to remain compatible with physiological pH. The protein structure was then minimized using Namd2 [39] and by keeping the backbone atoms fixed to preserve the resolved folding.

3.2. PLANTS Simulations

Concerning PLANTS simulations, ligand conformations and atomic charges were further optimized by semi-empirical PM7 method as implemented by MOPAC [40]. Docking simulations were performed by PLANTS, which is based on ant colony optimization (ACO). [40] Docking search was focused within an 8 Å radius sphere around the cocrystallized N3 inhibitor and, for each compound, 10 poses were generated and ranked by the ChemPLP scoring function with the speed equal to 1. PLANTS and MOPAC calculations were carried out by exploiting Warpengine, an in house developed system for distributed computing [41]. For the post-docking analyses, all generated poses having the ChemPLP score > 0 were discarded, and this induced the loss of 75 inactive compounds (as seen in Table 1).

3.3. LiGen Simulations

The geometrical docking procedure implemented in LiGen™, proprietary software developed by Dompé, was used for the reported docking simulations [42]. In detail, the docking search was focused within a 5.0 Å radius sphere around the co-crystallized ligand. The available void volume of the resulting pocket was defined by determining the free points within a 3D grid, which encompass the entire binding site. The free points are used by the docking procedures as well as to define the pharmacophore schemes. Specifically, the docking engine follows a specific workflow during which three docking scores are computed: first, the Pacman Score (PS) estimates a geometric fitting by evaluating the interaction between a ligand pose and the pocket, based on shape and volume complementarity. Then, the Chemical Score (CS), which encodes for the ligand binding interaction energy, is calculated by an in-house developed scoring function [43]. The last step involves a rigid body minimization of the docked ligand within the binding site, at the end of which a third score called the optimized chemical score (Csopt) is evaluated. All poses that do not fulfill geometric fitting or thresholds values of user-defined specific parameters are discarded, and this induced the loss of 744 compounds (among which were only six active molecules, as seen in Table 1).

With regard to pharmacophore analysis, the program implements three different probe atoms, based on the Tripos Force Field, to explore the binding pocket: (1) a positively charged sp3 nitrogen atom (ammonium cation), describing a hydrogen bond donor; (2) a negatively charged sp2 oxygen atom (as in a carboxyl group), representing a hydrogen bond acceptor; and (3) an sp3 carbon atom (methane), encoding for a hydrophobic group. The representative atom types can be modified, even though this selection produced the best outcomes in previous benchmarking analyses. For each free grid point, the binding energies between the probes and the protein atoms are evaluated by using an in house extended scoring function based on the work of Wang [39]. Every grid point will be identified as donor, acceptor, or hydrophobic according to which probe yields the best score.

The software then filters all the grid points to extract the key interaction sites in three steps. Firstly, the program averages the scores of all the grid points for each probe and selected those points having a score lower than the average. Based on these favorable grid points, LiGenPocket finds the pharmacophoric features. They are identified by clustering the neighbors' grid points, which are here defined as grid points with the same definition (donor, acceptor, or hydrophobic), falling at a distance less than 2.0 Å from a given point. The score averages of all points belonging to the so identified clusters are computed for each type of grid points, and those points, which show a score value lower than the average of their cluster, are discarded.

Secondly, the clusters of neighbor grid points that survive to the previous filtering process constitute the pharmacophoric features of the binding site. The geometric center of each cluster is thus defined as a pharmacophoric point. Lastly and for each pharmacophore element, the minimum distances between a ligand's atom and the closest compatible pharmacophoric point are calculated for each ligand. Each atom of the ligand is defined by a classification that parallels that of the used probes: four atom types are indeed considered and defined by a letter code (A, D, AD, and H are Acceptor, Donor, Acceptor and Donor, and Hydrophobic, respectively).

3.4. FRED Simulations

Ligand conformers were generated using OpenEye OMEGA [44]. Conformers with internal clashes and duplicates were discarded by the software, and the remaining ones were clustered based on of the root mean square deviation (RMSD). For this virtual screening, a maximum of 200 conformers per compound, clustered with an RMSD of 0.5 Å, was used. If the number of conformers generated exceeds the specified maximum, only the ones with the lowest energies are retained. For 1780 molecules, the generation of the rotamers was not possible due to stereochemistry issues and/or for the presence of large macrocycles, and they were removed from the library. The resulting library consisted of 11,755 molecules.

The target protein was processed using UCSF Chimera (v1.14). AMBER ff14SB was used to assign parameters to the standard residues, whereas the Antechamber module was used for the nonstandard residues. The charges for the nonstandard residues were calculated using the AM1-BCC method. The structure was minimized with 100 steps of gradient descent and 10 steps of conjugate descent, using the MMTK module. Rigid docking was then performed using OpenEye FRED [45] included in the OEDocking 3.4.0.2 suite (OpenEye Scientific Software, Santa Fe, NM. http://www.eyesopen.com). Each docked pose is scored using the Gaussian Shape scoring function. Finally, top scoring poses are converted into density fields to form the final shape potential field. The highest values in this field represent points where molecules can have a high number of contacts, without clashing into the protein structure. In its exhaustive search, FRED translates and rotates the structure of each conformer within the negative image of the active site and scores each pose. FRED first step has a default translational and rotational resolution of 1.0 and 1.5 Å, respectively. The 100 best scoring poses are then optimized with translational and rotational single steps of 0.5 and 0.75 Å, respectively, exploring all the 729 (six degrees of freedom with three positions = 36) nearby poses. The best scoring pose is retained and assigned to the compound. The binding poses were evaluated by using the Chemgauss4 scoring function implemented in OpenEye FRED [28]. For 24 molecules, the docking algorithm was unable to find a suitable binding pose, and these molecules were thus discarded from the analysis.

3.5. Glide Simulations

To perform docking experiments with Glide, the protein was preprocessed by the Protein Preparation Wizard from the Schrodinger Suite version 2019-4 with the default parameters [28]. The protonation states of each side chain were generated using Epik for pH = 7 ± 2 [36]. Protein minimization was performed using the OPLS3 force field [37]. All water molecules were removed. Glide software [29] was used for the docking calculations. Internal receptor grid boxes of 10 Å × 10 Å × 10 Å were defined and centered on the ligand atom position. The size of the outer binding box was determined by the ligand size (27 × 27 × 27 Å). A standard precision (SP) Glide docking was carried out, generating 20 poses per docked molecule. H-Bond constraints with D166, H163, and H164 were applied. Docking results were analyzed by Glide Docking score in the version 5.0 [46]. Here, Glide score was used to extract the best binding pose for each ligand. This is an empirical scoring function able to reproduce the trends of the binding affinity and is defined by the following equation:

$$GScore = a.vdW + b.Coul + Lipo + Hbond + Metal + Rewards + RotB + Site$$

where: vdW = van der Waals interaction energy; $Coul$ = Coulomb interaction energy; $Lipo$ = Lipophilic-contact plus phobic-attractive term; $HBond$ = Hydrogen-bonding term; $Metal$ = Metal-binding term (usually a reward); $Rewards$ = Various reward or penalty terms; $RotB$ = Penalty for freezing rotatable bonds; $Site$ = Polar interactions in the active site, and the coefficients of vdW and Coul are: a = 0.050, b = 0.150 for Glide 5.0 (the contribution from the Coulomb term is capped at −4 kcal/mol).

3.6. Rescoring Calculations

All the poses generated by the four docking programs were rescored by ReScore+ [47]. The computed scoring functions comprise (a) the various components of PLANTS [25] and XScore [24] scoring functions; (b) a set of scores computed by the VEGA suite, which encodes for polar and non-polar interaction energies [27]; (c) the MLP interactions scores for hydrophobic contacts [26]; (d) the recently proposed Contacts scores [18], which are simply based on several surrounding residues, and (e) the APBS score for evaluating ionic interactions [48]. Both the primary scores and the values from rescoring calculations were utilized to calculate binding and isomeric spaces as well as their combinations by applying a joining and a merging strategy (see below). For each considered scoring function, each explored space is defined by the following values: (1) the best scores including both the

lowest and the highest values (notice that the best value is not the lowest one for all scores); (2) the average score value; and (3) the score range and the standard deviation to encode for the spread of score values. For each ligand, all the generated poses were utilized to calculate the corresponding space parameters without exceptions.

In detail, the binding space was computed by averaging the computed scores for the poses of a given molecule/isomer. For molecules existing in multiple states, the space parameters corresponding to the isomer with the best primary score were considered. Similarly, the isomeric space was calculated by averaging the computed scores of the pose with the best primary score for all the isomers (clearly only for molecules existing in multiple states). In the so-called merged combination, the space parameters were calculated by averaging together the computed scores of all poses and all isomers. In the so-called joint combination, the consensus equations were developed by simultaneously considering the space parameters as computed for both binding and isomeric spaces. The descriptors for the binding and isomeric spaces were computed by using ad-hoc scripts of the VEGA suite of programs [27].

3.7. Consensus Analyses

The consensus analyses involved the primary scores and the scoring functions as computed by rescoring procedures. Notably, the analysis of the LiGen results also comprised the pharmacophoric distances as computed by this tool. The consensus analyses were performed by the EFO approach, which generates linear combinations of score values by exhaustively combining all possible variables and by optimizing a quality function based on both the early recognition (as encoded by the corresponding EF 1% values) and the entire ranking (as encoded by an asymmetry index applied to the distribution of the active molecules) [33].

By considering the high number of here analyzed descriptors along with the already included exhaustive search method, an incremental search algorithm was also implemented. In particular, given n descriptors in the input dataset, the equations with k variables are built by considering only the top ranked m equations with $k - 1$ variables (m is a user-defined parameter and is set by default equal to 30) and by combining them with the n descriptors avoiding repetitions. Therefore, the models to be evaluated are m $(n - k + 1)$ instead of the number of all possible combinations without repetitions, which are equal to n!/k! $(n - k)!$. A benchmark analysis by comparing the consensus models as generated by incremental and by exhaustive searches revealed that the former involves a performance decrease of about 10 % as assessed by the relative EF1% values when generating the three variable equations for the simplest case without space parameters (data not shown). Such a performance loss is seen as acceptable by considering that the incremental search algorithm allows the analysis of very extended datasets of descriptors and the development of consensus models including more than three variables. Furthermore, the reduced performances of the incremental algorithm equally affected all the here performed comparative analyses. This new search approach was implemented into a standalone and highly optimized version of EFO, which does not require the full installation of the VEGA program and can be freely downloaded at www.vegazz.net. In detail, 20 consensus models were generated for each analysis by combining all input variables without preliminary filtering processes and by always using the incremental search approach. The consensus equations were developed by including from one to five variables. The predictive power of the resulting equations was assessed by subdividing the dataset into training (70%) and test sets (30%) and repeating this task 10 times to minimize the randomness.

4. Conclusions

The study describes and compares a set of VS campaigns performed by using four different docking tools to repurpose an extended set of safe in man molecules as potential inhibitors of the SARS-CoV2 3CL-Pro enzyme. To assess the predictive performances of the here proposed docking strategies and due to the lack of known inhibitors for the considered

enzyme, the peculiar idea of the study is the exploitation of a training set composed of $\cong 500$ compounds that were reported as effective inhibitors (i.e., pIC50 > 5) of the SARS-CoV 3CL-Pro enzyme, a choice justified by the very high conservation degree between these two viral proteases.

All docking simulations were carried out by generating more than one pose per ligand and explicitly considering all possible isomers/states for those molecules existing in multiple states. In this way and after a rescoring analysis of all computed poses, the obtained results were utilized to calculate the descriptors of the resulting binding and isomeric space. The so-calculated score values for each utilized docking program were finally employed to develop consensus models by linearly combining them using the EFO approach.

Taken together, the obtained results allow for some concluding considerations, which can be summarized as follows:

(a) Apart from Fred, the performances of all other docking programs benefit from the inclusion of the binding space parameters as witnessed by an overall EF1% increase average equal to 18% with pIC50 > 6.

(b) The inclusion of the isomeric space parameters reveals, on average, a poorer beneficial effect and plays a clear role only with pIC50 > 5, a finding explainable considering the low number of actives existing in multiple states with pIC50 > 6.

(c) The merging combination of the two explored spaces is substantially ineffective, while the joint combination provides relevant enhancing effects, especially when analyzing the PLANTS and Glide results.

(d) The beneficial effects of the inclusion of space parameters markedly vary among the utilized docking engines; specifically, PLANTS and Glide appear to benefit from both spaces, the LiGen performances are positively affected only by the binding space, while Fred appears to be substantially insensitive to both spaces.

(e) Even though the primary scores play a relevant role in almost all generated predictive models, the rescoring procedures reveal a remarkably beneficial impact by increasing the performances of all used programs regardless of their sensitivity to space parameters.

(f) The enhancing effect of linearly combining diverse score values in the consensus models also varies among the docking programs with PLANTS and Fred showing the largest and the poorest overall effect, respectively.

(g) Although the primary objective of the study was to assess the role played by various post-docking procedures and the retrieved hits were not experimentally tested, it should be noted that the best ranked compounds include an encouraging number of heterogeneous molecules, the SARS-CoV-2 3CL-Pro inhibition activity of which was recently reported (see Table S8). This represents an indirect confirmation of the reliability of the reported simulations.

Since the SARS-CoV2 3CL-Pro is functionally active as a homodimer and many X-ray structures were recently resolved, the encouraging results here reported invite to repeat similar docking protocols by simultaneously considering both the monomers of different representative resolved structures to evaluate the enhancing effects exerted by binding and isomeric spaces for the resulting ensemble simulations. Not to mention that the same strategies could be also applied to explore representative frames coming from MD simulations.

Even though the employment of the known SARS-CoV 3CL-Pro inhibitors was justified by the very high conservation degree between these two enzymes, there is no doubt that the developed predictive models could be enhanced by utilizing true SARS-CoV 3CL-Pro inhibitors. One may figure out that the here described simulations could be exploited in a near future to develop increasingly performing predictive models as novel true inhibitors are identified.

To conclude, the here reported VS campaigns emphasize the generally beneficial effects of the applied post-docking procedures, even though their specific roles significantly

vary among the utilized pieces of software. These observed differences can be ascribed to the different implemented algorithms for docking search and scoring calculations, but they can also be due to the intrinsic geometrical and physicochemical features of the SARS-CoV-2 3CL-Pro binding pocket. This last consideration emphasizes that a robust assessment of the role of binding and isomeric spaces could require extensive benchmarking studies involving wide sets of diverse target proteins.

Supplementary Materials: The following are available online: Figure S1 Best EF1% values as obtained by the four tested docking programs using the common databases with pIC50 > 5 (S1A) and pIC50 > 6 (S1B); Figure S2 Trends of the frequency of the molecules shared at the same time by two, three or four rankings (S2A) or by specific pairs of ranking (S2B) when browsing the first half of the ranking positions (from 1 to 5000); Figure S3 Venn diagram showing the frequencies of the scaffolds detected within the Top500 molecules of the four computed rankings; Figure S4 Distribution of some selected physicochemical properties for the simulated molecules; Table S1 Occurrence of the various score values as observed in all the consensus models generated by using the PLANTS docking results; Table S2 Occurrence of the various score values as observed in all the consensus models generated by using the LiGen docking results; Table S3 Occurrence of the various score values as observed in all the consensus models generated by using the Fred docking results; Table S4 Occurrence of the various score values as observed in all the consensus models generated by using the Glide docking results; Table S5 Enrichment factors as obtained by analyzing the common dataset; Table S6 Occurrence of the various score values as observed in all the consensus models generated by all docking program; Table S7 Rankings as derived by using the best consensus models for the four utilized docking programs; Table S8 Common molecules as found within the top 500 positions of at least three rankings. All the generated consensus models for all performed analyses with the corresponding score values are available at http://www.exscalate4cov.network.

Author Contributions: A.R.B., G.R., C.M. and G.V. designed the study; C.M. prepared the screened database, C.M. and C.T. performed LiGen simulations; S.G. performed PLANTS simulations; A.P. developed the EFO approach; G.V. wrote the manuscript; S.A. performed Fred experiments; J.G. performed Glide experiments; S.A., J.G., B.J.P., F.M. and G.R. planned and validated Glide and Fred setup; A.R.B.: project administration and funding acquisition. All authors have read and agreed to the published version of the manuscript.

Funding: This research was conducted under the project "EXaSCale smArt pLatform Against paThogEns for Corona Virus–Exscalate4CoV" founded by the EU's H2020-SC1-PHE-CORONAVIRUS-2020 call, grant N. 101003551.

Institutional Review Board Statement: Not applicable.

Informed Consent Statement: Not applicable.

Data Availability Statement: Data available in a publicly accessible repository that does not issue DOIs. This data can be found here: http://www.exscalate4cov.network.

Conflicts of Interest: The authors declare no conflict of interest.

Sample Availability: Not available.

Abbreviations

3CL-Pro	3-chymostrypsin like Protease
ACO	ant colony optimization
AM1-BCC	Austin Model 1 (AM1) with bond charge correction (BCC)
APBS	Adaptive Poisson-Boltzmann Solver
CHARMM	Chemistry at Harvard Macromolecular Mechanics
CoV	Coronavirus, subfamily Coronavirinae and family Coronaviridae
Covid	Coronavirus Disease
CS	Chemical Score
CSopt	Optimized Chemical Score
CVFF	consistent-valence force field

EF	Enrichment Factors
EFO	Enrichment Factor Optimization
IC50	half maximal inhibitory concentration
MD	molecular dynamics
MERS	Middle-Est Respiratory Syndrome
MLP	Molecular Lipophilic Potential
MM-PBSA	molecular mechanics-Poisson–Boltzmann surface area
OPLS	Optimized Potentials for Liquid Simulations
PH	Pharmacophoric
PLIF	protein ligand interaction fingerprints
PLP	Piecewise Linear Potential
PM7	Parametric method 7
PS	Pacman Score
SARS	Severe Acute Respiratory Syndrome
SP	standard precision
VS	virtual screening

References

1. Zhu, N.; Zhang, D.; Wang, W.; Li, X.; Yang, B.; Song, J.; Zhao, X.; Huang, B.; Shi, W.; Lu, R.; et al. A Novel Coronavirus from Patients with Pneumonia in China, 2019. *N. Engl. J. Med.* **2020**, *382*, 727–733. [CrossRef]
2. Wang, C.; Horby, P.W.; Hayden, F.G.; Gao, G.F. A novel coronavirus outbreak of global health concern. *Lancet* **2020**, *395*, 470–473. [CrossRef]
3. Li, J.; Huang, D.Q.; Zou, B.; Yang, H.; Hui, W.Z.; Rui, F.; Yee, N.T.S.; Liu, C.; Nerurkar, S.N.; Kai, J.C.Y.; et al. Epidemiology of COVID-19: A systematic review and meta-analysis of clinical characteristics; risk factors; and outcomes. *J. Med. Virol.* **2020**. online ahead of print. [CrossRef]
4. Khailany, R.A.; Safdar, M.; Ozaslan, M. Genomic characterization of a novel SARS-CoV-2. *Gene Rep.* **2020**, *19*, 100682. [CrossRef] [PubMed]
5. Michel, C.J.; Mayer, C.; Poch, O.; Thompson, J.D. Characterization of accessory genes in coronavirus genomes. *Virol. J.* **2020**, *17*, 131. [CrossRef] [PubMed]
6. Chellapandi, P.; Saranya, S. Genomics insights of SARS-CoV-2 (COVID-19) into target-based drug discovery. *Med. Chem. Res.* **2020**, *29*, 1777–1791. [CrossRef]
7. Ashour, H.M.; Elkhatib, W.F.; Rahman, M.M.; Elshabrawy, H.A. Insights into the Recent 2019 Novel Coronavirus (SARS-CoV-2) in Light of Past Human Coronavirus Outbreaks. *Pathogens* **2020**, *9*, 186. [CrossRef] [PubMed]
8. Shyr, Z.A.; Gorshkov, K.; Chen, C.Z.; Zheng, W. Drug discovery strategies for SARS-CoV-2. *J. Pharmacol. Exp. Ther.* **2020**, *375*, 127–138. [CrossRef]
9. Jin, Z.; Du, X.; Xu, Y.; Deng, Y.; Liu, M.; Zhao, Y.; Zhang, B.; Li, X.; Zhang, L.; Peng, C.; et al. Structure of mpro from COVID-19 virus and discovery of its inhibitors. *Nature* **2020**, *582*, 289–293. [CrossRef]
10. Chuck, C.P.; Chow, H.F.; Wan, D.C.; Wong, K.B. Profiling of substrate specificities of 3C-like proteases from group 1, 2a, 2b, and 3 coronaviruses. *PLoS ONE* **2011**, *6*, e27228. [CrossRef]
11. Li, Q.; Kang, C. Progress in Developing Inhibitors of SARS-CoV-2 3C-Like Protease. *Microorganisms* **2020**, *8*, 1250. [CrossRef]
12. Gimeno, A.; Mestres-Truyol, J.; Ojeda-Montes, M.J.; Macip, G.; Saldivar-Espinoza, B.; Cereto-Massagué, A.; Pujadas, G.; Garcia-Vallvé, S. Prediction of Novel Inhibitors of the Main Protease (M-pro) of SARS-CoV-2 through Consensus Docking and Drug Reposition. *Int. J. Mol. Sci.* **2020**, *21*, 3793. [CrossRef] [PubMed]
13. Elmezayen, A.D.; Al-Obaidi, A.; Şahin, A.T.; Yelekçi, K. Drug repurposing for coronavirus (COVID-19): In silico screening of known drugs against coronavirus 3CL hydrolase and protease enzymes. *J. Biomol. Struct. Dyn.* **2020**, *26*, 1–13. [CrossRef] [PubMed]
14. Meyer-Almes, F.J. Repurposing approved drugs as potential inhibitors of 3CL-protease of SARS-CoV-2: Virtual screening and structure based drug design. *Comput. Biol. Chem.* **2020**, *88*, 107351. [CrossRef] [PubMed]
15. Olubiyi, O.O.; Olagunju, M.; Keutmann, M.; Loschwitz, J.; Strodel, B. High Throughput Virtual Screening to Discover Inhibitors of the Main Protease of the Coronavirus SARS-CoV-2. *Molecules* **2020**, *25*, 3193. [CrossRef]
16. Gao, L.Q.; Xu, J.; Chen, S.D. In Silico Screening of Potential Chinese Herbal Medicine Against COVID-19 by Targeting SARS-CoV-2 3CLpro and Angiotensin Converting Enzyme II Using Molecular Docking. *Chin. J. Integr. Med.* **2020**, *26*, 527–532. [CrossRef]
17. Macchiagodena, M.; Pagliai, M.; Procacci, P. Identification of potential binders of the main protease 3CLpro of the COVID-19 via structure-based ligand design and molecular modeling. *Chem. Phys. Lett.* **2020**, *750*, 137489. [CrossRef]
18. Vistoli, G.; Mazzolari, A.; Testa, B.; Pedretti, A. Binding Space Concept: A New Approach To Enhance the Reliability of Docking Scores and Its Application to Predicting Butyrylcholinesterase Hydrolytic Activity. *J. Chem. Inf. Model.* **2017**, *57*, 1691–1702. [CrossRef]
19. Vistoli, G.; Pedretti, A.; Testa, B. Assessing drug-likeness–what are we missing? *Drug Discov. Today* **2008**, *13*, 285–294. [CrossRef]

20. Mazzolari, A.; Gervasoni, S.; Pedretti, A.; Fumagalli, L.; Matucci, R.; Vistoli, G. Repositioning Dequalinium as Potent Muscarinic Allosteric Ligand by Combining Virtual Screening Campaigns and Experimental Binding Assays. *Int. J. Mol. Sci.* **2020**, *21*, 5961. [CrossRef]
21. Mazzolari, A.; Vistoli, G.; Testa, B.; Pedretti, A. Prediction of the Formation of Reactive Metabolites by A Novel Classifier Approach Based on Enrichment Factor Optimization (EFO) as Implemented in the VEGA Program. *Molecules* **2018**, *23*, 2955. [CrossRef] [PubMed]
22. Pedretti, A.; Mazzolari, A.; Gervasoni, S.; Vistoli, G. Rescoring and Linearly Combining: A Highly Effective Consensus Strategy for Virtual Screening Campaigns. *Int. J. Mol. Sci.* **2019**, *20*, 2060. [CrossRef]
23. Talarico, C.; Gervasoni, S.; Manelfi, C.; Pedretti, A.; Vistoli, G.; Beccari, A.R. Combining Molecular Dynamics and Docking Simulations to Develop Targeted Protocols for Performing Optimized Virtual Screening Campaigns on The hTRPM8 Channel. *Int. J. Mol. Sci.* **2020**, *21*, 2265. [CrossRef] [PubMed]
24. Wang, R.; Lai, L.; Wang, S. Further development and validation of empirical scoring functions for structure-based binding affinity prediction. *J. Comput.-Aided Mol. Des.* **2002**, *16*, 11–26. [CrossRef] [PubMed]
25. Korb, O.; Stützle, T.; Exner, T.E. Empirical scoring functions for advanced protein-ligand docking with PLANTS. *J. Chem. Inf. Model.* **2009**, *49*, 84–96. [CrossRef]
26. Vistoli, G.; Pedretti, A.; Mazzolari, A.; Testa, B. In silico prediction of human carboxylesterase-1 (hCES1) metabolism combining docking analyses and MD simulations. *Bioorg. Med. Chem.* **2010**, *18*, 320–329. [CrossRef]
27. Pedretti, A.; Mazzolari, A.; Gervasoni, S.; Fumagalli, L.; Vistoli, G. The VEGA suite of programs: A versatile platform for cheminformatics and drug design projects. *Bioinformatics* **2020**. online ahead of print. [CrossRef]
28. Sastry, G.M.; Adzhigirey, M.; Day, T.; Annabhimoju, R.; Sherman, W. Protein and ligand preparation: Parameters, protocols, and influence on virtual screening enrichments. *J. Comput. Aided Mol. Des.* **2013**, *27*, 221–234. [CrossRef]
29. Friesner, R.A.; Murphy, R.B.; Repasky, M.P.; Frye, L.L.; Greenwood, J.R.; Halgren, T.A.; Sanschagrin, P.C.; Mainz, D.T. Extra precision glide: Docking and scoring incorporating a model of hydrophobic enclosure for protein-ligand complexes. *J. Med. Chem.* **2006**, *49*, 6177–6196. [CrossRef]
30. Yang, S.; Chen, S.J.; Hsu, M.F.; Wu, J.D.; Tseng, C.T.; Liu, Y.F.; Chen, H.C.; Kuo, C.W.; Wu, C.S.; Chang, L.W.; et al. Synthesis, crystal structure, structure-activity relationships, and antiviral activity of a potent SARS coronavirus 3CL protease inhibitor. *J. Med. Chem.* **2006**, *49*, 4971–4980. [CrossRef] [PubMed]
31. Xie, X.; Muruato, A.E.; Zhang, X.; Lokugamage, K.G.; Fontes-Garfias, C.R.; Zou, J.; Liu, J.; Ren, P.; Balakrishnan, M.; Cihlar, T.; et al. A nanoluciferase SARS-CoV-2 for rapid neutralization testing and screening of anti-infective drugs for COVID-19. *Nat. Commun.* **2020**, *11*, 5214. [CrossRef]
32. Chiou, W.C.; Hsu, M.S.; Chen, Y.T.; Yang, J.M.; Tsay, Y.G.; Huang, H.C.; Huang, C. Repurposing existing drugs: Identification of SARS-CoV-2 3C-like protease inhibitors. *J. Enzyme Inhib. Med. Chem.* **2021**, *36*, 147–153. [CrossRef] [PubMed]
33. Jan, J.T.; Cheng, T.R.; Juang, Y.P.; Ma, H.H.; Wu, Y.T.; Yang, W.B.; Cheng, C.W.; Chen, X.; Chou, T.H.; Shie, J.J.; et al. Identification of existing pharmaceuticals and herbal medicines as inhibitors of SARS-CoV-2 infection. *Proc. Natl. Acad. Sci. USA* **2021**, *118*, e2021579118. [CrossRef] [PubMed]
34. Gupta, A.; Rani, C.; Pant, P.; Vijayan, V.; Vikram, N.; Kaur, P.; Singh, T.P.; Sharma, S.; Sharma, P. Structure-Based Virtual Screening and Biochemical Validation to Discover a Potential Inhibitor of the SARS-CoV-2 Main Protease. *ACS Omega.* **2020**, *5*, 33151–33161. [CrossRef]
35. Zhu, Y.; Xie, D.Y. Docking Characterization and in vitro Inhibitory Activity of Flavan-3-ols and Dimeric Proanthocyanidins Against the Main Protease Activity of SARS-Cov-2. *Front. Plant. Sci.* **2020**, *11*, 601316. [CrossRef] [PubMed]
36. Greenwood, J.R.; Calkins, D.; Sullivan, A.P.; Shelley, J.C. Towards the comprehensive, rapid, and accurate prediction of the favorable tautomeric states of drug-like molecules in aqueous solution. *J. Comput. Aided Mol. Des.* **2010**, *24*, 591–604. [CrossRef] [PubMed]
37. Harder, E.; Damm, W.; Maple, J.; Wu, C.; Reboul, M.; Xiang, J.Y.; Wang, L.; Lupyan, D.; Dahlgren, M.K.; Knight, J.L.; et al. OPLS3: A Force Field Providing Broad Coverage of Drug-like Small Molecules and Proteins. *J. Chem. Theory Comput.* **2016**, *12*, 281–296. [CrossRef]
38. Gervasoni, S.; Vistoli, G.; Talarico, C.; Manelfi, C.; Beccari, A.R.; Studer, G.; Tauriello, G.; Waterhouse, A.M.; Schwede, T.; Pedretti, A. A Comprehensive Mapping of the Druggable Cavities within the SARS-CoV-2 Therapeutically Relevant Proteins by Combining Pocket and Docking Searches as Implemented in Pockets 2.0. *Int. J. Mol. Sci.* **2020**, *21*, 5152. [CrossRef]
39. Phillips, J.C.; Hardy, D.J.; Maia, J.D.C.; Stone, J.E.; Ribeiro, J.V.; Bernardi, R.C.; Buch, R.; Fiorin, G.; Henin, J.; Jiang, W.; et al. Scalable molecular dynamics on CPU and GPU architectures with NAMD. *J. Chem. Phys.* **2020**, *153*, 044130. [CrossRef]
40. Stewart, J.J. Optimization of parameters for semiempirical methods VI: More modifications to the NDDO approximations and re-optimization of parameters. *J. Mol. Model.* **2013**, *19*, 1–32. [CrossRef]
41. Pedretti, A.; Mazzolari, A.; Vistoli, G. WarpEngine, a Flexible Platform for Distributed Computing Implemented in the VEGA Program and Specially Targeted for Virtual Screening Studies. *J. Chem. Inf. Model.* **2018**, *58*, 1154–1160. [CrossRef]
42. Beccari, A.R.; Cavazzoni, C.; Beato, C.; Costantino, G. LiGen: A high performance workflow for chemistry driven de novo design. *J. Chem. Inf. Model.* **2013**, *53*, 1518–1527. [CrossRef]

43. Beato, C.; Beccari, A.R.; Cavazzoni, C.; Lorenzi, S.; Costantino, G. Use of experimental design to optimize docking performance: The case of LiGenDock.; the docking module of LiGen.; a new de novo design program. *J. Chem. Inf. Model.* **2013**, *53*, 1503–1517. [CrossRef]

44. Hawkins, P.C.; Skillman, A.G.; Warren, G.L.; Ellingson, B.A.; Stahl, M.T. Conformer generation with OMEGA: Algorithm and validation using high quality structures from the Protein Databank and Cambridge Structural Database. *J. Chem. Inf. Model.* **2010**, *50*, 572–584. [CrossRef] [PubMed]

45. McGann, M. FRED pose prediction and virtual screening accuracy. *J. Chem. Inf. Model.* **2011**, *51*, 578–596. [CrossRef] [PubMed]

46. Halgren, T.A.; Murphy, R.B.; Friesner, R.A.; Beard, H.S.; Frye, L.L.; Pollard, W.T.; Banks, J.L. Glide: A new approach for rapid.; accurate docking and scoring. 2. Enrichment factors in database screening. *J. Med. Chem.* **2004**, *47*, 1750–1759. [CrossRef] [PubMed]

47. Pedretti, A.; Granito, C.; Mazzolari, A.; Vistoli, G. Structural Effects of Some Relevant Missense Mutations on the MECP2-DNA Binding: A MD Study Analyzed by Rescore+.; a Versatile Rescoring Tool of the VEGA ZZ Program. *Mol. Inform.* **2016**, *35*, 424–433. [CrossRef] [PubMed]

48. Baker, N.A.; Sept, D.; Joseph, S.; Holst, M.J.; McCammon, J.A. Electrostatics of nanosystems: Application to microtubules and the ribosome. *Proc. Natl. Acad. Sci. USA* **2001**, *98*, 10037–10041. [CrossRef]

Article

Structural Impacts of Drug-Resistance Mutations Appearing in HIV-2 Protease

Pierre Laville, Michel Petitjean and Leslie Regad *

Université de Paris, BFA, UMR 8251, CNRS, ERL U1133, Inserm, F-75013 Paris, France;
pierre.laville@u-paris.fr (P.L.); petitjean.chiral@gmail.com (M.P.)
* Correspondence: leslie.regad@u-paris.fr; Tel.: +33-1-57-27-82-72

Abstract: The use of antiretroviral drugs is accompanied by the emergence of HIV-2 resistances. Thus, it is important to elucidate the mechanisms of resistance to antiretroviral drugs. Here, we propose a structural analysis of 31 drug-resistant mutants of HIV-2 protease (PR2) that is an important target against HIV-2 infection. First, we modeled the structures of each mutant. We then located structural shifts putatively induced by mutations. Finally, we compared wild-type and mutant inhibitor-binding pockets and interfaces to explore the impacts of these induced structural deformations on these two regions. Our results showed that one mutation could induce large structural rearrangements in side-chain and backbone atoms of mutated residue, in its vicinity or further. Structural deformations observed in side-chain atoms are frequent and of greater magnitude, that confirms that to fight drug resistance, interactions with backbone atoms should be favored. We showed that these observed structural deformations modify the conformation, volume, and hydrophobicity of the binding pocket and the composition and size of the PR2 interface. These results suggest that resistance mutations could alter ligand binding by modifying pocket properties and PR2 stability by impacting its interface. Our results reinforce the understanding of the effects of mutations that occurred in PR2 and the different mechanisms of PR2 resistance.

Keywords: drug-resistance mutations; HIV-2 protease; structural characterization; induced structural deformations

Citation: Laville, P.; Petitjean, M.; Regad, L. Structural Impacts of Drug-Resistance Mutations Appearing in HIV-2 Protease. *Molecules* **2021**, *26*, 611. https://doi.org/10.3390/molecules26030611

Academic Editors: Marco Tutone and Anna Maria Almerico

Received: 18 December 2020
Accepted: 19 January 2021
Published: 25 January 2021

Publisher's Note: MDPI stays neutral with regard to jurisdictional claims in published maps and institutional affiliations.

1. Introduction

The human immunodeficiency virus (HIV) affects humans and causes the acquired immunodeficiency syndrome (AIDS). The treatment against HIV-1 infection corresponds to the same molecules that target four HIV proteins: the fusion protein, protease (PR), integrase, and reverse transcriptase. The same molecules are used to fight HIV-2 infection but HIV-2 is naturally resistant to most of these inhibitors [1–9]. Thus, it is important to develop new molecules that inhibit the HIV-2 replication, particularly against the HIV-2 protease (PR2).

PR is a homodimer that plays a major role in the virus maturation process: it hydrolyzes the viral polyproteins Gag and Gag-Pol, causing the development of immature virions. There are currently nine protease inhibitors (PIs) clinically recommended for treating HIV-1 infection [10]. These drugs bind to the PR catalytic site in the interface of the two monomers. This binding induces structural changes in the entire PR2, particularly in the flap region allowing the closing of binding pocket [10–12]. PR2 is naturally resistant to most of these PIs and only three of them are recommended for the treatment of HIV-2 infection: darunavir (DRV), saquinavir (SQV), and lopinavir (LPV) [1,10]. The natural resistance of PR2 is explained by amino-acid changes between PR1 and PR2 that induce structural changes in the entire structure [13]. Some of these structural changes, located in the binding pocket, modify properties and conformation of the PI-binding pocket and the internal interactions between PR2 and PIs [3,5,14–20]. Other structural changes, occurring

in the elbow and flap regions, alter the transition between the open and closed forms involved in ligand binding [13,21,22].

In addition to its natural resistance, many acquired resistance mutations appear in PR2. The identification of these mutations have been performed using genome sequencing studies of HIV-2 virus extracted from infected patients [2,4–6,8,23–28]. For example, it has been shown that the V47A, I50V, I54M, I82F, I84V, and L90M mutations lead to several PIs resistance [2,4,5,7,8]. Phenotypic susceptibility assays were used to confirm the resistance of some mutants [2,4,5,8,26,29]. For example, genotypic and phenotypic analyses showed that the I82F mutation causes resistance to indinavir (IDV) [26]. Furthermore, this mutation has been identified as causing hypersusceptibility to both DRV and SQV using phenotypic assays [2]. In addition, combinations of several mutations confer high resistance to several PIs [2,4,5,7,8,30]. For example, the I54M and I82F mutations induce resistance to all PIs and the V62A/L99F mutant is resistant to nelfinavir, IDV, and LPV [26]. Few studies have focused on the structural analysis of the impacts of drug-resistance mutations reported in PR2 because no tridimensional (3D) structure of PR2 mutant have been solved. These structural studies could help to understand the atomistic mechanism of resistance mutations. In our previous work, we performed the first structural analysis of the impact of 30 drug-resistant mutants of PR2 based on modeled structures. More precisely, we explored the consequences of drug-resistance mutations on PR2 structural asymmetry, an important property for ligand-binding and the target deformation [31]. Our findings suggested three possible resistance mechanisms of PR2: (i) mutations that induce structural changes in the binding pocket that could directly alter PI-binding, (ii) mutations that could impact the properties and conformation of the binding pocket by inducing structural changes in residues outside of the binding pocket but involved in interaction with pocket residues, and (iii) mutations that could modify the PR2 interface and its stability through structural changes in interface residues. These results have been based on PR2 backbone analysis. However, a better characterization of the structural impacts of drug-resistance mutations on PR2 structure including side-chain atoms could help in a better understanding of different proposed mechanisms.

In this study, we structurally analyzed a set of 31 drug-resistant mutants of PR2 that was updated relative to [31]. The 3D structure of each mutant was modeled and its structure was compared to the wild-type PR2 to locate structural rearrangements induced by drug-resistance mutations at backbone and side-chain atoms. The study reported that drug-resistance mutations could impact the flexbility of PR2 and the closing binding pocket, conformation and properties of PI-pocket and the composition and size of the PR2 interface.

2. Results

We studied the impact of 22 drug-resistance mutations on PR2 structure. These mutations appeared alone or in combination with others (two or three mutations per mutant) resulting in a set of 31 mutants, Figure 1A. First, mutant structures were modeled using FoldX software [32] and an energetic minimization step using the crystallographic structure of the wild-type PR2 (PDB code: 3EBZ [33]). Five 3D structures were built for each mutant to consider the different possible rotamers per amino acids as illustrated by Figure A1. This mutagenesis process resulted in a set of 155 mutant structures. The crystallographic structure of the wild-type PR2 (PDB code: 3EBZ [33]) was also energetically minimized with the protocol used for mutant structures. In the following, the minimized wild-type structure was referred as the wild-type structure and its structure was compared to the minimized mutant structures.

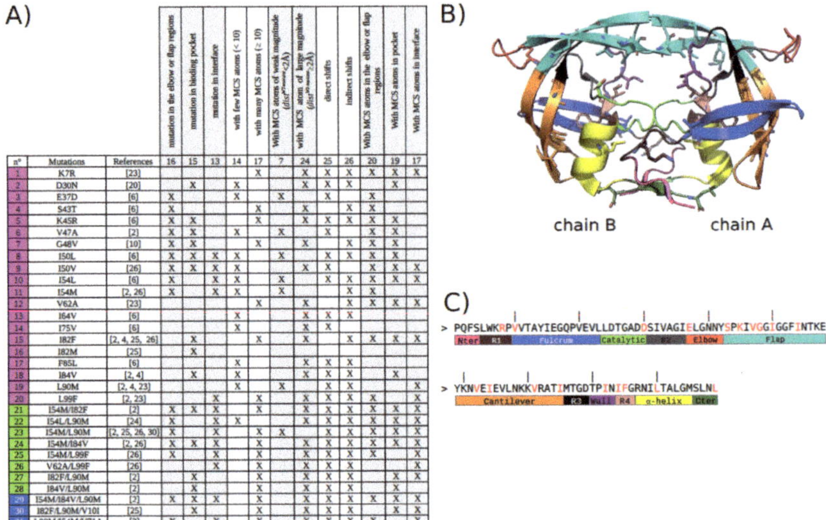

A)

n°	Mutations	References	mutation in the elbow or flap regions (16)	mutation in binding pocket (15)	mutation in interface (13)	with few MCS atoms (<10) (14)	with many MCS atoms (≥10) (17)	With MCS atoms of weak magnitude ($dist^{WT-mutant}<2$Å) (7)	with MCS atom of large magnitude ($dist^{WT-mutant}\geq 2$Å) (24)	direct shifts (25)	indirect shifts (26)	With MCS atoms in the elbow or flap regions (20)	with MCS atoms in pocket (19)	With MCS atoms in interface (17)
1	K7R	[23]					X		X	X	X	X	X	X
2	D30N	[20]		X		X			X	X	X		X	
3	E37D	[6]	X			X		X	X			X		
4	S43T	[6]	X				X		X		X	X		
5	K45R	[6]	X	X			X		X	X	X	X	X	
6	V47A	[2]	X	X		X		X		X		X	X	
7	G48V	[10]	X	X			X		X		X	X	X	
8	I50L	[6]	X	X	X	X		X		X	X	X	X	
9	I50V	[26]	X	X	X	X			X	X		X	X	X
10	I54L	[6]	X		X	X		X		X	X	X	X	X
11	I54M	[2, 26]	X		X	X		X			X	X		
12	V62A	[23]					X		X		X	X	X	X
13	I64V	[6]				X			X	X	X			
14	I75V	[6]				X			X	X				
15	I82F	[2, 4, 25, 26]		X			X		X		X	X	X	X
16	I82M	[25]		X										
17	F85L	[6]				X			X	X	X			
18	I84V	[2, 4]		X		X			X	X	X		X	
19	L90M	[2, 4, 23]				X		X	X	X				X
20	L99F	[2, 23]			X		X		X	X	X	X		X
21	I54M/I82F	[2]	X	X	X		X		X	X	X	X	X	X
22	I54L/L90M	[24]	X		X	X			X	X	X	X	X	X
23	I54M/L90M	[2, 25, 26, 30]	X		X		X	X	X	X	X	X	X	X
24	I54M/I84V	[2, 26]	X	X	X		X		X	X	X	X	X	X
25	I54M/L99F	[26]	X		X		X		X	X	X	X		X
26	V62A/L99F	[26]			X		X		X	X	X			X
27	I82F/L90M	[2]		X			X		X	X	X		X	X
28	I84V/L90M	[2]		X			X		X	X	X		X	
29	I54M/I84V/L90M	[2]	X	X	X		X		X	X	X	X	X	X
30	I82F/L90M/V10I	[25]		X			X		X	X	X		X	X
31	L90M/I54M/V71A	[2]	X		X		X		X	X	X	X		X

Figure 1. Description of the 31 drug-resistant mutants studied in this analysis. (**A**) Table listing the 31 drug-resistant mutants studied in this analysis. Single mutants are colored in magenta, double mutants in green and triple mutants in blue. (**B**) Location on PR2 structure of the 22 drug-resistance mutations included in the 31 mutant set. PR2 is represented in cartoon mode and colored according to the 13 regions defined in [34,35]. Mutations are represented in stick mode. (**C**) Amino acid sequence of PR2 presenting the limit of the 13 PR2 regions. All the mutated residues are colored in red in the sequence.

2.1. Identification of Atom Shifts in the Mutant-Structure Set

We first explored the structural deformations induced by mutations by locating shifted atoms in mutant structures relative to the wild-type structure. The shift of an atom was quantified by $dist^{WT-mutant}$, i.e., the distance between its positions in the wild-type and mutant structures. Figure A2 presents the distribution of $dist^{WT-mutant}$ of all atoms in the mutant structures set, except hydrogen atoms. These distances varied from 0.05 to 0.05, with 95% of atoms exhibited a $dist^{WT-mutant}$ smaller than 0.14 Å. These distances were summarized per structures by computing the RMSD between the wild-type and mutant structures (Figure A2). As expected, mutant structures exhibited close conformation than the wild-type, resulting in an average RMSD of 0.12 ± 0.09 Å. Only one mutant structure (one structure of the K7R mutant) exhibited a RMSD higher than 0.5 Å. Like in Liu et al., 2008 [36], we considered that a shift was significant if it had a magnitude higher than 0.3 Å because of uncertainties in the X-ray and mutant structures. Thus, to identify atom shifts induced by mutations, we retained the 2136 atoms with a $dist^{WT-mutant}$ higher than 0.3 Å in at least three structures of a given mutant. These selected atoms were denoted as "mutant-conserved shifted atoms" (MCS atoms) and the distribution of their $dist^{WT-mutant}$ is provided by Figure 2A. Most MCS atoms (70%) exhibited a shift of moderate magnitude (<1 Å). However, 17% of MCS atoms are of large magnitude (>2 Å), such as atom shifts detected in residue K69_A (Lysine 69 in chain A) in the L99F mutant that is illustrated in Figure 2B. Figure 2C presents another type of atom shift that corresponds to a flip of a ring of residue 3_B (residue 3 in chain B) in the five structures of the V62A/L99F mutant. This rearrangement type does not induce structural deformation and thus it does not seem to be linked to PI resistance.

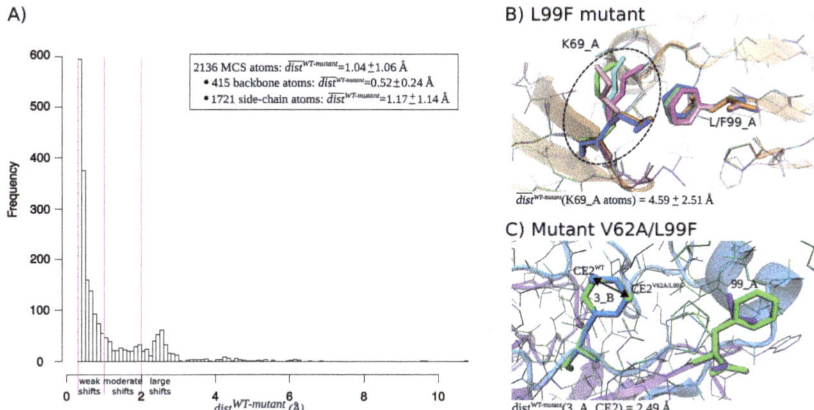

Figure 2. (**A**) Distribution of $dist^{WT-mutant}$ distances for MCS atoms extracted from the set of 155 mutant structures. Magenta lines corresponds to the cutoffs used to define a weak shift (0.3 Å $< dist^{WT-mutant} <$ 1 Å), moderate shifts (1 Å $\leq dist^{WT-mutant} <$ 2 Å), and large shifts ($dist^{WT-mutant} >$ 2 Å). (**B,C**) Illustration of atom shifts in the L99F (**A**) and V62A/L99F (**B**) mutants. (**B**) Superimposition of the five structures of the L99F mutant and the wild-type structure. Wild-type structure is colored in orange and represented in line and cartoon modes. The five structure of the mutant are represented in line mode and colored in magenta, cyan, blue, green, and pink. The L99F mutated residue is represented in stick mode. (**C**) Illustration of structural shift occuring at residue 3_B in the V62A/L99F mutant. The wild-type structure is represented in cartoon and line modes and colored according to its two chains: chain A is colored in purple and chain B is colored in marien blue. The mutant V62A/L99F structure is represented in lines and colored in green. The mutated residue 99_A and shifted residue 3_A are represented in stick mode. The arrow represents the $dist^{WT-mutant}$ of the CE atom of residue 3_B computed between the wild-type and the first structure of mutant V62A/L99F. $CE2^{V62A/L99F}$ and $CE2^{WT}$ correspond to atom CE of residue 3 of chain B in the V62A/L99F and wild-type structures and are represented in sphere mode.

As expected, side-chain atoms were overrepresented in the MCS atom set (Pearson's Chi-squared Test p-value = 3×10^{-28}) and they had larger $dist^{WT-mutant}$ value than backbone atoms (Student's t-test p-value of 6×10^{-95}). This means that drug-resistance mutations have more impacts on side-chain atoms than on backbone atoms. From the 2136 MCS atoms, 543 (=25%) are atoms of mutated residues. The shift of these atoms, named direct shifts, was a direct consequence of mutations. In contrast, 1593 (=75%) MCS atoms corresponded to indirect shifts, i.e., they occurred in non mutated residues, and their shifts resulted either from the intrinsic flexibility of atoms or from indirect impacts induced by the mutation through contacts between these atoms and mutated residues. Direct shifts had larger magnitude than indirect shifts, i.e., they exhibited an average higher $dist^{WT-mutant}$ distances than indirect shifts (Student's t-test p-value = 3×10^{-13}). To distinguish structural shifts resulting from mutation from those induced by flexibility, we detected MCS atoms in non minimized structures, i.e., mutant structures corresponded to the output of FoldX software. From this set of non-minimized structures, we located 883 MCS atoms, with 646 that were also detected as MCS atoms in the minimized structures set. This means that the shift of these MCS atoms observed in minimized structures (30% of MCS atoms) were the consequence of the mutation. As expected, these shifts, observed in minimized structures, occurred only on side-chain atoms and in or close to the mutated residue because FoldX software optimizes the configuration of only side chains in the vicinity of the mutated residue. The remained detected shifts were explained by the mutation and intrinsic flexibility of atoms.

Figure 3A shows that the entire PR2 structure was sampled by MCS atoms. A total of eight regions had few MCS atoms, i.e., less than 50 MCS atoms were detected in the

155 mutant structures (Figure 3). In contrast, some regions exhibited many MCS atoms, such as the cantilever and flap regions of the two chains and the α-helix region of chain B, with more than 150 MCS atoms. In addition, an assymetry between the two chains in terms of number of MCS atoms was observed. Indeed, chain B contained more MCS atoms than chain A (p-value of the Pearson's Chi-square test is of 5×10^{-24}). For example, the Nter, fulcrum, elbow, and R3 regions of chain A present few structural shifts, while they exhibited lots of deformations in chain B. Thus, even though mutations occurred in the two chains, they did not impact in the same way the two monomers.

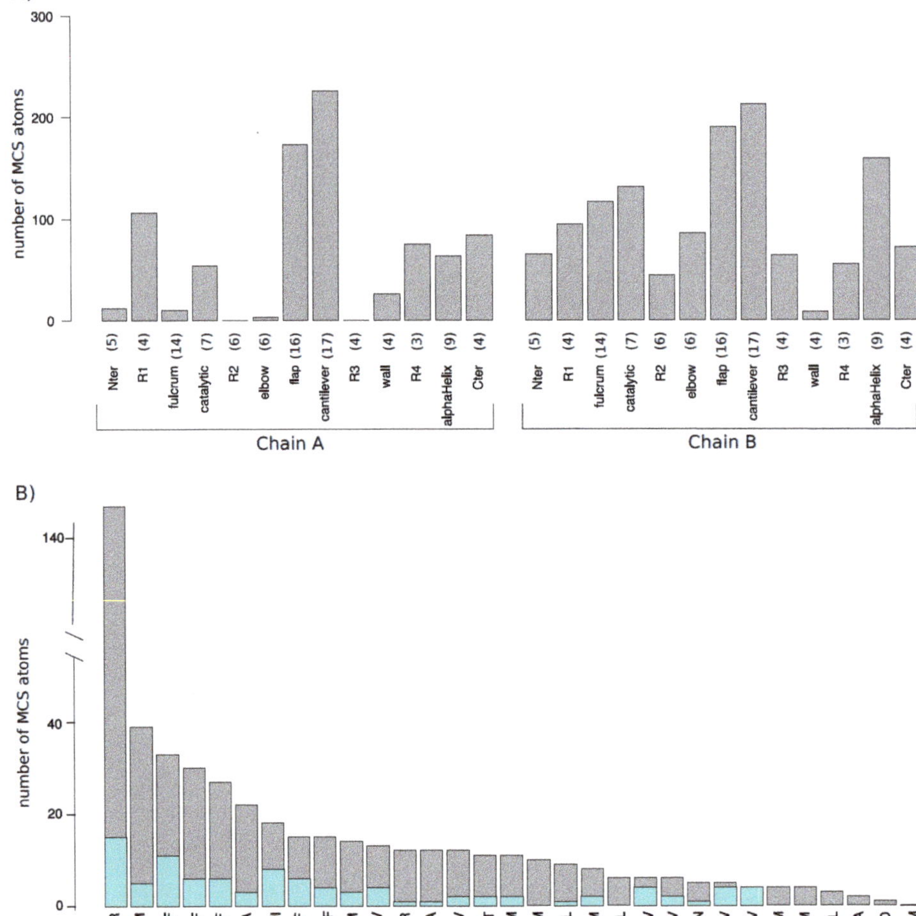

Figure 3. (**A**) Number of MCS atoms per mutant counted in the two PR2 chains. Each PR2 chain was divided in 13 regions according to [13]. Numbers in brackets indicate the size of each region in terms of amino acids. (**B**) Number of MCS atoms (grey) and MCS atoms with large magnitude ($dist^{WT-mutant} \geq 2$, cyan) per mutants. Mutants were sorted according to their number of MCS atoms.

Figure 3B presents the number of MCS atoms per mutants. On average, a mutant contained a moderate number of MCS atoms (16.26 ± 26.09). The I82M mutant was particular because it had no MCS atoms, revealing that the I82M mutation induced few structural changes in PR2 structure. In contrast, the K7R mutant was the mutant with the most MCS atoms (147 MCS located in 53 different residues as illustrated in Figure 4. A total of 14 mutants had less than 10 MCS atoms, indicating that these mutations induced few impacts on PR2 structure (Figure 1A). Except the I54L/L90M, all these mutants were single mutants. Although these mutants exhibited few deformations, 57% of them had shifts with a large magnitude ($dist^{WT-mutant} \geq 2$). For example, the D30N mutant induced 5 MCS atoms with one exhibiting large shift of 2.9 Å, while the L90M mutant caused four small shifts at residues 90_A/B and 97_B with a magnitude varying from 0.35 to 0.48 Å (Figures 3B and A3). In contrast, 17 mutants had many MCS atoms and all of these shifts had of large magnitude (Figure 3B). From these mutants, seven were single mutants, revealing that only one mutation could cause large deformations, such as those observed for the K7R and L99F mutants (Figures 4 and A3).

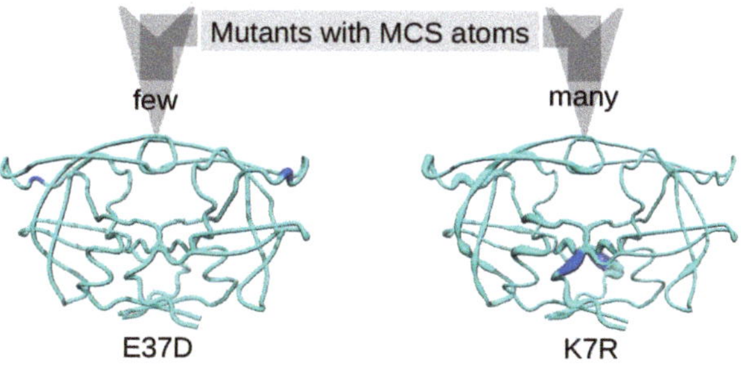

Figure 4. Location of the MCS atoms found in some mutants. PR2 is colored in cyan and represented in putty mode. The putty radius is relative to deformations induced by mutations: the higher the radius, the stronger the mutation-induced rearrangement. Mutated residues are colored in blue.

According to the location of MCS atoms, we differentiated three types of mutations (Figure 4). The first mutations corresponded to four mutations that induced structural rearrangements only in the mutated residues, i.e., having only direct impacts. For example, the I50V mutation caused structural changes in two atoms of residue 50 of the two chains, a residue involved in the binding pocket, the dimer interface and the flap region (Figure 4). The second type of mutations grouped five mutations impacting residues in their vicinity or further in structure, i.e., inducing indirect structural changes, such as the I82F mutations (Figure 1). Indeed, the I82F mutation induced many atoms rearrangements in five non mutated residues (8_A, 8_B, 21_B, 27_B, 49_B) with some of them of large magnitude. In this mutant structure, residue 8_A is located in the vicinity of mutated residue 82_A (located at less than 5.5 Å), while residues 27_B and 49_B that are

located at more than 6 Å of the mutated residue 82_A (Figure A8). The last type of mutations corresponded to mutations that induced both direct and indirect rearrangements (Figures 1 and 4). This mutation type grouped most of mutations. Figure 4 showed that the L99F mutation produced large shitfs in the mutated residues and also in its neighbor residues 68 and 69.

The location of MCS atoms (Figure A3) in mutant structures highlighted structural rearrangments located in important regions for PR2: in the elbow and flap regions that are implied in the PR2 deformations induced by ligand binding, in its pocket binding and in its interface. In the following, we explored the impacts of the studied resistance mutation in the PI-binding pocket and PR2 interface.

2.2. Impact of Mutations on the Properties of PI-Binding Pocket

From the 31 mutants, 15 had at least one mutation in the binding pocket (Figures 1A and A4). Except the I82M mutant, all these mutants presented MCS atoms in the pocket in the mutated or non mutated residue. Surprisingly, structural rearrangements in the binding pocket were also observed in five mutants without pocket mutations (K7R, I54L, V62A, I54L/L90M, I54M/L90M, Figure 1). A total of 36% of pocket atoms were deformed in at least one mutant, with an overrepresentation of side-chain atoms (Pearson's Chi-squared Test p-value $= 1 \times 10^{-3}$), see Figure A4.

To explore impacts of these mutations on the conformation and properties of the PI-binding pocket, PI-binding pockets were extracted from the 156 structures (1 wild-type and the 155 mutant structures). These 156 pockets were then classified according to their structural similarity quantified by pairwise RMSD (Figures A5 and 5). In addition, their volume, sphericity, and hydrophobicity values were compared to those of the wild-type pocket (Figure 6). First, the five structures of a given mutant were not always bundled in the classification or presented some variability in terms of descriptor values. This highlighted a structural diversity of the five structures of mutants. This is explained by the minimization effects and the fact that several rotamers were possible for some amino acids during the mutagenesis process as illustrated by Figure A1 for the K45R mutation. Except structures of the I82M, I54M, L90M, I54M/L99F, and I54M/V71A/L90M mutants, most structures of mutants without MCS in pocket were close to the wild-type pocket in the hierarchical classification and presented similar descriptor values than the wild-type pocket (Figures 5 and 6). Pocket of mutants with the K7R, I54M, I54L, I82F, and I84V mutations were the farthest to the wild-type pocket in the classification, meaning these mutations had the most impact on the pocket structure (Figure 5). Figure 6 showed that the K7R and I82F mutations also strongly modified pocket properties, like the K45R, V47A, G48V, I82M/F, and L90M mutations. More precisely, the V47A, I82F, and I82M mutations strongly decreased the pocket hydrophobicity. The I50V, I50L, V62A, and I84V mutations also caused a reduction of pocket hydrophobicity but with a weaker magnitude. The I50L and I82F mutations were also responsible of an increase of the pocket volume, in contrast to the I82M and I84V mutations that caused a reduction of the pocket volume. The G48V and I54M mutations increased the hydrophobicity of the pocket that was accompanied with a modification of the pocket size: the G48V mutation led to a reduction of the pocket volume in contrast to the I54M. An increase of pocket volume was also observed in the K7R, I54L, and L90M mutants with different magnitudes and a decrease of the pocket volume in the D30N and V47A mutants. The volume modification of the pocket of the K7R and D30N mutants was accompanied with an increase of the sphericity of the pocket. The K45R mutant was distinct because its five structure presented large diversity in terms of descriptor values Figure 6. Two of these pockets were bigger and less hydrophobic than the wild-type pocket while the three others were smaller, more hydrophobic and more spheric than the wild-type pocket.

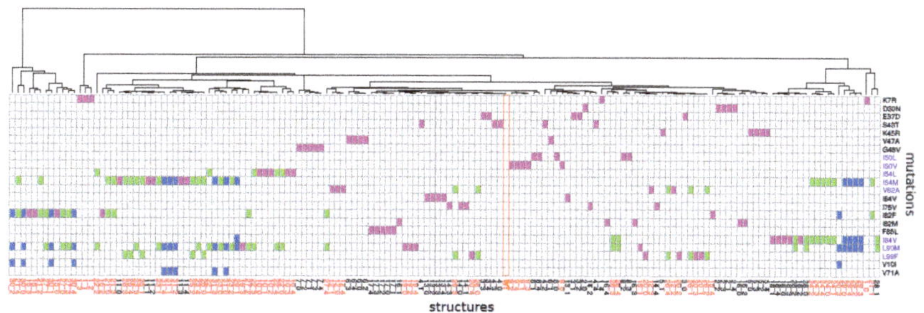

Figure 5. Hierarchical classification of wild-type and mutant pockets according to their conformational similarity quantified by pairwise RMSD computing using all pocket atoms. The table provides the description of each mutant structure in terms of mutations. Mutant structures are ranked according to their apparition in the classification. Single mutants are colored in magenta, double mutants in green and triple mutants in blue. The orange column locate the wild-type pocket. Mutations colored in purple correspond to mutations located in the binding pocket and mutant colored in red contain a MCS atom located in the binding pocket.

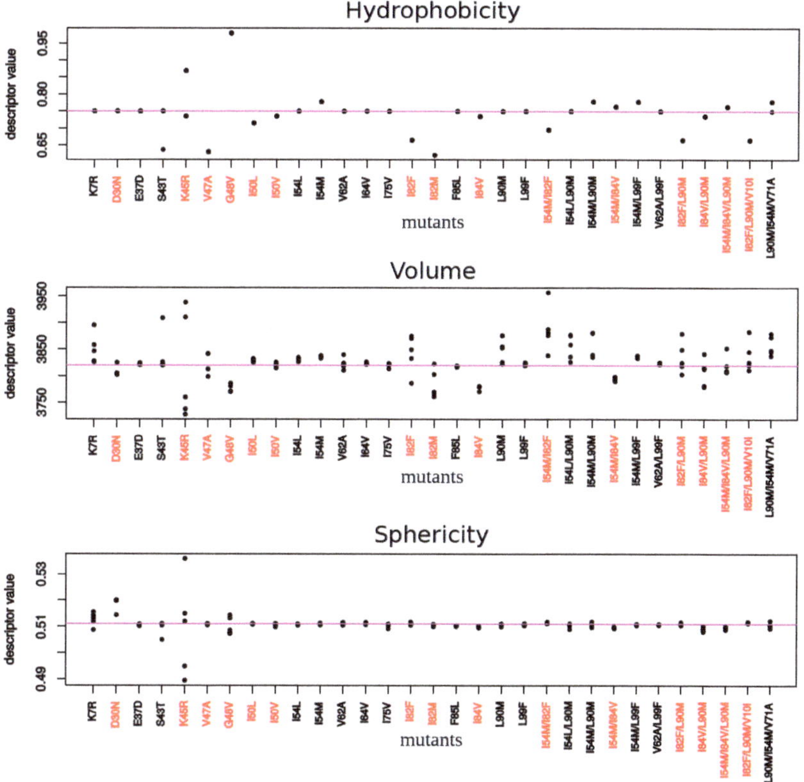

Figure 6. Hydrophobicity, volume, and sphericity of pockets extracted from each mutant structure. Red mutants correspond to mutants exhibiting at least one mutation in the binding. Each point corresponds to a mutant structure. For some mutants, less than five points appear meaning that several mutant structures exhibit same values for a given descriptor.

Figure A4 presents the MCS atoms occurring in the binding pocket in each mutant. We noted that some mutations induce structural rearrangements in residues important

for the PI binding. For example, the K7R, I50L/V and I54L mutations caused structural deformations in residues 25, 27, 30 that establish hydrogen bonds with PIs [18,33]. The D30N, K7R, I82F, I84V mutations led to atomic displacements in residues involved in van der Waals interactions with PI, such as in residues 23B, 27A, 28A, 30A, 49B, 48B, 82A and 84A.

Thus, the K7R, K45R, V47A, G48V, I50V/L, I54M, V62A, I82M/F, I84V, and L90M mutations could impact ligand binding by modifying pocket properties or the network of interactions with PIs.

2.3. Impact on Interface

From the 31 drug-resistant mutants, 13 had at least one mutation in the PR2 interface (Figure 1A). Except the I50L and I54M mutants, all these mutants contained MCS atoms in their interface. Five mutants without mutation in the interface presented structural deformations in their interface. To analyze the impact of these mutations on the PR2 interface, interfaces of the wild-type and mutant structures were extracted and compared in terms of amino acid composition and their size. To do so, a hierarchical classification of the 156 interfaces was computed according to their similarity in terms of interface composition (Figure 7). In addition, the Solvent Accessible Surface Area (SASA) value, measuring the interface size, of the two parts of the interface was computed for each mutant structure using NACCESS software [37] (Figure 8). Figure 7 showed that most structures without MCS atoms in interface were close to the wild-type interface in the classification, revealing that these mutations led weak changes in the PR2 interface. This was confirmed by the fact that these interfaces had similar size than the wild-type interface (Figure 8). Three mutants (I50L, I54M and I84V/L90M) without MCS atom in interface exhibited different interface composition relative to the wild-type. These differences in terms of interface composition led to an increase of the size of chain A interface in the I50L and I84V/L90M mutants. The G48V mutation was responsible of the presence in the interface of the two side-chain atoms of residue 48_A and the absence in the interface of one atom of residue 95_B and 99_B relative to the wild-type interface (Figure A6), resulting that the chain A interface of the mutant was larger and this of chain B was smaller than the wild-type interface (Figure 7). These differences in terms of interface observed in the I50L, I54M, and G48V mutants were explained by supplementary atoms in their interface induced by the mutation (Figure A6).

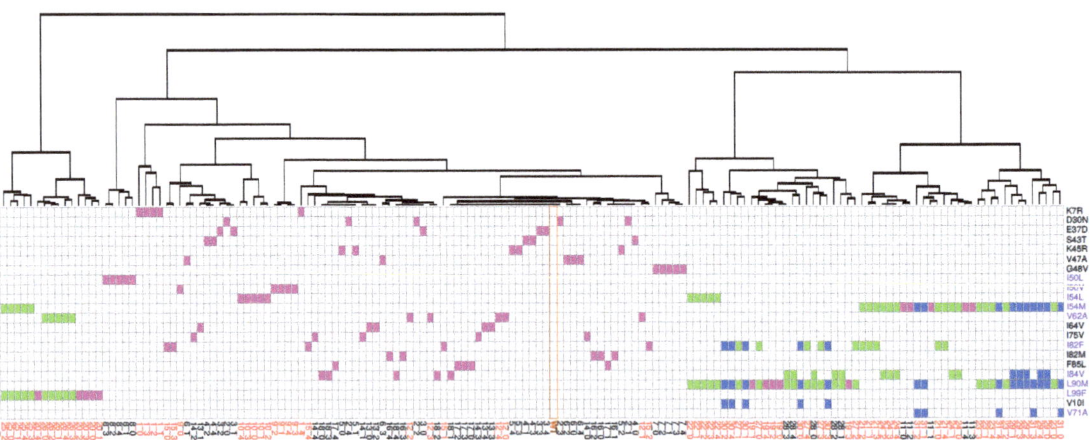

Figure 7. Classifications of the 156 structures according to their similitude of their interface. The table provides the description of each mutant structure in terms of mutations. Mutant structures are ranked according to their apparition in the classification. Single mutants are colored in magenta, double mutants in green and triple mutants in blue. The orange column locate the wild-type interface. Mutations colored in purple are invoved in the interface. Residues colored in red correspond to MCS atoms in the interface.

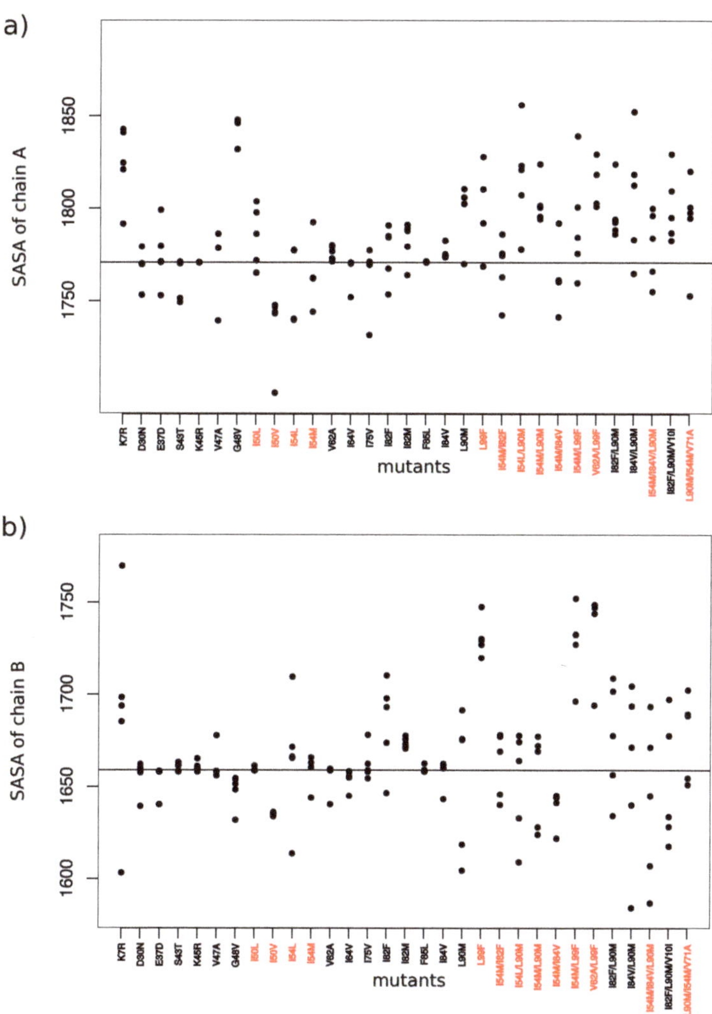

Figure 8. SASA values for interface of chain A (**a**) and chain B (**b**) for each mutant. Red mutants correspond to mutants having at least one mutation in PR2 interface.

From mutants having MCS atoms in the interface, the I50V, V62A, and I82F mutants corresponded to mutants inducing the less modifications in the PR2 interface (Figures 7 and 8). This was expected for the V62A mutant because only one MCS atom was observed in its interface. It was more surprising for the I50V and I82F mutants because large deformations were detected in interface-residues 50 and 8 of chains A and B, respectively. In contrast, the K7R, I54L, L90M, and L99F mutations alone or in combination with others contained many MCS atoms in the interface that strongly modified its composition and its size. The K7R, L90M, and L99F mutations caused an increase of the size of the two parts of PR2 interface, and the mutation I54L induced a weak increase of the size of the interface of chain B.

2.4. Impact of Combining Several Mutations Relative to Single Mutant

Figure 3 showed that most multiple mutants contained many MCS atoms with large magnitude, i.e., with $dist^{WT-mutant}$ distance higher than 2 Å. However, combining several

mutations did not significantly increase the average number of MCS atoms per mutant (Student's *t*-test *p*-value = 0.59, Figure 3B). Comparison of MCS atoms in the single and multiple mutants showed that several multiple mutants exhibited specific MCS atoms relative to the corresponding single mutants. For example, we highlighted a displacement in atoms of residue 50_B in the I54L and I54L/L90M mutants but the shift was larger in the double ($dist^{WT-mutant}$ for CD_50_B and CG1_50_B atoms >2.5 Å) than in single mutants ($dist^{WT-mutant}$ for CD_50_B and CG1_50_B atoms was of 0.69 and 0.62 Å in the I54L mutant). Combining several mutations could induce apparition or loss of MCS atoms relative to the corresponding mutants. For example, the combination of the I54M and I82F mutations caused structural shifts in pocket residues 50_A, 81_A/B, and 82_A/B, while no shift at these residues were observed in the I54M and I82F mutants (Figure A4). In contrast, the I82F mutant contained a large shift at residue 8_B, a residue involved in the PR2 interface and pocket, those were not found in the I54M/I82F mutants (Figure A4). These structural changes in the double mutant relative to the I82F mutant induced a weak decrease of the interface size and an increase of the pocket volume (Figure 6).

2.5. Impact of Using Different Structure Modeling Software

In this section, we explored the impact of using another structure-modeling software in the detection of structural rearrangements induced by mutations. To do so, we modeled the structure of the 31 mutants using the webserver Robetta, see Appendix G. These mutant structures were denoted as $mutant^{Robetta}$. For a better clarity of this section, the mutant structures built using our initial protocol (based on FoldX software and an energetic minimization step) were denoted as $mutant^{FoldX+Mini}$. First, we compared mutant structures generated with the wild-type structure (crystallographic structure), Figure A7A,B. We noted that the protocol based on FoldX software plus a minimization step led to a set of structures exhibiting a larger diversity than Robetta webserver. Then, the two mutant sets were compared by computing RMSD between $mutant^{Robetta}$ structures and the five structures of $mutant^{FoldX+Mini}$, denoted as $RMSD^{WT^*-mutant^{Robetta}}$. The two modeling protocols led to different structures with an average RMSD of 0.46 ± 0.04 between $mutant^{Robetta}$ and $mutant^{FoldX+Mini}$ structures (Figure A7C). We then compared the number of MCS atoms oberved in each structure of the two mutant sets, see Appendix G.3. Figure 9 presents the number of MCS atoms detected in each mutant. We noted that $mutant^{Robetta}$ structures exhibited substantially more MCS atoms than $mutant^{FoldX-mini}$ structures. This was explained by two reasons. First, the determination of the MCS atoms in $mutant^{FoldX-mini}$ structures was based on the comparison of the mutant structures and the minimized wild-type structures, while MCS atoms in $mutant^{Robetta}$ structures were detected by comparing mutant structures with the non minimized wild-type structure. The second reason was that an atom was detected as a MCS atom in the $mutant^{FoldX-mini}$ if it had a $dist^{WT-mutant}$ larger than 0.3 Å in at least three structures of the mutant. Figure 9 presents the number of atoms detected as MCS in both mutant sets. From the 504 MCS atoms per mutants detected in the $mutant^{FoldX+mini}$ set, 78 % were detected as MCS atoms in the $mutant^{Robetta}$ set. Although, the protocol based on FoldX software minimized the detection of structural rearrangements, the extracted structural deformations using this protocol was mainly found by a protocol based on another modeling software.

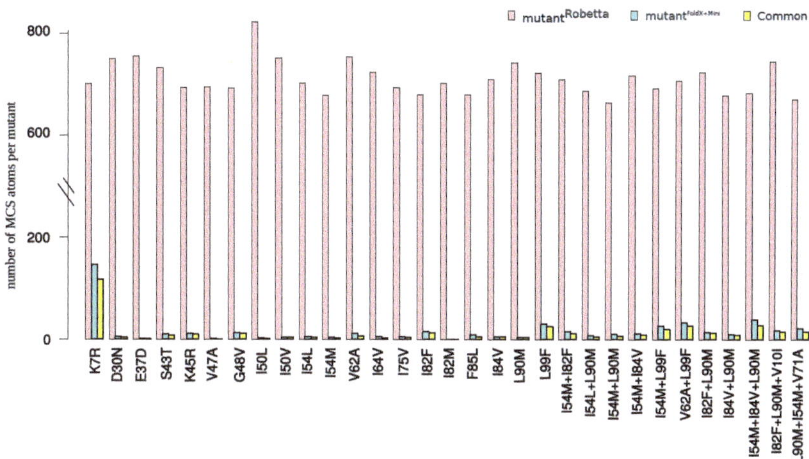

Figure 9. Number of MCS atoms in the *mutant*^{Robetta} set, in the *mutant*$^{FoldX+mini}$ set, and observed both in the two mutant sets.

3. Discussion

In this paper, we proposed a quantification and location of structural deformations at backbone and side-chain atoms of PR2 occurred in an update set of 31 drug-resistant mutants based on their modeled structures. We detected a set of atoms presenting a shift in at least one mutant structure relative to the wild-type structure. To identify structural rearrangements resulted from the mutation, we then retained only MCS atoms, i.e., atoms with a distance between its position in the mutant and wild-type structure higher than 0.3 Å in at least three of structures of a given mutant. The distance cutoff was set up to 0.3 Å to select only significant shifts and considering uncertainties in the X-ray structure, like in Liu et al. [36]. This step allowed us to detect on average 16.26 ± 26.09 MCS atoms per mutants. However, these two cutoffs could led to an under-estimation of detected structural shifts. This could explain the fact that no MCS atom was detected in the I82M mutant, while the binding pocket extracted from this mutant was smaller and less hydrophobic than the wild-type pocket. Indeed, using a distance cutoff of 0.1 Å, we counted 96 ± 107 MCS atoms per mutants and 17 MCS atoms in the I82M mutant.

The analysis of MCS atoms showed that they occurred in side-chain and backbone atoms in the mutated residues, in its vicinity or further in the structure. We noted that structural rearrangements in side-chain was more frequent and with larger magnitude than those observed in backbone atoms. Thus, drug-resistance mutations induced more deformations in side-chains than in backbone atoms. These results suggest that to combat against HIV drug resistance, it would be interesting to develop inhibitors that establish hydrogen interactions with backbone atoms. Favor backbone interactions between PI and PR is a strategy used for the design of DRV to avoid the detrimental effects of resistance mutations [38,39].

Our results showed that the studied drug-resistance mutations impacted all PR2 regions. However, even though mutations occurred in the two PR2 chains, we noted an assymetry in the impact of mutations in the two chains. This assymetric behavior of mutations is linked to the fact that PR2 is a structural asymmetric protein, i.e., its two chains exhibit different conformations in unbound and bound forms [11,12,35,40–43]. This structural asymmetry of PR2 could result from crystal packing, ligand binding, and intrinsic flexibility of PR2 [11], and may be involved in the structural changes of PR2, particularly upon ligand recognition and binding [11,12,41,42]. Our results are in agreement with previous findings that have showed that drug-resistance mutations could modify PR2 structural asymmetry [31]. Thus, as PR2 is an asymmetric protein, resistance mutations do not always have the same impact on the two chains.

The location onto PR2 structure of MCS atoms highlighted structural deformations that could be linked to resistance mechanisms. First, we observed, in 19 mutant structures, structural deformations in binding pocket residues as reported in Laville et al., 2020 [31]. Most of these mutations were located in the binding pocket, except the K7R, I54M and L90M mutations. Our findings revealed that the S43T, V47A, K45R, G48V, I82F, and I82M strongly modified pocket hydrophobicity. The two first decreased the hydrophobicity, while the four last increased it. The I50V, I50L, V62A, and I84V mutations induced also a modification of the pocket hydrophobicity but with a very weak magnitude. The K7R, I50L, I54M, I54L, I82F, and L90M mutations increased the pocket volume, while the D30N, V47A, G48V, I82M, and I84V mutations had the opposite effect. By comparing PR1 and PR2 binding pocket, we have previously observed that amino-acid changes occurring in pocket residues 31, 32, 46, 47, 76, and 82 increased the hydrophobicity of the binding pocket [19]. Chen et al. (2014) reported that these mutations have also an impact on the volume of the binding pocket [22]. In addition, the K7R, D30N, I50L/V, I54L, I82F, and I84V mutations seemed to have direct impact on the PI binding by causing structural rearrangements in pocket residues that establish hydrogen and van der Waals interactions with PIs. Secondly, our finding reported that mutations K7R, E37D, S43T, K45R, V47A, G48V, I50V, I50L, I54M, I54L, and V62A, occurring alone or in combination with others, induced directly or indirectly structural changes in the elbow and flap regions. These two regions are known to be important in the opening and closing mechanisms of the binding pocket during ligand binding [10]. Thus, these mutations could impact PI binding by modifying the flexibility and movement of the flap region upon PI binding. Thirdly, we reported that the K7R, I50L, I54L, G48V, L90M, and L99F mutations caused structural displacements that impacted the composition and size of the PR2 interface. This suggested that these mutations may alter the stability of PR2.

Several drug-resistant mutations were structurally studied in PR1 by comparing crystallographic structures of the wild-type and mutants in bound and unbound forms. These studies showed that drug-resistance mutations alter the conformation of flap residues and flap dynamics, modifying binding pocket properties and the interaction network with PI, and the PR2 stability. For example, it has been shown that some resistance mutations, such as the M46L, G48V, I50V, and I54V/M mutations, alter the conformation of flap residues and flap dynamics in PR1 [36,44–48]. The impact of resistance mutations on flap conformation had also observed in several multi-drug resistant mutants such as the PR20 (including 20 mutations), Flap+ (with L10I, G48V, I54V, and V82A mutations), and MDR-769 (with nine mutations) mutants having more opened flap region relative to the wild-type PR1 [49–52]. In addition, Shen et al., (2015) suggested that the E35D, M36I, and S37D mutations in the multi-drug resistant PRS17 mutant induce an increase of the flexibility of the flap region [53]. The impact of the amino-acid changes at residues 30, 50, 82, 84, and 90 on the pocket volume have been also observed in PR1 [48,49,52,54–57]. For example, the V82A and I84V mutations lead to an expansion of the active-site cavity [52,54], while the V82F and L90M mutations cause a volume reduction in the binding cavity [48,49,54,57]. The direct impact of mutations occurred in residues 30, 50, 82, and 84 on the network of PR-PI interactions were previously observed in PR1 [51,57–62]. The impact of the L90M mutation on the PR2 stability was previously observed in PR1 using urea denaturation experiment [63,64] or using sedimentation equilibrium analysis [65]. However, we noted several disagreements between findings obtained in PR1 and PR2. For example, Liu et al., (2008) observed structural deformations around the tip of the flap and 80 s loop (residues 78–82) in PR1 mutants G48V, I50V, I54V/M complexed with SVQ and DRV [36]. Our analysis of PR2 mutants detected displacements of atoms located around the flap tip in the I50V, I54M, and G48V mutants and in residue 53 in the G48V mutant. However, structural deformations in the 80 s loop is only found in the I54M mutant with a weak shift in backbone atom 79_B_0 ($dist^{WT-mutant}$ of 0.44 Å). An important structural rearrangement of the main-chain of residue 25 in the PR1 mutant L90M linking to the L90M resistance was observed in several studies [49,57]. Our result did not

highlight structural deformations at atoms of residue 25 in the L90M mutant of PR2; atoms of residue 25A/B exhibit an average $dist^{WT-mutant}$ of 0.07 ± 0.02 Å. In addition, we noted that the I84V and L90M mutations have not the same effect on the binding pocket in PR1 and PR2. Indeed, our findings showed that, in PR2, the I84V led to a reduction of pocket volume and the L90M mutation induced an augmentation of the pocket size in contrast in PR1 [48,49,52,54,57]. These disagreements in terms of impacts of drug-resistance mutations in PR1 and PR2 could be explained by the methodological differences of the two approaches. Indeed, our study was based on modeled structures, while PR1 studies used crystallographic structures. However, another reason that could explained these disagreements is the structural differences between PR1 and PR2 structures [13,18,19,40,42] leading to different PI-resistance profiles: PR2 is naturally resistant to six of the nine FDA (Food and Drug Administration)-approved PIs available for HIV-1 therapy[1,3,4]. More particularly, our previous work showed that the α-helix region (87–95), containing residue 90, presents different conformations in PR1 and PR2 and these structural differences could be partially explained by amino acid substitutions observed between the two PRs [13]. In addition, we previously showed that two PRs exhibit pockets with different properties: PR2 pockets are smaller and more hydrophobic than PR1 pockets [19,22]. These observation suggested that a same mutation could have different impacts on PR1 and PR2 structures.

In this study, we explored the impact of drug-resistance mutations on PR2 structure. To do so, we modeled 3D structures of mutant using as template the PR2 complexed with DRV (PDB code: 3EBZ [33]). However, it has been shown that drug-resistance mutations exhibit different sensitivities to the nine FDA-approved drugs [2,4,8,10,26]. For example, the I54M mutation lead to moderate resistance for indinavir, nelfinavir, tripanavir, DRV, and LPV, a high resistance for amprenavir, and is susceptible to SQV [26]. Thus, it would be interesting to consider different PIs and to cross information about the complete resistance profile of each mutant and the detected structural deformations. However, reliable resistance profile for PIs are difficult to collect for all mutants as few studies are available [3,4,26,29] and some of these studies have led to opposing results [10].

4. Materials and Methods

4.1. Data

In this study, we started from the list of the 30 drug-resistant mutants of PR2 studied in our previous study [31]. This mutant list was updated and a list of 31 drug-resistant mutants of PR2 was selected. This mutant set contains 22 different mutations (Figure 1) and there are 20 single mutants (with one mutation), 8 double mutants (with two mutations), and 3 triple mutants (with three mutations). These mutations sample the entire PR2 (Figure 1).

4.2. Structure Modeling

We modeled the 3D structure of each mutant using the protocol used in Laville et al., 2020 [31] based on FoldX suite [32] and Gromacs software [66]. This two-step protocol was applied to the wild-type crystallographic structure of the PR2 complexed with DRV (PDB code: 3EBZ [33]). First, the DRV ligand, metal atoms, and water molecules were removed from the crystallographic structure of PR2. Then, the `RepairPDB` command of the FoldX suite [32] was applied to the PR2 structure to reduce its energy. The in silico mutagenesis was then performed using the `BuildModel` command of the FoldX suite based on a side-chain rotamer library [32]. This step was performed five times to consider the several rotamers available for each amino acid [32] and generated five structures per mutant. An energetic minimization was then applied to the five modeled structures of a mutant using the protocol developed in our previous study [31]. We applied PROPKA software [67] to monoprotonate the oxygen atom OD2 of Aspartate 25 in chain B. Then, the system was solvated in a truncated octahedron box of explicit solvent (TIP3P water model) with a 12.0 Å buffer in each dimension and its charge was neutralized using chloride ions. The minimization of the system was performed using a two-step protocol using the force field AMBER ff99SB in GROMACS [66] by applying a steepest descent algorithm

combined with a conjugate gradient algorithm. A first step energy minimization allowed relaxing water molecules and counterions using a position harmonic restraining force of 100 kcal·mol^{-1}Å^{-2} on the heavy atoms of the protein. Then, restraints on protein atoms were removed using a second energy minimization step. The protocol used the particle mesh Ewald (PME) method to treat the long-range electrostatic interactions [68] and a cutoff distances of 10.0 Å for the long-range electrostatic and van der Waals interactions. This protocol was applied to the 31 mutants generating 155 mutant structures (5 structures per mutant). The minimization protocol was also applied to the wild-type structure and the minimized wild-type structure was named wild-type structure.

4.3. Identification of Shifted Atoms Induced by Mutations.

First, all shifted atoms in each mutant structure were extracted by comparing the position of atoms in the wild-type and mutant structures. To do so, we applied the method used in Perrier et al., 2019 [69]. Each mutant structure was superimposed onto the wild-type structure (the minimized wild-type structure) using PyMOL software [70]. Superimposition was based on all atoms. Hydrogen atoms were removed. Euclidean distances between the position of each atom in the mutant and wild-type structure were computed. These distances were denoted as $dist^{WT-mutant}$. Higher a $dist^{WT-mutant}$ value of an atom is, more the atom is shifted in the mutated structure relative to the wild-type structure. According to the modelization protocol, detected atom shifts resulted from the mutagenesis and minimization. To retain only significant structural rearrangements induced by resistance mutation, only atoms with a $dist^{WT-mutant}$ value higher than 0.3 Å were retained like in Liu et al., 2008 [36]. This distance cutoff of 0.3Å allowed selecting only significant structural displacements and removing uncertainties in the X-ray and mutant structures [36]. In addition, for each mutant, only shifted atoms observed in three of the five structures of the mutant were retained. These shifted atoms were named mutant-conserved shifted and noted MCS atoms.

4.4. Comparison of Wild-Type and Mutant Pockets

4.4.1. Pocket Estimation

From the 156 structures, we extracted the binding pocket. To consider the fact that the different known PI are structurally diverse and that resistance mutations induce resistance to one or several PI, we used the "common-ligand" approach to estimate pockets [11] that consisted to define the binding pocket as all atoms of PR2 capable to bind all co-crystallized ligands. To do so, a virtual ligand was built by superimposed all co-crystallized ligands extracted from a set of PRs. This virtual ligand was then placed in the query structure and its pocket was estimated as atoms located at least 4.5 Å of the virtual ligand. This protocol was applied on the 156 wild-type and mutant structures using the virtual ligand built from the PR set used in Triki et al., 2018 [11]. The pocket extracted from the wild-type structure was denoted as wild-type and those extracted from mutant structures were denoted as mutant pockets.

4.4.2. Comparison of the Conformation of Wild-Type and Mutant Pockets

To compare the conformation of the 156 pockets, we computed the root mean square deviation (RMSD) between each pocket pair (156 × 156). The calculation of RMSD was performed using PyMOL software [70] based on all pocket atoms. From these RMSD values, we computed a hierarchical classification of the pockets using the Ward method aggregation.

4.4.3. Comparison of the Properties of Wild-Type and Mutant Pockets

Each pocket was characterized by two geometrical descriptors (VOLUME_HULL and PSI), and one physicochemical descriptor (hydrophobicity_kyte) [71]. The VOLUME_HULL descriptor provides an estimation of the volume of a pocket. PSI measures the sphericity of a

pocket. These two descriptors were computed using PCI software [72]. `hydrophobicity_kyte` descriptor quantifies the hydrophobicity of a pocket.

4.5. Interface Comparison
4.5.1. Interface Extraction

The PPIC (Protein-Protein Interface Computation) program [73,74] was used to determine atoms involved in the interface of a structure. This program is parameter-free. It takes in input the 3D structure of a complex with two molecules (molecules or macromolecules) A and B. It defines the interface between the two molecules in two parts: interface of A and interface of B. Each interface part corresponds to the non-redundant set of all nearest neighbor atoms in one molecule of the atoms of other molecules. The extraction of neighbor atoms is performed using a simpler method of the Voronoï tessellation method [75]. Contrary to the Voronoï tessellation method, PPIC approach does not generate neighbors at long distances in the interface.

We used PPIC program to extract interface from the wild-type structure, denoted as wild-type interface, and the interface from the mutant structures, denoted as mutant interface.

4.5.2. Comparison of the Interface Composition

We compared the composition of the 156 interfaces by computed the Hamming distance, noted $D_{interface}$ between each interface pairs (Equation (1)).

$$D_{interface}(I1, I2) = N_{I1} + N_{I2} - 2 \times N_{I12} \tag{1}$$

with $I1$ and $I2$ the interfaces extracted from two structures, N_{I1} and N_{I2} the number of atoms of the interfaces $I1$ and $I2$, respectively, and N_{I12} the number of common atoms of the two interfaces $I1$ and $I2$.

Higher $D_{interface}(I1, I2)$ is, more dissimilar the two interfaces $I1$ and $I2$ are. To facilitate the interpretation of these distances, each $D_{interface}(I1, I2)$ value was normalized by the maximum number of atoms in the two interfaces ($N_{I1} + N_{I2}$) using Equation (2), resulting in the computation of the $D_{interface}(I1, I2)^{norm}$ value for each interface pair.

$$D_{interface}(I1, I2)^{norm} = \frac{D_{interface}(I1, I2)}{N_{I1} + N_{I2}}. \tag{2}$$

The $D_{interface}(I1, I2)^{norm}$ is ranking from 0 to 1, with a $D_{interface}(I1, I2)^{norm}$ equal to 0 means that the composition of the two interfaces $I1$ and $I2$ is identical.

Using the $D_{interface}(I1, I2)^{norm}$ values of each interface pairs, we computed a hierarchical classification of the 156 structures allowing to group structures according to their similarity in terms of interface composition. This classification was computed using the Ward method aggregation.

4.5.3. Comparison of the SASA of Interface

The SASA of the interface of each structure was determined using NACCESS software [37]. First the two monomers of each structure were separated. Then, the accessible surface area of all atoms of the 312 monomers was computed using NACCESS software [37]. The SASA values of the two parts of the interface (chains A and B) of a structure were obtained by computed the sum of the SASA of each atom detected as involved in the interface. The SASA value reflects the size of a interface.

5. Conclusions

In this study, we explored the impact of drug-resistance mutations reported in PR2. We first compared the modeled structure of 31 mutants with the wild-type PR2 structure to locate structural rearrangements induced by mutations. Secondly, we studied the impact of these deformations on the conformation and properties of the binding pocket and

on the interface. Our findings showed that one mutation could induce large structural deformations located in the mutated residue, in its vicinity or far in the structure. These structural deformations occur in both side-chain and backbone atoms, with on average more impact in the former. However, we revealed that resistance mutations do not always have the same behavior in the two monomers of PR2, which is link the structural asymmetry of this target.

The analysis of the location of structural rearrangements induced by resistance mutations provided clues to better understand resistance mechanisms. First, some mutations have a direct or indirect impacts on PI-binding. The K7R, V47A, G48V, I50L/V, I54L/M, V62A, I82F/M, I82M, I84V, and L90M induce structural rearrangements in the binding pocket that modify its conformation, volume and/or hydrophobicity. These changes in the binding pocket could have a negative effect on PI binding. In addition, some of these mutations (K7R, D30N, I50L/V, I54L, I82F, and I84V) have a direct impact on PI binding by causing structural displacements of residues establishing interactions with PI. Resistance mutations have also an impact on conformation of the elbow and flap regions, regions involving in the transition of the open and closed conformations of the PR2 upon ligand binding. Indeed, we reported that the K7R, E37D, S43T, K45R, V47A, G48V, I50V, I50L, I54M, I54L, and V62A led atom shifts in the elbow and flap regions that could modify the flexibility and movement of the flap region important in the binding-pocket closing. Thirdly, the K7R, G48V, I50L, I54L, L90M, and L99F mutations induce structural rearrangements in the PR2 interface that modify its composition and size that may alter the stability of PR2. Finally, we highlighted several residues that were never deformed in mutant structures.

In conclusion, this study explored the impact of a large number of PR2 resistance mutations on PR2 structure, particularly on its pocket binding and interface. Our results showed that some structural rearrangements induced by resistance mutations are located in important regions of PR2: the elbow and flap regions inducing in the PR2 deformation upon ligand binding, in the PI-binding pocket and in the interface of the two monomers. Some of these deformations modify the properties of binding pocket and the composition and size of the PR2 interface. These results suggest that the studied resistance mutations could alter PI binding by modifying the properties and flexibility of the pocket or the interaction between PR2 and PI and/or alter PR2 stability. Our results reinforce the resistance mechanisms proposed in our previous study [31] and lead to a better understanding of the effects of mutations that occurred in PR2 and the different mechanisms of PR2 resistance.

Author Contributions: L.R. conceive and designed the experiments; M.P. and L.R. supervised the project; P.L. and L.R. performed the experiments; P.L., M.P. and L.R. developed tools dedicated to analyses; P.L. and L.R. analyzed the data; L.R. wrote the paper. All authors reviewed the manuscript. All authors have read and agreed to the submitted version of the manuscript.

Funding: This work was supported by an ANRS Grant. PL is supported by a fellowship from the Ministère de l'Education Nationale de la Recherche et de Technologie (MENRT).

Institutional Review Board Statement: Not applicable.

Informed Consent Statement: Not applicable.

Data Availability Statement: The data presented in this study are openly available in https:// figshare.com/articles/dataset/Data_of_Laville_et_al_2021/13634147.

Acknowledgments: We are grateful to D. Flatters for helpful discussions and O. Taboureau for proofreading the manuscript. LR thanks the University of Paris and l'UFR SdV.

Conflicts of Interest: The authors declare no conflict of interest.

Abbreviations

The following abbreviations are used in this manuscript:

PR	Protease
PR2	HIV-2 Protease
PI	Protease inhibitor
MCS atoms	Mutant-Conserved Shifted atoms
RMSD	Root Mean Square Deviation
DRV	Darunavir
LPV	Lopinavir
SQV	Saquinavir
IDV	Indinavir
3D	Three-dimensional
SASA	Solvent Accessible Surface Area
PPIC	Protein-Protein Interface Computation

Appendix A

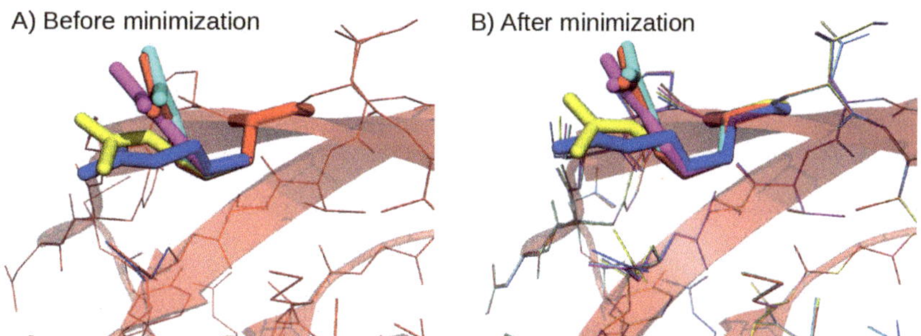

Figure A1. Illustration of the different rotamers for the arginine in position 45. Superimposition of the five structures of the K45R mutant before (**A**) and after the minimization process (**B**). Mutant structures are presented in line and cartoon mode. Residues R45 are presented in sticks.

Appendix B

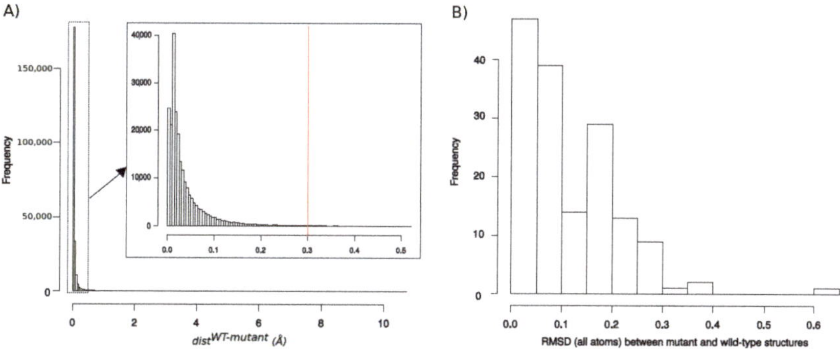

Figure A2. (**A**) Distribution of $dist^{WT-mutant}$ distances for each atom (hydrogen atoms were removed) in the set of 155 mutant structures. Red line corresponds the distance cutoff (0.3 Å) used to define an atom as shifted atom. (**B**) Distribution of RMSD (in Å) computed between wild-type and mutant structures. RMSD were computed using all atoms.

Appendix C

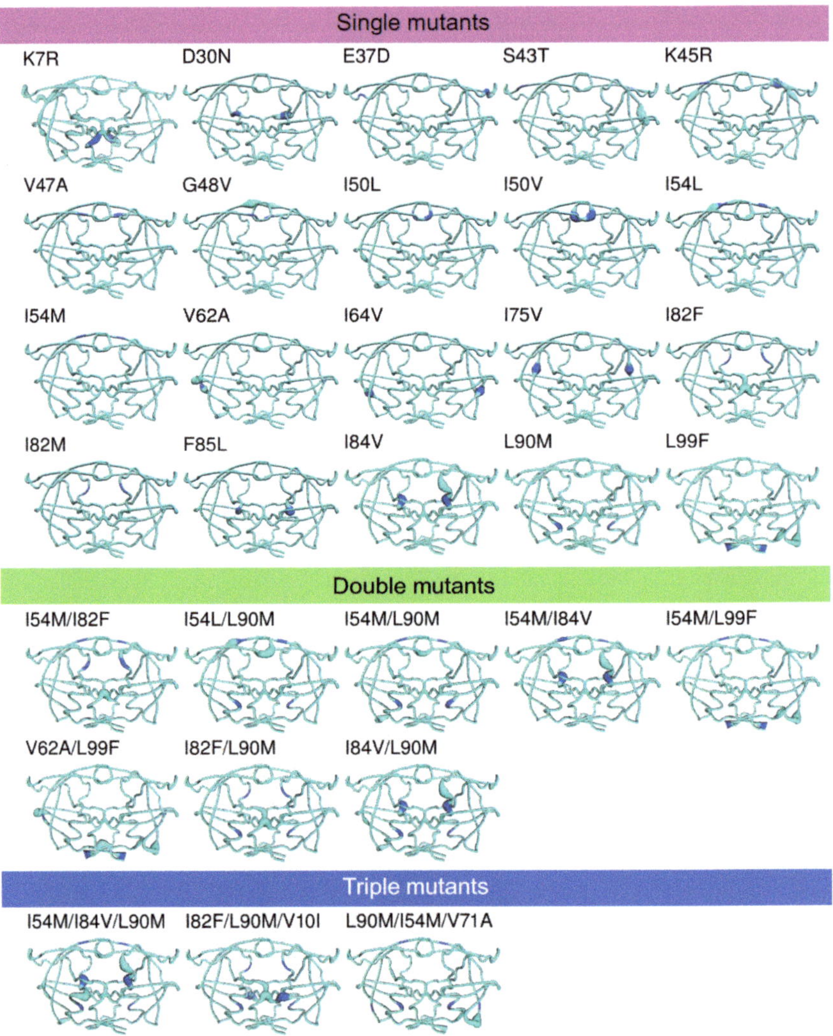

Figure A3. Location of the MCS atoms found in each mutant. PR2 is colored in cyan and represented in putty mode. The putty radius is relative to the deformation induced by mutations: the higher the radius, the stronger the mutation-induced rearrangement. Mutated residues are colored in blue.

Appendix D

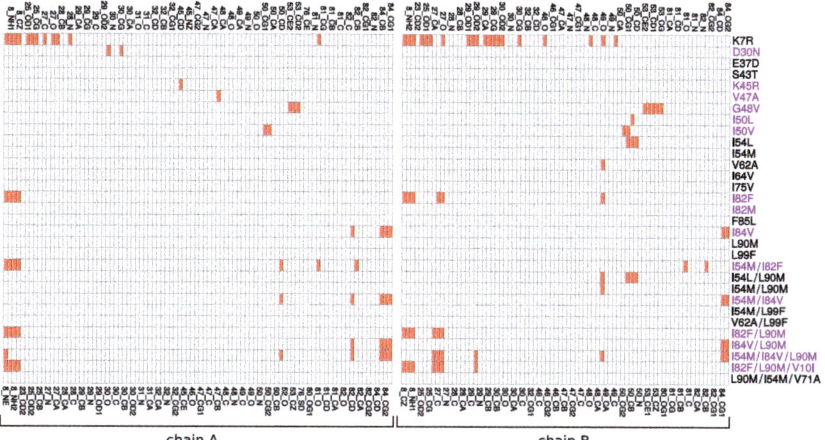

Figure A4. MCS atoms in the pocket. Mutants colored in purple present at least a mutation located in the binding pocket.

Appendix E

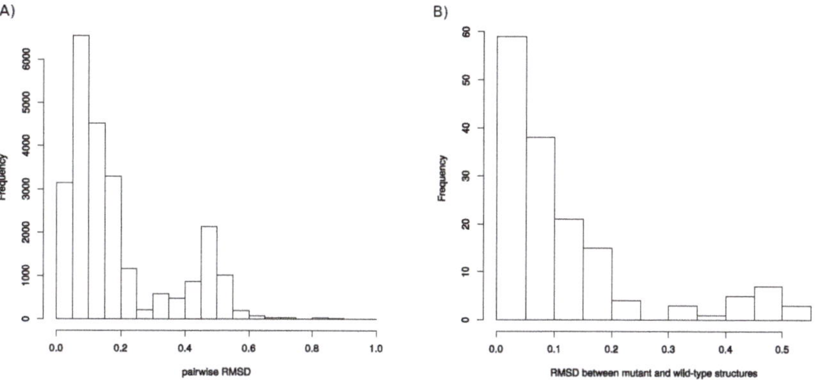

Figure A5. Comparison between pockets. (**A**) Distribution of RMSD computed between all pocket pairs. (**B**) Distribution of RMSD computed between the wild-type and mutant pockets.

Appendix F

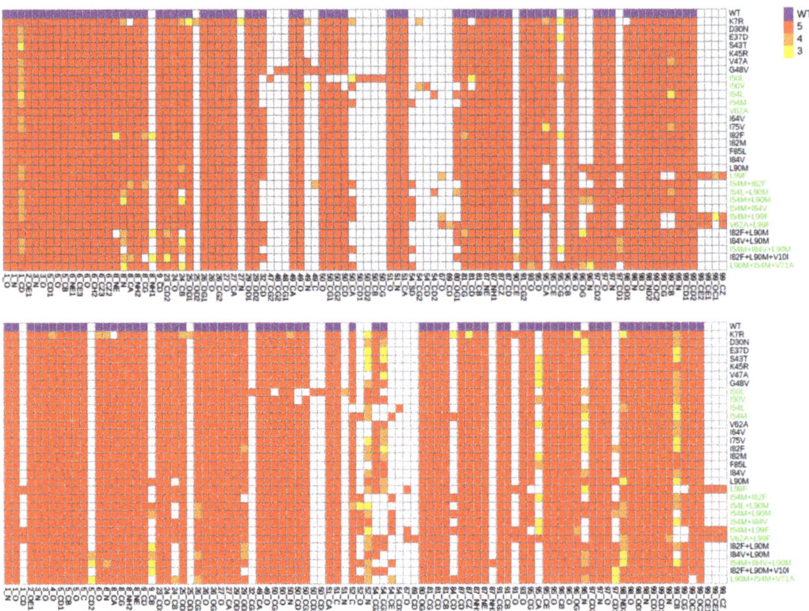

Figure A6. Number of mutant structure where a given atom is involved in the interface. Mutants colored in green present at least a mutation involved in the interface.

Appendix G

In this section, mutant structures built using our initial protocol (based on FoldX software and an energetic minimization step) were denoted as $mutant^{FoldX+Mini}$. The WT^* abbreviation described the crystallographic structure of PR2 complexed to the DRV that corresponded to the wild-type structure that was not minimized.

Appendix G.1. Protocol to Model Mutant Structures Using Robetta Software

Robetta webserver (https://robetta.bakerlab.org/) is a protein structure prediction service based on the RosettaCM method [76]. RosettaCM is a comparative modeling method that assembles structures using integrated torsion space-based and Cartesian space template fragment recombination, loop closure by iterative fragment assembly and Cartesian space minimization, and high-resolution refinement [76].

As in our initial modeling protocol, we used the PDB structure 3EBZ (PR2 complexed with DRV) as template. First, the DRV ligand, metal ions and water molecules were removed from the structure. Using this template, the structure of the 31 drug-resistant mutants was built using Robetta webserver with the "CM" option. Other parameters were set to defaults. This step resulted in a set of 31 mutant structures, named $mutant^{Robetta}$. The RMSD (based on all atoms) between the wild-type and $mutant^{Robetta}$ structures were computed using PyMoL software [70]. These RMSD values were refereed as $RMSD^{WT^*-mutant^{Robetta}}$.

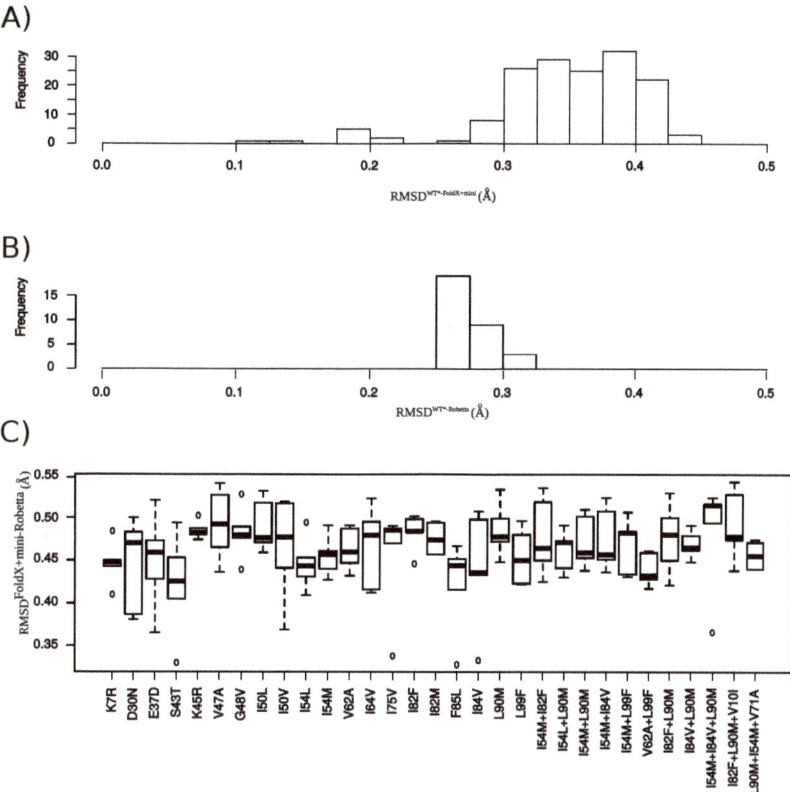

Figure A7. Comparison between $mutant^{FoldX+Mini}$ and $mutant^{Robetta}$ structures. (**A**) RMSD computed between the wild-type structure (not minimized, WT^* structure) and $mutant^{FoldX+Mini}$ structure. (**B**) RMSD computed between the wild-type structure (not minimized, WT^* structure) and $mutant^{Robetta}$ structure. (**C**) RMSD computed between each $mutant^{Robetta}$ structure and the five structure of $mutant^{FoldX+Mini}$.

Appendix G.2. Structural Comparison between the $Mutant^{FoldX+Mini}$ and $Mutant^{Robetta}$ Structures

First, we compared the two sets of PR2 mutants. For each mutant, its $mutant^{Robetta}$ structure was superimposed onto the five $mutant^{FoldX+Mini}$ mutants and the RMSD based on all atoms was computed using PyMoL software [70]. These RMSD were denoted as $RMSD^{FoldX+Mini-Robetta}$.

Appendix G.3. Detection of Structural Rearrangements in the Set of $Mutant^{Robetta}$ Structures

First, each $mutant^{Robetta}$ structure were superimposed onto the 3EBZ (PDB code) structure using PyMoL software [70]. Superimposition was based on all atoms. Euclidean distances between the position of each atom in the mutant and 3EBZ structure were computed. These distances were denoted as $dist^{WT^*-mutant^{Robetta}}$. An atom was considered as a MCS atom in a $mutant^{Robetta}$ structure if it had a $dist^{WT^*-mutant^{Robetta}}$ value higher than 0.3 Å.

Appendix H

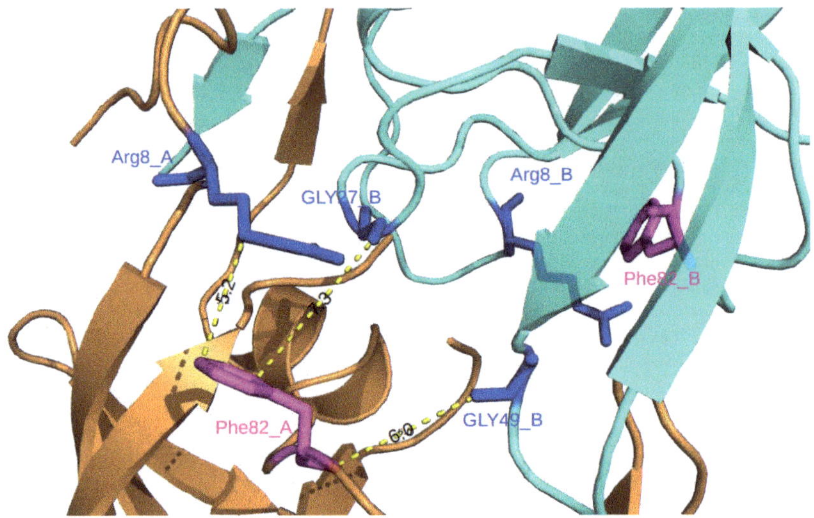

Figure A8. Location of residues exhibiting structural shifts in the I82F mutant. Chains A and B are presented in cartoon mode and colored in orange and blue, respectively. Mutated residues (82) are colored in magenta and presented in stick mode. Residues having shifted atoms are colored in blue and presented in stick mode.

References

1. Brower, E.; Bacha, U.M.; Kawasaki, Y.; Freire, E. Inhibition of HIV-2 protease by HIV-1 protease inhibitors in clinical use. *Chem. Biol. Drug. Des.* **2008**, *71*, 298–305, doi:10.1111/j.1747-0285.2008.00647.x.
2. Raugi, D.; Smith, R.; Ba, S.; Toure, M.; Traore, F.; Sall, F.; Pan, C.; Blankenship, L.; Montano, A.; Olson, J.; et al. Complex patterns of protease inhibitor resistance among antiretroviral treatment-experienced HIV-2 patients from senegal: Implications for second-line therapy. *Antimicrob. Agents Chemother.* **2013**, *57*, 2751–2760, doi:10.1128/AAC.00405-13.
3. Raugi, D.N.; Smith, R.A.; Gottlieb, G.S.; the University of Washington-Dakar HIV-2 Study Group. Four Amino Acid Changes in HIV-2 Protease Confer Class-Wide Sensitivity to Protease Inhibitors. *J. Virol.* **2016**, *90*, 1062–1069, doi:10.1128/JVI.01772-15.
4. Desbois, D.; Roquebert, B.; Peytavin, G.; Damond, F.; Collin, G.; Bénard, A.; Campa, P.; Matheron, S.; Chêne, G.; Brun-Vézinet, F.; et al. In vitro phenotypic susceptibility of human immunodeficiency virus type 2 clinical isolates to protease inhibitors. *Antimicrob. Agents Chemother.* **2008**, *52*, 1545–1548, doi:10.1128/AAC.01284-07.
5. Masse, S.; Lu, X.; Dekhtyar, T.; Lu, L.; Koev, G.; Gao, F.; Mo, H.; Kempf, D.; Bernstein, B.; Hanna, G.; et al. In vitro selection and characterization of human immunodeficiency virus type 2 with decreased susceptibility to lopinavir. *Antimicrob. Agents Chemother.* **2007**, *51*, 3075–3080, doi:10.1128/AAC.00146-07.
6. Cavaco-Silva, J.; Aleixo, M.; Van Laethem, K.; Faria, D.; Valadas, E.; Gonçalves, M.D.F.; Gomes, P.; Vandamme, A.; Cunha, C.; Camacho, R.J. Mutations selected in HIV-2-infected patients failing a regimen including atazanavir. *Antimicrob. Agents Chemother.* **2013**, *68*, 190–192, doi:10.1093/jac/dks363.
7. Bénard, A.; Damond, F.; Campa, P.; Peytavin, G.; Descamps, D.; Lascoux-Combes, C.; Taieb, A.; Simon, F.; Autran, B.; Brun-Vézinet, F.; et al. Good response to lopinavir/ritonavir-containing antiretroviral regimens in antiretroviral-naive HIV-2-infected patients. *AIDS* **2009**, *23*, 1171–1183, doi:10.1097/QAD.0b013e32832949f0.
8. Rodés, B.; Sheldon, J.; Toro, C.; Jiménez, V.; Alvarez, M.; Soriano, V. Susceptibility to protease inhibitors in HIV-2 primary isolates from patients failing antiretroviral therapy. *Antimicrob. Agents Chemother.* **2006**, *57*, 709–713, doi:10.1093/jac/dkl034.
9. Ntemgwa, M.; Toni, T.; Brenner, B.; Oliveira, M.; Asahchop, E.; Moisi, D.; Wainberg, M. Nucleoside and nucleotide analogs select in culture for different patterns of drug resistance in human immunodeficiency virus types 1 and 2. *Antimicrob. Agents Chemother.* **2009**, *53*, 708–715, doi:10.1128/AAC.01109-08.
10. Menéndez-Arias, L.; Alvarez, M. Antiretroviral therapy and drug resistance in human immunodeficiency virus type 2 infection. *Antivir. Res.* **2014**, *102*, 70–86, doi:10.1016/j.antiviral.2013.12.001.
11. Triki, D.; Cano Contreras, M.; Flatters, D.; Visseaux, B.; Descamps, D.; Camproux, A.; Regad, L. Analysis of the HIV-2 protease's adaptation to various ligands: characterization of backbone asymmetry using a structural alphabet. *Sci. Rep.* **2018**, *8*, 710, doi:10.1038/s41598-017-18941-3.

12. Ollitrault, G.; Fartek, S.; Descamps, D.; Camproux, A.; Visseaux, B.; Regad, L. Characterization of HIV-2 protease structure by studying its asymmetry at the different levels of protein description. *Symmetry* **2019**, *10*, 644, doi:doi:10.3390/sym10110644.
13. Triki, D.; Kermarrec, M.; Visseaux, B.; Descamps, D.; Flatters, D.; Camproux, A.; Regad, L. Exploration of the effects of sequence variations between HIV-1 and HIV-2 proteases on their three-dimensional structures. *J. Biomol. Struct. Dyn.* **2020**, *38*, 5014–5026, doi:10.1080/07391102.2019.1704877.
14. Gustchina, A.; Weber, I. Comparative analysis of the sequences and structures of HIV-1 and HIV-2 proteases. *Proteins Struct. Funct. Bioinform.* **1991**, *10*, 325–339, doi:10.1002/prot.340100406.
15. Sardana, V.; Schlabach, A.; Graham, P.; Bush, B.; Condra, J.; Culberson, J.; Gotlib, L.; Graham, D.E.A. Human Immunodeficiency Virus Type 1 Protease Inhibitors: Evaluation of Resistance Engendered by Amino Acid Substitutions in the Enzyme's Substrate Binding Site. *Biochemistry* **1994**, *33*, 2004–2010, doi:10.1021/bi00174a005.
16. Hoog, S.; Towler, E.; Zhao, B.; Doyle, M.; Debouck, C.; Abdel-Meguid, S. Human immunodeficiency virus protease ligand specificity conferred by residues outside of the active site cavity. *Biochemistry* **1996**, *35*, 10279–10286, doi:10.1021/bi960179j.
17. Tie, Y.; Boross, P.; Wang, Y.; Gaddis, L.; Hussain, A.; Leshchenko, S.; Ghosh, A.; Louis, J.; Harrison, R.; Weber, I. High resolution crystal structures of HIV-1 protease with a potent non-peptide inhibitor (UIC-94017) active against multi-drug-resistant clinical strains. *J. Mol. Biol.* **2004**, *23*, 341–352, doi:10.1016/j.jmb.2004.02.052.
18. Kovalevsky, A.; Louis, J.; Aniana, A.; Ghosh, A.; Weber, I. Structural evidence for effectiveness of darunavir and two related antiviral inhibitors against HIV-2 protease. *J. Mol. Biol.* **2008**, *384*, 178–192, doi:10.1016/j.jmb.2008.09.031.
19. Triki, D.; Billot, T.; Flatters, D.; Visseaux, B.; Descamps, D.; Camproux, A.; Regad, L. Exploration of the effect of sequence variations located inside the binding pocket of HIV-1 and HIV-2 proteases. *Sci. Rep.* **2018**, *8*, 5789, doi:10.1038/s41598-018-24124-5.
20. Hightower, M.; Kallas, E.G. Diagnosis, antiretroviral therapy, and emergence of resistance to antiretroviral agents in HIV-2 infection: A review. *Braz. J. Infect. Dis.* **2003**, *7*, 7–15, doi:10.1590/s1413-86702003000100002.
21. Kar, P.; Knecht, V.J. Origin of decrease in potency of darunavir and two related antiviral inhibitors against HIV-2 compared to HIV-1 protease. *Phys. Chem. B* **2012**, *116*, 2605–2614, doi:10.1021/JP211768N.
22. Chen, J.; Liang, Z.; Wang, W.; Yi, C.; Zhang, S.; Zhang, Q. Revealing origin of decrease in potency of darunavir and amprenavir against HIV-2 relative to HIV-1 protease by molecular dynamics simulations. *Sci. Rep.* **2014**, *4*, 6872, doi:10.1038/SREP06872.
23. Colson, P.; Henry, M.; Tourres, C.; Lozachmeur, D.; Gallais, H.; Gastaut, J.; Moreau, J.; Tamalet, C. Polymorphism and drug-selected mutations in the protease gene of human immunodeficiency virus type 2 from patients living in southern France. *J. Clin. Microbiol.* **2004**, *42*, 570–577, doi:10.1128/jcm.42.2.570-577.2004.
24. Storto, A.; Visseaux, B.; Bertine, M.; Le Hingrat, Q.; Collin, G.; Damond, F.; Khuong, M.; Blum, L.; Tubiana, R.; Karmochkine, M.; et al. ANRS HIV-2 CO5 cohort. Minority resistant variants are also present in hiv-2-infected antiretroviral-naive patients. *J. Antimicrob. Chemother.* **2018**, *73*, 1173–1176, doi:10.1093/jac/dkx530.
25. Damond, F.; Brun-Vezinet, F.; Matheron, S.; Peytavin, G.; Campa, P.; Pueyo, S.; Mammano, F.; Lastere, S.; Farfara, I.; Simon, F.; et al. Polymorphism of the human immunodeficiency virus type 2 (hiv-2) protease gene and selection of drug resistance mutations in hiv-2-infected patients treated with protease inhibitors. *J. Clin. Microbiol.* **2005**, *43*, 484–487, doi:10.1128/JCM.43.1.484-487.2005.
26. Ntemgwa, M.; Brenner, B.; Oliveira, M.; Moisi, D.; Wainberg, M. Variations in reverse transcriptase and RNase H domain mutations in human immunodeficiency virus type 1 clinical isolates are associated with divergent phenotypic resistance to zidovudine. *Antimicrob. Agents Chemother.* **2007**, *51*, 604–610, doi:10.1128/AAC.00646-07.
27. Larrouy, L.; Vivot, A.; Charpentier, C.; Bénard, A.; Visseaux, B.; Damond, F.; Matheron, S.; Chene, G.; Brun-Vezinet, F.; Descamps, D. ANRS CO5 HIV-2 Cohort. Impact of gag genetic determinants on virological outcome to boosted lopinavir-containing regimen in hiv-2-infected patients. *AIDS* **2013**, *27*, 69–80, doi:10.1097/QAD.0b013e32835a10d8.
28. Jallow, S.; Alabi, A.; Sarge-Njie, R.; Peterson, K.; Whittle, H.; Corrah, T.; Jaye, A.; Cotten, M.; Vanham, G.; McConkey, S.; et al. Virological response to highly active antiretroviral therapy in patients infected with human immunodeficiency virus type 2 (hiv-2) and in patients dually infected with hiv-1 and hiv-2 in the Gambia and emergence of drug-resistant variants. *J. Clin. Microbiol.* **2009**, *47*, 2200–2208, doi:10.1128/JCM.01654-08.
29. Mahdi, M.; Szojka, Z.; Mótyán, J.; Tözsér, J. Inhibition Profiling of Retroviral Protease Inhibitors Using an HIV-2 Modular System. *Viruses* **2015**, *71*, 6152–6162, doi:10.3390/v7122931.
30. Charpentier, C.; Visseaux, B.; Bénard, A.; Peytavin, G.; Damond, F.; Roy, C.; Taieb, A.; Chêne, G.; Matheron, S.; Brun-Vézinet, F.; et al. Transmitted drug resistance in French HIV-2-infected patients. *AIDS* **2013**, *27*, 1671–1684, doi:10.1097/QAD.0b013e32836207f3.
31. Laville, P.; Fartek, S.; Cerisier, N.; Flatters, D.; Petitjean, M.; Regad, L. Impacts of drug resistance mutations on the structural asymmetry of the HIV-2 protease. *BMC Mol. Cell Biol.* **2020**, *51*, 46, doi:10.1186/s12860-020-00290-1.
32. Guerois, R.; Nielsen, J.E.; Serrano, L. Predicting changes in the stability of proteins and protein complexes: A study of more than 1000 mutations. *J. Mol. Biol.* **2002**, *320*, 369–387, doi:10.1016/S0022-2836(02)00442-4.
33. Tie, Y.; Wang, Y.; Boross, P.; Chiu, T.; Ghosh, A.; Tozser, J.; Louis, J.; Harrison, R.; Weber, I. Critical differences in HIV-1 and HIV-2 protease specificity for clinical inhibitors. *Protein Sci.* **2012**, *21*, 339–350, doi:10.1002/pro.2019.
34. Sadiq, S.K.; de Fabritiis, G. Explicit solvent dynamics and energetics of HIV-1 protease flap opening and closing. *Proteins Struct. Funct. Bioinform.* **2010**, *78*, 2873–2885, doi:10.1002/prot.22806.

35. Triki, D.; Fartek, S.; Visseaux, B.; Descamps, D.; Camproux, A.; Regad, L. Characterizing the structural variability of HIV-2 protease upon the binding of diverse ligands using a structural alphabet approach. *J. Biomol. Struct. Dyn.* **2018**, *28*, 1–13, doi:10.1080/07391102.2018.1562985.

36. Liu, F.; Kovalevsky, A.; Tie, Y.; Ghosh, A.; Harrison, R.; Weber, I. Effect of flap mutations on structure of HIV-1 protease and inhibition by saquinavir and darunavir. *J. Mol. Biol.* **2008**, *381*, 102–115, doi:10.1016/j.jmb.2008.05.062.

37. Hubbard, S.; Thornton, J. NACCESS. In *Computer Program, Department of Biochemistry and Molecular Biology*; University College London: London, UK, 1993.

38. Koh, Y.; Nakata, H.; Maeda, K.; Ogata, H.; Bilcer, G.; Devasamudram, T.; Kincaid, J.; Boross, P.; Wang, Y.F.; Tie, Y.; et al. Novel bis-tetrahydrofuranylurethane-containing nonpeptidic protease inhibitor (PI) UIC-94017 (TMC114) with potent activity against multi-PI-resistant human immunodeficiency virus in vitro. *Antimicrob. Agents Chemother.* **2003**, *47*, 3123–3129, doi:10.1128/aac.47.10.3123-3129.2003.

39. Ghosh, A.K.; Dawson, Z.L.; Mitsuya, H. Darunavir, a conceptually new HIV-1 protease inhi-bitor for the treatment of drug-resistant HIV. *Bioorg. Med. Chem.* **2007**, *15*, 7576–7580, doi:10.1016/j.bmc.2007.09.010.

40. Tong, L.; Pav, S.; Pargellis, C.; Dô, F.; Lamarre, D.; Anderson, P. Crystal structure of human immunodeficiency virus (HIV) type 2 protease in complex with a reduced amide inhibitor and comparison with HIV-1 protease structures. *Proc. Natl. Acad. Sci. USA* **1993**, *90*, 8387–8391, doi:10.1073/pnas.90.18.8387.

41. Mulichak, A.M.; Hui, J.; Tomasselli, A.; Heinrikson, R.; Curry, K.; Tomich, C.; Thaisrivongs, S.; Sawyer, T.; Watenpaugh, K. The crystallographic structure of the protease from human immunodeficiency virus type 2 with two synthetic peptidic transition state analog inhibitors. *J. Biol. Chem.* **1993**, *268*, 13103–13109.

42. Tong, L.; Pav, S.; Mui, S.; Lamarre, D.; Yoakim, C.; Beaulieu, P.; Anderson, P. Crystal structures of HIV-2 protease in complex with inhibitors containing the hydroxyethylamine dipeptide isostere. *Structure* **1995**, *3*, 33–40, doi:10.1016/S0969-2126(01)00133-2.

43. Priestle, J.; Fassler, A.; Rösel, J.; Tintelnot-Blomley, M.; Strop, P.; Grütter, M. Comparative analysis of the X-ray structures of HIV-1 and HIV-2 proteases in complex with CGP 53820, a novel pseudosymmetric inhibitor. *Structure* **1995**, *3*, 381–389, doi:10.1016/S0969-2126(01)00169-1.

44. Kovalevsky, A.; Liu, F.; Leshchenko, S.; Ghosh, A.K.; Louis, J.; Harrison, R.; Weber, I. Ultra-high resolutioncrystal structure of HIV-1 protease mutant reveals two binding sites forclinical inhibitor TMC114. *J. Mol. Biol.* **2006**, *363*, 161–173, doi:10.1016/j.jmb.2006.08.007.

45. Zhang, H.; Wang, Y.; Shen, C.H.; Agniswamy, J.; Rao, K.; Xu, C.; Ghosh, A.; Harrison, R.; Weber, I. Novel P2 tris-tetrahydrofuran group in antiviral compound 1 (GRL-0519) fills the S2 binding pocket of selected mutants of HIV-1protease. *J. Med. Chem.* **2013**, *56*, 1074–1083, doi:10.1021/jm301519z.

46. Shen, C.; Wang, Y.; Kovalevsky, A.; Harrison, R.; Weber, I. Amprenavir complexes with HIV-1 protease andits drug-resistant mutants altering hydrophobic clusters. *FEBS* **2010**, *277*, 3699–3714, doi:10.1111/j.1742-4658.2010.07771.x.

47. Scott, W.; Schiffer, C. Curling of flap tips in HIV-1 protease as a mechanism for substrate entry and tolerance of drug resistance. *Structure* **2000**, *8*, 1259–1265, doi:10.1016/S0969-2126(00)00537-2.

48. Hong, L.; Zhang, X.; Hartsuck, J.; Tang, J. Crystal structure of an in vivo HIV-1 protease mutant in complex with saquinavir: Insights into the mechanisms of drug resistance. *Protein Sci.* **2000**, *9*, 1898–1904, doi:10.1110/ps.9.10.1898.

49. Agniswamy, J.; Louis, J.; Roche, J.; Harrison, R.; Weber, I. Structural Studies of a Rationally Selected Multi-Drug Resistant HIV-1 Protease Reveal Synergistic Effect of Distal Mutations on Flap Dynamics. *PLoS ONE* **2016**, *11*, e0168616, doi:10.1371/journal.pone.0168616.

50. Cai, Y.; Yilmaz, N.; Myint, W.; Ishima, R.; Schiffer, C. Differential Flap Dynamics in Wild-Type and a Drug Resistant Variant of HIV-1 Protease Revealed by Molecular Dynamics and NMR Relaxation. *J. Chem. Theory Comput.* **2012**, *8*, 3452–3462, doi:10.1021/ct300076y.

51. Zhang, Y.; Chang, Y.C.E.; Louis, J.; Wang, Y.; Harrison, Y.; Weber, I. Structures of darunavir-resistant HIV-1 protease mutant reveal atypical binding of darunavir to wide open flaps. *ACS Chem. Biol.* **2014**, *9*, 1351–1358, doi:10.1021/cb4008875.

52. Logsdon, B.; Vickrey, J.; Martin, P.; Proteasa, G.; Koepke, J.; Terlecky, S.R.; Wawrzak, Z.; Winters, M.A.; Merigan, T.C.; Kovari, L.C. Crystal structures of a multidrug-resistant humanimmunodeficiency virus type 1 protease reveal an expanded active-sitecavity. *J. Virol.* **2004**, *78*, 3123–3132, doi:10.1128/jvi.78.6.3123-3132.2004.

53. Shen, C.; Chang, Y.; Agniswamy, J.; Harrison, R.; Weber, I. Conformational variation of an extreme drug resistant mutant of HIV protease. *J. Mol. Graph. Model.* **2015**, *62*, 87–96, doi:10.1016/j.jmgm.2015.09.006.

54. Muzammil, S.; Armstrong, A.; Kang, L.; Jakalian, A.; Bonneau, P.; Schmelmer, V.; Amzel, L.; Freire, E. Unique thermodynamic response of tipranavir to human immunodeficiency virus type 1 protease drug resistance mutations. *J. Virol* . **2007**, *81*, 5144–5154.

55. Yeddi, R.; Proteasa, G.; Martinez, J.; Vickrey, J.; Martin, P.; Wawrzak, Z.; Liu, Z.; Kovari, I.; Kovaria, L. Contribution of the 80's loop of HIV-1 protease to the multidrug-resistant mechanism: crystallographic study of MDR769 HIV-1 protease variants. *Acta Crystallogr. D Biol. Crystallogr.* **2011**, *67*, 524–532, doi:10.1107/S0907444911011541.

56. Martin, P.; Vickrey, J.; Proteasa, G.; Jimenez Y.L. Wawrzak, Z.; Winters, M.; Merigan, T.; Kovari, L. Wide-open 1.3 Å structure of a multidrug-resistant HIV-1 protease as a drug target. *Structure* **2005**, *13*, 1887–1895, doi:10.1016/j.str.2005.11.005.

57. Kovalevsky, A.; Tie, Y.; Liu, F.; Boross, P.; Wang, Y.; Leshchenko, S.; Ghosh, A.; Harrison, R.; Weber, I.T. Effectiveness of Nonpeptide Clinical Inhibitor TMC-114 on HIV-1 Protease with Highly Drug Resistant Mutations D30N, I50V, and L90M. *J. Med. Chem.* **2006**, *49*, 1379–1387, doi:10.1021/jm050943c.

58. Prabu-jeyabalan, M.; Nalivaika, E.; King, N.; Schiffer, C. Viability of a drug-resistant human immunodeficiency virus type 1 protease variant: Structural insights for better antiviral therapy. *Society* **2003**, *77*, 1306–1315, doi:10.1128/JVI.77.2.1306-1315.2003.

59. Weber, I.; Agniswamy, J. HIV-1 Protease: Structural Perspectives on Drug Resistance. *Viruses* **2009**, *1*, 1110–1136, doi:10.3390/v1031110.

60. Meher, B.; Wang, Y. Interaction of I50V mutant and I50L/A71V double mutant HIV-protease with inhibitor TMC114 (darunavir): molecular dynamics simulation and binding free energy studies. *J. Phys. Chem.* **2012**, *116*, 1884–1900, doi:10.1021/jp2074804.

61. Mittal, S.; Bandaranayake, R.; King, N.; Prabu-Jeyabalan, M.; Nalam, M.; Nalivaika, E.; Yilmaz, N.; Schiffer, C. Structural and thermodynamic basis of amprenavir/darunavir and atazanavir resistance in HIV-1 protease with mutations at residue 50. *J. Virol.* **2013**, *87*, 4176–4184, doi:10.1128/JVI.03486-12.

62. Lockbaum, G.; Leidner, F.; Rusere, L.; Henes, M.; Kosovrasti, K.; Nachum, G.; Nalivaika, E.; Ali, A.; Yilmaz, N.; Schiffer, C. Structural Adaptation of Darunavir Analogues against Primary Mutations in HIV-1 Protease. *ACS Infect. Dis.* **2019**, *5*, 316–325, doi:10.1021/acsinfecdis.8b00336.

63. Mahalingam, B.; Louis, J.; Reed, C.; Adomat, J.; Krouse, J.; Wang, Y.; Harrison, R.; Weber, I. Structural and kinetic analysis of drug resistant mutants of HIV-1 protease. *Eur. J. Biochem.* **1999**, *263*, 238–245, doi:10.1046/j.1432-1327.1999.00514.x.

64. Mahalingam, B.; Wang, Y.; Boross, P.; Tozser, J.; Louis, J.; Harrison, R.; Weber, I. Crystal structures of HIV protease V82A and L90M mutants reveal changes in the indinavir-binding site. *Eur. J. Biochem.* **2004**, *271*, 1516–1524, doi:10.1111/j.1432-1033.2004.04060.x.

65. Xie, D.; Gulnik, S.; Gustchina, E.; Yu, B.; Shao, W.; Qoronfleh, W.; Nathan, A.; Erickson, J. Drug resistance mutations can effect dimer stability of HIV-1 protease at neutral pH. *Protein Sci.* **1999**, *8*, 1702–1707, doi:10.1110/ps.8.8.1702.

66. Abraham, M.; Murtolad, T.; Schulz, R.; Pálla, S.; Smith, J.; Hess, B.; Lindahl, E. Gromacs: High performance molecular simulations through multi-level parallelism from laptops to supercomputers. *SoftwareX* **2015**, *1–2*, 19–25, doi:10.1016/j.softx.2015.06.001.

67. Li, H.; Robertson, A.; Jensen, J.H. Very fast empirical prediction and rationalization of protein pK a values. *Proteins* **2005**, *61*, 704–721, doi:10.1002/prot.20660.

68. Darden, T.; York, D.; Pedersen, L. Particle mesh Ewald: An N log (N) method for Ewald sums in large systems. *J. Chem. Phys.* **1993**, *98*, 10089–10092, doi:10.1063/1.464397.

69. Perrier, M.; Castain, L.; Regad, L.; Todesco, E.; Landman, R.; Visseaux, B.; Yazdanpanah, Y.; Rodriguez, C.; Joly, V.; Calvez, V.; et al. HIV-1 protease, Gag and gp41 baseline substitutions associated with virological response to a PI-based regimen. *J. Antimicrob. Chemother.* **2019**, *74*, 1679–1692, doi:10.1093/jac/dkz043.

70. The PyMOL Molecular Graphics System, Version 2.0 Schrödinger, LLC. Available online: https://pymol.org/2/ (accessed on 20 January 2021).

71. Borrel, A.; Regad, L.; Xhaard, H.; Petitjean, M.; Camproux, A. PockDrug: A Model for Predicting Pocket Druggability That Overcomes Pocket Estimation Uncertainties. *J. Chem. Inf. Model.* **2015**, *55*, 882–895, doi:10.1021/ci5006004.

72. Cerisier, N.; Regad, L.; Triki, D.; Camproux, A.; Petitjean, M. Cavity versus ligand shape descriptors: Application to urokinase binding pockets. *J. Comput. Biol.* **2017**, *24*, 1134–1137, doi:10.1089/cmb.2017.0061.

73. Ozeel, V.; Perrier, A.; Vanet, A.; Petitjean, M. The Symmetric Difference Distance: A New Way to Evaluate the Evolution of Interfaces along Molecular Dynamics Trajectories; Application to Influenza Hemagglutinin. *Symmetry* **2019**, *11*, 662, doi:10.3390/sym11050662.

74. Laville, P.; Martin, J.; Launay, G.; Regad, L.; Camproux, A.; de Vries, S.; Petitjean, M. A non-parametric method to compute protein–protein and protein–ligands interfaces. Application to HIV-2 protease–inhibitors complexes. *bioRxiv* **2018**, doi:10.1101/498923.

75. Eppstein, D.; Paterson, M.; Yao, F. On nearest-neighbor graphs. *Discret. Comput. Geom.* **1997**, *17*, 263–282, doi:10.1007/PL00009293.

76. Song, Y.; DiMaio, F.; Wang, R.Y.R.; Kim, D.; Miles, C.; Brunette, T.; Thompson, J.; Baker, D. High-Resolution Comparative Modeling with RosettaCM. *Structure* **2013**, *21*, 1735–1742, doi:10.1016/j.str.2013.08.005.

Article

Multiple-Molecule Drug Design Based on Systems Biology Approaches and Deep Neural Network to Mitigate Human Skin Aging

Shan-Ju Yeh , Jin-Fu Lin and Bor-Sen Chen *

Laboratory of Automatic Control, Signal Processing and Systems Biology, Department of Electrical Engineering, National Tsing Hua University, Hsinchu 30013, Taiwan; m793281@gmail.com (S.-J.Y.); sweettofu531@gmail.com (J.-F.L.)
* Correspondence: bschen@ee.nthu.edu.tw

Abstract: Human skin aging is affected by various biological signaling pathways, microenvironment factors and epigenetic regulations. With the increasing demand for cosmetics and pharmaceuticals to prevent or reverse skin aging year by year, designing multiple-molecule drugs for mitigating skin aging is indispensable. In this study, we developed strategies for systems medicine design based on systems biology methods and deep neural networks. We constructed the candidate genomewide genetic and epigenetic network (GWGEN) via big database mining. After doing systems modeling and applying system identification, system order detection and principle network projection methods with real time-profile microarray data, we could obtain core signaling pathways and identify essential biomarkers based on the skin aging molecular progression mechanisms. Afterwards, we trained a deep neural network of drug–target interaction in advance and applied it to predict the potential candidate drugs based on our identified biomarkers. To narrow down the candidate drugs, we designed two filters considering drug regulation ability and drug sensitivity. With the proposed systems medicine design procedure, we not only shed the light on the skin aging molecular progression mechanisms but also suggested two multiple-molecule drugs for mitigating human skin aging from young adulthood to middle age and middle age to old age, respectively.

Keywords: skin aging; oxidative stress; aging progression mechanism; genome-wide genetic and epigenetic network (GWGEN); systems medicine design; multiple-molecule drug

Citation: Yeh, S.-J.; Lin, J.-F.; Chen, B.-S. Multiple-Molecule Drug Design Based on Systems Biology Approaches and Deep Neural Network to Mitigate Human Skin Aging. *Molecules* **2021**, *26*, 3178. https://doi.org/10.3390/molecules26113178

Academic Editors: Marco Tutone and Anna Maria Almerico

Received: 30 April 2021
Accepted: 24 May 2021
Published: 26 May 2021

Publisher's Note: MDPI stays neutral with regard to jurisdictional claims in published maps and institutional affiliations.

1. Introduction

Being the largest organ of the human body, the skin shows aging with biological age. Many people, especially female, like to spend money on cosmetics and pharmaceuticals regularly for preventing or reversing skin aging. Thus, changes in human skin caused by aging are important issues for both the pharmaceutical and cosmetic sectors worldwide [1]. Additionally, increasing life expectancy in developed countries reveals advancing age as the primary risk factor for numerous diseases [2]. Elder people tend to have dryness, itch, dyspigmentation, wrinkles, as well as benign and malignant tumors on skin. Under these worse conditions, they would feel sleep deprivation leading to having weakened immunity and getting infection. Hence, keeping our skin health promotes healthy aging [3]. Furthermore, identifying interventions, which are able to ameliorate skin aging progression, to delay, prevent or lessen age-related diseases is worth studying.

Human skin provides a primary protective barrier, routinely shielding us from allergens, microbes, and other environmental assaults, including solar ultraviolet (UV) irradiation, heat, infection, water loss, and injury. Skin aging is a complex process leading to the decrement of cutaneous functions and structures with time. Impaired epidermal barrier function, decline in resistance to infections and regenerative potential, and impairment of mechanical properties like loss of extensibility and elasticity are the essential biomarkers

of human skin ageing [4]. In general, skin aging can be regarded as two different processes. The first one is intrinsic aging, which is caused by biological age. The second one is extrinsic aging, which arises from solar UV exposure. The extrinsic factors contain the exposure under UV radiation and pollution, and poor nutrition resulting in alterations of DNA, RNA and protein in skin cells. The clinical manifestation of intrinsic aging is characterized by age spots, laxity, wrinkles, sagging, dryness, itchy, and the lower type I and III fibrillar collagens leading to dermal atrophy [5].

MicroRNAs (miRNAs) are a group of small noncoding RNAs, owning the post-transcriptional regulation ability to control gene expression negatively. Meanwhile, they are found to involve in many biological processes, such as epidermal development, proliferation, differentiation [6–8], inflammatory responses, immune regulation and wound healing in human skin [9,10]. Although we know that miRNAs might be a key player in the age-associated change, studies about age-related miRNAs in human skin remain limited [11]. As for long noncoding RNAs (lncRNAs), they are another type of noncoding RNA with >200 nucleotides. One review paper has summarized age-related lncRNAs and elucidated their roles in different aging process [12]. Since lncRNAs have versatile functions including gene regulation, chromatin structure modulation, genomic imprinting, cell growth and differentiation, and embryonic development, the dysregulated expression of lncRNAs may cause age-related diseases and disorders [13]. Recently, lncRNAs are regarded as potential targets for antiaging therapies [14]. Moreover, the well-known epigenetic modifications are DNA methylation and histone post-transcriptional modifications, including methylation, acetylation, ubiquitination, and phosphorylation. The accumulation of epigenetic alternations may not only contribute to skin aging but also promote malignant transformation [15,16]. H19, an epigenetic regulatory RNA, has been demonstrated to positively affect cell growth and proliferation and delay senescence [17]. With epigenetic silencing on LMNA, which is one of progeroid genes, we could observe a corresponding malignant transformation [18].

For the purpose of investigating skin aging process, researchers tried to identify the influence of microenvironment and epigenetic change on skin aging and put focus on some specific proteins, such as members of the collagen family, or cellular functions. However, the definitions of young skin and older skin are not fixed, that is, there is no definition of age range for young people and old people, respectively. Therefore, most studies compared young and old people with great differences in the research of skin aging progression. Although these studies proposed plenty of credible skin aging-associated theories and experimental results, the genomewide molecular progression mechanism of skin aging was unknown since the restriction of experimental methods and the attention to specific proteins and cellular functions. Moreover, although pharmacological interventions may prove to ameliorate the effect of aging on humans, the prohibitive expansion of treating healthy individuals in clinical trials over a long duration becomes a crucial difficulty in developing new drugs. On the contrary, repurposing drugs, which have been already approved for specific diseases, or those have been passed their safety tests but failed against their original indication, is more feasible than targeting aging itself with new drugs [19–21].

In recent years, pharmaceutical scientists put a lot of efforts into novel drug development based on the knowledge of existing drugs [22,23]. By performing in vitro search for drug discovery, researchers could identify interaction between drugs and targets (e.g., genes). However, due to the high cost and time consuming work, we could not conduct in vitro research most of the time. Instead, virtual screening in silico, selecting possible candidates first and verifying them in wet laboratory offer alternatives to us [24]. In general, docking simulation and machine learning method are considered to be two main approaches for in silico prediction of drug–target interaction [25]. For docking simulation, the process would be limited if the 3D structure of the protein is unknown [26]. To deal with this issue, chemogenomic methods, namely feature-based methods, transform drugs and targets into sets of descriptors (e.g., feature vector) allowing machine learning models to make prediction of drug–target interactions [27]. Chen at al. reviewed machine learning

methods and databases that used chemogenomic approaches for drug–target interaction prediction [28]. Except for traditional machine learning methods, the deep neural network has been employed in drug–target interaction prediction as well, such as deep belief neural networks [29], convolutional neural network [30], and multilayer perceptron [31,32].

In this study, we define different age intervals for each stage of skin aging. We build the candidate genome-wide genetic and epigenetic network (GWGEN) containing a candidate protein–protein interaction network (PPIN) and a candidate gene regulatory network (GRN) by big database mining. Moreover, it can be represented by a binary matrix. Assisted with microarray data of human skin, the false positives from the candidate GWGEN are pruned away by the system identification method and system order detection scheme. By doing so, we obtain real GWGENs of young-adult, middle-aged, and elderly skin aging as shown in Figures S1–S3, respectively. However, real GWGENs are still complex. Therefore, we further extract core GWGENs from real GWGWNs by the principal network projection (PNP) method. Based on the rank of projection values, we could obtain the core signaling pathways in respect of KEGG pathways to investigate skin aging molecular progression mechanisms for each stage of skin aging. To identify essential biomarkers in core signaling pathways, we refer to GenAge [33], which contains genes involved in human aging progression, and the Connectivity Map (CMap) [34] dataset to find overlap nodes being drug targets. To explore the potential candidate drugs toward our identified biomarkers, we trained a deep neural network of drug–target interactions in advance. By applying it, we could predict potential candidate drugs, which holds higher interaction probability to the identified biomarker. Afterwards, we designed two filters considering drug regulation ability and drug sensitivity to narrow down the candidate drugs. Consequently, we propose two potential multiple-molecule drugs i.e., niridazole, liothyronine, decitabine, pinacidil, and allantoin for mitigating skin aging from young adulthood to middle age; allantoin, diclofenac, mepyramine, resveratrol, and azathioprine for mitigating skin aging from middle age to old age.

2. Results

For the purpose of analyzing molecular progression mechanisms in human skin aging, extracting core signaling pathways from each core GWGEN becomes an essential issue. We defined three skin aging stages, including young-adult, middle-aged, and elderly human skin as shown in Figure 1. The research flowchart in Figure 2 shows how to construct the candidate GWGEN, real GWGENs, and core GWGENs so as to extract core singling pathways and investigate molecular progression mechanisms of human skin aging. By big database mining, the candidate GWGEN containing candidate PPIN and candidate GRN was constructed. With the help of the corresponding young-adult, middle-aged, and elderly skin microarray data, we applied system identification and system order detection methods to the candidate GWGEN for obtaining real GWGENs shown in Figures S1–S3 (Supplementary Materials), respectively. The Table 1 shows the number of nodes (e.g., proteins, TFs, miRNAs, and lncRNAs) as well as the edges standing for the interaction or regulation between two nodes for the candidate GWGEN and real GWGENs. According to Table 1, compared the nodes in candidate GWGEN to the nodes in real GWGENs, one could realize that the number of nodes diminish a lot in real GWGENs, reflecting that the false positives were removed successfully by the system order detection scheme. Since the real GWGENs were still too complicated to investigate molecular progression mechanisms of human skin aging, we applied principal network projection (PNP) method and selected the top-ranked 4000 nodes with significant projection values that could reflect 85% of the real GWGENs in three stages of skin aging to obtain core GWGENs (Figure 3a–c), respectively. In addition, for the genes in core GWGENs, we used the Database for Annotation, Visualization and Integrated Discovery (DAVID) Bioinformatics Resources version 6.8 to perform enrichment analyses for each stage of skin aging as shown in Tables S1–S3, respectively. Moreover, for investigating molecular progression mechanisms of skin aging conveniently, we denoted differential core signaling

pathways for young adult to middle-aged and middle-aged to elderly skin aging in respect of KEGG pathways, respectively. Based on skin aging molecular progression mechanisms, we identified essential biomarkers as drug targets for young-adult to middle-aged and middle-aged to elderly skin aging, respectively. Exploring candidate drugs toward our identified biomarkers, we trained a deep neural network of drug–target interaction in advance. We applied the trained model to predict the candidate drugs holding higher interaction probability with identified biomarkers. In order to narrow down the candidate drugs, we design two filters considering drug regulation ability and drug sensitivity. The more details will be discussed in the following sections.

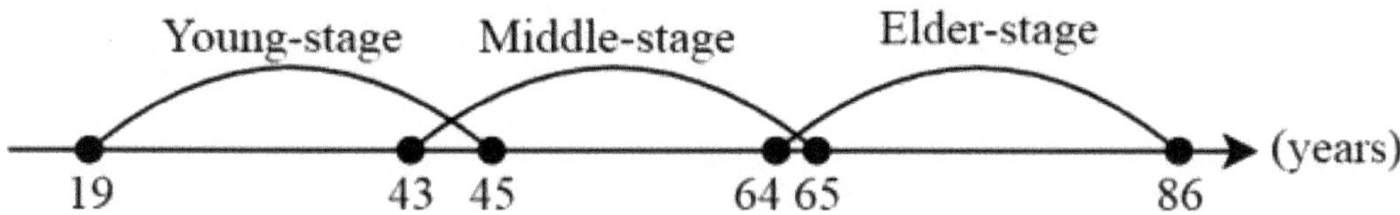

Figure 1. Skin aging stages. The figure denotes age intervals for each stage of skin. The young-adult, middle-aged, and elderly stage of skin are defined as 19 to 45 years old, 43 to 65 years old, and 64 to 86 years old, respectively.

Table 1. The table of the total number of nodes and edges in candidate GWGENs and identified real GWGENs for each stage of skin aging.

	Candidate GWGEN	Young-adult GWGEN	Middle-aged GWGEN	Elderly GWGEN
TFs	1851	464	491	383
TF—protein	429,829	11,715	10,888	10,234
TF—receptor	66,042	1824	1738	1639
TF—TF	27,892	513	468	223
TF—lncRNAs	1600	494	457	473
TF—miRNA	348	84	76	91
lncRNAs	666	593	522	582
lncRNAs—protein	19,520	1944	1902	2286
lncRNAs—TF	1443	88	90	84
lncRNAs—receptor	3018	293	288	366
miRNAs	126	111	99	107
miRNA—protein	45,613	4175	3997	4884
miRNA—receptor	6891	593	707	745
miRNA—TF	3028	782	680	849
Receptors	2388	2372	2366	2387
Proteins	15,347	13,195	13,219	13,213
PPIs	3,185,763	172,558	146,594	97,766
Total nodes	20,378	16,735	16,697	16,672
Total edges	3,790,987	195,063	167,885	119,640

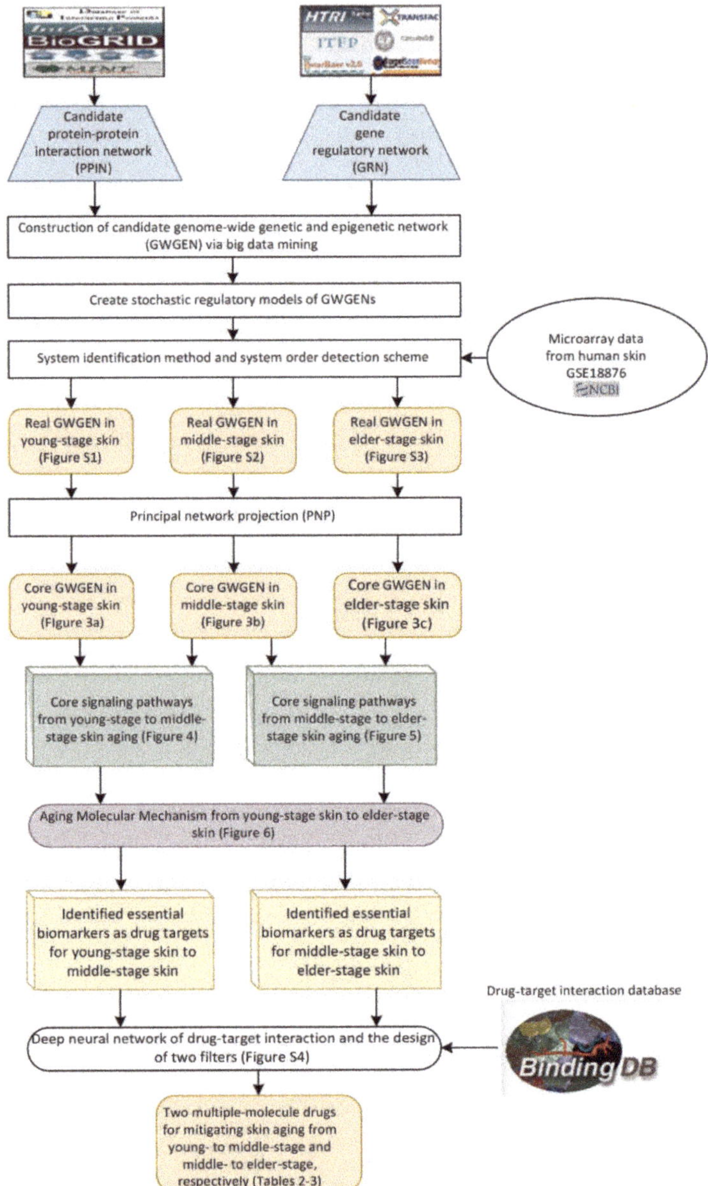

Figure 2. Reseasrch flowchart of systems medicine design for human skin aging. Flowchart of using systems biology methods to construct the candidate GWGEN, real GWGENs, core GWGENs, and core signaling pathways to find skin aging progression mechanisms for identifying essential biomarkers. After obtaining the essential biomarkers, we applied trained a deep neural network of drug–target interactions to predict the potential candidate drugs holding higher probability. To narrow down the candidate drugs, we considered drug regulation ability by querying the CMap dataset and drug sensitivity by referring to the sensitivity dataset from DepMap portal. Consequently, we proposed two multiple-molecule drugs to mitigate the skin aging from young-adult to middle-aged and middle-aged to elderly-stage.

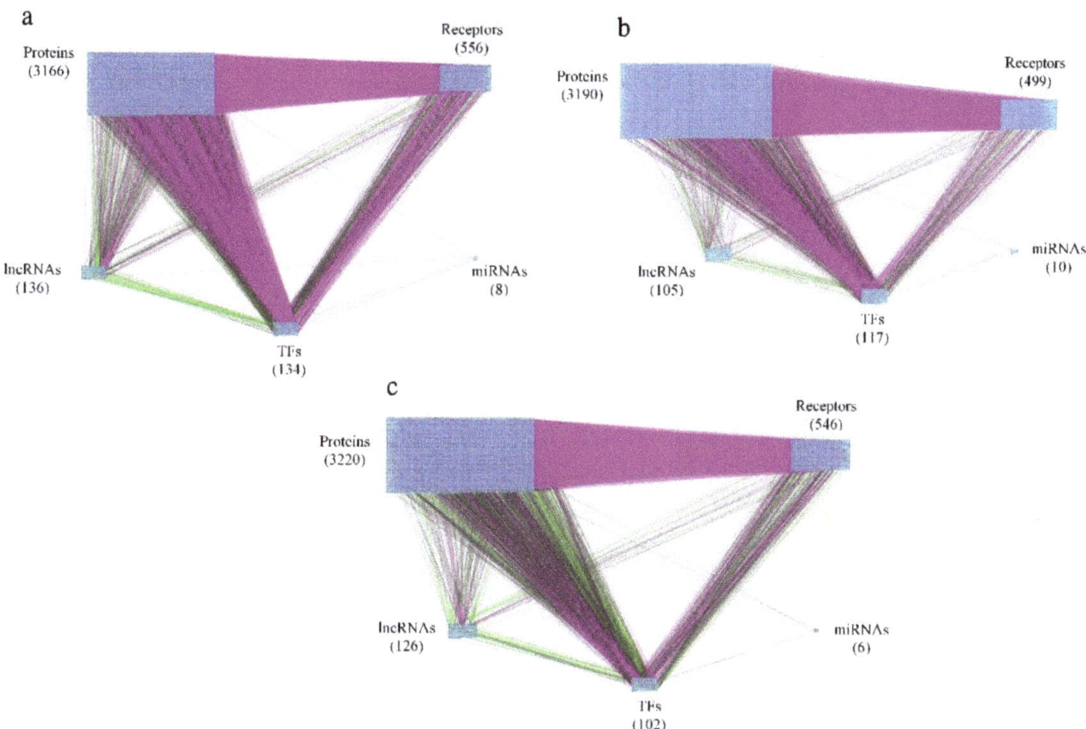

Figure 3. The core genomewide genetic and epigenetic networks (GWGEN). (**a–c**). (**a**) The core GWGEN of young-adult skin. The purple lines denote protein–protein interactions (PPIs). The green lines indicate transcriptional regulations by TFs and lncRNAs. The black lines represent post-transcriptional regulations by miNRAs. The total number of receptors, proteins, lncRNAs, TFs and miRNAs are 556, 3166, 136, 134 and 8, respectively. (**b**) The core GWGEN of middle-aged skin. The PPIs are in purple. The regulations from TFs and lncRNAs are in green. The black lines stand for the post-transcriptional regulations by miRNAs. The total number of receptors, proteins, lncRNAs, TFs and miRNAs are 499, 3190, 105, 117 and 10, respectively. (**c**) The core GWGEN of elderly-stage skin. The PPIs are shown in purple lines; regulations by TFs and lncRNAs are denoted in green; regulations from miRNAs are in black. The total number of receptors, proteins, lncRNAs, TFs and miRNAs are 546, 3220, 126, 102 and 6, respectively.

2.1. Differential Core Signaling Pathways from Young-Adult to Middle-Aged Skin Aging

The differential core signaling pathways from young-adult to middle-aged human skin were selected and analyzed as shown in Figure 4. According to our results, in core signaling pathways of young-adult skin aging only, receptor ESR1 receives microenvironment factor FASN to activate the TF SIRT6 through signaling transduction proteins PRR4 and LMNA. The TF SIRT6 could not only downregulate target gene *RBBP8*, which was modified by deacetylation, but also activate TF PARP1 to upregulate target gene *XRCC1* to promote cell proliferation and DNA repair in young-adult skin aging only. The receptor ESR1 also regulates TF JUN through signaling transduction proteins GOT1 and CHEK2 to regulate target gene *BRCA2* and *KPNA2*. TF JUN not only activates the target gene *BRCA2*, which was modified by phosphorylation, to promote DNA repair, but also downregulates the target gene *KPNA2* to promote cell proliferation, DNA repair and cell cycle in young-adult skin aging only.

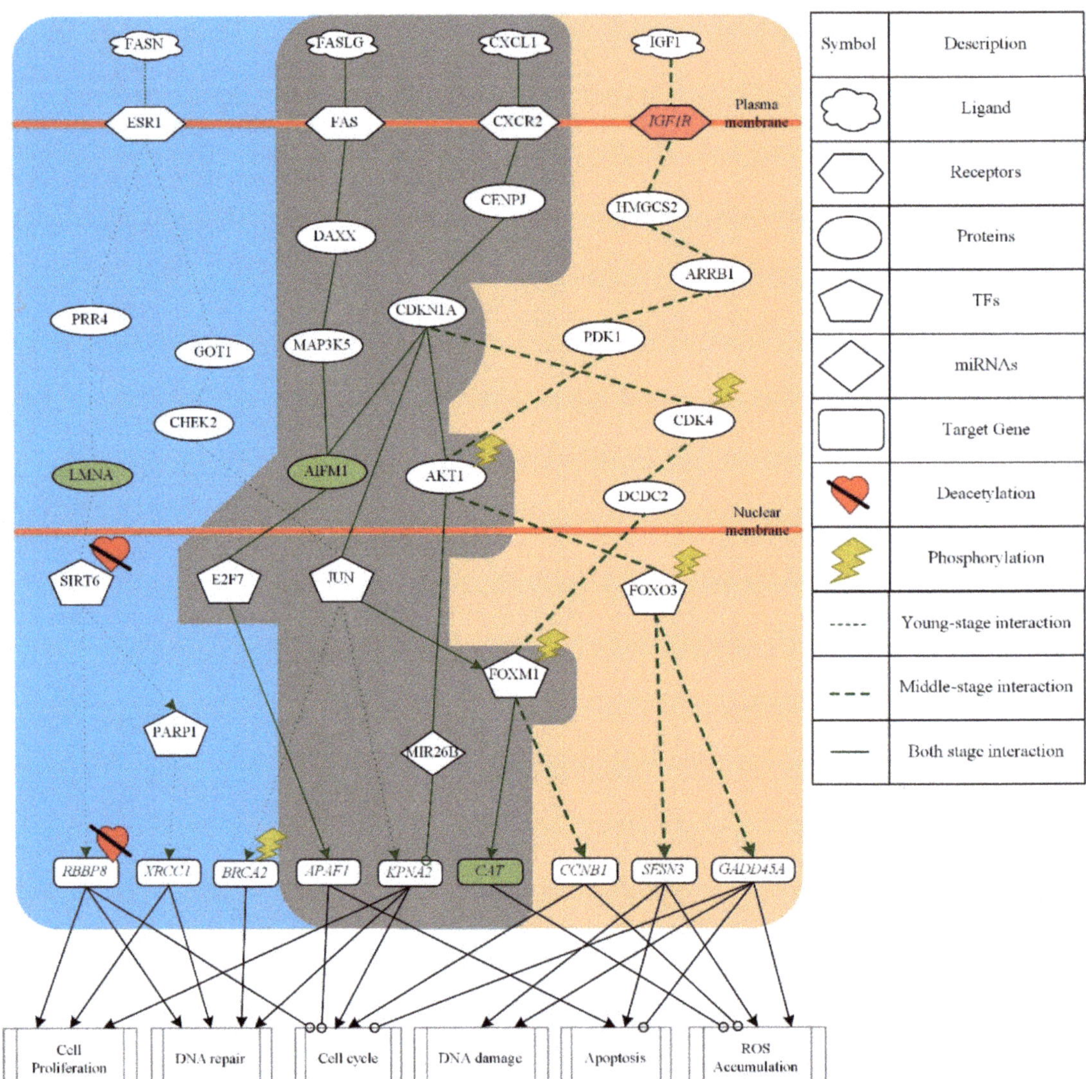

Figure 4. The core signaling pathways from young-adult to middle-aged skin aging. The green dot lines denote the signaling pathways in young-adult skin; the green dash lines represent the signaling pathways in middle-aged skin; the green solid lines indicate the signaling pathways in both stage; the green lines with arrow heads are upregulation (positive regulation); the green lines with circular heads are downregulation (negative regulation); the black solid lines with arrow heads mean activating cellular function; the black solid lines with circular heads mean inhibiting cellular function; the selected red target gene nodes indicate a higher gene expression in middle-aged skin compared with young-adult skin; the selected green target gene nodes indicate a lower gene expression in middle-aged skin compared with young-adult skin; the blue background shows young-adult skin; the brown background shows the overlap between young-adult and middle-aged skin; the skin color background shows middle-aged skin.

Next, in the core signaling pathways of both young-adult and middle-aged skin aging in Figure 4, the microenvironment factor FASLG was received by receptor FAS to activate TF E2F7 via signaling transduction proteins DAXX, MAP3K5 and AIFM1 to

downregulate target gene *APAF1* to promote apoptosis and cell-cycle arrest. In the next pathway, the receptor CXCR2 receive microenvironment factor CXCL1 to regulate TF E2F7, TF JUN, TF FOXM1 and miRNA MIR26B. First, the TF JUN was activated through signaling transduction proteins CENPJ and CDKN1A to activate TF FOXM1. As TF FOXM1, which was modified by phosphorylation, was activated, the target gene *CAT* was upregulated to inhibit ROS accumulation in both young-adult and middle-aged skin aging. The miRNA MIR26B was activated via signaling transduction proteins CENPJ, CDKN1A and AKT1 to inhibit target gene *KPNA2* so as to promote cell proliferation. It is noted that, protein AKT1 was modified by phosphorylation. The TF E2F7 was also regulated by protein AIFM1 to downregulate target gene *APAF1* in both young-adult and middle-aged skin aging.

In the core signaling pathways of middle-aged skin aging only, the TF FOXM1 was also activated through signaling transduction proteins CENPJ, CDKN1A, CDK4 and DCDC2 when receptor CXCR2 received microenvironment factor CXCL1 in middle-aged skin aging only as shown in Figure 4. With the activation of FOXM1, the target gene *CCNB1* was upregulated to promote cell-cycle arrest and inhibit ROS accumulation. For the next pathway in middle-aged skin aging only, the microenvironment factor IGF1 was received by receptor IGF1R to regulate TF FOXO3 via signaling transduction proteins HMGCS2, ARRB1, PDK1 and AKT1. Additionally, the protein AKT1 and TF FOXO3 were modified by phosphorylation. The TF FOXO3 downregulates not only target gene *SENS3* to promote DNA damage, apoptosis, and ROS accumulation, but also target gene *GADD45A* to promote DNA damage, apoptosis, and ROS accumulation and inhibit cell-cycle arrest in middle-aged skin aging only.

In summary, when the young-adult skin aging turned into middle-aged skin aging, DNA repair ability decreases and cell cycle starts to be arrested. Thereby, ROS accumulation increases and further promotes DNA damage and apoptosis in skin cells. Additionally, these molecular progression mechanisms from young-adult to middle-aged might potentially accelerate skin aging process in elderly skin aging. According to the core signaling analyses results and considering the overlap nodes between the GenAge and CMap datasets, we identified AIFM1, CAT, IGF1R, and LMNA as essential biomarkers for preventing skin aging from young adulthood to middle-age.

2.2. Differential Core Signaling Pathways from Middle-Aged to Elderly Skin Aging

The molecular progression mechanism based on differential core pathways from middle-aged skin to elderly skin aging is represented in Figure 5. In core pathways of middle-aged skin aging only, the TNF receptor superfamily member 1 alpha TNFRSF1A received microenvironment factor TNF to activate TF GATA2 through transduction proteins GABPA and STAT1 to upregulate target gene *MMP9* so as to inhibit collagen stability and skin homeostasis in middle-aged skin aging only. Note that STAT1 was modified by phosphorylation. In the next pathway, the microenvironment factor NGF was accepted by neurotrophic receptor tyrosine kinase1 NTRK1 and then transmitted the signal through transduction proteins EME1, HSPB1, NEDD9 and CPNE2 to activate TF ETS1. TF ETS1 could downregulate target gene *ERRFI1*, which was modified by hypermethylation, through activating miRNA MIR573 to promote homeostasis in middle-aged skin aging only.

Next, we focus on the core pathways of both middle-aged and elderly skin aging. In the first pathway, the receptor NTRK1 receives the microenvironment factor NGF and then transmits the signal through transduction proteins KPNA2, KAT5, CST2 and HRAS to activate TF GATA2. The protein KAT5 was modified by phosphorylation. After GATA2 was activated, target gene *BCL2* was upregulated to inhibit cell-cycle arrest and apoptosis in both middle-aged and elderly skin aging. For the second pathway, the receptor KIT could interact with the microenvironment factor KITLG to trigger TF AR through signaling transduction proteins FAM83H, HSPB1, PAX3 and H2AFB2. TF AR downregulated not only target gene *TYR* through triggering TF MITF to promote melanin synthesis, but also target gene *CDH1*, which was modified by phosphorylation, to promote cell-cycle,

apoptosis and DNA damage in both middle-aged and elderly skin aging. In the third pathway, receptor LRP1 could receive the microenvironment factor CYR61 to trigger the TF ETS1 via signaling transduction proteins RAMP1, MCM2, GEMIN4 and NOP56. In both middle-aged and elderly skin aging, the TF ETS1 could negatively regulate target gene *COL17A1* to promote melanin synthesis and inhibit collagen stability and skin homeostasis.

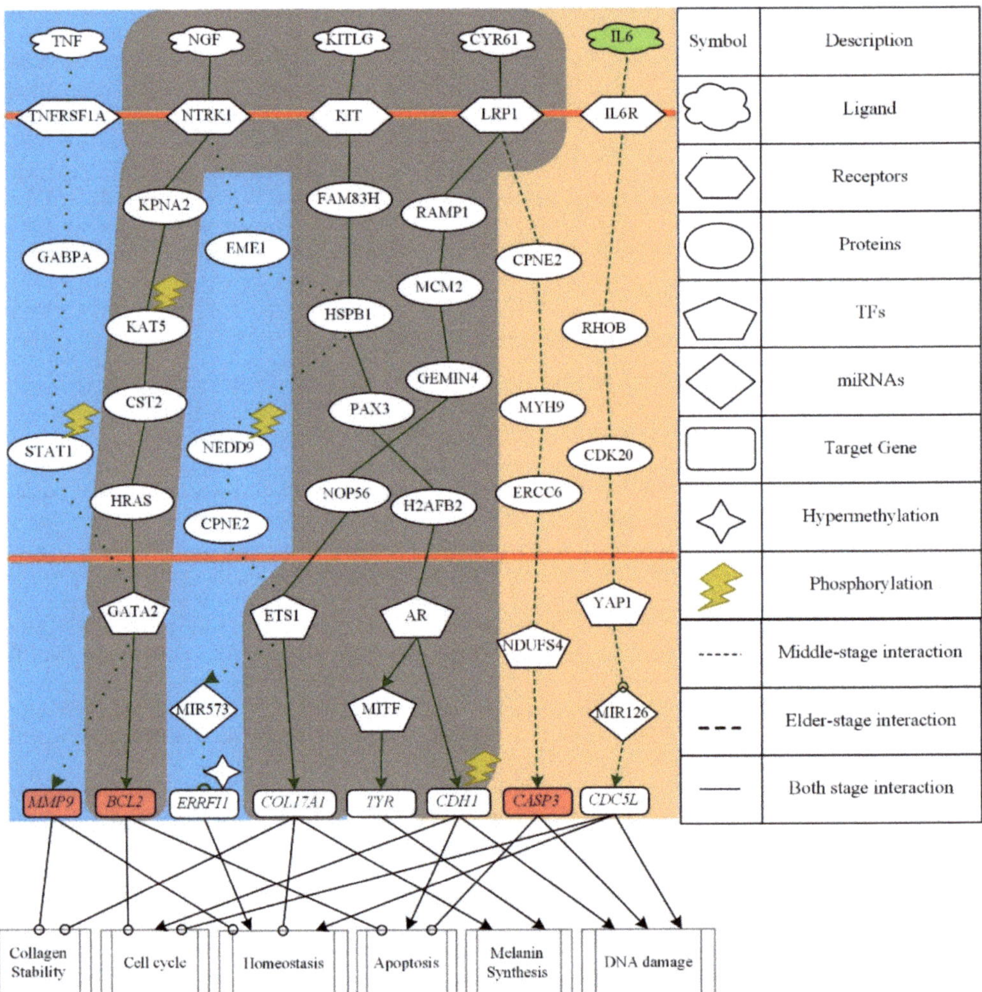

Figure 5. The core signaling pathways are obtained by projecting core GWGENs to KEGG pathways to investigate the aging progression mechanism from middle-aged to elderly skin aging. The green dotted lines denote the signaling pathways in middle-aged skin; the green dashed lines represent the signaling pathways in elderly skin; the green solid lines indicate the signaling pathways in both stages; the green lines with arrow heads are upregulation; the green lines with circle heads are downregulation; the black solid lines with arrow heads mean activating cellular function; the black solid lines with circle heads mean inhibiting cellular function; the selected red target gene nodes indicate a higher gene expression in elderly skin compared with middle-aged skin; the selected green target gene nodes indicate a lower gene expression in elderly skin compared with middle-aged skin; the blue background shows middle-aged skin; the brown background covers the overlap between middle-aged and elderly skin; the skin color background shows elderly skin.

In the core pathways of elderly skin aging only in Figure 5, CYR61/LRP1 could also trigger TF NDUFS4 through signaling transduction proteins CPNE2, MYH9 and ERCC6. The activated TF NDUFS4 might downregulate target gene *CASP3* to inhibit apoptosis and promote DNA damage. For another pathway, the microenvironment factor IL6 was accepted by receptor IL6R to trigger TF YAP1 through signaling transduction proteins RHOB and CDK20. In the elderly skin aging only, TF YAP1 activated target gene *CDC5L* through inhibiting MIR126 to inhibit cell cycle and promote skin homeostasis and DNA damage.

In summary, for skin aging molecular progression mechanisms from middle age to old age, we found that the promotion of cell cycle process, the inhibition of apoptosis, and the damage of DNA arose in elderly skin. Furthermore, skin homeostasis and collagen stability were destroyed to cause lower immunity and epidermal thinning, that is, the increment of wrinkles. According to core signaling analyses and considering the overlap nodes between the GenAge and CMap datasets, we identified MMP9, IL6, BCL2, and CASP3 as essential biomarkers for preventing skin aging from middle age to old age. Moreover, by extracting differential core signaling pathways from young-adult to elderly skin aging, some cellular dysfunctions including proliferation, DNA repair and damage, cell-cycle arrest, apoptosis, ROS accumulation, collagen stability, skin homeostasis, and melanin synthesis are induced in the skin aging process shown in Figure 6.

2.3. The Application of Deep Neural Network of Drug–Target Interaction Prediction and the Design of Two Filters Considering Drug Regulation Ability and Drug Sensitivity

To explore the drug–target interaction toward our identified biomarkers, we trained a deep neural network for drug–target interaction prediction. The design framework is shown in Figure S4. The interaction dataset used for training are from BindingDB [35]. In total, there are 80,291 known drug–target interactions between 38,015 drugs and 7292 proteins. The number of unknown drug–target interactions is 19,966,109, which is greater than the known drug–target interactions. Considering the class imbalance problem, we randomly chose the number of unknown interactions and made them the same size as known interactions. We trained the model using 70% of data, including 10% of data as the validation set. The remaining 30% of data was used as the testing set. To the data preprocessing before training the model, we performed feature scaling by standardization. Assisted with principal component analyses (PCA) for dimensionality reduction, we had 1000 out of 1359 features. For the architecture of deep neural network of drug–target interaction, we used Adam as an optimizer (learning rate = 0.003) with binary cross-entropy loss. The input layer had 1000 neurons followed by 512, 256, 128, and 64 neurons of hidden layers, respectively. The output layer has one neuron. Except for using sigmoid function to the output layer, we set a nonlinear activation function ReLU for each hidden layer. Moreover, the dropout 0.5, 0.4, 0.3, and 0.1 was applied to each hidden layer, respectively. For the trained deep neural network of drug–target interaction prediction, the training accuracy, validation accuracy, and testing accuracy were 95.469%, 93.230%, and 93.077%, respectively. From the perspective of the deep neural network framework application, we used it to predict the potential candidate drugs for our identified biomarkers. When the score of candidate drug approaches one, it would be selected. In other words, the higher the score, the higher probability of interacting between the candidate drug and the identified biomarker.

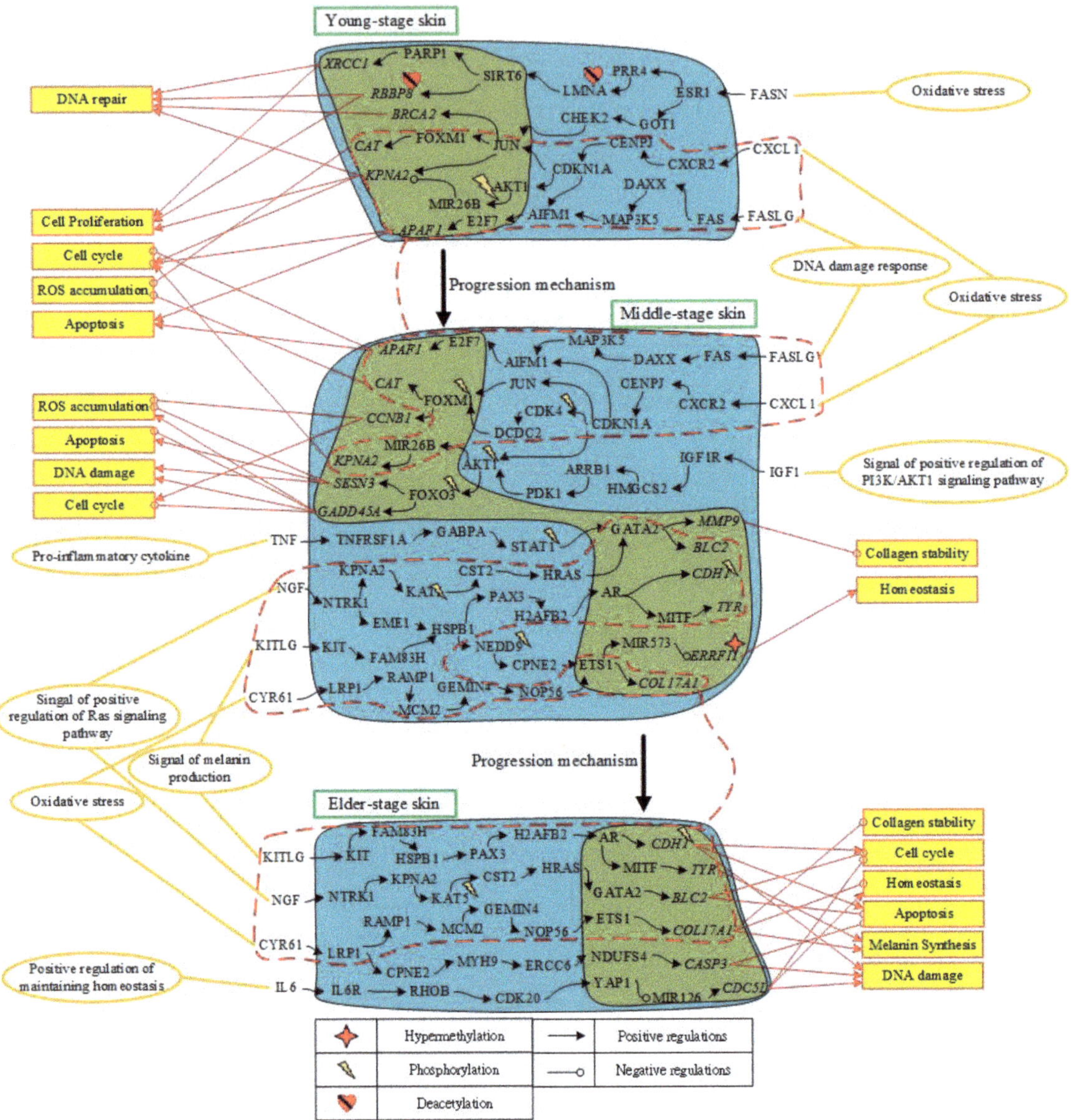

Figure 6. The overview of human skin aging molecular progression mechanisms from young-adult to middle-aged and then elderly skin aging. This figure summarizes the genetic and epigenetic progression mechanisms of skin aging in Figures 4 and 5. The upper horizontal part is the genetic and epigenetic progression mechanism from young-adult skin to middle-aged skin; the middle part indicates the genetic and epigenetic progression mechanisms from middle-aged skin to elderly skin; the red rectangle with orange background represents cellular functions; the yellow ellipse circles are microenvironment factors; the red dash lines surround the pathways and biomarkers that appear in two consecutive stages of skin; the black arrow lines represent the protein–protein interaction or transcriptional regulation; the black lines with circle head represent inhibit or downregulation; the red arrow lines represent the genes to induce cellular function; the red lines with circle head represent the genes to repress cellular function.

In order to narrow down the candidate drugs predicted by the deep neural network framework based on the identified biomarkers, we designed two filters considering drug regulation ability and drug sensitivity. With the help of the CMap dataset, we could know whether a gene was upregulated or downregulated after treating the small molecule compound. The abnormal up or down gene expression could be found by comparing the gene expression of identified biomarkers to the later skin stage. The goal for the first filter is to select candidate drugs, which could reverse the abnormal gene expression. Afterwards, we used the drug sensitivity dataset (PRISM Repurposing Primary Screen) to consider drug sensitivity. The second filter aims to find the drugs with around zero values implying that they would not influence the cell line too much since we are not going to kill or proliferate cells toward the skin corresponding cell line. Consequently, we proposed niridazole, liothyroninr, decitabine, pinacidil, and allantoin as a multiple-molecule drug for mitigating skin aging from young adulthood to middle age; and allantoin, diclofenac, mepyramime, resveratrol, and azathioprine as multiple-molecule drug for mitigating skin aging from middle age to old age. The drug targets with their corresponding drugs are shown in Tables S4 and S5.

3. Discussion

3.1. Investigating Skin Aging Molecular Progression Mechanisms by Differential Core Signaling Pathways from Young-Adult to Middle-Aged Human Skin Aging

In the first core pathway of young-adult skin aging only as shown in Figure 4, microenvironment factor fatty acid synthase FASN can promote cell proliferation, DNA repair, and cell-cycle arrest and interact with receptor ESR1 via crucial signaling transduction proteins PRR4, LMNA, GOT1, and CHEK2 to regulate TFs SIRT6, PARP1, and JUN. The signaling protein LMNA, which is an endogenous activator of TF SIRT6, could promote SIRT6-mediated downstream functions upon DNA damage. Moreover, protein LMNA could directly bind and activate TF SIRT6 toward histone deacetylation [36]. The TF SIRT6, which could control the longevity and regulation of DNA repair, could promote DNA repair and cell proliferation through the downregulation of the target gene *RBBP8*, which was mediated by deacetylation [37]. TF SIRT6 could also promote DNA repair under oxidative stress by activating TF PARP1 to upregulate target gene *XRCC1* [38]. TF PARP1 serves as a genomic caretaker by participating in several molecular mechanisms such as DNA repair and cell-cycle regulation. Therefore, PARP1 was considered as a longevity assurance and aging-promoting factor [39]. The target gene *XRCC1* upregulated by PARP1 was required for the viable and efficient repair for DNA single-strand breaks [40]. The TF JUN was also activated by the signaling transduction proteins GOT1 and CHEK2. The signaling transduction protein CHEK2 initiated by oxidative stress could regulate target gene *BRCA2* and *KPNA2* through interacting with TF JUN. In human cell, the serine kinase CHEK2 could induce the appropriate cellular response such as cell cycle checkpoint activation and DNA repair depending on the extent of damage, the cell type, and other factor. CHEK2 could participate in DNA repair by phosphorylating the target gene *BRCA2* through TF JUN [41]. The karyopherin alpha2 *KPNA2* expression had been reported to be induced in various proliferative skin disorders such as psoriasis and squamous cell carcinoma [42]. When the target gene *KPNA2* was downregulated by TF JUN, cell proliferation, cell cycle and DNA repair induced by CHEK2 were indirectly promoted.

In the core pathways of both young-adult and middle-aged skin aging, the microenvironment factor FASLG binds receptor FAS to regulate TF E2F7 through signaling transduction proteins DAXX, MAP3K5 and AIFM1. Responding to ROS, the microenvironment factor FASLG was activated, then binding to death receptor FAS to promote apoptosis pathway [43]. Signaling transduction protein MAP3K5 known as apoptosis signal-regulating kinase 1 (ASK1) could respond to oxidative stress and be activated [44]. Target gene *APAF1* is the core of the apoptosome, was activated by TF E2F7 to trigger the mitochondrial apoptotic pathway. Furthermore, target gene *APAF1* was also involved in the maintenance of genomic stability by the cell-cycle arrest response elicited upon DNA damage and promoted apoptosis [45]. For other pathways in both young-adult and middle-aged skin aging,

microenvironment factor CXCL1 bound the G-protein coupled receptor CXCR2 to activate signaling transduction protein CDKN1A through protein CENPJ. Protein CDKN1A known as CIP1, was a potent cyclin-dependent kinase inhibitor to regulate TFs E2F7, JUN, FOXM1, and miRNA MIR26B. First, protein CDKN1A transmits signal to AIFM1 so as to activate TF E2F7 to enhance apoptosis and cell-cycle arrest. TF JUN was also activated by protein CDNK1A to regulate FOXM1. Then TF FOXM1 upregulated target gene *CAT*, which was known as ROS detoxification enzyme and could defend the ROS accumulation [46]. After miRNA MIR26B inhibited the target gene *KPNA2*, the cell proliferation could be promoted [47].

In core pathways of middle-aged skin aging only, microenvironment factor CXCL1 also could regulate TF FOXM1 through signaling transduction proteins CENNPJ, CDKN1A, CDK4, and DCDC2 to trigger target gene *CCNB1* as shown in Figure 4. Cyclin dependent kinase 4 CDK4, which was modified by phosphorylation, is a positive regulator of cell cycle entry and can stabilize and activate FOXM1, thereby promote cell cycle and suppress the levels of reactive oxygen species [48]. TF FOXM1 also had been reported to be essential for proper cell cycle progression via activating cell cycle gene *CCNB1* for propelling specific cell cycle phase and inhibition ROS accumulation [49]. For another pathway, the microenvironment factor IGF1 was received by receptor IGF1R to activate FOXO3 via signaling transduction proteins HMGCS2, ARRB1, PDK1 and AKT1. The protein AKT1, which was modified by phosphorylation, could activate TF FOXO3. Protein phosphoinositide-dependent kinase PDK1 was one of the upstream kinases that activate AKT1. After AKT1, which is a key regulator of the PI3K/AKT1 signaling cascade controlling cell growth and survival, was activated and modified by phosphorylation, TF FOXO3 would be activated [50]. Moreover, it had been reported that the enhanced ROS production might further activate the signal of PI3K/AKT pathway, thus establishing a self-perpetuating cycle leading to further aging [51]. TF FOXO3, which was modified by phosphorylation, could downregulate target genes *SESN3* and *GADD45A* [52]. TF FOXO3 could decline ROS rescue pathway through downregulating the peroxiredoxin gene *SESN3*, which is responsible for the biphasic ROS accumulation. Therefore, FOXO3-induced ROS was increased and then accelerated for apoptosis and DNA damage [53]. Furthermore, phosphorylated FOXO3 also inhibited proapoptotic activity such as cell-cycle arrest by downregulating *GADD45A* [54]. The cause of the pleiotropic action of *GADD45* members, a decreased inducibility, might lead to far-reaching consequences such as DNA damage accumulation and disorder of cellular homeostasis and could eventually contribute to the aging process [55]. Therefore, we suggest that the downregulation of *GADD45A* also promotes ROS accumulation through cell-cycle arrest and the inhibition of proapoptotic activity.

3.2. Investigating Skin Aging Molecular Progression Mechanisms by Differential Core Signaling Pathways from Middle-Aged to Elderly Human Skin Aging

According to the core pathways of middle-aged skin aging only in Figure 5, the ligand TNF can inhibit collagen stability and skin homeostasis through receptor TNFRSF1A by transmitting the signal through significant signaling transduction proteins GABPA and STAT1 to TF GATA2. The proinflammatory cytokine tumor necrosis factor-alpha (TNF-A) inhibits collagen synthesis and enhances collagen degradation via increasing the production of target gene *MMP9*. It also increases the risk of cutaneous infections in the elderly by reducing skin immunity [56]. The activation of STAT1 is modified by phosphorylation. STAT1 has also been indicated as a potential target in the treatment of psoriasis, which is a chronic skin diseases [57]. TF GATA2 could upregulate target gene *MMP9* to digest collagen type IV, which is an important component of the basement membrane in skin [58].

For the next pathway, ligand NGF can promote skin homeostasis through receptor NTRK1 to transmit the significant signal via signaling transduction proteins EME1, HSPB1, NEDD9, and CPNE2 to upregulate TF ETS1. In human skin, proliferating keratinocytes release NGF in an increasing amount. Receptor NTRK1, known as tyrosine kinase receptor (TrkA) is the high-affinity receptor for NGF. At the skin level, NTRK1 could mediate NGF-induced keratinocyte proliferation [59]. Note that protein NEDD9 could be modified by

phosphorylation in human skin [60]. TF ETS1, which was regulated through the signaling pathway activated by ligand NGF, has been identified to be associate with skin aging [60]. The expression levels of MIR-573 were found to be lower in melanoma tissues and cell lines when compared to normal skin tissue. Moreover, MIR-573 reduction was demonstrated to be essential in melanoma initiation and progression [61]. Target gene *ERRFI1*, which was modified by hypermethylation, is required for proper epidermal homeostasis [62,63].

Focusing on core pathways in both middle-aged and elderly skin aging in Figure 5, the ligand NGF inhibits apoptosis and promotes cell-cycle arrest when received by receptor NTRK1 to activate TF GATA2 via signaling transduction proteins KPNA2, KAT5, CST2, and HRAS. Protein KAT5, which was modified by phosphorylation, has been presumed to serve as a potential biomarker for melanoma therapeutic target [64]. NGF can not only rescue human epidermal keratinocytes from spontaneous and UVB-induced apoptosis via NTRK1, but also protect keratinocytes from cell death via target gene *BCL2* family of apoptosis inhibitors [59]. Antiapoptotic function of target gene *BCL2* is regulated by phosphorylation. In addition, target gene *BCL2* could not only regulate cell cycle progression, but also act as an antioxidant that may regulate intracellular ROS. Expression of target gene *BCL2* has been observed to increase upon the induction of a senescence-like growth arrest or apoptosis by oxidative stress [65,66].

In the next core pathway of both middle-aged and elderly skin aging, the ligand KITLG promotes melanin synthesis, DNA damage, and inhibits cell-cycle arrest by modulating TFs AR and MITF via signaling transduction proteins FAM83H, HSPB1, PAX3, ATF5, and H2AFB2. The tyrosine kinase receptor KIT, its ligand KITLG, and TF MITF have been reported to play an important role of initiating and regulating signaling systems and transcription factors of melanin production. TF MITF also regulates melanocyte pigmentation by inducing target gene *TYR* [67]. Moreover, a previous study supposed that PAX3 and SOX10 could act together to induce the expression of MITF [68]. Target gene *CDH1*, which is downregulated by TF AR, has been reported to be regulated by phosphorylation [69]. It has been reported that cells lacking target gene *CDH1* have a shortened G1 phase, accumulate DNA damage, and undergo apoptosis [70].

In the final core pathways of both middle-aged and elderly skin aging, the ligand CYR61 could modify skin homeostasis and melanin synthesis through receptor LRP1 to transmit signal by signaling transduction proteins RAMP1, MCM2, GEMIN4 and NOP56 for upregulating TF ETS1. Responding to oxidative stress, CYR61 was elevated in the dermis of chronologically aged human skin, promoting aberrant collagen homeostasis by downregulating collagen members, the major structural protein in skin, to promote collagen degradation [71,72]. The loss of target gene *COL17A1* and MCM2 expression in advanced aged skin has been found to eventually cause epidermal thinning [73].

Focusing on the first core pathway of elderly skin aging only in Figure 5, through the signaling transduction starting from the ligand CYR61, TF NDUFS4 can promote apoptosis and DNA damage through signaling transduction proteins CPNE2, MYH9 and ERRC6. The ligand CYR61 interacting with receptor LRP1 has also been indicated to contribute to CCN1-induced ROS accumulation and CCN1/TNFA-induced apoptosis [74]. With the downregulating target gene *CASP3* by TF NDUFS4, senescence fibroblast can resist apoptosis death [75].

In the final core pathway of elderly skin aging only, the ligand IL6 could be accepted by receptor IL6R and then the significant signal is transmitted through signaling transduction proteins RHOB and CDK20 to activate TF YAP1. Proinflammatory cytokine IL6 has been suggested to be a biomarker of health status in the elderly [76]. TF YAP1 has been identified to play a physiological role in skin homeostasis, which can promote cell proliferation in the basal layer [77]. Knockdown of target gene *CDC5L* induces mitotic arrest and DNA damage [78].

3.3. The Genetic and Epigenetic Molecular Progression Mechanisms from Young-Adult to Elderly Human Skin Aging

The overview of overall skin aging molecular progression mechanisms is shown in Figure 6. Microenvironments trigger corresponding ligand signals to initiate some cellular dysfunctions affecting skin aging progression. Thus, core signaling pathways with the genetic and epigenetic modifications play a significant role in cellular dysfunctions of signaling transductions for each stage of skin aging.

In Figure 6, the core pathways of young-adult skin aging only, ligand FASN (oxidative stress) binds to receptor ESR1 to mediate two pathways. Responding to DNA damage signal to cause of oxidative stress, LMNA directly binds and activates TF SIRT6 toward histone deacetylation [36]. Activated SIRT6 promotes DNA repair cell-cycle and proliferation through the downregulating gene *RBBP8*, which was modified by deacetylation [37]. In addition, TF SIRT6 also promotes DNA repair and cell cycle under oxidative stress by activating TF PARP1 to upregulate target gene *XRCC1* [38]. Transduction protein CHEK2 also responds to oxidative stress from ligand FASN to activate TF JUN. TF JUN promotes DNA repair through phosphorylating target gene *BRCA2* [41]. Moreover, TF JUN could downregulate target gene *KPNA2* to promote cell proliferation, cell cycle and DNA repair [42].

In the core pathways of both young-adult and middle-aged skin aging, ligand CXCL1 (oxidative stress) binds to receptor CXCR2 to trigger protein CDKN1A. Activated protein CDKN1A not only upregulates target gene *CAT* to defend ROS accumulation through TF JUN and FOXM1, but also inhibits target gene *KPNA2* to promote cell proliferation by activating MIR26B [46,47]. Next, responding to ROS induced from DNA damage, the ligand FASLG interacts with FAS to initiate apoptosis pathway. Signaling transduction protein MAP3K5, activated by oxidative stress, triggers TF E2F7 to downregulate target gene *APAF1* and thereby involve in the maintenance of cell-cycle arrest upon DNA damage and promoting apoptosis [43–45]. Hence, in order to fight to the excessive accumulation of ROS upon decreasing the ability of DNA repair from young adulthood to middle-age, functions of apoptosis and cell-cycle arrest are raised.

In the core pathways of middle-aged skin aging only, the ligand CXCL1 (oxidative response) interacts with receptor CXCR2 and also activates signaling transduction protein CDK4. Phosphorylation of CDK4 positively regulates cell cycle entry and can stabilize and activate FOXM1 to upregulate target gene *CCNB1* to promote cell cycle phase and suppress the level of ROS [48,49]. Moreover, the ligand IGF1 (oxidative response) is received by receptor IGF1R to activate PI3K/AKT signaling pathway. Transduction protein AKT1 can activate TF FOXO3 through the modification by phosphorylation. Furthermore, TF FOXO3, which is modified by phosphorylation, downregulates genes *SESN3* and *GADD45A*. With the silence of target gene *SESN3*, ROS rescue pathway is declined, thus accelerating apoptosis with the increment of FOXO3-induced ROS. In addition, TF FOXO3 downregulates target gene *GADD45A* to promote ROS accumulation through cell-cycle arrest. The ligand TNF (proinflammatory cytokine) can inhibit collagen stability and skin homeostasis through activating TNFRSF1A and initiate the corresponding pathway. TF GATA2 was activated by phosphorylated transduction protein STAT1 to upregulate target gene *MMP9*. Increased gene *MMP9* can inhibit collagen synthesis and enhance collagen degradation. The ligand NGF (proliferating keratinocytes) interacts with receptor NTRK1 to activate TF ETS1, which was identified to be associative with skin aging [59,60]. MIR573 is activated by TF ETS1 and then negatively regulates gene *ERRFI1*. Downregulated gene *ERRFI1*, which was modified by hypermethylation, can maintain proper epidermal homeostasis [62,63].

In the core pathways of both middle-aged and elderly skin aging, the ligand NGF/NTRK1 (signal of positive regulation of Ras signaling pathway) activates TF GATA2 so as to regulate gene *BCL2* to protect keratinocytes from cell death [59]. Target gene *BCL2* not only triggers antiapoptotic function through the modification by phosphorylation, but also regulates cell cycle progression to act as an antioxidant of intracellular ROS [65,66]. The

ligand KITLG interacts with receptor KIT to promote melanin synthesis, DNA damage, and inhibits cell-cycle arrest by activating TF MITF through signaling transduction. TF MITF regulates melanocyte pigmentation by inducing gene *TYR* [67]. Due to the downregulation of gene *CDH1*, which is modified by phosphorylation, by TF AR, G1 phase is shortened to accumulate DNA damage and undergo apoptosis [69,70]. Next, the ligand CYR61 (oxidative stress) is received by receptor LRP1 to downregulate collagen members and promote collagen degradation [71,72]. After TF ETS1 is activated by signaling transduction, target gene *COL17A1* can be downregulated to destroy the balance between collagen stability and skin homeostasis and eventually cause epidermal thinning [73].

In the core pathways of elderly skin aging only, the ligand CYR61 (oxidative stress) interacts with receptor LRP1 so as to contribute to the CCN1-induced ROS accumulation. When target gene *CASP3* is downregulated by TF NDUFS4, senescence fibroblast could resist apoptosis death [74,75]. Moreover, since the ligand IL6 interacts with receptor IL6R to activate TF YAP1, which could promote cell proliferation in basal layer, it has been identified to play a physiological role in skin homeostasis [77]. Note that IL6 has been suggested as a biomarker of elderly health status [76]. After TF YAP1 downregulating target gene *CDC5L* through activating MIR126, functions of mitotic arrest and DNA damage were activated [78].

3.4. Two Multiple-Molecule Drugs Based on Identified Biomarkers to Mitigate Human Skin Aging

For mitigating the skin aging from young adulthood to middle age, we proposed a multiple-molecule drug including niridazole, liothyronine, decitabine, pinacidil, and allantoin. The drug targets were AIFM1, CAT, IGF1R, and LMNA as shown in Table 2. The black dot in Table 2 represents the proposed small molecules target to which identified biomarker (drug target). For instance, the niridazole has more potential to target to AIFM1 and CAT. Niridazole, an antiparasitic drug, could suppress delayed dermal hypersensitivity [79]. Studies have shown that combined therapy with liothyronine improved the treatment of hypothyroidism [80,81]. Decitabine, a DNA methyltransferase, induced changes in gene expression and cellular behavior associated with a regenerative response. Furthermore, wounds treated by decitabine were able to participate in regeneration [82]. Pinacidil is an effective antihypertensive drug for the treatment of mild to moderate essential hypertension [83]. In the meanwhile, according to the findings of one study, pinacidil may be utilized to prevent from UV-induced skin aging [84]. It is noted that allantoin, which is found in plants like chamomile, wheat sprouts, sugar beet, and comfrey, has been widely used in anti-aging serum [85,86]. Allantoin is also a well-known anti-irritating and hydrating agent as well as a peeling agent for skin [87,88].

Table 2. Drug targets and multiple-molecule drugs for preventing skin aging from young adulthood to middle-age.

Drug \ Target	AIFM1	CAT	IGF1R	LMNA
niridazole	●	●		●
liothyronine	●			
decitabine		●		
pinacidil			●	
allantoin			●	

●: Proposed small molecules target to the identified biomarkers (drug targets).

For mitigating the skin aging from middle-aged to elderly, we proposed a multiple-molecule drug consisting of allantoin, diclofenac, mepyramine, resveratrol, and azathioprine. The drug targets were MMP9, IL6, BCL2, and CASP3 as shown in Table 3. In Table 3, the black dot shows the drug target to each specific small molecule. For example, the drug target for allantoin are MMP9 and IL6. Diclofenac is a nonsteroidal anti-inflammatory drug.

It has been used to treat actinic keratoses developing in fair-skinned individuals with a history of overexposure to ultraviolet light [89]. To mepyramine, it works by preventing the action of histamine, which is a compound produced by the body when getting venom from insect bites [90]. Moreover, one study mentioned the stimulation from histamine would upregulate matrix metalloproteinase 9 (MMP9), which is also our proposed drug target for mepyramine [91]. Resveratrol is abundant in grape skin and seeds [92]. Responding to infection, stress, injury, bacteria or fungal infections, and UV-irradiation, it a popular ingredient in skincare products [93]. In the field of dermatology, azathioprine is an effective immunosuppressant that is extremely valuable in treating pemphigoid, generalized eczematous disorders, and actinic dermatitis [94]. Taken together, most of the proposed small-molecule compounds are approved by the U.S. Food and Drug Administration (FDA). Drug repurposing for identifying new uses of old drugs with the proposed systems biology approaches might provide the alternative way to find the effective drugs for mitigating skin aging.

Table 3. Drug targets and multiple-molecule drugs for preventing skin aging from middle age to old age.

Drug \ Target	MMP9	IL6	BCL2	CASP3
allantoin	●	●		
diclofenac	●			
mepyramine	●		●	
resveratrol			●	●
azathioprine			●	●

●: Proposed small molecules target to the identified biomarkers (drug targets).

3.5. The Limitations and Advantages to the Proposed Systems Medicine Design Procedure for Human Skin Aging

Gene expression has been widely used to infer other molecular type measures, such as proteomics, copy number variation, and mutation. In this study, we used human skin microarray data processed with cubic spline interpolation to help us construct GWGENs by system identification method via solving constrained linear least-squares estimation problem. After that, we computed Akaike's information criterion (AIC) for each gene to prune false positives. Increasing samples through data interpolation and computing AIC for detecting real systems order, we conquered the overfitting issue. Even though we applied AIC and performed the data interpolation for increasing sample size in each skin aging stage, it is noted that the estimated real GWGENs are near-optimum solutions but not unique solutions. Furthermore, we include basal level in protein, gene, miRNA, and lncRNA systems modeling. These terms imply the unknown interaction or epigenetic modification, and mutation. If we found a basal level change, which was higher than a threshold, we inferred the corresponding node was influenced by epigenetic modification or mutation. These findings have to be verified by a literature survey. Based on the progression molecular mechanisms in each skin aging stage, we could identify essential biomarkers. For exploring the drug–target interaction to our identified biomarkers, we trained a deep neural network of drug–target interaction in advance. In the drug–target data which we used to train the prediction model, if pairs have not been mentioned as known interactions in the BindingDB, we would assign them in the group of negative samples, meaning no interaction. However, the negative samples in our study do not mean without interaction. They might just be lack of experimental evidence or record nowadays. Although the proposed system medicine design procedure exists the aforementioned limitations, it still provides another viewpoint to shed the light on the human skin aging progression based on system level. Moreover, drug repurposing strategy, giving new uses for old drugs, has been used in this study. Most of the suggested small molecules are

approved by the FDA, which could shorten the time of clinical trials. Integrating systems biology approaches, deep learning framework and the design of two filters, we not only transferred biological knowledge into engineering interpretation but also applied them to drug discovery efficiently.

4. Materials and Methods

4.1. Overview of Systems Medicine Design Procedure of Human Skin Aging

In order to further understand skin aging molecular mechanisms from young adulthood to old age, we proposed a research flowchart as shown in Figure 2. At first, we collect several regulation and interaction databases including DIP [95], IntAct [96], BioGRID [97], BIND [98], MINT [99], HTRIdb [100], ITFP [101], Transfac [102], CircuitDB2 [103], and TargetScan [104] to construct the candidate GWGEN, which is composed of candidate protein-protein interaction network (PPIN) and candidate gene regulatory network (GRN). Moreover, the candidate GWGEN is a Boolean matrix. If two nodes have interaction, we would give one; if two nodes do not have interaction, we would give zero in it. With three-stage preprocessed microarray data, we then identify real GWGENs by system identification method and system order detection scheme. Since real GWGENs are still too complicated to investigate the skin aging progression mechanisms, we apply principal network projection (PNP) method to extract core GWGENs from real GWGENs based on the projection values. Subsequently, we denote the core signaling pathways in the style of KEGG pathways. According to the core signaling pathways, we investigate skin aging molecular mechanisms and identify essential biomarkers for young adulthood to middle age and middle age to old age, respectively. After that, we used the trained deep neural network of drug-target interaction to predict potential candidate drugs, which hold higher probability to have interactions with identified biomarkers. To narrow down the candidate drugs, we design two filters considering drug regulation ability and drug sensitivity by CMap [34] and PRISM Repurposing dataset [105]. Consequently, we propose two multiple-molecule drugs for slowing down human skin aging from young adulthood to middle age and from middle age to old age, respectively.

4.2. Data Preprocessing of Human Skin Microarray Data

We obtained human skin microarray data from GSE18876 containing the gene expression level of male skin. It included 50 ages in the range from 19 to 86 years old with 29,226 probes. One study has shown that *OR52N2*, *SIRT6*, *CPT1B*, *TUBAL3*, *COL1A1* and *MATN4* were significantly regulated with age. Furthermore, it also indicated that gene expressions of *OR52N2*, *SIRT6* and *CPT1B* increased with age and gene expressions of *TUBAL3*, *COL1A1* and *MATN4* decreased with age [106]. Therefore, we sketched the changes of gene expression levels of these typical genes. Based on this line graph and gene expression trend in aforementioned study, we defined young-adult, middle-aged and elderly skin as 19 to 45 years old, 43 to 65 years old and 64 to 86 years old, respectively. That is, the averages of gene expressions of *OR52N2*, *SIRT6* and *CPT1B* increased and the averages of gene expressions of *TUBAL3*, *COL1A1* and *MATN4* decreased from young adult stage to middle age, and then to old age in human male skin. In the estimation problem, one would easily face overfitting issue when the sample size is small and the feature size is big [107]. Hence, in this study, firstly, we increased the sample size to 500 for each skin aging stage by performing cubic spline data interpolation via *splin*, a MATLAB function [108–110]. Secondly, we utilized system order detection scheme by computing the AIC value to prune the false positives in the candidate GWGEN for finding the real GWGENs of the human skin aging systems. The more details would be discussed in the Section 4.5.

4.3. Dynamic Systems Modeling for the Candidate GWGEN

The candidate GWGWN consisting of PPIN and GRN. It is noted that GRN also includes miRNA regulation network and lncRNA regulation network. In the following

contents, we would take PPIN and GRN as an example, and the rest of them could be found in Supplementary Materials. The PPIs of human-protein i in the candidate PPIN can be described as a dynamic equation shown as below:

$$p_i(t+1) = p_i(t) + \sum_{j=1}^{I_i} \alpha_{ij}^P p_i(t)p_j(t) - \sigma_i^P p_i(t) + \lambda_i^P g_i(t) + \beta_i^P + \epsilon_i^P(t)$$
$$\text{, for } i = 1,\dots,I, \ -\sigma_i^P \leq 0 \text{ and } \lambda_i^P \geq 0. \tag{1}$$

where $p_i(t)$, $p_j(t)$, and $g_i(t)$ indicate the expression levels of the ith protein, the jth protein, and the ith gene at time t, respectively; α_{ij} denotes the interactive abilities between the ith protein with the jth protein in human skin cells; σ_i^P represents the degradation rate of the ith protein; λ_i^P indicates the translation effect from the corresponding mRNA to the ith protein; The basal level β_i^P signifies the regulations from other unknown regulators to the ith protein; I_i denotes the number of human proteins interacting with the ith protein in the candidate GWGENs; $\epsilon_i^P(t)$ signifies the noise of the ith protein owing to model uncertainty or measurement noise at time t.

The k gene in the candidate GRN can be represented as a dynamic equation in the following:

$$g_k(t+1) = g_k(t) + \sum_{i=1}^{I_k} a_{ki}^G p_i(t) - \sum_{r=1}^{R_k} b_{kr}^G g_k(t)m_r(t) + \sum_{\ell=1}^{L_k} c_{k\ell}^G o_\ell(t) - \mu_k^G g_k(t)$$
$$+\delta_k^G + \omega_k^G(t) \text{ for } k = 1,2,\dots,K, \ -b_{kr}^G \leq 0 \text{ and } -\mu_k^G \leq 0 \tag{2}$$

where $g_k(t)$, $p_i(t)$, $m_r(t)$, and $o_\ell(t)$ indicate the expression level of the kth gene, the ith transcription factor(TF), the rth miRNA and the ℓth lncRNA at time t, respectively; a_{ki}^G, $-b_{kr}^G$, and $c_{k\ell}^G$ represent the regulatory abilities of the ith TF, the repression ability of the rth miRNA, and the regulatory ability of the ℓth lncRNA on the kth gene, respectively; $-\mu_k^G$ signifies the degradation rate of the gene expression of the kth gene; The basal level δ_k^G denotes the regulations from other unknown regulators to the kth gene such as phosphorylation; $\omega_k^G(t)$ signifies the noise of the kth gene owing to model uncertainty or measurement noise at time t; I_k, R_k, and L_k mean the total number of TFs, miRNAs, and lncRNAs in the candidate GRN, respectively. Note that the biological regulatory mechanisms in skin cell in (2) involve TF transcription regulations by $\sum_{i=1}^{I_k} a_{ki}^G p_i(t)$, miRNA repressions by $-\sum_{r=1}^{R_k} b_{kr}^G g_k(t)m_r(t)$, lncRNA regulation by $\sum_{\ell=1}^{L_k} c_{k\ell}^G o_\ell(t)$, the mRNA degradation by $-\mu_k^G g_k(t)$, the basal level by δ_k^G, and the noise by $\omega_k^G(t)$. In this study, the effect of post-translational modification on skin aging is considered by the basal level term δ_k^G.

4.4. Systems Identification Approach in the Candidate GWGEN via Microarray Data

After systems modeling by Equations (1)–(4), we then perform the systems identification by solving the parameter estimation problems. The PPIN in Equation (1) can be rewritten in the following linear regression form:

$$p_i(t+1) = \begin{bmatrix} p_1(t)p_i(t) & \cdots & p_{I_i}(t)p_i(t) & g_i(t) & p_i(t) & 1 \end{bmatrix} \begin{bmatrix} \alpha_{i1}^P \\ \vdots \\ \alpha_{iI_i}^P \\ \lambda_i^P \\ 1-\sigma_i^P \\ \beta_i^P \end{bmatrix} + \epsilon_i^P(t) \tag{3}$$

$$= \psi_i^P(t)\theta_i^P + \epsilon_i^P(t) \ \text{ for } i = 1,2,\dots,I.$$

where $\psi_i^P(t)$, represents the regression vector that can be obtained from the microarray data and θ_i^P denotes the unknown parameter vector to be estimated for the ith protein in PPIN.

Furthermore, the Equation (3) of the ith protein can be augmented for Y_i time points shown as below:

$$\begin{bmatrix} p_i(t_2) \\ p_i(t_3) \\ \vdots \\ p_i(t_{Y_i+1}) \end{bmatrix} = \begin{bmatrix} \psi_i^P(t_1) \\ \psi_i^P(t_2) \\ \vdots \\ \psi_i^P(t_{Y_i}) \end{bmatrix} \theta_i^P + \begin{bmatrix} \epsilon_i^P(t_1) \\ \epsilon_i^P(t_2) \\ \vdots \\ \epsilon_i^P(t_{Y_i}) \end{bmatrix}, \; for \; i = 1,2,\ldots,I, \tag{4}$$

which can also be simplified by

$$P_i = \Psi_i^P \theta_i^P + E_i^P, \; for \; i = 1,2,\ldots,I \tag{5}$$

where

$$P_i = \begin{bmatrix} p_i(t_2) \\ p_i(t_3) \\ \vdots \\ p_i(t_{Y_i+1}) \end{bmatrix}, \; \Psi_i^P = \begin{bmatrix} \psi_i^P(t_1) \\ \psi_i^P(t_2) \\ \vdots \\ \psi_i^P(t_{Y_i}) \end{bmatrix}, \; E_i^P = \begin{bmatrix} \epsilon_i^P(t_1) \\ \epsilon_i^P(t_2) \\ \vdots \\ \epsilon_i^P(t_{Y_i}) \end{bmatrix}.$$

Therefore, the interaction parameters in the vector θ_i^P can be estimated by solving the following constrained least-squares estimation problem:

$$\hat{\theta}_i^P = \min_{\theta_i^P} \frac{1}{2} \| \Psi_i^P \theta_i^P - P_i \|_2^2, \; subject \; to \; A^P \theta_i^P \leq b^P, \tag{6}$$

where

$$A^P = \begin{bmatrix} 0 & \cdots & 0 & -1 & 0 & 0 \\ 0 & \cdots & 0 & 0 & 1 & 0 \end{bmatrix} \in \mathbb{R}^{2 \times (I_k+3)}, \; b^P = \begin{bmatrix} 0 \\ 1 \end{bmatrix}.$$

To estimate the interaction parameters in (1) by solving the parameter estimation problem in (6), we use an optimization toolbox function *lsqlin* in MATLAB. Simultaneously, we ensure the protein translation rate λ_i^P and the protein degradation rate $-\sigma_i^P$ to always be non-negative and non-positive value, respectively; that is, $\lambda_i^P \geq 0$ and $-\sigma_i^P \leq 0$.

Similarly, we rewrite the dynamic equation of GRN in the Equation (2) as the following linear regression form:

$$g_k(t+1) = [p_1(t) \; \cdots \; p_{I_k}(t) \; g_k(t)m_1(t) \; \cdots \; g_k(t)m_{R_k}(t) \quad o_1(t) \; \cdots \; o_{L_k}(t) \; g(t) \; 1] \begin{bmatrix} a_{k_1}^G \\ \vdots \\ a_{kI_k}^G \\ -b_{k_1}^G \\ \vdots \\ -b_{kR_k}^G \\ c_{k_1}^G \\ \vdots \\ c_{kL_k}^G \\ 1 - \mu_k^G \\ \delta_k^G \end{bmatrix} \tag{7}$$

$$+\omega_k^G(t) = \psi_k^G(t)\theta_k^G + \omega_k^G(t), \; for \; k = 1,2,\ldots,K$$

where $\psi_k^G(t)$, represents the regression vector that can be obtained from the microarray data and θ_k^G signifies the unknown parameter vector estimated for the kth gene in the GRN. Moreover, Equation (7) can be augmented for Y_k time points in the following form:

$$
\begin{bmatrix} g_k(t_2) \\ g_k(t_3) \\ \vdots \\ g_k(t_{Y_k+1}) \end{bmatrix} = \begin{bmatrix} \psi_k^G(t_1) \\ \psi_k^G(t_2) \\ \vdots \\ \psi_k^G(t_{Y_k}) \end{bmatrix} \theta_k^G + \begin{bmatrix} \omega_k^G(t_1) \\ \omega_k^G(t_2) \\ \vdots \\ \omega_k^G(t_{Y_k}) \end{bmatrix}, \text{ for } k = 1, 2, \ldots, K \qquad (8)
$$

Next, we simplify the Equation (8) as below:

$$
G_k = \Psi_k^G \theta_k^G + \Omega_k^G, \text{ for k} = 1, 2, \ldots, K \qquad (9)
$$

where

$$
G_k = \begin{bmatrix} g_k(t_2) \\ g_k(t_3) \\ \vdots \\ g_k(t_{Y_k+1}) \end{bmatrix}, \Psi_k^G = \begin{bmatrix} \psi_k^G(t_1) \\ \psi_k^G(t_2) \\ \vdots \\ \psi_k^G(t_{Y_k}) \end{bmatrix}, \Omega_k^G = \begin{bmatrix} \omega_k^G(t_1) \\ \omega_k^G(t_2) \\ \vdots \\ \omega_k^G(t_{Y_k}) \end{bmatrix}.
$$

Hence, the regulatory parameters in the vector θ_k^G can be estimated by solving the following constrained least-squares estimation problem:

$$
\hat{\theta}_k^G = \min_{\theta_k^G} \frac{1}{2} \| \Psi_k^G \theta_k^G - G_k \|_2^2, \text{ subject to } A^G \theta_k^G \leq b^G \qquad (10)
$$

where

$$
A^G = \begin{bmatrix} 0 & 0 & \cdots & 0 & 1 & 0 & \cdots & 0 & 0 & 0 & \cdots & 0 & 0 & 0 \\ 0 & 0 & \cdots & 0 & 0 & 1 & \cdots & 0 & 0 & 0 & \cdots & 0 & 0 & 0 \\ \vdots & \vdots & \ddots & \vdots & \vdots & \vdots & \ddots & \vdots & \vdots & \vdots & \ddots & \vdots & \vdots & \vdots \\ 0 & 0 & \cdots & 0 & 0 & 0 & \cdots & 1 & 0 & 0 & \cdots & 0 & 0 & 0 \\ 0 & 0 & \cdots & 0 & 0 & 0 & \cdots & 0 & 0 & 0 & \cdots & 0 & 1 & 0 \end{bmatrix} \in \mathbb{R}^{(R_k+1) \times (I_k+R_k+L_k+2)}
$$

$$
, b^G = \begin{bmatrix} 0 \\ \vdots \\ 1 \end{bmatrix}.
$$

By applying the function *lsqlin* in MATLAB optimization toolbox to solve the parameter estimation problem in Equation (10), we can estimate the regulatory parameters for GRN equation in Equation (2). Furthermore, we ensure that the miRNA repression rate $-b_{kr}^G$ to be a nonpositive value and the gene degradation rate $-\mu_k^G$ to be a nonpositive value for $k = 1, 2, \ldots K$ and $r = 1, 2, \ldots R_k$.

4.5. Pruning False Positives in Candidate GWGENs to Obtain Real GWGENs by System Order Detection Scheme

Due to the collected data, which we used for constructing the candidate GWGEN, come from different databases, the various experimental conditions and noises might result in getting many false-positive interactions and regulations after doing system identification. Thus, we have to apply system order detection scheme by computing AIC to detect the real system order of PPI model in Equation (1) and GRN model in Equation (2). According to Akaike's theory, the most accurate model has the smallest AIC value [111]. In other words, when the value of AIC achieves the minimum, the detected system order approaches to the real system order.

For PPI model in Equation (5), the AIC value of the ith protein can be defined in the following equation:

$$AIC_i^P(K_i) = \log\left\{\frac{1}{T_i}\left[P_i - \Psi_i^P\hat{\theta}_i^P\right]^T\left[P_i - \Psi_i^P\hat{\theta}_i^P\right]\right\} + \frac{2K_i}{T_i} \tag{11}$$

where $\hat{\theta}_i^P$ denotes the estimated interactive parameters of the ith protein from the solutions of the parameter estimation problem in Equation (6), and the covariance of estimated residual error is $(\varsigma_i^P)^2 = \frac{1}{T_i}\left[P_i - \Psi_i^P\hat{\theta}_i^P\right]^T\left[P_i - \Psi_i^P\hat{\theta}_i^P\right]$. In order to find out the real system order K_i^* of the ith protein in the PPI model so that $AIC_i^P(K_i^*)$, in Equation (11) can achieve the minimum value, we trade off the system order and the estimated residual error. By aforementioned system order detection method, PPIs with insignificant interaction abilities, which are out of K_i^*, could be regarded as false positives and be pruned away.

For the GRN model in Equation (9), AIC value of the kth gene can be defined as the following equation:

$$AIC_k^G(I_k, R_k, L_k) = \log\left\{\frac{1}{T_k}\left[G_k - \Psi_k^G\hat{\theta}_k^G\right]^T\left[G_k - \Psi_k^G\hat{\theta}_k^G\right]\right\} + \frac{(2I_k + R_k + L_k)}{T_k} \tag{12}$$

where $\hat{\theta}_k^G$ denotes the estimated regulatory parameters of the kth gene from the solutions of the parameter estimation problem in Equation (10), and the covariance of estimated residual error is $(\varsigma_k^G)^2 = \frac{1}{T_k}\left[G_k - \Psi_k^G\hat{\theta}_k^G\right]^T\left[G_k - \Psi_k^G\hat{\theta}_k^G\right]$. In order to find out the real system order I_k^*, R_k^*, and O_k^* of the kth gene in GRN so that $AIC_k^G(I_k^*, R_k^*, L_k^*)$, in (12) can achieve the minimum value, we trade off the system order and to estimate residual error. In this way, to kth gene, the gene regulations with insignificant regulatory abilities, which are out of I_k^*, R_k^*, and O_k^*, can be treated as false-positives and be pruned away from the candidate GRN. It is noted that we apply the same system order detection scheme on the miRNA model and the lncRNA model, which could be found in the Section S1.3 of Supplementary Materials.

After performing system identification and system order detection scheme, which pruned away the insignificant interactions and regulations in the candidate GWGEN, we eventually obtained the real GWGENs for three stage of human skin aging. However, it is still quite difficult to investigate the progression mechanisms of skin aging from these real GWGENs due to their high complexity. Here, we introduce the principal network projection (PNP) method to extract the core networks from the real GWGENs as core GWGENs to solve this issue. The details are described in the following section.

4.6. Extracting Core Networks from Real GWGENs by the Principal Network Projection Method

The PNP method is a network structure projection approach based on the principal singular values so as to reduce network dimension via deleting insignificant structures. In order to use the PNP method to extract the core networks from the real GWGENs, we have to construct a network matrix H consisting all of the estimated interactions and regulations in the real GWGEN (with the ith row denoting the interactions or regulations on the ith node, i.e., protein, gene, miRNA or lncRNA of real GWGEN) in the following formation:

$$
H = \begin{bmatrix}
\hat{\alpha}_{11} & \cdots & \hat{\alpha}_{1I} & 0 & \cdots & 0 & 0 & \cdots & 0 \\
\vdots & \hat{\alpha}_{ij} & \vdots & \vdots & 0 & \vdots & \vdots & 0 & \vdots \\
\hat{\alpha}_{I1} & \cdots & \hat{\alpha}_{II} & 0 & \cdots & 0 & 0 & \cdots & 0 \\
\hat{a}^G_{11} & \cdots & \hat{a}^G_{1I} & -\hat{b}^G_{11} & \cdots & -\hat{b}^G_{1R} & \hat{c}^G_{11} & \cdots & \hat{c}^G_{1Z} \\
\vdots & \hat{a}^G_{ki} & \vdots & \vdots & -\hat{b}^G_{kr} & \vdots & \vdots & \hat{c}^G_{kz} & \vdots \\
\hat{a}^G_{K1} & \cdots & \hat{a}^G_{KI} & -\hat{b}^G_{K1} & \cdots & -\hat{b}^G_{KR} & \hat{c}^G_{K1} & \cdots & \hat{c}^G_{KZ} \\
\hat{a}^M_{11} & \cdots & \hat{a}^M_{1I} & -\hat{b}^M_{11} & \cdots & -\hat{b}^M_{1R} & \hat{c}^M_{11} & \cdots & \hat{c}^M_{1Z} \\
\vdots & \hat{a}^M_{ri} & \vdots & \vdots & -\hat{b}^M_{rr} & \vdots & \vdots & \hat{c}^M_{rz} & \vdots \\
\hat{a}^M_{R1} & \cdots & \hat{a}^M_{RI} & -\hat{b}^M_{R1} & \cdots & -\hat{b}^M_{RR} & \hat{c}^M_{R1} & \cdots & \hat{c}^M_{RZ} \\
\hat{a}^L_{11} & \cdots & \hat{a}^L_{1I} & -\hat{b}^L_{11} & \cdots & -\hat{b}^L_{1R} & \hat{c}^L_{11} & \cdots & \hat{c}^L_{1Z} \\
\vdots & \hat{a}^L_{zi} & \vdots & \vdots & -\hat{b}^L_{zr} & \vdots & \vdots & \hat{c}^L_{zz} & \vdots \\
\hat{a}^L_{Z1} & \cdots & \hat{a}^L_{ZI} & -\hat{b}^L_{Z1} & \cdots & -\hat{b}^L_{ZR} & \hat{c}^L_{Z1} & \cdots & \hat{c}^L_{ZZ}
\end{bmatrix} \in \mathbb{R}^{(I+K+R+Z)\times(I+R+Z)} \tag{13}
$$

where $\hat{\alpha}_{ij}$ denotes the interactive abilities of the ith protein with the jth protein in the PPIN which could be obtained from $\hat{\theta}^P_i$ by solving parameter estimation problem in Equation (6) and pruning the false positives by AIC in Equation (11); $\hat{a}^G_{ki}$, $\hat{b}^G_{kr}$, and $\hat{c}^G_{kz}$ represent transcriptional regulative abilities from the ith TFs, the rth miRNAs and the zth lncRNAs onto the kth protein-coding genes, respectively, which could be obtained from $\hat{\theta}^G_k$ by solving parameter estimation problem in Equation (10) and pruning the false positives by AIC in (12); $\hat{a}^M_{ri}$, $\hat{b}^M_{rr}$, and $\hat{c}^M_{rz}$ indicate the transcriptional regulative abilities from the ith TFs, the rth miRNAs and the zth lncRNAs onto the rth miRNA's gene, respectively, which could be acquired from $\hat{\theta}^M_r$ by solving parameter estimation problem in Equation (S6) and pruning the false positives by AIC in Equation (S11); $\hat{a}^L_{zi}$, $\hat{b}^L_{zr}$, and $\hat{c}^L_{zz}$ indicate the transcriptional regulative abilities from the ith TFs, the rth miRNAs and the zth lncRNAs onto the zth lncRNA's gene, respectively, which could be acquired from $\hat{\theta}^L_z$ by solving parameter estimation problem in Equation (S10) and pruning the false positives by AIC in Equation (S12). It is noted that if interactions or regulations do not exist in the candidate GWGEN via big data mining or already have been pruned by AIC, the corresponding components in matrix H are padded with zero.

As the H have been constructed, we thereby extract the core GWGEN from the real GWGEN by the PNP method shown as below. At first, the combined network matrix H can be a factorization of the following singular value decomposition (SVD) form as below:

$$
H = U \times D \times V^T \tag{14}
$$

where $U \in \mathbb{R}^{(I+K+R+Z)\times(I+R+Z)}$, $V \in \mathbb{R}^{(I+R+Z)\times(I+R+Z)}$, and $D = \mathrm{diag}(d_1, \cdots, d_{I+R+Z})$. D is composed of $I + R + Z$ singular values of H and $d_1 \geq d_2 \geq \cdots \geq d_{I+R+Z}$. The eigen expression fraction E_h is defined in the following energy normalization:

$$
E_h = \frac{d_h^2}{\sum\limits_{h=1}^{I+R+Z} d_h^2} \tag{15}
$$

Then, we find out the minimum γ such that $\sum\limits_{h=1}^{\gamma} E_h \geq 0.85$. That is, top γ singular vectors of matrix H contain 85% core network structure of the real GWGEN from the energy point of view. Additionally, we define the projections of H to the top γ singular vectors of V as

$$
N_R(w,s) = h_{w,:} \times v^T_{:,s} \text{ for } w = 1, 2, \ldots, I^* + R^* + Z^* \text{ and } s = 1, 2, \ldots, \gamma \tag{16}
$$

where $h_{w:}$, and $v^T_{:,s}$ denote the wth row of H and the sth column of V, respectively. Subsequently, for the top γ right-singular vectors, we define the 2-norm projection value of proteins, genes, lncRNAs, and miRNAs (i.e., the nodes) in the real GWGEN as below:

$$D_R(w) = \left[\sum_{s=1}^{\gamma} [N_R(w,s)]^2 \right]^{1/2} \quad \text{for } w = 1, 2, \ldots, I^* + R^* + Z^* \text{ and } s = 1, 2, \ldots, \gamma \quad (17)$$

If the projection value, $D_R(w)$, approaches to zero for the wth node, it means that the wth node is almost independent to the principal network structure. That is, the larger the projection value is, the greater the contribution of the corresponding node to the core network is. By doing so, we can extract the core GWGEN by collecting nodes with large projection values from the real GWGENs and denote them in the KEGG pathway style to investigate the progression mechanisms of human skin aging.

4.7. Data Preprocessing for Training Deep Neural Network of Drug–Target Interaction in Advance

The drug–target interaction dataset comes from BindingDB [35]. The descriptors of drugs and targets are transformed by PyBioMed [112]. We install this package and import PyMolecule module and PyProtein module to transform drugs and targets into their descriptors under python 2.7 environment. The PyMolecule module in PyBioMed is responsible to compute the commonly used structural and physicochemical descriptors to be drug features. The drug features include constitutional and geometrical descriptors. Furthermore, the PyProtein module in PyBioMed is responsible for calculating the widely used descriptors, including structural and physicochemical properties of proteins and peptides from amino acid sequences, to be target features. Subseqently, concatenating the drug descriptor and the target descriptor, we describe properties of a drug and its target by a feature vector shown in (18). Moreover, the total number of drug features and target features are 363 and 996, respectively.

$$v_{drug-target} = [D, T] = [d_1, d_2, \ldots, d_I, t_1, t_2, \ldots, t_J] \quad (18)$$

where $v_{drug-target}$ indicates a feature vector of a drug-target pair; D denotes the feature vector of the drug; d_i indicates the ith drug feature; T represents the feature vector of the target; t_j is the jth target feature; I is the total number of drug features; J is the total number of target features. We conduct the same transformation for all the drug-target pairs to obtain their drug-target feature vectors.

Supplementary Materials: The following are available online, Figure S1: The real genome-wide genetic and epigenetic network (GWGEN) of young-stage skin, Figure S2: The real genome-wide genetic and epigenetic network (GWGEN) of middle-stage skin, Figure S3: The real genome-wide genetic and epigenetic network (GWGEN) of elder-stage skin, Figure S4: Deep neural network of drug-target interaction framework, Table S1: The pathway enrichment analysis of proteins through applying the DAVID in the core GWGEN of young-stage skin, Table S2: The pathway enrichment analysis of proteins through applying the DAVID in the core GWGEN of middle-stage skin, Table S3: The pathway enrichment analysis of proteins through applying the DAVID in the core GWGEN of elder-stage skin, Table S4: Drug targets with their corresponding small-molecule compounds, Table S5: Drug targets with their corresponding small-molecule compounds.

Author Contributions: Conceptualization, B.-S.C. and S.-J.Y.; methodology, S.-J.Y. and J.-F.L.; software, S.-J.Y. and J.-F.L.; validation, S.-J.Y. and J.-F.L.; formal analysis, S.-J.Y. and J.-F.L.; investigation, J.-F.L.; data curation J.-F.L.; writing—original draft preparation, S.-J.Y. and J.-F.L.; writing—review and editing, B.-S.C. and S.-J.Y.; visualization, J.-F.L.; supervision, B.-S.C.; funding acquisition, B.-S.C. All authors have read and agreed to the published version of the manuscript.

Funding: This research was funded by Ministry of Science and Technology grant number MOST 107-2221-E-007-112-MY3.

Institutional Review Board Statement: Not applicable.

Informed Consent Statement: Not applicable.

Data Availability Statement: The human skin data is from GSE18876 (https://www.ncbi.nlm.nih.gov/geo/query/acc.cgi?acc=GSE18876) (accessed on 19 May 2021). Drug sensitivity data is from depmap portal (https://depmap.org/portal/download/) (accessed on 19 May 2021).

Conflicts of Interest: The authors declare no conflict of interest.

References

1. Zhang, S.; Duan, E. Fighting against skin aging: The way from bench to bedside. *Cell Transplant.* **2018**, *27*, 729–738. [CrossRef]
2. Niccoli, T.; Partridge, L. Ageing as a risk factor for disease. *Curr. Biol.* **2012**, *22*, R741–R752. [CrossRef] [PubMed]
3. Blume-Peytavi, U.; Kottner, J.; Sterry, W.; Hodin, M.W.; Griffiths, T.W.; Watson, R.E.B.; Hay, R.J.; Griffiths, C.E.M. Age-associated skin conditions and diseases: Current perspectives and future options. *Gerontologist* **2016**, *56*, S230–S242. [CrossRef] [PubMed]
4. Batisse, D.; Bazin, R.; Baldeweck, T. Influence of age on the wrinkling capacities of skin. *Ski. Res. Technol.* **2002**, *8*, 148–154. [CrossRef]
5. Poljšak, B.D.R.G.; Godić, A. Intrinsic skin aging: The role of oxidative stress. *Acta Derm. Alp. Pannonica Adriat.* **2012**, *21*, 33–36.
6. Yi, R.; Poy, M.N.; Stoffel, M.; Fuchs, E. A skin microrna promotes differentiation by repressing 'stemness'. *Nature* **2008**, *452*, 225–229. [CrossRef] [PubMed]
7. Wei, T.; Orfanidis, K.; Xu, N.; Janson, P.; Ståhle, M.; Pivarcsi, A.; Sonkoly, E. The expression of microrna-203 during human skin morphogenesis. *Exp. Dermatol.* **2010**, *19*, 854–856. [CrossRef] [PubMed]
8. Hildebrand, J.; Rütze, M.; Walz, N.; Gallinat, S.; Wenck, H.; Deppert, W.; Grundhoff, A.; Knott, A. A comprehensive analysis of microrna expression during human keratinocyte differentiation in vitro and in vivo. *J. Investig. Dermatol.* **2011**, *131*, 20–29. [CrossRef] [PubMed]
9. Li, D.; Li, X.I.; Wang, A.; Meisgen, F.; Pivarcsi, A.; Sonkoly, E.; Ståhle, M.; Landén, N.X. Microrna-31 promotes skin wound healing by enhancing keratinocyte proliferation and migration. *J. Investig. Dermatol.* **2015**, *135*, 1676–1685. [CrossRef]
10. Fahs, F.; Bi, X.; Yu, F.-S.; Zhou, L.; Mi, Q.-S. New insights into micrornas in skin wound healing. *Iubmb Life* **2015**, *67*, 889–896. [CrossRef] [PubMed]
11. Smith-Vikos, T.; Slack, F.J. Micrornas and their roles in aging. *J. Cell Sci.* **2012**, *125*, 7–17. [CrossRef]
12. Kour, S.; Rath, P.C. Long noncoding rnas in aging and age-related diseases. *Ageing Res. Rev.* **2016**, *26*, 1–21. [CrossRef]
13. Yang, X.; Gao, L.; Guo, X.; Shi, X.; Wu, H.; Song, F.; Wang, B.J.P.O. A network based method for analysis of lncrna-disease associations and prediction of lncrnas implicated in diseases. *PLoS ONE* **2014**, *9*, e87797. [CrossRef]
14. Sousa-Franco, A.; Rebelo, K.; da Rocha, S.T.; Bernardes de Jesus, B. Lncrnas regulating stemness in aging. *Aging Cell* **2019**, *18*, e12870. [CrossRef] [PubMed]
15. Fraga, M.F. Genetic and epigenetic regulation of aging. *Curr. Opin. Immunol.* **2009**, *21*, 446–453. [CrossRef] [PubMed]
16. Esteller, M. Epigenetics in cancer. *N. Engl. J. Med.* **2008**, *358*, 1148–1159. [CrossRef]
17. Thorvaldsen, J.L.A.D.; Kristen, L.; Bartolomei, M.S. Deletion of the h19 differentially methylated domain results in loss of imprinted expression of h19 and igf2. *Genes Dev.* **1998**, *12*, 3693–3702. [CrossRef] [PubMed]
18. Fraga, M.F.; Esteller, M. Epigenetics and aging: The targets and the marks. *Trends Genet.* **2007**, *23*, 413–418. [CrossRef] [PubMed]
19. Partridge, L. Gerontology: Extending the healthspan. *Nature* **2016**, *529*, 154. [CrossRef]
20. Newman, J.C.; Milman, S.; Hashmi, S.K.; Austad, S.N.; Kirkland, J.L.; Halter, J.B.; Barzilai, N. Strategies and challenges in clinical trials targeting human aging. *J. Gerontol. Ser. A Biol. Sci. Med. Sci.* **2016**, *71*, 1424–1434. [CrossRef]
21. Jin, G.; Wong, S.T.C. Toward better drug repositioning: Prioritizing and integrating existing methods into efficient pipelines. *Drug Discov. Today* **2014**, *19*, 637–644. [CrossRef]
22. Keiser, M.J.; Setola, V.; Irwin, J.J.; Laggner, C.; Abbas, A.I.; Hufeisen, S.J.; Jensen, N.H.; Kuijer, M.B.; Matos, R.C.; Tran, T.B.; et al. Predicting new molecular targets for known drugs. *Nature* **2009**, *462*, 175–181. [CrossRef] [PubMed]
23. Langedijk, J.; Mantel-Teeuwisse, A.K.; Slijkerman, D.S.; Schutjens, M.H. Drug repositioning and repurposing: Terminology and definitions in literature. *Drug Discov. Today* **2015**, *20*, 1027–1034. [CrossRef]
24. Peska, L.; Buza, K.; Koller, J. Drug-target interaction prediction: A bayesian ranking approach. *Comput. Methods Programs Biomed.* **2017**, *152*, 15–21. [CrossRef] [PubMed]
25. Bagherian, M.; Sabeti, E.; Wang, K.; Sartor, M.A.; Nikolovska-Coleska, Z.; Najarian, K. Machine learning approaches and databases for prediction of drug–target interaction: A survey paper. *Brief. Bioinform.* **2020**, *22*, 247–269. [CrossRef] [PubMed]
26. Yamanishi, Y.; Araki, M.; Gutteridge, A.; Honda, W.; Kanehisa, M. Prediction of drug-target interaction networks from the integration of chemical and genomic spaces. *Bioinformatics* **2008**, *24*, i232–i240. [CrossRef] [PubMed]
27. Ezzat, A.; Wu, M.; Li, X.-L.; Kwoh, C.-K. Drug-target interaction prediction via class imbalance-aware ensemble learning. *BMC Bioinform.* **2016**, *17*, 509. [CrossRef] [PubMed]
28. Chen, R.; Liu, X.; Jin, S.; Lin, J.; Liu, J.J.M. Machine learning for drug-target interaction prediction. *Molecules* **2018**, *23*, 2208. [CrossRef]
29. Gao, K.Y.; Fokoue, A.; Luo, H.; Iyengar, A.; Dey, S.; Zhang, P. *Interpretable Drug Target Prediction Using Deep Neural Representation*; IJCAI: Stockholm, Sweden, 2018; pp. 3371–3377.

30. Lee, I.; Keum, J.; Nam, H.J.P.C.B. Deepconv-dti: Prediction of drug-target interactions via deep learning with convolution on protein sequences. *PLoS ONE* **2019**, *15*, e1007129. [CrossRef]
31. You, J.; McLeod, R.D.; Hu, P. Predicting drug-target interaction network using deep learning model. *Comput. Biol. Chem.* **2019**, *80*, 90–101. [CrossRef]
32. Tian, K.; Shao, M.; Wang, Y.; Guan, J.; Zhou, S. Boosting compound-protein interaction prediction by deep learning. *Methods* **2016**, *110*, 64–72. [CrossRef]
33. Tacutu, R.; Thornton, D.; Johnson, E.; Budovsky, A.; Barardo, D.; Craig, T.; Diana, E.; Lehmann, G.; Toren, D.; Wang, J.; et al. Human ageing genomic resources: New and updated databases. *Nucleic Acids Res.* **2018**, *46*, D1083–D1090. [CrossRef]
34. Lamb, J.; Crawford, E.D.; Peck, D.; Modell, J.W.; Blat, I.C.; Wrobel, M.J.; Lerner, J.; Brunet, J.-P.; Subramanian, A.; Ross, K.N.J.S. The connectivity map: Using gene-expression signatures to connect small molecules, genes, and disease. *Science* **2006**, *313*, 1929–1935. [CrossRef]
35. Gilson, M.K.; Liu, T.; Baitaluk, M.; Nicola, G.; Hwang, L.; Chong, J. Bindingdb in 2015: A public database for medicinal chemistry, computational chemistry and systems pharmacology. *Nucleic Acids Res.* **2015**, *44*, D1045–D1053. [CrossRef]
36. Ghosh, S.; Liu, B.; Wang, Y.; Hao, Q.; Zhou, Z. Lamin a is an endogenous sirt6 activator and promotes sirt6-mediated DNA repair. *Cell Rep.* **2015**, *13*, 1396–1406. [CrossRef]
37. Kaidi, A.; Weinert, B.T.; Choudhary, C.; Jackson, S.P. Human sirt6 promotes DNA end resection through ctip deacetylation. *Science* **2010**, *329*, 1348–1353. [CrossRef] [PubMed]
38. Mao, Z.; Hine, C.; Tian, X.; Van Meter, M.; Au, M.; Vaidya, A.; Seluanov, A.; Gorbunova, V. Sirt6 promotes DNA repair under stress by activating parp1. *Science* **2011**, *332*, 1443–1446. [CrossRef]
39. Mangerich, A.; Rkle, A. Pleiotropic cellular functions of parp1 in longevity and aging: Genome maintenance meets inflammation. *Oxidative Med. Cell. Longev.* **2012**, *2012*, 19. [CrossRef]
40. Brem, R.; Hall, J. Xrcc1 is required for DNA single-strand break repair in human cells. *Nucleic Acids Res.* **2005**, *33*, 2512–2520. [CrossRef]
41. Zannini, L.; Delia, D.; Buscemi, G. Chk2 kinase in the DNA damage response and beyond. *J. Mol. Cell Biol.* **2014**, *6*, 442–457. [CrossRef]
42. Umegaki-Arao, N.; Tamai, K.; Nimura, K.; Serada, S.; Naka, T.; Nakano, H.; Katayama, I. Karyopherin alpha2 is essential for rrna transcription and protein synthesis in proliferative keratinocytes. *PLoS ONE* **2013**, *8*, e76416. [CrossRef]
43. Redza-Dutordoir, M.; Averill-Bates, D.A. Activation of apoptosis signalling pathways by reactive oxygen species. *Biochim. Et Biophys. Acta Mol. Cell Res.* **2016**, *1863*, 2977–2992. [CrossRef]
44. Soga, M.; Matsuzawa, A.; Ichijo, H. Oxidative stress-induced diseases via the ask1 signaling pathway. *Int. J. Cell Biol.* **2012**, *2012*. [CrossRef]
45. Ferraro, E.; Pesaresi, M.G.; De Zio, D.; Cencioni, M.T.; Gortat, A.; Cozzolino, M.; Berghella, L.; Salvatore, A.M.; Oettinghaus, B.; Scorrano, L.; et al. Apaf1 plays a pro-survival role by regulating centrosome morphology and function. *J. Cell Sci.* **2011**, *124*, 3450. [CrossRef]
46. Lagouge, M.; Larsson, N.G. The role of mitochondrial DNA mutations and free radicals in disease and aging. *J. Intern. Med.* **2013**, *273*.
47. Tan, S.; Ding, K.; Li, R.; Zhang, W.; Li, G.; Kong, X.; Qian, P.; Lobie, P.E.; Zhu, T. Identification of mir-26 as a key mediator of estrogen stimulated cell proliferation by targeting chd1, greb1 and kpna2. *Breast Cancer Res.* **2014**, *16*, R40. [CrossRef]
48. Anders, L.; Ke, N.; Hydbring, P.; Choi, Y.J.; Widlund, H.R.; Chick, J.M.; Zhai, H.; Vidal, M.; Gygi, S.P.; Braun, P.; et al. A systematic screen for cdk4/6 substrates links foxm1 phosphorylation to senescence suppression in cancer cells. *Cancer Cell* **2011**, *20*, 620–634. [CrossRef]
49. Kwok, C.T.D.; Leung, M.H.; Qin, J.; Qin, Y.; Wang, J.; Lee, Y.L.; Yao, K.M. The forkhead box transcription factor foxm1 is required for the maintenance of cell proliferation and protection against oxidative stress in human embryonic stem cells. *Stem Cell Res.* **2016**, *16*, 651–661. [CrossRef]
50. Balasuriya, N.; McKenna, M.; Liu, X.; Li, S.S.C.; O'Donoghue, P. Phosphorylation-dependent inhibition of akt1. *Genes* **2018**, *9*, 450. [CrossRef]
51. Noh, E.-M.; Park, J.; Song, H.-R.; Kim, J.-M.; Lee, M.; Song, H.-K.; Hong, O.-Y.; Whang, P.H.; Han, M.-K.; Kwon, K.-B.; et al. Skin aging-dependent activation of the pi3k signaling pathway via downregulation of pten increases intracellular ros in human dermal fibroblasts. *Oxidative Med. Cell. Longev.* **2016**, *2016*, 6354261. [CrossRef]
52. Webb, A.E.; Brunet, A. Foxo transcription factors: Key regulators of cellular quality control. *Trends Biochem. Sci.* **2014**, *39*, 159–169. [CrossRef]
53. Hagenbuchner, J.; Kuznetsov, A.; Hermann, M.; Hausott, B.; Obexer, P.; Ausserlechner, M.J. Foxo3-induced reactive oxygen species are regulated by bcl2l11 (bim) and sesn3. *J. Cell Sci.* **2012**, *125*, 1191. [CrossRef] [PubMed]
54. Das, T.P.; Suman, S.; Alatassi, H.; Ankem, M.K.; Damodaran, C. Inhibition of akt promotes foxo3a-dependent apoptosis in prostate cancer. *Cell Death Dis.* **2016**, *7*, e2111. [CrossRef] [PubMed]
55. Moskalev, A.A.; Smit-McBride, Z.; Shaposhnikov, M.V.; Plyusnina, E.N.; Zhavoronkov, A.; Budovsky, A.; Tacutu, R.; Fraifeld, V.E. Gadd45 proteins: Relevance to aging, longevity and age-related pathologies. *Ageing Res. Rev.* **2012**, *11*, 51–66. [CrossRef] [PubMed]

56. Borg, M.; Brincat, S.; Camilleri, G.; Schembri-Wismayer, P.; Brincat, M.; Calleja-Agius, J. The role of cytokines in skin aging. *Climacteric* **2013**, *16*, 514–521. [CrossRef] [PubMed]

57. Hald, A.; Andrés, R.M.; Salskov-Iversen, M.L.; Kjellerup, R.B.; Iversen, L.; Johansen, C. Stat1 expression and activation is increased in lesional psoriatic skin. *Br. J. Dermatol.* **2013**, *168*, 302–310. [CrossRef] [PubMed]

58. Pittayapruek, P.; Meephansan, J.; Prapapan, O.; Komine, M.; Ohtsuki, M. Role of matrix metalloproteinases in photoaging and photocarcinogenesis. *Int. J. Mol. Sci.* **2016**, *17*, 868. [CrossRef] [PubMed]

59. Adly, M.A.; Assaf, H.; Hussein, M.R. Neurotrophins and skin aging. In *Textbook of Aging Skin*; Farage, M.A., Miller, K.W., Maibach, H.I., Eds.; Springer: Berlin/Heidelberg, Germany, 2017; pp. 515–527.

60. Zheng, M.; McKeown-Longo, P.J. Regulation of hef1 expression and phosphorylation by tgf-β1 and cell adhesion. *J. Biol. Chem.* **2002**, *277*, 39599–39608. [CrossRef]

61. Wang, H.; Chen, H.; Ma, M.; Wang, J.; Tang, T.; Ni, L.; Yu, J.; Li, Y.; Bai, B. Mir-573 regulates melanoma progression by targeting the melanoma cell adhesion molecule. *Oncol. Rep.* **2013**, *30*, 520–526. [CrossRef]

62. Ferby, I.; Reschke, M.; Kudlacek, O.; Knyazev, P.; Pantè, G.; Amann, K.; Sommergruber, W.; Kraut, N.; Ullrich, A.; Fässler, R.; et al. Mig6 is a negative regulator of egf receptor–mediated skin morphogenesis and tumor formation. *Nat. Med.* **2006**, *12*, 568–573. [CrossRef] [PubMed]

63. Raddatz, G.; Hagemann, S.; Aran, D.; Söhle, J.; Kulkarni, P.P.; Kaderali, L.; Hellman, A.; Winnefeld, M.; Lyko, F. Aging is associated with highly defined epigenetic changes in the human epidermis. *Epigenetics Chromatin* **2013**, *6*, 36. [CrossRef] [PubMed]

64. Chen, G.; Cheng, Y.; Tang, Y.; Martinka, M.; Li, G. Role of tip60 in human melanoma cell migration, metastasis, and patient survival. *J. Investig. Dermatol.* **2012**, *132*, 2632–2641. [CrossRef]

65. Deng, X.; Gao, F.; May, W.S. Bcl2 retards g1/s cell cycle transition by regulating intracellular ros. *Blood* **2003**, *102*, 3179–3185. [CrossRef] [PubMed]

66. Ryu, S.J.; Oh, Y.S.; Park, S.C. Failure of stress-induced downregulation of bcl-2 contributes to apoptosis resistance in senescent human diploid fibroblasts. *Cell Death Differ.* **2007**, *14*, 1020. [CrossRef]

67. D'Mello, S.A.N.; Finlay, G.J.; Baguley, B.C.; Askarian-Amiri, M.E. Signaling pathways in melanogenesis. *Int. J. Mol. Sci.* **2016**, *17*, 1144. [CrossRef] [PubMed]

68. Bondurand, N.; Pingault, V.; Goerich, D.E.; Lemort, N.; Sock, E.; Caignec, C.L.; Wegner, M.; Goossens, M. Interaction among sox10, pax3 and mitf, three genes altered in waardenburg syndrome. *Hum. Mol. Genet.* **2000**, *9*, 1907–1917. [CrossRef]

69. Zhou, Y.; Ching, Y.-P.; Chun, A.C.S.; Jin, D.-Y. Nuclear localization of the cell cycle regulator cdh1 and its regulation by phosphorylation. *J. Biol. Chem.* **2003**, *278*, 12530–12536. [CrossRef]

70. Harkness, T.A.A. Activating the anaphase promoting complex to enhance genomic stability and prolong lifespan. *Int. J. Mol. Sci.* **2018**, *19*, 1888. [CrossRef]

71. Quan, T.; Qin, Z.; Shao, Y.; Xu, Y.; Voorhees, J.J.; Fisher, G.J. Retinoids suppress cysteine-rich protein 61 (ccn1), a negative regulator of collagen homeostasis, in skin equivalent cultures and aged human skin in vivo. *Exp. Dermatol.* **2011**, *20*, 572–576. [CrossRef]

72. Quan, T.; Fisher, G.J. Role of age-associated alterations of the dermal extracellular matrix microenvironment in human skin aging: A mini-review. *Gerontology* **2015**, *61*, 427–434. [CrossRef]

73. Liu, N.; Matsumura, H.; Kato, T.; Ichinose, S.; Takada, A.; Namiki, T.; Asakawa, K.; Morinaga, H.; Mohri, Y.; De Arcangelis, A.; et al. Stem cell competition orchestrates skin homeostasis and ageing. *Nature* **2019**, *568*, 344–350. [CrossRef]

74. Juric, V.; Chen, C.-C.; Lau, L.F. Tnfα-induced apoptosis enabled by ccn1/cyr61: Pathways of reactive oxygen species generation and cytochrome c release. *PLoS ONE* **2012**, *7*, e31303. [CrossRef]

75. Marcotte, R.; Lacelle, C.; Wang, E. Senescent fibroblasts resist apoptosis by downregulating caspase-3. *Mech. Ageing Dev.* **2004**, *125*, 777–783. [CrossRef] [PubMed]

76. Maggio, M.; Guralnik, J.M.; Longo, D.L.; Ferrucci, L. Interleukin-6 in aging and chronic disease: A magnificent pathway. *J. Gerontol. Ser. A Biol. Sci. Med Sci.* **2006**, *61*, 575–584. [CrossRef] [PubMed]

77. Elbediwy, A.; Vincent-Mistiaen, Z.I.; Spencer-Dene, B.; Stone, R.K.; Boeing, S.; Wculek, S.K.; Cordero, J.; Tan, E.H.; Ridgway, R.; Brunton, V.G.; et al. Integrin signalling regulates yap and taz to control skin homeostasis. *Development* **2016**, *143*, 1674–1687. [PubMed]

78. Mu, R.; Wang, Y.B.; Wu, M.; Yang, Y.; Song, W.; Li, T.; Zhang, W.N.; Tan, B.; Li, A.L.; Wang, N.; et al. Depletion of pre-mrna splicing factor cdc5l inhibits mitotic progression and triggers mitotic catastrophe. *Cell Death Dis.* **2014**, *5*, e1151. [CrossRef] [PubMed]

79. Webster, L.T., Jr.; Butterworth, A.E.; Mahmoud, A.A.; Mngola, E.N.; Warren, K.S. Suppression of delayed hypersensitivity in schistosome-infected patients by niridazole. *N. Engl. J. Med.* **1975**, *292*, 1144–1147. [CrossRef]

80. Clyde, P.W.; Harari, A.E.; Getka, E.J.; Shakir, K.M. Combined levothyroxine plus liothyronine compared with levothyroxine alone in primary hypothyroidism: A randomized controlled trial. *JAMA* **2003**, *290*, 2952–2958. [CrossRef]

81. Escobar-Morreale, H.F.; Botella-Carretero, J.I.; Escobar del Rey, F.; Morreale de Escobar, G. Review: Treatment of hypothyroidism with combinations of levothyroxine plus liothyronine. *J. Clin. Endocrinol. Metab.* **2005**, *90*, 4946–4954. [CrossRef]

82. Aguilar, C.; Gardiner, D.M. DNA methylation dynamics regulate the formation of a regenerative wound epithelium during axolotl limb regeneration. *PLoS ONE* **2015**, *10*, e0134791. [CrossRef]

83. Friedel, H.A.; Brogden, R.N. Pinacidil. *Drugs* **1990**, *39*, 929–967. [CrossRef]

84. Cao, C.; Healey, S.; Amaral, A.; Lee-Couture, A.; Wan, S.; Kouttab, N.; Chu, W.; Wan, Y. Atp-sensitive potassium channel: A novel target for protection against uv-induced human skin cell damage. *J. Cell. Physiol.* **2007**, *212*, 252–263. [CrossRef]
85. Staiger, C. Comfrey: A clinical overview. *Phytother. Res.* **2012**, *26*, 1441–1448. [CrossRef]
86. Mazzulla, S.; Anile, D.; Sio, S.D.; Scaglione, A.; Seta, M.D.; Anile, A. In vivo evaluations of emulsion o/w for a new topical anti-aging formulation: Short-term and long-term efficacy. *J. Cosmet. Dermatol. Sci. Appl.* **2018**, *8*, 16.
87. Saito, M.L.; Oliveira, F. Confrei: Virtudes e problemas. *Rev. Bras. Farmacogn.* **1986**, *1*, 74–85. [CrossRef]
88. Kim, M.; Gu, M.J.; Lee, J.-G.; Chin, J.; Bae, J.-S.; Hahn, D. Quantitative analysis of bioactive phenanthrenes in dioscorea batatas decne peel, a discarded biomass from postharvest processing. *Antioxidants* **2019**, *8*, 541. [CrossRef] [PubMed]
89. Nelson, C.G. Diclofenac gel in the treatment of actinic keratoses. *Ther. Clin. Risk Manag.* **2011**, *7*, 207–211. [CrossRef] [PubMed]
90. Hägermark, O.; Strandberg, K.; Grönneberg, R. Effects of histamine receptor antagonists on histamine-induced responses in human skin. *Acta Derm. Venereol.* **1979**, *59*, 297–300.
91. Gschwandtner, M.; Purwar, R.; Wittmann, M.; Bäumer, W.; Kietzmann, M.; Werfel, T.; Gutzmer, R. Histamine upregulates keratinocyte mmp-9 production via the histamine h1 receptor. *J. Investig. Dermatol.* **2008**, *128*, 2783–2791. [CrossRef]
92. Galiniak, S.; Aebisher, D.; Bartusik-Aebisher, D. Health benefits of resveratrol administration. *Acta Biochim. Pol.* **2019**, *66*, 13–21. [CrossRef]
93. Hasan, M.; Bae, H. An overview of stress-induced resveratrol synthesis in grapes: Perspectives for resveratrol-enriched grape products. *Molecules* **2017**, *22*, 294. [CrossRef]
94. Patel, A.A.; Swerlick, R.A.; McCall, C.O. Azathioprine in dermatology: The past, the present, and the future. *J. Am. Acad. Dermatol.* **2006**, *55*, 369–389. [CrossRef] [PubMed]
95. Xenarios, I.; Rice, D.W.; Salwinski, L.; Baron, M.K.; Marcotte, E.M.; Eisenberg, D. Dip: The database of interacting proteins. *Nucleic Acids Res.* **2000**, *28*, 289–291. [CrossRef] [PubMed]
96. Kerrien, S.; Alam-Faruque, Y.; Aranda, B.; Bancarz, I.; Bridge, A.; Derow, C.; Dimmer, E.; Feuermann, M.; Friedrichsen, A.; Huntley, R.; et al. Intact–open source resource for molecular interaction data. *Nucleic Acids Res.* **2007**, *35*, D561–D565. [CrossRef] [PubMed]
97. Oughtred, R.; Chatr-aryamontri, A.; Breitkreutz, B.-J.; Chang, C.S.; Rust, J.M.; Theesfeld, C.L.; Heinicke, S.; Breitkreutz, A.; Chen, D.; Hirschman, J.; et al. Biogrid: A resource for studying biological interactions in yeast. *Cold Spring Harb. Protoc.* **2016**, *2016*, 080754. [CrossRef] [PubMed]
98. Bader, G.D.; Betel, D.; Hogue, C.W.V. Bind: The biomolecular interaction network database. *Nucleic Acids Res.* **2003**, *31*, 248–250. [CrossRef]
99. Chatr-aryamontri, A.; Ceol, A.; Palazzi, L.M.; Nardelli, G.; Schneider, M.V.; Castagnoli, L.; Cesareni, G. Mint: The molecular interaction database. *Nucleic Acids Res.* **2007**, *35*, D572–D574. [CrossRef]
100. Bovolenta, L.A.; Acencio, M.L.; Lemke, N. Htridb: An open-access database for experimentally verified human transcriptional regulation interactions. *BMC Genom.* **2012**, *13*, 405. [CrossRef]
101. Zheng, G.; Tu, K.; Yang, Q.; Xiong, Y.; Wei, C.; Xie, L.; Zhu, Y.; Li, Y. Itfp: An integrated platform of mammalian transcription factors. *Bioinformatics* **2008**, *24*, 2416–2417. [CrossRef]
102. Wingender, E.; Dietze, P.; Karas, H.; Knüppel, R. Transfac: A database on transcription factors and their DNA binding sites. *Nucleic Acids Res.* **1996**, *24*, 238–241. [CrossRef]
103. Friard, O.; Re, A.; Taverna, D.; De Bortoli, M.; Corá, D. Circuitsdb: A database of mixed microrna/transcription factor feed-forward regulatory circuits in human and mouse. *BMC Bioinform.* **2010**, *11*, 435. [CrossRef] [PubMed]
104. Agarwal, V.; Bell, G.W.; Nam, J.-W.; Bartel, D.P.J.E. Predicting effective microrna target sites in mammalian mrnas. *Elife* **2015**, *4*, e05005. [CrossRef] [PubMed]
105. Corsello, S.M.; Nagari, R.T.; Spangler, R.D.; Rossen, J.; Kocak, M.; Bryan, J.G.; Humeidi, R.; Peck, D.; Wu, X.; Tang, A.A.; et al. Non-oncology drugs are a source of previously unappreciated anti-cancer activity. *BioRxiv* **2019**, 730119.
106. Makrantonaki, E.; Brink, T.C.; Zampeli, V.; Elewa, R.M.; Mlody, B.; Hossini, A.M.; Hermes, B.; Krause, U.; Knolle, J.; Abdallah, M.; et al. Identification of biomarkers of human skin ageing in both genders. Wnt signalling—A label of skin ageing? *PLoS ONE* **2012**, *7*, e50393. [CrossRef] [PubMed]
107. Babyak, M.A.J.P.M. What you see may not be what you get: A brief, nontechnical introduction to overfitting in regression-type models. *Psychosom. Med.* **2004**, *66*, 411–421. [PubMed]
108. Lin, L.-H.; Lee, H.-C.; Li, W.-H.; Chen, B.-S. Dynamic modeling of cis-regulatory circuits and gene expression prediction via cross-gene identification. *BMC Bioinform.* **2005**, *6*, 258.
109. Chen, H.-C.; Lee, H.-C.; Lin, T.-Y.; Li, W.-H.; Chen, B.-S. Quantitative characterization of the transcriptional regulatory network in the yeast cell cycle. *Bioinformatics* **2004**, *20*, 1914–1927. [CrossRef]
110. Chen, B.-S.; Yang, S.-K.; Lan, C.-Y.; Chuang, Y.-J. A systems biology approach to construct the gene regulatory network of systemic inflammation via microarray and databases mining. *BMC Med. Genom.* **2008**, *1*, 46. [CrossRef]
111. Sakamoto, Y.; Ishiguro, M.; Kitagawa, G.J.D. The Netherlands: D. Reidel. Akaike information criterion statistics. *Dordr. Neth. D Reidel* **1986**, *81*, 26853.
112. Dong, J.; Yao, Z.-J.; Zhang, L.; Luo, F.; Lin, Q.; Lu, A.-P.; Chen, A.F.; Cao, D.-S. Pybiomed: A python library for various molecular representations of chemicals, proteins and dnas and their interactions. *J. Cheminform.* **2018**, *10*, 16. [CrossRef]

Sample Availability: Samples of the compounds are not available from the authors.

 *molecules*

Article

Identification of Potential COX-2 Inhibitors for the Treatment of Inflammatory Diseases Using Molecular Modeling Approaches

Pedro H. F. Araújo [1,2], Ryan S. Ramos [2], Jorddy N. da Cruz [2], Sebastião G. Silva [3], Elenilze F. B. Ferreira [1,2,4], Lúcio R. de Lima [2], Williams J. C. Macêdo [1,2,5], José M. Espejo-Román [6], Joaquín M. Campos [6] and Cleydson B. R. Santos [1,2,5,*]

[1] Graduate Program in Innovation Pharmaceutical, Federal University of Amapá, 68903-419 Amapá-AP, Brazil; pedro.henrique.fauro@gmail.com (P.H.F.A.); elenilze@yahoo.com.br (E.F.B.F.); williamsmacedo@yahoo.com.br (W.J.C.M.)

[2] Laboratory of Modeling and Computational Chemistry, Department of Biological and Health Sciences, Federal University of Amapá, 68902-280 Macapá-AP, Brazil; ryanquimico@hotmail.com (R.S.R.); jorddynevescruz@gmail.com (J.N.d.C.); luciorolima@gmail.com (L.R.d.L.)

[3] Campus Abaetetuba, Universidade Federal do Para, Ramal Manoel de Abreu, s/n-Mutirão, Abaetetuba, 68440-000 Pará, Brazil; profsebastiaogs@gmail.com

[4] Laboratory of Organic Chemistry and Biochemistry, University of State of Amapá, 68900-070 Macapá-AP, Brazil

[5] Laboratory of Molecular Modeling and Simulation System, Federal Rural University of Amazônia, Rua João Pessoa, 121, Capanema, 68700-030 Pará-PA, Brazil

[6] Department of Pharmaceutical Organic Chemistry, Faculty of Pharmacy, Biosanitary Institute of Granada (Ibs.GRANADA), Campus of Cartuja s/n, University of Granada, 18071 Granada, Spain; josemanuelespejo@correo.ugr.es (J.M.E.-R.); jmcampos@ugr.es (J.M.C.)

* Correspondence: breno@unifap.br

Academic Editor: Marco Tutone
Received: 25 August 2020; Accepted: 9 September 2020; Published: 12 September 2020

Abstract: Non-steroidal anti-inflammatory drugs are inhibitors of cyclooxygenase-2 (COX-2) that were developed in order to avoid the side effects of non-selective inhibitors of COX-1. Thus, the present study aims to identify new selective chemical entities for the COX-2 enzyme via molecular modeling approaches. The best pharmacophore model was used to identify compounds within the ZINC database. The molecular properties were determined and selected with Pearson's correlation for the construction of quantitative structure–activity relationship (QSAR) models to predict the biological activities of the compounds obtained with virtual screening. The pharmacokinetic/toxicological profiles of the compounds were determined, as well as the binding modes through molecular docking compared to commercial compounds (rofecoxib and celecoxib). The QSAR analysis showed a fit with R = 0.9617, R^2 = 0.9250, standard error of estimate (SEE) = 0.2238, and F = 46.2739, with the tetra-parametric regression model. After the analysis, only three promising inhibitors were selected, **Z-964**, **Z-627**, and **Z-814**, with their predicted pIC_{50} ($-\log IC_{50}$) values, **Z-814** = 7.9484, **Z-627** = 9.3458, and **Z-964** = 9.5272. All candidates inhibitors complied with Lipinski's rule of five, which predicts a good oral availability and can be used in in vitro and in vivo tests in the zebrafish model in order to confirm the obtained in silico data.

Keywords: *in silico*; COX-2 inhibitors; molecular modeling

1. Introduction

Inflammatory processes stand out for their pathophysiological aspect, as they are caused by pathogenic microorganisms, such as viruses, bacteria, fungi, and parasites that invade host cells to

reproduce, resulting in a complex and heterogeneous group of diseases that cause morbidity and mortality to those affected by them. At the same time, the need for specific attention to protect the integrity of human organisms from harmful or exogenous agents is emphasized [1].

Inflammatory mediation is regulated by the action of neutrophils, mast cells, eosinophils, macrophages, dendritic cells, and epithelial cells. It is a complex process involving vasodilation, chemotaxis, and increased permeability. Sensors called Toll-like receptors (TLR) recognize the products of pathogens, such as endotoxins and bacterial DNA that are located in the plasma membrane and in the endosomes, and they are capable of detecting intra and extracellular microorganisms [2].

When an injury occurs, platelets release complement proteins, where mast cells degranulate releasing histamine (vasodilation) and serotonin (cell permeability and diapedesis), where neutrophils are activated and migrate to the site of action induced by chemokines. Neutrophils phagocytose pathogenic organisms releasing mediators and attracting macrophages, increasing the release of pro-inflammatory mediators (prostaglandins and leukotrienes) and cytokines (interleukin 1 (IL-1), interleukin 6 (IL-6) and tumor necrosis factor (TNFα)). When cells are activated, the arachidonic acid (AA) of the cell membrane is converted by enzymes for the synthesis of prostaglandins and leukotrienes.

In one of the stages of the inflammatory process, prostaglandins, also called eicosanoids, are synthesized by triggering various stimuli that activate cell membrane receptors, when coupled with a regulatory protein that results in the activation of phospholipase A2 or with an increase in the concentration of Ca^{+2}. This type of enzyme hydrolyzes membrane phospholipids, consequently releasing the cascade of arachidonic acid, which is a substrate for the synthesis of physiopatological and inductive prostaglandins; see Figure 1 below [3].

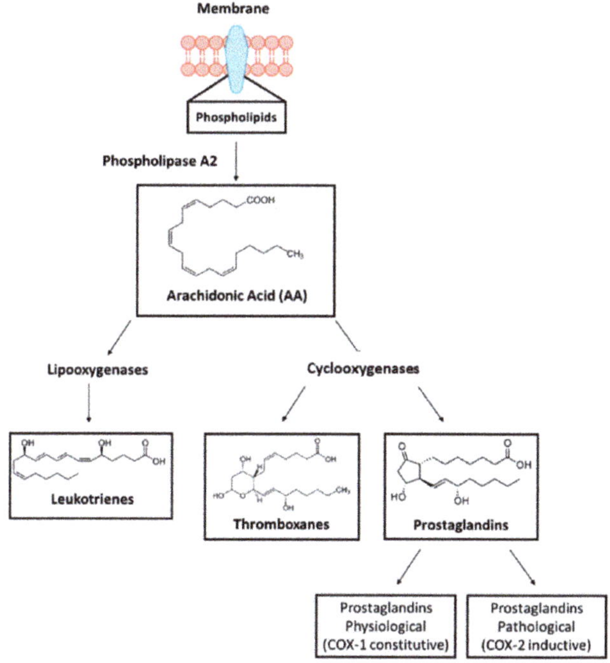

Figure 1. Cascade of arachidonic acid.

Prostaglandins-endoperoxide synthase (PTGS) are known as cyclooxygenases responsible for the synthesis of prostaglandins that are percussive biologically active molecules. Figure 1 shows that the conversion of AA into signaling molecules takes place in 2 moments. (i) The enzyme cyclooxygenase-1

(COX-1, PTGS1) catalyzes the addition of two free oxygen atoms to form the 1,2-dioxane bridge and a functional group to form prostaglandin G2 (endoperoxide (PGG2)). (ii) COX-2 (PTGS2) reduces the peroxide functional group to a secondary alcohol, forming prostaglandin H2 (PGH2). PGG2 and PGH2 are unstable, but they are precursors for the formation of other prostaglandins (PGE2, PGD2, PGF2), prostacyclin (PGl2), and thromboxane A2 (TXA2), which are commonly called eicosanoids.

Currently, it is known that two genes express two distinct, but similar, isoforms of COX. These enzymes catalyze the biosynthesis of prostaglandins and thromboxanes by reducing arachidonic acid; they are called COX-1 and COX-2, with similar general protein structures that essentially catalyze the same reaction. COX-1 is an enzyme considered to be constitutive; it is part of the homeostatic maintenance of various processes of the human organism, and it is present in most tissues, including stomach, kidney, and coronary arteries [4].

On the other hand, the COX-2 enzyme is considered inductive; that is, it expresses the inflammatory process in cells. The understanding of the COX inhibition process has allowed and still allows the development of new perspectives in relation to the therapeutic targets and the synthetic drugs produced, so that they are as selective and smooth as possible, corroborating their adverse effects [4,5]. Widely prescribed non-steroidal anti-inflammatory drugs (NSAIDs) are a class of drugs used to treat pain, fever, and inflammation. Since 1893, when acetylsalicylic acid was produced, NSAIDs have become the most accepted and prescribed drugs. However, it was not until 1971 that John Vane clarified the mechanism of action of these drugs: They inhibit cyclooxygenase (COX), preventing consequently the synthesis of prostaglandins [6].

Non-selective or traditional NSAIDs inhibit both isoforms (1 and 2), but COX-1 inhibition is the main cause of increased gastrointestinal bleeding or ulcer formation, abdominal pain, and dyspepsia, including indomethacin, naproxen, and ibuprofen. On the other hand, COX-2 inhibition, inducing pro-inflammatory processes, plays a role in pain relief with reduced gastrointestinal effects, which is usually expressed by the coxib class, among them rofecoxib, celecoxib, valdecoxib, and lumiracoxib [7].

The knowledge of the structures of COX-1 and COX-2 and their active sites constitute the fundamental basis for the development of more specific inhibitors for COX-2 and for the elaboration of studies of the structure–activity relationship of these products. During enzymatic activity, arachidonic acid binds to an Arg^{120} and to a Ser^{530}. An electron transfers Tyr^{385} to an oxidized heme, which is also bound within the enzyme, initiating the reaction cyclooxygenase. Several studies have attempted to elucidate how and where non-steroidal anti-inflammatory drugs act on cyclooxygenase to block prostaglandin synthesis. Within the hydrophobic channel of COX, an amino acid difference at position 523 (isoleucine in COX-1 and valine in COX-2) can be of critical importance in the selectivity of several drugs [8].

One of the main drugs used, with a selective effect to inhibit the pro-inflammatory effect of COX-2, was rofecoxib, which would have the potential for treatment without side effects such as ulcers and gastrointestinal problems. Studies by Bombardier et al. 2000 [5] reported that when compared to a 500 mg dose of naproxen per day for the same period, the incidence of efficacy was equivalent, but the administration of rofecoxib had less side effects such as bleeding gastric and duodenal ulcers. However, it was observed that the toxicological effect of rofecoxib, with a daily 50 mg dose for 9 consecutive months, doubled the acute myocardial infarction and strokes [8].

At the current stage of knowledge, in which the binding sites for specific inhibitors in COX-2 have already been described, and the three-dimensional protein structure of the enzyme is clearly established, the use of modern molecular modeling techniques should be able to triate new compounds of high affinity and specificity, but probably without the presence of the sulfonamide and sulfone groups of the second-generation compounds seen previously, thus representing the birth of a third generation of specific COX-2 inhibitors [9].

The process from a biological target project to a new drug discovery can take an average of 10 years or more, and the computational chemistry comes with an excellent direction in the rational

planning of drugs, with already countless cases of success involving or using computer simulations citing as an example the main factors: losartan, atorvastatin, and celecoxib.

Mathematical analyses accompany in silico studies in order to enable the reduction of costs and time to obtain positive results, observing the molecular structures and their possible affinities with the therapeutic target, using the quantitative structure–activity relationship (QSAR). This method aims to build parametric models for predicting inhibitory activity (IC_{50}) correlated with dependent variables such as physical–chemical, biological, and toxicological properties [10].

In these analyses, reduction filters are applied to the models used to predict inhibitory activity, correlating the molecule structures with their activity and toxicological potential, and comparing them with a positive control. In 2016, Brick et al. [10] applied the QSAR analysis to identify new antimalarial inhibitors from 1H-Imidazol-2-IL-Pyrimidine-4,6-Diamines, with reducing filters to eliminate descriptors that did not show correlation or information relevant to the process of statistical and toxicological analysis, beginning the screening with 107 compounds from the ZINC database and ending with four more promising compounds.

In this context, one of the main types of studies in progress by the scientific community is the in silico and in vivo studies of inhibitors for the inflammatory processes to search for new selective molecules. In parallel, the QSAR study (quantitative structure–activity relationships) uses multiparametric models that interrelate biological activity with the physical–chemical properties of selected molecules in order to predict their inhibitory capacity against the inflammatory mechanism [10]. Therefore, the objective of this work is the virtual screening of analogs of rofecoxib (Vioox®), based on pharmacophore and QSAR analysis, understanding approaches and molecular modeling techniques through free software that is easily accessible by the scientific community, in parallel with the prediction of pharmacokinetic properties and toxicity that show the possible effectiveness of the selected structures, according to the methodological scheme presented in Figure 2 (see more details in the Materials and Methods section).

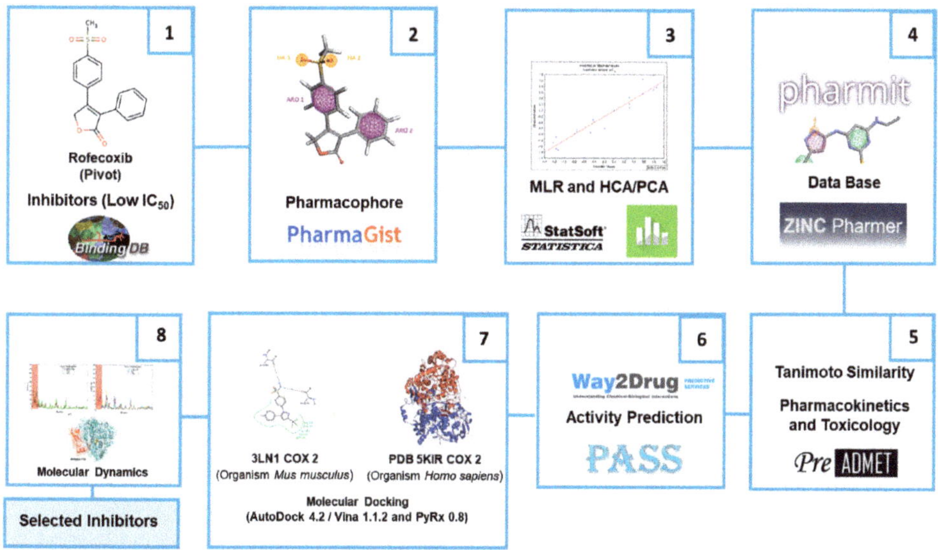

Figure 2. General scheme summarizing of the methodological steps.

2. Results and Discussion

2.1. Molecular Optimization and QSAR Analysis

The structure of rofecoxib was selected as a pivot given its potential to mitigate the gastrointestinal effects compared to other selective drugs. Although this drug has the unwanted effect of acute myocardial infarction, the objective is to detect essential pharmacophore characteristics through virtual screening so that the selected promising structures have the same effectiveness. On the other hand, this is one of the only structures that presents the complexed structure in the Protein Data Bank (PDB) (5KIR, https://www.rcsb.org/) for the *Homo sapiens* organism, bringing the results of an ideality in front of the human organism.

The 32 molecules (Rofecoxib as a pivot) for the analysis were selected in the BindingDB database (https://www.bindingdb.org/bind/index.jsp) obeying an increasing order of IC_{50}, with specific activity related to COX-2 and the *Homo sapiens* organism, in addition to not repeating inhibitory activity values, which could impact false-positive results through a straight line adjustment facilitated by of statistical analysis.

The molecular optimization values are shown in Table S1. The overlapping process was carried out by selecting molecules with the lowest energy value (PM3), since the optimization of molecular structures aims to bring the real structure closer to the energy minimum conformation, and with the observed experimental data, the optimization time quantification aims to elucidate the computational cost, as it is an expensive and time-consuming process [11].

Later, they were submitted to the PharmaGist software (https://bioinfo3d.cs.tau.ac.il/PharmaGist) for the extraction of physicochemical properties and construction of structure–activity relationship models (QSAR). The characteristics were analyzed with the aid of the Statistica® software, where the most relevant ones were used to predict the inhibitory activity as a function of the pIC_{50} value to decrease statistical inconsistencies. This software is capable of predicting the relationship between the inhibitor structure and its inhibitory activity, with a Pearson correlation cut-off of 0.4, obtaining a training set with n = 20 structures (methodology adopted by Santos, Cruz and Santos) [12–14]. Table 1 shows the selected descriptors. The atoms (ATM) characteristic presented the best correlation among all descriptors, with a value of 0.7651, allowing inferring that the number of atoms significantly interferes in the pIC_{50} responses of the selected molecules. However, it must be noted that the selected regression model is tetra-parametric, so the prediction analysis must take into account the contribution of each descriptor in the process of prediction of the inhibitory activity value, as is the case of aromatic characteristics (ARO) with a p value of 0.7358 and acceptors (ACC) with a p value of 0.6399, which also contributes to the prediction of the inhibitory activity.

This result can also be accompanied by the analysis of hierarchical cluster analysis (HCA) (Figure S1) performed with the aid of the Minitab® Trial software, allowing observation of the similarity between the physical–chemical characteristics and the inhibitory activity of the respective molecules, corroborating with the data obtained by Pearson's correlation. The characteristics of ATM and ARO show greater proximity to the predicted pIC_{50}. The descriptor ACC is inversely proportional to the predicted pIC_{50} value, which indicates that the presence of hydrophilic groups capable of establishing hydrogen-bonding interactions can increase the inhibitory potential of the selected structures.

The ATM characteristic may not be essential when analyzed individually; however, we observed that the greater the number of atoms present in a structure, the greater its volume and topological polar surface area (TPSA), which are both characteristics that are essential for good oral absorption of the medication in the body, consequently obeying the Lipinski rule. On the other hand, it is not the only relevant characteristic for the prediction of the values of inhibitory activity and must be corroborated with the other characteristics provided by the statistical analysis [15]. Figure S2 represents the PCA analysis for the selected molecules. It correlates its characteristics with the inhibitory activity; the compounds with the lowest activity are in red, and those with the best activity are in blue. Molecule **11** is displaced from the others because it presents values of hydrogen donors equal to

3, which is different from the others in the selected training group that present values of 0 and 1, showing statistically a decrease in the pIC_{50} values of the structures.

Table 1. Properties of the screened training molecules obtained from the PharmaGist server.

Inhibitor	Characteristics					
	ATM [a]	ARO [b]	DONN [c]	ACC [d]	pIC_{50}	IC_{50} (nM)
1 *	36	2	0	4	7.9208	12.00
2	45	4	1	2	10.0457	0.09
3	48	4	1	2	9.7695	0.17
4	50	2	1	3	9.6989	0.20
5	45	4	1	2	9.5528	0.28
6	42	4	1	2	9.4436	0.36
7	42	4	1	2	9.4317	0.37
8	42	4	1	2	9.2218	0.60
9	42	4	1	2	9.0969	0.80
10	42	3	1	3	9.0000	1.00
11	37	2	3	5	8.7447	1.80
12	44	3	1	5	8.4317	3.70
13	40	3	1	3	8.3979	4.00
14	42	3	0	4	8.3767	4.20
15	29	2	1	2	8.3000	5.00
16	41	3	1	4	8.2518	5.60
17	40	1	0	4	8.2218	6.00
18	36	1	0	4	8.0969	8.00
19	36	1	0	4	7.7212	19.00
20	33	2	0	3	7.6777	21.00
ATM	1.0000	0.5659	0.2027	−0.2280	0.7651	–
ARO		1.0000	0.3559	−0.6492	0.7358	–
DONN			1.0000	−0.0423	0.4743	–
ACC				1.0000	−0.6399	–

* Pivotal molecule. [a] Atoms; [b] Aromatics; [c] Donors; [d] Acceptors.

In addition, the values of the number of atoms provide a better forecast: molecule **11** (37 atoms) shows one of the lowest values for ATM, as well as **15, 18, 19,** and **20**. It is observed that for the most active molecules, the ATM characteristics are relevant to the value of inhibitory activity, shifting them to the most active side. All the most active structures have four aromatic groups, except for **molecule 4** which has only 2; however, its inhibitory activity is accentuated by the number of atoms in its structure (50).

In parallel, it must be understood that the predicted activity depends on the correlation between the four selected characteristics and their relative weight, and that the objective of the preliminary QSAR analysis is to investigate the most relevant characteristics among the data presented in the sampling of the structures already reported with inhibitory activity that are selective for COX-2. Figure S3 shows the analysis of HCA for the selected inhibitors. The HCA analysis gathers in hierarchical groups by similarity; the most active are represented in blue, and the least active ones are in red. It is observed that the data from the cluster followed the data obtained previously via QSAR and principal component analysis (PCA) analyses.

The tree-like dendrogram is seeking the structural similarities and the response to the inhibitory activity. It is noticed that **2** to **9** have four aromatic groups, with the exception of **4**, which has only two, but its activity is enhanced by the number of atoms of the structure. Figure 3 shows the structure of the eight most active molecules. The common group observed in these inhibitors is pyrrole, in addition to the methylsulfone group, with the exception of **4**. The presence of the methylsulfone group resembles the structure of rofecoxib and differs from the structures of the other coxibs (celecoxib and etoricoxibe), which have a sulfonamide functional group that is responsible for their toxicological characteristics. Pyrrole gives the appropriate lipophilic character to the molecule, which can help the molecule enter the COX-2 active channel [16].

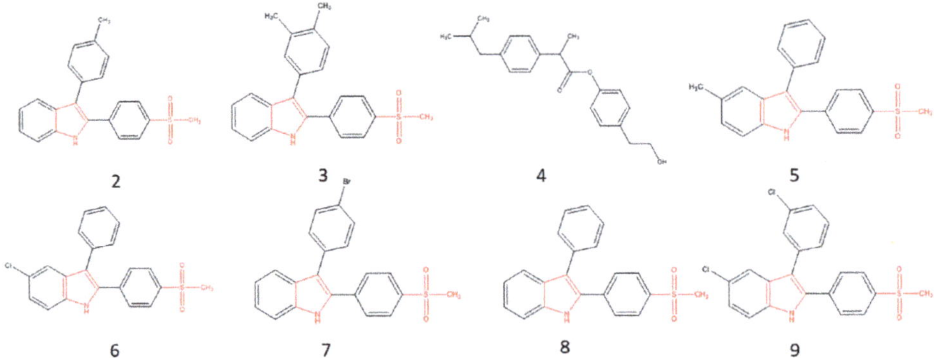

Figure 3. The most active molecules.

Table 2 shows the regression data of the descriptors used to verify the best model for predicting the inhibitory activity. A combination has been used to evaluate the statistical parameters and select the parametric prediction equation according to the best fit. It is observed that the physical-chemical parameters ATM, ACC, and ARO are significant for the final calculated pIC_{50} values. The best statistical parameters were obtained for the parametric tri and tetra models with R^2 values of 0.9599 and 0.9617, and variance ratios of 62.6373 and 46.2719, respectively. It is emphasized that the greater the number of equated scores, the greater the quality of the predictor model, although the values are statistically close [10].

Table 2. Parametric models and regression analysis values (R: Correlation coefficient; R^2: Correlation to square coefficient; $R_A{}^2$: Correlation coefficient to adjusted square. SEE: Standard estimated error; F: Variance ratio).

Mono Parametric						
Eq.	**Descriptors used**	**R**	**R^2**	**$R_A{}^2$**	**SEE**	**F**
1	ATM	0.7651	0.5854	0.5623	0.4804	25.4183
2	ARO	0.7358	0.5414	0.5159	0.5053	21.2522
3	ACC	0.6399	0.4095	0.3767	0.5733	12.4871
4	DONN	0.4743	0.2249	0.1819	0.6569	5.2251
Bi Parametric						
Eq.	**Descriptors used**	**R**	**R^2**	**$R_A{}^2$**	**SEE**	**F**
1	ATM + ACC	0.9022	0.8140	0.7921	0.3311	37.2030
2	ARO + ATM	0.8487	0.7203	0.6874	0.4060	21.8906
3	ATM + DONN	0.8316	0.6916	0.6554	0.4263	19.0684
4	ACC + DONN	0.7809	0.6099	0.5640	0.4795	13.2914
5	ARO + DONN	0.7701	0.5930	0.5452	0.4898	12.3893
6	ARO + ACC	0.7661	0.5869	0.5383	0.4934	12.0789
Tri Parametric						
Eq.	**Descriptors used**	**R**	**R^2**	**$R_A{}^2$**	**SEE**	**F**
1	ATM + ACC + DONN	0.9599	0.9215	0.9068	0.2217	62.6376
2	ARO + ATM + ACC	0.9053	0.8197	0.7859	0.3360	24.2481
3	ARO + ACC + DONN	0.8207	0.6736	0.6125	0.4521	11.0113
Tetra Parametric						
Eq.	**Descriptors used**	**R**	**R^2**	**$R_A{}^2$**	**SEE**	**F**
1	ATM + ACC + DONN + ARO	0.9617	0.9250	0.9050	0.2238	46.2719

Note that the parametric tetra prediction values were better adjusted with a correlation R = 0.9617 and standard error of estimate (SEE) = 0.2238, with a notable predictive capacity. Such alignment can be compared with the residual values found during the validation step of the equation, with $\Delta 4$ differences close to 0.2, which demonstrates the ability to predict the values of inhibitory activity. Table 3 analyzes the predicted models for the molecules, allowing inference of the difference (Δ = residual) between the pIC_{50} values found in an experimental way and the statistical prediction values for a defined parametric model and the equation was determined considering the highest statistical correlation.

Table 3. pIC_{50} values calculated using the prediction equations.

Molecule	pIC_{50}	Mono		Bi		Tri		Tetra	
		Eq. (1)	$\Delta 1$	Eq. (2)	$\Delta 2$	Eq. (3)	$\Delta 3$	Eq. (4)	$\Delta 4$
1 *	7.9208	8.2562	−0.3354	8.0305	−0.1097	7.7961	0.1247	7.7670	0.1538
2	10.0457	9.2561	0.7896	9.5495	0.4962	9.5773	0.4684	9.5693	0.4764
3	9.7695	9.5894	0.1801	9.8339	−0.0644	9.832	−0.0625	9.8414	−0.0719
4	9.6989	9.8116	−0.1127	9.6906	0.0083	9.6680	0.0309	9.7855	−0.0866
5	9.5528	9.2561	0.2967	9.5495	0.0033	9.5773	−0.0245	9.5693	−0.0165
6	9.4436	8.9228	0.5208	9.2651	0.1785	9.3226	0.1210	9.2972	0.1464
7	9.4317	8.9228	0.5089	9.2651	0.1666	9.3226	0.1091	9.2972	0.1345
8	9.2218	8.9228	0.2990	9.2651	−0.0433	9.3226	−0.1008	9.2972	−0.0754
9	9.0969	8.9228	0.1741	9.2651	−0.1682	9.3226	−0.2257	9.2972	−0.2003
10	9.0000	8.9228	0.0772	8.9322	0.0678	8.9888	0.0112	8.9925	0.0075
11	8.7447	8.3673	0.3774	7.7924	0.9523	8.5957	0.1490	8.6154	0.1293
12	8.4317	9.1450	−0.7133	8.4560	−0.0243	8.4910	−0.0593	8.4297	0.0020
13	8.3979	8.7006	−0.3027	8.7426	−0.3447	8.8190	−0.4211	8.8111	−0.4132
14	8.3767	8.9228	−0.5461	8.5993	−0.2226	8.3055	0.0712	8.2438	0.1329
15	8.3000	7.4785	0.8215	8.0327	0.2673	8.2189	0.0811	8.2529	0.0471
16	8.2518	8.8117	−0.5599	8.5045	−0.2527	8.5701	−0.3183	8.5297	−0.2779
17	8.2218	8.7006	−0.4788	8.4097	−0.1879	8.1357	0.0861	8.1972	0.0246
18	8.0969	8.2562	−0.1593	8.0305	0.0664	7.7961	0.3008	7.8344	0.2625
19	7.7212	8.2562	−0.5350	8.0305	−0.3093	7.7961	−0.0749	7.8344	−0.1132
20	7.6777	7.9229	−0.2452	8.0790	−0.4013	7.8752	−0.1975	7.8670	−0.1893
21 [a]	9.3010	9.0339	0.2671	9.0270	0.2740	8.7242	0.5768	8.7066	0.5944
22 [a]	8.8508	8.9228	−0.0720	9.2651	−0.4143	9.3226	−0.4718	9.2972	−0.4464
23 [a]	8.8239	9.3672	−0.5433	9.6443	−0.8204	9.3127	−0.4888	9.3508	−0.5269
24 [a]	8.6990	8.8117	−0.1127	8.8374	−0.1384	9.2534	−0.5544	9.2110	−0.5120
25 [a]	8.4815	9.0339	−0.5524	9.3599	−0.8784	9.0580	−0.5765	9.0787	−0.5972

	Prediction Equations
Mono	$pIC_{50} = 4.2566 + (0.1111 \times ATM)$
Bi	$pIC_{50} = 5.9493 + (0.0948 \times ATM) + (-0.3329 \times ACC)$
Tri	$pIC_{50} = 6.0749 + (0.0849 \times ATM) + (-0.3338 \times ACC) + (0.3495 \times DONN)$
Tetra	$pIC_{50} = 6.1250 + (0.0907 \times ATM) + (-0.3721 \times ACC) + (0.3766 \times DONN) + (-0.0674 \times ARO)$

* Pivotal molecule; [a] internal validation.

The best results in question were for **10** ($\Delta 4$ = 0.0075) and **12** ($\Delta 4$ = 0.0020) inhibitors, although **2** ($\Delta 4$ = 0.4764), **16** ($\Delta 4$ = −0.2779), and **18** ($\Delta 4$ = −0.2625) present residues greater than 0.2 in relation to the experimental data obtained. The margin of error (SEE = 0.2238) allows us to infer that the two may be within the desired perspective of residue, mainly, less than 1. The internal validation set demonstrated the detection of anomalous samples, which were excluded from the test set because they reduced the statistical correlation of the applied parametric method, with residues greater than 0.4, increasing the estimated error and deviating the correlation of the predicted values with the experimental data, which is justified by its exclusion from the test set initially. However, the reinclusion of these aims to determine the predictive capacity of the model, and results with residuals less than 1 are significant.

Figure 4 shows the projection of the data obtained in relation to the linear regression obtained, with a line adjustment of $R^2 = 0.9617$ of the tetra-parametric model, showing a good relationship between the experimental and predicted values.

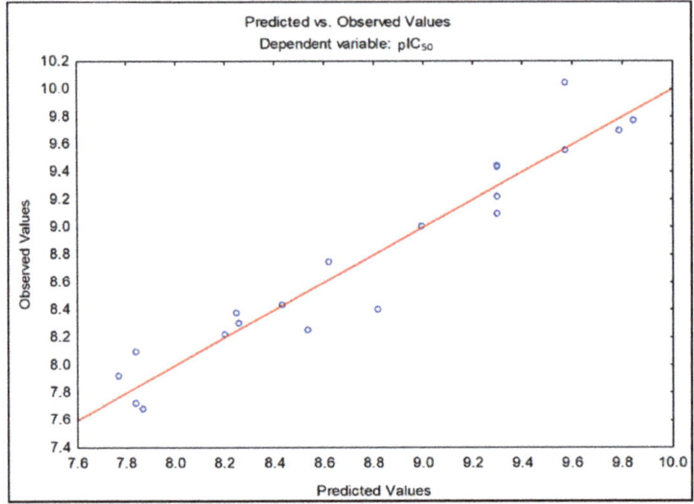

Figure 4. Linear correlation graph of the tetra-parametric model.

Table 4 shows the predicted values for the external validation set, applying the equations according to Table 4. For the validation test, a set between 20% and 30% of the total of the original set were used in the predictive model in order to prove its robustness [10]. It is observed that the values have good predictive quality for the molecules selected as an external set, with greater proximity for **26**, **27**, and **30**, with residues close to 0.1. This shows that the model has a significant correlation between the descriptors used.

Table 4. External validation set.

Molecule	pIC$_{50}$	Mono		Bi		Tri		Tetra	
		Eq. (1)	Δ1	Eq. (2)	Δ2	Eq. (3)	Δ3	Eq. (4)	Δ4
26	9.7447	9.3672	0.3775	9.3114	0.4333	9.6779	0.0668	9.6645	0.0802
27	9.2676	8.8117	0.4559	8.8374	0.4302	9.2534	0.0142	9.2110	0.0566
28	8.8508	8.9228	−0.0720	9.2651	−0.4143	9.3226	−0.4718	9.2972	−0.4464
29	8.6198	9.7005	−1.0807	8.2642	0.3556	8.2479	0.3719	8.1390	0.4808
30	8.2757	8.3673	−0.0916	8.1253	0.1504	8.2305	0.0452	8.1669	0.1088
31	8.1549	7.8118	0.3431	8.3171	−0.1622	8.4736	−0.3187	8.5250	−0.3701
Celecoxib	9.2839	8.7006	0.5833	8.4097	0.8742	8.4852	0.7987	8.4390	0.8449

2.2. Virtual Screening and Analysis of Pharmacokinetic and Toxicological Properties

After selecting the best inhibitors, these were used on the Protox II and Molinspiration servers to select the reduction filters; the compounds were directed to the ZincPharmer database through the Pharmit web server (http://pharmit.csb.pitt.edu/search.html) shown in Table S2, with the maximum and minimum values being selected in order to limit the promising structures to those within the pre-applied characteristics through the QSAR analysis, with a maximum limit of 2000 structures to be selected. The filters are applied on the online server, as well as the pharmacophore coordinates (below) elucidated for possible data reproduction and comparison with statistical analysis.

The pharmacophore structure obtained through the PharmaGist server is demonstrated, aligning the similarities of the twenty selected molecules (Table 5). It is observed that the characteristics of the pharmacophore follow the data obtained through statistical analysis, presenting two aromatic groups, two hydrophilic groups, and hydrogen receptors. Such pharmacophore characteristics are essential when compared to the central process molecule, which has two ARO groups and four ACC groups, allowing the tracking of molecules with physical and chemical characteristics closer to rofecoxib.

Table 5. Pharmacophore characteristics.

Pharmacophore Characteristics	Coordinates			
	x	y	z	Radius
Hydrogen Acceptor (HA1)	15.30	−10.69	−1.79	0.50
Hydrogen Acceptor (HA2)	17.52	−11.00	−2.88	0.50
Aromatic (ARO1)	17.28	−7.30	−1.54	1.10
Aromatic (ARO2)	21.57	−4.64	−1.16	1.10

On the other hand, in studies by Chakraborty, Sengupta, and Roy (2004) [17], linear multiple regression (RML) analyses were used to deduce statistically acceptable equations. The variation ratios were 0.675 for COX-1 and 0.842 for COX-2, observing three important pharmacophores groups: methyl sulfonyl portion, central phenyl ring, and terminal phenyl ring. These are relevant when compared to their affinity with the lipophilic channel present in the active sites of the enzymes, corroborating with the data obtained in the present study.

After the application of the reduction and selection filters of the 2000 compounds, they were submitted to similarity to Tanimoto to find out which ones are closer to the characteristics of the pivot molecule used in the process (Rofecoxib). The fifty eight (58) molecules were obtained with a Tanimoto index greater than 0.35 (see Table 6), which is a value that is considered reasonable for the application of toxicological and pharmacological prediction studies in silico, with three promising molecules being selected during these tests as reported below, subsequently applying the molecular coupling and molecular dynamics tests [18].

Table 6. Similarity studies of molecules using the Tanimoto Index.

Zinc	Tanimoto Index (T_i)		
	>0.25	>0.30	>0.35
Top Hits	1451	491	58

Table 7 shows the best results of the toxicological tests applied to the three inhibitors selected through the Tanimoto index and tests performed through the online server PreADMET (https://preadmet.bmdrc.kr/adme/) in order to screen those who present better absorption, distribution, and metabolism values besides limiting the possibilities of mutagenicity through toxicological tests. It is observed that molecules showed high LD_{50} values, with the exception of **Z-627**, but it presents good values for the absorption and distribution tests, contributing to its selection in the molecular docking tests. Carcinogenicity tests for rats and mice demonstrate a possibility of mutation for all the selected inhibitors; however, when compared to the control compound, it also observed this important side effect, and accordingly, it did not prevent the selection of these molecules. At the same time, the Ames test was used as a cut-off parameter between the most promising and those that would be excluded from the subsequent steps, where those that showed a positive result were eliminated from the process. This test assesses the possibility of mutagenicity of chemical compounds in media with a low histidine concentration, which allows the strains of *Salmonella typhimurium* to change and return to a prototypical state, which directly influences the carcinogenic response.

Table 7. Toxicological data of selected inhibitors.

Molecule	Toxicity				
	LD$_{50}$ (mg/kg) [a]	Toxicity Class [a]	Ames_test	Carcino-Mouse	Carcino-Rat
Rofecoxib	4500	V	Mutagen	Negative	Negative
Celecoxib	1400	IV	Mutagen	Negative	Negative
Z-814	2000	IV	Non-mutagen	Negative	Negative
Z-627	61	III	Non-mutagen	Negative	Negative
Z-964	5000	V	Non-mutagen	Negative	Negative

[a] Protox (http://tox.charite.de/protox_II) Class I: fatal if swallowed (LD$_{50}$ ≤ 5); Class II: fatal if swallowed (5 < LD$_{50}$ ≤ 50); Class III: toxic if swallowed (50 < LD$_{50}$ ≤ 300); Class IV: harmful if swallowed (300 < LD$_{50}$ ≤ 2000); Class V: may be harmful if swallowed (2000 < LD$_{50}$ ≤ 5000); Class VI: non-toxic (LD$_{50}$ > 5000).

The pharmacokinetic data for distribution are shown in Table 8. The plasma protein binding values (PPB) refer to the degree of binding of the inhibitors with the proteins present in the blood and C_{brain}/C_{blood} represents the permeability of the blood–brain barrier. Compounds with C_{brain}/C_{blood} values less than 1 do not have activity on the central nervous system (CNS).

Table 8. Distribution data of selected inhibitors.

Inhibitor	Distribution	
	C_{brain}/C_{Blood} [a]	PPB (%) [b]
Rofecoxib	0.0137	94.8036
Celecoxib	0.0272	91.0772
Z-814	0.9553	95.5607
Z-627	0.0344	85.3783
Z-964	0.4328	100.0000

[a] Permeability of the blood–brain barrier; [b] Plasma protein binding.

It is observed that **Z-964** shows 100% of binding with the plasma proteins, inferring the possibility of its bioaccumulation and a consequent increase in its half-life within the organism, since the unbound portion is metabolized, consequently is excreted, and the bound part is slowly released in order to maintain the balance of the medium. In parallel, **Z-627** showed an 85% binding, indicating that 15% of the fraction will not be bound, which increases the efficiency of diffusion and penetration into the cell membranes [19]. All selected compounds have no activity on the central nervous system, as they show values below one. In silico values for absorption are shown in Table 9.

Table 9. Absorption data for selected inhibitors.

Inhibitor	Absorption			
	P$_{Caco-2}$ (nm/sec) [a]	HIA (%) [b]	P$_{MDCK}$ (nm/sec) [c]	P-gp inhibition [d]
Rofecoxib	2.7294	98.2248	11.2740	Non
Celecoxib	0.4994	96.6870	45.0486	Inhibitor
Z-814	12.4185	98.6322	28.3061	Inhibitor
Z-627	0.6460	94.4692	1.4352	Non
Z-964	42.9100	95.5974	0.0517	Inhibitor

[a] Cell permeability; [b] Human intestinal absorption; [c] Cell permeability Maden Darby Canine Kidney; [d] in vitro P-glycoprotein inhibition.

The selected drug candidates showed high values for intestinal absorption (HIA> 94%), being one of the most important absorption, distribution, metabolism and excretion (ADME) properties [20]. The drug molecules are transported from the gastro-enteric tract to the blood circle and permeate the gastro-enteric membrane by various mechanisms, and among them, the activity of the P-glycoprotein must be taken into account. This P-glycoprotein is a common transporter in the intestinal penetration

of drugs, inferring in the hypothesis that the inhibitors **Z-964** and **Z-814**, because they present an in vitro inhibition of P-gp, decrease the efflux process through the passive permeability of the inhibitors, which is mediated by this protein. However, they have considerable absorption values when compared with those of the other molecules screened in this study.

The P_{MDKC} permeability value is significant for the **Z-814** inhibitor (28.3061 nm/sec), being higher than for the control compound. Values greater than two indicate a significant medication efflux. The compound **Z-964** showed a low permeability MDCK (0.0517 nm/sec) and **Z-627** approached the ideal (1.4352 nm/sec) [21,22]. In parallel, the Caco-2 permeability assay measures the flow rate of a compound through Caco-2 cell monolayers to predict the in vitro drug absorption, where values greater than two present drug efflux. Inhibitors **Z-814** (PCaco-2 = 12.4185 nm/sec) and **Z-964** (PCaco-2 = 42.9100 nm/sec) have higher values than rofecoxib (PCaco-2 = 2.7291 nm/sec), with the **Z-627** inhibitor (PCaco-2 = 0.6460 nm/sec) having a lower value; however, it is not a P-gp inhibitor, which can significantly interfere with intestinal absorption [22,23].

Table 10 demonstrates the predicted data for the biological activity of the selected inhibitors and compares the results against the selected controls rofecoxib and celecoxib, which were obtained from the PASS server (http://www.pharmaexpert.ru/passonline/). Celecoxib is used in everyday clinical practice, being part of the set of external validation and molecular docking of this research. The three selected inhibitors showed Pa > Pi values, indicating the possibility of activity in relation to the reported biological activities, mainly in terms of anti-inflammatory responses. **Z-627** has the best values for anti-arthritic (Pa = 0.985) and anti-inflammatory (Pa = 0.852) activities higher than controls (**Z-627** = 0.852, rofecoxib = 0.828, celecoxib = 0.663).

Table 10. Prediction of biological activity of selected inhibitors.

Inhibitor	Pa [a]	Pi [b]	Biological Activity	Pa [a]	Pi [b]	Adverse Effects
Rofecoxib	0.951	0.004	Antiarthritic	0.831	0.006	Extrapyramidal effect
	0.855	0.004	Non-Steroidal Anti-Inflammatory	0.640	0.006	Ulcer, peptic
	0.828	0.005	Anti-Inflammatory	0.531	0.040	Ulceration
	0.726	0.002	Cyclooxygenase-2 Inhibitor	0.385	0.024	Carcinogenic, group 1
Celecoxib	0.940	0.001	Cyclooxygenase Inhibitor	0.671	0.016	Pseudoporphyria
	0.936	0.001	Cyclooxygenase-2 Inhibitor	0.569	0.013	Ulcer, peptic
	0.809	0.007	Antiarthritic	0.498	0.024	Ulceration
	0.663	0.021	Anti-Inflammatory	0.385	0.134	Acidosis
Z-627	0.985	0.003	Antiarthritic	0.489	0.088	Extrapyramidal effect
	0.983	0.003	Antiosteoporotic	0.352	0.102	Visual acuity impairment
	0.852	0.005	Anti-Inflammatory	0.349	0.130	Fasciculation
	0.414	0.004	Cyclooxygenase Inhibitor	0.291	0.187	Interstitial nephritis
	0.192	0.016	Cyclooxygenase-2 Inhibitor	0.239	0.189	Ulcer, peptic
Z-814	0.821	0.006	Antiarthritic	0.761	0.015	Extrapyramidal effect
	0.569	0.024	Analgesic	0.747	0.037	Neutrophilic dermatoses
	0.527	0.049	Anti-Inflammatory	0.627	0.051	Hypercholesterolemic
	0.109	0.077	Cyclooxygenase Inhibitor	0.539	0.032	Adrenal contex hypoplasia
	0.084	0.065	Cyclooxygenase-2 Inhibitor	0.441	0.041	Ulcer
Z-964	0.689	0.007	Analgesic, Non-Opioid	0.428	0.106	Extrapyramidal effect
	0.688	0.018	Antiarthritic	0.383	0.083	Visual acuity impairment
	0.642	0.015	Analgesic	0.347	0.198	Nephrotic syndrome
	0.497	0.058	Anti-Inflammatory	0.313	0.166	Interstitial nephritis
	0.190	0.016	Cyclooxygenase-2 Inhibitor	0.309	0.207	Nephritis

[a] Pa = Possibility of activity; [b] Pi = Possibility of inactivity.

All the candidate inhibitors have the possibility of activity against COX-2, although it is below the value of the reference compounds. Nonetheless, it serves as a reference for possible activities that they may present during the in vivo tests to be performed. In addition, a prediction of adverse effects that they may have on the organisms was performed, verifying that all of them present a propensity of activity similar to the other selected control compounds (celecoxib and rofecoxib) in the case of extrapyramidal effects. However, they have a lower propensity for the emergence of gastrointestinal problems, such as ulcers, which is the main focus in the development of new selective

anti-inflammatory drugs, and the **Z-964** structure did not present the possibility of presenting such an adverse effect. Table 11 shows the physical–chemical data of the selected inhibitors.

Table 11. Cardiotoxic effects of selected molecules.

Inhibitor	Pa [a]	Pi [b]	Adverse Effect [c]	hERG [d]
Rofecoxib	0.561 0.547 0.345	0.031 0.036 0.305	Cardiac Failure Myocardial Infarction Hepatotoxicity	Medium Risk
Celecoxib	0.528 0.494	0.043 0.044	Cardiac Failure Myocardial Infarction	Medium Risk
Z-814	0.557 0.464	0.034 0.074	Myocardial Infarction Cardiac Failure	Low Risk
Z-627	0.516 0.479 0.710	0.041 0.066 0.096	Myocardial Infarction Cardiac Failure Hepatotoxicity	Medium Risk
Z-964	0.503 0.462	0.054 0.051	Cardiac Failure Myocardial infarction	High Risk

[a] Pa = Possibility of activity; [b] Pi = Possibility of inactivity; [c] Metatox web; [d] PreADMET.

Knowing the possibility that the structures present adverse cardiotoxic risks, the results were compared with the molecule still marketed in celecoxib, now considering its performance. However, it is observed that it presents risks of myocardial infarction and heart failure, which are analyzed through the Metatox (http://way2drug.com/mg2/gen_meta_all.php) and the hERG study (Human ether-a-go-go), through the PreADMET server, which refers to the blocking of the potassium channel, and that may cause cardiac collateral damage; see Table 11 below.

In view of the foregoing, this fact still did not allow its withdrawal from the market, as follow-up and adequate dosage reduce side effects and toxicological risks, which is a valid narrative for every drug currently sold; therefore, its cost–benefit must be evaluated. For now, it is seen that the molecules present a risk similar to or below the molecule withdrawn from the market (rofecoxb) and that which is still on the market (celecoxib), which does not preclude the possibility of being evaluated as candidates for specific COX-2 inhibitors. It is observed that the structures Z-814 and Z-627 present low and medium hERG risk, respectively, being better or equal to the molecules already commercialized, which makes their application as candidates for inhibitors of cyclooxygenase-2 possible. The Z-964 structure remains in the study due to the good results of bioavailability, in order to evaluate its experimental response in another study, as well as the others.

All candidate inhibitors present physical and chemical data within the acceptable range, showing no violation (Nv) or violating Lipinski's rule of five. This rule says that drugs with good oral bioavailability must obey four physicochemical parameters: molecular weight (MW) $\leq$ 500 g/mol, octanol/water partition coefficient (log P) $\leq$ 5, the number of hydrogen-bond donor groups (nHD) $\leq$ 5, and the number of hydrogen-bond acceptor groups (nHA) $\leq$ 10, see Table 12.

Table 12. Physicochemical data of the selected inhibitors.

Inhibitor	Properties							
	Milog P [a]	TPSA [b]	MW [c]	nHA [d]	nHD [e]	Nv [f]	Nrotb [g]	Volume
Rofecoxib	0.71	60.45	314.36	4	0	0	3	264.79
Celecoxib	3.61	77.99	381.38	5	2	0	4	298.65
Z-814	3.35	68.28	408.39	4	0	0	6	299.11
Z-627	1.38	75.27	330.41	5	2	0	4	286.58
Z-964	2.03	58.20	384.47	4	2	0	7	316.16

[a] Partition coefficient; [b] Topological Polar Surface Area; [c] Molecular Weight; [d] Number of Hydrogen Acceptors; [e] Number of Hydrogen Donors; [f] Number of violations; [g] Number of Rot bonds.

Table 12 shows that they have good absorption or permeability [24]. The pIC_{50} values (nM) were predicted for the selected molecules, see Table 13, according to the equations of Table 4, demonstrating acceptable values. The three selected inhibitors with Tanimoto index are shown in Figure 5.

Table 13. Predicted pIC_{50} values of the selected inhibitors and controls.

Inhibitor	Molecular Properties				Parametric QSAR Models ($pIC_{50} = -\log IC_{50}$)			
	ATM [a]	ARO [b]	DONN [c]	ACC [d]	Mono	Bi	Tri	Tetra
Rofecoxib	36	2	0	4	8.2562	8.0305	7.7961	7.7670
Celecoxib	40	3	1	4	8.7006	8.4097	8.4852	8.4390
Z-814	38	2	0	4	8.4784	8.2201	7.9659	7.9484
Z-627	41	2	2	3	8.8117	8.8374	9.2534	9.3458
Z-964	43	2	2	3	9.0339	9.0270	9.4232	9.5272

[a] Atoms; [b] Aromatics; [c] Donors; [d] Acceptors.

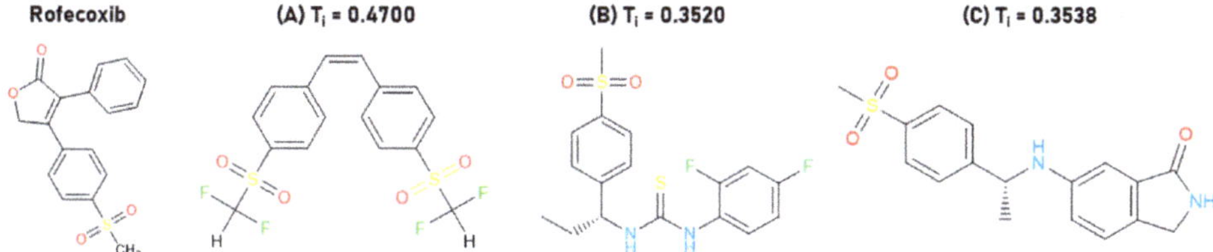

Figure 5. Selected inhibitors with Tanimoto index. (**A**) **Z-814**; (**B**) **Z-964**; (**C**) **Z-627**.

2.3. Molecular Docking

Figure 6 shows the poses calculated in relation to the deposited PDB complexes, with the deviation of the mean square root (RMSD) calculated at 0.91 Å for rofecoxib (RCX; 5KIR PDB code), 0.63 Å for celecoxib (**CEL**; 3LN1 PDB code) and 0.71 Å for indomethacin (**IMS**; 2YOE PDB code). Such a methodology provides alignment values for a maximum of 2 Å for the study of molecular docking, and accordingly, it validates the protocols used [25,26].

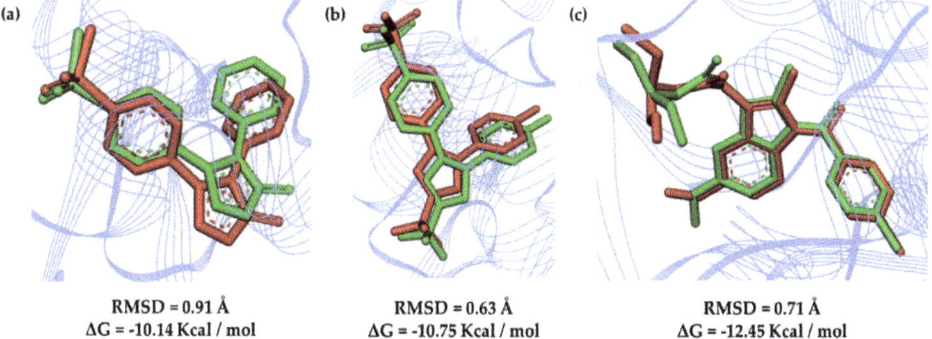

Figure 6. Overlapping poses of the crystallographic complexes (in green) with calculated poses (in red): (**a**) rofecoxib (**RCX**) for *Homo sapiens* (PDB 5KIR), (**b**) celecoxib (**CEL**) for *Mus musculus* (PDB 3LN1) and (**c**) indomethacin (**IMS**) *Ovis aries* (PDB 2OYE).

Figure 7 shows the interactions of the selected inhibitors with the control RCX in *Homo sapiens*. It is known that COX-1 and COX-2 have practically identical tertiary structures; however, the main difference between both is the replacement of Ile[434], His[513], and Val[434] residues in COX-1 by Val[434],

Arg[513], and Val[523] in COX-2, respectively. This allows an increase of approximately 25% of the active site that consists of a more accessible pocket with Arg[513] as a fundamental bonding site [16,27]. Figure 7a shows the main interactions of rofecoxib, within the pocket that provides a selective inhibition.

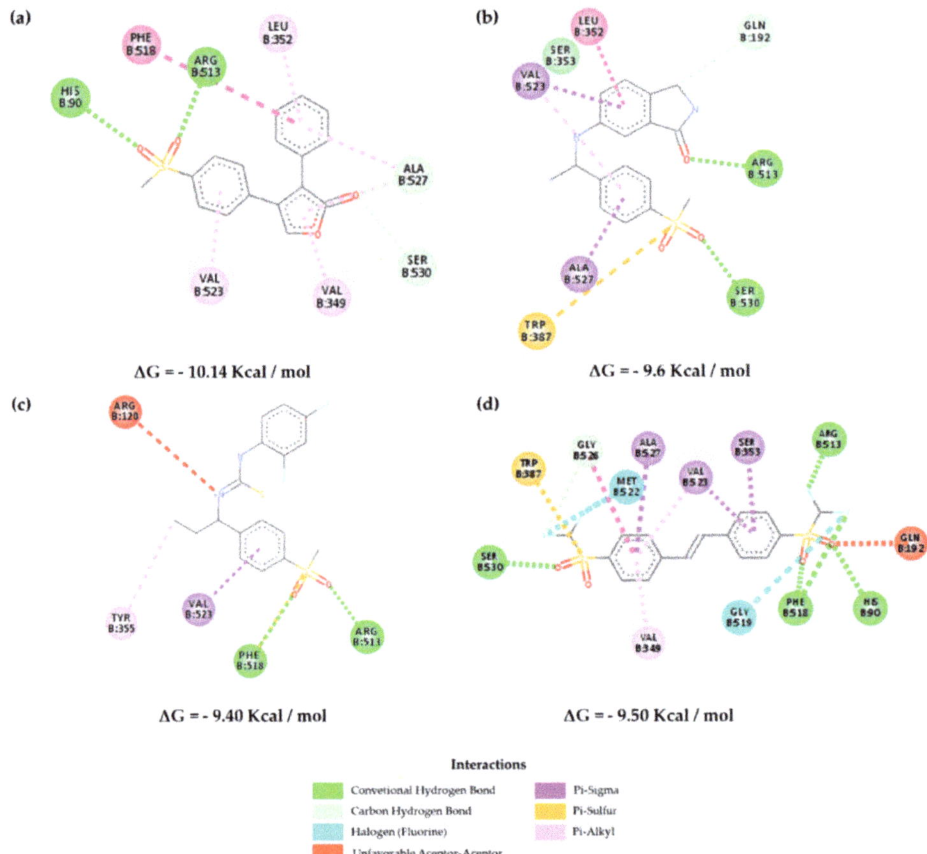

Figure 7. Interactions of the inhibitors (**a**) rofecoxib, (**b**) **Z-627**, (**c**) **Z-964**, and (**d**) **Z-814** with the active site of the structure of the Vioxx bound to the *Homo sapiens* cyclooxygenase-2 (COX-2, PDB 5KIR).

It is observed that the selected molecules **Z-814**, **Z-964**, and **Z-627** present a similarity of interactions with the amino acids of the hydrophobic region of the β leaf, Ser[530], and Val[523] for the former, and Ser[530], Phe[518], Val[523], and Leu[358] for the latter (Figure 7 and Table S3). The lipophilic channel of the enzyme is also constrained by the presence of Tyr[355] and Arg[513] on the enzyme surface, with the additional hydrogen-bond interaction between the Ala[527] and Val[523] phenolic group and the oxygen sulfone atoms of the structures.

On the other hand, in COX-2, there are some interactions that allow greater accessibility in the lipophilic channel in this isoform than in COX-1, which can be observed, indicating a greater ease of interaction with the active site of COX-2 via Phe[518]. This effect can also be translated by the negative Gibbs free energy required for the interaction to occur (Figure S6). The inhibitors selected may have an equivalent affinity in relation to the selected control compounds. The interaction with Ser[353] in **Z-814** demonstrates the possibility of a binding activity associated with low IC_{50} values [28–30].

In Figure 8 and Table S4, it is possible to verify the interactions of the selected inhibitors with the reference drug (**CEL**) against *Mus musculus*. Hydrophobic interactions are observed with Val[509],

Phe[504], Gly[512], Ser[339], and Leu[338], and hydrogen-bond interactions are observed with Gln[178], Phe[504], and Ser[339] for **CEL**. In parallel, we can observe the interactions for the selected inhibitors, where **Z-627** shows interactions with the hydrophobic residues Ser[339] and Val[509] as well as the control, and in addition, it presents a hydrogen-bond with Ala[513], showing selectivity [31,32].

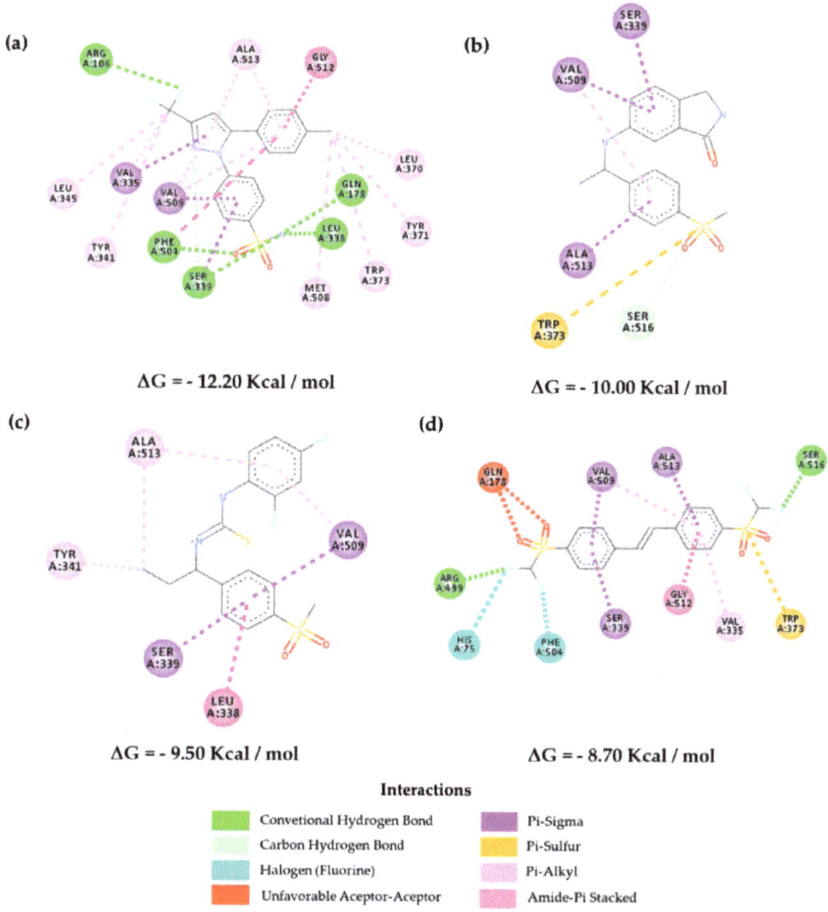

Figure 8. Interactions of inhibitors (**a**) celecoxib, (**b**) **Z-627**, (**c**) **Z-964**, and (**d**) **Z-814** with the active site of the *Mus musculus* COX-2 (PDB 3LN1).

On the other hand, the molecule **Z-964** shows greater interactions with Ala[513], Ser[339], and Val[509] in the lipophilic region present in the β-leaf of the enzyme and the hydrophilic residue Leu[338], which links to the sulfonic group of both inhibitors (sulfonamide for **CEL** and methylsulfone for **Z-964**). The **Z-814** molecule showed a lower affinity than the others, but it showed relative selectivity when it comes to the amino acid residues that are part of the interactions (Ser[339], Val[509], and Phe[504] (fluorine)) in the hydrophilic region of the molecule, which allows for interactivity in parallel with the CEL molecule. Furthermore, the data corroborate the QSAR analyses carried out when dealing with the connections with hydrogen acceptors, which are mainly influenced by the electronegativity of the selected structures. These interactions have already been observed in other studies, corroborating with the affinity data shown in Figure S5, in which **Z-967** shows an energy of 10.00 kcal/mol and **Z-964** shows an energy of 9.50 kcal/mol. These data are considered the most important ones [30,31].

Docking studies corroborate the preliminary QSAR results, as they consider that the presence of aromatic groups can influence the inhibitory activity of such molecules; nevertheless, chemical changes are necessary in order to decrease in the cytotoxic effect of the inhibitors when compared with the reference drugs, such as the replacement of the sulfonamide by a methylsulfone group (rofecoxib analogs) [28]. The QSAR analysis demonstrates a structure–activity relationship, as is the case with the characteristics **ARO, ACC**, and **DONN**, being closely linked with the possibilities of their interaction with the active enzyme site. Lipophilicity deals with an intrinsic relationship of the possibility of permeation and good oral availability, which was previously reported with obedience to the rule of five by Lipinski (logP $\leq$ 5) interacting with the side pocket of the enzyme [30,31].

The three structures were subjected to molecular coupling tests (Figure 9) to assess the possibility of being selective for COX-1 as well, which would determine the possibility of the appearance of undesirable adverse effects, such as gastrointestinal problems. They demonstrate a lower affinity possibility to the COX-2 enzyme as previously reported, with low bond energies (**Z-627** = −8.40 Kcal/mol, **Z-964** = −8.60 Kcal/mol, and **Z-814** = −6.80 Kcal/mol) compared to the selected control compounds and the indomethacin molecule (−10.70 Kcal/mol) deposited in the crystallographic structure of the PDB. The energy ratios (COX-2/COX-1 and COX-1/COX-2, as shown in Figure 10) were evaluated following an adaptation of the methodology adopted by Araújo and collaborators (2005) [7] that verified the influence of the most prescribed anti-inflammatory drugs of COX-2 on COX-1.

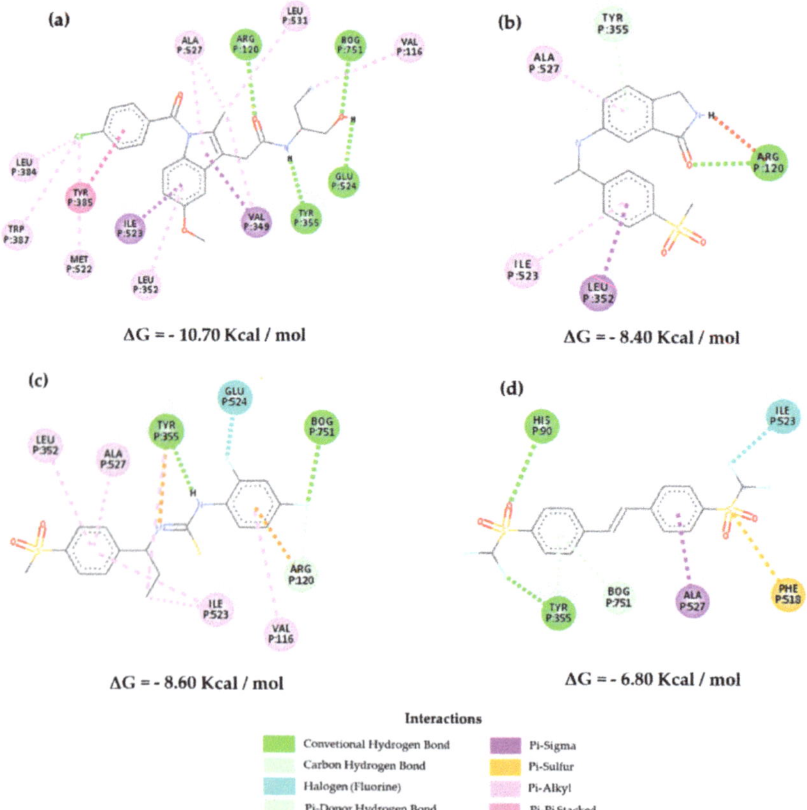

Figure 9. Interactions of the inhibitors (**a**) indomethacin, (**b**) **Z-627**, (**c**) **Z-964**, and (**d**) **Z-814** with the active site of the structure of the indomethacin complex to the *Ovis aries* COX-1 (PDB 2OYE).

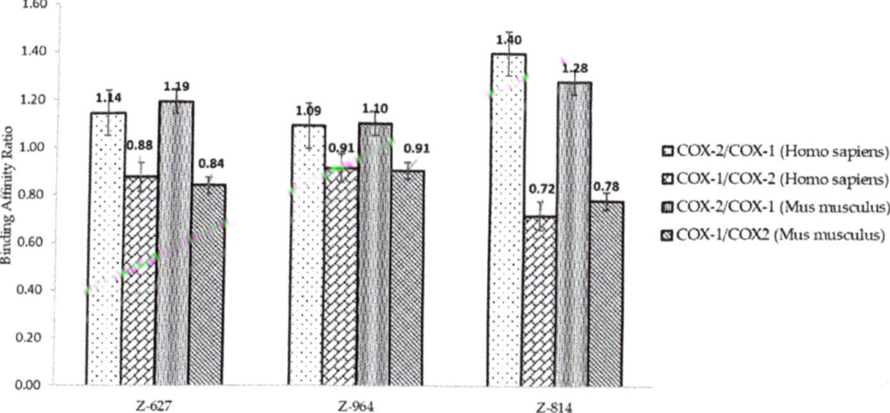

Figure 10. Binding affinity ratio of the selected structures.

It is observed that the activity ratio on COX-1 is lower when compared to COX-2, suggesting that the structures will not be highly selective for isoform 1, emphasizing that all the NSAIDs already prescribed have a small selectivity for this, which decreases the possibility of side effects [7]. On the other hand, the perspective of the advent of adverse effects can be compared in parallel with the prediction of biological PASS activity (Table 10), indicating a probability of few gastrointestinal effects. Furthermore, it is noted that selectivity in relation to COX-2 is given by the substitution of valine for an isoleucine in COX-1 in position 523, which in this case interacts with the phenolic ring of the selected structures. In addition, most inhibitors selective for COX-2 suggest not having free carboxylate groups, which contributes to this low affinity to isoform 1, and the high affinity is expressed by the interaction as the amino acid residue Arg[120] [33].

In this case, the **Z-627** structure presents this interaction relationship with the Arg[120] residue in the pyrroline portion of the structure, showing a possible structural rigidity and suggesting a possibility of structural modification in order to further limit the relationship estimate. However, the addition of the methylene group in the residue from Ile[523] indicates that interactions are restricted in access to the COX-1 side pocket, directly impacting the time-dependent competitive inhibition process in relation to COX-2 [34,35].

2.4. Structure–Activity Relationship of the Most Promising Molecules

The selected compounds (Figure 5) show a similar structure to that of the pivotal compound, having in their structures the methylsulfone group, showing no cytotoxic effect in relation to the sulfonamide group (**Z-627** and **Z-964**). Small substituents are the best, because they influence the volume of the molecule and possible van der Waals interactions with COX-2, which is a fact observed in docking studies. The introduction of fluorinated groups may show more significant activity.

According to Hayashi et al. 2012 [36], substituted analogs by acceptable hydrogen-bond groups potentiate the inhibitor activity. On the other hand, substitutions in the isoindoline nucleus can contribute to the inhibitor–enzyme stabilization, further demonstrating the fundamental role that the electrostatic and dipole–dipole interactions can play [37,38].

At the same time, the endocyclic nitrogen atoms included in five- or six-membered cycles such as pyrrole, pyridine, and pyrimidine, among others, may produce an increase in selectivity. The five-amino group in the isoindoline ring may favor the inhibitory activity of **Z-627** and, moreover, possible hydrogen-bond interactions through the methylsulfone group. The inhibition mechanism depends on the prostaglandin biosynthesis by means of arachidonic acid (AA), estimating that AA

fits into the channel cavity surrounded by amino acid residues with aromatic, aliphatic, and phenolic groups that establish several interactions.

Therefore, competitive or selective inhibitors bind to Val^{523} in COX-2, interfering with the arachidonic acid cascade and preventing the peroxidase action, as well as the formation of prostaglandins or thromboxanes (pro-inflammatory eicosanoids). In parallel with the studies carried out by Hayashi et al. 2012 [36], the best inhibitors **Z-627** and **Z-814** have two hydrogen-bond donors, as well as low values of TPSA and MW. For the three selected inhibitors we proposed theoretical synthetic routes—Supplementary Materials Figures S9–S11.

2.5. Molecular Dynamics Results and Affinity Energy

The studies of molecular dynamics simulations were carried out to understand more deeply the modes of interaction of the selected compounds with the target proteins. The results obtained through molecular dynamics simulations have served as support for the detailed evaluation of conformations over time observed in drug–receptor complexes [39–41]. Thus, understanding that the dynamics and changes in the movement of a protein are closely related to its biological function allows us to understand that the observation of these phenomena is extremely important. In this way, we carried out the investigation of the protein structure during the 100 ns of molecular dynamics simulations using the methods of root mean square deviations (RMSD) and root mean square fluctuations (RMSF). To plot the RMSD of the ligands, all the heavy atoms of the molecules were used, while to plot the RMSD and RMSF of the protein backbone, the Cα carbon atoms were used. In Figure 11, the graphs of the compounds that were bound to COX-2 of *Homo sapiens* are plotted, while in Figure 12, the RMSD plot of the complexes established with COX-2 of *Mus musculus* is displayed.

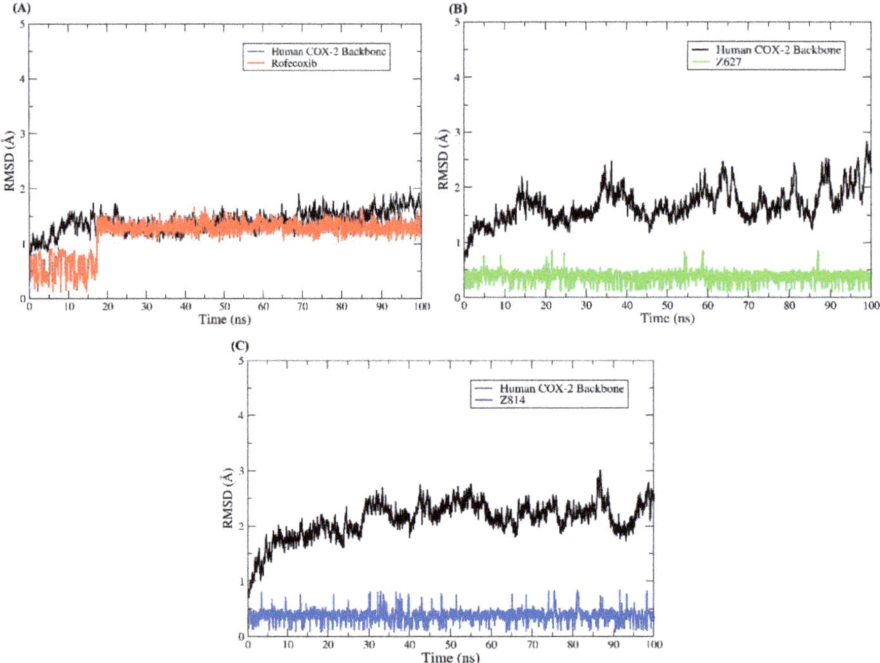

Figure 11. Root mean square deviations (RMSD) plot of complexes established with *Homo sapiens* COX-2. The protein backbone plot is colored black, but the ligand plots are colored in different ways. (**A**) RMSDs of the COX-2-rofecoxib system, (**B**) RMSDs of the COX-2-Z627 system, and (**C**) RMSDs of the COX-2-Z814 system.

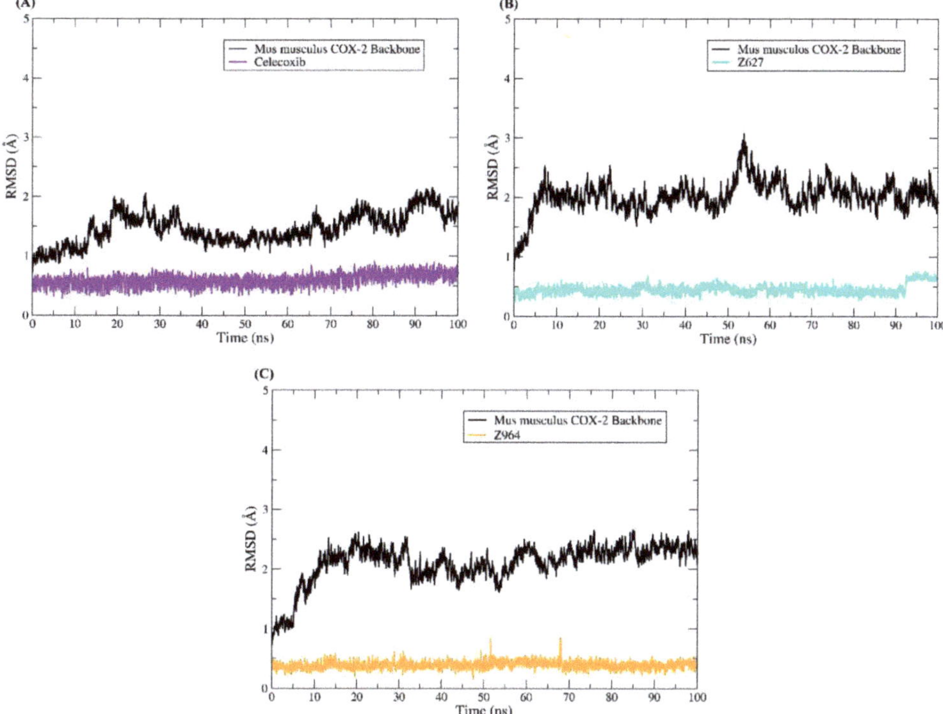

Figure 12. RMSD plot of complexes established with *Mus musculus* COX-2. The protein backbone plot is colored black, but the ligand plots are colored in different ways. (**A**) RMSDs of the COX-2-celecoxib system, (**B**) RMSDs of the COX-2-Z627 system, and (**C**) RMSDs of the COX-2-Z814 system.

Along the trajectories of MD simulations, COX-2 showed differences in the RMSDs of the complexes. The maximum Plator rising by the backbone RMSD was 3 Å, which was visualized in the COX-2-Z814 system, and the smallest fluctuation was observed in the COX-2-rofecoxib system. Despite the fluctuations displayed, this did not impair the interaction with the complexed ligands. It is important to note that the RMSD of the ligands showed low fluctuations and had a low RMSD value; this means that the ligands did not undergo drastic conformational changes after settling at the protein binding site.

Similar phenomena were observed for the complexes established between *Mus musculus* COX-2 and ligands. The backbone RMSD Plator was approximately 3 Å, and the ligands also remained in equilibrium throughout the 100 ns simulations, as observed in the RMSD graphs with small fluctuations.

The evaluations of the regions of the protein that obtained the greatest fluctuations along the trajectories of molecular dynamics were performed using the RMSF plot (see Figure 13). In general, the RMSF graphs showed a similar profile, even in the regions that suffered the greatest fluctuations. The greatest fluctuations were observed in the N-terminal portion of the protein.

This region is exposed to the solvent, being formed by alpha helices and beta leaves that are connected by long loop regions. Structurally, loops are the most flexible regions of the protein, so a region that exhibits many loops has a tendency to be flexible, as was observed in the RMSF plots displayed. Although this region is close to the active site of the ligands, its flexibility did not compromise the binding of the compounds, since all the ligands showed energy of favorable affinity with the protein, according to the molecular mechanics/generalized born surface area (MM-GBSA) results obtained. In addition, the fluctuation of this region did not affect the conformational stability of the ligand along

the trajectory of molecular dynamics, as the RMSD graphics of the ligands demonstrated that they remained stable along the trajectories without showing drastic changes in the RMSD plot.

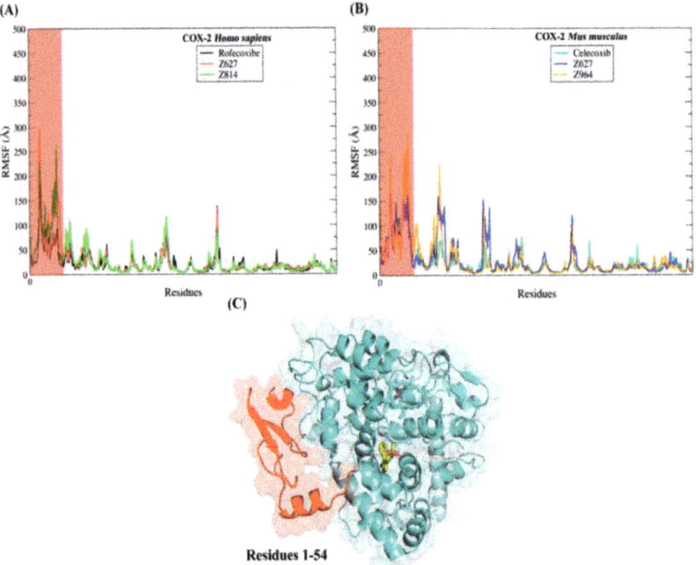

Figure 13. Root mean square fluctuations (RMSF) plot of the backbone of the proteins that established complexes with the compounds obtained by virtual screening. (**A**) RMSF for the COX-2 *Homo sapiens*; (**B**) RMSF for the COX-2 *Mus musculus*. (**C**) The region of the protein that has undergone the greatest fluctuations is highlighted in red.

In addition to structural analysis of the protein and ligands using RMSD and RMSF, we also evaluated whether the compounds are capable of interacting favorably with molecular targets. For this, we use the MM-GBSA method. The results obtained are summarized in Table 14.

Table 14. Affinity energy of COX-2 ligands systems.

Organism	Ligand	ΔG_{bind} (Kcal/mol)
	Rofecoxib	−48.15
Homo sapiens	**Z814**	−45.51
	Z627	−42.76
	Celecoxib	−47.78
Mus musculus	**Z627**	−41.63
	Z964	−44.27

All ligands have been shown to be able to interact favorably with COX-2. The selected compounds showed great affinity with COX-2 when we compared their values of affinity energy with the value obtained for the positive control of protein in the human body and Mus musculus. In the system established with human COX-2, compounds **Z-814** (ΔG_{bind} = −48.15 Kcal/mol) and **Z-627** (ΔG_{bind} = −45.51 Kcal/mol) showed binding affinity values similar to that obtained for rofecoxib (ΔG_{bind} = −42.76 Kcal/mol), which was the positive control.

Rofecoxib interacted through hydrogen bonds with the Arg[513] and His[90] residues, with an affinity energy of −2.14 and −1.82 Kcal/mol. Ligand **Z−814** established hydrogen bonds with His[90] and Tyr[385] with energy values of −1.53 and −1.48 Kcal/mol, while **Z−627** remained interacting with Phe[518] and

Ile^{517} with affinity values of −1.87 and −1.92 Kcal/mol. With the Mus musculus protein, the selected ligands, **Z–627** (ΔG_{bind} = −41.63 Kcal/mol) and **Z–964** (ΔG_{bind} = −44.27 Kcal/mol), also showed affinity values close to celecoxib (ΔG_{bind} = −47.78 Kcal/mol), which was used as a positive control. Celecoxib interacted with Phe^{504} and Ser^{339} through hydrogen bonds with affinity values of −1.42 and −1.95 Kcal/mol. Ligand **Z–627** interacted with Arg^{499} with an affinity of −1.81 Kcal/mol, while **Z–964** interacted with Phe^{504} with an affinity value of −1.68 Kcal/mol. The affinity energy values obtained with the MM-GBSA method, for the compounds selected by QSAR, were promising. This demonstrates that the selected substances can be considered as promising COX-2 inhibitors.

3. Materials and Methods

3.1. Selection of COX-2 Inhibitors

The molecule considered pivotal in the process was 4- (4-methylsulfonylphenyl)-3-phenyl-5*H*-furan-2-one (rofecoxib), which is known commercially as Vioxx®. It was taken from the BindingDB database (The Binding Database, https://www.bindingdb.org/bind/index.jsp) alongside twenty-four more molecules (Supplementary Material, Figure S7) to study the anti-inflammatory activity against COX-2 according to the literature data, following an increasing criterion of inhibitory activity, or IC$_{50}$ (Table 15). The molecules were aligned using the Discovery Studio® v. 4.0 program [42] for input on PharmaGist Web Server15 (http://bioinfo3d.cs.tau.ac.il/pharma/index.html) [43].

Table 15. Molecules selected in ascending order of IC$_{50}$.

Inhibitor	Molecule (PDB Code)	IC$_{50}$ (nM)	pIC$_{50}$	References
1 *	Rofecoxib (BDBM22369)	12.000	7.9208	[44]
2	BDBM50272105	0.090	10.0457	[45]
3	BDBM50272097	0.170	9.7695	[45]
4	BDBM50365267	0.200	9.6989	[4]
5	BDBM50272130	0.280	9.5528	[45]
6	BDBM50272111	0.360	9.4436	[45]
7	BDBM50272124	0.370	9.4317	[45]
8	BDBM50272106	0.600	9.2218	[45]
9	BDBM50272092	0.800	9.0969	[45]
10	BDBM50049011	1.000	9.0000	[46]
11	BDBM50189988	1.800	8.7447	[47]
12	BDBM50057618	3.700	8.4317	[48]
13	BDBM50151689	4.000	8.3979	[17]
14	BDBM50103310	4.200	8.3767	[49]
15	BDBM13066	5.000	8.3000	[50]
16	BDBM50297680	5.600	8.2518	[51]
17	BDBM50026234	6.000	8.2218	[52]
18	BDBM50332773	8.000	8.0969	[52]
19	BDBM50332765	19.000	7.7212	[52]
20	BDBM50336012	21.000	7.6777	[53]
21 [a]	BDBM50029613	0.500	9.3010	[54]
22 [a]	BDBM50272113	1.410	8.8508	[45]
23 [a]	BDBM50049030	1.500	8.8239	[46]
24 [a]	BDBM50272109	2.000	8.6990	[45]
25 [a]	BDBM50049013	3.300	8.4815	[46]
26 [b]	BDBM50272128	0.180	9.7447	[45]
27 [b]	BDBM50272090	0.540	9.2676	[45]
28 [b]	BDBM50272113	1.410	8.8508	[45]
29 [b]	BDBM50373566	2.400	8.6198	[55]
30 [b]	BDBM50057606	5.300	8.2757	[48]
31 [b]	BDBM50207446	7.000	8.1549	[56]
Celecoxib [b]	BDBM11639	0.520	8.4390	[45]

* Pivot; a Internal validation; b External validation.

3.2. Optimization of Molecular Structures and Determination of Pharmacophore Characteristics

The selected inhibitors were pre-optimized by means of Molecular Mechanics (MM+), followed by calculations of Austin Model 1 (AM1) and PM3 in the Hyper Chem 7® program (Table 2), with the lowest energy value used as a parameter of choosing the best model to carry out the construction of the pharmacophore hypothesis. Subsequently, the input was made to the PharmaGist Web Server 15 to determine the following characteristics: atoms (ATM), spatial characteristics (SF), characteristics (F), aromatic (ARO), hydrophobic (HYD), acceptor (ACC), and donor of hydrogen (DONN). The initial set presented 25 molecules, which were aligned according to the similarity with the selected pivot molecule, allowing the generation of pharmacophore models with the aid of the Discovery Studio® v. 4.0 program, following the methodology developed by us [10,12,14,57–59].

3.3. QSAR and PCA/HCA Studies

The inhibitory activity values were transformed into pIC_{50} ($-\log (IC_{50})$) in order to reduce the inconsistencies of the data obtained in an experimental way and homogenize the dataset, following the adopted methodological proposal [10,57,59]. In parallel, the importance of each pharmacophore descriptor was attributed—atoms (ATM), spatial characteristics (SF), characteristics (F), aromatic (ARO), hydrophobic (HYD), acceptor (ACC) and hydrogen donor (DONN); these were used for prediction in order to assess notoriety regarding the response to the pIC_{50} value through the Pearson correlation (p), using the software Statistica 7.0® and Minitab 19®, adapting the methodology adopted by Santos et al. 2015 and Ferreira et al. 2019 [12,59]. Pearson's coefficient (Equation (1)) measures the degree of linearity between two variables, assuming a value between +1 and −1. If one variable tends to increase while the others decrease, the value is negative. On the other hand, if both increase, the coefficient is positive. Moreover, $\overline{x}$ is the sample mean for the first variable; s_x is the standard deviation for the first sample; $\overline{y}$ is the sample mean for the second variable; s_y is the standard deviation for the second sample; and n is the column length.

$$\rho = \frac{\sum_{i=1}^{n}(x_i - \overline{x})(y_i - \overline{y})}{(n-1)s_x s_y} \tag{1}$$

The best pharmacophore descriptors were obtained considering the statistical quality relations of multiple linear regression (MLR), such as correlation coefficient (R), correlation coefficient squared (R^2), explained variance (adjusted R^2), standard error of estimate (SEE), and variance ratio (F), and they were transformed into parametric models for predicting the inhibitory activity at pIC_{50} values. The combinations were obtained using four parameters indicated by Pearson's correlation without repetition [12,59], according to Equation (2), where C = number of combinations, p = model type (p ≠ 0 and p = 4), and n = number of variables (n = 4).

$$Cp, n = \frac{n!}{p!(n-p)!} \tag{2}$$

For the prediction of the best model, in the internal validation stage, the random correlations between the descriptors and the inhibitory activity were measured to normalize the data obtained, applying the technique of detecting anomalous samples (outliers), in order to obtain a homogeneous set. This subset is considered as internal validation, for analysis of the prediction capacity of the selected model, comparing the data obtained during the two validations (internal and external; Figure S8). Principal component analysis (PCA) together with hierarchical cluster analysis (HCA) were applied in order to verify whether the model obtained corresponds to the degree of similarity, using Pearson's squared distance as a measurement parameter in the latter [60,61]. For the respective analyzes, Minitab v. 19® trial version was used.

3.4. Virtual Screening and Selection of Inhibitor Compounds

After selecting the best model via QSAR analysis, the selected molecules were superimposed to form a pharmacophore model. After inputing the pharmacophore, the search was performed within the ZINC database, selecting the 2000 most similar molecules, using the partition coefficient (log P), surface area (TPSA), number of atoms (Natoms), Molar Mass (MW), hydrogen acceptors (nHA), hydrogen donors (nHD), number of violations (Nv), number of revolutions (Nrotb), and volume, which were values as filters determined via Protox II servers (http://tox.charite.de/protox_II/) and molinspiration (https://www.molinspiration.com/cgi-bin/properties). The RMSD (Equation (3)) value was used as a reference parameter, which is the measure of the average distance between the atoms of the overlapping inhibitors, given in Angstroms, representing the quantitative similarity relationship between them. The lower the RMSD value, the better the model is compared to the target structure. δ_i^2 is the distance between atom i of any reference structure or the average position of N equivalent atoms.

$$\text{RMSD} = \sqrt{\frac{1}{N}\sum_{i=1}^{N}\delta_i^2} \tag{3}$$

Then, the Tanimoto test was performed via the BindingDB server. The similarity was determined according to the chemotype of the compounds screened with the pivotal molecule of the selection process to reduce and optimize the selection of compounds for determining pharmacokinetic and toxicological characteristics, using a cut-off index of 0.3, applying Equation (4) below [22].

$$J = \frac{M_{11}}{M_{01} + M_{10} + M_{11}} \tag{4}$$

where M_{11}—total number of attributes where A and B have a value of 1; M_{01}—total attributes where A is 0 and B is 1; M_{10}—total attributes where A is 1 and B is 0; M_{11}—total attributes where A and B have a value of 0.

3.5. Prediction of Toxicological and Pharmacokinetic Properties

Pharmacokinetic and toxicological studies were applied to inhibitors extracted from Pharmit via the ZincPharmer server. PreADMET v. 2.0 (https://preadmet.bmdrc.kr/) was used, which is an application based on a database on the web that is used for the prediction of ADME data (Absorption, Distribution, Metabolism, Excretion) with the following being selected: blood–brain barrier (BBB) penetration, in vitro permeability in Caco2 cells, human intestinal absorption (HIA), in vitro permeability of MDKC cells, in vitro P-glycoprotein inhibition, plasma protein binding (PPB), and Toxicological for Ames_Test, Carcinogenicity for Rats and Mice. The LD_{50} values were determined via Protox II servers (http://tox. charite.de/protox_II/) as well as toxicity class.

3.6. Prediction of Biological Activity of Selected Inhibitors

Activity predictions were made using the online PASS server (http://www.pharmaexpert.ru/ passonline), which allows you to predict the biological effects of compounds based on their formula using multilevel atom neighbors (VMA) descriptors, suggesting that the inhibitor's activity is expressed in terms of its chemical structure. Molecules with activities reported for anti-inflammatory and cyclooxygenase inhibitor effects were selected [25,61].

3.7. Molecular Docking

For this step, only the molecules with the best pharmacokinetic, toxicological, and biological parameters were selected for the study of molecular docking, in order to evaluate the interactions with selected inhibitors and the respective targets through the measurement of free energy interaction with amino acid residues and binding affinity. The crystallographic poses were extracted from the

web serve Protein Data Bank (PDB; https://www.rcsb.org/) for *Homo sapiens* with COX-2 complexed with the inhibitor rofecoxib having the code PDB 5KIR with a resolution of 2.697 Å, *Mus musculus* with COX-2 complexed with celecoxib having the PDB code 3LN1 with 2.40 Å resolution, and *Ovis aries* with COX-1 complexed with indomethacin having the PDB code 2OYE with 2.85 Å resolution; all structures were elucidated through X-Ray diffraction analyses.

Docking Study with AutoDock 4.2/Vina 1.1.2 via Graphical Interface PyRx (Version 0.8.30)

The selected inhibitors and proteins were prepared with the aid of Discovery Studio® 4.0 software, and the evaluation of the complexes with the ligand was evaluated using the AutoDock 4.2/Vina 1.1.2 software and the PyRx graphical interface version 0.8.30 (https://pyrx.sourceforge.io), with the standard exhaustiveness parameter of the software being the best conformation obtained through the analysis of the RMSD value. The validation protocol was based on the determination of the x, y, and z coordinates according to the average region of the active site; these values are observed in Table 16. The energy function score was used to evaluate the free binding energy (ΔG) of the interaction of the receptors with the ligands [25].

Table 16. Protocol data used in the validation of molecular docking.

Enzyme	Resolution	Inhibitor	Coordinates of the Grid Center	Grid Size (Points)
COX-2 (PDB code: 5KIR) *Homo sapiens*	2.400 Å	Rofecoxib	X = 23.214 Y = 41.353 Z = 3.517	36 x 26 y 28 z
COX-2 (PDB code: 3LN1) *Mus musculus*	2.697 Å	Celecoxib	X = 30.486 Y = −22.364 Z = -15.725	36 x 26 y 24 z
COX-1 (PDB code: 2OYE) *Ovis aries*	2.850 Å	Indomethacin (R)	X = 251.492 Y = 109.817 Z = −40.751	36 x 36 y 24 z

The calculation of binding affinity (ΔG) was also performed in order to compare the actual data obtained and the values predicted *in silico*, which was the same methodology adopted by Santos et al., 2020 [14], according to Equation (5).

$$\Delta G = -RT \ln K_i \tag{5}$$

where R (gas constant) is $1.987.10^{-3}$ kcal·mol^{-1}·K^{-1}, the temperature is 310 K for rofecoxib/celecoxib, and K_i is 310.10^{-9} M for rofecoxib and 340.10^{-9} M for celecoxib [28,32,52].

3.8. Molecular Dynamics Protocol

The initial structure for the system was obtained from molecular docking methods. The restrained electrostatic potential (RESP) protocol with the HF/6-31G* basis sets was applied to obtain the partial atomic charges of the atoms of each ligand [62–65]. The parameters of the ligand were constructed with the Antechamber module [66] using General Amber Force Field (GAFF) [67].

The amino acid protonation state was characterized using the PDB2PQR server [68]. The systems were built with the tLEaP module of the Amber 16 package [69–71]. The force field used to describe the protein in all simulations was ff14SB [72]. The protein–ligand system was solvated in an octahedron periodic box containing water molecules in the TIP3P model [73]. The partial charges were neutralized by adding counter-ions.

Energy minimization occurred in four stages. First, the water molecules and ions were optimized using 2000 cycles of the steepest descent and 3000 cycles of conjugate gradient. Then, the position of receptor-ligand hydrogen atoms was optimized using 4000 steps of the steepest descent algorithm and 3000 steps of the conjugate gradient. At the third stage, hydrogen atoms, water molecules, and ions were further optimized using 2500 steps of the steepest descent algorithm and 3500 steps of

the conjugate gradient. Finally, all atoms were minimized using 3000 steps of the steepest descent algorithm and three steps of the conjugate gradient.

Molecular dynamics simulations were performed at a constant volume by heating the systems up to 298 K. This heating was performed in five steps for a duration of 1 ns. After 100 ns, production runs were performed for each system.

The Particle Mesh Ewald method [74] was used for the calculation of the electrostatic interactions, and the bonds involving hydrogen atoms were restricted with the SHAKE algorithm—Restriction algorithm used to ensure that the distance between points of mass is maintained [75]. The temperature control was performed with the Langevin thermostat [76] within a collision frequency of 2 ps^{-1}.

3.8.1. Affinity Energy Calculations

To estimate the binding affinity (ΔG_{bind}), we used the molecular mechanics/generalized born surface area (MM-GBSA) methods [77–80]. The ΔG_{bind} was calculated according to the following equations:

$$\Delta G_{bind} = \Delta G_{complex} - \Delta G_{receptor} - \Delta G_{ligand} \tag{6}$$

$$\Delta G_{bind} = \Delta H - T\Delta S \approx \Delta E_{MM} + \Delta G_{solv} - T\Delta S \tag{7}$$

$$\Delta E_{MM} = \Delta E_{internal} + \Delta E_{ele} + \Delta E_{vdW} \tag{8}$$

$$\Delta G_{solv} = \Delta G_{GB} + \Delta G_{NP}. \tag{9}$$

The free energy of bonding (ΔG_{bind}) is the summation of the interaction energy of the gas phase among the protein–ligand (ΔE_{MM}), desolvation free energy (ΔG_{solv}), and system entropy ($-T\Delta S$). ΔE_{MM} is the result of the sum of internal energy ($\Delta E_{internal}$, sum of the energies of connection, angles and dihedral) electrostatic contributions (ΔE_{ele}), and the van der Waals term (ΔE_{vdW}). ΔG_{solv} is the sum of the polar (ΔG_{GB}) and non-polar (ΔG_{NP}) contributions. ΔG_{SA} was determined from the solvent accessible surface area (SASA) estimated by the linear combination of pairwise overlaps (LCPO) algorithm.

3.8.2. Per-Residue Free Energy Decomposition Analysis

Per-residue free energy decomposition was decomposed using the approach of MM/GBSA according to the following equation [14,81,82]:

$$\Delta G_{MM\text{-}GBSA} = \Delta E_{vdW} + \Delta E_{elec} + \Delta E_{pol} + \Delta E_{np}. \tag{10}$$

4. Conclusions

After the pharmacophore-based virtual screening, the QSAR analysis demonstrated a good line fit with $R^2 = 0.96$ and an equation with four main prediction parameters for pIC_{50}, ATM, ARO, ACC, and DON, where the ARO, ACC, and DON report the relationship with the three new and promising compounds selected and the pivot structure (rofecoxib). The development of the predetermined multiple linear regression model predetermined the pIC_{50} values for the selected compounds **Z-814** = 7.9484, **Z-627** = 9.3458, and **Z-964** = 9.5272. In database searches to evaluate possible applications that may have already been carried out, these substances are not used in specific biological activities (https://scifinder.cas.org/ and https://zinc.docking.org/).

The analyzes of toxicological prediction and bioavailability confirm the possibility of significant activity of the structures with a reduction of possible undesirable effects, of which **Z-627** was considered the most promising in view of all the tests applied via ADME analysis, without consequences for the CNS; this was corroborated with the main compounds selected. All selected compounds have the methyl sulfone group, unlike coxibs, which have the sulfonamide group. These three molecules do not present toxicological risks; they comply with the Lipinski rule of five, which provides for good oral availability, and PASS provides for a specific activity with a high probability of showing

promising anti-inflammatory activity, in addition to dim side effects in relation to the compound's selected controls. Molecular coupling tests demonstrate strong energy affinity with isoform 2 and low activity with isoform 1 through relationship analysis, which induces a possibility of minor side effects. Finally, zebrafish larvae should be analyzed to assess anti-inflammatory activity in the treatment of inflammatory disorders to confirm in silico results.

Supplementary Materials: The following are available online, Table S1: Energy values of the optimized molecules, Table S2: Filters applied according to the properties of the selected molecules, Table S3: Distance of Interactions for the structures for the PDB 5KIR, Table S4: Distance of Interactions for the structures for the PDB 3LN1, Table S5: Distance of Interactions for the structures for the PDB 2OYE, Figure S1: Dendrogram representing clustering of pharmacophores, Figure S2: Analysis of the main components for the sorted molecules. Scores (a) and Loading Graph (b), Figure S3: Dendrogram of selected molecules. More active (blue) and less active ones (red), Figure S4: Binding affinity results of compounds, including Vioxx bound (COX-2 – Homo sapiens), Figure S5: Binding affinity results of compounds, including celecoxib (COX-2 Mus musculus), Figure S6: Binding affinity results of compounds, including Indomethacin (COX-1 Ovis aries), Figure S7: Structures used in the molecular modeling, Figure S8: Structures used in the external validation set, Figure S9: Theoretical synthetic route for the preparation of compound A (Z-814), Figure S10: Theoretical synthetic route for the preparation of compound B (Z-964), Figure S11: Theoretical synthetic route for the preparation of compound C (Z-627).

Author Contributions: Conceptualization, P.H.F.A., W.J.C.M. and C.B.R.S.; methodology, P.H.F.A. and C.B.R.S..; software, R.S.R. and E.F.B.F.; validation, P.H.F.A., S.G.S., L.R.d.L., J.M.E.-R. and C.B.R.S,; formal analysis, P.H.F.A., R.S.R., J.N.d.C., J.M.C. and C.B.R.S.; investigation, P.H.F.A., R.S.R. and C.B.R.S..; resources, P.H.F.A., W.J.C.M., R.S.R. and C.B.R.S.; data curation, P.H.F.A., R.S.R. and C.B.R.S.; writing—original draft preparation, P.H.F.A. and C.B.R.S.; writing—review and editing, J.M.C and J.N.d.C.; visualization, P.H.F.A.; supervision, C.B.R.S.; project administration, C.B.R.S.; funding acquisition, J.M.C., C.B.R.S., P.H.F.A., W.J.C.M and E.F.B.F. All authors have read and agreed to the published version of the manuscript.

Funding: This research received no external funding.

Acknowledgments: To MEC/CAPES for the granting of development grants; UNIFAP/UEAP for financial assistance.

Conflicts of Interest: The authors declare no conflict of interest.

Abbreviations

AA—arachidonic acid, ACC—acceptors, ADME—absorption, distribution, metabolism and excretion, ARO—aromatic, ATM—atoms, C_{brain}/C_{blood}—permeability of the brain barrier, CEL—celecoxib, COX-1—cyclooxygenase 1, COX-2—cyclooxygenase 2, DONN—donors, HCA—hierarchal components analysis, HIA—human intestinal absorption, IC_{50}—inhibitory concentration, MDPI—Multidisciplinary Digital Publishing Institute, DOAJ—directory of open access journals, MilogP—partition coefficient, MW—molar weight, Natoms—number of atoms, nHA—number of hydrogen acceptors, nHD—number of hydrogen donors, Nrotb—number of rot bonds, Nv—number of violations, p—Pearson value, PCA—principal components analysis, Pcaco2—cell permeability, PDB—Protein Data Bank, P-gp inhibition—*In vitro* P-glecoprotein inhibition, P_{MDKC}—cell permeability Maden Darby canine kidney, PPB—plasma protein binding, QSAR—quantitative structure–activity relationships, RCX—rofecoxib, RMSD—deviation of the mean square root, T_i—Tanimoto Index, TPSA—topological polar surface area, Z-627—ZINC 170592627, Z-814—ZINC 33332814, Z-964—ZINC 225723964.

References

1. Medzhitov, R. Origin and physiological roles of inflammation. *Nature* **2008**, *454*, 428–435. [CrossRef]
2. Kim, M.H.; Son, Y.-J.; Lee, S.Y.; Yang, W.S.; Yi, Y.-S.; Yoon, D.H.; Yang, Y.; Kim, S.H.; Lee, D.; Rhee, M.H.; et al. JAK2-targeted anti-inflammatory effect of a resveratrol derivative 2,4-dihydroxy-N-(4-hydroxyphenyl)benzamide. *Biochem. Pharmacol.* **2013**, *86*, 1747–1761. [CrossRef]
3. Skinner, B.W.; Curtis, K.A.; Goodhart, A.L. Hypnotics and Sedatives. In *Side Effects of Drugs Annual*; Ray, S., Ed.; Elsevier: Indianapolis, IN, USA, 2018; Volume 40. [CrossRef]
4. Jain, S.; Tran, S.; El Gendy, M.A.; Kashfi, K.; Jurasz, P.; Velázquez-Martínez, C.A. Nitric Oxide Release Is Not Required to Decrease the Ulcerogenic Profile of Nonsteroidal Anti-inflammatory Drugs. *J. Med. Chem.* **2012**, *55*, 688–696. [CrossRef]
5. Bombardier, C.; Laine, L.; Reicin, A.; Shapiro, D.; Burgos-Vargas, R.; Davis, B.; Day, R.; Ferraz, M.B.; Hawkey, C.J.; Hochberg, M.C.; et al. Comparison of upper gastrointestinal toxicity of rofecoxib and naproxen in patients with rheumatoid arthritis. VIGOR Study Group. *N. Engl. J. Med.* **2000**, *343*, 1520–1528. [CrossRef]
6. Carvalho, W.A.; Carvalho, R.D.S.; Rios-Santos, F. Analgésicos inibidores específicos da ciclooxigenase-2: Avanços terapêuticos. *Braz. J. Anesthesiol.* **2004**, *54*, 448–464. [CrossRef] [PubMed]

7. Araujo, L.F.; Soeiro, A.D.M.; Fernandes, J.D.L.; Júnior, C.V.S. [Cardiovascular events: A class effect by COX-2 inhibitors]. *Arq. Bras. Cardiol.* **2005**, *85*, 222–229. [CrossRef] [PubMed]

8. Cryer, B.; Duboisø, A. The advent of highly selective inhibitors of cyclooxygenase—A review. *Prostaglandins Other Lipid Mediat.* **1998**, *56*, 341–361. [CrossRef]

9. Crofford, L.J. COX-1 and COX-2 tissue expression: Implications and predictions. *J. Rheumatol. Suppl.* **1997**, *49*, 15–19.

10. Birck, M.G.; Campos, L.J.; De Melo, E.B. Computacional Study Of 1 H -Imidazol-2-Yl-Pyrimidine-4,6-Diamines For Identification Of Potential Parent Compounds Of New Antimalarial Agents. *Quím. Nova* **2016**, *39*, 567–574. [CrossRef]

11. Munhoz, V.H.O.; Alcantara, A.F.C.; Pilo-Veloso, D. Conformational analysis by theoretical calculations of distinctin, an antimicrobial peptide isolated from Phyllomedusa distincta. *Quím. Nova* **2008**, *31*, 822–827. [CrossRef]

12. Santos, C.B.R.; Vieira, J.B.; Lobato, C.C.; Hage-Melim, L.I.S.; Souto, R.N.P.; Lima, C.S.; Costa, E.V.M.; Brasil, D.S.B.; Macêdo, W.J.D.C.; Carvalho, J.C.T. A SAR and QSAR Study of New Artemisinin Compounds with Antimalarial Activity. *Molecules* **2013**, *19*, 367–399. [CrossRef] [PubMed]

13. Cruz, J.V.; Serafim, R.B.; Da Silva, G.M.; Giuliatti, S.; Rosa, J.M.C.; Neto, M.F.A.; Leite, F.H.A.; Taft, C.A.; Silva, C.; Santos, C.B.R. Computational design of new protein kinase 2 inhibitors for the treatment of inflammatory diseases using QSAR, pharmacophore-structure-based virtual screening, and molecular dynamics. *J. Mol. Model.* **2018**, *24*, 225. [CrossRef] [PubMed]

14. Dos Santos, K.L.B.; Cruz, J.N.; Silva, L.B.; Ramos, R.S.; Neto, M.F.A.; Lobato, C.C.; Ota, S.S.B.; Leite, F.H.A.; Borges, R.S.; Silva, C.; et al. Identification of Novel Chemical Entities for Adenosine Receptor Type 2A Using Molecular Modeling Approaches. *Molecules* **2020**, *25*, 1245. [CrossRef]

15. E Clark, D. Rapid calculation of polar molecular surface area and its application to the prediction of transport phenomena. 1. Prediction of intestinal absorption. *J. Pharm. Sci.* **1999**, *88*, 807–814. [CrossRef] [PubMed]

16. Orlando, B.J.; Malkowski, M.G. Crystal structure of rofecoxib bound to human cyclooxygenase-2. *Acta Crystallogr. Sect. F Struct. Boil. Commun.* **2016**, *72*, 772–776. [CrossRef]

17. Chakraborty, S.; Sengupta, C.; Roy, K. Exploring QSAR with E-state index: Selectivity requirements for COX-2 versus COX-1 binding of terphenyl methyl sulfones and sulfonamides. *Bioorgan. Med. Chem. Lett.* **2004**, *14*, 4665–4670. [CrossRef]

18. Leão, R.P.; Cruz, J.V.; Da Costa, G.V.; Cruz, J.N.; Ferreira, E.F.B.; Silva, R.C.; De Lima, L.R.; Borges, R.S.; Dos Santos, G.B.; Santos, C.B.R. Identification of New Rofecoxib-Based Cyclooxygenase-2 Inhibitors: A Bioinformatics Approach. *Pharmaceuticals* **2020**, *13*, 209. [CrossRef]

19. Psimadas, D.; Georgoulias, P.; Valotassiou, V.; Loudos, G. Molecular Nanomedicine Towards Cancer: 111In-Labeled Nanoparticles. *J. Pharm. Sci.* **2012**, *101*, 2271–2280. [CrossRef]

20. Yan, A.; Wang, Z.; Cai, Z. Prediction of Human Intestinal Absorption by GA Feature Selection and Support Vector Machine Regression. *Int. J. Mol. Sci.* **2008**, *9*, 1961–1976. [CrossRef]

21. Wang, Q.; Rager, J.D.; Weinstein, K.; Kardos, P.S.; Dobson, G.L.; Li, J.; Hidalgo, I.J. Evaluation of the MDR-MDCK cell line as a permeability screen for the blood–brain barrier. *Int. J. Pharm.* **2005**, *288*, 349–359. [CrossRef]

22. Zhao, Y.H.; Le, J.; Abraham, M.H.; Hersey, A.; Eddershaw, P.J.; Luscombe, C.N.; Boutina, D.; Beck, G.; Sherborne, B.; Cooper, I.; et al. Evaluation of human intestinal absorption data and subsequent derivation of a quantitative structure–activity relationship (QSAR) with the Abraham descriptors. *J. Pharm. Sci.* **2001**, *90*, 749–784. [CrossRef]

23. Yazdanian, M.; Glynn, S.L.; Wright, J.L.; Hawi, A. Correlating partitioning and caco-2 cell permeability of structurally diverse small molecular weight compounds. *Pharm. Res.* **1998**, *15*, 1490–1494. [CrossRef] [PubMed]

24. Piccirillo, E.; Amaral, A.T.-D. BUSCA VIRTUAL DE COMPOSTOS BIOATIVOS: CONCEITOS E APLICAÇÕES. *Quím. Nova* **2018**, *41*, 662–677. [CrossRef]

25. Ramos, R.S.; Costa, J.D.S.; Silva, R.C.; Da Costa, G.V.; Rodrigues, A.B.L.; Rabelo Érica, M.; Souto, R.N.P.; Taft, C.A.; Silva, C.; Campos, J.M.; et al. Identification of Potential Inhibitors from Pyriproxyfen with Insecticidal Activity by Virtual Screening. *Pharmaceuticals* **2019**, *12*, 20. [CrossRef] [PubMed]

26. Mascarenhas, A.M.S.; De Almeida, R.B.M.; Neto, M.F.D.A.; Mendes, G.O.; Cruz, J.N.; Santos, C.B.R.; Botura, M.B.; Leite, F.H.A. Pharmacophore-based virtual screening and molecular docking to identify promising dual inhibitors of human acetylcholinesterase and butyrylcholinesterase. *J. Biomol. Struct. Dyn.* **2020**, 1–10. [CrossRef]

27. Garavito, R.M.; Malkowski, M.G.; DeWitt, D.L. The structures of prostaglandin endoperoxide H synthases-1 and -2. *Prostaglandins Other Lipid Mediat.* **2002**, *68*, 129–152. [CrossRef]

28. Magalhães, W.S.; Corrêa, C.M.; De Alencastro, R.B.; Nagem, T.J. Bases moleculares da ação anti-inflamatória dos ácidos oleanólico e ursólico sobre as isoformas da ciclo-oxigenase por docking e dinâmica molecular. *Quím. Nova* **2012**, *35*, 241–248. [CrossRef]

29. Fox, P.W.; Hershberger, S.L.; Bouchard, T.J. Genetic and enviromental contributions to the acquisition of a motor skill. *Nature* **1996**, *384*, 356–358. [CrossRef]

30. Limongelli, V.; Bonomi, M.; Marinelli, L.; Gervasio, F.L.; Cavalli, A.; Novellino, E.; Parrinello, M. Molecular basis of cyclooxygenase enzymes (COXs) selective inhibition. *Proc. Natl. Acad. Sci.* **2010**, *107*, 5411–5416. [CrossRef]

31. Inhibitor, S. Program Studi Biomedik, Fakultas Kedokteran Universitas Brawijaya, Malang, Indonesia Dosen Program Studi Pendidikan Biologi STKIP Pembangunan Indonesia Makassar 6. 2018. Available online: http://ojs.stkippi.ac.id/index.php/jip/article/view/128 (accessed on 19 August 2019).

32. Pham, V.C.; Shin, J.-S.; Choi, M.-J.; Kim, T.-W.; Lee, K.-J.; Kim, K.-J.; Huh, G.; Kim, J.-A.; Choo, D.J.; Lee, K.-T.; et al. Biological Evaluation and Molecular Docking Study of 3-(4-Sulfamoylphenyl)-4-phenyl-1H-pyrrole-2,5-dione as COX-2 Inhibitor. *Bull. Korean Chem. Soc.* **2012**, *33*, 721–724. [CrossRef]

33. Harman, C.A.; Turman, M.V.; Kozak, K.R.; Marnett, L.J.; Smith, W.L.; Garavito, R.M. Structural Basis of Enantioselective Inhibition of Cyclooxygenase-1 by S- -Substituted Indomethacin Ethanolamides. *J. Boil. Chem.* **2007**, *282*, 28096–28105. [CrossRef] [PubMed]

34. Gierse, J.K.; Koboldt, C.M.; Walker, M.C.; Seibert, K.; Isakson, P.C. Kinetic basis for selective inhibition of cyclo-oxygenases. *Biochem. J.* **1999**, *339*, 607–614. [CrossRef] [PubMed]

35. Walker, M.C.; Kurumbail, R.G.; Kiefer, J.R.; Moreland, K.T.; Koboldt, C.M.; Isakson, P.C.; Seibert, K.; Gierse, J.K. A three-step kinetic mechanism for selective inhibition of cyclo-oxygenase-2 by diarylheterocyclic inhibitors. *Biochem. J.* **2001**, *357*, 709–718. [CrossRef]

36. Hayashi, S.; Ueno, N.; Murase, A.; Nakagawa, Y.; Takada, J. Novel acid-type cyclooxygenase-2 inhibitors: Design, synthesis, and structure–activity relationship for anti-inflammatory drug. *Eur. J. Med. Chem.* **2012**, *50*, 179–195. [CrossRef] [PubMed]

37. Hayashi, S.; Nakata, E.; Morita, A.; Mizuno, K.; Yamamura, K.; Kato, A.; Ohashi, K. Discovery of {1-[4-(2-{hexahydropyrrolo[3,4-c]pyrrol-2(1H)-yl}-1H-benzimidazol-1-yl)piperidin-1-yl]cyclooctyl}methanol, systemically potent novel non-peptide agonist of nociceptin/orphanin FQ receptor as analgesic for the treatment of neuropathic pain: Design, synthesis, and structure–activity relationships. *Bioorgan. Med. Chem.* **2010**, *18*, 7675–7699. [CrossRef]

38. Hayashi, S.; Hirao, A.; Imai, A.; Nakamura, H.; Murata, Y.; Ohashi, K.; Nakata, E. Novel Non-Peptide Nociceptin/Orphanin FQ Receptor Agonist, 1-[1-(1-Methylcyclooctyl)-4-piperidinyl]-2-[(3R)-3-piperidinyl]-1H-benzimidazole: Design, Synthesis, and Structure–Activity Relationship of Oral Receptor Occupancy in the Brain for Orally Potent Antianxiety Drug(1, 2). *J. Med. Chem.* **2009**, *52*, 610–625. [CrossRef]

39. Silva, S.G.; Da Costa, R.A.; De Oliveira, M.S.; Cruz, J.N.; Figueiredo, P.L.B.; Brasil, D.D.S.B.; Nascimento, L.D.; Neto, A.D.J.C.; Junior, R.D.C.; Andrade, E.D.A. Chemical profile of Lippia thymoides, evaluation of the acetylcholinesterase inhibitory activity of its essential oil, and molecular docking and molecular dynamics simulations. *PLoS ONE* **2019**, *14*, e0213393. [CrossRef]

40. De Oliveira, M.S.; Cruz, J.N.; Silva, S.G.; Da Costa, W.A.; De Sousa, S.H.B.; Bezerra, F.W.F.; Teixeira, E.; Da Silva, N.J.N.; Andrade, E.D.A.; Neto, A.M.D.J.C.; et al. Phytochemical profile, antioxidant activity, inhibition of acetylcholinesterase and interaction mechanism of the major components of the Piper divaricatum essential oil obtained by supercritical CO_2. *J. Supercrit. Fluids* **2019**, *145*, 74–84. [CrossRef]

41. Costa, E.; Silva, R.; Espejo-Román, J.; Neto, M.D.A.; Cruz, J.; Leite, F.; Silva, C.; Pinheiro, J.; Macêdo, W.; Santos, C. Chemometric methods in antimalarial drug design from 1,2,4,5-tetraoxanes analogues. *SAR QSAR Environ. Res.* **2020**, 1–19. [CrossRef]

42. *Studio Discovery, Version 4.0*; Accelrys Software Inc.: San Diego, CA, USA, 2009.

43. Schneidman-Duhovny, D.; Dror, O.; Inbar, Y.; Nussinov, R.; Wolfson, H.J. PharmaGist: A webserver for ligand-based pharmacophore detection. *Nucleic Acids Res.* **2008**, *36*, W223–W228. [CrossRef]

44. Biava, M.; Porretta, G.C.; Poce, G.; De Logu, A.; Saddi, M.; Meleddu, R.; Manetti, F.; De Rossi, E.; Botta, M. 1,5-Diphenylpyrrole Derivatives as Antimycobacterial Agents. Probing the Influence on Antimycobacterial Activity of Lipophilic Substituents at the Phenyl Rings. *J. Med. Chem.* **2008**, *51*, 3644–3648. [CrossRef] [PubMed]

45. Gupta, A.K.; Gupta, R.A.; Soni, L.K.; Kaskhedikar, S.G. Exploration of physicochemical properties and molecular modelling studies of 2-sulfonyl-phenyl-3-phenyl-indole analogs as cyclooxygenase-2 inhibitors. *Eur. J. Med. Chem.* **2008**, *43*, 1297–1303. [CrossRef] [PubMed]

46. Huang, H.-C.; Li, J.J.; Garland, D.J.; Chamberlain, T.S.; Reinhard, E.J.; Manning, R.E.; Seibert, K.; Koboldt, C.M.; Gregory, S.A.; Anderson, G.D.; et al. Diarylspiro[2.4]heptenes as Orally Active, Highly Selective Cyclooxygenase-2 Inhibitors: Synthesis and Structure–Activity Relationships. *J. Med. Chem.* **1996**, *39*, 253–266. [CrossRef]

47. Navidpour, L.; Amini, M.; Shafaroodi, H.; Abdi, K.; Ghahremani, M.H.; Dehpour, A.R.; Shafiee, A. Design and synthesis of new water-soluble tetrazolide derivatives of celecoxib and rofecoxib as selective cyclooxygenase-2 (COX-2) inhibitors. *Bioorgan. Med. Chem. Lett.* **2006**, *16*, 4483–4487. [CrossRef] [PubMed]

48. Wesolowski, S.S.; Jorgensen, W.L. Estimation of binding affinities for celecoxib analogues with COX-2 via Monte Carlo-extended linear response. *Bioorgan. Med. Chem. Lett.* **2002**, *12*, 267–270. [CrossRef]

49. Habeeb, A.G.; Rao, P.N.P.; Knaus, E.E. Design and synthesis of 4,5-diphenyl-4-isoxazolines: Novel inhibitors of cyclooxygenase-2 with analgesic and antiinflammatory activity. *J. Med. Chem.* **2001**, *44*, 2921–2927. [CrossRef]

50. Stumpfe, D.; Bajorath, J. Exploring Activity Cliffs in Medicinal Chemistry. *J. Med. Chem.* **2012**, *55*, 2932–2942. [CrossRef]

51. Swarbrick, M.E.; Beswick, P.J.; Gleave, R.J.; Green, R.H.; Bingham, S.; Bountra, C.; Carter, M.C.; Chambers, L.J.; Chessell, I.P.; Clayton, N.M.; et al. Identification of [4-[4-(methylsulfonyl)phenyl]-6-(trifluoromethyl)-2-pyrimidinyl] amines and ethers as potent and selective cyclooxygenase-2 inhibitors. *Bioorgan. Med. Chem. Lett.* **2009**, *19*, 4504–4508. [CrossRef]

52. Wang, J.L.; Limburg, D.; Graneto, M.J.; Springer, J.; Hamper, J.R.B.; Liao, S.; Pawlitz, J.L.; Kurumbail, R.G.; Maziasz, T.; Talley, J.J.; et al. The novel benzopyran class of selective cyclooxygenase-2 inhibitors. Part 2: The second clinical candidate having a shorter and favorable human half-life. *Bioorgan. Med. Chem. Lett.* **2010**, *20*, 7159–7163. [CrossRef]

53. Xing, L.; Hamper, B.C.; Fletcher, T.R.; Wendling, J.M.; Carter, J.; Gierse, J.K.; Liao, S. Structure-based parallel medicinal chemistry approach to improve metabolic stability of benzopyran COX-2 inhibitors. *Bioorgan. Med. Chem. Lett.* **2011**, *21*, 993–996. [CrossRef]

54. Li, J.J.; Anderson, G.D.; Burton, E.G.; Cogburn, J.N.; Collins, J.T.; Garland, D.J.; Gregory, S.A.; Huang, H.-C.; Isakson, P.C. 1,2-Diarylcyclopentenes as Selective Cyclooxygenase-2 Inhibitors and Orally Active Anti-inflammatory Agents. *J. Med. Chem.* **1995**, *38*, 4570–4578. [CrossRef] [PubMed]

55. Orjales, A.; Mosquera, R.; López, B.; Olivera, R.; Labeaga, L.; Nunez, M.T. Novel 2-(4-methylsulfonylphenyl) pyrimidine derivatives as highly potent and specific COX-2 inhibitors. *Bioorgan. Med. Chem.* **2008**, *16*, 2183–2199. [CrossRef] [PubMed]

56. Blobaum, A.L.; Marnett, L.J. Molecular Determinants for the Selective Inhibition of Cyclooxygenase-2 by Lumiracoxib. *J. Boil. Chem.* **2007**, *282*, 16379–16390. [CrossRef] [PubMed]

57. Costa, J.D.S.; Costa, K.D.S.L.; Cruz, J.V.; Ramos, R.D.S.; Silva, L.B.; Brasil, D.D.S.B.; Silva, C.; Santos, C.B.R.; Macedo, W.J.D.C. Virtual Screening and Statistical Analysis in the Design of New Caffeine Analogues Molecules with Potential Epithelial Anticancer Activity. *Curr. Pharm. Des.* **2018**, *24*, 576–594. [CrossRef]

58. Cruz, J.V.; Neto, M.F.A.; Silva, L.B.; Ramos, R.S.; Costa, J.D.S.; Brasil, D.S.B.; Lobato, C.C.; Da Costa, G.V.; Bittencourt, J.A.H.M.; Silva, C.; et al. Identification of Novel Protein Kinase Receptor Type 2 Inhibitors Using Pharmacophore and Structure-Based Virtual Screening. *Molecules* **2018**, *23*, 453. [CrossRef]

59. Ferreira, E.F.B.; Silva, L.B.; Da Costa, G.V.; Costa, J.D.S.; Fujishima, M.A.T.; Leão, R.P.; Ferreira, A.L.S.; Federico, L.B.; Silva, C.; Campos, J.M.; et al. Identification of New Inhibitors with Potential Antitumor Activity from Polypeptide Structures via Hierarchical Virtual Screening. *Molecules* **2019**, *24*, 2943. [CrossRef]

60. Santos, C.B.R.; Braga, F.S.; Santos, C.F.; Costa, J.D.S.; De Melo, G.S.; De Mello, M.N.; Sousa, D.S.; Carvalho, J.C.T.; Socorro, D.D.; Brasil, B.; et al. Antimalarial Artemisinins Derivatives Study: Molecular Modeling and Multivariate Analysis (PCA, HCA, KNN, SIMCA and SDA). *J. Comput. Theor. Nanosci.* **2015**, *12*, 3443–3458. [CrossRef]

61. Ferreira, J.; Chaves, G.A.; Marino, B.L.B.; Sousa, K.P.A.; Souza, L.R.; Brito, M.F.B.; Teixeira, H.R.C.; Da Silva, C.H.T.P.; Santos, C.B.R.; Hage-Melim, L.I.S. Cannabinoid Type 1 Receptor (CB1) Ligands with Therapeutic Potential for Withdrawal Syndrome in Chemical Dependents ofCannabis sativa. *Chem. Med. Chem.* **2017**, *12*, 1408–1416. [CrossRef]

62. Wang, J.; Cieplak, P.; Kollman, P.A. How Well Does a Restrained Electrostatic Potential (RESP) Model Perform in Calculating Conformational Energies of Organic and Biological Molecules? *J. Comput. Chem.* **2000**, *21*, 1049–1074. [CrossRef]

63. Pinto, V.D.S.; Araújo, J.S.C.; Silva, R.C.; Da Costa, G.V.; Cruz, J.N.; Neto, M.F.A.; Campos, J.M.; Santos, C.B.R.; Leite, F.H.A.; Junior, M.C.S. In Silico Study to Identify New Antituberculosis Molecules from Natural Sources by Hierarchical Virtual Screening and Molecular Dynamics Simulations. *Pharmaceuticals* **2019**, *12*, 36. [CrossRef]

64. Costa, R.A.; Cruz, J.N.; Nascimento, F.C.A.; Silva, S.G.; Silva, S.O.; Martelli, M.C.; Carvalho, S.M.L.; Santos, C.B.R.; Neto, A.M.J.C.; Brasil, D.S.B. Studies of NMR, molecular docking, and molecular dynamics simulation of new promising inhibitors of cruzaine from the parasite Trypanosoma cruzi. *Med. Chem. Res.* **2018**, *28*, 246–259. [CrossRef]

65. Alves, F.S.; Rego, J.D.A.R.D.; Da Costa, M.L.; Da Silva, L.F.L.; Da Costa, R.A.; Cruz, J.N.; Brasil, D.D.S.B.; Brazil, D.D.S.B. Spectroscopic methods and in silico analyses using density functional theory to characterize and identify piperine alkaloid crystals isolated from pepper (Piper Nigrum L.). *J. Biomol. Struct. Dyn.* **2019**, *38*, 2792–2799. [CrossRef]

66. Wang, J.; Wang, W.; Kollman, P.A.; Case, D.A. Automatic atom type and bond type perception in molecular mechanical calculations. *J. Mol. Graph. Model.* **2006**, *25*, 247–260. [CrossRef] [PubMed]

67. Wang, J.; Wolf, R.M.; Caldwell, J.W.; Kollman, P.A.; Case, D.A. Development and testing of a general amber force field. *J. Comput. Chem.* **2004**, *25*, 1157–1174. [CrossRef] [PubMed]

68. Dolinsky, T.J.; Nielsen, J.E.; McCammon, J.A.; Baker, N.A. PDB2PQR: An automated pipeline for the setup of Poisson-Boltzmann electrostatics calculations. *Nucleic Acids Res.* **2004**, *32*, W665–W667. [CrossRef] [PubMed]

69. Case, D.A.; Cheatham, T.E., III; Darden, T.; Gohlke, H.; Luo, R.; Merz, K.M., Jr.; Onufriev, A.; Simmerling, C.; Wang, B.; Woods, R.J. The Amber biomolecular simulation programs. *J. Comput. Chem.* **2005**, *26*, 1668–1688. [CrossRef] [PubMed]

70. Cruz, J.N.; De Oliveira, M.S.; Silva, S.G.; Filho, A.P.D.S.S.; Pereira, D.S.; E Lima, A.H.L.; Andrade, E.H.D.A. Insight into the Interaction Mechanism of Nicotine, NNK, and NNN with Cytochrome P450 2A13 Based on Molecular Dynamics Simulation. *J. Chem. Inf. Model.* **2019**, *60*, 766–776. [CrossRef]

71. Santos, C.B.R.; Dos Santos, K.L.B.; Cruz, J.N.; Leite, F.H.A.; Borges, R.S.; Taft, C.A.; Campos, J.M.; Silva, C. Molecular modeling approaches of selective adenosine receptor type 2A agonists as potential anti-inflammatory drugs. *J. Biomol. Struct. Dyn.* **2020**, 1–13. [CrossRef]

72. Maier, J.A.; Martinez, C.; Kasavajhala, K.; Wickstrom, L.; Hauser, K.; Simmerling, C.L. ff14SB: Improving the Accuracy of Protein Side Chain and Backbone Parameters from ff99SB. *J. Chem. Theory Comput.* **2015**, *11*, 3696–3713. [CrossRef]

73. Jorgensen, W.L.; Chandrasekhar, J.; Madura, J.D.; Impey, R.W.; Klein, M.L. Comparison of simple potential functions for simulating liquid water. *J. Chem. Phys.* **1983**, *79*, 926–935. [CrossRef]

74. Darden, T.; York, D.; Pedersen, L. Particle mesh Ewald: AnN·log(N) method for Ewald sums in large systems. *J. Chem. Phys.* **1993**, *98*, 10089–10092. [CrossRef]

75. Ryckaert, J.-P.; Ciccotti, G.; Berendsen, H.J. Numerical integration of the cartesian equations of motion of a system with constraints: Molecular dynamics of n-alkanes. *J. Comput. Phys.* **1977**, *23*, 327–341. [CrossRef]

76. Izaguirre, J.A.; Catarello, D.P.; Skeel, R.D.; Wozniak, J.M. Langevin stabilization of molecular dynamics. *J. Chem. Phys.* **2001**, *114*, 2090–2098. [CrossRef]

77. Sun, H.; Li, Y.; Shen, M.; Tian, S.; Xu, L.; Pan, P.; Guan, Y.; Hou, T. Assessing the performance of MM/PBSA and MM/GBSA methods. 5. Improved docking performance using high solute dielectric constant MM/GBSA and MM/PBSA rescoring. *Phys. Chem. Chem. Phys.* **2014**, *16*, 22035–22045. [CrossRef]

78. Cruz, J.N.; Da Costa, K.S.; De Carvalho, T.A.A.; De Alencar, N.A.N. Measuring the structural impact of mutations on cytochrome P450 21A2, the major steroid 21-hydroxylase related to congenital adrenal hyperplasia. *J. Biomol. Struct. Dyn.* **2019**, *38*, 1425–1434. [CrossRef]

79. Da Costa, K.S.; Galúcio, J.M.; Da Costa, C.H.S.; Santana, A.R.; Carvalho, V.D.S.; Nascimento, L.D.D.; E Lima, A.H.L.; Cruz, J.N.; Alves, C.N.; Lameira, J. Exploring the Potentiality of Natural Products from Essential Oils as Inhibitors of Odorant-Binding Proteins: A Structure- and Ligand-Based Virtual Screening Approach To Find Novel Mosquito Repellents. *ACS Omega* **2019**, *4*, 22475–22486. [CrossRef]

80. Vale, V.V.; Cruz, J.N.; Viana, G.M.R.; Póvoa, M.M.; Brasil, D.D.S.B.; Dolabela, M.F. Naphthoquinones isolated from Eleutherine plicata herb: In Vitro antimalarial activity and molecular modeling to investigate their binding modes. *Med. Chem. Res.* **2020**, *29*, 487–494. [CrossRef]

81. Ramos, R.S.; Macêdo, W.J.C.; Costa, J.S.; Silva, C.H.T.D.P.D.; Rosa, J.M.C.; Da Cruz, J.N.; De Oliveira, M.S.; Andrade, E.H.D.A.; E Silva, R.B.L.; Souto, R.N.P.; et al. Potential inhibitors of the enzyme acetylcholinesterase and juvenile hormone with insecticidal activity: Study of the binding mode via docking and molecular dynamics simulations. *J. Biomol. Struct. Dyn.* **2019**, 1–23. [CrossRef]

82. Cruz, J.N.; Costa, J.F.S.; Khayat, A.S.; Kuca, K.; Barros, C.A.L.; Neto, A.M.J.C. Molecular dynamics simulation and binding free energy studies of novel leads belonging to the benzofuran class inhibitors of Mycobacterium tuberculosis Polyketide Synthase 13. *J. Biomol. Struct. Dyn.* **2018**, *37*, 1616–1627. [CrossRef]

Sample Availability: Samples of the compounds not available from the authors.

 *molecules*

Article

Computer-Driven Development of an in Silico Tool for Finding Selective Histone Deacetylase 1 Inhibitors

Hajar Sirous [1],*, Giuseppe Campiani [2], Simone Brogi [3],*, Vincenzo Calderone [3] and Giulia Chemi [2],†

1. Bioinformatics Research Center, School of Pharmacy and Pharmaceutical Sciences, Isfahan University of Medical Sciences, Isfahan 81746-73461, Iran
2. Department of Biotechnology, Chemistry and Pharmacy, DoE Department of Excellence 2018–2022, University of Siena, via Aldo Moro 2, 53100 Siena, Italy; campiani@unisi.it (G.C.); GChemi001@dundee.ac.uk (G.C.)
3. Department of Pharmacy, University of Pisa, via Bonanno 6, 56126 Pisa, Italy; vincenzo.calderone@unipi.it
* Correspondence: h_sirous@pharm.mui.ac.ir (H.S.); simone.brogi@unipi.it (S.B.); Tel.: +98-313-792-7065 (H.S.); +39-050-2219613 (S.B.)
† Present address: Wellcome Centre for Anti-Infectives Research, Drug Discovery Unit, Division of Biological Chemistry and Drug Discovery, University of Dundee, Dundee DD1 5EH, UK.

Academic Editors: Marco Tutone and Anna Maria Almerico
Received: 5 April 2020; Accepted: 20 April 2020; Published: 22 April 2020

Abstract: Histone deacetylases (HDACs) are a class of epigenetic modulators overexpressed in numerous types of cancers. Consequently, HDAC inhibitors (HDACIs) have emerged as promising antineoplastic agents. Unfortunately, the most developed HDACIs suffer from poor selectivity towards a specific isoform, limiting their clinical applicability. Among the isoforms, HDAC1 represents a crucial target for designing selective HDACIs, being aberrantly expressed in several malignancies. Accordingly, the development of a predictive in silico tool employing a large set of HDACIs (aminophenylbenzamide derivatives) is herein presented for the first time. Software Phase was used to derive a 3D-QSAR model, employing as alignment rule a common-features pharmacophore built on 20 highly active/selective HDAC1 inhibitors. The 3D-QSAR model was generated using 370 benzamide-based HDACIs, which yielded an excellent correlation coefficient value ($R^2 = 0.958$) and a satisfactory predictive power ($Q^2 = 0.822$; $Q^2_{F3} = 0.894$). The model was validated ($r^2_{ext\_ts} = 0.794$) using an external test set (113 compounds not used for generating the model), and by employing a decoys set and the receiver-operating characteristic (ROC) curve analysis, evaluating the Güner–Henry score (GH) and the enrichment factor (EF). The results confirmed a satisfactory predictive power of the 3D-QSAR model. This latter represents a useful filtering tool for screening large chemical databases, finding novel derivatives with improved HDAC1 inhibitory activity.

Keywords: 3D-QSAR; pharmacophore modeling; ligand-based model; HDACs; isoform-selective histone deacetylase inhibitors; aminophenylbenzamide

---

1. Introduction

Epigenetic defects in gene expression are well known in the onset and progression of cancer. In this context, pharmacological targeting of proteins of the cellular epigenetic machinery has provided opportunities for anti-cancer drug design [1,2]. Among epigenetic enzymes, histone deacetylases (HDACs) hold a fundamental role in regulating gene expression through histone post-translational modifications [3–5].

HDACs catalyze the removal of acetyl groups from the acetylated ε-amino termini of lysine residues located at the tails of the nucleosomal histones core. Histone deacetylation process leads to condensed

chromatin structure which concomitantly restricts the accessibility of related transcriptional factors to their target genes, thereby suppressing gene expression including tumor suppressor genes [6–8]. The abnormal regulation of this process culminates with the high expression level of HDACs. This event has been observed in the development of several human cancers. Consequently, effective inhibition of HDACs has recently gained importance as a valid therapeutic strategy to reverse aberrant epigenetic changes associated with cancer [9–11]. HDAC inhibitors (HDACIs) induce histone hyperacetylation and subsequent transcriptional re-activation of suppressed genes which are correlated with a variety of effects on tumor cells including apoptosis, differentiation, cell cycle arrest, inhibition of proliferation and cytostasis [12,13].

The HDAC family comprises 18 isoforms in mammalian cells which are categorized into four main classes (class I-IV) based on their structural and functional characteristics. HDACs belonging to class I (HDAC1–3 and 8), II (HDACs 4–7, 9 and 10) and IV (HDAC11) are all zinc-dependent metalloenzymes, while class III HDACs, also known as the sirtuins (SIRT1-7), requires NAD^+ as a cofactor for their catalytic function [6,14].

Extensive efforts over recent decades have led to the identification of four chemically diverse classes of HDACIs as potent antineoplastic agents including, hydroxamates, benzamides, cyclic peptides, and short-chain fatty acids [3]. The main breakthrough in developing these inhibitors was achieved by the US FDA approval of Vorinostat [15], Belinostat [16], Panobinostat (hydroxamate-based inhibitors) [17], Romidepsin (a cyclic peptide) [18] and Chidamide (a benzamide-based inhibitor) [19] for the treatment of lymphoma and myeloma. Moreover, several HDACIs such as Mocetinostat [20], Entinostat [21], Tacedinaline [22], Givinostat [23], and Abexinostat [24] are currently in clinical trials for treating various types of cancers. The structures of several approved and clinical HDACIs are shown in Figure 1.

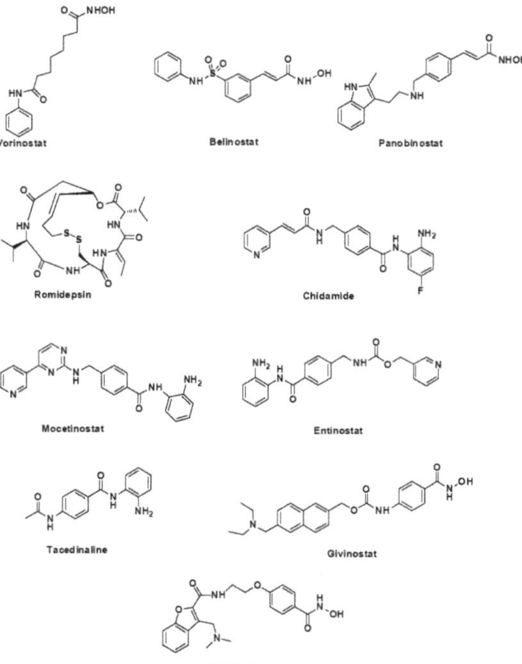

Figure 1. HDACIs approved by FDA and/or in clinical trials.

Despite these successes, the most known HDACIs target multiple HDAC isoforms and this poor selectivity represents a major drawback which limits their broad clinical utility [25,26]. Isoform-selective

HDACIs would offer superior therapeutic advantages due to limited off-target and undesirable effects, improved clinical efficacy and better tolerability [27,28]. Moreover, isoform-selective inhibitors would provide chemical tools to delineate the precise roles of individual HDAC isoforms in human diseases, including rare disorders [29]. Therefore, in recent years, identification of highly potent inhibitors with strict selectivity towards a specific isoform have caught more attention in the development of novel HDACIs for the epigenetic therapy [30–33]. In this context, due to the pivotal role of HDAC1 in the angiogenesis, proliferation, and survival of mammalian carcinoma cells, this isoform is particularly being sought as a preferred target for successful design of selective HDACIs [34–38].

A wide range of hydroxamic acid derivatives are known to be potent pan-inhibitors of several HDAC isoforms [39,40]. Accordingly, in recent years, there has been considerable interest in developing non-hydroxamate HDACIs with satisfactory selectivity towards a specific isoform. In this context, aminophenylbenzamide derivatives represent an important class of non-hydroxamate HDACIs owing to their potent HDAC inhibition along with desirable pharmacokinetic profile and excellent selectivity for HDAC class I enzyme. Furthermore, inhibitors containing an ortho-aminobenzamide typically exhibit relatively greater levels of selectivity for class I HDACs, particularly HDAC1 [35,41]. In addition, hydroxamic-derived inhibitors often suffer from some serious pharmacokinetic issues including poor metabolic stability, rapid clearance, undesirable oral absorption, and short half-life in plasma, whereas benzamides-based inhibitors show better metabolic stability and oral bioavailability [42–45]. For these solid reasons, renewed efforts are being directed towards the further exploration of innovative aminophenylbenzamide chemotypes as privileged and valuable scaffolds to develop isoform-selective HDACIs [46–48].

Recently, in silico techniques, including ligand-based methods such as pharmacophore modeling and three-dimensional quantitative structural activity relationship (3D-QSAR), have efficiently contributed to guide the discovery of novel bioactive molecules, with reduced costs in terms of money and time [49–51]. In fact, QSAR methods provide relationships between physicochemical properties of a series of compounds and their biological activities to obtain a reliable statistical model for predicting the activities of new chemical entities. The fundamental principle of the technique is that the change in structural properties determines modifications in biological activities of the compounds. In the classical QSAR approaches, affinities of ligands to their binding sites, inhibition constants, rate constants, and other biological data have been correlated with molecular properties including lipophilicity, polarizability, electronic and steric properties (Hansch analysis) or with structural features (Free-Wilson analysis). However, classical QSAR approach has only a limited utility for designing new molecules due to the lack of consideration of the 3D structure of the selected compounds. Accordingly, 3D-QSAR has emerged as a natural extension to the classical Hansch and Free-Wilson approaches, which exploits the three-dimensional properties of the ligands to predict their biological activities employing robust chemometric techniques such as partial least squares (PLS). The success of these methods can be attributed to several factors including identification of important features for the activity, rationalization of activity trends in molecules under study, prediction of the specific activity for a selected target or undesirable effects of new compounds. On the other hand, ligand-based methods have been used in virtual screening campaigns of chemical databases to find novel hits with improved potency and can be combined with other computational and experimental workflows to discover new potential drugs [52–57].

Until now, only limited number of 3D-QSAR studies for hydroxamate set of HDACIs have been reported [58–64]. However, to the best of our knowledge, no previous attempt has been made to seek the structural and chemical features of aminophenylbenzamide governing their HDAC inhibitory activities employing 3D-QSAR methodology along with a large set of compounds. Given the aforementioned therapeutic significance of this class of inhibitors, we developed and validated a 3D-QSAR model using a comprehensive set of previously reported benzamide derivatives as selective HDAC1 inhibitors. From this perspective, the software Phase, implemented in Maestro, was employed to explore a common-features pharmacophore hypothesis based on highly active ligands. This hypothesis was

then used as an alignment rule to derive a predictive 3D-QSAR model [65]. Such an in silico tool could aid not only in forecasting the HDAC1 inhibitory activity of newly designed chemical entities, but also offer a robust foundation for designing new selective HDACIs with increased binding affinities to HDAC1. Accordingly, the developed model could have a relevant implication in drug discovery campaign for searching isoform-selective HDACIs.

2. Results and Discussion

The application of 3D-QSAR methodology in the design of HDACIs has received little attention to date and only a few instances of field-based QSAR models (comparative molecular field analysis (CoMFA) and comparative similarity indices analysis (CoMSIA) methods) have been reported for hydroxamate-based inhibitors [58–64]. However, this approach has not been carried out for a large set of benzamide-based derivatives behaving as HDACIs. On the other hand, we have recently developed a series of predictive 3D-QSAR models for different purposes including the identification or rational design of new chemical entities for different targets [53,55], and the prediction of undesirable effects of novel molecules such as potential hERG K$^+$ channel related cardiotoxicity [57]. In all these cases, Phase was used to develop a computational tool using a pharmacophore-based alignment that links the information of pivotal functional groups of the ligands with their biological activity [65]. The fruitful results of the above-mentioned molecular modeling studies as well as therapeutic significance of benzamide chemotypes as valuable isoform-selective HDACIs inspired us to derive a pharmacophore-based 3D-QSAR model to be used as a screening filtering tool able to quantitatively predict the HDAC1 inhibitory activity of newly designed ligands.

2.1. Data Set Preparation

A comprehensive data set of 370 diverse HDACIs based on benzamide scaffold with functional biological activity expressed as IC$_{50}$ (see the Supplementary Materials for further details) with a range of HDAC1 inhibitory activities spanning five orders of magnitude (from 6.0 nM of compound **17** to 50 μM of compound **132**, Table S1) were selected from literature for developing a predictive 3D-QSAR model. Subsequently, an extensive conformational search for each ligand was performed employing MacroModel software (see experimental section for further details). Conformational analysis is crucial to enhance both the quality of the alignment for the molecules used to generate the 3D-QSAR model and the reliability of the in silico tool [53–57]. After the exhaustive conformational analysis of the selected ligands (Table S1), the generation of the 3D-QSAR model was started.

2.2. Pharmacophore Modeling and 3D-QSAR Model Generation

As a first step to develop the 3D-QSAR model, 20 most active compounds with IC$_{50}$ values ≤ 10 nM included in the data set (ligands **1–20**, Figure 2, Table S1) were considered to find out common pharmacophore hypotheses that were subsequently scored and ranked by the software Phase. This means that the highly active compounds possess common features that are responsible for the activity exploited by a 3D pharmacophore hypothesis. Therefore, a pharmacophore hypothesis provides a rational picture of primary chemical features of ligands responsible for HDAC1 inhibitory activity and therefore can be used as a reliable alignment rule for the 3D-QSAR model development.

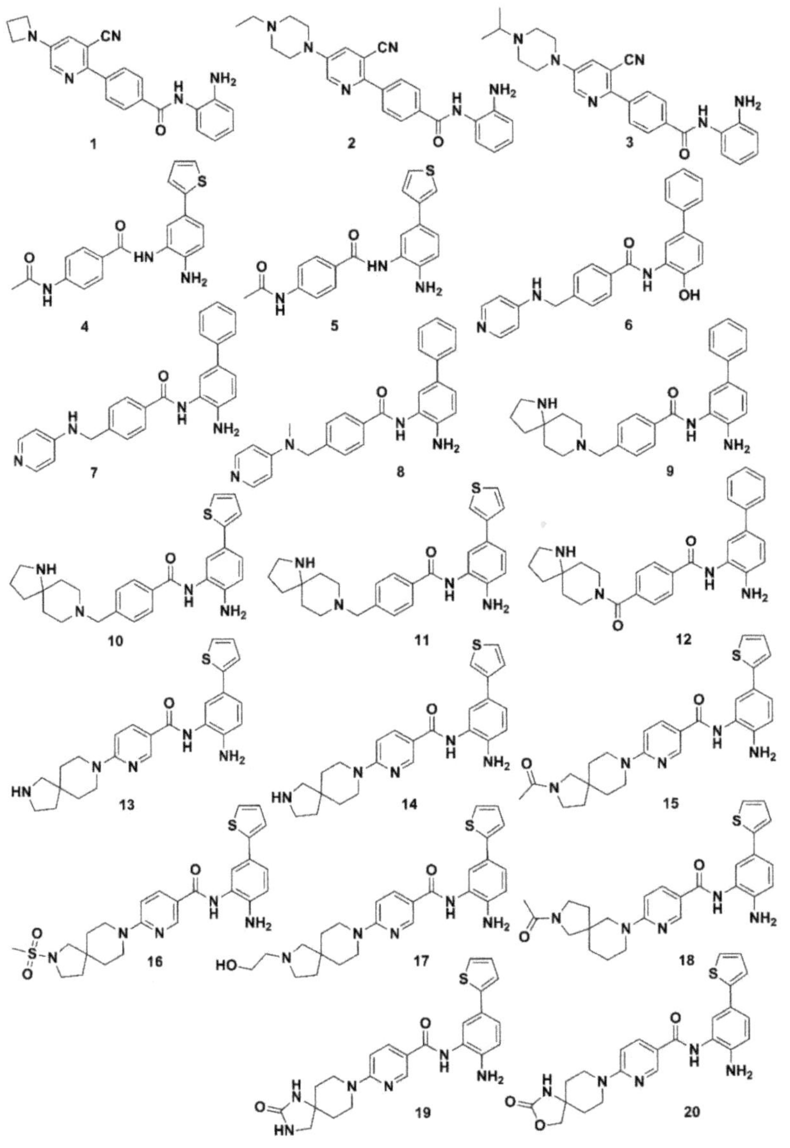

Figure 2. Chemical structure of highly active compounds against HDAC1 (IC_{50} comprised between 0.004 µM and 0.01 µM) used for generating a common-features pharmacophore.

To have optimal combination of sites or features shared by the most active ligands, the minimum and maximum number of site points were set on 5. This means that we selected 5 as maximum features to include in the pharmacophore models. Among the 26 common pharmacophore hypotheses generated by the software Phase, only those models which showed superior alignment with the active compounds were identified by calculating the survival score. The survival scoring function of Phase module identifies the best candidate hypothesis from the generated models and offers an overall ranking of all the hypotheses. The scoring algorithm includes contributions from the alignment of site points and vectors, volume overlap, selectivity, number of ligands matched, relative conformational

energy, and activity. To identify pharmacophore models with more active and less inactive features, all models were mapped to inactive compounds and scored. If inactives score well, the hypothesis could be invalid because it does not discriminate between actives and inactives. Therefore, adjusted survival score was calculated by subtracting the inactive score from the survival score of these pharmacophores.

After the scoring, the model ADDRR, herein referred to ADDRR hypothesis, with the maximum adjusted survival score (3.769) and lowest relative conformational energy, was selected as the top-ranked hypothesis among the generated 3D model hypotheses. The different scoring parameters for the selected hypothesis (ADDRR) were provided in Table 1. The 3D spatial arrangement of all features with inter-feature distance constraints of ADDRR are presented in Figure 3. As shown in this figure, the hypothesis was characterized by the five main features: one hydrogen-bond acceptor (A), two hydrogen-bond donors (D), and two aromatic rings (R).

Table 1. The different scoring parameters for the best pharmacophore hypothesis, matching all 20 highly active compounds used.

HYPO ID	Survival	Survival—Inactive	Site	Vector	Volume	Selectivity	Matches	Energy	Activity	Inactive
ADDRR	3.769	1.841	0.97	0.999	0.798	1.578	20	0.006	2.097	1.928

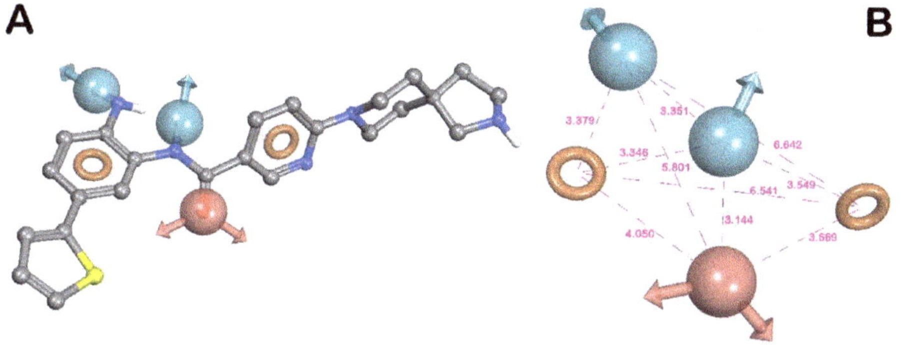

Figure 3. (**A**) Superposition of highly active compound **13** and ADDRR hypothesis. (**B**) ADDRR hypothesis and its inter-feature distances. Features are as follows: H-bond acceptor = A, red vector; H-bond donor = D, blue vectors; aromatic feature = R, orange ring (pictures were generated by means of Maestro software).

Figure 3A depicts one of the most active ligands in the set (compound **13**, Table S1), mapped onto the ADDRR pharmacophore. As depicted in the mentioned figure, compound **13** thoroughly fits all features of the pharmacophore model, underlining the previous findings on structural components required for interacting with the HDAC1 binding site [66–68]. As illustrated in Figure 3, the carbonyl oxygen of the benzamide group served as a hydrogen-bond acceptor (HBA) feature, while two hydrogen-bond donor (HBD) features were mapped to the protons of the 2-aminophenyl NH_2 and amide NH. Furthermore, out of two aromatic features, one was mapped to the phenyl ring of the 2-aminophenyl. The other aromatic feature was mapped to the pyridine ring of the nicotinamide moiety. This hypothesis was well corroborated by accepted common pharmacophore model for HDACIs comprising the zinc binding group (ZBG), a linker and a cap group as established by computational and biophysical studies reported for HDAC1 inhibitors [66–70]. Based on the current study, the aniline groups of the benzamide-based inhibitors are served as ZBG, coordinating the catalytic zinc ion in the HDAC1 active site. Moreover, it is well known that H-bonds formation with key residues of HDAC1 active site are commonly found for the ZBG of this class of HDAC1 inhibitors. In particular, in addition to the zinc ion coordination, protons of the 2-aminophenyl NH_2 group could also establish hydrogen bonds with His140 and His141, while the carbonyl oxygen of the benzamide portion could also form another hydrogen binding interaction with hydroxyl group of

Tyr303. It has been reported that NH of amide could offer the appropriate HBD vector to address the Gly149 through H-bond formation. The presence of two aromatic features capable of participating in π-π stacking interactions with hydrophobic residues Phe150, Tyr204, Phe205, and Tyr303, represents another important requisite for further stabilization of ligand binding. These residues were located in a long and narrow hydrophobic tube-like channel and thus interaction with them allow tubular access of ligand into active site [66,67,69,70]. Accordingly, the above-mentioned ADDRR hypothesis imparts the key features of ligand for providing the relevant interactions with the HDAC1 active site.

The ADDRR pharmacophore hypothesis was then employed as alignment rule to derive the 3D-QSAR model. In this step, the compounds were randomly divided into training (70%) and test sets (30%) taking into account that the response range was well-covered in both sets (Table S1). This choice was made to warrant the inclusion of the positive information originating from 70% of the compounds enclosed in the training set (corresponding to 259 compounds), for the development of the computational tool. Moreover, the compounds kept in the test set (30%, 111 compounds) ensures an appropriate assessment of the predictive power of the generated model through an exhaustive internal validation. The atom-based version of Phase's 3D-QSAR workflow was preferred to the pharmacophore-based one. Such a choice allowed us to take into account contributions associated with all the important structural features other than pharmacophore for HDAC1 inhibitory activity such as the steric clashes. To enhance the model accuracy and evade overfitting phenomenon, models containing one up to seven factors were generated for the studied data set. Statistical parameters for each model are provided in Table 2. Model featuring seven factors was preferred and selected because it better performed in comparison with other models. The reliability of the selected model is justified by the fact that all statistical parameters were in acceptable range. In this regard, the correlation and cross-validated correlation coefficients ($R^2 = 0.958$ and $Q^2 = 0.822$, respectively) of the selected model along with the Pearson R-value (R-Pearson = 0.915) were extremely satisfactory, indicating a close correspondence between estimated and experimental IC_{50} values. Moreover, the high Fisher ratio (F = 822.1) suggested a statistically significant regression model, which was further supported by the small value of the variance ratio (P = 4.377×10^{-169}), an indication of a high degree of confidence. Finally, the small values of the standard deviation and the root-mean-square error (SD: 0.178 and RMSE: 0.281, respectively) also provided indication about the robustness of the developed computational model. Moreover, the Q^2_{F3} value clearly indicates that the 3D-QSAR model with seven factors is robust.

Table 2. 3D-QSAR statistical parameters of the seven Latent Variables (LVs) Phase-derived sets of models.

LVs	R^{2a}	SD^b	F^c	P^d	$RMSE^e$	Q^{2f}	$Q^2_{F3}{}^g$	$R\text{-Pearson}^h$
1	0.3408	0.6978	132.9	4.699×10^{-22}	0.5199	0.3886	0.635	0.6747
2	0.6273	0.5257	215.5	1.344×10^{-55}	0.3979	0.6420	0.791	0.8282
3	0.7620	0.4209	272.2	3.629×10^{-79}	0.3615	0.7045	0.823	0.8579
4	0.8775	0.3025	455.0	1.704×10^{-144}	0.3514	0.7207	0.833	0.8690
5	0.9159	0.2512	551.0	9.327×10^{-134}	0.2971	0.8004	0.881	0.9003
6	0.9433	0.2067	698.6	6.590×10^{-134}	0.2865	0.8143	0.890	0.9100
7	0.9582	0.1778	822.1	4.377×10^{-169}	0.2808	0.8217	0.894	0.9152

[a] R^2: value of r^2 of the regression. [b] SD: standard deviation of the regression. [c] F: variance ratio. [d] P: significance level of variance ratio. [e] RMSE: root-mean-square error in the test set predictions. [f] Q^2: value of Q^2 for the predicted activities. [g] Q^2_{F3}: value of Q^2_{F3} for the predicted activities calculated as reported in Materials and Methods section. [h] R-Pearson: correlation between the predicted and observed selectivity index values for the test set.

A scatter plot of experimental versus predicted activities was generated to assess the results (Figure 4). Based on this plot, the IC_{50} values were reliably predicted for both training and test set molecules (Table S1). This plot along with the aforementioned statistical features clearly imply the significance of the approach and indicate a QSAR model with a robust predictive power.

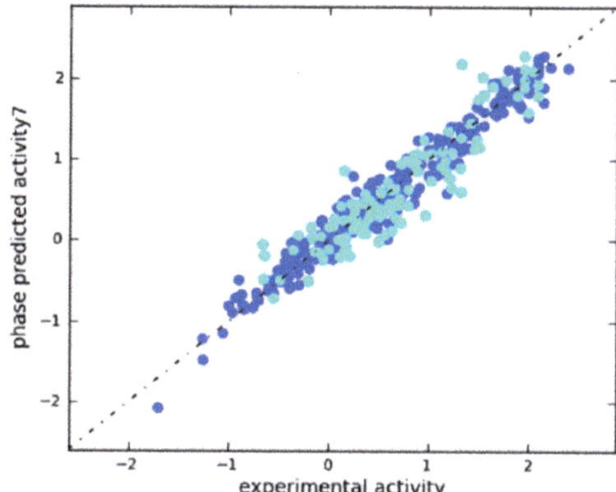

Figure 4. Scatter plot for the predicted (Phase-predicted activity) and the observed (experimental activity) pIC$_{50}$ values (µM) as calculated by the 3D-QSAR model applied to the training set (blue) and test set (cyan) compounds.

Three-dimensional aspects obtained from the QSAR model were visualized using 3D plots of the crucial volume elements occupied by ligands. Such plots allow the visual analysis of important features of ligand structures along with their contributions to the biological activity. The 3D plot representation of the whole model, superimposed to the highly (**3, 10**, and **13**), moderate (**22, 213**, and **239**), and less active derivatives (**218, 279**, and **299**), is depicted in Figure 5. In this illustration, the blue and red cubes indicate the positive and negative coefficients, respectively. In fact, blue cubes refer to ligand regions in which the specific feature is important for better activity, whereas the red cubes are indicative of a particular structural feature or functional group which is not essential for the activity or is likely to decrease the activity. Cubes with small positive and negative coefficients, which therefore did not greatly affect the activity, were filtered out by setting a 1.50×10^{-2} coefficient threshold. Remarkably, compounds **10** and **13** (Figure 5A,B, respectively) as well as other highly active ligands, mainly lodge in the blue regions, while the less active compounds such as **218** and **299** (Figure 5G,I, respectively) largely resides on the red regions. Moreover, regarding some compounds with moderate activity and generally all compounds with limited activity, we also observed a significant inability to match all the pharmacophore features, according to the decrease of inhibitory potencies.

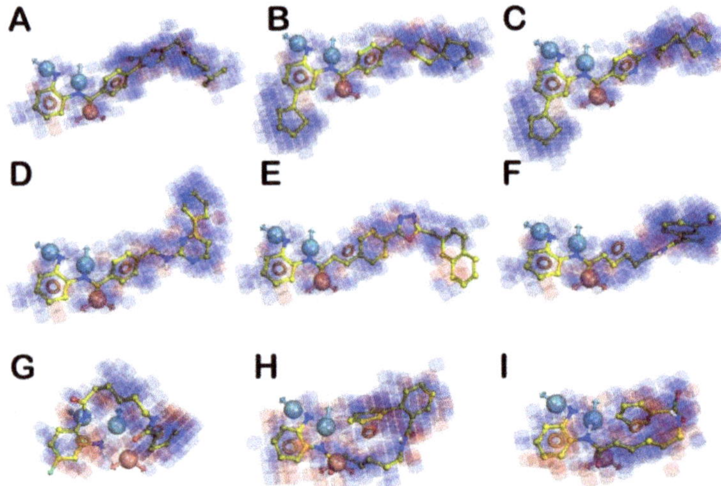

Figure 5. (**A–C**) Superposition of highly active compounds **3**, **10**, and **13**, respectively with the 3D-QSAR model. (**D–F**) Superposition of moderate active compounds **22** (Mocetinostat), **213**, and **239**, respectively with the 3D-QSAR model. (**G–I**) Superposition of less active compounds **218**, **279**, and **299**, respectively with the 3D-QSAR model. The picture was generated by means of Maestro software (Schrödinger, LLC, New York, NY, USA, 2015).

2.3. In Silico 3D-QSAR Model Validation

2.3.1. Validation Using External Test Set

After the generation of the 3D-QSAR model, a preliminary in silico validation was performed using an external test set selected from the literature that have not been used for generating the computational model. This set was composed of 113 compounds with different inhibitory activities against HDAC1 (ranging from 5.8 nM to 1140 nM; Table S2 in the Supplementary Materials). As reported in Table S2, our model was satisfactorily efficient in estimating the HDAC1 inhibitory activity of compounds included in the external test set. In the scatter plot depicted in Figure 6, the experimental and predicted pIC_{50} values of these compounds are also displayed, offering a reasonable correlation coefficient ($r^2_{ext\_ts} = 0.794$). This result provided further confirmation that the correlation shown by the model is not accidental.

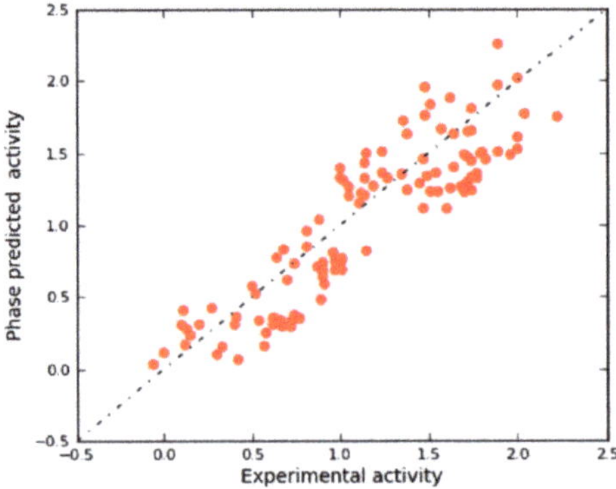

Figure 6. Scatter plot for the predicted (Phase-predicted activity) and observed (experimental activity) pIC_{50} values (μM) as calculated by the 3D-QSAR model with 7 factors applied to the external test set.

2.3.2. Validation Using Decoy Set and Receiver-Operating Characteristic (ROC) Curve Approach

For a further validation and to assess the performance of the developed 3D-QSAR model, we employed a validation method based on generation of decoys set. This procedure is usually employed to evaluate the capability of in silico tools such as 3D-QSAR models to discriminate between active or inactive molecules [71–75]. Starting from highly active compounds (ligands 1–20 in Table S1), 86 additional compounds with good activity against HDAC1 (cutoff IC_{50} < 35 nM; Tables S1 and S2) were selected from the training, test and external validation sets for a total of 106 compounds (Table S3) from which decoys were generated. For this set of active ligands, DUD-E server generated 5764 decoys. After an appropriate minimization and conformational search of decoys, we have combined them with the active molecules (referred as A in Figure 7A) for a total of 5870 compounds (referred as D in Figure 7A) that were then subjected to a virtual screening using the developed 3D-QSAR model. Interestingly, the results of this evaluation supported the validity of the proposed model. Analysis of the database screening results (Figure 7A) indicated a trend in which inactive compounds fail to completely satisfy all the pharmacophore features, thus making their predicted activity very poor or absent. In contrast, the 3D-QSAR model was reasonably efficient in the estimation of HDAC1 inhibitory activity of active compounds.

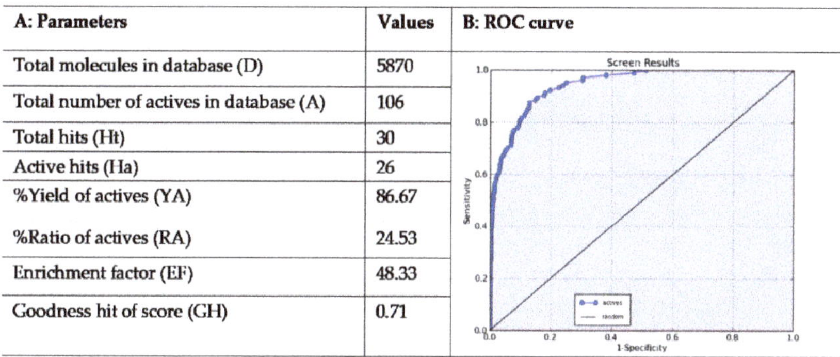

A: Parameters	Values	B: ROC curve
Total molecules in database (D)	5870	
Total number of actives in database (A)	106	
Total hits (Ht)	30	
Active hits (Ha)	26	
%Yield of actives (YA)	86.67	
%Ratio of actives (RA)	24.53	
Enrichment factor (EF)	48.33	
Goodness hit of score (GH)	0.71	

Figure 7. (**A**) EF and GH scores obtained by the application of 3D-QSAR model in a database screening; and (**B**) receiver-operating characteristic (ROC) curve generated from database screening.

According to the screening results (Figure 7A), the top 30 ranked compounds were considered to be hits (Ht). This cutoff value could represent a suitable number of molecules (about 1% of database) to be purchased after a virtual screening campaign. Remarkably, among Ht, 26 (Ha) compounds belonged to the set of 106 known HDAC1 inhibitors. Furthermore, this qualitative analysis was well supported by the calculation of some statistical parameters such as EF and GH score (see Materials and Methods section for calculation details). In this regard, the calculated EF was 48.33, which implies that it could be about 48.33 times more probable to select active compounds from the hit list compared with random selection from the complete database. The estimated GH score value of 0.71, larger than 0.5, indicates a great reliability of the model (Figure 7A). This suggests that the developed computational model can serve as efficient tool in virtual screening studies to find out novel chemical entities behaving as selective HDAC1 inhibitors.

The applicability of the proposed 3D-QSAR model was further evaluated by means of the receiver-operating characteristic (ROC) curve. The ROC curve approach is a well-recognized metric used as an objective way to assess the balance between model sensitivity (capability to discover true positives) and specificity (capability to avoid false positives) [55,57,76,77]. For this purpose, 5870 compounds employed in the previous validation step, were ranked according to their predicted activity values as estimated by the 3D-QSAR model. The output of the ROC curve provided a score for appraising the overall performance of the model. In particular, the closer the ROC score is to 1.0, the

better is the model at discriminating active from inactive compounds. ROC curve analysis of our in silico model yielded a satisfactory Area Under the Curve (AUC) score of 0.94 (Figure 7B), providing additional evidence about the predictivity of the developed 3D-QSAR model.

3. Materials and Methods

3.1. Hardware and Software Specifications

All computational tasks in this study were carried out using molecular modeling package from Schrödinger suite 2015 (Schrödinger, Inc., LLC, New York, NY, USA) installed on an Intel(R) Xeon(R) CPU E5-2620 v2 @ 3.30 GHz, 64 GB RAM with 12 processors, and a 2GB graphics card of NVIDIA Quadro K2200 running Ubuntu 10.04 LTS (long-term support) as operating system. Access to the Schrödinger modules as well as the capability to organize and analyze data was provided by Maestro as a portal interface of Schrödinger [78].

3.2. Ligands and Data Set Preparation

A comprehensive set of HDAC1 inhibitors characterized by the 2-aminophenylbenzamide scaffold with known IC_{50} values that vary over a wide range was collected from the literature [66,67,79–92] and the bindingDB database [93]. The selection criterion for the compounds to be included in the set was that their HDAC1 inhibition was evaluated using the same fluorescent assay based on the fluorogenic substrate Fluor-de-Lys. This inclusion criterion allowed us to obtain a homogeneous set of compounds regarding their biological evaluation. This step is crucial to develop a predictive model since the data selection is pivotal for adding the correct information to a software for developing computational models. The 3D structures of all ligands were built using the builder panel in the Maestro. For the molecules possessing known stereochemistry, the absolute configuration was specified during the drawing of the compounds. All structures were treated by LigPrep module of Schrödinger suite 2015 [94] in order to generate the most probable ionization state at the cellular pH value (7.4 ± 0.2) as reported by us [95–97]. Moreover, the OPLS-AA_2005 force field was used for optimization, which produces the lowest energy conformer of the ligand [98]. The prepared ligands were then submitted to MacroModel software [99] in order to obtain an exhaustive conformational analysis using the OPLS-AA_2005 as force field. The solvent effects are simulated employing the analytical Generalized-Born/Surface-Area (GB/SA) model [100], and no cutoff for non-bonded interactions was selected. Molecular energy minimizations were performed using Polak–Ribiere conjugate gradient (PRCG) method with 2000 maximum iterations and 0.001 gradient convergence threshold. The conformational searches were carried out by employing MCMM (Monte Carlo Multiple Minimum) torsional sampling method. Automatic setup with 21 kJ/mol (5.02 kcal/mol) in the energy window for saving structure and a 0.5 Å cutoff distance for redundant conformers was used.

3.3. 3D-QSAR Model Generation

The software package Phase 4.2 [101], implemented in Maestro suite, was used to generate pharmacophore hypotheses and 3D-QSAR models for HDAC1 inhibitors based on 2-aminophenylbenzamide scaffold. Given a set of molecules with high affinity for a particular protein target, this software uses fine-grained conformational sampling and a range of scoring techniques to identify a common-features pharmacophore hypothesis, which conveys 3D structural characteristics that are critical for the activity. Pharmacophore feature sites for the molecules were specified by a set of features well-defined in Phase as hydrogen-bond acceptor (A), hydrogen-bond donor (D), hydrophobic group (H), negatively charged group (N), positively charged group (P) and aromatic ring (R). No user-defined feature was employed for the present study. The ligands prepared as reported in the previous step, were imported into the "develop common pharmacophore hypotheses" panel of Phase with their respective biological activity values. Twenty active compounds (Figure 2 and in Table S1 in the Supplementary Materials) possessing highly inhibitory potency against HDAC1 were

selected for generating the pharmacophore hypotheses. Common-features pharmacophore hypotheses were identified, scored, and ranked by means of conformational analysis and tree-based partitioning techniques. In the score hypotheses step, common pharmacophores are examined, and a scoring procedure is applied to identify the pharmacophore from each surviving n-dimensional box that yields the best alignment of the active set ligands. This pharmacophore provides the means to explain how the active molecules bind to the receptor.

The best ranked pharmacophore model obtained by Phase (ADDRR, shown in Figure 3A superimposed to aminophenylnicotinamide analogue **13**), consisted of five features: one hydrogen-bond acceptor (A), two hydrogen-bond donors (D), and two aromatic functions (R). The inter-feature distances (Figure 3B) were measured by using the site measurements tool implemented in the software Phase. This pharmacophore was used as alignment rule for further 3D-QSAR analysis. All the molecules used for the QSAR studies (Table S1) were aligned to the selected pharmacophore hypothesis. In the present study, we set a pIC_{50} threshold for the selection of active and inactive ligands. These pIC_{50} values were also used as the dependent variable in the 3D-QSAR calculations. In particular, compounds that showed an IC_{50} comprised between 5 and 50 μM were considered to be inactive ligands. Moderate inhibitors were considered compounds with IC_{50} between 10 nM and 5 μM, while compounds possessing an $IC_{50} \leq 10$ nM were assigned as potent inhibitors of HDAC1 and consequently as actives during 3D-QSAR model generation. Remarkably, to avoid possible faults arising from the inclusion in the set of molecules with uncertain activity, only molecules with experimentally definite inhibitory activity have been selected to develop the in silico model. Atom-based QSAR models were developed for ADDRR hypothesis using 259 compounds in the training set (370 compounds were randomly divided 70% in the training and 30% in the test set) and a grid spacing of 0.5 Å. QSAR models were generated by means of PLS method. An internal validation was achieved employing leave-n-out (LnO) technique as specified in Phase user manual (Phase, version 4.2, User Manual, Schrödinger press, LLC, New York, NY, 2015). As reported by Todeschini et al. the internal validation results generally expressed in terms of Q^2 metrics should be amended introducing Q^2_{F3} metrics for the internal validation of the QSAR models [102]. For this purpose, we calculated these metrics (Table 2) employing the formula reported below (Equation (1)).

$$Q^2_{F3} = 1 - \frac{\sum_{i=1}^{n_{OUT}}(y_i - \hat{y}_{i/i})^2 / n_{OUT}}{\sum_{i=1}^{n_{TR}}(y_i - \overline{y}_{TR})^2 / n_{TR}} \tag{1}$$

where y_i is the experimental response of the ith object, $\hat{y}_{i/i}$ is the predicted response when the ith object is not in the training set, n_{TR} and n_{OUT} are the number of training and prediction objects, respectively, and $\overline{y}_{TR}$ is the average value of the training set experimental responses. Moreover, to avoid overfitting/underfitting phenomena, we considered 7 factors that is an appropriate for the number of selected compounds. In fact, although there is no limit on the maximum number of factors, but as a general rule, we stopped adding factors when the standard deviation of regression is approximately equal to the experimental error (calculated as median error among the selected compounds).

3.4. In Silico 3D-QSAR Model Validation

After the generation of the 3D-QSAR model, a preliminary in silico validation was performed using a large external test set of compounds (113 molecules) selected from the literature [83,84,89,103–106] (Table S2 in the Supplementary Materials) that have not been used for generating and cross validating the model. These compounds were prepared by using Maestro, LigPrep, and MacroModel, adopting the same procedure for preparing the molecules used to derive the model. Moreover, to further assess that the chosen model with 7 factors better performs with respect to the other Phase-derived models, we applied the validation method employing the external test set to all the generated QSAR models (Table 2). This workflow established that the model with 7 factors is the best performing model of the series in predicting the activity of the external test set with a correlation coefficient $r^2_{ext\_ts} = 0.794$

(Figure 6) (LVs 1, $r^2_{ext\_ts}$ = 0.421; LVs 2, $r^2_{ext\_ts}$ = 0.698; LVs 3, $r^2_{ext\_ts}$ = 0.657; LVs 4, $r^2_{ext\_ts}$ = 0.712; LVs 5, $r^2_{ext\_ts}$ = 0.735; LVs 6, $r^2_{ext\_ts}$ = 0.787; Figures S1–S6, respectively).

Further validation of the model was done by enrichment study using decoy test [107]. For this purpose, the Enhanced (DUD-E) web server [108] was employed to generate a set of useful decoys generated from a collection of 106 active compounds from three sources: 1) active compounds used to develop the pharmacophore model, 2) other compounds with good activity against HDAC1 used in 3D-QSAR studies and 3) the most active compounds of the external test set. This collection consisted of 106 active compounds with $IC_{50} \leq 35$ nM (Table S3). For this set of active ligands, the DUD-E server provided 5764 inactive ligands (redundant structures in the output files were deleted) from a subset of the ZINC database filtered using the Lipinski's rules for drug-likeness, for a total of 5870 compounds (5764 inactives plus 106 actives). Each of these inactive decoys was selected to bear a resemblance to the physicochemical properties of the reference ligand but differ from it in terms of 2D structure (e.g., large difference of Tanimoto coefficient between decoys and active molecules). Although largely used, the approach based on decoys sets presents some limitations (i.e., the decoy sets often span a small, synthetically feasible subset of molecular space and are restricted in physicochemical similarity compared with actives). After the generation, the decoys sets were downloaded as 126 smiles files and imported into Maestro and submitted to LigPrep application to properly convert smiles into 3D structures as well as for removing potential erroneous structures. Subsequently, to perform a minimization and a conformational search of the obtained structures MacroModel program was employed (same parameters for ligands preparation were applied). A single file containing conformers of active molecules and decoys was created and submitted to Phase software for predicting the inhibitory activity of database against HDAC1 using the developed 3D-QSAR model and employing "search for matches" option. After decoys generation and activity evaluation, the Güner–Henry score, i.e., goodness of hit list (GH) and enrichment factor (EF) values were estimated by Equations (2) and (3), respectively.

$$EF \ = \ \frac{Ha/Ht}{(A/D)} \tag{2}$$

$$GH \ = \ \left\{ \frac{Ha * (3A + Ht)}{4HtA} \right\} * \left[1 - \frac{(Ht - Ha)}{(D - A)} \right] \tag{3}$$

where *Ht* represents the total number of compounds in the hit list found by virtual screening, *Ha* is the total actives found by virtual screening considering the top 30-ranked position (positions comprise within the cutoff value). The total number of compounds (*Ht*) might represent the number of molecules to purchase after a virtual screening protocol and almost the 1% of the considered database (*D*). *A* represents the total of the active derivatives enclosed in the database, and *D* stands for the total number of molecules existing in the set. The range of *GH* score varies from 0 to 1. The *GH* score 0 means a null model, while the *GH* score 1 denotes generation of an ideal model. Moreover, the % yield of actives (% *YA*) and % ratio of actives (% *RA*) were evaluated by Equations (4) and (5), respectively.

$$\%YA \ = \ \left[\left(\frac{Ha}{Ht} \right) * 100 \right] \tag{4}$$

$$\%RA \ = \ \left[\left(\frac{Ha}{A} \right) * 100 \right] \tag{5}$$

Moreover, to assess the predictive power of the 3D-QSAR model, a ROC was employed through an Enrichment Calculator (enrichment.py) script [55–57,76]. The mentioned script calculates the enrichment metrics, including area under the receiver-operating characteristic curve (AUC), from virtual screening by means of the output structure file and a list of known active molecules. The output of the screening protocol, using active molecules and decoys, consisted of a list of molecules ranked by the predicted activity from the top-predicted molecules as estimated by the 3D-QSAR model. These ranking data along with a list file of active molecules were submitted to the enrichment.py application.

4. Conclusions

The present study describes the generation of a ligand-based pharmacophore model (ADDRR) for a subset of 20 highly active aminophenylbenzamide derivatives reported as selective HDAC1 inhibitors by employing the software Phase implemented in the Schrödinger molecular modeling suite. With the aid of pharmacophore-based alignment rule, a meaningful 3D-QSAR model was derived and validated employing of the QSAR models a large set of benzamide-based HDAC1 inhibitors (training set, test set, and an external test set for a total of 483 molecules) by using PLS analysis. The main objective of this approach was to develop an in-house computational tool for the prediction of HDAC1 inhibitory activity during the design of innovative aminophenylbenzamide chemotypes as privileged therapeutic scaffold in the isoform-selective HDACIs research. The validation outcomes confirmed that the proposed 3D-QSAR model is endowed with satisfactory predictive power taking into account favorable structural requirements responsible for HDAC1 inhibitory activity. This aspect has been computationally investigated since the selectivity is implicit in the template molecules; however prospective validation is needed to exploit the performance of the model. In fact, the developed 3D-QSAR model can be used for rationally designing novel and selective HDACIs. Moreover, based on the computational investigation, the developed model possesses a rationale for virtual screening campaign, with huge potential in isoform-selective HDACIs drug discovery, and it can effectively provide a set of guidelines for the design and optimization of novel derivatives with greater activity towards HDAC1.

Supplementary Materials: The following are available online, Table S1: Experimental (Observed column) and predicted (Predicted column) activity pIC_{50} for compounds used for developing the 3D-QSAR model., Table S2: Experimental (Observed column) and predicted (Predicted column) activity pIC_{50} for compounds included in the external test set, Table S3: Active compounds used for generating a decoys set, Figures S1–S6: Scatter plots for all the model generated by Phase.

Author Contributions: Conceptualization, H.S. and S.B.; methodology, H.S., G.C. (Giulia Chemi) and S.B.; software, H.S., G.C. (Giulia Chemi) and S.B.; validation, H.S., G.C. (Giulia Chemi) and S.B.; formal analysis, H.S., G.C. (Giulia Chemi), G.C. (Giuseppe Campiani), V.C. and S.B.; investigation, H.S., G.C. (Giulia Chemi); data curation, H.S., G.C. (Giulia Chemi), G.C. (Giuseppe Campiani), V.C. and S.B.; writing—original draft preparation, H.S. and S.B.; writing—review and editing, H.S., G.C. (Giulia Chemi), S.B., G.C. (Giuseppe Campiani), V.C.; supervision, S.B. All authors have read and agreed to the published version of the manuscript.

Funding: This research was funded by the Bioinformatics Research Center in Isfahan University of Medical Sciences (Iran) grant number 298137 to H.S.

Acknowledgments: The authors thank Lorenzo Chemi for the technical assistance during the calculation of Q^2_{F3}.

Conflicts of Interest: The authors declare no conflict of interest.

References

1. Conway, S.J.; Woster, P.M.; Greenlee, W.J.; Georg, G.; Wang, S. Epigenetics: Novel Therapeutics Targeting Epigenetics. *J. Med. Chem.* **2016**, *59*, 1247–1248. [CrossRef]
2. Yoo, C.B.; Jones, P.A. Epigenetic therapy of cancer: Past, present and future. *Nat. Rev. Drug Discov.* **2006**, *5*, 37–50. [CrossRef]
3. Bertrand, P. Inside HDAC with HDAC inhibitors. *Eur. J. Med. Chem.* **2010**, *45*, 2095–2116. [CrossRef]
4. Falkenberg, K.J.; Johnstone, R.W. Histone deacetylases and their inhibitors in cancer, neurological diseases and immune disorders. *Nat. Rev. Drug Discov.* **2014**, *13*, 673–691. [CrossRef]
5. Paris, M.; Porcelloni, M.; Binaschi, M.; Fattori, D. Histone deacetylase inhibitors: From bench to clinic. *J. Med. Chem* **2008**, *51*, 1505–1529. [CrossRef]
6. de Ruijter, A.J.; van Gennip, A.H.; Caron, H.N.; Kemp, S.; van Kuilenburg, A.B. Histone deacetylases (HDACs): Characterization of the classical HDAC family. *Biochem. J.* **2003**, *370*, 737–749. [CrossRef]
7. Kouzarides, T. Chromatin modifications and their function. *Cell* **2007**, *128*, 693–705. [CrossRef]
8. Rodriquez, M.; Aquino, M.; Bruno, I.; De Martino, G.; Taddei, M.; Gomez-Paloma, L. Chemistry and biology of chromatin remodeling agents: State of art and future perspectives of HDAC inhibitors. *Curr. Med. Chem.* **2006**, *13*, 1119–1139. [CrossRef]

9. Bolden, J.E.; Peart, M.J.; Johnstone, R.W. Anticancer activities of histone deacetylase inhibitors. *Nat. Rev. Drug Discov.* **2006**, *5*, 769–784. [CrossRef]

10. Marks, P.A.; Dokmanovic, M. Histone deacetylase inhibitors: Discovery and development as anticancer agents. *Expert Opin. Investig. Drugs* **2005**, *14*, 1497–1511. [CrossRef]

11. Minucci, S.; Pelicci, P.G. Histone deacetylase inhibitors and the promise of epigenetic (and more) treatments for cancer. *Nat. Rev. Cancer* **2006**, *6*, 38–51. [CrossRef]

12. Brittain, D.; Weinmann, H.; Ottow, E. Recent Advances in the Medicinal Chemistry of Histone Deacetylase Inhibitors. *Annu. Rep. Med. Chem.* **2007**, *42*, 337–348. [CrossRef]

13. Miller, T.A. Patent status of histone deacetylase inhibitors. *Expert Opin. Ther. Pat.* **2005**, *14*, 791–804. [CrossRef]

14. Gregoretti, I.; Lee, Y.-M.; Goodson, H.V. Molecular Evolution of the Histone Deacetylase Family: Functional Implications of Phylogenetic Analysis. *J. Mol. Biol.* **2004**, *338*, 17–31. [CrossRef]

15. Duvic, M.; Talpur, R.; Ni, X.; Zhang, C.; Hazarika, P.; Kelly, C.; Chiao, J.H.; Reilly, J.F.; Ricker, J.L.; Richon, V.M.; et al. Phase 2 trial of oral vorinostat (suberoylanilide hydroxamic acid, SAHA) for refractory cutaneous T-cell lymphoma (CTCL). *Blood* **2006**, *109*, 31–39. [CrossRef]

16. Lee, H.Z.; Kwitkowski, V.E.; Del Valle, P.L.; Ricci, M.S.; Saber, H.; Habtemariam, B.A.; Bullock, J.; Bloomquist, E.; Li Shen, Y.; Chen, X.H.; et al. FDA Approval: Belinostat for the Treatment of Patients with Relapsed or Refractory Peripheral T-cell Lymphoma. *Clin. Cancer Res.* **2015**, *21*, 2666–2670. [CrossRef]

17. Garnock-Jones, K.P. Panobinostat: First Global Approval. *Drugs* **2015**, *75*, 695–704. [CrossRef]

18. VanderMolen, K.M.; McCulloch, W.; Pearce, C.J.; Oberlies, N.H. Romidepsin (Istodax, NSC 630176, FR901228, FK228, depsipeptide): A natural product recently approved for cutaneous T-cell lymphoma. *J. Antibiot.* **2011**, *64*, 525–531. [CrossRef]

19. Qiao, Z.; Ren, S.; Li, W.; Wang, X.; He, M.; Guo, Y.; Sun, L.; He, Y.; Ge, Y.; Yu, Q. Chidamide, a novel histone deacetylase inhibitor, synergistically enhances gemcitabine cytotoxicity in pancreatic cancer cells. *Biochem. Biophys. Res. Commun.* **2013**, *434*, 95–101. [CrossRef]

20. Garcia-Manero, G.; Assouline, S.; Cortes, J.; Estrov, Z.; Kantarjian, H.; Yang, H.; Newsome, W.M.; Miller, W.H.; Rousseau, C.; Kalita, A.; et al. Phase 1 study of the oral isotype specific histone deacetylase inhibitor MGCD0103 in leukemia. *Blood* **2008**, *112*, 981–989. [CrossRef]

21. Knipstein, J.; Gore, L. Entinostat for treatment of solid tumors and hematologic malignancies. *Expert Opin. Investig. Drugs* **2011**, *20*, 1455–1467. [CrossRef]

22. Gediya, L.K.; Belosay, A.; Khandelwal, A.; Purushottamachar, P.; Njar, V.C.O. Improved synthesis of histone deacetylase inhibitors (HDIs) (MS-275 and CI-994) and inhibitory effects of HDIs alone or in combination with RAMBAs or retinoids on growth of human LNCaP prostate cancer cells and tumor xenografts. *Bioorg. Med. Chem.* **2008**, *16*, 3352–3360. [CrossRef] [PubMed]

23. Finazzi, G.; Vannucchi, A.M.; Martinelli, V.; Ruggeri, M.; Nobile, F.; Specchia, G.; Pogliani, E.M.; Olimpieri, O.M.; Fioritoni, G.; Musolino, C.; et al. A phase II study of Givinostat in combination with hydroxycarbamide in patients with polycythaemia vera unresponsive to hydroxycarbamide monotherapy. *Br. J. Haematol.* **2013**, *161*, 688–694. [CrossRef] [PubMed]

24. Evens, A.M.; Balasubramanian, S.; Vose, J.M.; Harb, W.; Gordon, L.I.; Langdon, R.; Sprague, J.; Sirisawad, M.; Mani, C.; Yue, J.; et al. A Phase I/II Multicenter, Open-Label Study of the Oral Histone Deacetylase Inhibitor Abexinostat in Relapsed/Refractory Lymphoma. *Clin. Cancer Res.* **2015**, *22*, 1059–1066. [CrossRef]

25. Bruserud, O.; Stapnes, C.; Ersvær, E.; Gjertsen, B.; Ryningen, A. Histone Deacetylase Inhibitors in Cancer Treatment: A Review of the Clinical Toxicity and the Modulation of Gene Expression in Cancer Cells. *Curr. Pharm. Biotechnol.* **2007**, *8*, 388–400. [CrossRef]

26. Subramanian, S.; Bates, S.E.; Wright, J.J.; Espinoza-Delgado, I.; Piekarz, R.L. Clinical Toxicities of Histone Deacetylase Inhibitors. *Pharmaceuticals* **2010**, *3*, 2751–2767. [CrossRef]

27. Karagiannis, T.C.; El-Osta, A. Will broad-spectrum histone deacetylase inhibitors be superseded by more specific compounds? *Leukemia* **2006**, *21*, 61–65. [CrossRef]

28. Roche, J.; Bertrand, P. Inside HDACs with more selective HDAC inhibitors. *Eur. J. Med. Chem.* **2016**, *121*, 451–483. [CrossRef]

29. Brindisi, M.; Saraswati, A.P.; Brogi, S.; Gemma, S.; Butini, S.; Campiani, G. Old but Gold: Tracking the New Guise of Histone Deacetylase 6 (HDAC6) Enzyme as a Biomarker and Therapeutic Target in Rare Diseases. *J. Med. Chem.* **2020**, *63*, 23–39. [CrossRef]

30. Bieliauskas, A.V.; Pflum, M.K.H. Isoform-selective histone deacetylase inhibitors. *Chem. Soc. Rev.* **2008**, *37*, 1402. [CrossRef]

31. Guha, M. HDAC inhibitors still need a home run, despite recent approval. *Nat. Rev. Drug Discov.* **2015**, *14*, 225–226. [CrossRef] [PubMed]

32. Ononye, S.N.; van Heyst, M.; Falcone, E.M.; Anderson, A.C.; Wright, D.L. Toward isozyme-selective inhibitors of histone deacetylase as therapeutic agents for the treatment of cancer. *Pharm. Pat. Anal.* **2012**, *1*, 207–221. [CrossRef] [PubMed]

33. Zhang, L.; Han, Y.; Jiang, Q.; Wang, C.; Chen, X.; Li, X.; Xu, F.; Jiang, Y.; Wang, Q.; Xu, W. Trend of Histone Deacetylase Inhibitors in Cancer Therapy: Isoform Selectivity or Multitargeted Strategy. *Med. Res. Rev.* **2015**, *35*, 63–84. [CrossRef] [PubMed]

34. Glaser, K.B.; Li, J.; Staver, M.J.; Wei, R.-Q.; Albert, D.H.; Davidsen, S.K. Role of Class I and Class II histone deacetylases in carcinoma cells using siRNA. *Biochem. Biophys. Res. Commun.* **2003**, *310*, 529–536. [CrossRef]

35. Hu, E.; Dul, E.; Sung, C.-M.; Chen, Z.; Kirkpatrick, R.; Zhang, G.-F.; Johanson, K.; Liu, R.; Lago, A.; Hofmann, G.; et al. Identification of Novel Isoform-Selective Inhibitors within Class I Histone Deacetylases. *J. Pharmacol. Exp. Ther.* **2003**, *307*, 720–728. [CrossRef]

36. Kawai, H.; Li, H.; Avraham, S.; Jiang, S.; Avraham, H.K. Overexpression of histone deacetylase HDAC1 modulates breast cancer progression by negative regulation of estrogen receptor? *Int. J. Cancer* **2003**, *107*, 353–358. [CrossRef]

37. Rikimaru, T.; Taketomi, A.; Yamashita, Y.-I.; Shirabe, K.; Hamatsu, T.; Shimada, M.; Maehara, Y. Clinical Significance of Histone Deacetylase 1 Expression in Patients with Hepatocellular Carcinoma. *Oncology* **2007**, *72*, 69–74. [CrossRef]

38. Senese, S.; Zaragoza, K.; Minardi, S.; Muradore, I.; Ronzoni, S.; Passafaro, A.; Bernard, L.; Draetta, G.F.; Alcalay, M.; Seiser, C.; et al. Role for Histone Deacetylase 1 in Human Tumor Cell Proliferation. *Mol. Cell Biol.* **2007**, *27*, 4784–4795. [CrossRef]

39. Khan, N.; Jeffers, M.; Kumar, S.; Hackett, C.; Boldog, F.; Khramtsov, N.; Qian, X.; Mills, E.; Berghs, S.C.; Carey, N.; et al. Determination of the class and isoform selectivity of small-molecule histone deacetylase inhibitors. *Biochem. J.* **2007**, *409*, 581–589. [CrossRef]

40. Shen, S.; Kozikowski, A.P. Why Hydroxamates May Not Be the Best Histone Deacetylase Inhibitors-What Some May Have Forgotten or Would Rather Forget? *ChemMedChem* **2016**, *11*, 15–21. [CrossRef]

41. Beckers, T.; Burkhardt, C.; Wieland, H.; Gimmnich, P.; Ciossek, T.; Maier, T.; Sanders, K. Distinct pharmacological properties of second generation HDAC inhibitors with the benzamide or hydroxamate head group. *Int. J. Cancer* **2007**, *121*, 1138–1148. [CrossRef] [PubMed]

42. Bonfils, C.; Kalita, A.; Dubay, M.; Siu, L.L.; Carducci, M.A.; Reid, G.; Martell, R.E.; Besterman, J.M.; Li, Z. Evaluation of the Pharmacodynamic Effects of MGCD0103 from Preclinical Models to Human Using a Novel HDAC Enzyme Assay. *Clin. Cancer Res.* **2008**, *14*, 3441–3449. [CrossRef] [PubMed]

43. Chou, C.J.; Herman, D.; Gottesfeld, J.M. Pimelic Diphenylamide 106 Is a Slow, Tight-binding Inhibitor of Class I Histone Deacetylases. *J. Biol. Chem.* **2008**, *283*, 35402–35409. [CrossRef] [PubMed]

44. Kelly, W.K.; Richon, V.M.; O'Connor, O.; Curley, T.; MacGregor-Curtelli, B.; Tong, W.; Klang, M.; Schwartz, L.; Richardson, S.; Rosa, E.; et al. Phase I clinical trial of histone deacetylase inhibitor: Suberoylanilide hydroxamic acid administered intravenously. *Clin. Cancer Res.* **2003**, *9*, 3578–3588.

45. Nagaoka, Y.; Maeda, T.; Kawai, Y.; Nakashima, D.; Oikawa, T.; Shimoke, K.; Ikeuchi, T.; Kuwajima, H.; Uesato, S. Synthesis and cancer antiproliferative activity of new histone deacetylase inhibitors: Hydrophilic hydroxamates and 2-aminobenzamide-containing derivatives. *Eur. J. Med. Chem.* **2006**, *41*, 697–708. [CrossRef]

46. Moradei, O.M.; Mallais, T.C.; Frechette, S.; Paquin, I.; Tessier, P.E.; Leit, S.M.; Fournel, M.; Bonfils, C.; Trachy-Bourget, M.-C.; Liu, J.; et al. Novel Aminophenyl Benzamide-Type Histone Deacetylase Inhibitors with Enhanced Potency and Selectivity. *J. Med. Chem.* **2007**, *50*, 5543–5546. [CrossRef]

47. Suzuki, T.; Miyata, N. Non-hydroxamate Histone Deacetylase Inhibitors. *Curr. Med. Chem.* **2005**, *12*, 2867–2880. [CrossRef]

48. Wang, D.-F.; Helquist, P.; Wiech, N.L.; Wiest, O. Toward Selective Histone Deacetylase Inhibitor Design: Homology Modeling, Docking Studies, and Molecular Dynamics Simulations of Human Class I Histone Deacetylases. *J. Med. Chem.* **2005**, *48*, 6936–6947. [CrossRef]

49. Chemi, G.; Brogi, S. Breakthroughs in Computational Approaches for Drug Discovery. *J. Drug Res. Dev.* **2017**, *3*. [CrossRef]
50. Verma, J.; Khedkar, V.; Coutinho, E. 3D-QSAR in Drug Design—A Review. *Curr. Top. Med. Chem.* **2010**, *10*, 95–115. [CrossRef]
51. Yang, S.-Y. Pharmacophore modeling and applications in drug discovery: Challenges and recent advances. *Drug Discov. Today* **2010**, *15*, 444–450. [CrossRef]
52. Cherkasov, A.; Muratov, E.N.; Fourches, D.; Varnek, A.; Baskin, I.I.; Cronin, M.; Dearden, J.; Gramatica, P.; Martin, Y.C.; Todeschini, R.; et al. QSAR Modeling: Where Have You Been? Where Are You Going To? *J. Med. Chem.* **2014**, *57*, 4977–5010. [CrossRef] [PubMed]
53. Brogi, S.; Corelli, F.; Di Marzo, V.; Ligresti, A.; Mugnaini, C.; Pasquini, S.; Tafi, A. Three-dimensional quantitative structure–selectivity relationships analysis guided rational design of a highly selective ligand for the cannabinoid receptor 2. *Eur. J. Med. Chem.* **2011**, *46*, 547–555. [CrossRef] [PubMed]
54. Brogi, S.; Papazafiri, P.; Roussis, V.; Tafi, A. 3D-QSAR using pharmacophore-based alignment and virtual screening for discovery of novel MCF-7 cell line inhibitors. *Eur. J. Med. Chem.* **2013**, *67*, 344–351. [CrossRef] [PubMed]
55. Zaccagnini, L.; Brogi, S.; Brindisi, M.; Gemma, S.; Chemi, G.; Legname, G.; Campiani, G.; Butini, S. Identification of novel fluorescent probes preventing PrP Sc replication in prion diseases. *Eur. J. Med. Chem.* **2017**, *127*, 859–873. [CrossRef] [PubMed]
56. Brogi, S.; Brindisi, M.; Joshi, B.P.; Sanna Coccone, S.; Parapini, S.; Basilico, N.; Novellino, E.; Campiani, G.; Gemma, S.; Butini, S. Exploring clotrimazole-based pharmacophore: 3D-QSAR studies and synthesis of novel antiplasmodial agents. *Bioorg. Med. Chem. Lett.* **2015**, *25*, 5412–5418. [CrossRef] [PubMed]
57. Chemi, G.; Gemma, S.; Campiani, G.; Brogi, S.; Butini, S.; Brindisi, M. Computational Tool for Fast in silico Evaluation of hERG K+ Channel Affinity. *Front. Chem.* **2017**, *5*. [CrossRef]
58. Liu, B.; Lu, A.J.; Liao, C.Z.; Liu, H.B.; Zhou, J.J. 3D-QSAR of Sulfonamide Hydroxamic Acid HDAC Inhibitors. *Acta Phys. Chim. Sin.* **2005**, *21*, 333–337. [CrossRef]
59. Guo, Y.; Xiao, J.; Guo, Z.; Chu, F.; Cheng, Y.; Wu, S. Exploration of a binding mode of indole amide analogues as potent histone deacetylase inhibitors and 3D-QSAR analyses. *Bioorg. Med. Chem.* **2005**, *13*, 5424–5434. [CrossRef]
60. Juvale, D.C.; Kulkarni, V.V.; Deokar, H.S.; Wagh, N.K.; Padhye, S.B.; Kulkarni, V.M. 3D-QSAR of histone deacetylase inhibitors: Hydroxamate analogues. *Org. Biomol. Chem.* **2006**, *4*, 2858. [CrossRef]
61. Ragno, R.; Simeoni, S.; Valente, S.; Massa, S.; Mai, A. 3-D QSAR Studies on Histone Deacetylase Inhibitors. A GOLPE/GRID Approach on Different Series of Compounds. *J. Chem. Inf. Model.* **2006**, *46*, 1420–1430. [CrossRef] [PubMed]
62. Chen, Y.; Li, H.; Tang, W.; Zhu, C.; Jiang, Y.; Zou, J.; Yu, Q.; You, Q. 3D-QSAR studies of HDACs inhibitors using pharmacophore-based alignment. *Eur. J. Med. Chem.* **2009**, *44*, 2868–2876. [CrossRef] [PubMed]
63. Wang, D.-F.; Wiest, O.; Helquist, P.; Lan-Hargest, H.-Y.; Wiech, N.L. QSAR Studies of PC-3 cell line inhibition activity of TSA and SAHA-like hydroxamic acids. *Bioorg. Med. Chem. Lett.* **2004**, *14*, 707–711. [CrossRef] [PubMed]
64. Xie, A.; Liao, C.; Li, Z.; Ning, Z.; Hu, W.; Lu, X.; Shi, L.; Zhou, J. Quantitative Structure-Activity Relationship Study of Histone Deacetylase Inhibitors. *Anti-Cancer Agents Med. Chem.* **2004**, *4*, 273–299. [CrossRef] [PubMed]
65. Dixon, S.L.; Smondyrev, A.M.; Knoll, E.H.; Rao, S.N.; Shaw, D.E.; Friesner, R.A. PHASE: A new engine for pharmacophore perception, 3D QSAR model development, and 3D database screening: 1. Methodology and preliminary results. *J. Comput. Aided Mol. Des.* **2006**, *20*, 647–671. [CrossRef] [PubMed]
66. Abdizadeh, T.; Kalani, M.R.; Abnous, K.; Tayarani-Najaran, Z.; Khashyarmanesh, B.Z.; Abdizadeh, R.; Ghodsi, R.; Hadizadeh, F. Design, synthesis and biological evaluation of novel coumarin-based benzamides as potent histone deacetylase inhibitors and anticancer agents. *Eur. J. Med. Chem.* **2017**, *132*, 42–62. [CrossRef]
67. Hamblett, C.L.; Methot, J.L.; Mampreian, D.M.; Sloman, D.L.; Stanton, M.G.; Kral, A.M.; Fleming, J.C.; Cruz, J.C.; Chenard, M.; Ozerova, N.; et al. The discovery of 6-amino nicotinamides as potent and selective histone deacetylase inhibitors. *Bioorg. Med. Chem. Lett.* **2007**, *17*, 5300–5309. [CrossRef]
68. Sixto-López, Y.; Bello, M.; Correa-Basurto, J. Insights into structural features of HDAC1 and its selectivity inhibition elucidated by Molecular dynamic simulation and Molecular Docking. *J. Biomol. Struct. Dyn.* **2018**, *37*, 584–610. [CrossRef]

69. Brindisi, M.; Cavella, C.; Brogi, S.; Nebbioso, A.; Senger, J.; Maramai, S.; Ciotta, A.; Iside, C.; Butini, S.; Lamponi, S.; et al. Phenylpyrrole-based HDAC inhibitors: Synthesis, molecular modeling and biological studies. *Future Med. Chem.* **2016**, *8*, 1573–1587. [CrossRef]

70. Brindisi, M.; Senger, J.; Cavella, C.; Grillo, A.; Chemi, G.; Gemma, S.; Cucinella, D.M.; Lamponi, S.; Sarno, F.; Iside, C.; et al. Novel spiroindoline HDAC inhibitors: Synthesis, molecular modelling and biological studies. *Eur. J. Med. Chem.* **2018**, *157*, 127–138. [CrossRef]

71. Braga, R.C.; Andrade, C.H. Assessing the Performance of 3D Pharmacophore Models in Virtual Screening: How Good are They? *Curr. Top. Med. Chem.* **2013**, *13*, 1127–1138. [CrossRef] [PubMed]

72. Dror, O.; Schneidman-Duhovny, D.; Inbar, Y.; Nussinov, R.; Wolfson, H.J. Novel Approach for Efficient Pharmacophore-Based Virtual Screening: Method and Applications. *J. Chem. Inf. Model.* **2009**, *49*, 2333–2343. [CrossRef] [PubMed]

73. Krishna, S.; Singh, D.K.; Meena, S.; Datta, D.; Siddiqi, M.I.; Banerjee, D. Pharmacophore-Based Screening and Identification of Novel Human Ligase I Inhibitors with Potential Anticancer Activity. *J. Chem. Inf. Model.* **2014**, *54*, 781–792. [CrossRef] [PubMed]

74. Sakkiah, S.; Thangapandian, S.; John, S.; Lee, K.W. Pharmacophore based virtual screening, molecular docking studies to design potent heat shock protein 90 inhibitors. *Eur. J. Med. Chem.* **2011**, *46*, 2937–2947. [CrossRef]

75. Thangapandian, S.; John, S.; Sakkiah, S.; Lee, K.W. Pharmacophore-based virtual screening and Bayesian model for the identification of potential human leukotriene A4 hydrolase inhibitors. *Eur. J. Med. Chem.* **2011**, *46*, 1593–1603. [CrossRef] [PubMed]

76. Triballeau, N.; Acher, F.; Brabet, I.; Pin, J.-P.; Bertrand, H.-O. Virtual Screening Workflow Development Guided by the "Receiver Operating Characteristic" Curve Approach. Application to High-Throughput Docking on Metabotropic Glutamate Receptor Subtype 4. *J. Med. Chem.* **2005**, *48*, 2534–2547. [CrossRef]

77. Zhao, W.; Hevener, K.E.; White, S.W.; Lee, R.E.; Boyett, J.M. A statistical framework to evaluate virtual screening. *BMC Bioinform.* **2009**, *10*. [CrossRef]

78. *Maestro*; Version 10.1; Schrödinger, LLC: New York, NY, USA, 2015.

79. Andrews, D.M.; Gibson, K.M.; Graham, M.A.; Matusiak, Z.S.; Roberts, C.A.; Stokes, E.S.; Brady, M.C.; Chresta, C.M. Design and campaign synthesis of pyridine-based histone deacetylase inhibitors. *Bioorg. Med. Chem. Lett.* **2008**, *18*, 2525–2529. [CrossRef]

80. Methot, J.L.; Chakravarty, P.K.; Chenard, M.; Close, J.; Cruz, J.C.; Dahlberg, W.K.; Fleming, J.; Hamblett, C.L.; Hamill, J.E.; Harrington, P.; et al. Exploration of the internal cavity of histone deacetylase (HDAC) with selective HDAC1/HDAC2 inhibitors (SHI-1:2). *Bioorg. Med. Chem. Lett.* **2008**, *18*, 973–978. [CrossRef]

81. Kattar, S.D.; Surdi, L.M.; Zabierek, A.; Methot, J.L.; Middleton, R.E.; Hughes, B.; Szewczak, A.A.; Dahlberg, W.K.; Kral, A.M.; Ozerova, N.; et al. Parallel medicinal chemistry approaches to selective HDAC1/HDAC2 inhibitor (SHI-1:2) optimization. *Bioorg. Med. Chem. Lett.* **2009**, *19*, 1168–1172. [CrossRef]

82. Methot, J.L.; Hamblett, C.L.; Mampreian, D.M.; Jung, J.; Harsch, A.; Szewczak, A.A.; Dahlberg, W.K.; Middleton, R.E.; Hughes, B.; Fleming, J.C.; et al. SAR profiles of spirocyclic nicotinamide derived selective HDAC1/HDAC2 inhibitors (SHI-1:2). *Bioorg. Med. Chem. Lett.* **2008**, *18*, 6104–6109. [CrossRef]

83. Li, Y.; Wang, Y.; Xie, N.; Xu, M.; Qian, P.; Zhao, Y.; Li, S. Design, synthesis and antiproliferative activities of novel benzamides derivatives as HDAC inhibitors. *Eur. J. Med. Chem.* **2015**, *100*, 270–276. [CrossRef] [PubMed]

84. Li, Y.; Zhou, Y.; Qian, P.; Wang, Y.; Jiang, F.; Yao, Z.; Hu, W.; Zhao, Y.; Li, S. Design, synthesis and bioevalution of novel benzamides derivatives as HDAC inhibitors. *Bioorg. Med. Chem. Lett.* **2013**, *23*, 179–182. [CrossRef] [PubMed]

85. Siliphaivanh, P.; Harrington, P.; Witter, D.J.; Otte, K.; Tempest, P.; Kattar, S.; Kral, A.M.; Fleming, J.C.; Deshmukh, S.V.; Harsch, A.; et al. Design of novel histone deacetylase inhibitors. *Bioorg. Med. Chem. Lett.* **2007**, *17*, 4619–4624. [CrossRef] [PubMed]

86. Andrews, D.M.; Stokes, E.S.; Carr, G.R.; Matusiak, Z.S.; Roberts, C.A.; Waring, M.J.; Brady, M.C.; Chresta, C.M.; East, S.J. Design and campaign synthesis of piperidine- and thiazole-based histone deacetylase inhibitors. *Bioorg. Med. Chem. Lett.* **2008**, *18*, 2580–2584. [CrossRef]

87. Hirata, Y.; Hirata, M.; Kawaratani, Y.; Shibano, M.; Taniguchi, M.; Yasuda, M.; Ohmomo, Y.; Nagaoka, Y.; Baba, K.; Uesato, S. Anti-tumor activity of new orally bioavailable 2-amino-5-(thiophen-2-yl)benzamide-series histone deacetylase inhibitors, possessing an aqueous soluble functional group as a surface recognition domain. *Bioorg. Med. Chem. Lett.* **2012**, *22*, 1926–1930. [CrossRef]

88. Kiyokawa, S.; Hirata, Y.; Nagaoka, Y.; Shibano, M.; Taniguchi, M.; Yasuda, M.; Baba, K.; Uesato, S. New orally bioavailable 2-aminobenzamide-type histone deacetylase inhibitor possessing a (2-hydroxyethyl)(4-(thiophen-2-yl)benzyl)amino group. *Bioorg. Med. Chem.* **2010**, *18*, 3925–3933. [CrossRef]

89. Mahboobi, S.; Dove, S.; Sellmer, A.; Winkler, M.; Eichhorn, E.; Pongratz, H.; Ciossek, T.; Baer, T.; Maier, T.; Beckers, T. Design of chimeric histone deacetylase- and tyrosine kinase-inhibitors: A series of imatinib hybrides as potent inhibitors of wild-type and mutant BCR-ABL, PDGF-Rbeta, and histone deacetylases. *J. Med. Chem.* **2009**, *52*, 2265–2279. [CrossRef]

90. Rajak, H.; Kumar, P.; Parmar, P.; Thakur, B.S.; Veerasamy, R.; Sharma, P.C.; Sharma, A.K.; Gupta, A.K.; Dangi, J.S. Appraisal of GABA and PABA as linker: Design and synthesis of novel benzamide based histone deacetylase inhibitors. *Eur. J. Med. Chem.* **2012**, *53*, 390–397. [CrossRef]

91. Valente, S.; Trisciuoglio, D.; De Luca, T.; Nebbioso, A.; Labella, D.; Lenoci, A.; Bigogno, C.; Dondio, G.; Miceli, M.; Brosch, G.; et al. 1,3,4-Oxadiazole-containing histone deacetylase inhibitors: Anticancer activities in cancer cells. *J. Med. Chem.* **2014**, *57*, 6259–6265. [CrossRef]

92. Rusche, J.R.; Peet, N.P.; Hopper, A.T. Compositions Including 6-Aminohexanoic Acid Derivatives as HDAC Inhibitors. U.S. Patent 9,265,734, 23 February 2016.

93. Liu, T.; Lin, Y.; Wen, X.; Jorissen, R.N.; Gilson, M.K. BindingDB: A web-accessible database of experimentally determined protein-ligand binding affinities. *Nucleic Acids Res.* **2007**, *35*, D198–D201. [CrossRef] [PubMed]

94. *LigPrep*; Version 3.3; Schrödinger, LLC: New York, NY, USA, 2015.

95. Sirous, H.; Chemi, G.; Campiani, G.; Brogi, S. An integrated in silico screening strategy for identifying promising disruptors of p53-MDM2 interaction. *Comput. Biol. Chem.* **2019**, *83*, 107105. [CrossRef] [PubMed]

96. Sirous, H.; Chemi, G.; Gemma, S.; Butini, S.; Debyser, Z.; Christ, F.; Saghaie, L.; Brogi, S.; Fassihi, A.; Campiani, G.; et al. Identification of Novel 3-Hydroxy-pyran-4-One Derivatives as Potent HIV-1 Integrase Inhibitors Using in silico Structure-Based Combinatorial Library Design Approach. *Front. Chem.* **2019**, *7*, 574. [CrossRef] [PubMed]

97. Sirous, H.; Fassihi, A.; Brogi, S.; Campiani, G.; Christ, F.; Debyser, Z.; Gemma, S.; Butini, S.; Chemi, G.; Grillo, A.; et al. Synthesis, Molecular Modelling and Biological Studies of 3-hydroxypyrane- 4-one and 3-hydroxy-pyridine-4-one Derivatives as HIV-1 Integrase Inhibitors. *Med. Chem.* **2019**, *15*, 755–770. [CrossRef] [PubMed]

98. Jorgensen, W.L.; Maxwell, D.S.; Tirado-Rives, J. Development and Testing of the OPLS All-Atom Force Field on Conformational Energetics and Properties of Organic Liquids. *J. Am. Chem. Soc.* **1996**, *118*, 11225–11236. [CrossRef]

99. *MacroModel*; Version 10.7; Schrödinger, LLC: New York, NY, USA, 2015.

100. Still, W.C.; Tempczyk, A.; Hawley, R.C.; Hendrickson, T. Semianalytical treatment of solvation for molecular mechanics and dynamics. *J. Am. Chem. Soc.* **1990**, *112*, 6127–6129. [CrossRef]

101. *Phase*; Version 4.2; Schrödinger, LLC: New York, NY, USA, 2015.

102. Todeschini, R.; Ballabio, D.; Grisoni, F. Beware of Unreliable Q2! A Comparative Study of Regression Metrics for Predictivity Assessment of QSAR Models. *J. Chem. Inf. Model.* **2016**, *56*, 1905–1913. [CrossRef]

103. Frechette, S.; Leit, S.; Woo, S.H.; Lapointe, G.; Jeannotte, G.; Moradei, O.; Paquin, I.; Bouchain, G.; Raeppel, S.; Gaudette, F.; et al. 4-(Heteroarylaminomethyl)-N-(2-aminophenyl)-benzamides and their analogs as a novel class of histone deacetylase inhibitors. *Bioorg. Med. Chem. Lett* **2008**, *18*, 1502–1506. [CrossRef]

104. Marson, C.M.; Matthews, C.J.; Atkinson, S.J.; Lamadema, N.; Thomas, N.S. Potent and Selective Inhibitors of Histone Deacetylase-3 Containing Chiral Oxazoline Capping Groups and a N-(2-Aminophenyl)-benzamide Binding Unit. *J. Med. Chem.* **2015**, *58*, 6803–6818. [CrossRef]

105. Zhu, Y.; Chen, X.; Ran, T.; Niu, J.; Zhao, S.; Lu, T.; Tang, W. Design, synthesis and biological evaluation of urea-based benzamides derivatives as HDAC inhibitors. *Med. Chem. Res.* **2017**, *26*, 2879–2888. [CrossRef]

106. Harrington, P.; Kattar, S.; Miller, T.A.; Stanton, M.G.; Tempest, P.; Witter, D.J. 4-Carboxybenzylamino Derivatives as Histone Deacetylase Inhibitors. U.S. Patent 8,389,553, 5 March 2013.

107. Kirchmair, J.; Markt, P.; Distinto, S.; Wolber, G.; Langer, T. Evaluation of the performance of 3D virtual screening protocols: RMSD comparisons, enrichment assessments, and decoy selection—What can we learn from earlier mistakes? *J. Comput. Aided Mol. Des.* **2008**, *22*, 213–228. [CrossRef] [PubMed]
108. Mysinger, M.M.; Carchia, M.; Irwin, J.J.; Shoichet, B.K. Directory of Useful Decoys, Enhanced (DUD-E): Better Ligands and Decoys for Better Benchmarking. *J. Med. Chem.* **2012**, *55*, 6582–6594. [CrossRef] [PubMed]

 *molecules*

Article

The Study on the hERG Blocker Prediction Using Chemical Fingerprint Analysis

Kwang-Eun Choi, Anand Balupuri and Nam Sook Kang *

Graduate School of New Drug Discovery and Development, Chungnam National University, 99 Daehak-ro, Yuseong-gu, Daejeon 34134, Korea; hwendiv@naver.com (K.-E.C.); balupuri@cnu.ac.kr (A.B.)
* Correspondence: nskang@cnu.ac.kr; Tel.: +82-42-821-8626

Academic Editors: Marco Tutone and Anna Maria Almerico
Received: 6 May 2020; Accepted: 2 June 2020; Published: 4 June 2020

Abstract: Human ether-a-go-go-related gene (hERG) potassium channel blockage by small molecules may cause severe cardiac side effects. Thus, it is crucial to screen compounds for activity on the hERG channels early in the drug discovery process. In this study, we collected 5299 hERG inhibitors with diverse chemical structures from a number of sources. Based on this dataset, we evaluated different machine learning (ML) and deep learning (DL) algorithms using various integer and binary type fingerprints. A training set of 3991 compounds was used to develop quantitative structure–activity relationship (QSAR) models. The performance of the developed models was evaluated using a test set of 998 compounds. Models were further validated using external set 1 (263 compounds) and external set 2 (47 compounds). Overall, models with integer type fingerprints showed better performance than models with no fingerprints, converted binary type fingerprints or original binary type fingerprints. Comparison of ML and DL algorithms revealed that integer type fingerprints are suitable for ML, whereas binary type fingerprints are suitable for DL. The outcomes of this study indicate that the rational selection of fingerprints is important for hERG blocker prediction.

Keywords: hERG toxicity; drug discovery; fingerprints; machine learning; deep learning

1. Introduction

The human ether-a-go-go related gene (hERG) or KCNH2 gene encodes a voltage-gated potassium channel known as the hERG channel. This channel plays a key role in cardiac action potential repolarization. Reduced function of hERG causes potential action prolongation and increases the risk for potentially fatal ventricular arrhythmia, torsades de pointes. Therefore, preclinical hERG testing is essential in the drug discovery process to avoid cardiac toxicity [1]. Recently, many drugs such as astemizole, terfenadine, cisapride, thioridazine, grepafloxacin and sertindole were withdrawn from the market due to undesired cardiotoxicity effects [2,3]. Nowadays, the US Food and Drug Administration (FDA) demands in vitro hERG assay of lead compounds prior to clinical trials [4]. The side effects of unexpected hERG channel binding by drug candidates are a major challenge in the drug discovery process [5]. The development of an accurate prediction model for hERG channel blockers is crucial in the early stages of drug discovery and development.

Although the electron microscopy structure of the membrane protein hERG is known [6], its X-ray crystal structure is not available. Thus, structure-based hERG blocker prediction is challenging. However, a few structure-based hERG blocker predictions were attempted with homology modeling using structures of the related potassium ion channels as templates [7]. Several researchers have applied a ligand-based drug design approach to identify hERG blockers. They used various machine learning (ML) algorithms such as naïve Bayes (NB) [8], support vector machine (SVM) [9], random forest (RF) [10] and k-NN [11]. Sun et al. reported an NB model with a receiver operating characteristic (ROC) value of 0.87 on the basis of 1979 compounds [12]. Jia et al. applied SVM methods and

atom-type descriptors without fingerprints on 1043 compounds to develop a model with accuracy of 0.94 [9]. Yap et al. developed a model with accuracy of 0.97 based on 310 compounds using similar methodology [13]. Marchese Robinson et al. built an RF model with Matthews correlation coefficient (MCC) of 0.83 using a dataset of 368 compounds. They used the extended-connectivity fingerprint (ECFP_4) for developing the model [14]. Kim et al. developed a model with accuracy of 0.96 using the RF algorithm on a dataset of 293 compounds [15]. Chavan et al. reported a k-NN model with accuracy of 0.55 on the basis of 1967 compounds [11]. The deep learning (DL) approach with artificial neural networks (ANN) has also been used to predict hERG blockers [16,17]. Cai et al. applied DL to 7889 compounds and obtained accuracy of 0.93 and an area under the receiver operating characteristic curve (AUC) value of 0.97 for the best model [17]. Zhang et al. applied DL to 1871 compounds and developed a model with accuracy of 0.78 [18]. Several research papers have reported the application of ML or DL techniques on hERG blockers. However, smaller datasets were used in the previous publications as compared to the recent papers utilizing the DL approach. Datasets for hERG blockers have grown in recent years.

Fingerprints used in the former ligand-based drug design studies on hERG as well as other targets were mostly binary types representing 0 or 1 [19–22]. Fingerprints describe chemical substructures as numerical category data. The number of a binary type fingerprint is usually restricted to 1024 bits [23]. The integer type fingerprints describe chemical structures in more detail with various substructures. They produce a large number of category data types useful for ML. Although integer type fingerprints indicate diversity of chemical substructures, they are limited to commercial software such as the Pipeline Pilot (PP) module of Discovery Studio (DS) software (BIOVIA, San Diego, CA, USA) [24]. In the present study, we calculated integer as well as binary type fingerprints for a dataset of hERG blockers. Integer type fingerprints were computed using PP and they included extended-connectivity fingerprints (ECFP_2, ECFP_4 and ECFP_6) and functional-class fingerprints (FCFP_2, FCFP_4 and FCFP_6). Binary type chemistry development kit (CDK) fingerprints (standard, extended and graph) were computed using the R package. We used both types of fingerprints for ML and DL. Our results show that rational selection of fingerprints is important for hERG blocker prediction. We have discussed the advantages and disadvantages of integer and binary type fingerprints in ML and DL.

2. Results

2.1. Dataset Splitting

As discussed in the methodology section, a dataset consisting of hERG blockers was divided into the training and test sets in the ratio 4:1. The training and test sets consisted of 3991 and 998 compounds, respectively. Principle component analysis (PCA) was performed to verify the diversity of the chemical space of the dataset. PCA analysis with eight common descriptors, including partition coefficient (AlogP), molecular weight (MW), hydrogen-bond donor (HBD), hydrogen-bond acceptor (HBA), rotatable bond number (RBN), number of rings (Num Rings), number of aromatic rings (Num Arom Rings) and molecular fractional polar surface area (MFPSA), showed high chemical diversity of the compounds within the training and test sets (Figure 1). Generally, PCA descriptors are indirect numeric type representations of chemical structures. However, fingerprints representing chemical substructures are not numeric type. They are Boolean type indicating existence or nonexistence (1 or 0) of a unique fingerprint. Therefore, additional integer and binary type fingerprints were computed to validate the division of the dataset. As shown in Figure 2, integer type PP fingerprints (ECFP_6 and FCFP_6) and binary type CDK fingerprints (standard and extended) showed similar fingerprint frequency for the training and test sets. Binary type CDK fingerprints displayed relatively lower fingerprint frequency than integer type PP fingerprints. This is due to the limited size of the binary type CDK fingerprints (1024 bits). PCA analysis and fingerprint frequency calculations validate the appropriate splitting of the dataset into training and test sets.

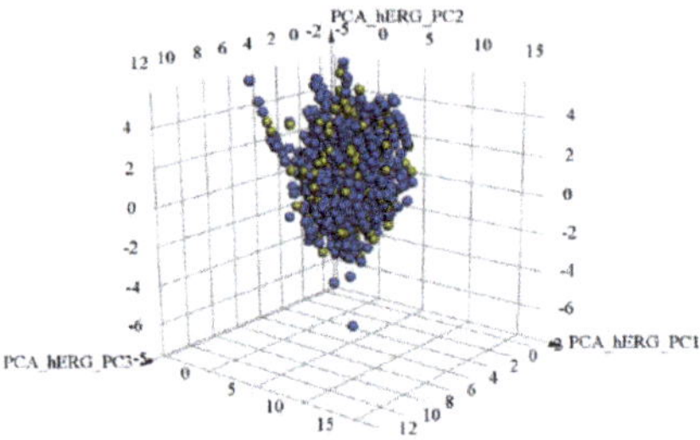

Figure 1. Principle component analysis (PCA) for the training and test sets of hERG blockers. Independent variables included partition coefficient (AlogP), molecular weight (MW), hydrogen-bond donor (HBD), hydrogen-bond acceptor (HBA), rotatable bond number (RBN), number of rings (Num Rings), number of aromatic rings (Num Arom Rings) and molecular fractional polar surface area (MFPSA). Blue and yellow spheres indicate training and test sets, respectively.

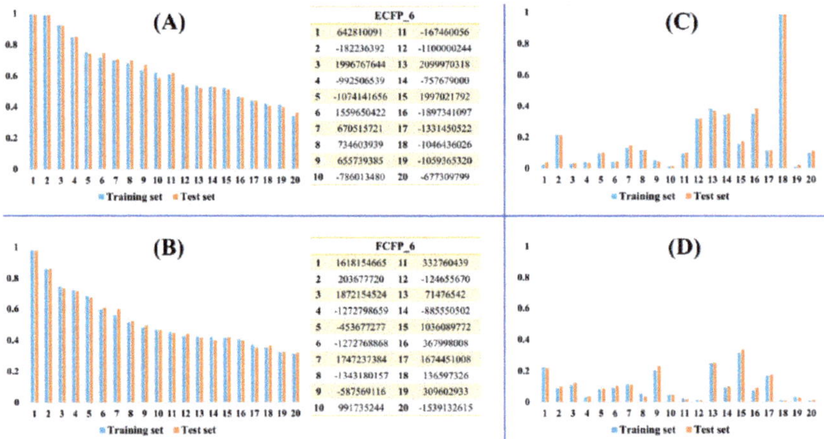

Figure 2. Fingerprint frequency of the training and test sets of hERG blockers. X-axis denotes fingerprint identifier whereas Y-axis denotes the fingerprint frequency. Representative integer type PP fingerprints: (**A**) ECFP_6 and (**B**) FCFP_6. Representative binary type chemistry development kit (CDK) fingerprints: (**C**) standard and (**D**) extended. For validating the appropriate split of the dataset into training and test sets, the top 20 of PP integer fingerprint identifiers were selected as representatives depending on the inclusion order. Furthermore, the top 20 of CDK binary fingerprint identifiers were chosen as representatives depending on the binary sequence (1024 bits). Fingerprint frequency was calculated as the observed number of unique fingerprint identifiers/the total number of chemical data (training set = 3991 and test set = 998).

2.2. Model Building and Evaluation

The training set comprising 3991 compounds was used to build the models and the test set consisting of 998 compounds was used to evaluate the performance of the developed models. Different integer and binary type fingerprints were utilized for model building (Figure 3 and Supplementary materials Tables S1 and S2). All control models which lacked fingerprints and included only descriptors (AlogP, MW, HBD, HBA, RBN, Num Rings, Num Arom Rings and MFPSA) as features showed average predictive accuracy (Q) of 0.77 (ML = 0.79 and DL = 0.75) and an average area under the receiver operating characteristic curve (AUC) value of 0.82 (ML = 0.82 and DL = 0.81). The addition of fingerprints in the features improved the performance of the models. All models including integer type PP fingerprints (ECFP_2, FCFP_2, ECFP_4, FCFP_4, ECFP_6 and FCFP_6) displayed average Q

and AUC values of 0.87 (ML = 0.86 and DL = 0.88) and 0.92 (ML = 0.91 and DL = 0.93), respectively. Integer type fingerprints were converted to binary type using the "Convert Fingerprint" module of PP. The conversion of integer type to binary type fingerprints decreased the performance of the models. All models including converted binary type PP fingerprints demonstrated average Q and AUC values of 0.77 (ML = 0.75 and DL = 0.81) and 0.81 (ML = 0.77 and DL = 0.84), respectively. The open-source based CDK fingerprints (standard, extended and graph) are originally binary type and cannot be converted to integer type. Models with original binary type CDK fingerprints showed a comparatively higher performance than converted binary type PP fingerprints. All models, including binary type CDK fingerprints, exhibited average Q and AUC values of 0.86 (ML = 0.83 and DL = 0.89) and 0.90 (ML = 0.87 and DL = 0.93), respectively. Integer type PP fingerprints exhibited a higher performance than original binary type CDK fingerprints for ML. However, integer type PP fingerprints demonstrated a slightly lower performance than original binary type CDK fingerprints for DL. Compared to integer type PP fingerprints and original binary type CDK fingerprints, converted binary type PP fingerprints showed lower performance for both ML and DL. Comparison of different algorithms revealed that the RF algorithm produced better models than others. Models derived using the RF algorithm exhibited average Q and AUC values of 0.90 and 0.95, respectively, for both integer type PP fingerprints as well as original binary type CDK fingerprints. Furthermore, the RF algorithm produced the best model with FCFP_2 integer type PP fingerprint. This model showed Q and AUC values of 0.91 and 0.95, respectively.

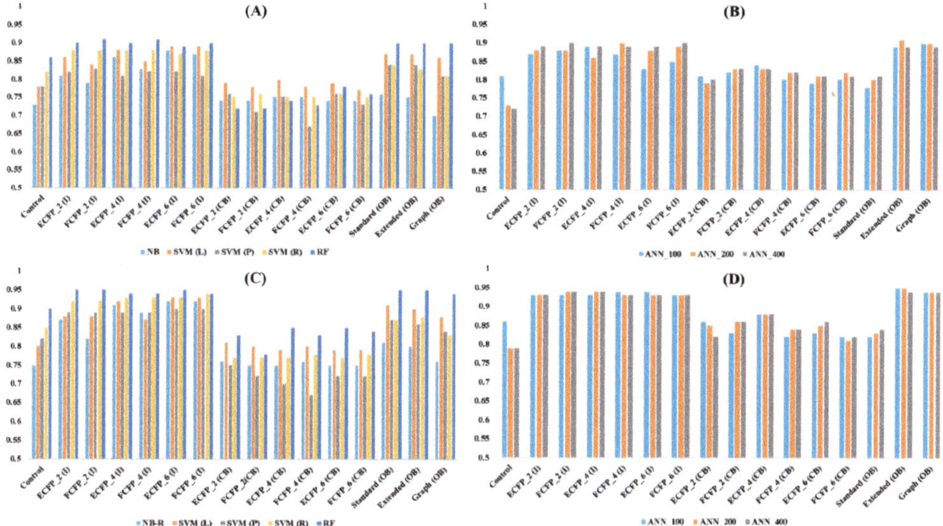

Figure 3. Models built using integer and binary type fingerprints for hERG blockers. X-axis denotes fingerprint types whereas Y-axis denotes the performance. Control lacking fingerprint only include descriptors, I: integer type PP fingerprints, CB: converted binary type PP fingerprints, OB: original binary type CDK fingerprints, NB: naïve Bayes, SVM: support vector machine (L: linear, P: polynomial, R: radial), RF: random forest, ANN: artificial neural network (layer size 100, 200, and 400) for deep learning. (**A**,**B**) Accuracy value for the models using machine learnings (**A**) and deep learning (**B**). (**C**,**D**) AUC value for machine learnings (**C**) and deep learning (**D**).

2.3. External Validation

Two external sets were used for further validation of the models. As discussed in the methodology section, external set 1 (Ex-1) from ChEMBL database and external set 2 (Ex-2) from recent publications [25–29] consisted of 263 and 47 hERG blockers, respectively. Model predictions for Ex-1 and Ex-2 using integer

and binary type fingerprints are summarized in Figure 4 and Tables S3–S6 (Supplementary materials), respectively. The control models exhibited an average Q value of 0.79 (ML = 0.80 and DL = 0.78) for Ex-1. The Ex-1 showed average Q values of 0.87 (ML = 0.86 and DL = 0.88), 0.80 (ML = 0.77 and DL = 0.82) and 0.86 (ML = 0.83 and DL = 0.88) for integer type PP fingerprints, converted binary type PP fingerprints and original binary type CDK fingerprints, respectively. Ex-1 displayed higher predictive accuracy for both integer type PP fingerprints and original binary type CDK fingerprints than for the control models. However, integer type PP fingerprints produced comparatively higher predictive accuracy than original binary type CDK fingerprints. Converted binary type PP fingerprints showed predictive accuracy similar to the control models. Integer type PP fingerprints exhibited higher predictive accuracy than original binary type CDK fingerprints for ML. However, integer type PP fingerprints exhibited predictive accuracy similar to the original binary type CDK fingerprints for DL. Control models exhibited an average Q value of 0.76 (ML = 0.79 and DL = 0.73) for Ex-2. The Ex-2 showed average Q values of 0.80 (ML = 0.80 and DL = 0.80), 0.71 (ML = 0.71 and DL = 0.70) and 0.70 (ML = 0.72 and DL = 0.68) for integer type PP fingerprints, converted binary type PP fingerprints and original binary type CDK fingerprints, respectively. Compared to the control models, Ex-2 displayed higher predictive accuracy for integer type PP fingerprints while it showed lower predictive accuracy for original binary type CDK fingerprints and converted binary type PP fingerprints. Integer type PP fingerprints exhibited higher predictive accuracy for both ML and DL as compared to the original binary type CDK fingerprints and converted binary type PP fingerprints.

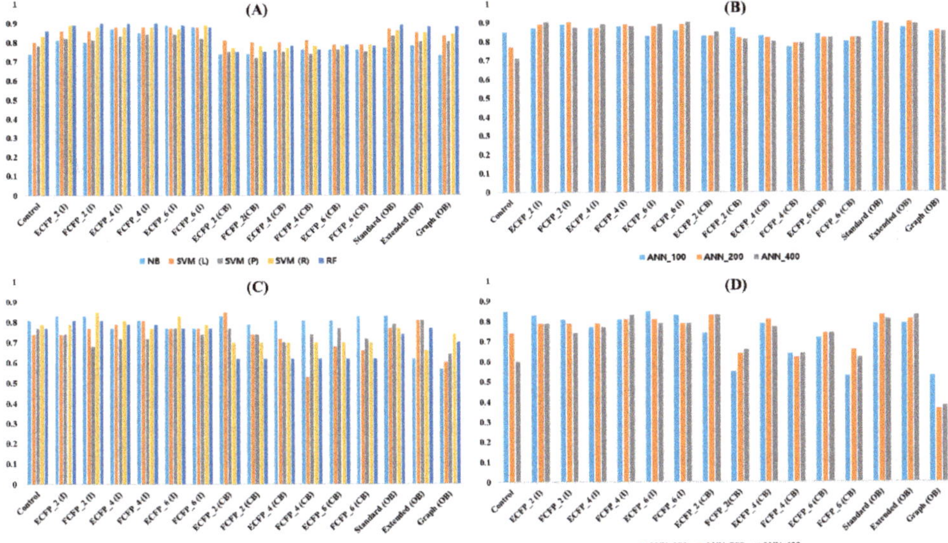

Figure 4. Model prediction to external sets. X-axis denotes fingerprint types whereas Y-axis denotes the performance for accuracy. Control lacking fingerprint include only descriptors, I: integer type PP fingerprints, CB: converted binary type PP fingerprints, OB: original binary type CDK fingerprints, NB: naïve Bayes, SVM: support vector machine (L: linear, P: polynomial, R: radial), RF: random forest, ANN: artificial neural network (layer size 100, 200, and 400) for deep learning. (**A,B**) Ex-1 prediction for machine learnings (**A**) and deep learning (**B**). (**C,D**) Ex-2 prediction for machine learnings (**C**) and deep learning (**D**).

In accordance with the model-building results, converted binary type PP fingerprints reduced the predictive accuracy of the models. Original binary type CDK fingerprints exhibited comparatively higher predictive accuracy than converted binary type PP fingerprints for Ex-1. However, Ex-2 displayed slightly lower predictive accuracy for original binary type CDK fingerprints as compared to converted binary type PP fingerprints. This might be due to the small data size of the Ex-2

(47 compounds). Further parameters for external validation of the developed models are provided in Tables S3–S6 (Supplementary materials). These included true positive (TP), true negative (TN), false positive (FP), and false negative (FN). It can be seen in the supplementary tables that TN predictions were higher than FN. This might be because compounds in the whole dataset as well as the external sets were intended to develop as inhibitors for other drug targets but they also inhibited hERG, to produce toxicity. During model building, the RF algorithm displayed the same average Q value of 0.90 for both integer type PP fingerprints as well as original binary type CDK fingerprints. However, the RF model predictions for Ex-1 and Ex-2 using integer type PP fingerprints were found to be better than original binary type CDK fingerprints. The RF models with integer type PP fingerprints showed average Q values of 0.89 and 0.79 for Ex-1 and Ex-2, respectively. The RF models with original binary type CDK fingerprints exhibited average Q values of 0.88 and 0.74 for Ex-1 and Ex-2, respectively. The best model that was obtained using the RF algorithm and FCFP_2 integer type PP fingerprint showed Q values of 0.90 and 0.81 for Ex-1 and Ex-2, respectively.

3. Discussion

The unwanted blockage of the hERG channel by drug candidates could lead to fatal cardiotoxicity. Thus, it is essential to screen compounds for activity on hERG channels early in the drug discovery process to decrease the risk of a drug candidate failing in preclinical safety studies. Due to the unavailability of the crystal structure of the hERG channel, researchers mainly use the ligand-based drug design approach to identify hERG blockers. Several ML and DL approaches have been applied. However, fingerprints used in the majority of the previous studies were mostly binary type [30,31]. These fingerprints are limited in size and generally restricted to 1024 bits. The integer type fingerprints describe chemical structures in more detail, but they are limited to commercial software. Fingerprints of the integer type have not been explored much for the prediction of hERG blockers or for the prediction of other drug target inhibitors. In this study, we computed both integer and binary type fingerprints for a dataset of hERG blockers and evaluated various ML (NB, SVM, RF and Bagging) and DL (ANN) algorithms. A training set of 3991 compounds was used to develop QSAR models. The performance of the developed models was evaluated using a test set of 998 compounds. Models were further validated using two external sets (Ex-1: 263 compounds and Ex-2: 47 compounds).

The overall results showed that addition of fingerprints to the descriptors improved the performance of the models. Models with integer type PP fingerprints displayed slightly better performance than models with binary type CDK fingerprints. Conversion of integer type PP fingerprints to binary type PP fingerprints reduced the performance of the models. Except for small-sized Ex-2, models with binary type CDK fingerprints showed better performance than converted binary type PP fingerprints. Comparison of different algorithms revealed that models built by RF outperformed those built by other algorithms. This is in agreement with a previous study that reported high performance of the RF models [17]. The RF algorithm demonstrated a similar performance for both integer type PP fingerprints as well as original binary type CDK fingerprints during model building. However, the RF models with integer type PP fingerprints showed comparatively better predictions for both the external sets than the RF models with original binary type CDK fingerprints. The best model was obtained using the RF algorithm and FCFP_2 integer type PP fingerprint. Comparison of ML and DL algorithms revealed that ML models with integer type PP fingerprints demonstrated better performance and predictions than ML models with binary type CDK fingerprints. On the other hand, DL models performed slightly better with binary type CDK fingerprints as compared to DL models with integer type PP fingerprints. However, DL models exhibited similar predictions for both integer type PP fingerprints and binary type CDK fingerprints for Ex-1. In the case of small-sized Ex-2, ML and DL models showed better predictions with integer type PP fingerprints as compared to ML and DL models with binary type CDK fingerprints.

The outcomes of the study suggested that integer type fingerprints improved the performance and predictive ability of QSAR models. Although integer type fingerprints could be applied to both

ML and DL, these fingerprints did not improve the predictive accuracy of DL models significantly. Moreover, they required long computation time due to the large number of features. Numerous fingerprint identifiers caused memory problems in some ANN packages such as "nnet". Due to memory problems, integer type fingerprints needed to be converted to binary type for DL. Our results demonstrated that conversion of fingerprints from integer to binary type reduced the performance and the predictive ability of the models in both ML and DL. Compared to integer type fingerprints, original binary type fingerprints produced DL models with slightly better performance. Furthermore, the predictive ability of DL models with binary type fingerprints was comparable to DL models with integer type fingerprints. Accordingly, binary type fingerprints are recommended for DL. Binary type fingerprints were suitable for both ML and DL due to their limited size. However, binary type fingerprints produced ML models with comparatively lower performance and predictive ability than ML models with integer type fingerprints. Consequently, integer type fingerprints are recommended for ML. In conclusion, rational selection of fingerprints is important for hERG blocker prediction.

4. Materials and Methods

4.1. Dataset

A dataset consisting of 5252 compounds with hERG inhibition values (IC_{50} and K_i) was obtained from the ChEMBL database [32]. The reported K_i values were converted into IC_{50} values. In accordance with previous studies, compounds with $IC_{50} \leq 1$ μM and $IC_{50} > 10$ μM were classified as active and inactive compounds, respectively [15,16]. Active and inactive compounds were defined as 1 and 0, respectively. Prior to splitting the dataset into training and test sets, 263 compounds (5% of the dataset) were extracted by random selection as Ex-1 using DS 2019 software (BIOVIA, San Diego, CA, USA). The remaining dataset with 4989 compounds was randomly partitioned into training and test sets in the ratio 4:1. The training and test sets consisted of 3991 and 998 compounds, respectively. In addition to the ChEMBL dataset, 47 hERG blockers reported in recent research papers were collected as Ex-2 from recent publications [25–29]. Active and inactive compounds for the external set 2 (Ex-2) were defined in the same way as discussed for the ChEMBL dataset (Table 1). The training set was used for the model building whereas the test and external sets (Ex-1 and Ex-2) were used for model evaluation.

Table 1. Dataset used in the present study.

Source	Set	Compounds	Active Compounds	Inactive Compounds
ChEMBL database (Total compounds: 5252)	Training set	3991	1201	2790
	Test set	998	278	720
	Ext-1	263	67	196
Recent research papers (Total compounds: 47)	Ext-2	47	18	29

4.2. Fingerprint Calculation

The PP integer type fingerprints were computed using the "molecular fingerprint" module of PP. The "atom abstraction" option was set to atom type and functional class for calculating ECFP and FCFP fingerprints, respectively. The "maximum distance" option was set to 2, 4 and 6 for generating different types of ECFP and FCFP fingerprints (ECFP_2, FCFP_2, ECFP_4, FCFP_4, ECFP_6 and FCFP_6). Integer type fingerprints were converted to binary type fingerprints using the "Convert Fingerprint" module of PP. CDK binary type fingerprints (standard, extended and graph) were computed with "rcdk" package [33] of R (version 3.5.2, R Core Team, Vienna, Austria) [34]. The "get.fingerprint" function was used for the calculation of the fingerprints. The fingerprint frequency was determined for validating the appropriate split of the dataset into training and test sets. Fingerprint frequency was

calculated as the observed number of unique fingerprint identifiers/the total number of chemical data (training set = 3991 and test set = 998).

4.3. Model Building

In this study, we evaluated one DL algorithm: ANN, and four ML algorithms: NB, SVM, RF and bagging. DS 2019 software was used for NB and bagging because it supports only these two algorithms. R package was employed for NB, SVM, RF, ANN. NB is available in both R and DS packages. NB-DS computes only integer type fingerprints but it is possible to calculate both integer as well as binary type fingerprints using NB-R. We utilized both the packages to evaluate NB. The "create Bayesian model" protocol of DS was used for NB-DS. The laplacian value was set to the default value of 1. The "create recursive partitioning model" protocol of DS was employed for decision-tree based bagging such as RF. The number of trees was set to 100. The "e1071" package [35] of R was utilized for NB and SVM. NB models were created using the "naïveBayes" function. The Laplace parameter was set to 1. SVM models were developed using the "svm" function. The linear, polynomial and radial methods were implemented for SVM. While implementing, linear, polynomial and radial kernel functions were selected. The "randomForest" package [36] of R was used for a decision-tree based RF. The number of trees was set to 100. The "h2o" package [37] of R was utilized for ANN with three hidden layers. The "Rectifier activation" function was employed, and the number of iterations was set to 20. The layer size parameter was set to 100, 200, and 400 for three hidden layers. Descriptors were calculated using DS. These included AlogP, MW, HBD, HBA, RBN, Num Rings, Num Arom Rings and MFPSA. Except for the control models, all developed models included fingerprints. Control models comprised only descriptors.

4.4. Model Evaluation

Model performance was evaluated in terms of predictive accuracy (Q), the area under the receiver operating characteristic curve (AUC), true positive (TP), true negative (TN), false positive (FP) and false negative (FN). The "calculate molecular property" protocol of DS 2019 software and "predict" function of R package were used for the model evaluation.

Supplementary Materials: The following are available online. Table S1. Models built using integer type fingerprints. Table S2. Models built using binary type fingerprints. Table S3. Model prediction for external set 1 using integer type fingerprints. Table S4. Model prediction for external set 1 using binary type fingerprints. Table S5. Model prediction for external set 2 using integer type fingerprints. Table S6. Model prediction for external set 2 using binary type fingerprints. The supplementary excel data for the compound list and detailed model prediction data used in this research for hERG blocker prediction.

Author Contributions: Conceptualization, N.S.K.; methodology, K.-E.C.; software, K.-E.C.; validation, N.S.K., K.-E.C. and A.B.; formal analysis, K.-E.C. and A.B.; investigation, K.-E.C.; resources, N.S.K.; data curation, K.-E.C.; writing—original draft preparation, K.-E.C.; writing—review and editing, N.S.K. and A.B.; visualization, K.-E.C.; supervision, N.S.K.; project administration, N.S.K.; funding acquisition, N.S.K. All authors have read and agreed to the published version of the manuscript.

Funding: This research was supported by Basic Science Research Program through the National Research Foundation of Korea (NRF) funded by the Ministry of Science, NRF-2020R1A2C100691511 and NRF-2016M3A9E4947695.

Conflicts of Interest: The authors declare no conflict of interest.

References

1. Priest, B.; Bell, I.M.; Garcia, M. Role of hERG potassium channel assays in drug development. *Channels* **2008**, *2*, 87–93. [CrossRef]
2. Lee, H.-M.; Yu, M.-S.; Kazmi, S.R.; Oh, S.Y.; Rhee, K.-H.; Bae, M.-A.; Lee, B.H.; Shin, D.-S.; Oh, K.-S.; Ceong, H. Computational determination of hERG-related cardiotoxicity of drug candidates. *BMC Bioinform.* **2019**, *20*, 250. [CrossRef] [PubMed]

3. Laverty, H.; Benson, C.; Cartwright, E.; Cross, M.; Garland, C.; Hammond, T.; Holloway, C.; McMahon, N.; Milligan, J.; Park, B. How can we improve our understanding of cardiovascular safety liabilities to develop safer medicines? *Br. J. Pharmacol.* **2011**, *163*, 675–693. [CrossRef] [PubMed]

4. Cantilena, L., Jr.; Koerner, J.; Temple, R.; Throckmorton, D. OIII-A-1: FDA evaluation of cardiac repolarization data for 19 drugs and drug candidates. *Clin. Pharmacol. Ther.* **2006**, *79*, P29. [CrossRef]

5. Fernandez, D.; Ghanta, A.; Kauffman, G.W.; Sanguinetti, M.C. Physicochemical features of the HERG channel drug binding site. *J. Biol. Chem.* **2004**, *279*, 10120–10127. [CrossRef] [PubMed]

6. Wang, W.; MacKinnon, R. Cryo-EM structure of the open human ether-à-go-go-related K+ channel hERG. *Cell* **2017**, *169*, 422–430.e10. [CrossRef]

7. Saxena, P.; Zangerl-Plessl, E.-M.; Linder, T.; Windisch, A.; Hohaus, A.; Timin, E.; Hering, S.; Stary-Weinzinger, A. New potential binding determinant for hERG channel inhibitors. *Sci. Rep.* **2016**, *6*, 24182. [CrossRef]

8. Liu, L.-l.; Lu, J.; Lu, Y.; Zheng, M.-y.; Luo, X.-m.; Zhu, W.-l.; Jiang, H.-l.; Chen, K.-x. Novel Bayesian classification models for predicting compounds blocking hERG potassium channels. *Acta Pharmacol. Sin.* **2014**, *35*, 1093. [CrossRef]

9. Jia, L.; Sun, H. Support vector machines classification of hERG liabilities based on atom types. *Bioorg. Med. Chem* **2008**, *16*, 6252–6260. [CrossRef]

10. Song, M.; Clark, M. Development and evaluation of an in silico model for hERG binding. *J. Chem. Inf. Model.* **2006**, *46*, 392–400. [CrossRef]

11. Chavan, S.; Abdelaziz, A.; Wiklander, J.G.; Nicholls, I.A. A k-nearest neighbor classification of hERG K+ channel blockers. *J. Comput. Aided Mol. Des.* **2016**, *30*, 229–236. [CrossRef] [PubMed]

12. Sun, H. An accurate and interpretable Bayesian classification model for prediction of hERG liability. *Chem. Med. Chem.* **2006**, *1*, 315–322. [CrossRef] [PubMed]

13. Yap, C.; Cai, C.; Xue, Y.; Chen, Y. Prediction of torsade-causing potential of drugs by support vector machine approach. *Toxicol. Sci.* **2004**, *79*, 170–177. [CrossRef] [PubMed]

14. Marchese Robinson, R.L.; Glen, R.C.; Mitchell, J.B. Development and comparison of hERG blocker classifiers: Assessment on different datasets yields markedly different results. *Mol. Inform.* **2011**, *30*, 443–458. [CrossRef]

15. Kim, J.H.; Chae, C.H.; Kang, S.M.; Lee, J.Y.; Lee, G.N.; Hwang, S.H.; Kang, N.S. The predictive QSAR model for hERG inhibitors using Bayesian and random forest classification method. *Bull. Korean Chem. Soc.* **2011**, *32*, 1237. [CrossRef]

16. Polak, S.; Wiśniowska, B.; Ahamadi, M.; Mendyk, A. Prediction of the hERG potassium channel inhibition potential with use of artificial neural networks. *Appl. Soft Comput.* **2011**, *11*, 2611–2617. [CrossRef]

17. Cai, C.; Guo, P.; Zhou, Y.; Zhou, J.; Wang, Q.; Zhang, F.; Fang, J.; Cheng, F. Deep Learning-Based Prediction of Drug-Induced Cardiotoxicity. *J. Chem. Inf. Model.* **2019**, *59*, 1073–1084. [CrossRef]

18. Zhang, Y.; Zhao, J.; Wang, Y.; Fan, Y.; Zhu, L.; Yang, Y.; Chen, X.; Lu, T.; Chen, Y.; Liu, H. Prediction of hERG K+ channel blockage using deep neural networks. *Chem. Biol. Drug Des.* **2019**, *94*, 1973–1985. [CrossRef]

19. Drwal, M.N.; Siramshetty, V.B.; Banerjee, P.; Goede, A.; Preissner, R.; Dunkel, M. Molecular similarity-based predictions of the Tox21 screening outcome. *Front. Environ. Sci.* **2015**, *3*, 54. [CrossRef]

20. Myint, K.-Z.; Wang, L.; Tong, Q.; Xie, X.-Q. Molecular fingerprint-based artificial neural networks QSAR for ligand biological activity predictions. *Mol. Pharm.* **2012**, *9*, 2912–2923. [CrossRef]

21. Myint, K.Z.; Xie, X.-Q. *Artificial Neural Networks*; Springer: New York, NY, USA, 2015; pp. 149–164.

22. Fan, T.; Sun, G.; Zhao, L.; Cui, X.; Zhong, R. QSAR and classification study on prediction of acute oral toxicity of N-nitroso compounds. *Int. J. Mol. Sci.* **2018**, *19*, 3015. [CrossRef] [PubMed]

23. Fernández-de Gortari, E.; García-Jacas, C.R.; Martinez-Mayorga, K.; Medina-Franco, J.L. Database fingerprint (DFP): An approach to represent molecular databases. *J. Cheminformat.* **2017**, *9*, 9. [CrossRef]

24. Rogers, D.; Hahn, M. Extended-connectivity fingerprints. *J. Chem. Inf. Model.* **2010**, *50*, 742–754. [CrossRef] [PubMed]

25. Zhang, X.; Liu, T.; Wang, B.; Gao, Y.; Liu, P.; Li, M.; Du, L. Astemizole-based turn-on fluorescent probes for imaging hERG potassium channel. *Med. Chem. Comm.* **2019**, *10*, 513–516. [CrossRef] [PubMed]

26. Yu, Y.; Wu, Z.; Shi, Z.-C.; He, S.; Lai, Z.; Cernak, T.A.; Vachal, P.; Liu, M.; Liu, J.; Hong, Q. Accelerating the discovery of DGAT1 inhibitors through the application of parallel medicinal chemistry (PMC). *Bioorgan. Med. Chem. Lett.* **2019**, *29*, 1380–1385. [CrossRef] [PubMed]

27. Li, L.; Okumu, A.; Dellos-Nolan, S.; Li, Z.; Karmahapatra, S.; English, A.; Yalowich, J.C.; Wozniak, D.J.; Mitton-Fry, M.J. Synthesis and anti-staphylococcal activity of novel bacterial topoisomerase inhibitors with a 5-amino-1, 3-dioxane linker moiety. *Bioorgan. Med. Chem. Lett.* **2018**, *28*, 2477–2480. [CrossRef]

28. Bagdanoff, J.T.; Chen, Z.; Acker, M.; Chen, Y.-N.; Chan, H.; Dore, M.; Firestone, B.; Fodor, M.; Fortanet, J.; Hentemann, M. Optimization of Fused Bicyclic Allosteric SHP2 Inhibitors. *J. Med. Chem.* **2019**, *62*, 1781–1792. [CrossRef]

29. Zhang, X.; Liu, T.; Li, Q.; Li, M.; Du, L. Aggregation-Induced Emission: Lighting Up hERG Potassium Channel. *Front. Chem.* **2019**, *7*, 54. [CrossRef]

30. Xuan, S.; Liang, H.; Wang, Z.; Yan, A. Classification of blocker and non-blocker of hERG potassium ion channel using a support vector machine. *Sci. China Chem.* **2013**, *56*, 1413–1423. [CrossRef]

31. Du, F.; Babcock, J.J.; Yu, H.; Zou, B.; Li, M. Global analysis reveals families of chemical motifs enriched for hERG inhibitors. *PLoS ONE* **2015**, *10*, e0118324. [CrossRef]

32. Mendez, D.; Gaulton, A.; Bento, A.P.; Chambers, J.; De Veij, M.; Félix, E.; Magariños, M.P.; Mosquera, J.F.; Mutowo, P.; Nowotka, M. ChEMBL: Towards direct deposition of bioassay data. *Nucleic Acid Res.* **2019**, *47*, D930–D940. [CrossRef] [PubMed]

33. Guha, R. Chemical informatics functionality in R. *J. Stat. Softw* **2007**, *18*, 1–16. [CrossRef]

34. R Core Team. *R: A Language and Environment for Statistical Computing*; R Foundation for Statistical Computing: Vienna, Austria, 2018; Available online: https://www.R-project.org/.

35. David, M.; Evgenia, D.; Kurt, H.; Andreas, W.; Friedrich, L. e1071: Misc Functions of the Department of Statistics, Probability Theory Group (Formerly: E1071), TU Wien. R Package Version 1.7-3. 2019. Available online: https://CRAN.R-project.org/package=e1071 (accessed on 31 May 2020).

36. Liaw, A.; Wiener, M. Classification and regression by randomForest. *R News* **2002**, *2*, 18–22.

37. Erin, L.; Navdeep, G.; Spencer, A.; Anqi, F.; Arno, C.; Cliff, C.; Tom, K.; Tomas, N.; Patrick, A.; Michal, K.; et al. h2o: R Interface for the 'H2O' Scalable Machine Learning Platform. R Package Version 3.28.1.2. 2020. Available online: https://github.com/h2oai/h2o-3 (accessed on 31 May 2020).

Sample Availability: Samples of the compounds are not available from the authors.